# Technology Transfer and Business Enterprise

# The International Library of Critical Writings in Business History

*Series Editor:* Geoffrey Jones
Professor of Business History,
University of Reading

# Technology Transfer and Business Enterprise

*Edited by*

## David J. Jeremy

*Reader in Business History*
*Faculty of Management and Business*
*Manchester Metropolitan University*

An Elgar Reference Collection

Published by
Edward Elgar Publishing Limited
Gower House
Croft Road
Aldershot
Hants GU11 3HR
England

Edward Elgar Publishing Company
Old Post Road
Brookfield
Vermont 05036
USA

**British Library Cataloguing in Publication Data**
Technology Transfer and Business
Enterprise. – (International Library of
Critical Writings in Business History;
No.9)
I. Jeremy, David J.  II. Series
338.9

**Library of Congress Cataloguing in Publication Data**
Technology transfer and business enterprise / edited by David J.
Jeremy.
    p.  cm. — (An Elgar reference collection) (International
library of critical writings in business history; 9)
    Includes bibliographical references and index.
    1. Technology transfer—Economic aspects.  2. Technology transfer—
Economic aspects—Case studies.  I. Jeremy, David J.  II. Series.
III. Series: International library of critical writings in business
history; 9.
HC79.T4T458  1993
338.9'26—dc20
                                                                93–35787
                                                                CIP

ISBN 1 85278 727 9

Printed in Great Britain by Galliard (Printers) Ltd, Great Yarmouth

# Contents

# Acknowledgements

The editor and publishers wish to thank the following who have kindly given permission for the use of copyright material.

Tetsuo Abo for excerpt: Tetsuo Abo (1992), 'Japanese Motor Vehicle Technologies abroad in the 1980s', in David J. Jeremy (ed.), *The Transfer of International Technology: Europe, Japan and the USA in the Twentieth Century*, 167–90.

American Economic Association for article: Kenneth J. Arrow (1969), 'Classificatory Notes on the Production and Transmission of Technological Knowledge', *American Economic Review*, **LIX** (2), May, 29–35.

Basil Blackwell Ltd for article: Christine MacLeod (1992), 'Strategies for Innovation: The Diffusion of New Technology in Nineteenth-Century British Industry', *Economic History Review*, **XLV** (2), 285–307.

Berg Publishers Ltd for excerpt: Kristine Bruland (1991), 'The Norwegian Mechanical Engineering Industry and the Transfer of Technology, 1800–1900', in *Technology Transfer and Scandinavian Industrialisation*, 229–67.

Butterworth-Heinemann Ltd, Oxford, UK, for article: David J. Jeremy (1982), 'Immigrant Textile Machine Makers along the Brandywine, 1810–1820', *Textile History*, **13** (2), Autumn, 225–48.

Cambridge University Press for article: Mira Wilkins (1974), 'The Role of Private Business in the International Diffusion of Technology', *Journal of Economic History*, **XXXIV** (1), March, 166–88.

Elsevier Science Publishers B.V. for article: Bruce Kogut and Sea Jin Chang (1991), 'Technological Capabilities and Japanese Foreign Direct Investment in the United States', *Review of Economics and Statistics*, **LXXIII** (3), August, 401–13.

Harvard Business School for excerpt: David J. Jeremy (1977), 'Damming the Flood: British Government Efforts to Check the Outflow of Technicians and Machinery, 1780–1843', *Business History Review*, **LI** (1), Spring, 1–34.

Haworth Press, Inc. for excerpt: Philippe Gugler and John H. Dunning (1993), 'Technology-Based Cross-Border Alliances', in R. Culpan (ed.), *Multinational Strategic Alliances*, 123–65.

Historical Association for article: J.R. Harris (1976), 'Skills, Coal and British Industry in the Eighteenth Century', *History*, **61** (202), June, 167–82.

Kent State University Press for article: Nathan Rosenberg (1972), 'Factors Affecting the Diffusion of Technology', *Explorations in Economic History*, **10** (1), Fall, 3–33.

Macmillan Ltd for excerpt: Sanjaya Lall (1985), *Multinationals, Technology and Exports: Selected Papers*, 114–30.

Macmillan Publishing Company for excerpt: Everett M. Rogers with F. Floyd Shoemaker (1971), *Communication of Innovations: A Cross-Cultural Approach*, 2nd edn., 18–40, 174–96.

MIT Press for excerpt: David J. Jeremy (1981), 'Conclusion: Some Perspectives on the Translantic Diffusion of Early Industrial Textile Technologies', in *Transatlantic Industrial Revolution: The Diffusion of Textile Technologies between Britain and America, 1790–1830s*, 252–62, 335–6.

MIT Press Journals for article: Raymond Vernon (1966), 'International Investment and International Trade in the Product Cycle', *Quarterly Journal of Economics*, **LXXX** (2), May, 190–207.

National Institute of Economic and Social Research for article: L. Nabseth (1974), 'Summary and Conclusions', in L. Nabseth and G.F. Ray (eds.), *The Diffusion of New Industrial Processes: An International Study*, Vol. 29 of *Economic and Social Studies*, 294–315.

Oxford University Press, Inc. for excerpt: Linsu Kim (1991), 'Pros and Cons of International Technology Transfer: A Developing Country's View', in Tamir Agmon and Mary Ann Von Glinow (eds.), *Technology Transfer in International Business*, 223–39.

K.G. Saur Verlag for excerpt: Reinhard Neebe (1993), 'Technology Transfer and Foreign Trade in the Early Years of the Federal Republik of Germany', in *German Yearbook on Business History 1989–92*, 133–55.

Sidgwick & Jackson for excerpt: Jay Tuck (1986), 'CoCom – The Toothless Watchdog: What are We Doing?', *High-Tech Espionage: How the KGB Smuggles NATO's Strategic Secrets to Moscow*, 180–206.

Geoffrey Tweedale for excerpt: Geoffrey Tweedale (1992), 'Aspects of the Anglo-American Transfer of Computer Technology: The Formative Years, 1930s–1960s', in David J. Jeremy (ed.), *The Transfer of International Technology: Europe, Japan and the USA in the Twentieth Century*, 90–117.

United Nations for excerpt: 'Transnational Corporations, Technology and Growth' (1992), in United Nations, Department of Economic and Social Development, Transnational Corporations and Management Division, *World Investment Report, 1992: Transnational Corporations as Engines of Growth*, 131–62.

University of Chicago Press for articles: Timo Myllyntaus (1991), 'The Transfer of Electrical Technology to Finland, 1870–1930', *Technology and Culture*, **32** (2/1), April, 293–317; Russell I. Fries (1975), 'British Response to the American System: The Case of the Small-Arms Industry after 1850', *Technology and Culture*, **16** (3), July, 377–403; John H. Jensen and Gerhard Rosegger (1978), 'Transferring Technology to a Peripheral Economy: The Case of Lower Danube Transport Development, 1856–1928', *Technology and Culture*, **19** (4), October, 675–702; Arnold Krammer (1981), 'Technology Transfer as War Booty: The U.S. Technical Oil Mission to Europe, 1945', *Technology and Culture*, **22** (1), January, 68–103.

Western Economic Association International for article: M.L. Burstein (1984), 'Diffusion of Knowledge-Based Products: Applications to Developing Economies', *Economic Inquiry*, **XXII**, October, 612–33.


Every effort has been made to trace all the copyright holders but if any have been inadvertently overlooked the publishers will be pleased to make the necessary arrangement at the first opportunity.

In addition the publishers wish to thank the Library of the London School of Economics and Political Science, the Photographic Unit of the University of London Library, and the Marshall Library, Cambridge University, for their assistance in obtaining these articles.

# Introduction

Sharing technology[1] is widely recognized as a principal means of relieving world poverty. It is especially crucial for developing countries. Creating the capacity to achieve economic growth will raise standards of living and help to reduce the proportion of the world's population (some 20% or 800 million people in the early 1980s) who die prematurely. All this was brought to the West's attention by the Brandt Commission on International Development Issues in 1980, with the publication of their paperback *North–South: A Programme for Survival* (London: Pan Books). Furthermore, the benefits of sharing technology between industrial leaders and followers do not flow in one direction only. The experience of Japan since the late nineteenth century demonstrates that. However, the hard fact that technology transfer does not occur evenly or smoothly across the international economy raises questions that have troubled and challenged both academics and policy makers, and will continue to do so as long as the global disparities between rich and poor remain. For example, can we discern any repeating, and thus potentially repeatable, patterns in the diffusion of technology? What circumstances have explained successful transfers? What has constituted appropriate technology? What part has the state played in promoting or denying transfers? How significant have entrepreneurs and business organizations been? This collection of essays aims to bring an historical perspective to the theory of technology transfer and the entrepreneurial roles of individuals, the state and the business organization in the transfer process.

## I  Models of Technology Transfer

Technology transfer is rarely, if ever, a disembodied technological matter. The movement of people or capital is a necessary, though not always sufficient, concomitant of the movement of technology. International trade has been an important matrix for technology transfer. Consequently, the economists' trade theories have provided important contributions to models (explanations) of technology transfer. David Ricardo's early nineteenth century theory of comparative advantage (demonstrating that economies benefit by specializing in activities at which they are relatively better than their neighbours) explained why all might profit in a free trade international economy. Hecksher and Ohlin in the 1930s and 1940s traced the sources of comparative advantage to relative proportions of the factors of production in differing countries.[2] However, none built technology transfer into their theories until Raymond Vernon did so in 1966.

In his seminal article (Chapter 1), Vernon related the international spread of innovations to international trade and a cyclical (and hence an historical) process. In an advanced economy (like Britain in the early nineteenth century or the United States in the mid-twentieth century) there were unique opportunities to market innovations meeting wants at high levels of income and produced by high labour costs – for example, luxury textiles in Britain or skyscraper elevators in the USA. At this first stage in the model labour costs are high initially because of the lack of standardization. As demand grows, so products and production methods become more standardized and labour costs fall. Knowledge of the new goods and methods spreads internationally and foreign demand emerges in advanced economies, met at first by exports from the pioneer economy.

When will a second stage arrive, when pioneer producers make direct foreign investments and thus transplant their new technology? Vernon, drawing on classical economics theory, suggests this will happen when the average costs of production abroad are lower than the marginal costs of production plus transport costs of exports from the pioneer manufacturer. In establishing foreign plants, knowledge and communication between originating entrepreneurs and executives and their foreign markets become crucial, because average costs of production abroad are initially unknown. However, lower labour costs abroad suggest that foreign production and the diffusion of technology to other advanced countries will be successful. Over time, methods of production are refined in the advanced economies receiving the new technology and eventually imports from the pioneer economy of high-income, labour-saving products have been replaced by similar exports to it from advanced rival imitators.

A third stage arrives when fully standardized products and production are transferred to less developed countries, again in pursuit of low labour costs, import substitution and exports to capture world markets. In this manner, the modern cotton textile manufacture spread from the West to the Far East in the late nineteenth century.

While Vernon's work explains very broadly and in the long run how international trade may promote technology transfer, his model cannot suffice. Nothing is said about the adaptation of new technology to suit foreign conditions, economic or social. Neither does the rational economic model adequately explain why technology transfer has failed or been delayed. Nor does it throw much light on the actual process of diffusion. For these matters, further understanding has come from empirical work by economists, sociologists and historians.

In a number of works Nathan Rosenberg, an economist, has probed interactions between economic growth and technical change. His focus on technology diffusion is exemplified in Chapter 2. At its heart is the debate about the importance of technological change to economic development. Economists and economic historians had played down the role of technology, seeing it not just as part of the 'residual' (that part of economic growth unexplained by economic causes), but as less significant than institutional factors – here, like privateering in explaining the retarded adoption of late sixteenth century Dutch fast ship design. Rosenberg raised the possibility that technical change could be much slower and more far-reaching than economists allowed. This turned attention to the process of technology diffusion where Rosenberg has emphasized the gradual, cumulative aspects of inventive activity (in contrast to Schumpeter's leaps forward), the importance of machine-making skills and 'complementarities' (technical constraints on the primary innovation).

The possibility of non-economic factors coming into play in the process of technology transfer brought fresh importance to the sociologists' much older research on the diffusion of innovations. This was summarized by Everett Rogers and Floyd Shoemaker, who found numerous reports for and against many of the generalizations about the transfer of innovations being tested by anthropologists, agricultural economists, educationalists, geographers and many other specialists. However, some propositions met overwhelming support, and from these Rogers and Shoemaker produced a typology of the diffusion process (Chapter 3). Summarized as S-M-C-R-E (source, message, channel, receiver, effects), this model emphasized the role of channels of communication, change agents and the S-curve of adoption over time. How did this fit with the economists' models? Kenneth Arrow, another economist, highlighted the conflict in his influential article on the production and transmission of technological knowledge (Chapter 4). In essence, he argued, 'the economists are studying the demand for information by potential innovators and sociologists the problems in the supply of communication channels.' (Chapter 4, p. 102). But was it possible to unravel the two that easily? What if, for instance, social systems (like the patent system) helped to alter demand, or differing standards of living (shaped by both economic and non-economic factors) contributed to the international migration flows carrying skilled technical migrants and bearers of new technology?

The complexities of the technology transfer process drove investigators in the 1970s and 1980s back to empirical studies in the search for models. Four very different approaches can be found in the remaining articles in this section (Chapters 5–8). Nabseth reported a case study investigation of the diffusion between firms in different countries of new industrial processes introduced between 1945 and the early 1970s (Chapter 5). In explaining the speed of adoption, profitability and expected relative advantage stand out, as might be expected. The United Kingdom's widely-observed record of a slow rate of adoption was attributed in the steel industry (where it was compared to the United States and Sweden) to uncertainty created by successive nationalization and denationalization programmes, and to the survival of small and medium-sized firms that did not maximize profits (in the sense of equating marginal cost to marginal revenue). Attitudes of management to new technology were also crucial to diffusion between firms.

Another approach was that of the historian. Though inhibited by the imperfections of the evidence, historians have the freedom to be eclectic, to bring together data from all directions, whether economic, sociological or technical. Reflecting concerns about the spread of early industrialization out from Britain, several scholars like John Harris (see Chapter 9) and David Jeremy have written a series of detailed case studies. Harris has examined the flows of technology from Britain to France over a range of industries. Jeremy's work (Chapter 6) has been confined to the textile industries and the movement of technologies between Britain and the USA before the 1840s. He found that comparatively few individual artisans, rather than the mass of migrants, were the key channels of transatlantic transfer; that local capitalists in the receptor economy were needed to establish the new technology in industrial settings; and that the imported technologies would be modified by relative factor (land, capital and labour) proportions and prices on one hand, and social systems (like the employment and status of New England mill girls) on the other.

The early nineteenth century conditions in which technology transfer occurred differed vastly from those of the late twentieth century. In the interim, human knowledge has grown in quantity many times over; knowledge of technological matters has spread from the largely empirically-based mechanical and (elementary) chemical kinds to the scientifically-based nuclear, electronic and genetic sorts. Communications, which in the eighteenth century required weeks to take knowledge between continents, now permit immediate person-to-person contact with anyone else on the globe linked by the international telephone network. Advanced business organizations, then mostly confined to the family partnership operating a single plant and producing a single good or service, are now multinational constellations of hundreds and even thousands of interconnected groups encircling the globe. Given these and other prodigious changes in technical and social systems, what new models have emerged to explain technology transfer in the late twentieth century?

The extent of the reported findings about modern transfers of technology in the context of business may be glimpsed in the sources to the articles reprinted in Chapters 8 and 23, much of it written by economists. One of the most prolific of the economists has been John H. Dunning. He has concentrated on the activities of multinational companies and registered their evolution into transnational corporations (TNCs) which 'operate through a global network of related enterprises.' (Chapter 8, p. 173). To explain why these TNCs exchange technology with one another he calls on games theories, transaction costs economics and organizational decision theory, and produces his own OLI paradigm (linking a firm's ownership to the country's location and the firm's hierarchical control to explain why technology sharing, or any other form of direct foreign investment would be chosen) (Chapter 8).

In this section on models of technology diffusion, M.L. Burstein's article (Chapter 7) has been included for several reasons. It confronts the difficulty, neglected by business historians and historians of technology, of the decay of profitability as knowledge is diffused. It deals with a very modern problem associated with the spread of computer hardware and software, and it typifies the sort of techniques used by quantitative economists studying technology transfer. In essence, Burstein argues that modern innovations need to be 'bundled' or sold in a cluster to avoid the loss of markets and profits to free-riding rivals.

## II  Agencies of Technology Transfer

While the remaining articles in this collection do deal with explanations of various aspects of technology transfer (e.g. speed of transfer, degree of failure to transfer, etc.), they have been chosen primarily to show how transfers have been achieved. The focus is on the carriers and frameworks. To illustrate the persistence of patterns over time and place and industry, the articles range in subject from the late eighteenth to the late twentieth century, from Europe to America to the Far East, and from textiles to motor vehicles and computers.

### A  Artisans and Professionals

During the period of the classical industrial revolution (say, 1760–1830), the flows of technology from leader (Britain) to follower nations presented many patterns which have been repeated since. The clearest has been that intangibles matter as much as, sometimes more than, tangibles. Harris's article (Chapter 9) demonstrates that non-verbal craft skills were essential to coal-fuel industrial technologies. Without these craft skills, embodied in a range of (initially, British) workmen and managers, rival French and Swedish industrialists either had great difficulty in acquiring, or could not acquire, new iron-working techniques. The same story could be told about other industries.[3] Jeremy's essay (Chapter 10) on British textile machinists in one sub-region of the USA during the early national period shows also that while key workmen might be few in number, they could soon transmit their skills to indigenous craftsmen sharing the same language, cultural values and social systems, including the Western empirical, mechanical engineering paradigm (in Thomas Kuhn's sense, of an intellectual framework within which the 'normal science' of the day operates[4]). In these articles, while Harris has looked at operators and technicians, Jeremy has closely examined machine makers. Not all conveyors of new technology lived up to their claims. Where they did, they brought best-practice designs and, as numbers of 'new-knowledge' artisans grew, competition led to falling prices.

It might be assumed that the arrival of a new paradigm, the shift from an empirical to a scientific basis for technological knowledge, might have seen the locus of new technology move from the technician's brain-eye-hand to objective forms, like formulae, engineering drawings, and the like. Kristine Bruland's research (Chapter 11) finds that the Norwegian mechanical engineering industry in the nineteenth century heavily relied on its own engineers travelling abroad, either for training or for industrial site visits, and on foreign contacts to learn the latest techniques. In numbers, itinerant Norwegian engineers were much more significant than immigrant engineers from Britain, Germany and elsewhere. Timo Myllyntaus (Chapter 12) discovered something similar in Finland. The early start and development of electrification there was due to a number of factors, one of which was a positive attitude on the part of entrepreneurs and society towards the securing and adoption of innovations. In this the foreign training or experience of electrical engineers became a prime instrument. Nor has the role of the individual as technology carrier diminished in the mid- and late twentieth century. Geoff Tweedale's account of the Anglo-American transfer of computer technology (Chapter 13) illustrates that while profit and war might be the prime motivators for exchanging or purloining advances in computer knowledge, international movements and meetings between the few at the forefront were important means of transatlantic transfer. A growing literature on the production and diffusion of high technology in the micro-electronic industry has stressed the role of individuals operating in flexible business structures, whether in Silicon Valley, California or along Route 128 around Boston.[5]

While the articles reprinted here have probed the nature of intangible skills and attitudes in varying degrees of detail, two points seem clear. First, the cluster of intangibles that are of primary importance belong to technicians and engineers. They are the machine builders, the system builders. Knowledge of foreign innovations first arrives with them, whether they are immigrant or indigenous engineers. From their workshops and offices the new best practices, adapted to suit local conditions and needs, move outwards into the host economy.[6] Second, in trying to conceptualize the process of transfer, it is possible to view the creation of a reservoir of machine- and system-building skills as a preliminary stage in the successful diffusion of a new industrial technology, or a new technological paradigm. This fits on one hand with Vernon's cyclical theory, and on the other with the observed stages noted by Jeremy for textile technologies crossing to the USA and by Nakaoka, Yuzawa and Uchida for the spread of textile, railway and electrical technologies to Japan in the late nineteenth century.[7] Once the reservoir of skills is in place then, within the same engineering paradigm, it will be possible to import 'normal technological models' from more advanced countries and to transfer the cumulative advances they embody simply by the techniques of reverse engineering. This is exactly what the South Koreans have done across a wide range of industries (see Chapter 20).

The process of technology transfer via individual emigrant artisans and professionals has not been, and cannot be, as predictable or planned as transfer through organizations. To some extent, migration waves responded to detectable economic cycles and social conditions causing 'pushes' in the country of origin and 'pulls' in the country of destination. However, knowledge about distant parts was often very incomplete and uncertain. On the other hand, European emigrants have tended to be young male adults, newly trained, open to new ideas and with long working lives ahead.[8] Whether considering the hundreds who took new technology to the USA early in the nineteenth century (see Chapters 6 and 10), the 8,500 Westerners temporarily hired to help modernize Meiji Japan,[9] or the scientists and engineers who went from Europe to the USA in the 'brain drain' of the 1950s and 1960s,[10] the processes of personal adjustment to a new society would inevitably release, extinguish or reduce the individual's potential as a technology carrier.[11] The literature on transatlantic migrations also shows that more than a quarter of all emigrants eventually returned home.

## B  The Role of the State *

Since the 1960s the debate about the sources of and routes towards industrialization has produced much discussion on the role of the state in promoting economic growth. Alexander Gerschenkron argued that the state in a backward economy might take the place of the entrepreneur. Not only would the state provide large amounts of capital to buy new technology from abroad and then establish infant industry plants, it would also choose appropriate technologies, usually of a large-scale and capital-intensive kind to compensate for shortages in the quantity and quality of its own skilled labour force. Heavy or capital goods industries would be favoured.[12]

Examples of state interventions in backward, 'follower' countries came from Russia, Germany and Japan. In the debate, Gerschenkron was responding to W.W. Rostow who had proposed a linear, rather than a multi-path, stage model. The critical stage in Rostow's model (which was heavily based on the British case) was 'take-off'. Prominent among developments at this phase were a release of resources from agriculture, an accumulation of social overhead capital (transport systems, especially), the emergence of financial institutions and dynamic entrepreneurs. This constellation of conditions would move a traditional, subsistence economy into the accelerating capital accumulation that distinguished the developmental, industrializing state.[13] With these two models in mind we might expect the state to be indifferent towards the transfer of technology when its own economy was far ahead of international competition, as in Britain before 1860, in the USA between the early and mid-twentieth century, and in Japan since the 1980s. Everywhere else among rapidly developing nations the state should have been deeply engaged in securing new, capital goods technology to assist in bringing it level with international competitors.

Nothing so simple transpired. In Britain during the classic period of her industrial revolution manufacturing interests pressured the state into taking drastic legal measures against the outflow of technicians and machinery, as David Jeremy details (Chapter 14). Only the practical difficulties of sealing off the economy and islands of Britain from cultural neighbours and economic rivals frustrated these governmental measures. Nor did America's technological leadership up to the late twentieth century lead to governmental complacency about technology transfer. The scale, complexity and destructive power of military technology, combined with global conflicts between democratic and totalitarian (Right and Left) political systems, ensured that. Arnold Krammer shows (Chapter 17) that one of the rewards of military victory energetically sought by the United States government in 1944–45 in the immediate wake of its advancing tanks and troops was the capture of new technology developed by Nazi Germany.

Here, too, is a case of the spread of new technical knowledge by the military machine, and clearly outside market mechanisms.

Governmental protection of new technology developed in the USA since 1945 has had something of a mixed record if Jay Tuck's journalistic account (Chapter 19) is a reliable adumbration of work yet to be done by the historians. One side of Reinhard Neebe's article (Chapter 18) documents how in post-war Germany, the United States government for a decade suppressed the technological capacity of West Germany. As for Japan, exports of new technology entail 'guidance' from the Ministry of International Trade and Industry (MITI) and, if investment grants and technical assistance are needed (as with technology transfers to China before the late 1980s), agencies of the Ministry of Finance will be involved.[14]

The mixed record of the state in securing technology transfer for follower economies is illustrated in the three articles by Russell Fries, John Jensen, and Gerhard Rosegger and Linsu Kim. Fries (Chapter 15) pinpoints a reverse flow situation. During the half century before the Civil War the United States developed the so-called 'American System' of manufacture: a cluster of specialized machine tools and measuring techniques which permitted a hitherto-unknown degree of interchangeability (and productivity) in the production of consumer durable goods, from rifles and clocks to sewing machines, bicycles and typewriters.[15] A new paradigm in production technology had arrived. First faced with the American System at the Great Exhibition of 1851, and the prospects of becoming a technological follower, the British government might be expected to take steps to catch up. Because the new system had particular applicability to military small arms manufacture, where it had first appeared, there was a double motive for securing its transfer and adoption. Fries shows that while the British government did establish a replica of an American System small arms factory and refused to place military orders with private contractors unless they supplied interchangeable weapons, the government made little or no impact on the technology used to meet civilian demand for sporting rifles or African muskets. Peculiar and powerful market forces delayed technology transfer.

Jensen and Rosseger's article on transport in the lower Danube basin (Chapter 16) stresses the importance both of an international commission and of nationalism in effecting technology transfer (with foreign engineers implementing the international, national and private wills to achieve economic development). In the very recent case of South Korea, Kim (Chapter 20) conclusively demonstrates the crucial role of government policy in securing direct foreign investment and technology transfer from abroad. His essay also represents a management specialist's analysis of technology transfer.

To explain all the observed dimensions of technology transfer and interactions with business we have to consider the nature of the state. Chalmers Johnson has noted 'that the innumerable things a state does can be arranged in rough rank order according to its priorities, and that a state's first priority will define its essence.'[16] A state may be committed to any number of first priorities: defence, revolution, religion, welfare, regulation, development. And, to cite Johnson again, the state's first priority may change over time. Japan has been a remarkably consistent development state, since the Meiji Restoration of 1867–68 at least. Modernization was seen as the way to achieve economic and military parity with the Western powers. Technology transfer therefore became a process of the highest importance. Of the 8500 foreigners employed in Japan in the Meiji period (1867–1912), 800 were directly employed by the national government (see Yasumuro, Note 9). The role of the Ministry of International

Trade and Industry closely orchestrating economic growth in Japan, since 1954 especially, further underlines Japan's goal of being a development state. In contrast stood the ideological state whose persecution of dissenters induced minority emigrations. The consequences for technology transfer, as with the French Huguenots of the sixteenth century or the German Jews of the 1930s, could backfire badly on ideological regimes.[17]

## C   Private Business and Multinational Enterprise

How long have private firms engaged in the international transfer of technology? Which businesses have the greatest vested interest in diffusing innovations? How have private businesses facilitated or obstructed technology transfer? What problems have they encountered?

Firms engaged in industrial espionage, labour piracy and the acquisition of technical knowledge as far back as there have been groups of individuals in economic competition. The research of Eric Robinson, A.E. Musson, John Harris, David Jeremy and others has revealed a lively exchange of men and machines between Britain and Europe and the United States (or its predecessor colonies) in the eighteenth century.[18] Individuals might carry new technology between continents in a seemingly haphazard fashion, carried by tides of migration, but the implantation of new technology in a host country needed more than the presence of mechanics or engineers. The entrepreneur and the firm he (usually he) organized were almost always the essential concomitants for the establishment of new methods of production or the production of new products.

The first article in this section, by Christine MacLeod (Chapter 21), considers the diffusion of innovations between firms in nineteenth century Britain, and questions the emphasis placed on machine makers by Rosenberg. While underscoring the importance of engineering firms, first explored by Rosenberg (Chapter 2) and supported by Bruland (Chapter 11), she tests the hypothesis that machine users would be more likely than machine builders to restrict the diffusion of their innovations. From a range of cases she argues that user-inventors could achieve the best diffusion rates by means of demonstration effects and by licensing other users and thus stimulating competition. Builder-inventors, the machine makers, facilitated diffusion by eroding secretiveness and creating channels of information; by making cumulative improvements (as emphasized by Rosenberg); and by spreading the use of machine tools which though slow to achieve interchangeability in Britain, did raise standards of accuracy. It might be added that machine builders could and did slow down diffusion by refusing to adopt interchangeable screw threads. Variant screw threads ensured that repair work came back to them.[19]

The role of multinational firms in technology transfer was first tackled among business historians by Mira Wilkins (Chapter 22). She identified four company-level agencies of transfer: the export of goods; the registration of patents; the despatch of technical knowledge; and direct foreign investment. Based on her pioneering work on multinational enterprises (MNEs), her analysis still holds good. Fresh weight to the importance of technology in multinational business has been given by Michael Porter of Harvard Business School. In seeking explanations for growth differentials between nations and industries he isolates factor conditions, demand conditions, related industries, and firm strategy, structure and rivalry.[20] But it is in their creation and application of new technology that multinational firms can leave rivals behind, or themselves be left behind. 'Of all the things that can change the rules of competition, technological change is the most prominent.'[21]

In at least two respects, however, the scene has changed over the past 20 years. First, in many instances, the MNEs of the 1970s have become the transnational corporations (TNCs) of the 1990s. This has posed new levels of competition and organizational complexity. Global strategies have to be pursued. Strategic alliances between TNCs are formed to reduce high risks and the rising costs of R&D and rapid obsolescence. The national allegiance on which much of Porter's theory rests may be giving way among corporate strategy-makers to a supra-national allegiance to the firm, the TNC. Certainly negotiations between the state (national, federal or, as with the European Economic Community, quasi-federal) and TNCs increasingly exercise the minds of bureaucrats, state and corporate. As Dunning and others have shown, the new technologies of the 1980s lay in the direction of information technology (computers, robotics and lasers) and biotechnology (food processing, chemicals, pharmaceuticals and genetic engineering) and led to the formation of TNC hierarchical galaxies, 'with a pivotal group of firms (which control key technologies), being surrounded by satellite suppliers and customers.'[22] Flexible manufacturing systems, down to domesticated robots, have become best practice.

Second, while technology transfer may be largely in the hands of TNCs, it now occurs within two distinct relationships in the international economy. There is transfer between *developed* country A and *developed* country B. There is also transfer between *developed* country A and *developing* country D. Two sorts of concern have therefore arisen about technology transfer. There are questions about the spread of technology outwards from Japan, which in per capita GNP has edged ahead of the USA. Does it lead to deindustrialization? Do Japanese plants in the USA or the UK lose their design and related manufacturing capacities, and simply become 'screwdriver plants'? Is employment shifted abroad? Or, will Japanese management techniques revive flagging Western industrial systems? Is it in fact possible to transfer Japanese technology, given the intangibles? There are also questions about the transfer of technology to the developing world. While some Asian countries like South Korea are hard on the heels of Japan, what can be done to transfer technology to parts of Africa or the India subcontinent which seem interminably gripped in poverty? What is the appropriate technology that could be introduced? What risks and incentives face TNCs in slowly developing countries? These and other questions relating to economic development, in which technology transfer *per se*, is bound up, come to mind. Most are beyond the scope of this essay.

Clues to answers to some of these questions are found in the United Nations report of 1992 (Chapter 23). Several emphases must be highlighted here. First, there is the awareness that technology consists of hardware (capital equipment), software (like specifications and computer programs), and services (the human skills required in engineering, management and marketing). Second, that innovation now comes largely from corporations themselves, rather than academic institutions and research laboratories. Third, while much innovation has originated in a TNC's home country, foreign subsidiaries and affiliates have been responsible for modifying it to suit local circumstances. Fourth (and this echoes Mira Wilkins), technology acquisition has come through publications, trade, foreign direct investment (through wholly-owned foreign affiliates and joint ventures), and through non-equity links (like patents, licences, contractual agreements and strategic alliances).

The means and impact of transferring technology between advanced industrial nations in the late twentieth century have been investigated by economists in a large literature, of which the article by Bruce Kogut and Sea Jin Chang (Chapter 24) is representative. They show that Japanese direct investment, like that from other foreign countries, was targeted on industries intensive in R&D capability and high growth potential. Tetsuo Abo's article (Chapter 25) is particularly interesting in that it highlights the culture-embodied elements in Japan's process technologies, and illustrates one of the barriers to technology transfer identified by Wilkins.

The United Nations report (Chapter 23) might have more clearly distinguished transfers of technology to developing countries, the South of the Brandt Report, from transfers between developed economies. In the mid-1980s, Dunning reckoned that no more than a quarter of MNEs' foreign direct investment went to developing countries.[23] Not surprisingly, imports of capital equipment are registered as a prime determinant of the productive capacities of developing countries. In the 1980s, in absolute terms, Africa has fallen back and Asia and the Pacific leapt ahead. Sanjaya Lall (Chapter 26) makes the important distinction between know-how and know-why (basic research) technical knowledge. He then argues that multi-national corporations operating in India have spread know-how to local firms, via the MNCs' Indian affiliates, through linkages and competition; but have been much less willing to part with know-why capacity. From the literature it would seem that the call, made in the 1970s by E.F. Schumacher, for a 'greener' intermediate technology (more labour-intensive, small-scale, flexible and less costly) for slowly developing countries, has been unheeded.[24]

Little has been said about barriers to technology transfer carried out by firms, whether small or global. Mira Wilkins (Chapter 22) identifies nine kinds of barrier. Summarized they amount to economic, technological and cultural barriers. All three kinds have to be overcome for transfer to be effective.

## Conclusion

The transfer of technology can take place outside the commercial setting of capitalist enterprise, for example, between primitive cultures, between military organizations, or between non-profit agencies. Approximate patterns which it follows in some of those situations have been extensively traced by sociologists. The international diffusion of technology between competitive economies is a much more complex matter. It is simultaneously a cyclical, a spatial and a multi-dimensional phenomenon. It is *cyclical* in several senses. In the long term it is contingent on the comparative advantages between nations and the competitive advantages developed by firms across them. The models of Vernon and Porter have explanatory power at this first level. There are other dynamics feeding into the cyclical, however. One is the changing scientific and technological paradigm which has shifted 'normal science' and 'new technology' from eighteenth century mechanical engineering empirically discovered to modern science-based information and bio-technology. Another movement within the large international economic cycle is a rough stages-of-diffusion process, starting with the artisans and pro-fessionals who take new knowledge, then find backers in the host economy, and finally see modifications made and transmitted back to the originating economy. International transfer of technology is also *spatial*. There is a physical movement of skilled people, hardware, software and services between countries and organizations. Differences in geology, climate, time, language, organization, political system, culture, have to be taken into account. Technology transfer increasingly happens because it is engineered by business organizations, but the international scale and the capitalist setting mean that it is also a *multi-dimensional* phenomenon. Economic and ideological contexts provide motivation for transfer. Cultural systems (government and politics, law, education, language, underlying values, for example) facilitate the movement of new technical knowledge. Technological knowledge, generated for the most part within firms, but also in academic settings, imposes its own limitations on ease of transfer. Paradoxically, while the complete process of successful technology transfer requires a set of layered and interdependent conditions, like the fibrous strands in a rope, it is still possible (as cases of military and industrial espionage testify) for a single individual to take the crucial steps that will lead to effective technology transfer.

The overall lesson from the articles reprinted here is that technology transfer almost always occurs because of economic motives, but economic models do not fully explain the process. Economic conditions may largely explain long-run trends, but the interventions of population migration, the state and the business organization (increasingly the last), combined with cultural differentials and changes in the technology itself, are necessary inputs in any calculations about transferring technology.

## Notes

1. In this essay I am using the terms 'sharing technology', 'technology transfer' and 'technology diffusion' interchangeably to denote the complex process under discussion. While transfer can occur between countries, regions, firms or individuals, I am almost always considering transfer between countries.
2. James Foreman-Peck, *A History of the World Economy: International Economic Relations since 1850*, Brighton: Wheatsheaf Books, 1983, pp. 37–45.
3. For example, a disassembled spinning mule sent to Philadelphia in the 1780s left local mechanics uncomprehending and after four years they sent it back to Britain. See David J. Jeremy, 'British Textile Technology Transmission to the United States: The Philadelphia Region Experience, 1770–1820', *Business History Review*, **47** (1973).
4. Thomas Kuhn, *The Structure of Scientific Revolutions*, Chicago: Chicago University Press, 1970. This is elaborated into a model of culture change in the context of early industrialization by Anthony F.C. Wallace in his *Rockdale: The Growth of an American Village in the Early Industrial Revolution*, New York: Alfred Knopf, 1978.
5. See, for example, Edward B. Roberts, *Entrepreneurs in High Technology: Lessons from MIT and Beyond*, New York: Oxford University Press, 1991.
6. The importance of machine shops in diffusing new techniques was first registered by Rosenberg (see below).
7. See Tetsuro Nakaoka, 'The Transfer of Cotton Manufacturing Technology from Britain to Japan'; Takeshi Yuzawa, 'The Transfer of Railway Technologies from Britain to Japan, with Special Reference to Locomotive Manufacture'; and Hoshimi Uchida, 'The Transfer of Electrical Technologies from the United States and Europe to Japan, 1869–1914', all in David J. Jeremy (ed.), *International Technology Transfer: Europe, Japan and the USA, 1700–1914*, Aldershot: Edward Elgar, 1991.
8. Dudley Baines, *Emigration from Europe, 1815–1930*, London: Macmillan, 1991.
9. Ken'ichi Yasumuro, 'Engineers as Functional Alternatives to Entrepreneurs in Japanese Industrialisation', in Jonathan Brown and Mary B. Rose (eds.), *Entrepreneurship, Networks and Modern Business*, Manchester: Manchester University Press, 1993.
10. Neebe (Chapter 18); George Louis Payne, *Britain's Scientific and Technological Manpower*, Stanford University Press, Stanford, CA: 1960.
11. See Charlotte Erickson, *Invisible Immigrants: The Adaptation of English and Scottish Immigrants*

*in Nineteenth Century America*, Coral Gables, Florida: University of Miami Press, 1972.

12. Alexander Gerschenkron, *Economic Backwardness in Historical Perspective*, Cambridge, MA: Harvard University Press, 1962.

13. W.W. Rostow, *The Stages of Economic Growth*, Cambridge: Cambridge University Press, 1960. Models of economic growth are usefully compared in Clive Trebilcock, *The Industrialisation of the Continental Powers, 1780–1914*, London: Longman, 1981.

14. See Martha Caldwell Harris, 'Technology Transfer and Sino-Japanese Relations', in Tamir Agmon and Mary Ann Von Glinow (eds.), *Technology Transfer in International Business*, New York: Oxford University Press, 1991.

15. See Nathan Rosenberg (ed.), *The American System of Manufactures*, Edinburgh: Edinburgh University Press, 1969; Merritt Roe Smith, *Harpers Ferry Armory and the New Technology: The Challenge of Change*, Ithaca, NY: Cornell University Press, 1977; David A. Hounshell, *From the American System to Mass Production, 1800–1932: The Development of Manufacturing Technology in the United States*, Baltimore: Johns Hopkins University Press, 1984; Donald R. Hoke, *Ingenious Yankees; The Rise of the American System of Manufactures in the Private Sector*, New York: Columbia University Press, 1990; Carolyn C. Cooper, *Shaping Invention: Thomas Blanchard's Machinery and Patent Management in Nineteenth-Century America*, New York: Columbia University Press, 1991.

16. Chalmers Johnson, *MITI and the Japanese Miracle: The Growth of Industrial Policy, 1925–1975*, Stanford, CA: Stanford University Press, 1982, p. 305.

17. See Warren C. Scoville, 'The Huguenots and the Diffusion of Technology', *Journal of Political Economy*, **60**, 1952.

18. See, for example, A.E. Musson and Eric Robinson, *Science and Technology in the Industrial Revolution*, Manchester: Manchester University Press, 1969; Barrie M. Ratcliffe (ed.), *Great Britain and Her World, 1750–1914*, Manchester: Manchester University Press, 1975; David J. Jeremy, *Transatlantic Industrial Revolution*, Cambridge, MA: MIT Press, 1981.

19. A standard was proposed in Britain by Joseph Whitworth in 1841 and was adopted by most large firms by 1860, but the age of steel then demanded a new more finely-pitched thread. In the USA four alternatives were still available in the late nineteenth century. See L.T.C. Rolt, *Tools for the Job: A History of Machine Tools to 1950*, London: HMSO, 1986; Bruce Sinclair, 'At the Turn of the Screw: William Sellers, the Franklin Institute, and a Standard American Thread', *Technology and Culture*, **10**, 1969.

20. Michael E. Porter, *The Competitive Advantage of Nations*, London: Macmillan, 1990, p. 71.

21. Michael E. Porter, *Competitive Advantage: Creating and Sustaining Superior Performance*, New York: Free Press, 1985, p. 164.

22. See, for example, John H. Dunning, 'Multinational Enterprises and Industrial Restructuring in the UK', *Lloyds Bank Review*, **158**, October 1985.

23. Dunning *ibid*., 1985, p. 7.

24. E.F. Schumacher, *Small Is Beautiful: A Study of Economics as if People Mattered*, London: Abacus, 1973.

# Part I
# Theory and Process

# [1]

## INTERNATIONAL INVESTMENT AND
## INTERNATIONAL TRADE
## IN THE
## PRODUCT CYCLE *

RAYMOND VERNON

Anyone who has sought to understand the shifts in international trade and international investment over the past twenty years has chafed from time to time under an acute sense of the inadequacy of the available analytical tools. While the comparative cost concept and other basic concepts have rarely failed to provide some help, they have usually carried the analyst only a very little way toward adequate understanding. For the most part, it has been necessary to formulate new concepts in order to explore issues such as the strengths and limitations of import substitution in the development process, the implications of common market arrangements for trade and investment, the underlying reasons for the Leontief paradox, and other critical issues of the day.

As theorists have groped for some more efficient tools, there has been a flowering in international trade and capital theory. But the very proliferation of theory has increased the urgency of the search for unifying concepts. It is doubtful that we shall find many propositions that can match the simplicity, power, and universality of application of the theory of comparative advantage and the international equilibrating mechanism; but unless the search for better tools goes on, the usefulness of economic theory for the solution of problems in international trade and capital movements will probably decline.

The present paper deals with one promising line of generalization and synthesis which seems to me to have been somewhat neglected by the main stream of trade theory. It puts less emphasis upon comparative cost doctrine and more upon the timing of innovation, the effects of scale economies, and the roles of ignorance and uncertainty in influencing trade patterns. It is an approach

* The preparation of this article was financed in part by a grant from the Ford Foundation to the Harvard Business School to support a study of the implications of United States foreign direct investment. This paper is a by-product of the hypothesis-building stage of the study.

## INVESTMENT AND TRADE                                                      191

with respectable sponsorship, deriving bits and pieces of its inspiration from the writings of such persons as Williams, Kindleberger, MacDougall, Hoffmeyer, and Burenstam-Linder.[1]

Emphases of this sort seem first to have appeared when economists were searching for an explanation of what looked like a persistent, structural shortage of dollars in the world. When the shortage proved ephemeral in the late 1950's, many of the ideas which the shortage had stimulated were tossed overboard as prima facie wrong.[2] Nevertheless, one cannot be exposed to the main currents of international trade for very long without feeling that any theory which neglected the roles of innovation, scale, ignorance and uncertainty would be incomplete.

### LOCATION OF NEW PRODUCTS

We begin with the assumption that the enterprises in any one of the advanced countries of the world are not distinguishably different from those in any other advanced country, in terms of their access to scientific knowledge and their capacity to comprehend scientific principles.[3] All of them, we may safely assume, can secure access to the knowledge that exists in the physical, chemical and biological sciences. These sciences at times may be difficult, but they are rarely occult.

It is a mistake to assume, however, that equal access to scientific principles in all the advanced countries means equal probability of the application of these principles in the generation of new products. There is ordinarily a large gap between the knowledge of a scientific principle and the embodiment of the principle in

1. J. H. Williams, "The Theory of International Trade Reconsidered," reprinted as Chap. 2 in his *Postwar Monetary Plans and Other Essays* (Oxford: Basil Blackwell, 1947); C. P. Kindleberger, *The Dollar Shortage* (New York: Wiley, 1950); Erik Hoffmeyer, *Dollar Shortage* (Amsterdam: North-Holland, 1958); Sir Donald MacDougall, *The World Dollar Problem* (London: Macmillan, 1957); Staffan Burenstam-Linder, *An Essay on Trade and Transformation* (Uppsala: Almqvist & Wicksells, 1961).

2. The best summary of the state of trade theory that has come to my attention in recent years is J. Bhagwati, "The Pure Theory of International Trade," *Economic Journal*, LXXIV (Mar. 1964), 1–84. Bhagwati refers obliquely to some of the theories which concern us here; but they receive much less attention than I think they deserve.

3. Some of the account that follows will be found in greatly truncated form in my "The Trade Expansion Act in Perspective," in *Emerging Concepts in Marketing*, Proceedings of the American Marketing Association, December 1962, pp. 384–89. The elaboration here owes a good deal to the perceptive work of Se'ev Hirsch, summarized in his unpublished doctoral thesis, "Location of Industry and International Competitiveness," Harvard Business School, 1965.

a marketable product. An entrepreneur usually has to intervene to accept the risks involved in testing whether the gap can be bridged.

If all entrepreneurs, wherever located, could be presumed to be equally conscious of and equally responsive to all entrepreneurial opportunities, wherever they arose, the classical view of the dominant role of price in resource allocation might be highly relevant. There is good reason to believe, however, that the entrepreneur's consciousness of and responsiveness to opportunity are a function of ease of communication; and further, that ease of communication is a function of geographical proximity.[4] Accordingly, we abandon the powerful simplifying notion that knowledge is a universal free good, and introduce it as an independent variable in the decision to trade or to invest.

The fact that the search for knowledge is an inseparable part of the decision-making process and that relative ease of access to knowledge can profoundly affect the outcome are now reasonably well established through empirical research.[5] One implication of that fact is that producers in any market are more likely to be aware of the possibility of introducing new products in that market than producers located elsewhere would be.

The United States market offers certain unique kinds of opportunities to those who are in a position to be aware of them.

First, the United States market consists of consumers with an average income which is higher (except for a few anomalies like Kuwait) than that in any other national market — twice as high as that of Western Europe, for instance. Wherever there was a chance to offer a new product responsive to wants at high levels of income, this chance would presumably first be apparent to someone in a position to observe the United States market.

Second, the United States market is characterized by high unit labor costs and relatively unrationed capital compared with practically all other markets. This is a fact which conditions the demand for both consumer goods and industrial products. In the case of consumer goods, for instance, the high cost of laundresses contributes to the origins of the drip-dry shirt and the home washing machine. In the case of industrial goods, high labor cost leads to the early

---

4. Note C. P. Kindleberger's reference to the "horizon" of the decision-maker, and the view that he can only be rational within that horizon; see his *Foreign Trade and The National Economy* (New Haven: Yale University Press, 1962), p. 15 *passim.*

5. See, for instance, Richard M. Cyert and James G. March, *A Behavioral Theory of the Firm* (Englewood Cliffs, N.J.: Prentice-Hall, 1963), esp. Chap. 6; and Yair Aharoni, *The Foreign Investment Decision Process,* to be published by the Division of Research of the Harvard Business School, 1966.

development and use of the conveyor belt, the fork-lift truck and the automatic control system. It seems to follow that wherever there was a chance successfully to sell a new product responsive to the need to conserve labor, this chance would be apparent first to those in a position to observe the United States market.

Assume, then, that entrepreneurs in the United States are first aware of opportunities to satisfy new wants associated with high income levels or high unit labor costs. Assume further that the evidence of an unfilled need and the hope of some kind of monopoly windfall for the early starter both are sufficiently strong to justify the initial investment that is usually involved in converting an abstract idea into a marketable product. Here we have a reason for expecting a consistently higher rate of expenditure on product development to be undertaken by United States producers than by producers in other countries, at least in lines which promise to substitute capital for labor or which promise to satisfy high-income wants. Therefore, if United States firms spend more than their foreign counterparts on new product development (often misleadingly labeled "research"), this may be due not to some obscure sociological drive for innovation but to more effective communication between the potential market and the potential supplier of the market. This sort of explanation is consistent with the pioneer appearance in the United States (conflicting claims of the Soviet Union notwithstanding) of the sewing machine, the typewriter, the tractor, etc.

At this point in the exposition, it is important once more to emphasize that the discussion so far relates only to innovation in certain kinds of products, namely to those associated with high income and those which substitute capital for labor. Our hypothesis says nothing about industrial innovation in general; this is a larger subject than we have tackled here. There are very few countries that have failed to introduce at least a few products; and there are some, such as Germany and Japan, which have been responsible for a considerable number of such introductions. Germany's outstanding successes in the development and use of plastics may have been due, for instance, to a traditional concern with her lack of a raw materials base, and a recognition that a market might exist in Germany for synthetic substitutes.[6]

6. See two excellent studies: C. Freeman, "The Plastics Industry: A Comparative Study of Research and Innovation," in *National Institute Economic Review*, No. 26 (Nov. 1963), p. 22 *et seq.*; G. C. Hufbauer, *Synthetic Materials and the Theory of International Trade* (London: Gerald Duckworth, 1965). A number of links in the Hufbauer arguments are remarkably similar to

194        *QUARTERLY JOURNAL OF ECONOMICS*

Our hypothesis asserts that United States producers are likely to be the first to spy an opportunity for high-income or labor-saving new products.[7] But it goes on to assert that the first producing facilities for such products will be located in the United States. This is not a self-evident proposition. Under the calculus of least cost, production need not automatically take place at a location close to the market, unless the product can be produced and delivered from that location at lowest cost. Besides, now that most major United States companies control facilities situated in one or more locations outside of the United States, the possibility of considering a non-United States location is even more plausible than it might once have been.

Of course, if prospective producers were to make their locational choices on the basis of least-cost considerations, the United States would not always be ruled out. The costs of international transport and United States import duties, for instance, might be so high as to argue for such a location. My guess is, however, that the early producers of a new product intended for the United States market are attracted to a United States location by forces which are far stronger than relative factor-cost and transport considerations. For the reasoning on this point, one has to take a long detour away from comparative cost analysis into areas which fall under the rubrics of communication and external economies.

By now, a considerable amount of empirical work has been done on the factors affecting the location of industry.[8] Many of these studies try to explain observed locational patterns in conventional cost-minimizing terms, by implicit or explicit reference to labor cost and transportation cost. But some explicitly introduce problems of communication and external economies as powerful locational forces. These factors were given special emphasis in the analyses which were a part of the New York Metropolitan Region Study of the 1950's. At the risk of oversimplifying, I shall try to summarize what these studies suggested.[9]

some in this paper; but he was not aware of my writings nor I of his until after both had been completed.

7. There is a kind of first-cousin relationship between this simple notion and the "entrained want" concept defined by H. G. Barnett in *Innovation: The Basis of Cultural Change* (New York: McGraw-Hill, 1953) p. 148. Albert O. Hirschman, *The Strategy of Economic Development* (New Haven: Yale University Press, 1958), p. 68, also finds the concept helpful in his effort to explain certain aspects of economic development.

8. For a summary of such work, together with a useful bibliography, see John Meyer, "Regional Economics: A Survey," in the *American Economic Review*, LIII (Mar. 1963), 19–54.

9. The points that follow are dealt with at length in the following publications: Raymond Vernon, *Metropolis, 1985* (Cambridge: Harvard Uni-

In the early stages of introduction of a new product, producers were usually confronted with a number of critical, albeit transitory, conditions. For one thing, the product itself may be quite unstandardized for a time; its inputs, its processing, and its final specifications may cover a wide range. Contrast the great variety of automobiles produced and marketed before 1910 with the thoroughly standardized product of the 1930's, or the variegated radio designs of the 1920's with the uniform models of the 1930's. The unstandardized nature of the design at this early stage carries with it a number of locational implications.

First, producers at this stage are particularly concerned with the degree of freedom they have in changing their inputs. Of course, the cost of the inputs is also relevant. But as long as the nature of these inputs cannot be fixed in advance with assurance, the calculation of cost must take into account the general need for flexibility in any locational choice.[1]

Second, the price elasticity of demand for the output of individual firms is comparatively low. This follows from the high degree of production differentiation, or the existence of monopoly in the early stages.[2] One result is, of course, that small cost differences count less in the calculations of the entrepreneur than they are likely to count later on.

Third, the need for swift and effective communication on the part of the producer with customers, suppliers, and even competitors is especially high at this stage. This is a corollary of the fact that a considerable amount of uncertainty remains regarding the ultimate dimensions of the market, the efforts of rivals to preempt that market, the specifications of the inputs needed for production, and the specifications of the products likely to be most successful in the effort.

All of these considerations tend to argue for a location in which communication between the market and the executives directly concerned with the new product is swift and easy, and in which a wide

versity Press, 1960), pp. 38–85; Max Hall (ed.), *Made in New York* (Cambridge: Harvard University Press, 1959), pp. 3–18, 19 *passim*; Robert M. Lichtenberg, *One-Tenth of a Nation* (Cambridge: Harvard University Press, 1960), pp. 31–70.

1. This is, of course, a familiar point elaborated in George F. Stigler, "Production and Distribution in the Short Run," *Journal of Political Economy*, XLVII (June 1939), 305, *et seq*.

2. Hufbauer, *op. cit.*, suggests that the low price elasticity of demand in the first stage may be due simply to the fact that the first market may be a "captive market" unresponsive to price changes; but that later, in order to expand the use of the new product, other markets may be brought in which are more price responsive.

variety of potential types of input that might be needed by the production unit are easily come by. In brief, the producer who sees a market for some new product in the United States may be led to select a United States location for production on the basis of national locational considerations which extend well beyond simple factor cost analysis plus transport considerations.

## THE MATURING PRODUCT [3]

As the demand for a product expands, a certain degree of standardization usually takes place. This is not to say that efforts at product differentiation come to an end. On the contrary; such efforts may even intensify, as competitors try to avoid the full brunt of price competition. Moreover, variety may appear as a result of specialization. Radios, for instance, ultimately acquired such specialized forms as clock radios, automobile radios, portable radios, and so on. Nevertheless, though the subcategories may multiply and the efforts at product differentiation increase, a growing acceptance of certain general standards seems to be typical.

Once again, the change has locational implications. First of all, the need for flexibility declines. A commitment to some set of product standards opens up technical possibilities for achieving economies of scale through mass output, and encourages long-term commitments to some given process and some fixed set of facilities. Second, concern about production cost begins to take the place of concern about product characteristics. Even if increased price competition is not yet present, the reduction of the uncertainties surrounding the operation enhances the usefulness of cost projections and increases the attention devoted to cost.

The empirical studies to which I referred earlier suggest that, at this stage in an industry's development, there is likely to be considerable shift in the location of production facilities at least as far as internal United States locations are concerned. The empirical materials on international locational shifts simply have not yet been analyzed sufficiently to tell us very much. A little speculation, however, indicates some hypotheses worth testing.

Picture an industry engaged in the manufacture of the high-income or labor-saving products that are the focus of our discussion. Assume that the industry has begun to settle down in the United States to some degree of large-scale production. Although the first

3. Both Hirsch, *op. cit.*, and Freeman, *op. cit.*, make use of a three-stage product classification of the sort used here.

## INVESTMENT AND TRADE 197

mass market may be located in the United States, some demand for the product begins almost at once to appear elsewhere. For instance, although heavy fork-lift trucks in general may have a comparatively small market in Spain because of the relative cheapness of unskilled labor in that country, some limited demand for the product will appear there almost as soon as the existence of the product is known.

If the product has a high income elasticity of demand or if it is a satisfactory substitute for high-cost labor, the demand in time will begin to grow quite rapidly in relatively advanced countries such as those of Western Europe. Once the market expands in such an advanced country, entrepreneurs will begin to ask themselves whether the time has come to take the risk of setting up a local producing facility.[4]

How long does it take to reach this stage? An adequate answer must surely be a complex one. Producers located in the United States, weighing the wisdom of setting up a new production facility in the importing country, will feel obliged to balance a number of complex considerations. As long as the marginal production cost plus the transport cost of the goods exported from the United States is lower than the average cost of prospective production in the market of import, United States producers will presumably prefer to avoid an investment. But that calculation depends on the producer's ability to project the cost of production in a market in which factor costs and the appropriate technology differ from those at home.

Now and again, the locational force which determined some particular overseas investment is so simple and so powerful that one has little difficulty in identifying it. Otis Elevator's early proliferation of production facilities abroad was quite patently a function of the high cost of shipping assembled elevator cabins to distant locations and the limited scale advantages involved in manufacturing elevator cabins at a single location.[5] Singer's decision to invest in Scotland as early as 1867 was also based on considerations of a sort sympathetic with our hypothesis.[6] It is not unlikely that the

4. M. V. Posner, "International Trade and Technical Change," *Oxford Economic Papers*, Vol. 13 (Oct. 1961), p. 323, *et seq.* presents a stimulating model purporting to explain such familiar trade phenomena as the exchange of machine tools between the United Kingdom and Germany. In the process he offers some particularly helpful notions concerning the size of the "imitation lag" in the responses of competing nations.
5. Dudley M. Phelps, *Migration of Industry to South America* (New York: McGraw-Hill, 1963), p. 4.
6. John H. Dunning, *American Investment in British Manufacturing Industry* (London: George Allen & Unwin, 1958), p. 18. The Dunning book

overseas demand for its highly standardized product was already sufficiently large at that time to exhaust the obvious scale advantages of manufacturing in a single location, especially if that location was one of high labor cost.

In an area as complex and "imperfect" as international trade and investment, however, one ought not anticipate that any hypothesis will have more than a limited explanatory power. United States airplane manufacturers surely respond to many "noneconomic" locational forces, such as the desire to play safe in problems of military security. Producers in the United States who have a protected patent position overseas presumably take that fact into account in deciding whether or when to produce abroad. And other producers often are motivated by considerations too complex to reconstruct readily, such as the fortuitous timing of a threat of new competition in the country of import, the level of tariff protection anticipated for the future, the political situation in the country of prospective investment and so on.

We arrive, then, at the stage at which United States producers have come around to the establishment of production units in the advanced countries. Now a new group of forces are set in train. In an idealized form, Figure I suggests what may be anticipated next.

As far as individual United States producers are concerned, the local markets thenceforth will be filled from local production units set up abroad. Once these facilities are in operation, however, more ambitious possibilities for their use may be suggested. When comparing a United States producing facility and a facility in another advanced country, the obvious production-cost differences between the rival producing areas are usually differences due to scale and differences due to labor costs. If the producer is an international firm with producing locations in several countries, its costs of financing capital at the different locations may not be sufficiently different to matter very much. If economies of scale are being fully exploited, the principal differences between any two locations are likely to be labor costs.[7] Accordingly, it may prove wise for the international firm to begin servicing third-country markets from the new location. And if labor cost differences are large enough to offset transport

is filled with observations that lend casual support to the main hypotheses of this paper.

7. Note the interesting finding of Mordecai Kreinin in his "The Leontief Scarce-Factor Paradox," *The American Economic Review*, LV (Mar. 1965), 131–39. Kreinin finds that the higher cost of labor in the United States is not explained by a higher rate of labor productivity in this country.

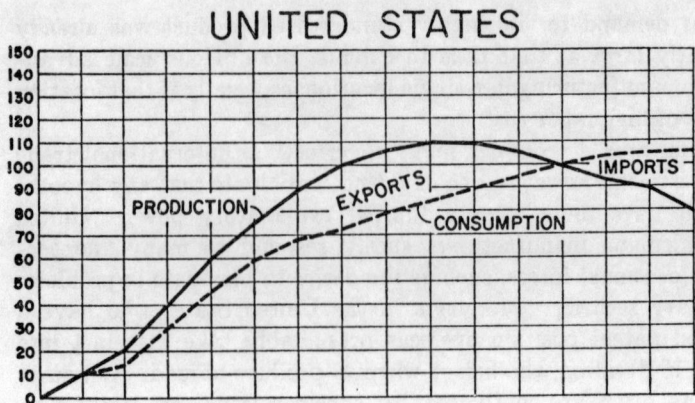

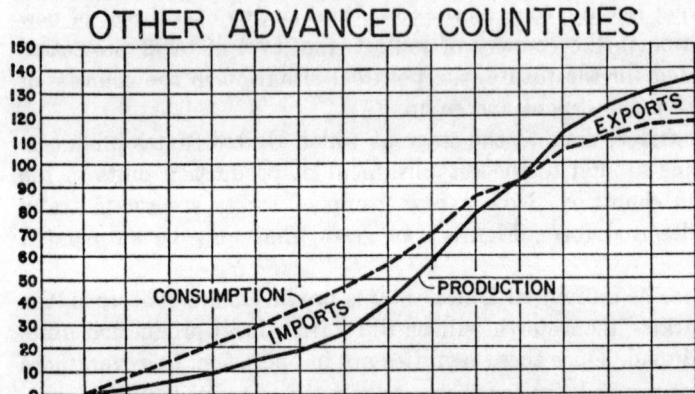

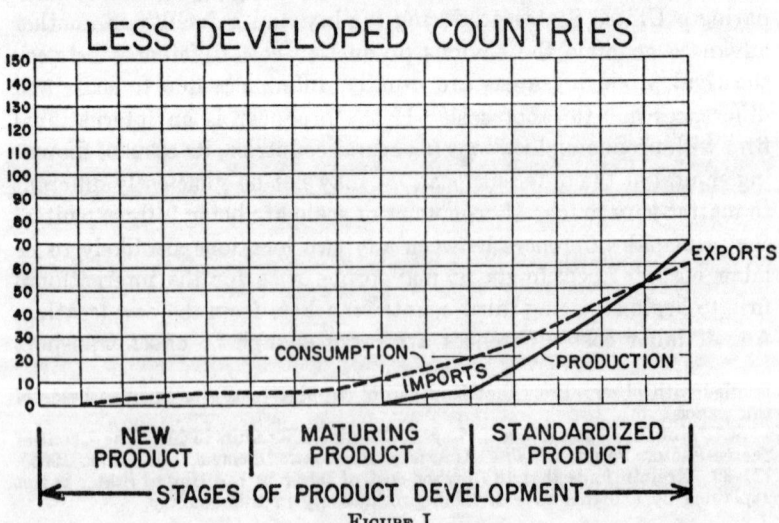

FIGURE I

costs, then exports back to the United States may become a possibility as well.

Any hypotheses based on the assumption that the United States entrepreneur will react rationally when offered the possibility of a lower-cost location abroad is, of course, somewhat suspect. The decision-making sequence that is used in connection with international investments, according to various empirical studies, is not a model of the rational process.[8] But there is one theme that emerges again and again in such studies. Any threat to the established position of an enterprise is a powerful galvanizing force to action; in fact, if I interpret the empirical work correctly, threat in general is a more reliable stimulus to action than opportunity is likely to be.

In the international investment field, threats appear in various forms once a large-scale export business in manufactured products has developed. Local entrepreneurs located in the countries which are the targets of these exports grow restive at the opportunities they are missing. Local governments concerned with generating employment or promoting growth or balancing their trade accounts begin thinking of ways and means to replace the imports. An international investment by the exporter, therefore, becomes a prudent means of forestalling the loss of a market. In this case, the yield on the investment is seen largely as the avoidance of a loss of income to the system.

The notion that a threat to the status quo is a powerful galvanizing force for international investment also seems to explain what happens after the initial investment. Once such an investment is made by a United States producer, other major producers in the United States sometimes see it as a threat to the status quo. They see themselves as losing position relative to the investing company, with vague intimations of further losses to come. Their "share of the market" is imperiled, viewing "share of the market" in global terms. At the same time, their ability to estimate the production-cost structure of their competitors, operating far away in an unfamiliar foreign area, is impaired; this is a particularly unsettling state because it conjures up the possibility of a return flow of products to the United States and a new source of price competition, based on cost differences of unknown magnitude. The uncertainty can be reduced by emulating the pathfinding investor and by investing in the same area; this may not be an optimizing investment

8. Aharoni, *op. cit.*, provides an excellent summary and exhaustive bibliography of the evidence on this point.

### INVESTMENT AND TRADE      201

pattern and it may be costly, but it is least disturbing to the status quo.

Pieces of this hypothetical pattern are subject to empirical tests of a sort. So far, at any rate, the empirical tests have been reassuring. The office machinery industry, for instance, has seen repeatedly the phenomenon of the introduction of a new product in the United States, followed by United States exports,[9] followed still later by United States imports. (We have still to test whether the timing of the commencement of overseas production by United States subsidiaries fits into the expected pattern.) In the electrical and electronic products industry, those elements in the pattern which can be measured show up nicely.[1] A broader effort is now under way to test the United States trade patterns of a group of products with high income elasticities; and, here too, the preliminary results are encouraging.[2] On a much more general basis, it is reassuring for our hypotheses to observe that the foreign manufacturing subsidiaries of United States firms have been increasing their exports to third countries.

It will have occurred to the reader by now that the pattern envisaged here also may shed some light on the Leontief paradox.[3] Leontief, it will be recalled, seemed to confound comparative cost theory by establishing the fact that the ratio of capital to labor in United States exports was lower, not higher, than the like ratio in the United States production which had been displaced by competitive imports. The hypothesis suggested in this paper would have the United States exporting high-income and labor-saving products in the early stages of their existence, and importing them later on.[4] In the early stages, the value-added contribution of industries engaged in producing these items probably contains an

---

9. Reported in U.S. Senate, Interstate and Foreign Commerce Committee, *Hearings on Foreign Commerce*, 1960, pp. 130–39.

1. See Hirsch, *op. cit.*

2. These are to appear in a forthcoming doctoral thesis at the Harvard Business School by Louis T. Wells, tentatively entitled" International Trade and Business Policy."

3. See Wassily Leontief, "Domestic Production and Foreign Trade: The American Capital Position Re-examined," *Proceedings of the American Philosophical Society*, Vol. 97 (Sept. 1953), and "Factor Proportions and the Structure of American Trade: Further Theoretical and Empirical Analysis," *Review of Economics and Statistics*, XXXVIII (Nov. 1956).

4. Of course, if there were some systematic trend in the inputs of new products — for example, if the new products which appeared in the 1960's were more capital-intensive than the new products which appeared in the 1950's — then the tendencies suggested by our hypotheses might be swamped by such a trend. As long as we do not posit offsetting systematic patterns of this sort, however, the Leontief findings and the hypotheses offered here seem consistent.

unusually high proportion of labor cost. This is not so much because the labor is particularly skilled, as is so often suggested. More likely, it is due to a quite different phenomenon. At this stage, the standardization of the manufacturing process has not gotten very far; that is to come later, when the volume of output is high enough and the degree of uncertainty low enough to justify investment in relatively inflexible, capital-intensive facilities. As a result, the production process relies relatively heavily on labor inputs at a time when the United States commands an export position; and the process relies more heavily on capital at a time when imports become important.

This, of course, is an hypothesis which has not yet been subjected to any really rigorous test. But it does open up a line of inquiry into the structure of United States trade which is well worth pursuing.

### THE STANDARDIZED PRODUCT

Figure I, the reader will have observed, carries a panel which suggests that, at an advanced stage in the standardization of some products, the less-developed countries may offer competitive advantages as a production location.

This is a bold projection, which seems on first blush to be wholly at variance with the Heckscher-Ohlin theorem. According to that theorem, one presumably ought to anticipate that the exports of the less-developed countries would tend to be relatively labor-intensive products.

One of the difficulties with the theorem, however, is that it leaves marketing considerations out of account. One reason for the omission is evident. As long as knowledge is regarded as a free good, instantaneously available, and as long as individual producers are regarded as atomistic contributors to the total supply, marketing problems cannot be expected to find much of a place in economic theory. In projecting the patterns of export from less-developed areas, however, we cannot afford to disregard the fact that information comes at a cost; and that entrepreneurs are not readily disposed to pay the price of investigating overseas markets of unknown dimensions and unknown promise. Neither are they eager to venture into situations which they know will demand a constant flow of reliable marketing information from remote sources.

If we can assume that highly standardized products tend to have a well-articulated, easily accessible international market and

INVESTMENT AND TRADE                    203

to sell largely on the basis of price (an assumption inherent in the definition), then it follows that such products will not pose the problem of market information quite so acutely for the less-developed countries. This establishes a necessary if not a sufficient condition for investment in such industries.

Of course, foreign investors seeking an optimum location for a captive facility may not have to concern themselves too much with questions of market information; presumably, they are thoroughly familiar with the marketing end of the business and are looking for a low-cost captive source of supply. In that case, the low cost of labor may be the initial attraction drawing the investor to less-developed areas. But other limitations in such areas, according to our hypothesis, will bias such captive operations toward the production of standardized items. The reasons in this case turn on the part played in the production process by external economies. Manufacturing processes which receive significant inputs from the local economy, such as skilled labor, repairmen, reliable power, spare parts, industrial materials processed according to exacting specification, and so on, are less appropriate to the less-developed areas than those that do not have such requirements. Unhappily, most industrial processes require one or another ingredient of this difficult sort. My guess is, however, that the industries which produce a standardized product are in the best position to avoid the problem, by producing on a vertically-integrated self-sustaining basis.

In speculating about future industrial exports from the less-developed areas, therefore, we are led to think of products with a fairly clear-cut set of economic characteristics.[5] Their production function is such as to require significant inputs of labor; otherwise there is no reason to expect a lower production cost in less-developed countries. At the same time, they are products with a high price elasticity of demand for the output of individual firms; otherwise, there is no strong incentive to take the risks of pioneering with production in a new area. In addition, products whose production process did not rely heavily upon external economies would be more obvious candidates than those which required a more elaborate industrial environment. The implications of remoteness also would be critical; products which could be precisely described by standardized specifications and which could be produced for inventory without fear of obsolescence would be more relevant than those

5. The concepts sketched out here are presented in more detail in my "Problems and Prospects in the Export of Manufactured Products from the Less-developed Countries," U.N. Conference on Trade and Development, Dec. 16, 1963 (mimeo.).

which had less precise specifications and which could not easily be ordered from remote locations. Moreover, high-value items capable of absorbing significant freight costs would be more likely to appear than bulky items low in value by weight. Standardized textile products are, of course, the illustration par excellence of the sort of product that meets the criteria. But other products come to mind such as crude steel, simple fertilizers, newsprint, and so on.

Speculation of this sort draws some support from various interregional experiences in industrial location. In the United States, for example, the "export" industries which moved to the low-wage south in search of lower costs tended to be industries which had no great need for a sophisticated industrial environment and which produced fairly standardized products. In the textile industry, it was the grey goods, cotton sheetings and men's shirt plants that went south; producers of high-style dresses or other unstandardized items were far more reluctant to move. In the electronics industry, it was the mass producers of tubes, resistors and other standardized high-volume components that showed the greatest disposition to move south; custom-built and research-oriented production remained closer to markets and to the main industrial complexes. A similar pattern could be discerned in printing and in chemicals production.[6]

In other countries, a like pattern is suggested by the impressionistic evidence. The underdeveloped south of Italy and the laggard north of Britain and Ireland both seem to be attracting industry with standardized output and self-sufficient process.[7]

Once we begin to look for relevant evidence of such investment patterns in the less-developed countries proper, however, only the barest shreds of corroboratory information can be found. One would have difficulty in thinking of many cases in which manufacturers of standardized products in the more advanced countries had made significant investments in the less-developed countries with a view of exporting such products from those countries. To be sure, other

6. This conclusion derives largely from the industry studies conducted in connection with the New York Metropolitan Region study. There have been some excellent more general analyses of shifts in industrial location among the regions of the United States. See e.g., Victor R. Fuchs, *Changes in the Location of Manufacturing in the United States Since 1929* (New Haven: Yale University Press, 1962). Unfortunately, however, none has been designed, so far as I know, to test hypotheses relating locational shifts to product characteristics such as price elasticity of demand and degree of standardization.

7. This statement, too, is based on only impressionistic materials. Among the more suggestive, illustrative of the best of the available evidence, see J. N. Toothill, *Inquiry into the Scottish Economy* (Edinburgh: Scottish Council, 1962).

## INVESTMENT AND TRADE    205

types of foreign investment are not uncommon in the less-developed countries, such as investments in import-replacing industries which were made in the face of a threat of import restriction. But there are only a few export-oriented cases similar to that of Taiwan's foreign-owned electronics plants and Argentina's new producing facility, set up to manufacture and export standard sorting equipment for computers.

If we look to foreign trade patterns, rather than foreign investment patterns, to learn something about the competitive advantage of the less-developed countries, the possibility that they are an attractive locus for the output of standardized products gains slightly more support. The Taiwanese and Japanese trade performances are perhaps the most telling ones in support of the projected pattern; both countries have managed to develop significant overseas markets for standardized manufactured products. According to one major study of the subject (a study stimulated by the Leontief paradox), Japanese exports are more capital-intensive than is the Japanese production which is displaced by imports; [8] this is what one might expect if the hypothetical patterns suggested by Figure I were operational. Apart from these cases, however, all that one sees are a few provocative successes such as some sporadic sales of newsprint from Pakistan, the successful export of sewing machines from India, and so on. Even in these cases, one cannot be sure that they are consistent with the hypothesis unless he has done a good deal more empirical investigation.

The reason why so few revelant cases come to mind may be that the process has not yet advanced far enough. Or it may be that such factors as extensive export constraints and overvalued exchange rates are combining to prevent the investment and exports that otherwise would occur.

If there is one respect in which this discussion may deviate from classical expectations, it is in the view that the overall scarcity of capital in the less-developed countries will not prevent investment in facilities for the production of standardized products.

There are two reasons why capital costs may not prove a barrier to such investment.

First, according to our hypotheses, the investment will occur in industries which require some significant labor inputs in the production process; but they will be concentrated in that subsector of the

8. M. Tatemoto and S. Ichimura, "Factor Proportions and Foreign Trade: the Case of Japan," *Review of Economics and Statistics*, XLI (Nov. 1959), 442–46.

industry which produces highly standardized products capable of self-contained production establishments. The net of these specifications is indeterminate so far as capital-intensiveness is concerned. A standardized textile item may be more or less capital-intensive than a plant for unstandardized petro-chemicals.

Besides, even if the capital requirements for a particular plant are heavy, the cost of the capital need not prove a bar. The assumption that capital costs come high in the less-developed countries requires a number of fundamental qualifications. The reality, to the extent that it is known, is more complex.

One reason for this complexity is the role played by the international investor. Producers of chemical fertilizers, when considering whether to invest in a given country, may be less concerned with the going rate for capital in that country than with their opportunity costs as they see such costs. For such investors the alternatives to be weighed are not the full range of possibilities calling for capital but only a very restricted range of alternatives, such as the possibilities offered by chemical fertilizer investment elsewhere. The relevant capital cost for a chemical fertilizer plant, therefore, may be fairly low if the investor is an international entrepreneur.

Moreover, the assumption that finance capital is scarce and that interest rates are high in a less-developed country may prove inapplicable to the class of investors who concern us here.[9] The capital markets of the less-developed countries typically consist of a series of water-tight, insulated, submarkets in which wholly different rates prevail and between which arbitrage opportunities are limited. In some countries, the going figures may vary from 5 to 40 per cent, on grounds which seem to have little relation to issuer risk or term of loan. (In some economies, where inflation is endemic, interest rates which in effect represent a negative real cost are not uncommon.)

These internal differences in interest rates may be due to a number of factors: the fact that funds generated inside the firm usually are exposed to a different yield test than external borrowings; the fact that government loans are often floated by mandatory levies on banks and other intermediaries; and the fact that funds borrowed by governments from international sources are often re-

---

9. See George Rosen, *Industrial Change in India* (Glencoe, Ill.: Free Press, 1958). Rosen finds that in the period studied from 1937 to 1953, "there was no serious shortage of capital for the largest firms in India." Gustav F. Papanek makes a similar finding for Pakistan for the period from 1950 to 1964 in a book about to be published.

## INVESTMENT AND TRADE                    **207**

loaned in domestic markets at rates which are linked closely to the international borrowing rate, however irrelevant that may be. Moreover, one has to reckon with the fact that public international lenders tend to lend at near-uniform rates, irrespective of the identity of the borrower and the going interest rate in his country. Access to capital on the part of underdeveloped countries, therefore, becomes a direct function of the country's capacity to propose plausible projects to public international lenders. If a project can plausibly be shown to "pay its own way" in balance-of-payment and output terms at "reasonable" interest rates, the largest single obstacle to obtaining capital at such rates has usually been overcome.

Accordingly, one may say that from the entrepreneur's viewpoint certain systematic and predictable "imperfections" of the capital markets may reduce or eliminate the capital-shortage handicap which is characteristic of the less-developed countries; and, further, that as a result of the reduction or elimination such countries may find themselves in a position to compete effectively in the export of certain standardized capital-intensive goods. This is not the statement of another paradox; it is not the same as to say that the capital-poor countries will develop capital-intensive economies. All we are concerned with here is a modest fraction of the industry of such countries, which in turn is a minor fraction of their total economic activity. It may be that the anomalies such industries represent are systematic enough to be included in our normal expectations regarding conditions in the less-developed countries.

\*     \*     \*     \*     \*

Like the other observations which have preceded, these views about the likely patterns of exports by the less-developed countries are attempts to relax some of the constraints imposed by purer and simpler models. Here and there, the hypotheses take on plausibility because they jibe with the record of past events. But, for the most part, they are still speculative in nature, having been subjected to tests of a very low order of rigorousness. What is needed, obviously, is continued probing to determine whether the "imperfections" stressed so strongly in these pages deserve to be elevated out of the footnotes into the main text of economic theory.

HARVARD GRADUATE SCHOOL OF BUSINESS ADMINISTRATION

# [2]

# Factors Affecting the Diffusion of Technology

NATHAN ROSENBERG

*The University of Wisconsin*

I

The rate at which new techniques are adopted and incorporated into the productive process is, without doubt, one of the central questions of economic growth. New techniques exert their economic impact as a function of the rate at which they displace older techniques and the extent to which the new techniques are superior to the old ones. Although we are a still a very long way from being able to assess the exact role of technological change—as distinct from all other factors—in generating the rise in resource productivity which is at the heart of the growth process, it is, I think, clear that the contribution of technological change itself will have to be established through the study of diffusion. Only in this way can we develop a closer understanding of the rate at which new techniques, once invented, have been translated into events of economic significance.

Although these remarks are, I believe, sufficiently uncontroversial, it is a striking historiographical fact that the serious study of the diffusion of new techniques is an activity no more than fifteen years old.[1] Even today, if we focus upon the most critical events of the industrial revolution, such as the introduction of new techniques of power generation and the climactic events in metallurgy, our ignorance of the rate at which new techniques were adopted, and the factors accounting for these rates is, if not total, certainly no cause for professional self-congratulation. Much of the history of the past

I wish to acknowledge the helpful comments and criticism received on an earlier version of this paper by the participants in the economic history workshops at the University of Chicago, Indiana University, and the University of Wisconsin. I am also grateful to Peter Lindert, who was particularly helpful at a later stage.

[1] For an admirable earlier study, see Marc Bloch's "Avenement et conquetes du moulin a eau," *Annales d'histoire economique et sociale*, 7 (1935), 538–563. Bloch provides a masterly analysis, turning primarily on changing legal and economic conditions as they affected the availability of servile labor, of the lag of an entire millennium between the invention of the watermill and its widespread adoption.

two centuries or so has been written by scholars with impressive credentials for technological history and its minutiae, but with a more limited appreciation of the economic consequences of technological events. Thus, H. W. Dickinson, in his classic account of the history of the steam engine states, with respect to the consequences of the improvements which Watt had affected, that by the end of the eighteenth century, "The textile industry had been transformed from handicraft into a machine industry."[2] Although Dickinson wrote with a commanding authority on purely technological matters, it is necessary to record that he was almost a half century off the mark with respect to his economic history. By the end of the eighteenth century a suitable technology simply did not exist for the application of power to fully mechanized spinning or weaving operations.[3]

If we turn to the recent works of economic historians for further assistance on an issue as central to industrialization as the diffusion of steam power in the nineteenth century, we are offered some illumination but not nearly as much, surely, as we are entitled to expect.[4] Habakkuk states that "steam did not begin to play an im-

---

[2] H. W. Dickinson, *A Short History of the Steam Engine* (Cambridge: Cambridge University Press, 1938), p. 89.

[3] Even in cotton textiles, the most fully mechanized of the textile branches, full mechanization was achieved only in the second quarter of the nineteenth century. The protracted agony of the hand-loom weavers was being acted out in the 1830s and 1840s as the improved power loom was widely adopted, and it was only in the latter decade that the number of power-loom weavers exceeded the number of hand-loom weavers—although the proportion of *output* accounted for by the hand-loom weavers was, of course, far smaller than the proportion which they constituted of the weaving labor force. Even so, there were an estimated 40,000 cotton hand-looms still at work in 1850. In woolen textiles, which was slower to mechanize, hand-loom weavers persisted in and around Leeds, the most highly mechanized woolen center, during the 1850s. J. H. Clapham, *An Economic History of Modern Britain*, (Cambridge: Cambridge University Press, 1962), I Ch. XIV; Phyllis Deane and W. A. Cole, *British Economic Growth 1688–1959* (Cambridge: Cambridge University Press, 1962), pp. 191–192. See also Maurice Dobb, *Studies in the Development of Capitalism* (London: Routledge and Kegan Paul, 1946), pp. 263–265.

[4] For evidence on the limited impact of Watt's steam engine on the broad range of British industries by 1800, a full quarter century after first practical success had been achieved, see John Lord, *Capital and Steam Power* (London: P. S. King and Son, 1923), Ch. 8. However, since Boulton and Watt's engine was widely pirated and Newcomen engines remained in use, Lord's figures should not be taken as representing the total amount of

## Factors Affecting the Diffusion of Technology                    5

portant role in powering the British economy until the 1830s and 1840s, and was not massively applied until the 1870s and 80s. Even as late as 1870 less than a million horsepower was generated by steam in the factories and workshops of Great Britain."[5] Although Habakkuk does not indicate the source, these assertions are clearly based on Mulhall's estimates which included a figure of 900,000 horsepower for the fixed steam-power of Great Britain in 1870.[6] Mulhall was, of course, a great statistical pioneer, but his estimates were necessarily crude and, in some cases, no more than informed guesses. It is, nevertheless, symptomatic of the limited research conducted on the subject of technological diffusion so far that our knowledge of the diffusion of steam power has not been advanced, in the twentieth century, substantially beyond Mulhall's venerable *Dictionary of Statistics*. This late nineteenth-century work is still the last word on the subject. Our knowledge of the sequence of events at the purely technical level remains far greater than our knowledge of the *translation* of technical events into events of economic significance.

The present paper attempts to take some steps toward closing the gap between the technical and the economic realms of discourse. In doing so, I am hopeful that we will eventually end up with a better understanding of the nature of the linkages between these two realms. Such an improved understanding should enable us to get a better handhold on the timing of the diffusion process and thereby make it possible to formulate and to test more precise hypotheses concerning the spread of new technologies. This paper does not attempt to examine the whole range of factors influencing diffusion. Rather, it concentrates on certain supply side considerations. Needless to say, alterations in relative factor and commodity prices also affect the rate of diffusion of new technologies—but that is obviously

---

steam power in use in Great Britain. See A. E. Musson and E. Robinson, "The Early Growth of Steam Power," *Economic History Review*, 2nd series, 9 (April 1959), 418–439, and J. R. Harris, "The Employment of Steam Power in the 18th Century," *History*, 52 (1967), 133–148.

[5] H. J. Habakkuk, *American and British Technology in the 19th Century* (Cambridge: Cambridge University Press, 1962), pp. 184–185. See also A. E. Musson and Eric Robinson, *Science and Technology in the Industrial Revolution* (Manchester: University of Manchester Press, 1969), p. 72, and David Landes, *The Unbound Prometheus* (Cambridge: Cambridge University Press, 1969), p. 221.

[6] M. G. Mulhall, *The Dictionary of Statistics* (London: G. Routledge and Sons, Ltd., 1892), p. 545.

so. We argue here, not so obviously, that, factor and commodity prices aside, the rate at which new technologies replace old ones will depend upon the speed with which it is possible to overcome an array of supply side problems. These problems have not been uniformly resistant to efforts to overcome them, but have proven to be of varying degrees of intractability. This is the basic justification for the focus of this paper.

## II

*The Continuity of Inventive Activity*

That the diffusion of inventions is an essentially economic phenomenon, the timing of which can be largely explained by expected profits, is by now well established. The extended labors of Griliches and Mansfield have clearly demonstrated the power and scope of purely economic explanations in the diffusion of individual inventions. However, if one examines the history of the diffusion of many inventions, one cannot help being struck by two characteristics of the diffusion process: its apparent overall slowness on the one hand and the wide variations in the rates of acceptance of different inventions, on the other.[7] It is argued in this paper that a better understanding of the timing of diffusion is possible by probing more deeply at the technological level itself, where it may be possible to identify factors accounting for both the general slowness as well as wide variations in the rate of diffusion.

How slow is slow? When we speak of diffusion as being relatively slow, we are obviously implying some sort of dating procedure as well as expressing a comparative or absolute judgment. It should be noted at the outset that whether inventions are measured as diffusing rapidly or slowly depends in large part upon the selection of dates. If one dates the steam engine from the achievements of Newcomen around the first decade of the eighteenth century rather than from the work of Watt in the last third of the eighteenth century, as is commonly the case, one gets a much slower rate of diffusion. But on the basis of criteria commonly employed, the case for dating the steam engine from Newcomen is a perfectly compelling one. His atmospheric steam engine was not only technically workable, it was commercially feasible as well. To be sure, it experienced great heat loss in its operation and was a voracious consumer of

[7] See, for example, E. Mansfield, "Technical Change and the Rate of Imitation," *Econometrica*, 29 (October 1961), 741–766.

fuel, but it nevertheless survived the market test and was widely used in the eighteenth century, primarily for pumping water out of mines. One very cautious study has identified 60 Newcomen engines for the period 1712–1733 and no less than 300 for the years 1734–1781.[8] Moreover, even in 1800, by which time Watt's patents had all expired, Newcomen engines not only continued to be used but, due apparently to their low construction and maintenance costs and long life expectancy, still continued to be *built*. One might almost be tempted to say of James Watt that he was "just an improver," although such a statement would be comparable to saying of Napoleon that he was just a soldier or of Bach that he was just a court musician. That is to say, Watt's improvements on the steam engine transformed it from an instrument of limited applicability at locations peculiarly favored by access to cheap fuel, to a generalized power source of much wider significance. Nevertheless, even if we date the steam engine from Watt's accomplishments of the 1760s and 1770s, it still took a full century of improvement and design change before this new power source surpassed water power in manufacturing and displaced the sail on ocean-going vessels.

The essential point to be grasped here is that inventive activity is, itself, best described as a gradual process of accretion, a cumulation of minor improvements, modifications, and economies, a sequence of events where, in general, continuities are much more important than discontinuities. Even where it is possible to identify major inventions which seem to represent entirely new concepts and therefore genuine discontinuities, sharp and dramatic departures from the past, there are usually pervasive technological as well as economic forces at work which tend to slow down and to flatten out the impact of such inventions in terms of their contribution to raising resource productivity. Thus, even the big technological breakthroughs which are associated with such names as Darby, Watt, Cort, and Bessemer, usually have much more gently declining slopes of cost reduction flowing from their technical contributions than the historical literature would lead us to expect.

The fact is that the period which is looked at as encompassing the diffusion of an invention is usually much more than that: It is a period when critical inventive activity (what Usher called "secondary inventions") and essential design improvements and modifications are still going on. Although we might be tempted to dismiss

[8] J. R. Harris, "Employment of Steam Power," p. 147.

this later work as much less important than the initial technological breakthrough, there is no good *economic* reason for this attitude, for it is precisely this later work which first establishes commercial feasibility and therefore shapes the possibilities for diffusion. We need to approach this whole area of research with a clearer appreciation of the continuum of inventive activity, running from initial conceptualization (the "Eureka! I have it!" stage) to establishment of technical feasibility (invention) to commercial feasibility (innovation) to subsequent diffusion. By concentrating our attention upon the sharp discontinuities associated with major inventions, we are misrepresenting the manner in which the gradual growth in the stock of useful knowledge is transformed into improvements in resource productivity.

Although we find it a convenient verbal shorthand to speak of the "displacement" of one technique by another, the historical process is often one of a series of smaller and highly tentative steps. Thus, there were several intermediate stages in the displacement of sail by steam: Steamships were at first fully rigged and long continued to carry at least auxiliary sails, particularly as an insurance against breakdowns (which were not infrequent) on long ocean voyages, whereas sailing ships, in their twilight days, were often furnished with auxiliary engines.

The inadequate conception of invention as an intermittent and discontinuous phenomenon has been shared by the historian and the economist. The historian finds that it immensely simplifies the writing of chronological history if particular events—in this case inventions—can be precisely pinpointed in time, just like the Great Fire of London, the accession of the House of Hanover to the British throne, the Treaty of Paris or the abolition of the Corn Laws. The economist, for his part, finds it enormously convenient for analytical purposes to distinguish between movements along existing isoquants in response to alterations in factor prices, and shifts in the isoquants themselves due to the intermittent phenomenon of technological change. Both disciplines, therefore, have in the past found it convenient to focus upon technological change as a discontinuous phenomenon which can be precisely located in historical time. But by breaking into the continuum of inventive activity in this way, we are, subtly but inevitably, distorting our approach to the diffusion process. Once an invention has been "made," after all, the natural expectation is that all that remains in the sequence is for it to be adopted. Any delay becomes a "lag."

*Factors Affecting the Diffusion of Technology*            9

However, in viewing the problem in this way, we are very much underestimating the technological and economic importance of the subsequent "improvements." We are engaging in a sort of conceptual foreshortening which distorts our view of later events. For we are led to treat the period after the conventional dating of an invention as one where a fairly well-established technique is awaiting adoption whereas, in fact, highly significant technological and economic adaptations are typically waiting to be made. It is this same foreshortening of perspective which greatly increases our general impression of the slowness of diffusion.

Thus, Enos has studied the length of the lag between invention and innovation for 35 important innovations and has subjected his results to statistical analysis. In his sample the arithmetic mean for the interval between invention and innovation is 13.6 years.[9] Enos defines the date of an invention as "the earliest conception of the product in substantially its commercial form" and dates an innovation from "the first commercial application or sale."[10] It is difficult, however, in view of his definition of invention, to know just what significance to attach to the size of these lags. In some cases the date selected for invention corresponds fairly closely to a time when the technical problems were reasonably well solved. In other cases his choice of dates seems to be based more closely upon "earliest conception," when many important technical problems still remained to be solved. For example, Enos places the invention of the cotton-picker in 1889 and its innovation in 1942—a lag of 53 years. It was in 1889 that Angus Campbell first made use, on an experimental basis, of a spindle-type picker, but the machine was far from constituting a satisfactory picker. As Jewkes and his co-authors had pointed out, "the machine left cotton on the ground and caused damage to the bolls and blooms."[11] It will be readily conceded that these are serious deficiencies in a device for picking cotton (even though one can conceive of possible price relationships where the adoption of such a machine would have been economically rational). Numerous important technical problems therefore remained to be

---

[9] John L. Enos, "Invention and Innovation in the Petroleum Refining Industry," in *The Rate and Direction of Inventive Activity* (Princeton: University Press, 1962), p. 309. Most of Enos's information is drawn from J. Jewkes, D. Sawers, and R. Stillerman, *The Sources of Invention* (London: St. Martin's Press, 1958).

[10] *Ibid.*, p. 308.

[11] Jewkes et al., *Sources*, p. 283

solved after 1889. Similar complaints can be raised over many of the other dates employed in the Enos study—such as the 79-year interval for the fluorescent lamp (invention dated 1859), the 56-year interval for the gyro-compass (invention dated 1852), or the 22-year interval for television (invention dated 1919).[12] In these and in many other cases, much critical and indispensable inventive activity remained to be performed. The length of the interval, in other words, was at least partly—and in many cases primarily—due to the time required for carrying out *further* inventive activity. It is therefore extremely difficult to know what significance ought to be attached to Enos's numerical findings concerning these "lags."

## III

*Improvements in Inventions after Their First Introduction*

It follows from the conception of invention adopted in this paper that most inventions are relatively crude and inefficient at the date when they are first recognized as constituting a new invention. They are, of necessity, badly adapted to many of the ultimate uses to which they will eventually be put; therefore, they may offer only very small advantages, or perhaps none at all, over previously-existing techniques. Diffusion under these circumstances will necessarily be slow because the clear superiority of the new technique over the old has not yet been established or, perhaps, because the new technique or process alters the quality of the final product in unfortunate or unpredictable ways. Thus, as John Nef has shown, the emergence of a coal-using technology in England was seriously impeded in the seventeenth century by the fact that the use of the mineral fuel damaged the final product—as in the case of glass-making and the drying of malt for the brewing industry.[13] Indeed, it was out of this attempt to maintain high quality of final product while using coal as a fuel that major advances were made in furnace design and the technique of coking was eventually developed.[14] More important, the very slow diffusion of the coke-smelting of iron after Abraham Darby's first success in 1709 was due in part to the

[12] Enos, *Inventive Activity*, p. 307.

[13] John Nef, "The Progress of Technology and the Growth of Large-Scale Industry in Great Britain, 1540–1640," *Economic History Review*, 5 (1934), 15–18.

[14] Beer made of malt that had been dried by raw coal was, apparently, practically undrinkable.

Factors Affecting the Diffusion of Technology          11

fact that "coke pig-iron produced a bar inferior in tensility and ductility to that made from charcoal pig: it was 'cold-short' and unsuitable for the production of wares of quality."[15] Consequently the use of coke pig-iron was confined to the much smaller, cast iron branch of the iron industry. The adoption of a new technique, then, is often limited by imperfections in the product which, in turn, are only gradually overcome or bypassed. Such problems connected with quality control were long a source of persistent difficulties in the use of iron and steel. The inability to achieve a rigid and precise control of the quality and therefore of the performance characteristics of the metal was a major handicap in the application of ferrous materials to a range of industrial uses. Such control came only with the introduction of the Bessemer and open-hearth methods, since these made it possible to control the carbon content within very narrow limits.[16]

[15] T. S. Ashton, *Iron and Steel in the Industrial Revolution* (Manchester: University of Manchester Press, 1924), p. 35. Furthermore, as Ashton points out: "It seems very probable that the production of sound coke-iron was not a sudden creation but the result of many trials in which failure and success alternated. The inventory taken at Coalbrookdale in 1718 shows that there was an accumulation of 'sculls,' or defective iron, which were sold at a low price to a neighboring forgemaster. Such sculls may have been produced frequently in the early days of the process, and though it is beyond question that marketable iron was produced every year, it might have been difficult for Darby himself to say exactly when the problem had reached a final solution" (p. 33).

[16] The frequent inferiority and, therefore, slow diffusion of new inventions in their early stages of development is strikingly apparent in military history. The English long-bow swept the field at Agincourt in 1415, long after the first introduction of gunpowder and cannon into Europe. Although Europeans sought to develop an effective field artillery in the fifteenth century, they did not succeed in overcoming the technical problems involved until the first half of the seventeenth century. Until that time such weapons were of very limited effectiveness, aside from their use in sieges. Their limited mobility and slow rate of fire enabled them to be easily overcome by massed charges. See C. Cipolla, *Guns, Sails and Empires* (New York: Random House, 1965), Ch. 1 and Epilogue.

The Texas Rangers, in spite of all the guns and powder in their armories, could not establish a decisive superiority over the fierce Comanches under the special circumstances of mounted combat until the availability of the six shooter. For an absorbing account, see W. P. Webb, *The Great Plains* (Boston: Ginn and Co., 1931), pp. 167–179. Before the availability of multi-shot weapons, Webb concludes that the Indians were at a distinct advantage. "In the first place, the Texans carried at most three shots; the Comanche carried twoscore or more arrows. It took the Texan a minute to

If it is true that inventions in their early forms are often highly imperfect and constitute only slight improvements over earlier techniques, it also follows that the pace at which subsequent improvements are made will be a major determinant of the rate of diffusion.[17] Indeed it may very well be the case that such improvements will reduce total costs by an amount greater than the reduction in costs of the initial invention over the older techniques which it eventually replaced. Mak and Walton argue that, on the Louisville-New Orleans route, "The introduction of the steamboat, 1815–20, led to a significant fall in real freight costs, but the absolute as well as the relative decline in real freight rates was greatest during the period of improvement, 1820–60."[18] Not all of the improvement in productivity, of course, was attributable to technological change. In addition to the increase in cargo carrying capacity per measured ton and the extension of the navigation season, which *were* primarily

---

reload his weapon; the Indian could in that time ride three hundred yards and discharge twenty arrows. The Texan had to dismount in order to use his rifle effectively at all, and it was his most reliable weapon; the Indian remained mounted throughout the combat. Apparently the one advantage possessed by the white man was a weapon of longer range and more deadly accuracy than the Indian's bow, but the agility of the Indian and the rapidity of his movements did much to offset this advantage" (p. 169).

[17] This is not to suggest that a continuous rate of improvement in a technology implies some continuous rate of its diffusion. When a technology is still at a very primitive stage in its development, even substantial reductions in cost may have little effect upon diffusion. On the other hand as a new technology reaches the cost levels of competing methods, relatively small additional cost reductions may bring it below critical threshold levels and thus lead to rapidly accelerating rates of diffusion. For a rigorous application of a threshold function in the study of diffusion, see Paul David, "The Mechanization of Reaping in the Ante-Bellum Midwest," in *Industrialization in Two Systems*, ed., H. Rosovsky (New York: John Wiley and Sons, 1966), pp. 3–28. David's threshold function, it should be noted, relates to farm size, and he demonstrates, how the rising relative cost of harvest labor lowered that farm size, leading finally to the rapid introduction of the reaper in the Midwest during the 1850s.

[18] James Mak and Gary Walton, "Steamboats and the Great Productivity Surge in River Transportation," *Journal of Economic History*, 32 (September 1972), p. 625. They add: "We qualify this conclusion to the extent that the productivity gains from steam power, which are reflected in the fall in rates, 1815–20, may have been understated somewhat because of slow entry or limited competition in this early period. Nevertheless, there were 17 vessels in operation on Western rivers in 1817, and 69 were in operation by 1820. It seems reasonable to assume that the initial impact of the steam engine had occurred by 1820" (footnote 15, p. 625).

*Factors Affecting the Diffusion of Technology*                    *13*

the result of technological changes, there were significant reductions in cargo collection times and passage times, which were not. Nevertheless, it is clear that the overall increase in total factor productivity associated with this major transportation innovation came in the years *following* its initial introduction.[19]

A similar conclusion is reached by Enos in his study of technical progress in the petroleum refining industry in the twentieth century. Enos examined the introduction of four major new processes in petroleum refining: thermal cracking, polymerization, catalytic cracking, and catalytic reforming. In measuring the benefits for each new process he distinguished between the "alpha phase"—or cost reductions which occur when the new process is introduced—and the "beta phase"—or cost reductions flowing from the subsequent improvements in the new process. Enos finds that the average annual cost reductions generated by the beta phase of each of these innovations considerably exceeds the average annual cost reductions generated by the alpha phase (4.5% as compared to 1.5%). On this basis he asserts: "The evidence from the petroleum refining industry indicates that improving a process contributes even more to technological progress than does its initial development."[20]

One final, general point needs to be made in concluding this section. It seems to be extraordinarily difficult to visualize and to anticipate the uses to which an invention will be put. Railroads were originally thought of as essentially feeders to canals and other forms of water transportation. In the early days of radio at the turn of the century, it was regarded primarily as a supplement to wire com-

[19] A cumulation of minor design changes on the steamboat had the effect of substantially increasing the length of the navigation season for each steamboat size class. By steadily reducing the draft in relation to tonnage and cargo carrying capacity, steamboat designers and builders brought about major improvements in the productivity of capital by enabling steamboats to operate a longer portion of the year. Thus, Mak and Walton state that, as a rough average, "The navigation season was extended from approximately six months, before 1830, to about nine months, during the last half of the ante-bellum period." (Mak and Walton, "Steamboats," p. 634.) See also Louis Hunter, *Steamboats on the Western Rivers* (Cambridge: Harvard University Press, 1949), Ch. 2 and pp. 219-225.

[20] John L. Enos, "A Measure of the Rate of Technological Progress in the Petroleum Refining Industry," *Journal of Industrial Economics*, 6 (June 1958), 180. See also the same author's "Invention and Innovation in the Petroleum Refining Industry," in *The Rate and Direction of Inventive Activity* (Princeton: Princeton University Press, 1962), pp. 299-321.

munication services, to be used only where wire was not practicable —as in certain isolated locations or for ships at sea.[21] Finally, even so versatile an inventor as Thomas Edison is said to have thought that the phonograph would be useful principally to record the wishes of old men on their death beds.

It is easy to sneer at such failures of vision and poverty of imagination—especially since, in history, we always have the immense advantage of knowing how the story ended. Nevertheless, past experience suggests that the prediction of how a given invention will fit into the social system, the uses to which it will be put, and the alterations it will generate, are all extraordinarily difficult intellectual exercises.[22] Such difficulties, in turn, must have played an important role in slowing down the pace of diffusion.

Even when an invention genuinely contains important elements of novelty, there is a strong tendency to conceptualize it in terms of the traditional or familiar. Thus the transition to a new technique is often slowed by the extreme difficulty of breaking away from the old forms and embracing the different logic of a new technique or principle.[23]

[21] This limited conception of the potential of the radio in turn slowed the pace at which it was developed because, as a result, "most of the original research on the development of wireless communication was initially oriented toward the relatively simple task of transmitting impulses for telegraphic communications." Frank Lynn, *An Investigation of the Rate of Development and Diffusion of Technology in Our Modern Industrial Society*, in *The Employment Impact of Technological Change*, 6 vols. (Washington, D.C.: U.S.G.P.O., 1966), II, p. 68.

[22] In a closely related context, Simon Kuznets has pointed to the almost congenital pessimism of professional judgments on the possibilities for technological change over the years. "Experts are usually specialists skilled in, and hence bound to, traditional views; and they are, because of their knowledge of one field, likely to be cautious and unduly conservative. Hertz, a great physicist, denied the practical importance of shortwaves, and others at the end of the nineteenth century reached the conclusion that little more could be done on the structure of matter. Malthus, Ricardo, and Marx, great economists, made incorrect prognoses of technological change at the very time that the scientific bases for these changes were evolving. On the other hand, imaginative tyros like Jules Verne and H. G. Wells seemed to sense the potentialities of technological change. It is well to take cognizance of this consistently conservative bias of experts in evaluating the hypothesis of an unlimited effective increase in the stock of knowledge and in the corresponding potential of economic growth" (Simon Kuznets, *Economic Growth and Structure* [New York: W. W. Norton and Co., 1965], p. 89).

[23] Marx saw this point clearly: "To what an extent the old forms of the instruments of production influenced their new forms at first starting, is

*Factors Affecting the Diffusion of Technology* 15

IV

*The Development of Technical Skills Among Users*

Closely associated with this gradual improvement in the innovation itself is the development of the human skills upon which the use of the new technique depends in order to be effectively exploited. There is, in other words, a learning period the length of which will depend upon many factors, including the complexity of the new techniques, the extent to which they are novel or rely on skills already available or transferable from other industries, etc.[24] There

---

shown by, amongst other things, the most superficial comparison of the present powerloom with the old one, of the modern blowing apparatus of a blast-furnace with the first inefficient mechanical reproduction of the ordinary bellows, and perhaps more strikingly than in any other way, by the attempts before the invention of the present locomotive, to construct a locomotive that actually had two feet, which after the fashion of a horse, it raised alternately from the ground. It is only after considerable development of the science of mechanics, and accumulated practical experience, that the form of a machine becomes settled entirely in accordance with mechanical principles, and emancipated from the traditional form of the tool that gave rise to it" (Karl Marx, *Capital* [New York: Modern Library Edition, n.d.], p. 418). Similarly on water, abortive attempts were made to imitate nature. One such attempt—Lord Stanhope's ill-fated paddle steamer—has been preserved in the sad lines of the poet, T. Baker (fl. 1837–1857):

> Lord Stanhope hit upon a novel plan
> Of bringing forth this vast Leviathan
> (This notion first Genevois' genius struck);
> His frame was made to emulate the duck;
>
> Webb'd feet had he, in Ocean's brine to play;
> With whale-like might he whirl'd aloft the spray;
> But made with all this splash but little speed;
> Alas! the duck was doom'd not to succeed!

The Steam-Engine, Canto IV

As reprinted in D. B. Wyndham Lewis and Charles Lee, *The Stuffed Owl: An Anthology of Bad Verse* (London: J. M. Dent and Sons 1952), pp. 193–194.

[24] In this sense, the *sequence* of events in history becomes very important in explaining the experiences of individual industries. The problems encountered and solved in industry A often turn out to provide valuable externalities in the form of knowledge, techniques and labor skills which become available to industries B, C, and D. Thus, Usher has argued that the industrial revolution in England owed much to the technical skills which had been developed by generations of craftsmen in the production of

is abundant evidence from a variety of sources showing sustained reductions in real labor costs per unit of output in situations where labor was employed in a plant using unchanged facilities. Indeed, the phenomenon is sufficiently well established that it has come to be known as the "Horndal Effect," after the Swedish steelworks where output per manhour was observed to increase at about 2% per year for fifteen years in spite of the fact that the plant and production techniques remained unchanged. The phenomenon has been further documented in several industries, most notably air-frame production, machine tools, shipbuilding, and textiles.[25]

---

clocks and watches. Machines which had been developed in this trade "stand out as the most conspicuous examples of instruments of precision. The lessons learned by the craftsmen of these trades formed the basis for the development of the engineering sciences in the late eighteenth century and the early nineteenth century. These timekeepers presented a substantial array of notable devices for the control of motion. These devices involved all the primary problems of geared mechanisms. The marine chronometer required delicate adjustments to the expansion and contraction of metals during small changes of temperature. The pendulum clocks presented important problems in the theory of dynamics. The development of the pendulum clock rested upon a complete mathematical treatment of the forces operating in a pendulum. The escapements of both clocks and watches called for considerable refinements in the design of gear teeth, and the problems received full mathematical treatment in the course of the eighteenth century. Much of the work done for Arkwright on the spinning machine was entrusted to a clockmaker. George Stephenson learned much of his mechanics by repairing and studying clocks. The rapid development of the engineering sciences after Watt's inventions was largely due to the extensive mathematical treatment of the problems of dynamics involved in the construction of these small instruments of precision" (W. Bowden, M. Karpovich and A. P. Usher, *An Economic History of Europe Since 1750* [New York: American Book Co., 1937], p. 308). See also A. E. Musson and E. Robinson, "The Origins of Engineering in Lancashire," *Journal of Economic History*, 20 (June 1960), especially 219–222. Musson and Robinson conclude "that clock-, if not watch-, makers, and above all clock-tool makers, were in very great demand for textile-machine making and contributed materially to the early growth of engineering" (p. 222).

[25] A. Alchian, "Reliability of Progress Curves in Airframe Production," *Econometrica*, 31 (October 1963), 679–692; Werner Hirsch, "Firm Progress Ratios," *Econometrica*, 24 (April 1956), 136–143; Leonard Rapping, "Learning and World War II Production Functions," *Review of Economics and Statistics*, 47 (1965), 81–86; Paul David, "Learning by Doing and Tariff Protection: A Reconsideration of the Case of the Ante-Bellum U.S. Cotton Textile Industry," *Journal of Economic History*, 30, (September 1970), 521–601; Kenneth Arrow, "The Economic Implications of Learning by Doing," *Review of Economic Studies*, 29 (June 1962), 155–173. According

Factors Affecting the Diffusion of Technology         17

While the existence of learning curves within the framework of an *established* technology is well recognized, the role of learning experiences in accounting for the gradual improvements of *new* technologies and their slow diffusion has not received much attention. Since it takes time to acquire such skills, it will also take time to establish the superior efficiency of a new technique over existing ones. The point is nicely illustrated with respect to the adoption of Henry Cort's puddling process:

> One of the most important problems associated with puddling in its early years was that of training a labor force that could produce high quality bar iron with the process. The ironmasters who initially adopted puddling had to train workers in the use of a process that was not only new, but was also somewhat of a "mystery" to everyone, including Cort. An efficient puddler was a workman who could not only do the strenuous labor of moving masses of iron in and out of the puddling furnaces, but could also develop a "feel" for the process itself. He had to learn to determine from the color of the flames in the puddling furnace and the texture of the molten metal when the pig iron was fully decarburized or had "come to nature," i.e., when the carbon and other impurities had been sufficiently removed. Puddling was a backbreaking job that also required a great deal of judgment and experience was probably the best teacher. The development of a highly skilled labor force was crucial to the success of the puddling process and the lack of such a labor force was probably the greatest single impediment to its rapid adoption.[26]

The adoption of the steam engine afforded innumerable ex-

---

to Hirsch, the U.S. Air Force "for quite some time had recognized that the direct labor input per airframe declined substantially as cumulative airframe output went up. The Stanford Research Institute, and the RAND Corporation initiated extensive studies in the late forties, and the early conclusions were that, in so far as World War II airframe data were concerned, doubling cumulative airframe output was accompanied by an average reduction in direct labor requirements of about 20%. This meant that the average labor requirement after doubling quantities of output was about 80% of what it had been before. Soon the aircraft industry began talking about the 'eighty per cent curve'" (Hirsch, p. 136). It is possible, of course, that cost reductions which have been attributed to learning by doing have actually been due to other factors which have not been correctly identified, especially in cases where learning by doing has been defined as a residual. See John Chipman, "Induced Technical Change and Patterns of International Trade," in *The Technology Factor in International Trade*, ed. Raymond Vernon (New York: National Bureau of Economic Research, 1970), pp. 95–98.

[26] Charles K. Hyde, "Technological Change in the British Iron Industry, 1700–1860," unpublished doctoral dissertation, University of Wisconsin, pp. 112–113. See also Landes, p. 92.

amples of the importance as well as the slow accumulation of know-how, which is essential to the successful operation of a new technology. The length of life and the frequency of breakdowns of steam engines required that overloading be scrupulously avoided, but this in turn involved the accumulation of experience concerning optimum loads. Again, the life expectancy of engine boilers required careful attention to such matters as pressure levels and appropriate feed-water arrangements—indeed, when these matters were *not* attended to, the life of a boiler was likely to be abruptly, and sometimes disastrously, terminated.[27]

The manner in which these new technical skills are acquired is relevant to the speed of the diffusion process in another way. Many of the technical skills are acquired through direct, on-the-job participation in the work process. Since these include a large component of uncodified skills (or know-how), such skills were not readily transferable through formal education or the printed word, but required the movement of qualified personnel. Where and to the extent that this was so, it placed a serious constraint upon the speed of geographic diffusion.[28]

V

*The Development of Skills in Machine-making*

The next portion of my argument deals with the development of the skills involved not in *using* the new techniques but in developing the skills and facilities in machine-making itself. This involves the broadest questions of industrial organization and specialization and lies at the very heart of the industrialization process. Successful invention and successful *diffusion* of inventions in industrializing economies has required, above all, a growth in the capacity to devise, to adapt, and of course, to produce at low cost, machinery which has been made suitable to highly specialized end uses. Before a new invention can join the family of technical options genuinely available to the economy, elaborate arrangements must sometimes be made.

[27] Loss of life was especially fearful aboard steamboats on western rivers, and this loss of life led to an early assertion of the investigatory and regulatory activites of the federal government. See John G. Burke, "Bursting Boilers and Federal Power," *Technology and Culture*, 7 (Winter 1966).

[28] Nathan Rosenberg, "Economic Development and the Transfer of Technology: Some Historical Perspectives," *Technology and Culture*, 11 (October 1970), 550–575.

Factors Affecting the Diffusion of Technology          19

The mere conceptualization of a solution may be, and often has been, very far removed in calendar time from the availability of a method which is technologically workable, much less commercially feasible.

It is an often-told tale in the history of inventions that they have to sit on the shelves long after their initial conceptualization because of the absence of the appropriate mechanical skills, facilities, and design and engineering capacity required to translate them into a working reality. This is why the emergence of a capital goods industry, with a sophisticated knowledge of metallurgy and the capacity to perform reliable precision work in metals, was so critical to industrialization in its eighteenth- and nineteenth-century form. The desperate and unsuccessful improvisations which otherwise had to be resorted to is perfectly captured in the picture of James Watt stuffing soaked rags in the gaps between his pistons and cylinders in an effort to prevent loss of steam, until Wilkinson's boring mill finally provided him with reasonably accurate cylinders. The commercial practicability of Watt's steam engine with its separate condensing chamber really dates from 1776, the year in which Wilkinson's boring mill, invented in 1774, became available for the preparation of his cylinders.[29]

It might be said of Watt that he was singularly lucky in having a cannon-maker such as Wilkinson nearby, but the essential point, surely, is that his conceptualizations were not so far in advance of the technical capacities of the metal-workers of his day as to render his ideas unfeasible.[30] Such was the lot of da Vinci whose notebooks are crowded with machinery sketches far in advance of the technical

[29] K. R. Gilbert, "Machine Tools," in Charles Singer et al., *A History of Technology* (London: Oxford University Press, 1958), IV, p. 421. See also S. Smiles, *Industrial Biography* (London: John Murray 1908), pp. 178–182. As Landes points out about the separate condenser, Watt "saved the energy that had previously been dissipated in reheating the cylinder at each stroke. This was the decisive breakthrough to an 'age of steam' not only because of the immediate economy of fuel (consumption per output was about a fourth that of the Newcomen machine) but even more because this improvement opened the way to continuing advances in efficiency that eventually brought the steam-engine within reach of all branches of the economy and made of it a universal prime mover" (Landes, *Unbound Prometheus*, p. 102).

[30] At a much earlier date, the cannon makers had borrowed important techniques, especially that of casting in bronze, from the makers of church bells. Such are the vagaries—and ironies—of the industrial learning process. Cipolla, pp. 23, 25.

skills of early sixteenth-century Florence or Europe. Breech-loading cannon had been made as early as the sixteenth century but could not be fired in reasonable safety (to the user at least!) until precision in metal-working made it possible to produce an airtight breech and properly-fitting case. Christopher Polham, a Swede, devised many techniques for the application of machinery to the quantity production of metal and metal products, but could not successfully implement his conceptions with the power sources and clumsy wooden machinery of the first half of the eighteenth century. Charles Babbage had already conceived of the main features of the modern computer over a hundred years ago and had incorporated these features into his "analytical engine," a project for which he actually received a large subsidy from the British Government. Babbage's failure to complete his ingenious scheme was due to the inability of contemporary British metal-working to deliver the components which were indispensable to the machine's success.[31]

People like da Vinci and Babbage were, to use the popular phrase, "far ahead of their time." But, to give the phrase some operational content, it is the state of development of the capital goods industries, more than any other single factor, which determines whether and to what extent an invention is ahead of its time. Each important invention goes through a gestation period of varying length, while the capital goods industries adapt themselves to the special needs and requirements of the new technique. Therefore the pace of technical advance in the user industry may depend critically upon events in the capital goods sector. This process of problem-solving and accommodation is central to a better understanding of the timing of technical change and the rate of diffusion of new inventions.[32] For it is the speed with which performance characteristics are improved, techniques modified to meet the needs of specialized users, and the price of the invention gradually reduced, which determine its acceptability among an increasingly widening circle of potential users.

[31] S. H. Hollingdale and G. C. Tootill, *Electronic Computers* (London: Pelican, 1965), p. 46 and Ch. 2 and 3. Babbage, it is interesting to note, had borrowed the system of punched cards from the Jacquard loom—where they had been used to control the introduction of threads in weaving brocade. See Charles Babbage, *Passages from the Life of a Philosopher* (London: Longman, Green, Longman, Roberts, and Green, 1864), pp. 116–118.

[32] Nathan Rosenberg, "Technological Change in the Machine Tool Industry, 1840–1910," *Journal of Economic History*, 23 (December 1963).

*Factors Affecting the Diffusion of Technology*        *21*

VI

*Complementarities*

A further element significantly affecting the timing of the diffusion process, and one where the capital goods sector also plays an
important role, lies in the complementarity in productive activity
between different techniques. That is to say, a given invention, however promising, often cannot fulfill anything like its potential unless
*other* inventions are made relaxing or bypassing constraints which
would otherwise hamper its diffusion and expansion. It is for this
reason that a single technological breakthrough hardly ever constitutes a complete innovation. Before the productivity-increasing
benefits of any single breakthrough can be realized, many other ac-
commodations need to be made. The expansion of a productive
activity runs into a series of new constraints or bottlenecks. As one
bottleneck is overcome, others eventually assert themselves and need
to be expanded. Although these bottlenecks can often be overcome
by committing more resources to a particular activity, frequently
inventive activity is called for. In both cases, however, time-consuming procedures are involved which hold back the further expansion and diffusion of the new technique.

The history of American railroads in the second half of the
nineteenth and the early twentieth centuries provides compelling
evidence that the growth in productivity was the product of many
subsequent inventions, none of which was available in the early
years of railroad building—say 1840. The growth in productivity
was, to begin with, very great. It has been calculated that the incremental expenses required for meeting the railroad demands of 1910
traffic loads with the technology available back in 1870 would have
amounted to about $1.3 billion. The cumulation of small innova-
tions and relatively modest individual design changes brought about,
between 1870 and 1910 alone, a more than tripling of freight car
capacities with only a small increase in dead weight, and a more than
doubling of locomotive force with the introduction of more powerful
engines. The greater loads and greater speeds made possible by the
improved rolling stock could not have been achieved, however,
without several other significant inventions: the control of train
movements through use of the telegraph, block signalling, air
brakes, automatic couplers, and the substitution of steel rails for
iron. Not all these inventions were equally significant in reducing
costs, nor were they, as a result, adopted with equal speed. Air

brakes and automatic couplers (first employed in 1869 and 1873 respectively) were coolly received and were eventually adopted only after the passage of national legislation.[33] Steel rails, however, in spite of their considerably higher price, were rapidly adopted. Steel rails were first used by the Pennsylvania Railroad in the early 1860s, and by 1890 they accounted for 80% of all track mileage.[34] The critical importance of steel rails to the growth in railroad productivity was that they were far more durable, lasting more than ten times as long as iron rails, and that they were capable of bearing far heavier loads than iron rails without breaking. Indeed, the old iron rails were simply incapable of supporting the 1910 locomotives, and would have crushed under their average weight of 70 tons.[35]

The argument made here with respect to complementarities in railroad operation could, if space permitted, be expanded to encompass other classes of invention not so far mentioned, such as bridgebuilding. Bridgebuilding in both America and Great Britain underwent drastic changes in structure, design, and materials as the railroad network expanded and confronted engineers with problems and requirements respecting such matters as strength, rigidity, and fire resistance for which there was absolutely no precedent in the pre-railroad age. The need to provide bridges suitable to the requirements of the railroads led, in Great Britain, to a systematic study of iron as a building material. The outcome of this study was a major advance in knowledge concerning the structural properties of iron —resistance of beams and plates, strength of girders, compression and tensile strengths, etc.[36] The knowledge thus acquired soon had a wide range of applications wherever iron was used as a building

[33] A. Fishlow, "Productivity and Technological Change in the Railroad Sector, 1840-1910," in Studies in Income and Wealth, no. 30, *Output, Employment and Productivity in the United States after 1800* (New York: National Bureau of Economic Research 1966), pp. 635, 641.

[34] As Fishlow points out, the main advantage of the air brake and automatic coupler was increased speed. "Greater speed in itself is not an unmixed blessing, however. Unless engine capacity is not being fully utilized, higher speeds can be attained only by the sacrifice of load. What the air brake and automatic coupler really did, therefore, was to allow a greater element of choice in train operation, permitting higher speed when it was more desirable than larger loads." *Ibid.*, p. 636.

[35] *Ibid.*, pp. 635, 639-640.

[36] "Report of the Commissioners Appointed to Inquire into the Application of Iron to Railway Structures," *Parliamentary Papers*, 1849, Vol. 29; William Fairbairn, *Useful Information for Engineers*, London, 1860, vol. II, pp. 223-228.

Factors Affecting the Diffusion of Technology              23

material—in shipbuilding, multistoried buildings, cranes, steam engines, etc.

The present discussion is merely illustrative of a much larger class of complementarities which exert significant effects upon the pattern of diffusion. The argument could readily be duplicated almost endlessly from different sectors of industry and agriculture.[37]

## VII

*Improvements in "Old" Technologies* ·

The discussion so far has examined a variety of factors which have the effect of slowing down the diffusion of new techniques. We have considered several classes of reasons why the full productivity-increasing effects of a new technique may take a great deal of time to assert themselves. There is, however, an additional powerful explanation for such slowness which has received surprisingly little attention. That is, the "old" technology continues to be improved after the introduction of the "new," thus postponing even further the time when the old technology is clearly outmoded. Yet, curiously, it is a very general practice among historians to fix their attention upon the story of the new method as soon as its technical feasibility has been established and to terminate all interest in the old. The result, again, is to sharpen the belief in abrupt and dramatic discontinuities in the historical record.

A closer look at this record discloses not only the persistence of

[37] The contemporary experience with the introduction of new, high-yielding rice and wheat varieties in Southeast Asia forcefully exemplifies the argument advanced here. It is currently being discovered that the adoption of the high-yielding rice varieties generates a whole new series of requirements with respect to fertilizer use, water management, harvesting, processing (drying, storing, milling, grading), and disease and insect control. Although many of these problems can be dealt with by conventional means, others cannot, e.g., double-cropping calls for low-cost harvesting, threshing, and other machinery of a kind not existing ten years ago. Much attention is currently being devoted to the development of such machinery at the International Rice Research Institute at Los Banos, The Philippines, where the new rice varieties were first developed.

The really distinctive feature of the new rice varieties is that they have been genetically designed so that they are highly fertilizer-responsive. Indeed, without large doses of fertilizer the new varieties are no more productive than the old ones. A continued diffusion of the new varieties would seem to call for cost-reducing innovations in the provision of fertilizer and the entire range of complementary inputs. Better still would be the development of new seed varieties not *requiring* these complementary inputs.

old technologies in places where location or resource availability provided special advantages, but often major improvements in these technologies long after they were supposed to have expired. Thus, the slowness with which the stationary steam engine, as noted earlier, established itself as a new power source in the first half of the nineteenth century, was due in part to important improvements which continued to be made in design and construction of water wheels. Many of these improvements centered around the introduction of iron as a building material, and men like William Fairbairn could achieve international reputations well into the nineteenth century as builders of water wheels.[38] In fact, early steam engines were commonly used to supplement the action of a water mill by pumping the water from the lower mill pond back to the upper pond—thus enabling it to run over the water wheel many times. Needless to say, this was a cumbersome and inefficient arrangement. Nevertheless, until the development of the rotative steam engine which converted the oscillation of the beam into rotary motion, it was an essential expedient since the steam engine prior to that was really no more than a water pump.[39]

An additional fillip was given to the utilization of water after 1840 as a result of the introduction of the water turbine, which further reduced the cost of water power.[40] In America, where early industry had been heavily concentrated in New England, a region with highly favorable water power locations, water power was probably the main power source for manufacturing until well into the second half of the nineteenth century. Even as late as 1869, steam power accounted for barely over one half of primary-power capacity in U.S. manufacturing—51.8% as compared to 48.2% for water.[41] In

[38] William Pole, *The Life of Sir William Fairbairn, Bart* (London: Longmans, Green and Co., 1877), and William Fairbairn, *Treatise on Mills and Millwork* (London: Longman; Part I, 1861 and Part II, 1863).

[39] R. L. Hills, *Power in the Industrial Revolution* (Manchester: Manchester University Press 1970), Ch. 8. Hills also points out that the waterwheel was long preferred to the steam engine in spinning because "it was essential to have as regular speed as possible. Until some method of automatically controlling the revolutions had been found, the waterwheel with its greater steadiness was preferable" (*Ibid.*, p. 175).

[40] Victor Clark, *History of Manufactures in U.S.* (New York: McGraw-Hill, 1929), II, pp. 407-408. Although of French origin, much of the practical development of the turbine was performed in America.

[41] Allen H. Fenichel, "Growth and Diffusion of Power in Manufacturing, 1839-1910," in Studies in Income and Wealth no. 30, *Output, Employment and Productivity in the United States after 1800* (New York: National

## Factors Affecting the Diffusion of Technology 25

general it may be said that steam power tended to be adopted earliest in locations where water power sources were scarce and where fuel was either abundant or easily transportable.

Even so "primitive" a power source as the windmill, which might have been regarded as a certain early casualty to the steam engine, experienced a considerable growth in at least one English county. According to Finch, the number of windmills in Kent employed in the grinding of corn grew from 95 to 239 between 1769 and the 1840s.[42] By the 1860s steam had decisively established its superiority over wind in cornmills, and windmills continued to be operated only so long as they provided quasi-rents to their owners. Jevons, writing in the middle of the 1860s, pointed out: "Wind-cornmills still go on working until they are burnt down, or go out of repair; they are then never rebuilt, but their work is transferred to steam-mills."[43]

In iron and steel, where the location of particular resource inputs and their chemical properties played an extremely important role (especially because of the intimate connection between the presence of such chemicals as sulfur and phosphorus and a poor quality final product), the introduction into the United States of the modern, mineral-based British technology was long delayed. Furthermore, the old and the new technology coexisted for long periods of time after the new technology was finally introduced. As late as 1840 almost 100% of all pig iron in the U.S. was still being produced with charcoal. Although the proportion subsequently declined rapidly as anthracite and, later, bituminous coal were introduced,[44] the tonnage of pig iron produced by charcoal continued to rise through the 1880s and reached its alltime annual peak in 1890.[45] The most interesting point here for present purposes is that during the 1840s and 1850s, when the new mineral fuel tech-

---

Bureau of Economic Research, 1966), pp. 443–478, Appendix B. In absolute terms waterpower capacity continued to grow through the first decade of the twentieth century (Table A-1).

[42] William Coles-Finch, *Watermills and Windmills* (London: C. W. Daniel, 1933), pp. 136–137.

[43] W. S. Jevons, *The Coal Question* (London: Macmillan and Co., 1865), p. 136.

[44] "The proportion of pig iron made with charcoal declined from close to 100 per cent in 1840 to about 45 per cent in the middle 1850s ... and to about 25 per cent at the close of the Civil War" (Peter Temin, *Iron and Steel in 19th Century America* [Cambridge: MIT Press, 1964], p. 82).

[45] *Ibid.*, Table C-2.

nology was being introduced, it was primarily the reduction in the demand for charcoal iron which accounted for the relative decline in the charcoal sector as compared to the mineral fuel sector. Indeed, Fogel and Engerman actually conclude that, between 1842 and 1858, the growth in total factor productivity in the "backward" charcoal sector probably exceeded the growth in total factor productivity in the "modern" anthracite sector.[46]

My point so far has been that one of the reasons new technologies seem to displace old ones slowly is that the old technologies continue to improve. But the point needs to be seen within a larger framework, for there is often an intimate connection between innovations, on the one hand, and improvements in older technologies on the other. That is to say, innovations often appear to *induce* vigorous and imaginative responses on the part of industries for which they are providing close substitutes. What is being suggested here is a possible lack of symmetry in the manner in which business firms respond to alterations in their profit prospects. The imminent threat to a firm's profit margins which are presented by the rise of a new competing technology seems often in history to have served as a more effective agent in generating improvements in efficiency than the more diffuse pressures of intra-industry competition. Indeed, such asymmetries may be an important key to a better understanding of the workings of the competitive process, even though they find no place at present in formal economic theory.

Thus it has often been asserted that, by the 1850s, the iron hull cargo steamship had displaced the sailing ship and that Britain built its worldwide trading empire on the new vessel. In fact, while the complex problems of designing an efficient steamship, with its iron hull, engines, screw propellers, and very high fuel requirements, were being worked out, the wooden sailing ship also underwent a series of drastic changes. Builders of sailing ships responded to the competition of iron and steam by a number of imaginative changes in hull design, including the use of iron itself in a "composite" hull —wood placed on an iron skeleton. They adopted a range of labor-saving machinery to reduce crew requirements. According to Graham:

[46] Robert Fogel and Stanley Engerman, "A Model for the Explanation of Industrial Expansion During the 19th Century: With an Application to the American Iron Industry," as reprinted in *The Reinterpretation of American Economic History*, Robert Fogel and Stanley Engerman, (New York: Harper and Row, 1971), pp. 159–162.

Although the steam ship had successfully wedged its way into the overseas trade, mainly by carrying passengers and subsidized mails, the evolving sailing ship of the 1860's and 1870's—faster than its predecessors, with double the space for cargo in proportion to tonnage, and manned and navigated by about one-third the number of men—retained on broad oceans a predominance almost as marked as that of the screw steamer in the coastal and neighbouring waters of Europe. Even when the opening of the Suez Canal in 1869 reduced the longest gap between coaling stations from some 5,000 to 2,000 miles, although the China trade was eventually lost, most of the traffic to the Bay of Bengal, the East Indies, South America or Australia, was still conducted by the sailing ship which continued to be the more economical carrier for the greatly expanding trade in bulky commodities, such as iron and coal, the jute and rice of India and Burma, the wool of Australia, the nitrate fertilizer of Chile and the wheat of California.[47]

Even the rapid growth of the steamship fleet in the 1870s offered some consolation to the sailing ship: "it was coal more than any other article that brought a new lease of life to the commercial sailing ship in the latter days of her ascendancy. As the cheapest coal carrier, as well as the cheapest warehouse in the world, the sailing ship became the chief replenisher of overseas coaling bases and depots."[48] Furthermore, some of the design changes which improved the performance of the sailing ship were also made possible by steam, in this case by the steam tugboat which, "taking them in and out of harbor, relieved the windjammers of need for handiness, enabling greater length and fine lines, and enabling guaranteed sailings out of a harbor."[49]

[47]C. S. Graham, "The Ascendancy of the Sailing Ship, 1850–1885," *Economic History Review*, 2nd series, 9 (August 1956), 81. For a careful examination of the economic impact of technological changes in the steamship, see Charles K. Harley, "The Shift from Sailing Ships to Steamships, 1850–1890: a Study in Technological Change and its Diffusion," in *Essays on a Mature Economy: Britain after 1840*, ed. Donald N. McCloskey (London: Methuen, 1971), pp. 215–231.

[48]*Ibid.*, p. 84

[49]S. C. Gilfillan, *Inventing the Ship* (Chicago: Follett Publishing Co., 1935), p. 157. It is not unusual for the "new" technology to extend the life of the "old" by providing it with some form of externality. Thus, the arrival of the steamboat on western rivers brought about significant reductions in labor costs in flatboat operation by providing flatboatmen with a speedy form of upriver transportation—a trip which had previously been both very slow and costly. See E. Haites and J. Mak, "Ohio and Mississippi River Transportation, 1810–1860," *Explorations in Economic History*, 8 (Winter 1970–1971), fn. 36, and Mak and Walton, "Steamboats," p. 19.

By the 1880s the sailing ship finally lost its dominance even on long distance hauls to steamships, which were now equipped with high pressure compound engines and a range of superior components provided by the recent breakthroughs in steel-making technology—including the very important boiler plates and boiler tubes, so essential to high pressure and fuel economy.[50] The sailing ship of the 1880s was far superior to its predecessor of 1850 or so, and it seems plausible to attribute this improvement to the strong competition of steam.[51] Obviously one cannot assert this with authority, because we do not know what the sailing ship of the 1880s would have been like in the absence of such intertechnological competition. But it seems like a reasonable conjecture for which there is analogous evidence in the experience of other industries. Thus, technological competition recently appears to have been a powerful force among materials producers (where, for example, the increasing competition from aluminum seems to have led to the setting up of product-research and engineering laboratories in the steel industry), among suppliers of transportation services, and among the major kinds of fuel. Not only does this form of competition generate economically-beneficial consequences; it also plays a significant role in explaining the rate of diffusion of some new techniques.

[50] See Douglass North, "Ocean Freight Rates and Economic Development, 1750–1913," *Journal of Economic History*, 18, (December 1958); Douglass North, *Growth and Welfare in the American Past* (Englewood-Cliffs, N.J.: Prentice-Hall, 1966), Ch. 9; and Gary Walton, "Productivity Change in Ocean Shipping After 1870: A Comment," *Journal of Economic History*, 30 (June 1970). North states: "Although the steamship substituted for the sailing ship in passenger travel as early as 1850, it did not substitute for the sailing ship in the carriage of bulk goods in ocean transportation until much later. Indeed, as late as 1880 most of the goods carried in ocean transportation were going by sail, and the changeover from sail to steam did not occur in most of the long-haul routes in the world until the very end of the nineteenth century . . ." (*Growth and Welfare*, p. 110).

[51] The sailing ship adapted to its more specialized role as a long-distance carrier of bulk cargoes by modifying its sails and rigging so as to reduce crew requirements. In part this was done by "abolishing the lightest sails, broadening the upper yards, and cutting in two the largest sails until furling could replace almost all reefing" (S. C. Gilfillan, p. 160). The American merchant fleet was slower than the British in substituting steam for sail. As late as 1913 almost 20% of the gross tonnage of American merchant vessels still consisted of sailing vessels. *Historical Statistics of the U.S., Colonial Times to 1957* (Washington, D.C.: U.S. Printing Office, 1960), p. 444.

## VIII

*Diffusion and its Institutional Context*

This paper has discussed, at some length, several categories of technological considerations which have influenced the pace of diffusion. Needless to say, the treatment has been suggestive rather than exhaustive; in fact, the number of variables—social, legal, and institutional as well as economic and technological—which might retard the diffusion process is virtually limitless. Nevertheless, it is important that an effort be made to maintain conceptual clarity among these categories because our understanding of the process of long-term economic growth is influenced, in important ways, by this conceptual apparatus. Ever since Abramovitz and Solow opened up the problem of "The Residual," economists have been attempting to sort out the contributions of various factors to economic growth and, particularly, to measure the contribution of technological change as distinguished from all other possible factors. Whereas the entire residual was for some time uncritically attributed to technological change (although not by Abramovitz or Solow) a later, more discriminating approach has attempted to isolate other factors—changes in organization, improvements in the quality of the labor force, etc.—and to measure their separate contributions.[52] In this difficult but essential process of "cutting technological change down to size," however, there is a danger of going too far, by assigning an independent and separate role to factors which really exert their effects upon the growth of productivity by retarding or accelerating the rate of technological diffusion.

A recent example of the position I am criticizing is North's

---

[52] North threw out the following challenge several years ago: "I would hazard the speculation that if we ever did the research necessary to get some crude idea of the magnitudes involved, we would discover that improved economic organization was as important as technological change in the development of the Western world between 1500 and 1830. I mean by this, improvements in the factor and product markets, reduction in impediments to efficient resource allocation, and economies of scale. Moreover, the complementarity between physical and human capital in the development, application, and spread of technological change requires equal analytical attention before we can begin to make sense on this subject. Clearly, we need to overhaul our view of the whole process by which the Western world developed in the last five or six centuries" (Douglass North, "The State of Economic History," *American Economic Review, Papers and Proceedings,* 55 [May 1965], 87–88.

otherwise admirable study of "Sources of Productivity Change in Ocean Shipping, 1600–1850."[53] North states that his objective is to "identify as precisely as possible those sources of productivity usually lumped into the general category of technological change." And, he adds: "The conclusion which emerges from the study is that a decline in piracy and an improvement in economic organization account for most of the productivity change observed."[54]

With a portion of North's argument there is no disagreement whatever. North makes an important contribution to our understanding of productivity growth during the period by demonstrating that organizational and marketing improvements were highly significant. He shows that, in the tobacco trade before the Revolutionary War, the increased number of round trips per year was due, not to increased speed resulting from technological change, but rather to the introduction of a system of factors and an increasing centralization of inventories. These organizational improvements, by making it easier and much quicker for ships to secure cargoes, substantially reduced the ratio of port time to sea time, and thereby sharply increased the quantity of freight which could be carried by a given stock of ships.[55]

The other portion of North's argument, involving an attempt to downgrade the contribution of technological change to the growth of productivity in ocean shipping, is more questionable. North finds that, before 1800, the most important source of the great reduction in crew size requirements was the decline in piracy and privateering. His "downgrading" of the role of technical change proceeds as follows:

> One can ask at this point ... the extent to which technical changes in shipping and in ship construction account for the changes in manning requirements and, indeed, in observed ship speed. There is no doubt that the ship of the nineteenth century was in striking contrast to the ship of the early seventeenth century. Except for one crucial point it could be argued that smaller crews were made feasible precisely by technological improvements in sail and rigging. The obstacle to this argument is that by 1600 the Dutch had developed a ship, the flute, which cost less to construct than existing ships, had a tons-per-man ratio similar to that of nineteenth-century ships on the Atlantic route, was at least as fast as existing ships, and could be (and was) constructed of 500–600 tons burden. While the design was copied and

[53] *Journal of Political Economy*, 76 (September–October 1968), 953–970.
[54] *Ibid.*, p. 953.
[55] *Ibid.*, pp. 960–963.

modified over the next two centuries, the essential economic characteristics were not basically altered. The enigma to be explained, therefore, is why the flute (or ships of similar design) took so long to spread to all the commodity routes in the world, once it had entered the Baltic route and the English coal trade in the first half of the seventeenth century. The answer lies in the very nature of the flute and its great advantages in that it was lightly built, frequently carried no armament, was easy to sail, and had simple rigging. These characteristics had all come about because the Dutch enjoyed a large-volume bulk trade in the Baltic, where piracy had already been eliminated. Only as privateering was driven from other seas and as improvements took place in market organization was it possible to put into general active service ships designed exclusively for the carrying trade.[56]

The trouble with this paragraph is that the diffusion process has been completely lost from view. A superior technology in the form of the Dutch flute existed by 1600, but security considerations long confined its adoption to a small portion of ocean trade. As piracy and privateering were suppressed in the course of the eighteenth century, the superior vessel was widely adopted on new routes with the expected rise in productivity. But, if all this is so, the elimination of piracy and privateering emerge as factors which influence shipping productivity only as intervening variables: i.e., the threat which they posed to the security of shipping was responsible for *the very slow diffusion of a superior technology*—"the flute (or ships of similar design.)" There seems to be general agreement that, in the *absence* of the security threat, the flute design would have been adopted much earlier.[57] North, however, in his legitimate concern with deflating the overblown spectre of technological change, gives the impression—doubtless unintended—that it was scarcely of any significance whatever in the period with which he is concerned. It goes unmentioned in his final sentence, which states: "The conclusion one draws is that the decline of piracy and privateering and the development of markets and international trade shared honors as primary

[56]*Ibid.*, p. 964.
[57]Thus Walton states, with respect to colonial shipping: "As the obstacles of piracy and similar hazards were eliminated, specialized cargo-carrying vessels possessing the input characteristics of the flyboat were adopted. In the process, the costs of shipping were substantially reduced, which had a favorable impact on the development of a trading Atlantic community" (Gary Walton, "Obstacles to Technical Diffusion in Ocean Shipping, 1675-1775," *Explorations in Economic History*, 8 (Winter 1970-1971), p. 136.

factors in the growth of shipping efficiency over this two-and-a-half
century period."[58]

The interpretation which North has placed upon his historical
account has been restated by Fogel and Engerman in their volume,
*The Reinterpretation of American Economic History*, where North's
article has been reprinted. Fogel and Engerman regularly refer to
North's article as showing that factor productivity did not rise as a
result of *new* inventions.

> In the case of ocean shipping, Douglass North ... found that a rapid
> and protracted increase in total factor productivity took place despite
> the absence of a single major *new* invention. According to North the
> rise of efficiency was due largely to the change in the proportion of
> large ships in the Atlantic fleet. This diffusion of large ships was set
> off, not by *new* technological knowledge, but by a change in institu-
> tional conditions.[59]

And, earlier: "Thus, *new* equipment plays virtually no role in
Douglass North's explanation of the 50 per cent fall in the cost of
ocean transportation that he finds for the 250-year period between
1600 and the middle of the nineteenth century."[60]

But if a superior ship designed specifically for improved cargo-
carrying capacity had been developed by 1600, it is no verbal quibble
to say that the improvements in ocean shipping productivity due to
the eventual adoption of this design should correctly be regarded as
belonging to the category of technological change. The portion of
North's paper dealing with piracy is not an explanation of productivity growth which is *independent* of technological change, although
it is frequently made to sound that way. Rather, it is a cogent and
forceful explanation for the very slow *diffusion* of a major techno-

---

[58] North, "Sources of Productivity Change," p. 967.

[59] Robert Fogel and Stanley Engerman, *The Reinterpretation of American Economic History* (New York: Harper and Row, 1971), p. 206, emphasis
added.

[60] *Ibid.*, p. 5, emphasis added. Also, p. 100: "North argues that most of
these changes were due to improved organization rather than new equipment. ... He holds that no *new* technological knowledge was required for the
switch from fleets of predominantly small ships to fleets of predominantly
large ones. ... What then explains the dominance of the small over the large
ship in the seventeenth and eighteenth centuries and then the rapid shift
toward large ships between 1800 and 1860? North again finds the answer
not in *new* technological knowledge, but in institutional change. He argues
that the elimination of piracy made it feasible to build large, light vessels for
the exclusive purpose of carrying cargoes." [Emphasis added.]

logical innovation. What seems to be at issue here—and this emerges with particular clarity in the writing of Fogel and Engerman—is that benefits attributable to technological change are being arbitrarily confined to recent or "new" developments. But there is no obvious reason why the productivity-increasing effects of technological change should be confined to changes of recent vintage. Surely the essential point, on which all would agree, is that the productivity of any technology is never independent of its institutional context and therefore needs to be studied within that context. North's paper should be interpreted as a striking demonstration of this point, for he shows how this institutional context can account for the very slow diffusion of a superior shipping technology.

## IX

The several arguments of this paper all add up to a perspective and a program for research rather than a sharply defined set of conclusions. A variety of reasons have been advanced for believing that a new technique establishes its advantages over old ones only slowly, and it has been argued that the apparent slowness of the *diffusion* of technologies is linked to this process and needs to be studied in relation to it. In spite of the occasional appearance of inventions which seem to be spectacular for their *technological* novelty, the *economic* impact of such inventions is much more diffuse and gradual. Their introduction into the texture of the economy is more accurately—if less dramatically—viewed as occurring along a gradual downward slope of real costs rather than as a Schumpeterian gale of creative destruction. At the same time, it is perfectly apparent that the question posed earlier has not been answered: How slow is slow? (How fast is fast?). But it should be clear by now that I would not worry excessively about that failure so long as we can advance our understanding of the reasons for the *actual* historical pace of technological diffusion. Once that pace has been established, we can each go our separate ways in deciding what we choose to regard as fast or slow.

# [3]

Excerpts from Everett M. Rogers with F. Floyd Shoemaker, *Communication of Innovations: A Cross-Cultural Approach*, 18–40, 174–96.

## Elements in the Diffusion of Innovations

Crucial elements in the diffusion of new ideas are (1) the *innovation* (2) which is *communicated* through certain *channels* (3) over *time* (4) among the members of a *social system*.* It is the element of time which distinguishes diffusion from other types of communication research. As pointed out previously in this chapter, diffusion research deals only with messages that are *new* ideas.

The four elements of diffusion differ only in nomenclature from the essential elements of most general communication models. For example, Aristotle proposed a very simple model of oral communication consisting of the speaker, the speech, and the listener. Laswell described all communication as dealing with *"who says what, through what channels* of communication, to *whom* with what . . . *results"* (Smith and others, 1946, p. 212). The S–M–C–R model, cited earlier, consists of (1) source, (2) message, (3) channel, and (4) receivers, to which we might add (5) the *effects* of communication. Obviously, this S–M–C–R–E communication model (Berlo, 1960) corresponds closely to the elements

*These four elements are similar to those listed by Katz and others (1963) as essential in any diffusion study: (1) the *acceptance*, (2) over *time*, (3) of some specific *item*—an idea or practice, (4) by individuals, groups, or other *adopting units*, linked to (5) specific *channels* of communication, (6) to a *social structure*, and (7) to a given system of values or *culture*. We do not include element 1 as a separate item among our four, as we see acceptance or adoption as an effect of communication (our element 2). We collapse Katz and others' (1963) elements 4, 6, and 7 in our fourth element, because they make up various aspects of the social system. No single chapter in the present volume deals solely with the effects of culture on diffusion, but in various chapters we shall discuss specific interfaces of culture and diffusion. Examples are cultural consequences of diffusion (Chapter 11), how an innovation's cultural compatibility affects its rate of adoption (Chapter 4), and the influence of traditional and modern norms on diffusion (Chapter 1).

of diffusion (Figure 1-2): (1) the receivers are members of a social system, (2) the channels* are the means by which the innovation spreads, (3) the message is a new idea, (4) the source is the origin of the innovation (an inventor, scientist, change agent, opinion leader, and the like), and (5) the effects are changes in knowledge, attitude, and overt behavior (adoption or rejection) regarding the innovation.

## The Innovation

An *innovation* is an idea, practice, or object perceived as new by an individual. It matters little, so far as human behavior is concerned, whether or not an idea is "objectively" new as measured by the lapse of time since its first use or discovery. It is the perceived or subjective newness of the idea for the individual that determines his reaction to it. If the idea seems new to the individual, it is an innovation.

"New" in an innovative idea need not be simply new knowledge. An innovation might be known by an individual for some time (that is, he is aware of the idea**), but he has not yet developed a favorable or unfavorable attitude toward it, nor has he adopted or rejected it. The "newness" aspect of an innovation may be expressed in knowledge, in attitude, or regarding a decision to use it.

Every idea has been an innovation sometime. Any list of innovations must change with the times. Black Panthers, computers, micro-teaching, birth control pills, chemical weed sprays, LSD, heart transplants, and laser beams might still be considered innovative ideas at this writing, but the reader in North America will probably find many of these items adopted or even discontinued at the time of reading. This list also illustrates the great variety

---

*The source and channel are seldom distinguished in most diffusion research, which is usually conducted by obtaining recall data from receivers via personal interviews. These receivers can report the communication channels from which they directly obtained information about the innovation, but they may be unaware of the original source of the new idea (in the case of technological innovations, this original source is ultimately a scientist). Some receivers may recognize the immediate source of information about the innovation when the channels are interpersonal, as in the case of a farmer-receiver obtaining information via interpersonal channels from a change agent, such as an agricultural extension agent. But generally the equivalent of the communication channel element in diffusion research is both the source and channel in the S–M–C–R–E model.

**There is, of course, more to knowledge about an innovation than simply awareness of it. Also important as a basis for effective decision making (adoption or rejection) about an innovation is the degree of knowledge about how properly to use the idea.

**Figure 1-2
Elements in the diffusion of innovations
and the S-M-C-R-E communication model are similar.**

| Elements in the S-M-C-R-E Model: | Source | — | Message | — | Channel | — | Receiver | — | Effects |
|---|---|---|---|---|---|---|---|---|---|
| Corresponding elements in the diffusion of innovations: | Inventors, scientists, change agents, or opinion leaders | | Innovation (Perceived attributes, such as relative advantage, compatibility, etc.) | | Communication channels (Mass media or interpersonal) | | Members of a social system | | Consequences over time 1. Knowledge 2. Attitude change (persuasion) 3. Behavioral change (adoption or rejection) |

of material products, ideological beliefs, social movements, and so on that qualify as innovations. Among the wide range of innovations that have been analyzed by diffusion researchers are a new speech form among oil drillers (Boone, 1949), nuclear warfare among nations (Beaton and Maddox, 1962), a rumor aboard a submarine (Allingham, 1964), and snowmobilies among the Lapps (Pelto and others, 1969).

It should not be assumed that the diffusion and adoption of all innovations is necessarily desirable. In fact, we shall review studies of harmful and uneconomical innovations* that are generally not desirable for either the individual or his social system.**

## Symbolic Versus Action Innovation-Decisions

Most but not all of the new ideas analyzed in this book are of a material, technological variety,*** consisting of an object as well as an idea. An innovation may have two components: (1) an *idea* component, and (2) an *object* component (that is, the material or physical product aspect of the idea). All innovations must have the ideational component, of course, but many do not have a physical referent. One criterion for classifying innovations is whether or not the innovation has an object component associated with it.

Innovations with only an idea component cannot be adopted in a sense that can be physically observed. Adoption is essentially a *symbolic* decision in this case. In contrast, innovations that also have an object component invoke an *action* adoption.**** Examples of innovations requiring symbolic decisions are Communist ideology, news events, and rumors.

---

*We have relatively few such studies in comparison to the number of inquiries about "desirable" innovations. One illustration is Francis' (1960) investigation of the adoption of a bogus and costly item of farm equipment, the oats incubator. There are also a few studies of the *discontinuance* of "undesirable" innovations, like Graham and Gibson's (1967) study of the cessation of cigarette smoking for health reasons.

**Further, the same innovation may be desirable for one adopter in his situation but undesirable for another potential adopter in a different situation.

***When a more restrictive definition of innovation is needed, it can be preceded by an appropriate adjective such as "technological" or "material." Actually, most innovations studied in past research have been both technological and material. We do not limit our meaning of technological to material objects only. For example, the idea of assembly line organization in a factory is an example of social technology. Technology can be broadly defined as a design for instrumental action.

****Further, some innovations cannot (or should not) be adopted immediately. Change agents are seeking, in these cases, only to persuade their clients to be willing (and perhaps able) to adopt without actually doing so. An example is the efforts by the U.S. Office of Civil Defense to inform the U.S. public about the nature of fallout shelters and to convince them

### Characteristics of Innovations

It should not be assumed, as it often has in the past, that all innovations are equivalent units of analysis. This is a gross oversimplification. Whereas it may take an innovation like modern math only five or six years to reach complete adoption, another new idea such as team teaching may require several decades to reach widespread use. The several characteristics of innovations, as sensed by the receivers, contribute to their different rate of adoption.

1. *Relative advantage* is the degree to which an innovation is perceived as better than the idea it supersedes. The degree of relative advantage may be measured in economic terms, but often social prestige factors, convenience, and satisfaction are also important components. It matters little whether the innovation has a great deal of "objective" advantage. What does matter is whether the individual *perceives* the innovation as being advantageous. The greater the perceived relative advantage of an innovation, the more rapid its rate of adoption.

2. *Compatibility* is the degree to which an innovation is perceived as being consistent with the existing values, past experiences, and needs of the receivers. An idea that is not compatible with the prevalent values and norms of the social system will not be adopted as rapidly as an innovation that is compatible. The adoption of an incompatible innovation often requires the prior adoption of a new value system. An example of an incompatible innovation is the use of the IUCD (intra-uterine contraceptive device) in countries where religious beliefs discourage use of birth control techniques.

3. *Complexity* is the degree to which an innovation is perceived as difficult to understand and use. Some innovations are readily understood by most members of a social system; others are not and will be adopted more slowly. For example, the rhythm method of family planning is relatively complex for most peasant housewives to comprehend because it requires understanding human reproduction and the monthly cycle of ovulation. For this reason, attempts to introduce the rhythm method in village India have been much less successful than campaigns to diffuse the loop, a type of IUCD, which is a much less complex idea in the eyes of the receiver. In general those new ideas requiring little additional learning investment on the part of the receiver will be

to use the shelters if the need should ever occur. But there is no intent by the Civil Defense officials to bring about actual use of the shelters. A similar illustration of an *anticipatory innovation-decision* was that made in Iowa in 1968 where the extension service conducted an active campaign to inform farmers about how to control the corn pest that was currently a major problem in the neighboring state of Nebraska, in anticipation of the future need for such knowledge.

adopted more rapidly than innovations requiring the adopter to develop new skills and understandings.

4. *Trialability*\* is the degree to which an innovation may be experimented with on a limited basis. New ideas which can be tried on the installment plan will generally be adopted more quickly than innovations which are not divisible. Ryan and Gross (1943) found that not one of their Iowa farmer respondents adopted hybrid seed corn without first trying it on a partial basis. If the new seed could not have been sampled experimentally, its rate of adoption would have been much slower. Essentially, an innovation that is trialable represents less risk to the individual who is considering it.

5. *Observability*\*\* is the degree to which the results of an innovation are visible to others. The easier it is for an individual to see the results of an innovation, the more likely he is to adopt. For example, a technical assistance agency in Bolivia introduced a new corn variety in one town. Within two years the local demand for the seed far exceeded the supply. The farmers were mostly illiterate, but they could easily observe the spectacular results achieved with the new corn and were thus persuaded to adopt. In the United States a rat poison that killed rats in their holes diffused very slowly among farmers because its results were not visible.

The five attributes just described are not a complete list, but they are the most important characteristics of innovations, past research indicates, in explaining rate of adoption.

Given that an innovation exists and that it has certain attributes, communication between the source and the receivers must take place if the innovation is to spread beyond its inventor. Now we turn our attention to this second element in diffusion.

## Communication Channels

*Communication* is the process by which messages are transmitted from a source to a receiver. In other words communication is the transfer of ideas

---

\*This attribute of an innovation was referred to as "divisibility" in an earlier version of this book (Rogers, 1962b, p. 131). We prefer the convention of "trialability" because it implies a somewhat broader meaning, including the notion of psychological trial (see Chapter 4 of this book).

\*\*This attribute was referred to as "communicability" in the earlier edition of this book (Rogers, 1962b, p. 132), but we prefer the present convention because of its more precise meaning (see Chapter 4).

*Elements of*
*Diffusion*                                                                    **24**

from a source with a viewpoint of modifying the behavior of receivers. A communication *channel* is the means by which the message gets from the source to the receiver.

As we stated earlier, diffusion is a subset of communication research that is concerned with new ideas. The essence of the diffusion process is the human interaction by which one person communicates a new idea to one or several other persons. At its most elementary level, the diffusion process consists of (1) a new idea, (2) individual A who has knowledge of the innovation, (3) individual B who is not yet aware of the new idea, and (4) some sort of communication channel connecting the two individuals. The nature of the social relationships between A and B determines the conditions under which A will or will not tell B about the innovation, and further, it influences the effect that the telling has on individual B.

The communication channel by which the new idea reaches B is also important in determining B's decision to adopt or reject the innovation. Usually the choice of communication channel lies with A, the source, and should be made in light of (1) the purpose of the communication act, and (2) the audience to whom the message is being sent. If A wishes simply to inform B about the innovation, *mass media channels** are often the most rapid and efficient, especially if the number of Bs in the audience is large. On the other hand, if A's objective is to persuade B to form a favorable attitude toward the innovation, an *interpersonal channel*** is more effective.

Therefore, the source should choose between mass media and interpersonal channels on the basis of the receiver's stage in the innovation–decision process, whether at the knowledge or persuasion stage. This brings us to discussion of a third element in diffusion, time.

## Over Time

Time is an important consideration in the process of diffusion.*** The time dimension is involved (1) in the innovation-decision process by which an individual passes from first knowledge of the innovation through its adoption

---

*Mass media channels are all those means of transmitting messages that involve a mass medium, such as radio, television, film, newspapers, magazines, and so on, which enable a source of one or a few individuals to reach an audience of many.
**Interpersonal channels are those that involve a face-to-face exchange between two or more individuals.
***"Time is the key to diffusion research" (Katz and others, 1963).

Elements in the Diffusion
of Innovations                                                   25

or rejection, (2) in the innovativeness of the individual, that is, the relative earliness-lateness with which an individual adopts an innovation when compared with other members of his social system, and (3) in the innovation's rate of adoption in a social system, usually measured as the number of members of the system that adopt the innovation in a given time period.


## The Innovation-Decision Process

The *innovation-decision process* is the mental process through which an individual passes from first knowledge of an innovation to a decision to adopt or reject and to confirmation of this decision. Many diffusion researchers have conceptualized a cumulative series of five stages in the process: (1) from awareness (first knowledge of the new idea), (2) to interest (gaining further knowledge about the innovation), (3) to evaluation (gaining a favorable or unfavorable attitude toward the innovation), (4) to small-scale trial, (5) to an adoption or rejection decision.* We prefer to conceptualize four main functions or steps in the process: (1) knowledge, (2) persuasion, (3) decision, and (4) confirmation.** The *knowledge function* occurs when the individual is exposed to the innovation's existence and gains some understanding of how it functions. The *persuasion function* occurs when the individual forms a favorable or unfavorable attitude toward the innovation. The *decision function* occurs when the individual engages in activities which lead to a choice to adopt or reject the innovation. The *confirmation function* occurs when the individual seeks reinforcement for the innovation-decision he has made, but he may reverse his previous decision if exposed to conflicting messages about the innovation.

An example is presented in Figure 1-3 to clarify the meaning of the innovation-decision process and to show the importance of the time dimension. Mr. Skeptic, an Iowa farmer, first learned of hybrid seed corn from an agricultural extension agent in 1935 (the knowledge function). However, he was not convinced to plant hybrid corn on his own farm until 1937, after he had discussed the innovation with several neighbors (the persuasion function).

---

*These five stages were utilized in the previous version of this book (Rogers, 1962b).
**We feel this four-stage process is an improvement over the traditional "adoption process," which it replaces, because our present model of the innovation-decision process makes provision for rejection as well as adoption decisions and allows for post-decision communication behavior which usually reinforces the original decision, but may lead to its reversal. The present model is conceptually linked to the notions of decision making, the learning process, and dissonance reduction.

*Elements of
Diffusion*                                                                          26

Skeptic purchased a small sack of hybrid seed in 1937 and by 1939 was planting 100 percent of his corn acreage in hybrids. When did he adopt hybrid corn ?

Skeptic adopted in 1939 when he decided to continue full scale use of the innovation (decision function). *Adoption* is a decision to make full use of a new idea as the best course of action available.\* The *innovation-decision period* is the length of time required to pass through the innovation-decision process; in

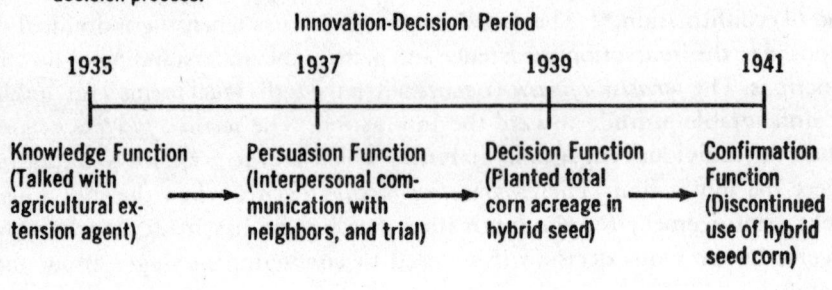

**Figure 1-3
The innovation-decision period
from knowledge to decision and confirmation.**

The innovation-decision process is the mental process through which an individual passes from first knowledge of an innovation to a decision to adopt or reject and to later confirmation of this decision. There are four functions in this process: (1) knowledge, (2) persuasion (attitude formation and change), (3) decision (adoption or rejection), and (4) confirmation. For the farmer depicted in this diagram, confirmation of his adoption decision continued until 1941, when conflicting messages reached him, and he decided to discontinue the innovation. The innovation-decision period is the length of time required to pass through the innovation-decision process.

**Innovation-Decision Period**

| 1935 | 1937 | 1939 | 1941 |
|------|------|------|------|

| Knowledge Function (Talked with agricultural extension agent) | Persuasion Function (Interpersonal communication with neighbors, and trial) | Decision Function (Planted total corn acreage in hybrid seed) | Confirmation Function (Discontinued use of hybrid seed corn) |

the present instance it lasted four years.\*\* The innovation decision can also take a negative turn; that is, the final decision can be *rejection*, a decision not to adopt an innovation.

The last function in the innovation-decision process is confirmation, a stage at which the receiver seeks reinforcement for the adoption or rejection

\*This definition, based upon Zaltman (1964, p. 24), differs slightly in wording from that utilized in the previous edition of the present book (Rogers, 1962b, p. 17), where adoption was defined as a decision to continue full-scale use of an innovation.

\*\*For practical reasons, the length of the innovation-decision period is usually measured from first knowledge until the decision to adopt (or reject), although in a strict sense it should perhaps be measured to the time of confirmation. The trouble with this latter procedure is that the confirmation function may continue over an indefinite time period.

decision he has made. Occasionally, however, conflicting and contradictory messages reach the receiver about the innovation, and this may lead to discontinuance on one hand or later adoption (after rejection) on the other. In the case of Mr. Skeptic, a decision was made to *discontinue* use of the innovation after previously adopting it. Farmer Skeptic became dissatisfied with hybrid seed and discontinued its use in 1941, when he again planted all of his corn acreage in open-pollinated seed. Discontinuances occur for many other reasons, including replacement of the innovation with an improved idea. Discontinuances occur only after the individual has fully adopted the idea.

### Innovativeness and Adopter Categories

If Skeptic adopted hybrid seed in 1939 and the average farmer in his community adopted in 1936, Skeptic is less innovative than the average member of his system. *Innovativeness* is the degree to which an individual is relatively earlier in adopting new ideas than the other members of his system.* Rather than describing Mr. Skeptic as "less innovative than the average member of his social system," it is handier and more efficient to refer to him as being in the "late majority" adopter category. This shorthand notation saves words and contributes to clearer understanding, for diffusion research shows clearly that each of the adopter categories has a great deal in common. If Skeptic is like most others in the late majority category, he is below average in social status, makes little use of mass media channels, and secures most of his new ideas from peers via interpersonal channels. In a similar manner, we shall present a concise word-picture of each of the other four adopter categories in this book (Chapter 5). *Adopter categories* are the classifications of members of a social system on the basis of innovativeness. The five adopter categories used here are: (1) innovators, (2) early adopters, (3) early majority, (4) late majority, and (5) laggards.

Obviously, the measure of innovativeness and the classification of the system's members into adopter categories are based upon the relative time at which an innovation is adopted.

### Rate of Adoption

There is a third specific way in which the time dimension is involved in the diffusion of innovations. *Rate of adoption* is the relative speed with which an

*By "relatively earlier" we mean earlier in terms of actual time of adoption, rather than whether the individual *perceives* he adopted the innovation relatively earlier than others in his system.

innovation is adopted by members of a social system. This rate of adoption is usually measured by the length of time required for a certain percentage of the members of a system to adopt an innovation. Therefore, we see that rate of adoption is measured using an innovation or a system, rather than an individual, as the unit of analysis. Innovations that are perceived by receivers as possessing greater relative advantage, compatibility, and the like have a more rapid rate of adoption (as we pointed out in a previous section of this chapter).

There are also differences in the rate of adoption for the same innovation in different social systems. Generally, diffusion research shows that systems typified by modern, rather than traditional, norms will have a faster rate of adoption. How do we classify systems as to modern or traditional? What is a social system?

## Among Members of a Social System

A *social system* is defined as a collectivity of units which are functionally differentiated and engaged in joint problem solving with respect to a common goal. The members or units of a social system may be individuals, informal groups, complex organizations, or subsystems. The social system analyzed in a diffusion study may consist of all the peasants in a Latin American village, students at a university, high schools in Thailand, medical doctors in a large city, or members of an aborigine tribe. Each unit in a social system can be functionally differentiated from every other member. All members cooperate at least to the extent of seeking to solve a common problem or to reach a mutual goal. It is this sharing of a common objective that binds the system together.

It is important to remember that diffusion occurs within a social system, because the social structure of the system affects the innovation's diffusion patterns in several ways. The social system constitutes a set of boundaries within which innovations diffuse. In this section we shall deal with the following topics: How the social structure affects diffusion, the effect of traditional and modern norms on diffusion, the roles of opinion leaders and change agents, and types of innovation-decisions. All these issues involve interfaces between the social system and the diffusion process that occurs within it.

### Social Structure and Diffusion

To the extent that the members in a social system are differentiated, structure then exists within the system. Social structure develops through the

*Elements in the Diffusion*
*of Innovations*                                                          29

arrangement (such as in an hierarchical fashion) of the statuses or positions in a system. A formal organization such as a government agency has a well-developed formal social structure consisting of titled positions, giving those in a higher ranked status the right to give orders to those of lesser rank and to expect the orders to be carried out. Even an informal grouping has some degree of structure inherent in the interpersonal relationships among its members, determining who interacts with whom and under what circumstances. Naturally, both formal and informal social structures have an effect on human behavior and how it changes in response to communication stimuli.

Diffusion and social structure are complexly interrelated.

1. *The social structure acts to impede or facilitate the rate of diffusion and adoption of new ideas through what are called "system effects."*\* The basic notion of system effects is that the norms, social statuses, hierarchy, and so on of a social system influence the behavior of individual members of that system. *System effects* are the influences of the system's social structure on the behavior of the individual members of the social system.

In the case of innovation diffusion, one can conceptualize an individual's innovation behavior as explained by two types of variables: (1) the *individual's* personality, communication behavior, attitudes, and so on, and (2) the nature of his *social system*. Which type is more important?

Several investigations point out the importance of system influences. Van den Ban (1960) studied the effects of traditional and modern norms (for a sample of Wisconsin townships) on the innovativeness of farmers. Although such individual characteristics as a farmer's education, size of farm, and net worth were positively related to his innovativeness, the township norms were even better predictors of farmer innovativeness. Van den Ban concluded that a farmer with a high level of education, on a large farm, and with a high net worth, but residing in a township with traditional norms, adopted fewer farm innovations than if he had a lower level of education and a smaller farm in a township where the norms were modern.

Qadir (1966) conducted a somewhat parallel inquiry in twenty-six Filipino rural neighborhoods. He found that in modern systems with a social climate favorable to the adoption of innovations, even individuals lacking much education, mass media exposure, or modern attitudes, acted in an innovative manner. This finding suggests Generalization 1-1: *System effects* (such as

---

\*Sometimes system effects are synonymously referred to as "compositional effects" (Davis and others, 1961), as "contextual effects," or as "structural effects" (Blau, 1957 and 1960; Campbell and Alexander, 1965; and Tannenbaum and Bachman, 1964). None of these studies, however, looked at system effects on diffusion behavior.

system norms, the composite educational level of one's peers, and the like) *may be as important in explaining individual innovativeness as such individual characteristics as education, cosmopoliteness, and so on.**

2. *Diffusion may also change the social structure of a system.* Some new ideas are "restructuring" innovations in that they change the structure of the social system itself. The adoption of a village development council changes the village social structure by adding a new set of statuses. The initiation of a research and development unit within an industrial firm and the departmentalization of a public school are also restructuring innovations. In many instances the restructuring affects the rate of future innovation diffusion within the system.**

A system's social structure and the way in which innovations diffuse in that system bear a close but subtly intertwining relationship. As Katz (1961) remarked, "It is as unthinkable to study diffusion without some knowledge of the social structures in which potential adopters are located as it is to study blood circulation without adequate knowledge of the structure of veins and arteries." Therefore, a system's social structure affects diffusion, and vice versa.

### System Norms and Diffusion

We have just pointed out that a system's norms affect an individual's innovation-adoption behavior. *Norms* are the established behavior patterns

---

*Since system effects have only recently been recognized as particularly important, we need carefully designed investigations of the role of these effects in the diffusion of innovations. Further support for this proposition comes from Saxena's (1968) study of system effects on farmer innovativeness in Indian villages, and from Davis' (1968) companion investigation in Nigeria. In both studies, system (village) variables explained a portion of the variance in individual innovativeness not also explained by individual variables (e.g., literacy, mass media exposure, and so on). The specific research studies which support or do not support each of the generalizations in this book are listed in Appendix A.

**In fact one strategy for speeding the rate of diffusion of new ideas is to restructure the social system. Such an approach is generally more appropriate and perhaps easier to accomplish when the system is a formal organization, than when it is a relatively more informal system such as a peasant village. Changing the norms of such a system might be a very difficult feat for a change agent, at least in the short range.

Those of highest status and power in a system, the elite, are obviously in a position to serve as "gatekeepers" in controlling the flow of innovations into the system from external sources. They are more likely to favor the introduction of functioning innovations, which do not threaten to disturb the status quo of the system's social structure, than restructuring innovations. Perhaps an illustration of this point is the oligarchic leaders of Latin American nations who promote technological innovations in agriculture and marketing, but oppose such restructuring innovations as land reform and overhauling of the tax system. This viewpoint will be discussed further in Chapter 11.

*Elements in the Diffusion
of Innovations* *31*

for the members of a given social system.\* They define a range of tolerable behavior and serve as a guide or a standard for the members of a social system.

A system's norms can be a barrier to change, as was shown in our example of water-boiling in a Peruvian community. Such resistance to new ideas is often found in norms relating to food. In India, for example, sacred cows roam the countryside while millions of people are undernourished. Pork cannot be consumed by Moslems. Polished rice is eaten in most of Asia and the United States, even though whole rice is more nutritious.

In additonal to influencing the original adoption or rejection of an innovation, norms also influence the manner in which an innovation will be integrated into the existing way of life of the receivers, that is, the consequences of the innovation. When horses were introduced into the Shoshone culture, the Indians readily accepted them for they had prior experience with horses, which they had stolen from pioneers for food (Harris, 1940). Although U.S. Bureau of Indian Affairs agents had intended that the horses be used for transportation, the Indians ate them!

Most of the researches which provide evidence of the influence of norms on diffusion have suffered from methodological shortcomings such as the limitations of their case study nature. Only one or two social systems\*\* are usually included in these analyses. Needed are investigations of a large sample of comparable systems, so that we can have some assurance of the generalizability of the results. In such investigations it is necessary to conceptualize and operationalize the traditional-modern dimension of system norms, which is itself a serious methodological task.

## Traditional and Modern Norms

We conceptualize system norms that are most relevant for innovation diffusion as either traditional or modern. These two kinds of norms are *ideal*

---

\*There have been two schools of thought among sociologists as to the meaning of norm. The neo-positivists (such as Lundberg, Chapin, and Dodd) defined norms as a *standard behavior* represented by such measures of central tendency in a distribution as a mean, median, or mode. The social actionists defined norm as a group *expectation* for a certain type of behavior. This argument between the "what is" versus the "what ought to be" meanings of norm has subsided recently with a tendency toward a more operational definition of norm that may reflect either a standard or an expectation for behavior, or both.
\*\*Illustrative of such investigations in which a pair of contrasting social systems (such as rural communities, schools, or colleges), one with modern and one with traditional norms, are compared in terms of their social structure and the communication behavior of their members are: Eibler (1965), Leuthold (1965), Lionberger and Chang (1965), Campbell and Holik (1960), van den Ban (1963b), Davis (1965), Rogers and van Es (1964), and Rogers with Svenning (1969).

*types*, conceptualizations based on observations of reality and designed to facilitate comparisons. Ideal types do not necessarily exist empirically but may be constructed by abstracting to a logical extreme the characteristics of the behavior under analysis. Developed purely for methodological reasons, ideal types provide a framework for anlysis. They are ideal not in the sense that they describe what ought to be, but rather in the sense that they logically accentuate some dimension of analysis. Since early times social scientists have conceptualized opposing pairs of ideal types, which are called "polar types," for the purposes of analyzing behavior occurring within social systems. Our conception of traditional and modern norms is a synthesis, at least in part, of various aspects of these previous typologies.*

A number of synonyms may be used to describe modern norms: Innovative progressive, developed, scientific, rational, and so on. The crucial dimension is that individuals in social systems with modern norms view change favorably, predisposing them to adopt new ideas more rapidly than individuals in traditional systems. Traditional social systems can be characterized by:

1   Lack of favorable orientation to change.
2   A less developed or "simpler" technology.
3   A relatively low level of literacy, education, and understanding of the scientific method.
4   A social enforcement of the status quo in the social system, facilitated by affective personal relationships, such as friendliness and hospitality, which are highly valued as ends in themselves.
5   Little communication by members of the social system with outsiders. Lack of transportation facilities and communication with the larger society reinforces the tendency of individuals in a traditional system to remain relatively isolated.
6   Lack of ability to empathize or to see oneself in others' roles, particularly the roles of outsiders to the system. An individual member in a system with traditional norms is not likely to recognize or learn new social relationships involving himself; he usually plays only one role and never learns others.

In contrast, a modern social system is typified by:
1   A generally positive attitude toward change.

*These ideal types include, for example, the *Gemeinschaft* and *Gesellschaft* of Töennies, Weber's rational and traditional types, Merton's local and cosmopolitan, and the sacred and secular types of Becker. The traditional and modern ideal types are based most directly on the work of Parsons (1951, pp. 101ff), Parsons and Shils (1951, pp. 80ff), Redfield (1956), Weber (1947, pp. 115–116), Wolf (1955), and particularly Lerner (1958).

*Elements in the Diffusion*
*of Innovations*                                                 33

2   A well developed technology with a complex division of labor.
3   A high value on education and science.
4   Rational and businesslike social relationships rather than emotional and affective.
5   Cosmopolite perspectives, in that members of the system often interact with outsiders, facilitating the entrance of new ideas into the social system.
6   Empathic ability on the part of the system's members, who are able to see themselves in roles quite different from their own.

In summary, a social system with modern norms is more change oriented, technologically developed, scientific, rational, cosmopolite, and empathic. A traditional system embodies the opposite characteristics.*

There is one danger in attempting to fit our thinking into the framework of ideal types: There is a tendency to overemphasize the extent of the differences. Traditional and modern ideal types are actually the end points of a continuum on which actual social system norms may range. We should not forget that the norms of most systems are distributed between the extremes that we have described.

One should not conclude from our discussion that traditional norms are necessarily undesirable. In many cases, tradition lends stability to a social system that is undergoing rapid change and is in danger of disorganization.** Modern systems have their own unique drawbacks, including slums, pollution of water and air, alienation, neuroses, and an almost endless list of social problems rooted in the consequences of "progress."

The reader should remember that it is possible for an individual to be a

*Three main methods have been utilized to measure modern and traditional norms in diffusion research: (1) the average innovativeness of the system's members, (2) their attitudes toward innovators, and (3) key informants' ratings (Rogers, 1962b, pp. 67–70). None of these measures of system norms is above methodological criticism, and future research efforts should be directed toward developing improved measures. Nevertheless, existing measures provide a rough indication of a system's norms, in that it is usually possible to say that the norms of one social system are relatively more traditional or modern than those of another system (or systems).

**Perhaps an ideal rate of change is somewhat less than that which would lead to disequilibrium within the system. *Disequilibrium* occurs when the rate of change is too rapid to permit the system to adjust. A system is said to be in a state of *dynamic equilibrium* when change is occurring at a rate commensurate with the system's ability to cope with it. *Stable equilibrium* occurs when there is almost no change in the structure or functioning of the social system. The latter state would occur on a traffic circle without vehicles; a dynamic equilibrium has cars moving around it. Disequilibrium would occur if there were one car too many and all traffic stopped. For a more detailed discussion of ideal rates of change, see Chapter 11.

*Elements of*
*Diffusion*                                                                 *34*

member of more than one social system and that the norms of these different
systems may vary on the traditional-modern continuum. And if the norms of
the systems to which the individual belongs are widely divergent, he is likely
to experience cross-pressures in making innovation decisions. For instance,
a school teacher who is continuing his part-time graduate education in a
university where new ideas are constantly discussed is likely to experience
conflict when he attempts to introduce these innovations into the traditional
school system where he teaches.

Not only is the traditionalism-modernism of a social system's norms
important in predicting individual diffusion behavior, but also the *commitment*
of the individual to the social system affects his conformity to its norms. An
innovative teacher in a traditional school may be relatively unaffected by the
norms because the local school is not important as a reference group to him.
Thus, an individual's integration into a social system,* as well as the nature of
the system's norms, need to be studied in order to fully explain his adoption
behavior.**

### Opinion Leaders and Change Agents

In this section we have discussed the influence of the social structure on
the members' diffusion behavior. Now we turn to the different roles that
individuals play in a social system and the effect of these roles on diffusion
patterns. Specifically, we shall look at two roles: Opinion leaders and change
agents.

Very often the most innovative member of a system is perceived as a
deviant from the social system, and he is accorded a somewhat dubious status
of low credibility by the average members of the system. His role in diffusion
(especially in persuading others about the innovation) is therefore likely to be
limited. On the other hand there are members of the system who function in

*Both Yadav (1967) and Guimarães (1968) found that a high degree of communication
integration characterized more modern peasant villages. *Communication integration* is the
degree to which the units in a social system are interconnected by interpersonal communication channels. It seems logical that more integrated systems should be more modern (and
their members relatively more innovative) because once a new idea enters these systems, it
will spread quickly to all members of the systems.
**Jamias (1964) found that Michigan dairy farmers who were more dogmatic (that is, whose
belief systems were more rigidly compartmentalized) conformed more closely to social
system norms than did farmers who were less dogmatic. This finding suggests that perhaps
personality variables such as dogmatism may affect the degree to which system norms
influence individual behavior.

*Elements in the Diffusion
of Innovations* 35

the role of opinion leader. They provide information and advice about innovations to many others in the system.

*Opinion leadership* is the degree to which an individual is able to informally influence other individuals' attitudes or overt behavior in a desired way with relative frequency.* It is a type of informal leadership, rather than being a function of the individual's formal position or status in the system. Opinion leadership is earned and maintained by the individual's technical competence, social accessibility, and conformity to the system's norms.** Several researches indicate that when the social system is modern, the opinion leaders are quite innovative; but when the norms are traditional, the leaders also reflect this norm in their behavior. By their close conformity to the system's norms, the opinion leaders serve as an apt model for the innovation behavior of their followers.

In any system, naturally, there may be both innovative and also more traditional opinion leaders. These influential persons can lead in the promotion of new ideas, or they can head an active opposition. In general, when opinion leaders are compared with their followers, we find that they (1) are more exposed to all forms of external communication, (2) are more cosmopolite, (3) have higher social status, and (4) are more innovative (although the exact degree of innovativeness depends, in part, on the system's norms).

Opinion leaders are usually members of the social system in which they exert their influence. In some instances individuals with influence in the social system are professionals representing change agencies external to the system. A *change agent* is a professional who influences innovation-decisions in a direction deemed desirable by a change agency. He usually seeks to obtain the adoption of new ideas, but he may also attempt to slow down diffusion and prevent the adoption of what he believes are undesirable innovations. Change agents often use opinion leaders within a given social system as lieutenants in

---

*Thus, our definition of opinion leadership implies a leadership-followership *relation* between two or more people, rather than an abstract attribute of an individual leader. We follow Merton (1957, p. 415) in the notion that opinion leadership has not occurred unless the opinions or overt behavior of the followers is different from what it would have been if the opinion leader had not interacted with others in the system. Our present definition of opinion leadership also differs somewhat from that expressed in the previous edition of the present book (Rogers, 1962b, p. 16), which we now feel did not adequately convey the notion that opinion leaders may be either active or passive (or both) in influencing their followers. That is, opinion leaders may be sought by, or they may seek, their followers.
**These criteria for opinion leadership correspond roughly to those suggested by Katz (1967).

their campaigns of planned change. There is research evidence that opinion leaders can be "worn-out" by change agents who overuse them.* Opinion leaders may be perceived by their peers as too much like the change agents; thus, the opinion leaders lose their credibility with their former followers.

### Types of Innovation-Decisions

The social system has yet another important kind of influence on the diffusion of new ideas. Innovations can be adopted or rejected by individual members of a system or by the entire social system. The relationships between the social system and the decision to adopt innovations may be described in the following manner:

1. *Optional decisions* are made by an individual regardless of the decisions of other members of the system. Even in this case, the individual's decision is undoubtedly influenced by the norms of his social system and his need to conform to group pressures. The decision of an individual to begin wearing contact lenses instead of eye glasses, an Iowa farmer's decision to adopt hybrid corn, and a housewife's adoption of birth control pills are examples of optional decisions.

2. *Collective decisions* are those which individuals in the social system agree to make by consensus. All must conform to the system's decision once it is made. An example is fluoridation of a city's drinking water. Once the community decision is made, the individual has little practical choice but to adopt fluoridated water. It does indeed "take two to tango" (Katz, 1962), once the partners have agreed to dance.

3. *Authority decisions* are those forced upon an individual by someone in a superordinate power position, such as a supervisor in a bureaucratic organization. The individual's attitude toward the innovation is not the prime factor in his adoption or rejection; he is simply told of and expected to comply with the innovation-decision which was made by an authority. Few research studies have yet been conducted of this type of innovation-decision, which must be very common in an organizational society such as the U.S. today. In all authority decisions we must distinguish between (1) the decision maker, who is one (or more) individual(s), and (2) the adopter or adopters, who carry out the decision. In the case of optional and collective decisions these two roles (of deciding and adopting) are performed by the same individual(s).

---

*This point will be discussed in detail in Chapters 6 and 7.

These three types of innovation-decisions range on a continuum from optional decisions (where the adopting individual has almost complete responsibility for the decision), through collective decisions (where the adopter has some influence in the decision), to authority decisions (where the adopting individual has no influence in the innovation decision). Collective and authority decisions are probably much more common than optional decisions in formal organizations,* such as factories, public schools, or labor unions, in comparison with other fields like agriculture and medicine where innovation-decisions are usually optional.

Generally, the fastest rate of adoption of innovations results from authority decisions (depending, of course, on whether the authorities are traditional or modern). In turn, optional decisions can be made more rapidly than the collective type. Although made most rapidly, authority decisions are more likely to be circumvented and may eventually lead to a high rate of discontinuance of the innovation.** Where change depends upon compliance under public surveillance, it is not likely to continue once the surveillance is removed.

The type of innovation-decision for a given idea may change or be changed over time. Automobile seat belts, during the early years of their use, were installed in private autos largely as optional decisions. Then in the 1960s many states began to require by law installation of seat belts in all new cars. In 1968 a federal law was passed to this effect. An optional innovation-decision then became a collective decision.***

There is yet a fourth type of innovation-decision which is essentially a sequential combination of two or more of the three types**** we have just discussed. *Contingent decisions* are a choice to adopt or reject which can be made only after a prior innovation-decision. An individual member of a social system is free to adopt or not to adopt a new idea only after his system's innovation-decision. A teacher cannot adopt or reject the use of an overhead

*We shall focus on collective innovation-decisions in Chapter 9 and on authority decisions in Chapter 10.

**Therefore, authority innovation-decisions often result in a rapid rate of adoption but in a relatively low quality decision, that cannot effectively be put into action, at least over an extended period of time.

***But in another sense an optional decision was still required by the automobile driver or passenger to use the belts, that is, to fasten them when getting in the seat. The collective decision in 1968 led to a rapid (100 percent) installation of the belts but not to a parallel increase in the use of the safety devices.

****Of course there are also many innovation-decisions which are difficult to categorize in that they fall between these three types; nevertheless, the three types are heuristically distinct.

projector in his classroom until the school system has decided to purchase one; at that point the teacher can decide to use or reject the overhead projector. In the Punjab State of India hybrid corn adoption is a contingent decision because hybrid corn requires a growing season two weeks longer than open-pollinated varieties, and villagers release their cattle to roam for forage across the unfenced fields once their corn is harvested. One can readily imagine the difficulty of making an optional decision to adopt hybrid corn in the Punjab without a prior collective decision by the entire village. One can also imagine contingent decisions in which the first decision is of an authority sort followed by optional or collective decisions. The distinctive aspect of contingent decision making is that two(or more) tandem decisions are required; either of the decisions may be optional, collective, or authority.

## Summary

A main theme in this book is that communication is essential for social change.

*Social change* is the process by which alteration occurs in the structure and function of a social system. We suggest three sequential stages in the process of social change: (1) *invention*, the process by which new ideas are created or developed, (2) *diffusion*, the process by which these new ideas are communicated to the members of a given social system, and (3) *consequences*, the changes that occur within the social system as a result of the adoption or rejection of the innovation.

Change is either immanent or contact. *Immanent change* occurs when members of a social system with little or no external influence create and develop a new idea (that is, invent it), and then it spreads within the system. *Contact change* occurs when sources external to the social system introduce a new idea; contact change may be either selective or directed. *Selective contact change* results when members of a social system are exposed to external influences and adopt or reject a new idea from that source on the basis of their needs. *Directed contact change*, or planned change, is caused by outsiders who, on their own or as representatives of change agencies, intentionally seek to introduce new ideas in order to achieve goals they have defined. Much change that occurs today is directed, and this variety is the main concern of the present book.

## Summary 39

*Communication* is the process by which messages are transferred from a source to a receiver. Essential elements in the communication process are the source, message, channels, and receivers. Diffusion is essentially a special type of communication concerned with the spread of messages that are *new* ideas. A certain degree of risk is usually associated with the reception of innovations, and this leads to somewhat different behaviors on the part of the individual than if he were receiving routine ideas.

Another distinctive aspect of diffusion as a subfield of communication is that heterophily is most often present between source and receiver. *Heterophily* is the degree to which pairs of individuals who interact are different in certain attributes, such as beliefs, values, education, social status, and the like. The opposite of heterophily is *homophily*, the degree to which pairs of individuals who interact are similar in certain attributes. Generally, most human communication takes place between individuals who are homophilous, a situation that leads to more effective communication. Therefore, the extent of source-receiver heterophily, which is often present in the diffusion of innovations, leads to special problems in securing effective communication.

The main elements in the diffusion of new ideas are: (1) the *innovation*, (2) which is *communicated* through certain *channels*, (3) *over time*, (4) among the members of a *social system*. An *innovation* is an idea, practice, or object perceived as new by an individual. The characteristics of an innovation, as perceived by the members of a social system, determine its rate of adoption. Five attributes of innovations are: (1) relative advantage, (2) compatibility, (3) complexity, (4) trialability, and (5) observability.

*Communication channels* are the means by which a message gets from a source to a receiver. Mass media channels are more effective in creating knowledge of innovations, whereas interpersonal channels are more effective in forming and changing attitudes toward the new idea.

Time is involved in diffusion in (1) the innovation-decision process, (2) innovativeness, and (3) an innovation's rate of adoption. The *innovation-decision process* is the mental process through which an individual passes from first knowledge of an innovation to a decision to adopt or reject, and to later confirmation of this decision. We conceptualize four functions in this process: (1) knowledge, (2) persuasion, (3) decision, and (4) confirmation. *Adoption* is a decision to make full use of a new idea as the best course of action. *Rejection* is a decision not to adopt an innovation. *Discontinuance* is a decision to cease use of an innovation after previously adopting it. Discontinuance, then, is essentially adoption of an innovation, followed by rejection.

*Innovativeness* is the degree to which an individual is relatively earlier in adopting new ideas than other members of his social system. We specify five *adopter categories*, classifications of the members of a social system on the basis of innovativeness: (1) innovators, (2) early adopters, (3) early majority, (4) late majority, and (5) laggards. *Rate of adoption* is the relative speed with which an innovation is adopted by members of a social system.

A *social system* is a collectivity of units which are functionally differentiated and engaged in joint problem solving with respect to a common goal. It is important to remember that diffusion occurs within a social system, because the system's social structure can have an important influence on the spread of new ideas.

Social structure consists of the statuses or positions in a social system and how these statuses are arranged, such as in an hierarchical fashion. The social structure of a system acts to impede or facilitate the rate of diffusion and adoption of new ideas through what are called "system effects." The norms, social statuses, hierarchy, and so on of a social system influence the behavior of individual members of that system. Evidence from several researches indicates that the system effects may be as important in explaining individual innovativeness as education, cosmopoliteness, and the like. Diffusion may also change the social structure of a system in the sense that many innovations are of a restructuring nature.

*Norms* are the established behavior patterns for the members of a social system. Two ideal types of norms can be distinguished: Traditional and modern. A system with modern norms is more change oriented, technologically developed, scientific, rational, cosmopolite, and empathic, whereas a traditional system is the opposite of these.

*Opinion leadership* is the degree to which an individual is able to informally influence other individuals' attitudes or overt behavior in a desired way with relative frequency. A *change agent* is a professional person who attempts to influence innovation-decisions in a direction that he feels is desirable.

We consider three main types of innovation-decisions: (1) *optional decisions*, which are made by an individual regardless of the decisions of other members of the system; (2) *collective* decisions, which individuals in the social system agree to make by consensus, and (3) *authority* decisions, which are forced upon an individual by someone in a superordinate power position. These three types of decisions may, of course, occur in a sequential order so that an optional decision cannot be made until a prior collective decision has been made; these *contingent innovation-decisions* are a choice to adopt or reject which can be made only after a prior innovation-decision.

# 5                                           *Adopter*

The innovator makes enemies of all those who prospered under
the old order, and only lukewarm support is forthcoming from
those who would prosper under the new . . . because men are
generally incredulous, never really trusting new things unless
they have tested them by experience.

NICCOLO MACHIAVELLI (1961, p. 51)


A slow advance in the beginning, followed by rapid and uniformly accelerated progress, followed again by progress that
continues to slacken until it finally stops: These are the three
ages of real social beings which I call inventions. . . . If taken
as a guide by the statistician and by the sociologist, [they]
would save many illusions.         GABRIEL TARDE (1903, p. 127)

# Categories

.�— —.. ——— .—.—. —. — . .—.        —.—. .— — . ——. ——— .—. .. . ...

$A$LL individuals in a social system do not adopt an innovation at the same time. Rather, they adopt in an ordered time sequence, and they may be classified into adopter categories on the basis of when they first begin using a new idea. We could describe each individual adopter in a social system in terms of his time of adoption, but this would be tedious work. It is much easier and more meaningful to describe adopter categories,* each containing individuals with a similar degree of innovativeness. There is much practical usefulness for change agents if they can identify potential innovators and laggards in their client audience and utilize different change strategies with each such subaudience.

We know more about innovativeness, the degree to which an individual is relatively earlier in adopting new ideas than other members of his social system, than any other concept in diffusion research (Chapter 2). The expressed, short-term goal of most change agencies is to facilitate the adoption of innovations by their clients. Because increased innovativeness is the objective of change agencies, it has become the main dependent variable in the diffusion research these change agencies sponsor. A further reason for the prime focus on innovativeness in diffusion research, especially in less developed countries, is that innovativeness is the best single indicator of

---

*\*Adopter categories* are the classifications of the members of a social system on the basis of innovativeness.

modernization* (Rogers with Svenning, 1969, p. 292). Innovativeness indicates behavioral change, the ultimate goal of modernization programs, rather than cognitive or attitudinal change.

This chapter suggests one method of categorizing adopters and demonstrates the usefulness of this technique with research findings from both more developed and less developed nations. We shall discuss the normality of adopter distributions, the method of classifying adopters, characteristics of adopter categories, and predicting innovativeness.

# Innovativeness and Adopter Categories

## The Need to Standardize Categories

Titles of adopter categories are about as numerous as diffusion researchers themselves. The inability of diffusion researchers to agree on common semantic ground in assigning terminology has led to a plethora of adopter descriptions. The most innovative individuals have been termed progressists, high-triers, experimentals, lighthouses, advance scouts, and ultradopters. Least innovative individuals have been called drones, parochials, and diehards. The fertile disarray of adopter categories and methods of categorization, illustrated by the adopter categories, emphasizes the need for standardization. How can a reader compare research findings about adopter categories until there is standardization of both the nomenclature and the classification system? Fortunately, one method of adopter categorization has gained a dominant position in recent years, one based upon the S-shaped curve of adoption.

## S-Curve of Adoption Over Time

It is the time variable which allows researchers to classify adopter categories and to plot diffusion curves.** Research has generally shown that the

---

*\*Modernization* is defined as the process by which individuals change from a traditional way of life to a more complex, technologically advanced, and rapidly changing style of life. This is best indicated by an individual's actual use of new ideas in agriculture, health, family living, and other fields.
\*\*As such, an adopter distribution is one type of diffusion curve, which represents the number of knowers or adopters of an innovation per unit of time. In the present book we utilize

*Innovativeness and*
*Adopter Categories*                                                    *177*

adoption of an innovation follows a normal, bell-shaped curve when plotted over time on a frequency basis. If the cumulative number of adopters is plotted, the result is an S-shaped curve. Figure 5-1 shows that the same adoption data can be represented by either a bell-shaped (frequency) or an S-shaped (cumulative) curve.

the term "adopter distribution," rather than "diffusion curve," for the sake of greater precision.

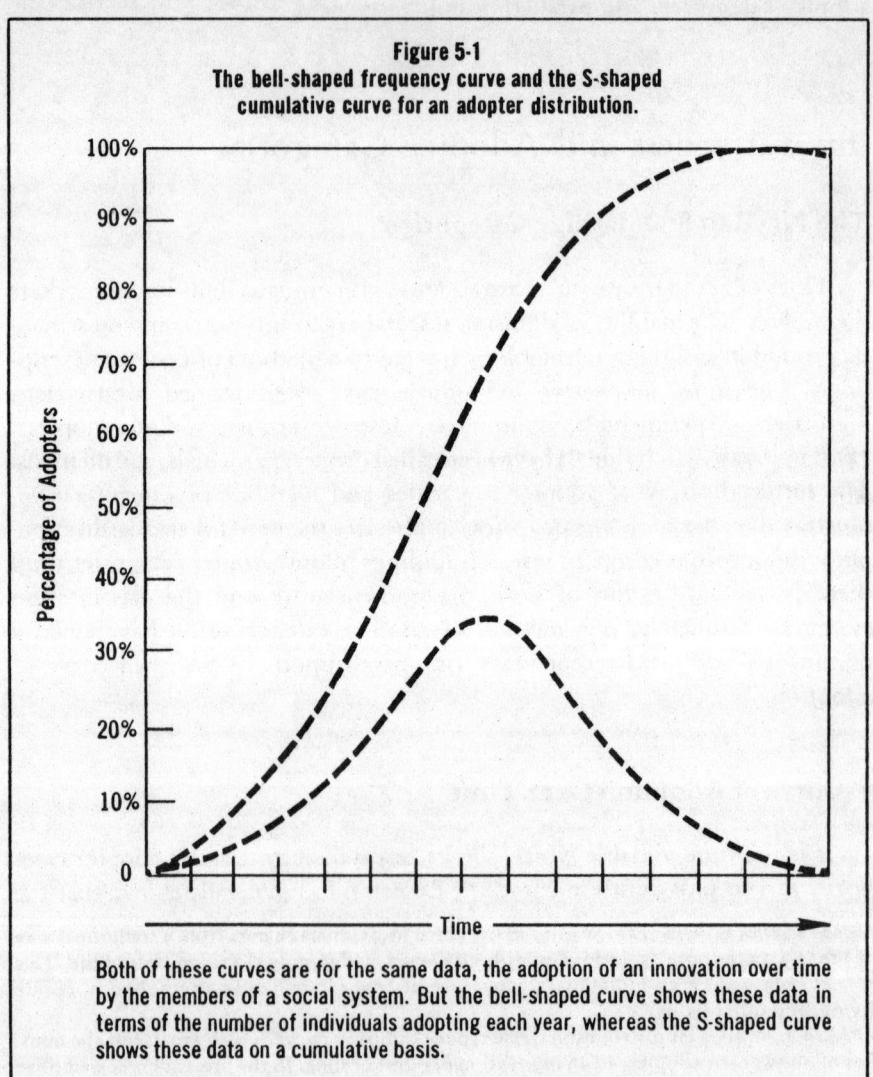

**Figure 5-1**
**The bell-shaped frequency curve and the S-shaped**
**cumulative curve for an adopter distribution.**

Both of these curves are for the same data, the adoption of an innovation over time by the members of a social system. But the bell-shaped curve shows these data in terms of the number of individuals adopting each year, whereas the S-shaped curve shows these data on a cumulative basis.

*Adopter
Categories*                                                                     *178*

The S-shaped adopter distribution rises slowly at first when there are few adopters in a time period. Then it accelerates to a maximum when half of the individuals in the system have adopted. It then increases at a gradually slower rate as the few remaining individuals finally adopt. This S-shaped curve is normal. Why?

## *Learning Curves*

Psychological research indicates that individuals learn a new skill, or bit of knowledge, or set of facts through a learning process which, plotted over time, follows a normal curve. When an individual is confronted with a new situation in the psychologist's laboratory, he makes many errors at the beginning. After a series of trials, the errors decrease until learning capacity has been reached. When plotted, these data yield a curve of increasing gains at first and later become a curve of decreasing gains. The gain in learning per trial is proportionate to (1) the product of the amount already learned, and (2) the amount remaining to be learned before the limit of learning is reached. It should be pointed out, however, that the S-shaped learning curve was not selected by psychologists because of any formal learning theory but simply because it resulted from learning experiments. Thus, their intellectual basis was empirical rather than conceptual. The learning curve provides reason to expect adopter distributions to be normal.* If a social system is substituted for the individual in the learning curve, it seems reasonable to expect that experience with the innovation is gained as each successive member in the social system adopts it. Each adoption in the social system is in a sense equivalent to a learning trial by an individual.

## *Diffusion Effect*

Another reason for expecting normal adopter distributions is the *diffusion effect*, defined as the cumulatively increasing degree of influence upon an individual within a social system to adopt or reject an innovation. This influence results from the increasing rate of knowledge and adoption or rejection of the innovation in the system.** Adoption of a new idea is the

*It has been found that many human traits are normally distributed, whether the trait is a physical characteristic, such as weight or height, or a behavioral trait, such as intelligence or the learning of information. Hence, a variable such as degree of innovativeness might be expected to be normally distributed also.

**A more detailed discussion of the diffusion effect was presented in Chapter 4.

result of human interaction. If an innovation is introduced in a social system, there are theoretical grounds for expecting the number of adoptions over time to be normally distributed. If the first adopter of the innovation discusses it with two other members of the social system, and these two adopters each pass the new idea along to two peers, the resulting distribution follows a binomial expansion.* This mathematical function follows a normal shape when plotted.

Of course, several of the assumptions underlying this hypothetical example are seldom found in reality. For instance, members of a social system do not have completely free access to interact with one another. Status barriers, geographical location, and other variables affect diffusion patterns. The diffusion effect begins to level off after half of the individuals in a social system have adopted, because each new adopter finds it increasingly difficult to tell the new idea to a peer who has not yet adopted, for such nonknowers become increasingly scarce.

## Testing Adopter Distributions for Normality

It has generally been found that *adopter distributions follow a bell-shaped curve over time and approach normality.*** There are useful implications to be found for a standard method of adopter categorization, as we shall soon see.

Eight adopter distributions for single innovations tested by Rogers (1958b) were bell-shaped and all approached normality, although half of those tested were found to deviate significantly from normality (Table 5-1). Similar evidence is provided by Bose (1964b), who found the adoption of crop chemicals was normally distributed in each of seven Indian villages, even though these villages differed widely in their norms, size, and social make-up. Further, Andrus (1965) found that innovativeness scores (based on the adoption of new household products by a national sample of 11,000 U.S. families) formed a bell-shaped, but not exactly a normal, curve. Table 5-1 shows that the distribution of innovativeness scores for samples of Iowa and Ohio farmers are normal.

*A pattern that is similar to that of an unchecked infectious epidemic. Bailey (1957, pp. 29–37, 155–159) provides epidemiological models for this biological phenomenon that could usefully be applied to the diffusion of innovations.

**DeFleur (1966) suggests that the curve of dicontinuance may also approach normality, although there is yet no research evidence on this point.

*Adopter*
*Categories*                                                                 180

## Table 5-1   Normality of Adopter Distributions

| INNOVATION (OR INNOVATIVENESS SCORES) | NORMALITY OF ADOPTER DISTRIBUTION | RESEARCH STUDY |
|---|---|---|
| 1  2,4-D weed spray in Iowa (all adopters) | Normal | Beal and Rogers (1960) |
| 2  2,4-D weed spray in Iowa (beginning farmers excluded) | Normal | Beal and Rogers (1960) |
| 3  Antibiotics in Iowa (all adopters) | Not normal[b] | Beal and Rogers (1960) |
| 4  Antibiotics in Iowa (beginning farmers excluded) | Not normal[a] | Beal and Rogers (1960) |
| 5  Hybrid corn (Iowa) | Not normal[b] | Ryan (1948) |
| 6  Hybrid corn (Virginia) | Not normal[b] | Dimit (1954) |
| 7  2,4-D weed spray (Ohio) | Normal | Rogers (unpublished data) |
| 8  Warfarin rat poison (Ohio) | Normal | Rogers (unpublished data) |
| 9  Adoption of farm innovations scores (Iowa) | Normal | Rogers (1958b) |
| 10  Adoption of farm innovations scores (Ohio) | Normal | Rogers (1958b) |

[a]Deviation from normality is significant at the 5 percent level.
[b]Deviation from normality is significant at the 1 percent level.
*Source:* Rogers, 1958b; used by permission.

## A Method of Adopter Categorization

A researcher seeking standardization of adopter categories faces three problems: (1) determining the number of adopter categories to conceptualize, (2) deciding on the portion of the members of a system to include in each category, and (3) determining the method, statistical or otherwise, of defining the adopter categories.

There is no question, however, about the criterion for adopter categorization. It is innovativeness, the degree to which an individual is relatively earlier in adopting new ideas than other members of his social system.* Innovative-

*One of the major problems in measuring innovativeness is the inaccuracy of recall by some individuals and for some innovations. In one study which collected data both by recall and from records, it was found that doctors' statements tended to make them appear more up-to-date than did druggists' prescription records (Menzel, 1957). On the other hand, Havens (1962b) found that Ohio dairy farmers were quite accurate in reporting from memory their date of adoption of bulk milk tanks. Havens checked the adoption dates with records of the farmers' milk purchasers. The different results from the two studies may be due to the type of innovation, the latter requiring a substantial cash outlay as well as a change in behavior.

ness is a "relative" dimension, in that one either has more or less of it than others in a social system. Innovativeness is a continuous variable, and partitioning it into discrete categories is only a conceptual device, much like dividing the continuum of social status into upper, middle, and lower classes.

Before describing a method of adopter categorization, it is important to specify the characteristics which a set of categories should possess. Ideally, categories should: (1) be *exhaustive,* or include all the respondents of the sample, (2) be *mutually exclusive,* or exclude from any other category a respondent who appears in one category, and (3) be derived from *one classificatory principle* (Jahoda and others, 1951, p. 264).

We have previously demonstrated that adopter distributions closely approach normality. This is important because the normal frequency distribution has several characteristics which may be used in classifying adopters. One of these characteristics or parameters is the mean $(\bar{\chi})$, or average, of the sample. Another parameter of a distribution is the standard deviation (sd), a measure of dispersion about the mean. The standard deviation explains the average amount of variance on either side of the mean for a sample.

These two statistics, the mean $(\bar{\chi})$ and the standard deviation (sd), can be used to divide a normal adopter distribution into categories. If vertical lines are drawn to mark off the standard deviations on either side of the mean, the curve is divided into categories in a way that results in a standardized percentage of respondents in each category. Figure 5-2 shows the normal frequency distribution divided into five adopter categories: (1) innovators, (2) early adopters, (3) early majority, (4) late majority, and (5) laggards. These five adopter categories and the approximate percentage of individuals included in each are located on the adopter distribution in Figure 5-2.

The area lying to the left of the mean time of adoption minus two standard deviations includes the first 2.5 percent of the individuals to adopt an innovation—the *innovators.* The next 13.5 percent to adopt the new idea are included in the area between the mean minus one standard deviation and the mean minus two standard deviations; they are labeled *early adopters.* The next 34 percent of the adopters, called *early majority,* are included in the area between the mean date of adoption and minus one standard deviation. Between the mean and one standard deviation to the right of the mean are located the next 34 percent to adopt the new idea, the *late majority.* The last 16 percent are called *laggards.*

This method of adopter classification is probably the most widely used in current diffusion studies. However, as can be observed, it is not a symmetrical classification in that there are three adopter categories to the left of the mean and only two to the right. One solution would be to break laggards into two categories, such as early and late laggards, but laggards seem to be a fairly homogeneous category. Similarly, innovators and early adopters could be combined into a single class to achieve symmetry, but their quite different characteristics mark them as two distinct categories.

Another difficulty in our method of adopter classification is incomplete adoption, which occurs for innovations that have not reached 100 percent use at the time of their study. This means that our five-fold classification scheme is not completely exhaustive. But the problem of incomplete adoption or nonadoption is eliminated when a series of innovations is combined into a composite innovativeness scale.

Three principles of categorization were suggested earlier in this section. Innovativeness as a criterion fulfills each of these requirements. The five adopter categories are exhaustive (except for nonadopters), mutually exclusive, and are derived from one classification principle. The method of adopter categorization just described is the most widely used in diffusion research today.

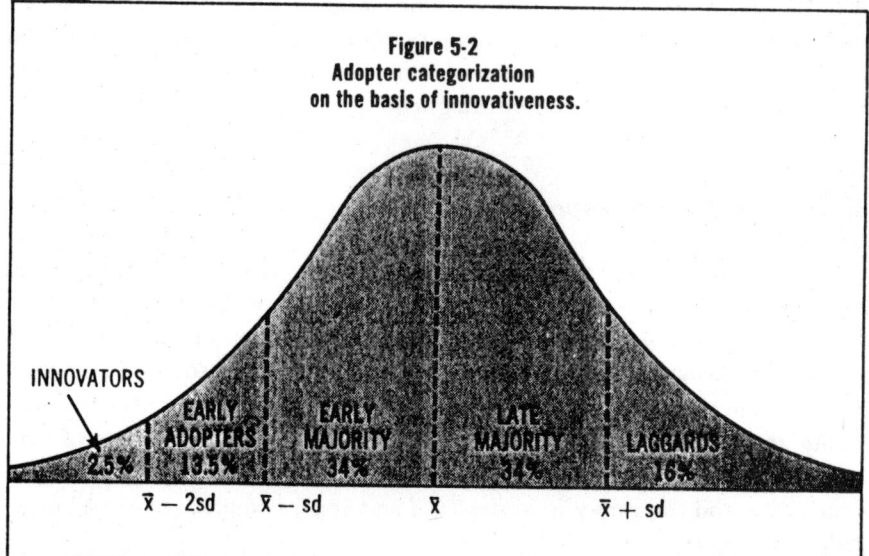

**Figure 5-2**
**Adopter categorization**
**on the basis of innovativeness.**

INNOVATORS

EARLY ADOPTERS | EARLY MAJORITY | LATE MAJORITY | LAGGARDS
2.5% | 13.5% | 34% | 34% | 16%

$\bar{x} - 2sd$      $\bar{x} - sd$            $\bar{x}$                $\bar{x} + sd$

The innovativeness dimension, as measured by the time at which an individual adopts an innovation or innovations, is continuous. However, this variable may be partitioned into five adopter categories by laying off standard deviations from the average time of adoption.

# Adopter Categories as Ideal Types

The five adopter categories set forth in this chapter are ideal types. *Ideal types* are conceptualizations that are based on observations of reality and designed to make possible comparisons. The traditional and modern norms described in Chapter 1 are examples of ideal types. The function of ideal types is to guide research efforts and serve as a framework for the synthesis of research findings.

Actually, there are no pronounced breaks in the innovativeness continuum between each of the five categories. Ideal types are not simply an average of all the observations about an adopter category. Exceptions to the ideal types must be found. If no exceptions or deviations are located, ideal types would not be necessary. Ideal types are based on abstractions from empirical cases and are intended as a guide for theoretical formulations and empirical investigations. However, they are not a substitute for these investigations.

We now present a thumbnail sketch of the dominant characteristics and subcultural values of each adopter category, which will be followed by more detailed generalizations. Actually, there are few adequate investigations completed of the values of each adopter category; hence, the following section is an abstraction from a variety of studies not aimed specifically at determining value differences among adopter categories.

## Innovators: Venturesome

Observers have noted that venturesomeness is almost an obsession with innovators. They are eager to try new ideas. This interest leads them out of a local circle of peers and into more cosmopolite social relationships. Communication patterns and friendships among a clique of innovators are common, even though the geographical distance between the innovators may be great. Being an innovator has several prerequisites. These include control of substantial financial resources to absorb the possible loss due to an unprofitable innovation and the ability to understand and apply complex technical knowledge.

The salient value of the innovator is venturesomeness. He desires the hazardous, the rash, the daring, and the risky. The innovator also must be willing to accept an occasional setback when one of the new ideas he adopts proves unsuccessful.

## Early Adopters: Respectable

Early adopters are a more integrated part of the local social system than are innovators. Whereas innovators are cosmopolites, early adopters are localites. This adopter category, more than any other, has the greatest degree of opinion leadership in most social systems. Potential adopters look to early adoptors for advice and information about the innovation. The early adopter is considered by many as "the man to check with" before using a new idea. This adopter category is generally sought by change agents to be a local missionary for speeding the diffusion process. Because early adopters are not too far ahead of the average individual in innovativeness, they serve as a role model for many other members of a social system.* The early adopter is respected by his peers. He is the embodiment of successful and discrete use of new ideas. And the early adopter knows that he must continue to earn this esteem of his colleagues if his position in the social structure is to be maintained.

## Early Majority: Deliberate

The early majority adopt new ideas just before the average member of a social system. The early majority interact frequently with their peers, but leadership positions are rarely held by them. The early majority's unique position between the very early and the relatively late to adopt makes them an important link in the diffusion process.

The early majority may deliberate for some time before completely adopting a new idea. Their innovation-decision is relatively longer than that of the innovator and the early adopter. "Be not the last to lay the old aside, nor the first by which the new is tried," might be the motto of the early majority. They follow with deliberate willingness in adopting innovations, but seldom lead.

## Late Majority: Skeptical

The late majority adopt new ideas just after the average member of a social system. Adoption may be both an economic necessity and the answer to increasing social pressures. Innovations are approached with a skeptical and cautious air, and the late majority do not adopt until most others in their

*It will be pointed out in Chapter 6 that the degree of opinion leadership possessed by each adopter category depends in part on the social system's norms.

social system have done so. The weight of system norms must definitely favor the innovation before the late majority are convinced. They can be persuaded of the utility of new ideas, but the pressure of peers is necessary to motivate adoption.

## Laggards: Traditional

Laggards are the last to adopt an innovation. They possess almost no opinion leadership. They are the most localite in their outlook of all adopter categories; many are near isolates. The point of reference for the laggard is the past. Decisions are usually made in terms of what has been done in previous generations. This individual interacts primarily with others who have traditional values. When laggards finally adopt an innovation, it may already have been superseded by another more recent idea which the innovators are already using. Laggards tend to be frankly suspicious of innovations, innovators, and change agents. Their tradition direction slows the innovation-decision process to a crawl. Adoption lags far behind knowledge of the idea. Alienation from a too-fast-moving world is apparent in much of the laggard's outlook. While most individuals in a social system are looking to the road of change ahead, the laggard has his attention fixed on the rear-view mirror.

# Characteristics of Adopter Categories

From content analyses of research publications in the Diffusion Documents Center at Michigan State University, we have gleaned over 3,000 findings relating various independent variables to innovativeness.* Research findings on the characteristics of adopter categories are summarized as generalizations under the following headings: (1) socioeconomic status, (2) personality variables, and (3) communication behavior.

## Socioeconomic Characteristics

Generalization 5-1: *Earlier adopters are no different from later adopters in age*. There is inconsistent evidence about the relationship of age and innovativeness; about half of the 228 studies on this subject show no relationship,

*Nearly 60 percent of the relationships produced by the content analysis have innovativeness as the dependent variable.

*Adopter*

*Categories* 186

20 percent show that earlier adopters are younger, and 30 percent indicate they are older.*

Generalization 5-2: *Earlier adopters have more years of education than do later adopters.*

Generalization 5-3: *Earlier adopters are more likely to be literate than are later adopters.*

Generalization 5-4: *Earlier adopters have higher social status than later adopters.*** Status is indicated by such variables as income, level of living, possession of wealth, occupational prestige, self-perceived identification with a social class, and the like. But however measured, about two-thirds of such inquiries find a positive relationship of status with innovativeness.

Generalization 5-5: *Earlier adopters have a greater degree of upward social mobility than later adopters.* Although definitive empirical support is lacking, our evidence suggests that earlier adopters are not only of higher status but are on the move in the direction of still higher levels of social status. In fact, they may be using the adoption of innovations as one means of getting there.

Generalization 5-6: *Earlier adopters have larger sized units ( farms, and so on) than later adopters.*

Generalization 5-7: *Earlier adopters are more likely to have a commercial (rather than a subsistence) economic orientation than are later adopters.* A subsistence orientation is typified by a traditional peasant who produces only for his own consumption and not for sale. Greater innovativeness comes with the advent of a commercial orientation in which farm products are raised for market.

Generalization 5-8: *Earlier adopters have a more favorable attitude toward credit (borrowing money) than later adopters.*

Generalization 5-9: *Earlier adopters have more specialized operations than later adopters.*

The social characteristics of earlier adopters thus mark them as better educated, of higher social status, and the like. They are wealthier, more

---

*Rogers (1962b, pp. 173–174) reanalyzed Gross' (1942) original data to demonstrate there are wider differences in age between adopter categories when age at the time of *adoption* of hybrid seed was used, rather than age at the time of *interview*.

**Not only do earlier adopters have higher social status, but they may also have a higher degree of "status inconsistency," the degree to which an individual's various status dimensions, such as income, education, and occupation, are interrelated. Thus, an individual with high status inconsistency might be relatively high in education and in occupational prestige but low in income. Wells and MacLean (1962) found that more innovative Michigan farmers had higher status consistency, contrary to our expectations. This study was tentative in nature but suggests a lead for further research on status inconsistency and innovativeness.

specialized, and have larger sized units. Wealth and innovativeness appear to go hand-in-hand.* Do innovators innovate because they are rich, or are they rich because they innovate? The answer to this cause-and-effect question cannot be answered on the basis of available correlational data. However, there is adequate reason why wealth and innovativeness vary together. Greatest profits go to the first to adopt; therefore, the innovator gains a financial advantage through his innovations. Some new ideas are costly to adopt and require large initial outlays of capital. Only the wealthy units in a social system may be able to adopt these innovations. The innovators become richer and the laggards become poorer through this process. Because the innovator is the first to adopt, he must take risks that can be avoided by later adopters. Certain of the innovator's new ideas are likely to fail. He must be wealthy enough to absorb the loss from these occasional failures. It should be pointed out that although wealth and innovativeness are highly related, economic factors do not offer a complete explanation of innovative behavior (or even approach doing so). For example, although agricultural innovators tend to be wealthy, there are many rich farmers who are not innovators.

## Personality Variables

Personality variables associated with innovativeness have not yet received their share of research attention, perhaps because of difficulties of measuring these dimensions in field interviews.**

Generalization 5-10: *Earlier adopters have greater empathy than later adopters*. Empathy is the ability of an individual to project himself into the role of another person. This ability is an important quality for the innovator, who must be able to think counterfactually, be imaginative, and take the roles of heterophilous others in order to communicate effectively with them.

Generalization 5-11: *Earlier adopters are less dogmatic than later adopters*. Dogmatism is a variable representing a relatively closed belief system, a set of beliefs that are strongly held. The highly dogmatic person would not welcome new ideas; he prefers to hew to the past in a closed manner.

---

*Cancian (1967) argues that the relationship of wealth and innovativeness may be curvilinear because of the intervening variable of perceived risk.
**Harp (1960) feels that the inclusion of personality variables in analyses of innovativeness will contribute little. He states that if other sociological variables are included in investigations of innovativeness, the effect of ". . . personality may disappear." This is, of course, an empirical question, yet to be fully answered.

*Adopter*
*Categories*                                                      *188*

Generalization 5-12: *Earlier adopters have a greater ability to deal with abstractions than later adopters.* Innovators must be able to adopt a new idea largely on the basis of abstract stimuli, such as are received from the mass media. But later adopters can observe the innovation in the here-and-now of a peer's operation. Therefore, they need less ability to deal with abstractions.

Generalization 5-13: *Earlier adopters have greater rationality than later adopters.* Rationality is use of the most effective means to reach a given end.

Generalization 5-14: *Earlier adopters have greater intelligence than later adopters.*

Generalization 5-15: *Earlier adopters have a more favorable attitude toward change than later adopters.*

Generalization 5-16: *Earlier adopters have a more favorable attitude toward risk than later adopters.*

Generalization 5-17: *Earlier adopters have a more favorable attitude toward education than later adopters.*

Generalization 5-18: *Earlier adopters have a more favorable attitude toward science than later adopters.* Because most innovations are the products of scientific research, it is logical than innovators should be more favorably inclined toward science.

Generalization 5-19: *Earlier adopters are less fatalistic than later adopters.* Fatalism is the degree to which an individual perceives a lack of ability to control his future. How can a change agent convince a client to adopt innovations that will control the size of his family and give him better health and a higher level of living when the client believes that his future is determined by fate?

Generalization 5-20: *Earlier adopters have higher levels of achievement motivation than later adopters.* Achievement motivation is a social value which emphasizes a desire for excellence in order for an individual to attain a sense of personal accomplishment.

Generalization 5-21: *Earlier adopters have higher aspirations (for education, occupations, and so on) than later adopters.*

## Communication Behavior

Generalization 5-22: *Earlier adopters have more social participation than later adopters.*

Generalization 5-23: *Earlier adopters are more highly integrated with the social system than later adopters.* Communication integration is the degree to which the units in a social system are interconnected by interpersonal communication channels.

Generalization 5-24: *Earlier adopters are more cosmopolite than later adopters.*
The innovators' reference groups are more likely to be outside rather than
within their social system. They travel widely and are involved in matters
beyond the boundary of their local system. For instance, as shown in Chapter
2, Iowa hybrid corn innovators traveled more often to urban centers like
Des Moines than did the average farmer (Ryan and Gross, 1943). Medical
doctors who innovated a new drug attended more out-of-town professional
meetings than noninnovators (Coleman and others, 1966). Industrial and
educational organizations that are markedly innovative are more likely to hire
consultants, a cosmopolite influence. Thai peasant innovators were found to
visit Bangkok more frequently (Goldsen and Rallis, 1957, pp. 25–28). Chaparro
(1955) concluded that Costa Rican innovators were amazingly cosmopolite.
Over 84 percent had visited the U.S., 62 percent had traveled to Europe, and
67 percent had visited Mexico.

Generalization 5-25: *Earlier adopters have more change agent contact than later
adopters.*

Generalization 5-26: *Earlier adopters have greater exposure to mass media
communication channels than later adopters.*

Generalization 5-27: *Earlier adopters have greater exposure to interpersonal
communication channels than later adopters.*

Generalization 5-28: *Earlier adopters seek information about innovations more
than later adopters.*

Generalization 5-29 *Earlier adopters have greater knowledge of innovations than
later adopters.*

Generalization 5-30: *Earlier adopters have a higher degree of opinion leadership
than later adopters.* Although we find that innovativeness and opinion leadership are positively related, we know that the degree to which these two
variables are related depends in part on the norms of the system. In a modern
system opinion leaders are more likely to be innovators than in traditional
systems (Chapter 6).

Generalization 5-31: *Earlier adopters are more likely to belong to systems with
modern rather than traditional norms, than are later adopters.*

Generalization 5-32: *Earlier adopters are more likely to belong to well integrated
systems than are later adopters.* The internal "trickle-down" of new ideas in a
well integrated system is faster, enabling the members of such systems to learn
about new ideas more rapidly.

In summary, we see that most of these independent variables in the thirty-
two generalizations are positively related to innovativeness (Figure 5-3). This
means that innovators will score higher on these variables than laggards. For

instance, Rogers with Svenning (1969, p. 300) found that in traditional Colombian villages the innovators averaged thirty trips a year to cities whereas the laggards averaged only 0.3 trips.

A few variables, such as dogmatism and fatalism, are negatively related (Figure 5-3), and opinion leadership is greatest for early adopters, at least in most systems.

Thus, a set of characteristics of each adopter category emerges from past diffusion research. The important differences among these categories suggest that change agents might utilize somewhat different strategies of change with

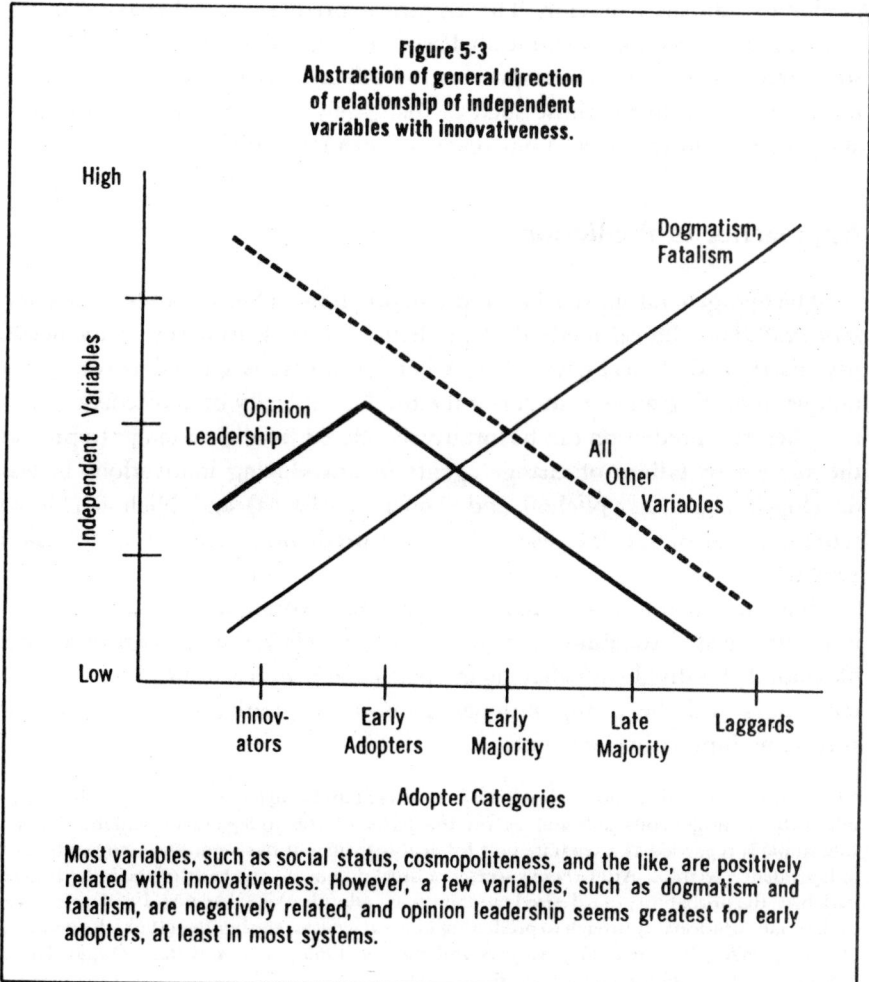

**Figure 5-3
Abstraction of general direction
of relationship of independent
variables with innovativeness.**

Most variables, such as social status, cosmopoliteness, and the like, are positively related with innovativeness. However, a few variables, such as dogmatism and fatalism, are negatively related, and opinion leadership seems greatest for early adopters, at least in most systems.

each. Thus, one might appeal to innovators to adopt an innovation because it was soundly tested and developed by credible scientists, but this approach would not be effective with laggards.

# Predicting Innovativeness

One of the goals of social science is to provide an empirical basis for predicting human behavior. The empirical prediction of behavior is not meaningful unless it is theoretically based and logically consistent. Although such prediction is of clear import to the scientist, it is perhaps even more useful and relevant for those such as change agents, development planners, and administrators, whose immediate concern is action.

## Approaches to Prediction

The two general approaches to predicting human behavior are *clinical* and *statistical*. The clinical method of prediction is used primarily in medicine, psychiatry, and clinical psychology where each case is viewed as somewhat unique, requiring an assessment of the total complexity of antecedent symptoms before a prediction can be intuitively offered. In their attempt to predict the success or failure of change agents in introducing innovations in less developed countries, Niehoff and Anderson (1964a) and Niehoff (1966a) represent one of the few uses of clinical prediction methods in diffusion research.

Statistical methods of prediction are more commonly used to predict innovativeness. Probability (or stochastic) models are used to forecast the likelihood of individuals behaving in a particular manner. Multiple correlation techniques and the configurational method* are perhaps most useful for predicting innovative behavior.

*The configurational method of prediction consists of dividing a sample of respondents into relatively homogeneous subsamples on the basis of the independent variables. Each subsample is regarded as a separate unit for analysis, since it has a unique configuration of independent variables. After these successive breakdowns on the basis of the independent variables, the probability of a desired outcome (e.g., adoption) is calculated. For illustrations of the configurational approach to predicting innovativeness, see Findley (1968), Rogers and Havens (1962a), Bonilla (1964), Rogers and Ramos (1965), Ross and Bang (1965), Keith (1968a), Keith (1968b), Herzog and others (1968).

## Multiple Correlation Approach

*Multiple correlation* is a statistical procedure designed to analyze and explain the variance in a dependent variable in components due to the effects of various independent variables. The goal of the multiple correlation approach is to predict a maximum of the variance in the dependent variable, which in the present case is innovativeness. Table 5-2 shows a summary of multiple correlation analyses of innovativeness. We notice a general trend for the percentage of explained variance in innovativeness to increase with the passing of years until in the mid-1960s up to about 80 percent of the variance in innovativeness was explained. This may be partly attributable to the advent of computer data analysis which allows the inclusion of a greater number of independent variables in these analyses. Further, a greater variety of independent variables were included in these studies; economic and social psychological dimensions are evidenced in the recent studies, along with variables indicating social structural aspects.

One advantage of the multiple correlation approach is that it discloses the degree to which each independent variable is related to innovativeness, while controlling the effects of all other independent variables. This yields an indicant of the novel contribution of each independent variable in explaining innovativeness. For instance, we have just reviewed evidence in this chapter that social status and cosmopoliteness are both related to innovativeness. But we also generally find that status and cosmopoliteness are correlated. The relationship of cosmopoliteness to innovativeness may be due to the relationship of both with social status. Multiple correlation methods help us to untangle the complex webs of interrelationships among our independent variables as they relate to innovativeness.

Despite the fact that some multiple correlation analyses have explained up to 80 percent of the variance in innovativeness, most attempts are much less successful. Further research is needed to raise the level of predictive accuracy. Further, most of the studies reviewed in Table 5-2 are actually "postdiction" rather than prediction in that they did not validate their predictions on a future sample of similar respondents.

# Simulation of Innovation Diffusion

## Computer Simulation

The investigations of diffusion researchers have traditionally been bound by their research tools to examinations of slices or cross-sections of the

**Table 5-2  Summary of Multiple Correlation Analyses of Innovativeness**

| | INVESTIGATOR | RESPONDENTS | PERCENTAGE OF VARIANCE IN INNOVATIVENESS EXPLAINED (%) | NUMBER OF INDEPEN-DENT VARIABLES UTILIZED |
|---|---|---|---|---|
| 1 | Copp (1956) | Kansas farmers | 50.0 | 5 |
| 2 | Fleigel (1956) | Wisconsin farmers | 32.0 | 6 |
| 3 | Copp (1958) | Wisconsin farmers | 52.0 | 4 |
| 4 | Rogers (1957a) | Iowa farmers | 17.0 | 5 |
| 5 | Armstrong (1959) | Kentucky farmers | 42.1 | 3 |
| 6 | Ramsey and others (1959) | New York dairy farmers | 9.6 | 4 |
| 7 | Hobbs (1960) | Iowa farmers | 29.7 | 7 |
| 8 | Sizer and Porter (1960) | West Virginia farmers | 25.9 | 4 |
| 9 | Straus (1960) | Wisconsin farmers | 33.6 | 3 |
| 10 | Kimball (1960) | Michigan farm families | 25.0 | 6 |
| 11 | McMillion (1960) | Large farmers in New Zealand | 39.9 | 5 |
| 12 | Rogers and Havens (1961b) | Ohio farmers | 56.4 | 5 |
| 13 | Flinn (1961) | Truck growers in 7 Ohio communities | 56.6 | 4 |
| 14 | Cohen (1962) | New Jersey families | 54.8 | 3 |
| 15 | Rogers and Havens (1962a) | Ohio farmers | 64.1 | 5 |
| 16 | Deutschmann and Fals Borda (1962b) | Colombian farmers | 56.3 (and 68.9 when using 27 variables) | 8 |
| 17 | Junghare (1962) | Farmers in India | 23.8 | 7 |
| 18 | Madigan (1962a) | Heads of households and other males in the Philippines | 17.1 | 3 |
| 19 | Neill (1963) | Ohio farmers | 40.5 | 6 |
| 20 | Havens (1963a) | Colombian farmers | 47.3 | 3 |
| 21 | Flinn (1963) | Truck growers in Ohio | 64.1 | 5 |
| 22 | Jain (1965) | Farmers in Canada | 50.3 | 7 |
| 23 | Haring (1965) | Wisconsin farmers | 50.2 | 34 |
| 24 | Andrus (1965) | U.S. consumers | 41.0 | 21 |
| 25 | Rogers (1966a) | Colombian farmers in five communities | From 24.1 to 39.0 | 6 |
| 26 | Morgan and others (1966) | U.S. household heads | 16.0 | 5 |
| 27 | Beal and Sibley (1966) | Guatemalan Indian farmers | 78.0 (42.0 when using 6 variables) | 51 |
| 28 | Moulik and others (1966) | Farmers in India | 81.0 | 4 |
| 29 | Whittenbarger and Maffei (1966) | Colombian farmers | 44.4 | 5 |
| 30 | Ramos (1966a) | Colombian farmers | 12.9 | 9 |
| 31 | Singh (1966b) | Indian farmers | 63.5 | 6 |
| 32 | Wish (1967) | Retail food stores in Puerto Rico | 87.5 | 35 |
| 33 | Chattopadhyay and Pareek (1967) | Indian peasants | 59.0 | 3 |
| 34 | Herzog and others (1968a) | Brazilian peasants | 43.0 | 13 |
| 35 | Roy and others (1968) | Indian farmers | 50.0 | 15 |
| 36 | Ascroft and others (1969) | Nigerian peasants | 42.0 | 13 |

process at one point in time. Methodological limits have necessitated slow-motion analyses which hold a slice of the process stationary while observing the dynamics of diffusion. Now with the flexible time considerations provided by the computer, it is possible to fuse the stationary analysis with the continuing process and capture the important variables in action. This can be done with the technique of computer simulation.

The result of computer simulation is the reproduction of the social process that one seeks to mimic.* If the simulated process does not correspond to reality data, then one knows that he must adjust his model or set of rules governing the simulated process.

Torsten Hägerstrand, a quantitative geographer at the Royal University of Lund, Sweden, is the father of diffusion simulation research. His work on computer simulation began in the early 1950s but was published only in Swedish. For many years the "paper curtain" of language barriers prevented the diffusion of his work to U.S. researchers. From the mid-1960s, however, his work has been carried forward in a series of investigations by quantitative geographers. Examples of simulations are the diffusion of deep well drilling in Colorado (Bowden, 1965a) and agricultural innovations in Colombia (Hanneman, 1969). These studies and others like them suggest that computer simulation of diffusion holds promise as a means to explore the complexities of the process as it unfolds over time. This potential, however, has not been very fully realized.

## Simulation Training Games

Not only does diffusion simulation have an important research potential, but it may also be useful as a training device. A simulation game called "Change Agent" has been developed in the Department of Communication at Michigan State University. This game, which requires about one hour to play, asks the participant to take the role of a change agent in a peasant village of 100 families. His objective is to choose optimally among various strategies of change so as to obtain as close as possible to 100 percent adoption of an innovation by the villagers. For example, the change agent can choose to locate the important opinion leaders in the village (at a "cost" of twenty days of effort each) or he can broadcast a radio program about the innovation (which requires ten days to prepare). The game's scoring system converts each of the

*Probably one of the best known computer simulations was the prediction of voting behavior in the 1960 national presidential election (Pool and Abelson, 1961).

player's decisions into the percentage of clients who adopt the innovation as a result. The player can even choose to obtain feedback as to the percentage adoption of the innovation that he has obtained at any given point during the playing of "Change Agent" (at a cost of thirty days for hypothetically surveying his peasants).

The advantage of training games like "Change Agent" is that they teach future change agents the realistic nature of their job before they are actually in it. Games seem especially important in creating interest and involvement in learning more about the diffusion of innovations.

# Summary

*Adopter categories* are classifications of individuals within a social system on the basis of innovativeness, the degree to which an individual is relatively earlier in adopting new ideas than other members of his system. A variety of categorization systems and titles for adopters have been used in past research studies. This chapter suggests a possible conceptual standardization of these categories.

The research of sociologists, learning psychologists, and students of the diffusion effect provide theoretical reasons for expecting adopter categories to be normal. The *diffusion effect* is the cumulatively increasing degree of influence upon an individual within a social system to adopt or reject an innovation. This influence results from the increasing rate of knowledge and adoption or rejection of the innovation in the system. Adopter distributions tend to follow an S-shaped curve over time and approach normality.

The continuum of innovativeness can be partitioned into five adopter categories (innovators, early adopters, early majority, late majority, and laggards) on the basis of two characteristics of a normal distribution, the mean and the standard deviation. These five categories are an arbitrary classification system.

The suggested adopter categories are ideal types, conceptualizations based on observations and designed to institute comparisons. Dominant values of each category are: innovators—venturesome; early adopters—respectable; early majority—deliberate; late majority—skeptical; and laggards—traditional. The relatively earlier adopters in a social system tend to have more education, a higher social status, more upward social mobility, larger units, a commercial rather than a subsistence orientation, a favorable attitude toward credit, and more specialized operations. Earlier adopters also have greater empathy, less dogmatism, greater ability to deal with abstractions, greater rationality, and more favorable attitudes toward change, risk, education, and science. They are less fatalistic and have higher achievement motivation scores and higher aspirations for their children. Earlier adopters have more social participation, are more highly integrated with the system, are more cosmopolite, have more change agent contact, have more exposure to both mass media and interpersonal channels, seek information more, have higher knowledge of innovations, and have more opinion leadership. They usually belong to systems with modern norms and to well integrated systems.

Social scientists have used two approaches to predicting innovative human behavior: Clinical and statistical analysis. Multiple correlation attempts to explain the maximum variation in the dependent variable resulting from several independent variables.

A relatively new research tool is the computer, which has been used to simulate diffusion processes. But diffusion simulation is more than a predictive device; it has great value as a training tool for change agents.

# [4]

## CLASSIFICATORY NOTES ON THE PRODUCTION AND TRANSMISSION OF TECHNOLOGICAL KNOWLEDGE

*By* Kenneth J. Arrow
*Harvard University*

### I. *Introduction*

Analysis of production functions over the last twelve years has suggested strongly that (*a*) a major proportion of the increase in per capita income cannot be explained by increases in the capital-labor ratio, and (*b*) production functions differ strongly among nations and indeed among regions. Since a production function is defined relative to a given body of technological knowledge, (*a*) implies that technological knowledge has been growing over time and (*b*) that technological knowledge varies over countries.

An economist could just leave the analysis at that, asserting that the causes which determine the amount of technological knowledge at any one time and place lie as much outside his province as the tastes which determine consumption patterns. But in fact, we know that significant quantities of resources are being expended by profit-making institutions on research and development. From the studies of Griliches [4] and Mansfield [8, Part IV], we know that the diffusion of technological knowledge, at least within a given economy, is partly governed by profitability considerations. Hence, it is suggested, we must regard the body of technological knowledge as the result as well as the cause of economic changes.

Economists have had a further, more detailed, preoccupation: that of the bias of technological change. A production function, if it shifts due to increased knowledge, can shift in many ways. In a two-factor model, Hicks spoke of "capital-saving" and "laborsaving" innovations and raised the question whether the bias is itself induced by economic considerations. Fellner [3] convincingly argued that factor prices, per se, should have no tendency to cause bias, since the aim of the entrepreneur is to minimize the total of costs. Kennedy [5] then sought to carry the Fellner analysis further by introducing explicitly the trade-offs between different possible biases in innovation. Kennedy, like most other current writers, has tended to replace the Hicks classification by introducing a more explicit and restricted model of technological change, namely, factor augmentation:

$$(1) \qquad Y = F[A(t)\,K,\, B(t)\,L],$$

where $Y$ is output, $K$ is capital, $L$ is labor, and $A(t)$ and $B(t)$ are the total augmentations of capital and labor, respectively. Then there is postulated a "transformation function" for knowledge, in the form of a trade-off among the rates of growth of $A$ and $B$ and research expenditures.

Models of this type and others (e.g., Arrow [1]) have the natural motivation of using the well-tried tools of production and distribution theory in what appears to be a related field. "Knowledge," as reflected in the variables $A$ and $B$, appears as an input to physical output, and we then need to supplement the ordinary production relation with an additional relation determining these newly defined inputs.

It is the suggestion of this paper that such models, though they may well be useful descriptively (however, this remains to be tested empirically), do not capture the essential features of the creation and transmission of knowledge. Technological progress is in the first instance the reduction in uncertainty. The product of a research and development effort is an observation on the world which reduces its possible range of variation. The outputs of different research projects are qualitatively different; there is no gain in acquiring the same information twice. The production of knowledge is thus basically different in character from the production of goods, where successive items can be qualitatively identical. The research and development process indeed has quantitative implications for factor productivity, particularly if it is measured at a relatively aggregated level, but these implications are a far from exhaustive description of the research and development process itself.

Research and development is thus intimately connected with the problems of uncertainty reduction which have been the objects of research in mathematical statistics and information theory. The problem of the transmission of knowledge—why different individuals and nations do not have access to equal bodies of knowledge —even more clearly requires analysis of qualitative distinctness of different items of information. Here the disciplines of information and communication theory, learning theory in psychology, and diffusion theory in sociology may be brought to bear.

The remainder of this paper will be devoted to elaborations and exemplifications of these themes. It cannot be claimed that any usable model has yet emerged, but it is hoped that the remarks here will be useful in further developments.

## II. *The Production of Knowledge*

Knowledge arises from deliberate seeking, but it also arises from observations incidental on other activities. Haavelmo, Kaldor, and I (see Arrow [1] for references) have all stressed that the activities

of production and investment may lead to increases in productivity without any identifiable allocation of resources to that end. An illuminating special case of increased knowledge is the discovery of natural resources. Exploration is analogous to research; but when a community is becoming settled, the members, in the course of their ordinary activities, frequently continue to make unexpected discoveries. Sutter established an agricultural community; his employee, Marshall, saw some yellow flecks in the mill race.

A more general formulation including both research and learning by doing can be formulated and will, I think, be useful. The Bayesian language will be used (for an excellent introduction, see Raiffa [11, Chaps. 1–5]); however, any sensible analysis of uncertainty will lead to equivalent formulations. Let the term "activity" be used, as usual for any process described by inputs and outputs, but we are particularly interested in the case where the outputs are not known with certainty. It can easily happen that the outcomes for different activities will be dependent random variables in the sense of a subjective probability. The case most interesting from the present point of view is that where there is an underlying unknown parameter upon which the probability distributions of outcomes for the different activities depend; then observing the outcome of one activity changes the *a posteriori* distribution of outcomes of the other. This need not happen of course; if the outcome of an activity is known with certainty a priori, then observing the outcome cannot change the probabilities of outcome of any other activity. More generally, if there is statistical independence between the outcomes of activities, the probability of distribution of the outcomes of one is unaffected by observations on the outcomes of another. Thus, if I really believe a pair of dice is fair, observing

## THEORY OF INNOVATION

any outcome is of no use in predicting a subsequent one. But if I suspect bias and I express my suspicions by an appropriate subjective probability distribution over the possible outcomes, then observing an outcome certainly does change my subsequent expectations in accordance with Bayes's theorem.

Thus, an activity will in general have two valuable consequences: the physical outputs themselves and the change in information about other activities.

In many cases, one or the other effect predominates. The classical research situation is one in which the actual output (e.g., of nylon) is of negligible importance compared to the information gain —*a posteriori* the probability that a substance with the properties of nylon can be produced is now 1, whereas a priori it may have been a small figure. In the cases of learning by doing the opposite holds; the motivation for engaging in the activity is the physical output, but there is an additional gain, which may be relatively small, in information which reduces the cost of further production.

Once these polar cases are presented, it becomes clear that intermediate cases are possible, and in fact they not only occur but are, I would hold, frequent. In fact, the bulk of research and development expenditures are actual steps in the production process—design, engineering, tooling, and manufacturing and marketing start-up costs (see Mansfield [7, p. 106]). Each stage involves uncertainties with regard to costs and, at the end, with regard to demand. At each stage, then, something is learned with regard to the probability distribution of outcomes for future repetitions of the activity. At the same time, the physical outputs are expected to be directly valuable.

The problem of the optimal choice of a sequence of activities where both the physical outcome and the information acquired are relevant to the choice is in principle a problem of statistical decision theory, but in fact only fragmentary results have been attained. Analytically, the question is difficult, and, perhaps not surprisingly, the results that have been attained can hardly be stated simply, though definite methods of computing the optimal solutions exist. (In my judgment, this will be an increasingly common situation in economic theory; broad general theorems of the kind we admire can usually only be found under undesirably restrictive conditions. What theory can imply in a broad range of cases is a computing algorithm. To test theory, then, we need econometric evidence or at least well-informed quantitative judgments as inputs to the computing process.) Some beginnings of a microeconomic theory of research and development are found in the work of T. Marschak, Glennan and Nelson (see [10, Chaps. 1, 2 and 5]).

Building a macroeconomic theory on microeconomic foundations is not an easy task; indeed, there is not one completely successful example. But in the case of the production of knowledge, the task is apparently much harder. To proceed from an individual to a collective theory that consumption depends on income or output on inputs involves incompletely justified steps; but it is at least reasonably clear what the aggregates themselves are. Information at the individual level is describable either as the actual outcome of a particular activity or as a whole conditional distribution over states of nature, with the conditions being the actual outcomes. Such a probability distribution is hard to describe in any simple way, and aggregating this information over individuals is even harder.

Given the probabity distribution at any moment of time, individuals make production decisions. The actual outputs are not known with certainty, but one might plau-

32                        AMERICAN ECONOMIC ASSOCIATION

sibly take their expected value and regard its relation to the inputs as the production function. This will shift over time as additional information is acquired. The current models in the growth field assume (*a*) that the effect of a given body of information on the production function can be summarized in a few parameters and (*b*) that these same parameters summarize the possibilities for acquiring new information. The first assumption is undoubtedly essential if one is to get anywhere. But the second is on the face of it very misleading. Consider an economy with labor as the only factor of production and operating under constant returns to scale. Then the effect of any given information structure on the production function is completely summarized in the output-labor ratio. But a given output-labor ratio may correspond to very different information structures and therefore very different potentialities for future productivity increases. One might have been based on very thorough exploration of limited areas of investigation, with little further room for increased knowledge, while the other might have investigated more widely and have potential productivity gains available at little additional cost. Securities analysts use such considerations in their evaluations; and historical examples of this difference are frequently cited.

One might be tempted to consider as an alternative measure of aggregate knowledge Shannon's measure of information. There are of course a host of practical difficulties in applying this measure to technological knowledge. But unfortunately it does not seem to be a correct measure for either the supply or the demand side. The irrelevance of Shannon's measure to the demand price for knowledge was shown by J. Marschak [9]. To take an extreme case, consider two states of the world which are indistinguishable with regard to technology; e.g., distinguished only with

regard to the number of craters on the dark side of the moon. Any discrimination between these two states would be an increase in information, but surely not one relevant to productivity. On the supply side, similarly, the cost of achieving a given increase in information may differ according to the particular observations made.

### III. *Transmission of Knowledge*

The observer of the outcome of an activity can be supposed to form new probability judgments; but how does this affect the information structures of others? The transmission of the observation or of the revised probability judgments must take place over channels which have a limited capacity and are therefore costly. Though the language is borrowed from communications theory, the really limited channels are human minds, not telegraphs or printed words.

Even for the individual there is a problem of channel capacity. To transform his a priori into *a posteriori* probabilities as the result of observations which have taken time, he must remember his a priori probabilities; but memory is a channel for transmission between points of time and is notoriously limited in capacity and subject to error. Natural memory can of course be supplemented with artificial aids —books, files, computer memories—but ultimately there remains a capacity limitation. As a result, even for an individual, the transformation of probabilities and therefore the acquisition of information is less than ideal. Winter in an unpublished paper has argued that memory limitation accounts for the results of the famous Humphreys experiment which has been so crucial in modern learning theory: An event $E$ occurs with probability $p > \frac{1}{2}$ on each of a succession of trials, with independence between trials. The individual is asked before each trial to predict

## THEORY OF INNOVATION                                         33

whether or not $E$ will occur; he is not told the value of $p$ but must in effect infer it from observations. If his aim is to maximize the expected number of successes, the rational strategy would be always to predict $E$, at least after the process has been observed long enough to establish almost certainly that $p > \frac{1}{2}$. But in fact the subjects wind up predicting sometimes $E$ and sometimes not $E$, and the frequency with which $E$ is predicted approaches very closely indeed to $p$. This might be explained by the assumption that they only remember a limited number of previous observations and thus never can be very well assured that it is best to predict $E$. (To be sure, many other explanations have been advanced in the literature.)

There have been studies by both economists and sociologists on the diffusion of innovations; the relation between the two types of studies resembles nothing so much as the parable of the blind men and the elephant. While Griliches and Mansfield stress the profitability of the investment and the risks involved, the sociologists (see, e.g., Rogers [12] and, for a more theoretical treatment, Coleman [2, Chap. 17]) are concerned with the nature of the channel connecting the adopters of an innovation with potential followers. While mass media play a major role in alerting individuals to the possibility of an innovation, it seems to be personal contact that is most relevant in leading to its adoption. Thus, the diffusion of an innovation becomes a process formally akin to the spread of an infectious disease.

There is nothing irreconcilable in the two viewpoints: in effect, the economists are studying the demand for information by potential innovators and sociologists the problems in the supply of communication channels. Different communication channels have different costs (or, equivalently, different capacities), where these costs in-

clude the ability of the sender to "code" the information and the recipient to "decode" it.

The understanding of transmission of knowledge is of especial importance in two of the key socioeconomic problems of our time: ($a$) international inequalities in productivity, and ($b$) the failure of the educational system in reducing income inequality. The two problems have a considerable formal similarity. If one nation or class has the knowledge which enables it to achieve high productivity, why is not the other acquiring that information? That a nation or class has a consistently high productivity implies a successful communication system within the nation or class, so the problem turns on the differential between costs of communication within and between classes.

(It is not intended to assert that difficulties in communication of information are necessarily the sole source of total factor productivity differences. There can be withholding of information to perpetuate monopoly positions; and both in foreign trade and in economic relations between the races, income differentials can arise from exploitation; see Krueger [6]. There might also be genetic factors. But the available evidence certainly suggests that communication problems are a major and perhaps predominant source of productivity and income differentials.)

It may be worthwhile speculating on some of the causes of these differential channel costs. Note first that all the sociological work on diffusion has put great stress on personal contact. Mass media may provide some overall alertness to change, but, except for the most alert and daring innovators, it is the example and advice of those known personally that are apparently most potent in securing acceptance of innovation. As Coleman stresses, however, personal contacts are by no means randomly distributed in the population, and the manner in which any trait (in particular, an item of information) is diffused throughout the system is correspondingly altered. To cite one case, the diffusion of use of new drugs was higher among physicians practicing in pairs than among those practicing singly.

I believe these facts cast light on the basic factors at work in the difficulties of transmission of knowledge across nations and through the educational system. Two general principles may be suggested: (1) A channel has greater capacity if the receiver regards it as more reliable; this is why personal contacts are frequently so important. (2) To a large extent, channels of communication serve purposes other than the diffusion of innovations—friendship, convenience—and the direction of diffusion may be dictated by factors in addition to profitability. Thus international channels, which typically have less other purposes to serve, are more expensive. In particular, and this may be very important, personal contacts across nations are obviously much less than within.

In amplification of (1), there is one special case of unreliability from the receiver's point of view that deserves some attention: the inability of the receiver to understand the message. As already remarked, every piece of information can be regarded as transmitted in a code and can only be used if decoded. In the first instance, a language itself is a code, and the sheer difficulty of translation perhaps can be underestimated. (The inability of English-speaking economists to learn from their French, German, and Italian colleagues is notorious.) There are problems in nonverbal forms of communication. When the British in World War II supplied us with the plans for the jet engine, it took ten months to redraw them to conform to American usage. More subtly, as several gifted observers of the educational scene have observed, there are class and

racial differences in the meaning of words, not so much in the literal denotation, but in the connotations and associations, and in the significance of nonverbal behavior. In the complicated interplay of messages between teacher and student, the unreliabilities of communication can lead to extreme inefficiencies.

## IV. *Exponential Growth and Stationary Bias*

To conclude, we might briefly consider two aspects of current growth models which are inconsistent with models of technological progress as information-seeking and transmitting.

One is the universal tendency to take some form of rate of growth of productivity as a variable of the system. Typically the model ends up in a quasi-stationary state with a constant rate of productivity growth. Now no known form of adaptive model, with probabilities modified by experience, leads to such a result; on the contrary, such models invariably lead to asymptotes in which output is constant. There is a limit to what can be learned even with infinitely many opportunities. Actually, with respect to the very long run, such a conclusion seems very reasonable. You cannot get something for nothing, ever, and it seems unreasonable to suppose that, by waiting a sufficient length of time, you can get any given output for arbitrarily small input. Eternal exponential technological growth is just as unreasonable as eternal exponential population growth.

But of course exponential technological growth does have the advantage of being consistent with the observed facts; if anything, the observed rate of growth of total factor productivity is increasing. I can only conjecture that, as in the case of population, the true law is something like the logistic curve, but we are still in the early phases, which resemble the exponential.

## THEORY OF INNOVATION    35

At the moment, that is, each new item of information opens up the opportunity for acquiring additional items in a constant ratio (in some sense), but eventually the scarcity of additional information will become apparent.

The other aspect is the Kennedy invention-possibility curve, the trade-off between rates of augmentation of different factors for a given research budget. In the first place, there is, of course, no warrant for assuming that technological progress is factor-augmenting; pieces of knowledge about the workings of the world need not be associated in any simple way with different factors of production. In the second place, as already remarked, in the long run, it is unlikely that rates of augmentation are the right variables to enter into any such relation. But most important, there is no warrant for the stability of the curve over time, however it is described. The curve of efficient possible innovations at any time must surely shift with the particular nature of the information acquired, a nature not summed up in productivity measures of any kind. My conjecture is that this is one place where theory should not presume to take the place of history. The set of opportunities for innovation at any one moment are determined by what the physical laws of the world really are and how much has already been learned and is therefore "accidental" from the viewpoint of economics. My guess is that economic factors have little to do with bias in technological progress (though they may have a good deal to do with its magnitude). European desire for spices in the late fifteenth century

may have had a good deal to do with motivating Columbus' voyages, but the brute, though unknown, facts of geography determined what in fact was their economic results.

### REFERENCES

1. K. J. ARROW, "The Economic Implications of Learning by Doing," *Rev. of Econ. Studies,* June, 1962, pp. 155–73.
2. J. S. COLEMAN, *Introduction to Mathematical Sociology* (Free Press, 1964).
3. W. FELLNER, "Does the Market Direct the Relative Factor-Saving Effects of Technological Progress?" in *The Rate and Direction of Inventive Activity* (Princeton, 1962), pp. 171–88.
4. Z. GRILICHES, "Hybrid Corn: An Exploration in the Economics of Technological Change," *Econometrica,* Oct., 1957, pp. 501–22.
5. C. KENNEDY, "Induced Bias in Innovation and the Theory of Distribution," *Econ. J.,* Sept., 1964, pp. 541–47.
6. A. O. KRUEGER, "The Economics of Discrimination," *J.P.E.,* Oct., 1963, pp. 481–86.
7. E. MANSFIELD, *The Economics of Technological Change* (Norton, 1968).
8. ———, *Industrial Research and Technological Innovation* (Norton, 1968).
9. J. MARSCHAK, "Remarks on the Economics of Information," in *Contributions to Scientific Research in Management* (Western Data Processing Center, Univ. of California, 1960), pp. 79–98.
10. T. MARSCHAK, T. K. GLENNAN, JR., AND R. SUMMERS, *Strategy for R&R* (New York, Springer, 1967).
11. H. RAIFFA, *Decision Analysis* (Addison-Wesley, 1968).
12. E. ROGERS, *Diffusion of Innovations* (Free Press, 1962).

# [5]

Excerpt from L. Nabseth and G.F. Ray (eds), *The Diffusion of New Industrial Processes: An International Study*, 294–315.

# CHAPTER 11

# SUMMARY AND CONCLUSIONS

### By L. Nabseth, IUI [1]

## INTRODUCTION

In the Introduction to this book we stressed the importance of diffusion of new technical processes as a decisive factor in raising efficiency and productivity in industry. Under special circumstances, even if a new process was used only by its innovator in his home country, it might still be of great economic importance. But it is mostly through its diffusion to other companies, both intra-nationally and internationally (including to foreign subsidiaries)[2] that a new process becomes really significant. A free flow of ideas and information between firms and countries is normally essential if a process innovation is to be of economic importance on the international scene. Cultures and societies which suppress such a flow are at a disadvantage in their efforts to increase productivity, and processes developed in secret, for instance in defence work, will probably be diffused much slower than other new technologies. But even in countries like those studied in this book, where information on new technology is fairly freely accessible, there is a real problem of delay in the spread of information on new processes. The theory that such information diffuses very rapidly among companies in different countries was not substantiated in the two cases where we obtained detailed information, numerically controlled machine tools and special presses. We shall take up this point again later.

This book is focused on the diffusion in some countries of a few new

---

[1] This chapter is based primarily on the different chapters of this book. But other results from the investigation have been presented elsewhere, and to some extent they are taken into account. These are: M. Breitenacher, 'Innovation und Imitation fördern den technischen Fortschritt', *Wirtschaftskonjunktur (Berichte des IFO-Instituts für Wirtschaftsforschung)*, vol. 21, no. 3, July 1969; Davies, 'The clay brick industry and the tunnel kiln'; A. Gebhardt, 'Der Tunnelofen in der Ziegelindustrie – Beispiel einer neuen Technologie', *IFO-Schnelldienst*, vol. 24, no. 27, July 1971; L. Nabseth, 'The diffusion of innovations in Swedish industry', paper given to the International Economic Association conference at St Anton, 1971; G. F. Ray, 'The diffusion of new technology', also 'Ergebnisse von Diffusionsuntersuchungen in Europa' in O. Hatzold (ed.), *Innovation in der Wirtschaft*, Munich, IFO, 1970 and 'New technology and enterprise decisions' in Z. Román (ed.) *Progress and Planning in Industry*, Budapest, Akadémiai Kiadó, 1972; H. Schedl, 'Geringe Verbreitung schützenloser Webmaschinen in der Baumwollindustrie', *IFO-Schnelldienst*, vol. 23, no. 40, October 1970; Smith, 'The weaving of cotton and allied textiles in Great Britain'.

[2] For a discussion of the importance of the multinational firm for the diffusion of new technology see OECD, *Gaps in Technology. Analytical Report*, Paris, 1970.

SUMMARY AND CONCLUSIONS        295

processes introduced since World War II. When the project started, we thought it would be easy to find examples of important new technologies and that the only problem would be to choose between the many alternatives. As it turned out, however, it proved quite difficult for engineers in different industries to find examples of new technologies with major effects. For those working close to the production line, technological change seems to go much more in steps than by leaps, or perhaps one could say that it often seems to go by a few leaps and then many small steps. (Float glass is an example in this book.) This general conclusion is based on more material than that presented here, since our attempts to find good examples of new processes were quite extensive. The importance of smooth and continuous progress is also stressed in the introduction to chapter 6, even though basic oxygen steel was considered at first to be one of the clearer cases of a major 'leap forward' in technology.

In the Introduction it was stated that the aims of the investigation were:

(a) to assess the scope and extent of the diffusion of selected new processes in the countries covered, and to establish the international differences in the level and speed of their diffusion;

(b) to study the factors which affect the speed of diffusion;

(c) to make some attempt to account for the differences between countries.

The obvious question to ask in this Summary is: did the group succeed in its intentions?

### INTERNATIONAL DIFFERENCES IN THE DIFFUSION OF SELECTED NEW PROCESSES

We were interested only in new processes, or, from the sellers' point of view, new products in the producer-goods industries. New consumer-goods were not considered; one obvious reason being that quite a number of studies of the international diffusion of consumer products, for instance, television sets, cars and man-made fibres, are already available. Studies of new processes, especially in the industrial sector, are fewer and mostly relate to American conditions.[1] Our investigations provided a simple explanation for this: it is more difficult to study producer- than consumer-goods. There are quite good statistics of stocks

[1] Griliches, 'Hybrid corn'; Mansfield, *Industrial Research and Technological Innovation*; J. E. Tilton, *International Diffusion of Technology: the case of semi-conductors*, Washington (DC), The Brookings Institution, 1971.

of the latter, but the same is not true of production equipment, with the one exception of the iron and steel industry among the cases considered here.

For this study, it was decided to obtain information directly from firms, partly because this was thought to improve the possibility of explaining differences in diffusion patterns between countries, partly because it seemed the only way to get the information required, and partly because we were also interested in differences between companies (for instance between firms of different sizes) and in comparisons with other studies.[1] We used some other sources, such as trade associations and individual experts, and found that for our kind of economic research, it was easier to get data from firms in small countries than in large ones.[2] Probably this is because in smaller countries there are fewer companies to contact and it is easier to convince them of the value of the research project. For some big countries, notably Germany, Italy and the United States, the information presented on, for instance, special presses, is so restricted that one must admit the diffusion diagrams are not really representative. This does not, of course, mean that the information is useless. On this point one general conclusion seems to be that producers or licence-holders of the processes can provide as reliable data on diffusion as the firms which use the process (special presses, float glass and shuttleless looms were good cases in point).

One reason for restricting the new techniques considered to processes introduced since the war[3] was that in many countries earlier data were unobtainable. But it meant that in most cases diffusion is still going on, and will continue during the seventies and possibly the eighties as well. In one respect, this represents a limitation on the data and analysis; on the other hand, there is value – certainly novelty value – in studying diffusion at an early stage, as it is for instance in continuous casting. For other processes, such as shuttleless looms and basic oxygen steel, diffusion is well under way or almost complete. In fact the part of the diffusion process analysed for continuous casting is regarded only as a preliminary development stage in basic oxygen steel. This lack of complete data also caused problems in the econometric calculations: for instance, for special presses and shuttleless looms arbitrary introduction dates had to be assigned to non-adopters so that all the available information could be used.

But even if, ideally, one had all the data required, problems would still arise in presenting diffusion patterns for different countries. Total

---

[1] In Mansfield's studies the only comparisons are between processes and firms.

[2] Some members of our research team think, however, that this is more a question of the technology studied than of the size of country.

[3] The tunnel kiln is an exception; it originated earlier, although its wider application has also been post-war.

SUMMARY AND CONCLUSIONS　　　　　297

output in a country produced by the new process, the increase in capacity that is equipped with the process, the number of firms, plants, or machines using the process, can all be stated, but the data ought to relate to the questions to be answered. From a macro-economic point of view, one might argue that it is the influence of diffusion on total production that is of primary interest – whether a process is used intensively in a few firms or more extensively in many. On the other hand, the decision to introduce new processes must be made by some-one, usually managements of firms or plants, so that, to analyse in depth the decision process in relation to new techniques, data from firms or even individual plants are needed. Furthermore, company data are required to understand how the diffusion process might be influenced, for instance speeded up, although, if the object is simply to forecast the diffusion pattern, some type of curve fitting to aggregate data might be sufficient.

The real problem in presenting diffusion diagrams, however, is not the numerator but the choice of a meaningful denominator: what is a reasonable basis of comparison?[1] In studying the diffusion of a new technique, in the paper or steel industry for instance, the data can be related either to the total production of paper or steel, or to the number of firms in the industry. But problems at once arise: either the new technique may never be suitable for certain types of paper or steel, or it may improve over time, so that while it is unsuitable for parts of the production or some types of firms initially, it will be suitable later on. Examples of the first case are special presses, which have never been suitable for some types of paper, and tunnel kilns, which cannot be used for certain grades of clay. The basic oxygen process illustrates the second case, since initially it could not be adapted for big plants, for producing specialised steels, or for processing high-phosphoric ores. A third problem arises with numerically controlled machine tools, which can be used to produce parts of many different products that are, how-ever, not clearly definable, and a fourth (relating both to the numerator and the denominator) is how to fix a starting date for commercial operation of a new process which has been improved over a long period, but initially could be used only in some types of plant. When is it possible to say that the process has really become an innovation in the Schumpeterian sense of the word?

These problems can be solved in different ways, all more or less arbitrary. In the chapters on continuous casting and tunnel kilns, a ' technological ceiling ' was assumed for each country, but the problem of defining the ceiling remained. For the diffusion charts to provide a

---

[1] See also G. F. Ray, 'On defining diffusion of new techniques', *Business Economist*, vol. 4, Summer 1972, pp. 82–8.

definition, either the process must have been in use for a very long time, or some sort of theoretical definition is needed. Another solution is to ignore that part of the production or production equipment which is definitely unsuitable for the new process. This was done for special presses and tunnel kilns, but, of course, here too there is an arbitrary element.

Yet another way of handling the problem is to use Mansfield's method of counting only those firms that have already introduced the new process.[1] Only fairly well established processes can then be studied, and those firms that prefer to retain the old process, possibly because it too has been improved, are not taken into account, thus precluding deeper analysis of the procedure behind the choice of new processes. Examples of competition between a new process and an improved version of the old can be found in tunnel kilns (described in chapter 5), in basic oxygen steel (chapter 6) and in continuous cooking in pulp production.[2]

Even if all the firms that have not introduced the new technique are eliminated from the analysis, the problem remains of how to analyse the case where there is a choice of innovations. This has affected many companies in the computer field, and there are examples of it in the studies of new methods of steel-plate marking and cutting in ship-building.[3]

In general, comparisons of diffusion patterns between processes and between countries require the utmost care. The arbitrary assumptions which are unavoidable mean that conclusions based on casual inspection can easily be quite wrong. In the literature of diffusion, sigmoid curves, representing the development through time of the number of firms using the new technology, or the volume of production or capacity, have been widely used, but two distinct questions remain. The first is simply whether such curves give a good statistical fit to the observed data. (When the diffusion process is manifestly incomplete and the number of observations from the past is small, one may not be able to distinguish with any confidence between the first part of a sigmoid curve, the beginning of an exponential expansion and a straight line.) Secondly, even if one is satisfied that the curve is a good pictorial representation of the facts, how should this particular shape be interpreted? The problem is especially acute when, as in the present context, the measure of diffusion is a ratio for which the appropriate denominator is difficult to find. The analogy from epidemic diseases sometimes used

---

[1] Mansfield, *Industrial Research and Technological Innovation*.

[2] Nabseth, 'The diffusion of innovations in Swedish industry'.

[3] Ray, 'The diffusion of new technology', pp. 72–6; also briefly mentioned in chapter 2 above.

SUMMARY AND CONCLUSIONS       299

in consumer expenditure theory did not seem very helpful.[1] Moreover, in some cases presenting the data in this form might be misleading: it could well be that, in the early stages, overriding technical reasons allowed a new process to start rapidly in one country although it remained quite inappropriate in another.

### FACTORS AFFECTING THE SPEED OF DIFFUSION

At the beginning of this project, a common theoretical and methodological approach to the problem of diffusion of new technology was sought. It is obvious from the preceding chapters that a strictly 'common' approach was not found. As mentioned in the Introduction, this was less a failure to agree on a standard analysis than a question of differences in the techniques studied and in the empiricial material available. Furthermore, it may very well be that such a standardisation would in fact have hidden important explanatory factors in the diffusion of some processes. New processes are very different – in their relation to old processes, in their own improvement over time, and in other respects – and too much of a common approach might very well conceal these differences. Nevertheless, there are sufficient similarities in the analyses of the different processes to provide a basis for comparisons between them as well as within them.

Of the different stages of diffusion within a firm – first information, awareness, consideration and adoption – it is primarily the adoption stage that has been analysed in this study, although some material on information and awareness is also included. In the adoption stage, a few explanatory variables common to nearly all the processes were found – relative advantage or profitability, other economic variables, institutional circumstances and management attitudes. Before considering their influence, however, the data on 'first information' must be examined.

### First information

For the individual firm, information about a new technique is a precondition to adoption. From this point of view, it would seem as relevant to study the diffusion of the information as of the actual adoption. On the other hand, it might be argued that knowledge about new processes spreads fairly rapidly in and between industrialised countries, so that understanding the diffusion of production is not much help in understanding the diffusion of adoption. The bigger and the more important an innovation, the more valid this argument seems to be, but here we must remember what has been said before about the

---

[1] See Rogers, *Diffusion of Innovations*.

difficulty of deciding when an invention has really become an innovation (for instance, in the case of oxygen steel this differed between different types of plant within the industry). Certain information might therefore be very important to some people within an industry, but much less interesting to others. Furthermore, in industries like textiles, paper, bricks and brewing, there are many small firms with limited possibilities of obtaining, let alone making use of, information about new processes. Taking all these things into account, it seems quite probable that the diffusion of information about a new technique is a time-consuming process. Carter and Williams in their study of British industry, conclude that 'the backward firm may not hear of an idea for several years after it is first made known'.[1] Of course, the data presented here are not precise; it is difficult to recall what exactly happened 15 years ago, even if – as in many cases – some of those who introduced the new process are still with the company.

In this study, data on first information were gathered for numerically controlled machine tools and for special presses. Some Swedish data were also obtained for other new processes.[2] These confirm the hypothesis that the diffusion of information about new techniques is quite a slow process. Timelags of about ten years between the first and the last firm are not uncommon. For the majority of firms, however, this is shortened to about five years, which means there are a few firms in each country which lag behind the others. One hypothesis often advanced is that information is diffused more rapidly than adoption.[3] This is not supported by our material, taking into account all the processes studied. The interim report (chapter 2) found that the later a process is introduced in a country, the more rapid the diffusion process will be. This may perhaps be because, during the long time preceding adoption, the process is improved and so spreads more rapidly. A similar hypothesis could be advanced regarding information about new processes, but is is not well supported by the data. Possibly, once started, information diffuses more quickly in small countries, like Austria and Sweden, than in large ones, like Germany and the United Kingdom. This is not so for all the processes studied, but the weakness of the data must be remembered here.

Another interesting question in this connection is whether big firms get information about new processes earlier than small firms. In table 11.1 which covers Germany, Sweden and the United States, this hypothesis gets some support in numerically controlled machine tools and

---

[1] Carter and Williams, *Industry and Technical Progress*, ch. 16. We must, however, bear in mind that this study deals with British conditions. It is uncertain if these results can be extrapolated to the other countries studied.

[2] Nabseth, 'The diffusion of innovations in Swedish industry.'

[3] Rogers, *Diffusion of Innovations*, p. 108.

rather weaker support in special presses. If, however, the American firms (which nearly all have over 1,500 employees, and obtained information in the period 1963–5) are excluded, then the hypothesis gets strong support from German and Swedish data on special presses.

The best sources of information on special presses seem to be the same in the different countries – technical journals, manufacturers and licence-holders of the process. The important question is whether or not

Table 11.1. *Size of firm and date of first information*

|  | First information obtained | | |
|---|---|---|---|
|  | Before 1960 | 1960 or later | Total |
| NC machine tools | | | |
| Employing less than 1000 | 22 | 21 | 43 |
| Employing 1000 or more | 30 | 8 | 38 |
| Total | 52 | 29 | 81 |
| Special presses | | | |
| Employing less than 1000 | 10 | 69 | 79 |
| Employing 1000 or more | 7 | 28 | 35 |
| Total | 17 | 97 | 114 |
| Special presses excluding US firms | | | |
| Employing less than 1000 | 10 | 69 | 79 |
| Employing 1000 or more | 7 | 11 | 18 |
| Total | 17 | 80 | 97 |

SOURCE: inquiry.

information on a new technique and its adoption are related (there is some indication that they are, especially in the Swedish material). This could, of course, have important policy implications: improving information channels to firms would then speed up diffusion of the process. On the other hand, obtaining and evaluating information is sometimes costly to a firm, so that the problem of being informed is to some extent a question of management attitudes. If one does not want to be early in introducing new processes, it is otiose to be well-informed about innovations in the field.

*Profitability or relative advantage*

In all discussions about the diffusion of new technology, the profitability or relative advantage of the new process in relation to the old stands out as an explanatory variable. But it is by no means clear what the causal relationship ought to be. With perfect foresight and perfect capital markets one could say that, as long as the internal rate of return on a new process exceeded a certain level (due regard being taken, of course, of the capital equipment in use), a firm ought to introduce the technique in question as soon as possible. It would then be possible to distinguish only between users and non-users of a process; no theory could be advanced about the order of introduction among firms.

In fact, we know that, since the capital market is imperfect, firms' opportunity costs differ.[1] A high internal rate of return for one firm may very well be considered a rather low rate for another, at least in the short term, so that firms are likely to differ in their speed of application of a new process. A hypothesis about the order of introduction among firms requires that the opportunity costs for different companies be known, but this information could not be obtained.

Secondly, new processes are not introduced with perfect foresight; on the contrary, it is the very essence of new technical processes, especially in the early stages, that the outcome of their introduction is uncertain. Will the process work according to plan or not? From this point of view it seems quite natural that the higher the *ex ante* internal rate of return of a new process the more likely is its introduction, since, *ceteris paribus*, it then offers a high safety margin for uncertainty.[2] The lower the *ex ante* internal rate of return of a process (above a certain level) the longer it may have to wait for introduction, so that the margin of risk is diminished. To a first approximation this type of reasoning lies behind the profitability discussions in the different chapters of this book.

However, one general conclusion seems to be that calculating the profitability of a new process is more difficult than is usually acknowledged in studies on the subject. For some processes, for instance numerically controlled machines and continuous casting, profitability turned out very difficult to calculate *ex post*, and even more difficult *ex ante*. This does not mean that firms do not try to estimate the relative advantage of a new process, but rather that their calculations are very subjective. Of course, new investments always involve uncertainties, but they are probably greater in the introduction of new technologies than, for instance, in straightforward replacement. It follows that profitability calculations for new processes are very much linked with management attitudes, especially when experience of the technology is scarce and perhaps contradictory.

It appears from the technical descriptions in the different chapters that the introduction of a new process is regarded as a unique event by each firm. The very presentation of a profitability calculation therefore

---

[1] This may also be the case even when capital markets are perfect.

[2] Strictly speaking a firm must be assumed to combine the rate of return with some probability distribution for the actual outcome.

involves subjective elements, with some managements stressing the risks and uncertainties, others the benefits that can be gained. An excellent example of this uncertainty problem turned up quite fortuitously in the relationship between continuous casting and the basic oxygen process. Profitability calculations for continuous casting took into account the use of oxygen steel, but very few firms' cost calculations for the oxygen process explicitly considered any extra advantages from combining it with other new processes. This would certainly be very difficult in analysis of the kind presented here, and it could be argued that even the most far-sighted management could not have foreseen, in the early 1960s when many oxygen steel decisions were being made, the drastic changes and improvements in continuous casting that were to occur in the late 1960s. Nevertheless, it is the function of management to foresee the indirect advantages or disadvantages of a certain course of action.

There are many examples given of the difficulties encountered in using a new technology, especially by the first firms. They may be 'leaders' in introducing the process but 'followers' when it comes to its profitability. Of course, a 'follower' in introduction is not necessarily a 'leader' in benefits. On these problems, however, the investigation succeeded in obtaining little empirical evidence.

Another serious difficulty for empirical investigations is that the profitability of a new process for a firm changes over time, not only because processes are improved (old processes as well as new), but also because factor prices change. Tunnel kilns are a good example: they can use either coal or oil, the latter giving a superior technical performance, so that the expected price ratio between oil (or gas) and coal is important; it has been changing over time. In spite of such difficulties, which give the term 'rational behaviour' a rather subtle meaning when it comes to new technology, the profitability aspect has been considered in one way or other in all the chapters. A general conclusion seems to be that profitability is an important factor in explaining the diffusion of new processes, not only in distinguishing between users and non-users, but also in explaining diffusion among firms and within them. This agrees with results obtained elsewhere.[1] But although profitability is a significant variable in most calculations, it is more difficult to say anything about its importance relative to other factors in explaining diffusion of new technology. This is not surprising in view of the calculation difficulties and all the proxies for profitability that have to be used. The most elaborate profitability calculations in this

---

[1] For a discussion of other findings see E. Mansfield, 'Determinants of the speed of application of new technology', paper given to the International Economic Association conference at St Anton, 1971.

study were made for special presses, where profitability explains more of the variation among firms than elsewhere, although some important elements may still have been omitted from the calculations.

When trying to pick out proxies for profitability, it is natural that the various authors should have relied heavily on conventional investment theory. Introducing a new technical process normally involves investment; the application of gibberellic acid in malting is the only process studied which requires negligible capital investment for its introduction.[1] Two relevant variants of investment theory are the accelerator principle and the vintage approach.[2] According to the former it is an increase in demand which calls for investment in new productive capacity. This may also be supposed to lead to more rapid introduction of new processes, since they are very often superior to the old. (However, the accelerator principle may also lead to increased use of outdated capital equipment, which could complicate the econometric analysis.) This principle gets very firm support in our material, notably for oxygen steel and special presses, but to some extent also for tunnel kilns, shuttle-less looms and float glass. For instance, in new paper machines special presses are cheaper to introduce than the old type and this, of course, leads to rapid diffusion in firms and countries which are investing heavily in new paper capacity.

The vintage approach implies that, with given demand, new processes will be introduced more rapidly in firms with old than with new capital equipment, simply because replacement of equipment is more urgent for the former. This theory gets some, rather weak, support from our data. For example, the float glass process was probably delayed in France by the expensive new glass capacity that had been installed just before the new process became available.[3] But the vintage theory may be reversed in some new processes. For instance, it is quite clear that special presses are more profitable in newer paper machines. Consequently firms and countries with many old paper machines would be expected to have a slower rate of diffusion than those with newer machines, and this hypothesis was borne out by the data. Another variant of the vintage approach is illustrated by the use of gibberellic acid in malting. As already mentioned, this requires no

---

[1] Lack of *human* capital might be an obstacle to diffusion of this process.

[2] For a more detailed discussion of the accelerator principle see for instance D. J. Smyth, 'Empirical evidence on the accelerator principle', *Review of Economic Studies*, vol. 31, June 1964, pp. 185–202; or R. C. O. Matthews, *The Trade Cycle*, Cambridge University Press, 1958. For a more detailed discussion of vintage models see for instance L. Johansen, *Production Functions*, Amsterdam, North-Holland, 1972, and Salter, *Productivity and Technical Change*.

[3] This particular decision, however, perhaps ought not to be considered in isolation, but in the context of the policy of a multinational company. The same company built their first float glass factory in Italy.

SUMMARY AND CONCLUSIONS                    305

new investment, but the extra capacity it gives at one stage of production would be useless unless the remaining parts of the process were also increased in capacity. Profitability calculations may thus differ between firms according to the situation in stages of the process not directly connected with this innovation.

A major element in profitability calculations are the relative prices different firms have to pay for factors of production. This element differs not only between firms in the different countries, but also between firms in the same country, although the latter was more difficult to take into account. Compared with old processes, new ones may save capital, labour, raw material, or some combination of all three; sometimes they imply increased use of one or more factors of production, for instance capital. Thus differences between firms in relative factor prices have been used in this study to explain varying diffusion rates between different countries. The changes in the factors of production used that the processes would imply are not always clear from the descriptions in the different chapters, but usually it seems fairly plain whether they are positive or negative, and then the profitability variables also seem to contribute to the explanations of different rates of diffusion. In oxygen steel, for instance, differences in scrap prices or availability between countries have a significant explanatory value. The use of coal and the production of fletton bricks in the United Kingdom make tunnel kilns less profitable there than in other countries, and this shows up in the diffusion patterns. Numerically controlled machines and shuttleless looms, which are, apart from their other advantages, very labour-saving, seem to have been adopted more rapidly in firms and countries with a relatively high wage level, such as the United States and Sweden.

In considering the profitability calculations of a new technology, the problem of disposing of increased output is an important factor; one major conclusion from the case studies is that introducing new processes very often increases capacity. For special presses this is a major advantage, and the same is true of shuttleless looms and gibberellic acid; but one production line with the float glass process makes more glass than the whole consumption of Austria. This expansion introduces another element of uncertainty which underlines the previous point concerning the subjectivity of any profitability calculations. With the formula used for special presses, for example, differences between British and Swedish firms seem at first sight rather odd. This process would pay for itself in a year or less in many firms in both countries, but whilst Swedish firms have introduced the process, many British firms have not. A possible explanation is that British firms are less certain they can sell the extra output than Swedish firms, nearly all of which are exporters. In float glass the very large capacity created may often act as a disincentive; when a combined Danish–Swedish plant chose the Pittsburg process instead, the main reason probably was that the expected output of float glass was considered too large to sell easily. Gibberellic acid is of no use to a maltster if the demand for malt is already satisfied. It was found that vertically integrated firms introduced shuttleless looms more rapidly and to a greater extent than non-integrated firms, possibly because the former are more certain of finding outlets for the increased output. Similarly, heavy foreign competition was found to act as an impediment to diffusion, and this could be another factor making for uncertainty in expected demand for the increased output.

In contrast with the accelerator principle, whereby the investment in new plant is decided by an actual or expected change in demand, here new investment *entails* additional capacity, and the question is whether, by price reduction or other means, the firm can dispose profitably of the extra output. This may require gaining market-shares at the expense of others, which raises the question of what constitutes rational behaviour in an uncertain world. If, for instance, Japanese firms started investing heavily in steel capacity while oxygen steel was still in its introductory stage, and, by adapting the process to their special conditions, they obtained a competitive advantage over American firms and invaded the American market, is it still defensible to say (as we tend to as economists) that American firms acted rationally in waiting until the process was well developed and appropriate for their own production facilities before investing in it? Another example is found in continuous casting. Austrian and British firms, which started very early experimenting with this new technology, thought it was best suited to special steel production and followed this line in their development work. Development work started somewhat later by American, German and Swedish firms and directed to application in large-scale production of commercial steels turned out after all to be on the right track, and the 'late comers' gained competitive advantages over the 'pioneers'. Were the pioneers then 'irrational'?

*Other economic variables*
The introduction of a new process usually involves investment; the financing of that investment suggests itself as a factor explaining differences in diffusion behaviour between firms. Of course, if the internal rate of return on an investment is high, it will pay to borrow money if internal funds are not available. But there are at least two problems. First, capital markets may be imperfect, so that it is not possible to borrow more than a certain amount. Secondly, as mentioned

SUMMARY AND CONCLUSIONS                    307

before, the risk factor is important in investment in new technology, especially when experience is scarce. As in investment in research and development, a firm may be reluctant to borrow money for this purpose. The implication is that greater willingness to invest in new processes might be expected if they could be financed out of the firm's own resources. This hypothesis was tested in different ways as shown in the various chapters, but the results were frequently limited by lack of information. They are most clear-cut in oxygen steel, for which it is shown that an increased cash flow leads to higher investment, and this in turn means a more rapid diffusion of the new technology. Special presses give some support to the hypothesis, although it is rather difficult to interpret these results. It is not surprising that the hypothesis is better supported in explaining investment in expensive new equipment, such as that for the basic oxygen process, than in relatively cheap machines, such as special presses. Weavers in Austria, the United Kingdom and Italy often mentioned lack of capital as a major reason for being non-users of shuttleless looms. In Britain, small firms introduced cheaper water and jet looms quite extensively instead of the more expensive Sulzer looms. Capital availability probably also influenced when the float glass process was introduced by some firms.

The importance of international contacts may also indicate the influence of financial resources: firms with close international contacts are to a much larger extent users of shuttleless looms than firms without such contacts, which may, or course, have effects other than purely financial ones on the diffusion of new technology.

It is often said that large firms are more willing to introduce new technology than smaller ones, and this argument has been used to justify the concentration of production. One reason for large firms being more willing to introduce new processes rapidly could lie on the financial side, if, as has, been argued, their ability to acquire the necessary finance is greater. Another argument takes account of the risk factor. The failure of a new process in a small company might be a catastrophe; in a large firm, it might be just a small loss to be set off against profits in other fields. A further advantage for large firms may be that they can afford a management qualified to evaluate the advantages of new technology; small firms may have to wait upon the experience of others.

Other investigations indicate that firms must be of a certain size to respond rapidly to new technical processes.[1] This does not mean, however, that they must be very big in any absolute sense; furthermore, the threshold size seems to vary for different technologies. We acquired extensive data on the size of firms, which showed considerable variation,

[1] Mansfield, 'The speed of response of firms to new techniques.'

308                    NEW INDUSTRIAL PROCESSES

so that it was possible to compare behaviour as to new technology for different size groups. The results seem largely to support the findings elsewhere. It is by no means inevitable that large firms should be the first to introduce a new process; on the contrary, there are many examples in our material of smaller firms taking the lead, for instance, in oxygen steel and continuous casting, also in shuttleless looms.[1] But we must not interpret this as showing differences in attitude or bureaucratic slowness in the big firms; the technologies of oxygen steel and continuous casting were first developed for smaller firms, larger firms had no particular profitability advantage in the new processes until they had been developed further. Of course, this raised a further question of why these technologies were first developed in smaller firms, which has not been fully answered in this study. Numerically controlled machines are another example of the aspect just mentioned. Here there is no correlation between size of firm and date of introduction, but there is a definite tendency for the number of machines per thousand employees to be higher in smaller than in larger firms. This, however, might very well be because large firms have a different pattern of production, with longer series to which these machines are not as well suited. The only case where we found a definite tendency for larger firms to introduce the process earlier was in special presses, where this was quite significant in the econometric calculations. However, it might equally well have been caused by the profitability of this new technology: larger firms have more paper machines than small firms, which probably means that, on average, each large firm has both more new and more old machines than each small firm. As special presses are more profitable on newer machines, large firms are likely to show an automatic lead over small ones when considering only the date of *first* introduction. Again, many small firms are non-users of shuttleless looms (although, as just mentioned, some small firms were early users); this might be an example of a threshold size effect. Another case is gibberellic acid, where small firms may not have the resources to employ people qualified to handle the new technology.

*Institutional circumstances*

Quite a few differences between firms in different countries can be explained by institutional differences in the countries concerned, which may take a variety of forms. Some have an effect only for a limited period, whereas others are more fundamental. A very good example of the first is the successive nationalisation and denationalisation of the

---

[1] We must, however, be aware of possible differences in the definition of the term 'commercial application'. A pilot plant in a big American firm would probably be defined as commercial in Austria or Sweden.

British steel industry during the whole post-war period. This led to a climate of uncertainty and relatively low investment for some years and, in turn, influenced British investment in oxygen steel and continuous casting, possibly being one reason for the slow diffusion of these processes in the United Kingdom. A more permanent difference between the countries concerned is found in connection with gibberellic acid. In Germany the process is prohibited, and the legal situation in Italy is unclear, which inevitably influences diffusion, or the lack of it, in these countries.

Another institutional difference between countries lies in the structure of industry. Numerically controlled machines were, to start with, extremely well suited for the aircraft industry. The United States and the United Kingdom had large aircraft industries which clearly contributed to the rapid spread of these machines, whereas in the other countries studied this industry was much smaller or negligible. Countries with a comparatively large number of small firms in an industry seem often to experience slower diffusion of a new process than where the industry is more concentrated. The many small firms in the cotton-type weaving industry in the United Kingdom and in Italy are examples in this study. Another aspect of the same thing is the importance of industrial and research associations in a country; strong associations and high membership ratios might speed up the diffusion rate, as seems to have been the case with special presses in Sweden.

*Management attitudes*

Although the explanatory variables mentioned so far account for much of the difference in behaviour between firms, it is clear, not least from the econometric calculations, that the unexplained residuals are also quite large. To some extent no doubt, these residuals represent measurement errors, and errors in the variables, which may not measure what was intended, for instance profitability. But another way of looking at the problem is to say that there are differences in management attitudes towards new technology. Other things being equal, some firms are more willing to take risks on new processes than others. This would not be at all surprising, taking into account the difficulties mentioned earlier of making profitability calculations.

It is not too difficult for an aggressive management (in the sense of chapter 6) to present favourable *ex ante* profitability calculations to its board of directors, just as it is not difficult for a less aggressive management to make the calculations less optimistic. Inevitably fundamental problems of management ability (and management luck) enter into all discussions of diffusion of new technology. Rational behaviour in this context is indeed a difficult concept. Other studies also have concluded that differences in attitudes towards new technology affect the rate of diffusion of new processes.[1]

In some chapters of this study we tried to take explicit account of attitudes, although fully aware of the pitfalls in measuring them. These attempts support the hypothesis that management attitudes do differ between firms. The most elaborate analysis is in chapter 6, where it cannot be said to be altogether a success, but at least it suggests one way of attacking the problem and probably could be elaborated. A more direct method in the same chapter, using productivity measures as a gauge of management aggressiveness, gave slightly better results, but of course it is questionable whether productivity differences between firms are really good proxies for differences in attitudes towards new technology.

In chapter 4 it is shown that the same firms tend to be early or late in the introduction of all new processes, not just the one studied, and that this applies to most countries. This result is confirmed in another paper.[2] Some firms seem to want to be among the pioneers, others like to watch their experiences. This theory received further support from discussions with sellers of some of the capital equipment used. Nevertheless the question remains of why certain firms want to be the first; improved measures of aggressiveness might help to answer this question.

Some of the proxies used for profitability or for other economic variables could also be regarded as proxies for management attitudes. Vertically integrated firms, which introduced shuttleless looms earlier and more extensively than non-integrated firms, might have done so as the result of different risk and profitability conditions, but another interpretation could be that under given conditions vertical integration is a sign of progressiveness on the part of management. In that case integration could stand as a proxy for management attitudes.

We also learned that the introduction of new technology often means big changes in structure and administrative practices, changes that some organisations are more willing to accept than others. A good example is numerically controlled machines, the introduction of which is usually accompanied by radical changes in organisation. Big firms have an advantage here in that they can more easily take the risks connected with introducing numerical control, but more research is needed on, for instance, the sources and training of managers, and on organisational structure, to understand what types of firms most easily adapt to new processes.

[1] See for instance Carter and Williams, *Industry and Technical Progress*, especially ch. 10 and ch. 16.

[2] Nabseth, 'The diffusion of innovations in Swedish industry.'

EXPLAINING DIFFUSION DIFFERENCES BETWEEN COUNTRIES

One reason for undertaking this study was to try and explain differences between the participating countries in diffusion patterns of new processes. In the United Kingdom it has often been said that the British are good at basic research and inventing new things, but bad at developing these inventions into innovations and making use of innovations developed initially in other countries. For the NIESR this was one reason for the whole project: they wanted to test this hypothesis. We have tried in one way or another to find factors which can at least partially explain the differences between countries for all the processes considered, but it would be an exaggeration to claim that we succeeded in this task. The analysis is not always up to desirable scientific standards, partly because of the poor quality of the primary data. On the other hand, we have succeeded in finding a number of reasons why diffusion patterns differ between the countries concerned. These may not comprise the full explanation, or even most of it, but we have at least started to account for the differences, and this is more than is yet to be found in the international literature on the subject.

In chapter 2 it was found that, according to the indicators used, no country among those studied had an outstanding general lead in introducing new techniques. In all subsequent chapters we have tried to include American data, which are sometimes rather poor, but if they can be accepted the conclusion of the interim report seems to stand. The United States has not turned out a 'leader' in the sense used in chapter 2. In some cases, for instance numerically controlled machines, diffusion has been much quicker in the United States, whereas in others, oxygen steel, continuous casting and special presses, this is not the case. The implications of 'leading' and 'following' in the introduction of new technology are by no means clear. Simply to count diffusion percentages for firms at different points of time, or to estimate the percentage of output produced by the process in question, does not of itself yield significant results. The diffusion patterns must be adjusted for economic and institutional differences among countries to give more meaningful content to the words 'leader' and 'follower', for example, as indicators of progressive or non-progressive managements. But, as pointed out many times, this is a difficult task when management attitudes and profitability calculations are as intertwined as in decisions on new technology. In comparing the progressiveness of management between industries and countries, it is easier to compare some of the outcomes of behaviour, like profits, profit margins and output growth, than 'inputs' like the use of new technology.

If a government, for some reason, wants to influence diffusion patterns it must, however, know the effects of changes in economic and institutional circumstances on management behaviour. If the uncertainty created by successive nationalisation and denationalisation of the British iron and steel industry had not been engendered, and in consequence the diffusion of oxygen steel and continuous casting had been quicker in that industry than it actually was, then the diffusion patterns for the techniques chosen – in the sense used in the interim report – would have been outstanding in the United Kingdom compared with the other countries. This would imply a denial of the initial hypothesis that British firms lag behind in the use of new technology, but on the other hand, in some instances the data, if they can be accepted at face value, show rather peculiar behaviour by British firms, especially regarding special presses. This is a process that has a very short pay-off period for many companies, yet many British firms have still not introduced the technology.[1] These firms are certainly small by British standards, but not by international standards; nor is there any lack of information about the new process. The data support a hypothesis which needs further research, that, in the United Kingdom after the war, there was room for quite a few small and medium-sized firms that did not maximise profits in a neo-classical sense – competition from abroad was not keen enough to weed them out. In Sweden, with among other things lower tariff barriers, this was not so. Out of a random sample of new processes introduced since the war, Sweden, being a small country, would not be expected to be the innovator of more than a few. But it appears that a new process, once started in another country, spreads quickly in Sweden if experience seems promising; the relatively large foreign trade and heavy foreign competition,[2] together with the close contacts between Swedish firms (in associations and research work, for example) lead to the rapid introduction of new technology. Continuous casting, tunnel kilns and automatic transfer lines in car production are cases in point. Rapid diffusion of such new technologies has been a necessity for firms working in a country paying the highest wages in Europe.

If we consider the various processes in more detail,[3] we find that they differ in the factors which explain the differences in their diffusion

---

[1] One essential point that has been raised in relation to the pay-off calculations is the proportion of integrated to non-integrated paper production. Some British experts maintain that this is important and that, with due regard to this fact, pay-off values for British firms would be increased. In Sweden, on the other hand, it is maintained that the facts cannot explain differences in behaviour between firms in the two countries.

[2] Notwithstanding what is said on p. 306, above, this means that we do not know with certainty whether heavy foreign competition acts mostly as a stimulus or as a hindrance in the diffusion of new technology.

[3] These remarks apply to six processes only, excluding gibberellic acid and float glass because of their special features.

patterns between countries. Differences in wage levels are considered the most important factor in explaining the lead of the United States and Sweden in the rate of diffusion of numerically controlled machines. Similarly, their slow diffusion in Italy and Austria is thought to be influenced primarily by low wages. This is, of course, a clear profitability consideration. Another important factor has been the size of the aircraft industry in the United States and Great Britain. Early information about this technology might have had some effect both in the United States and in Sweden.

The rapid diffusion of special presses in Sweden is strongly affected by the fact that this was where they were invented and innovated (also independently in the United States), which meant a rapid spread of information about the process. Furthermore, the process seems to have been highly profitable, partly because of investment in new machinery, partly because exports provided a good outlet for the increased capacity created. Such a situation did not prevail in the United Kingdom, where the consumption of paper increased at a much slower rate. Heavy foreign competition and the necessity of buying pulp from abroad may also have held back the process in the United Kingdom, but some differences in management attitudes must probably also be taken into account. In Italy, large investment in new paper machines and a rapid increase in paper consumption explain much of the more rapid diffusion than in Austria and Germany. In Austria a rather old stock of paper machines seems also to have hampered diffusion of the process. The most curious pattern is found in the United States. The process was invented and innovated there too, but still diffusion of both information and adoption was very slow. A slow growth in paper consumption and low investment activity might account for this; furthermore, much recent expansion and investment by American paper companies has taken place in Canada.

The diffusion of shuttleless looms seems to have been most rapid in Germany and the United States. In both countries the large size of firms, their vertical integration, and the possibilities for shift-work probably made this technology profitable from the beginning. Furthermore, import competition has probably not been as severe in Germany as in Sweden, the United Kingdom and, maybe, Italy. In the United Kingdom and Italy the large number of small firms has probably also impeded diffusion; many of them could not afford the expensive investment. In Sweden the large size of firms has probably had a positive influence on the diffusion rate, and the heavy import competition has meant that they have been forced either to invest in new technology or to leave the market.

As to oxygen steel, the main reason why this process diffused much more rapidly in Austria than in any other country seems to have been that the technology was invented and developed there, encouraged by the production structure of the Austrian iron and steel industry and its precarious situation after the war. When the technology had developed sufficiently to be usable in big plants, diffusion started on a more rapid scale in countries like Germany, the United Kingdom and the United States, where plants were big. A faster increase in steel output in Germany than in the United Kingdom gave a more rapid diffusion in the former country. Diffusion in the United States was probably discouraged to some extent by readily available scrap, whereas diffusion in the United Kingdom may have suffered from the uncertainties of nationalisation. In Sweden the relative importance of special steel meant a less rapid diffusion than elsewhere. The Italian position, however, cannot be explained from the data; steel production increased very sharply in that country during the 1960s, but electric production seems to have had more attraction than the basic oxygen process.

The early development of continuous casting of steel started fairly independently in Austria, Germany and the United Kingdom, and rapid and similar diffusion might have been expected in all three countries. However, in Austria and also the United Kingdom, diffusion was delayed in the 1960s by attempts to develop the technology for the production of special steels, which turned out rather a failure. In Sweden and the United States, and later in Germany, development concentrated on its application to large-scale production of ordinary steels, which proved more successful. A fairly uniform output, either produced by small plants with electric furnaces or by larger plants using oxygen converters, offered advantages for the introduction of this technology. From this point of view the plant structure in Austria, Germany and Sweden favoured the process, especially when compared with the British and American structure. In Germany, where the pioneer steelworks are situated, a very rapid diffusion started during the later 1960s.

Brick production has increased most rapidly in Austria and Italy. This ought to have stimulated the diffusion of tunnel kilns in these countries. Brick consumption *per capita* is also much higher in these countries and in the United Kingdom than in Sweden and the United States. In Austria the diffusion of the process has been comparatively rapid, but not in Italy, where the high capital costs of tunnel kilns have meant instead widespread further modification of the Hoffmann kiln. In the United Kingdom, where the first tunnel kiln was built as early as 1902, the use of coal and the unique raw material of fletton bricks have resulted in slow diffusion. Germany and Sweden have both had rather rapid diffusion associated with a decrease in brick production;

SUMMARY AND CONCLUSIONS                    315

keen competition seems to have made it necessary for firms wanting to survive to install the tunnel kiln.

### WERE WE STARTING THE PROJECT NOW

In a project like this which has extended over a long period – more than five years – much experience has been gained which may be of help to others entering the same field of research. We can ask ourselves: what would we do differently if we were starting the project now?

In the beginning we discussed whether we should try to get information about diffusion patterns for a large number of new processes, or whether we should confine ourselves to just a few, to be studied in depth. Taking into account all the difficulties described in interpreting diffusion rates, the line we chose of selecting a limited number of processes seems to have been the right one, given the resources available. But the attempt to get most of the data required from individual firms rapidly reached a limit of strongly diminishing returns. In some countries it was just too difficult. If we were to start again, we would probably rely much more on information from other sources, such as producers and sellers of new equipment and licence-holders, and have it confirmed by a few companies. Establishing good contacts with such producers seems to be a better approach, and obviously the more one can rely on published data the better, although when it comes to new processes inevitably such data will be scarce indeed. Another approach would have been to go deeper into a limited number of companies, but then one would probably have had to give up the idea of making any kind of industry-wide or country-wide generalisations.

As to methodology, this study was undertaken in two consecutive stages, a procedure which was dictated by financial considerations. If there had been no such restriction, it would have been better if we could have given more time at the beginning to discussions on how to analyse our material. Then it might have been easier to compare diffusion of the various processes in detail.

Finally, it must be remembered that in this study six different research institutes in six countries were involved. This has stimulated much fruitful discussion and provided each institute with a more rounded appreciation of the complexity of the diffusion process than might otherwise have been possible. But it has also to be said that, despite all its advantages, international cooperation does slow down the progress of the research itself.

# [6]

Excerpt from David J. Jeremy, *Transatlantic Industrial Revolution: The Diffusion of Textile Technologies Between Britain and America, 1790–1830s*, 252–62, 335–6.

# 14

## Conclusion: Some Perspectives on the Transatlantic Diffusion of Early Industrial Textile Technologies

The westward transatlantic diffusion of textile technologies had startling economic and social results. Few who watched the first fruits of transfer, a hand-powered carding machine and an eighty-spindle jenny, laboriously operated on a rattling carriage in Philadelphia's 1788 Independence Day parade could have imagined what these omens of a machine age portended.[1] Yet within twenty years or so, the United States boasted nearly 100,000 cotton mill spindles. Between 1810 and 1820 this figure more than tripled and then it more than tripled again in the 1820s.[2] "I suppose," a Rhode Island cousin wrote in 1826 to the machinist Aza Arnold, then at Great Falls, New Hampshire, "you are building factories & filling them with Machinery at a great rate—so that I should hardly know your village were I to land in it from a balloon."[3] By 1831 perhaps 700 firms had over 1.2 million cotton spindles and 33,500 looms (mostly power-driven) in water-driven mills on never-failing streams, chiefly east of the Alleghenies and mostly between New Hampshire in the north and Maryland in the south. In the early 1830s, fixed capital investment in the United States cotton industry, computed at nearly $45 million, was over a third of the size of the fixed capital investment in the contemporary British cotton industry.[4] This rate of growth, accomplished within forty years, was an astonishing testimony to the success of technological diffusion.

Americans were not merely imitators. Although their first experiments in factory spinning closely copied British models, the involvement of Boston's leading business interests in the research and development stage of power loom weaving led to a series of innovations in the organization and design of cotton manufacturing equipment that disquieted rival British manufacturers as early as 1826.[5] These innovations made the Waltham-type coarse-goods cotton factory at least 10 percent more efficient in throughput of materials than its British counterpart.

Modified technology permitted, and indeed was partly the product of, modified factory discipline. Before it soured in the late 1830s, the benevolent paternalism of the Lowell corporations made their name an international byword for enlightened industrialism. Between the late 1820s and the Civil War Lowell was to the New World what New Lanark was to the Old, making a visit to Lowell nearly as obligatory for foreign tourists as a sight of Niagara Falls.[6] Dickens in 1842 admired the cleanliness, vigor, dress, and deportment of the Lowell operatives, and especially their musical and literary talents, evidenced in the boardinghouses' "joint stock pianos," circulating libraries, and the girls' own periodical, the *Lowell Offering*.[7] Although the meaning of the Lowell mill and boardinghouse disciplines continues to intrigue historians, there can be no doubt that, to the beholder, the contrast between conditions in the Massachusetts mills and British cotton factories was sharp.[8] Dickens, writing as reporter rather than novelist, saw the difference as "between the Good and Evil, the living light and deepest shadow."[9]

In Rhode Island and other states to the south, the technology and organization of cotton factories remained closer to English models. As a Pennsylvania state investigating committee discovered in 1837, a fifth of factory employees were children under the age of twelve; adults and children alike worked eleven to fourteen hours a day, six days a week; and "no particular attention is paid to the education or morals of the children, by the employers."[10]

253

In the American woolen manufacture, the results of technology diffusion were less spectacular, not least because the woolen fiber inhibited technical change. Even so, diffusion led to a measure of technical and economic transformation. American woolen men visiting England's clothing districts in the mid-1820s disdainfully noted the persistence of billies, small jennies, and hand looms. Condenser cards, large jennies, jacks, and power looms distinguished the bigger, best-practice American woolen factories, especially those between the Hudson and Merrimack valleys, where woolen and cotton manufacturing were often contiguous or shared the same managements.[11] Glimpses of the spirit of change and improvement pervading American woolen manufacture in the 1820s occasionally surface in the record. One New Hampshire mill manager told his directors in 1826, "The adoption of the new system of loco-motive machinery throughout our buildings and the enlargement of our Dye House, now nearly equal to another Factory, will justify the expenditure of the time and money devoted to these objects."[12] Condenser cards and power looms formed part of this "loco-motive" production line. For an aggregate estimate of economic change, the absence of fixed capital investment data in the contemporary British woolen industry denies an international comparison. However, by 1820, when half of the American woolen industry was located in New York, Connecticut, and Massachusetts, the whole American woolen manufacture was just under a quarter of the size of the American cotton industry in fixed capital investment and between a tenth and a sixth in spindleage. (See appendix D.)

254

In attempting to understand the complex process of technology diffusion that lay behind America's early industrial transformation, this study reaches a number of conclusions about the circumstances hindering and promoting transfer and about conditions inducing technical modifications. To begin, the investigation lends definition to the roles played by immigrant artisans in transatlantic transfer. Most obviously, the artisan emerges as the preeminent technology carrier in this period. The finding hardly comes as a surprise. Requiring new skills, the new technologies appeared in published verbal forms only belatedly and then incompletely. However other grounds for emphasizing the importance of the artisan become plain when obstacles to transfer are considered. In Britain, mill-level secretiveness, resistance to the publication of technical knowledge, and a chaotic search system in the London patent repositories all elevated the knowledge and skill of key industrial workers in the diffusion process. Even the severe but ineffective laws against artisan emigration underscored the value of skilled workers to foreign rivals. The gathering pace of inventive activity, and with it innovation, the increasing complexity of machinery employed in textile processing, and the incremental nature of technical change together implied that overseas imitators would place heavy reliance on experienced British artisans.

As technology carrier, the skilled worker was not without problems. Conflicting British regional traditions in manufacturing and machine making, the British pursuit of higher-quality textile products, resistance to technical change in some trades, and widespread attitudes of secretiveness, to say nothing of union militancy in the 1820s, conspired

*Conclusion*

against a wholly sanguine view of artisans as technology carriers. For these reasons, technical publications and other rational vehicles of technology diffusion, like international exhibitions and manufacturers' conferences, would extensively supplant migrant artisans later in the nineteenth century.[13]

The contribution of artisans in diffusing the four textile technologies to America exhibited no simple repeating pattern. At the first stage, the conveyance of technical potential to the United States, immigrant workers proved indispensable in the transfer of cotton factory spinning and woolen machine carding and spinning. But with cotton power loom weaving and calico printing, American visitors to Britain largely displaced the immigrant artisan as carrier. The difference reflects partly the intensifying of American business interests after 1807 and partly the state of the technologies. Power loom weaving caught American attention before the completion of mechanization, forcing Americans to share in research and development costs, while calico printing comprised a complex group of processes combining partially understood industrial chemistry with design and mechanics, requiring more thorough investigation by American imitators. In the establishment of prototype mills and technological models, immigrant artisans predominated where the technology necessitated skills acquired only by lengthy learning periods: for several machines in woolen manufacturing (billy, jenny, gig, cloth shear), for mule spinning and mill managing in cotton yarn production, and for calico printing skills and printery superintendence in cotton finishing. With internal diffusion in the United States, immigrants seemed most conspicuous before 1807, when the process primarily rested on kin groups and trade networks. Slater's impact, for example, owed much to this situation. With increases in the scale of production, associated with power loom weaving, capital goods firms became more important vehicles of diffusion, as the Locks and Canals Company at Lowell demonstrated. Selling agents serving numbers of manufacturing firms also disseminated technical information among their clients. Nevertheless a few immigrants, possessing skills that eluded mechanization or rapid learning, figured in internal diffusion. The Scholfield family in woolen manufacturing and the dozens of Lancashire printers who moved among the New England printworks in the 1820s were cases in point.

In an assessment of the role of British textile workers as agents of technology transfer, a clear distinction must be made between individual artisans and artisans in aggregate. The initial conveyance across the Atlantic of new information and skills arose from the emigration of a relatively small number of British workers, perhaps a few dozen, many of whom shared in setting up prototype mills and to some extent in the process of internal diffusion. On the other hand, the mass of textile workers reaching the United States from Britain present a very different picture. Quantitative analyses dispel any notion that they, collectively, represented the new British textile technologies. At any time between 1770 and 1831, data show, at least 74 percent of immigrant textile operatives were weavers, almost certainly hand loom weavers. The proportion of operatives with new industrial skills rose from 11 percent of those immigrating (and staying with their trades) on the eve of the War of 1812 to 17.5 percent (of

3,632 new immigrants) in the 1820s. However, numbers of woolen manufacturers arriving annually rose appreciably between the War of 1812 and the 1820s (from 7 to nearly 24) and even more so did numbers of general machine makers (from 10 to 100). Yet a substantial level of transience was suggested by the overall 60 percent of workers in their teens and twenties and a similar proportion unaccompanied by family.

255

The immigrants who shared in transatlantic transfer did not conform to the pattern displayed by immigrants in aggregate. Besides having industrial experience, most of the technology carriers were much older men. John Murray, who set up an early though not very successful power loom, was in his mid-forties. So too was John D. Prince, the calico printery manager. The Thorp brothers who introduced roller calico printing in 1809 were in their thirties when they left England. And the Scholfields were in their mid-thirties when they set out for Boston. Slater, certainly, had just reached his majority when he sailed from London, and John McDonald, who set up an early power loom, was in his twenties. Not all the technology carriers were unaccompanied when they set out for America. The Scholfields went in a family group. It might be concluded that greater experience and emotional stability promoted successful transatlantic transfer. All that can certainly be said is that the successful artisan carriers of the technologies under review did not conform to the typical picture of the British textile immigrant in the United States: young, unaccompanied males possessing preindustrial skills. These conclusions underscore the importance of studying technological migrations at both the individual and aggregate levels.[14]

256

The implications of this aggregate immigration pattern are intriguing. Did the superfluity of hand loom weavers support the persistence of traditional technologies and thereby slow industrial growth? The question is hard to answer, even for Philadelphia, renowned for its hand loom weavers until the Civil War. All that can be said now is that Philadelphia was not an especially preferred destination for immigrant British weavers in the period 1770–1831; neither was the Philadelphia region, in comparison with New England or New York, short of skilled industrial immigrants from Britain's textile industries. On this basis, any delay that Philadelphia experienced in industrializing cannot be attributed to immigrant weavers, though a detailed local study could well alter this conclusion.

The significance for American industrialization of the supply of skilled industrial immigrants also invites further exploration. A comparison between numbers of new immigrants linked with all textile trades and numbers added to the expanding American cotton and woolen industries in the 1820s discloses large shortfalls at operative, managerial, and machine-making levels. At the most, fresh immigration could have met about a quarter of the demand for new operatives in the two American industries during this decade. At the managerial level, the presence of manufacturers may overstate the potential contribution of immigration to the two industries' labor expansion of the 1820s. Including manufacturers, immigration could supply about half the new demand for cotton mill managers; excluding them, it could meet less than 5 percent. In the case of woolen managers, prospects were brighter for Americans. Assuming that clothiers

*Conclusion*

were manufacturers and not just finishers, current immigration could have provided nearly 70 percent of new American demand. Current immigration of machine makers could have furnished perhaps 40 percent of new demand for American textile machinery makers. The relative abundance of immigrant machine makers (1,001) compared with skilled industrial operatives (under 600) may well help to explain the low wages of American machine makers, which Rosenberg hailed as "truly startling."[15] The closeness between American and British machine makers' wage levels and therefore the lower labor costs in building American capital goods presented an additional reason for Americans to prefer capital-intensive equipment.

These estimates, for the moment at least, are the best available, but they are also imperfect. Nevertheless, in view of the conservative nature of my quantitative estimates, I maintain that new immigration could at most supply 25 to 50 percent of the growing demand for textile workers in the United States during the 1820s. By this date American manufacturers had crossed the threshold into independence of Britain's technical know-how. And this seems quite consistent with my earlier finding that before 1807, individual immigrant artisans secured textile technology transfer at the first three stages of diffusion and thereafter figured primarily at the introduction and prototype stages. Their diminishing role in the internal diffusion stage simply parallels industrial growth. By the 1820s the cotton and woolen industries expanded, in part because the general reservoir of technical knowledge and skills in the United States had both spread beyond the confines of the immigrant community and risen in level. Only technologies new in

Britain (and America), or older ones embodying hard-won, nonmechanized skills, needed the presence of British immigrants in the 1820s.

Even were my figures perfect, this quantitative bird's-eye view could not be the complete picture. For example, some trades in Britain that seemed applicable to the United States proved inappropriate because of the diverging courses of British and American innovation; English mule spinners and weaving overseers made this discovery at Dover, New Hampshire, in the 1820s. The hazard of overemphasizing statistics will become evident in local studies pursuing the economic and social adaptation of immigrant workers. Charlotte Erickson's *Invisible Immigrants* demonstrates the kind of detailed career profiles and contrasting British and American local contexts that can be assembled to investigate the motivations, movements, and adaptation of immigrant Englishmen. From this work, it is clear that insecurity, loneliness, illness, and early death awaited some proportion of immigrants; skilled industrial immigrants, plagued by these misfortunes, saw their economic potential crumble. Machine maker James Standring, machine engraver Thomas Lonsdale, and mule spinner James Leard, all ruined by alcoholism, illustrate the kind of destructive effect that noneconomic influences could wreak on promising industrial immigrants. Although this study throws new light on the role of artisans in the diffusion of textile technologies to early industrial America, much local work remains to be done.

Another conclusion relating to the channels of technology transfer also confirms the familiar: industrial espionage inevitably accompanies industrial competition, and no amount of public or

*Some Perspectives*

private protection can eliminate the borrowing or
piracy. Artisans slipped through the British customs net because it lacked an infallible mechanism
for identifying occupations and because an individual emigrant's intentions could never be positively
established. Yet it does seem that Britain's prohibitory laws raised technology acquisition costs for
foreign borrowers. Insurance on smuggled equipment added at least 30 to 45 percent to purchase
prices in the 1820s, though this halved by the late
1830s. Circuitous export routes, additional warehousing costs, and disassembly and repackaging
charges probably raised total smuggling costs in the
1820s nearer to 50 percent of machinery retail
prices. Since the most astute industrial spies "always take the easiest and cheapest course," Americans naturally imported relatively small amounts of
equipment from Britain.[16] They included key parts
or even whole machines when design, quality, or
price was better than that obtainable in the United
States, and American visitors evinced a keen interest in Britain's best-practice technology. In official
applications for machinery export orders made to
the British Board of Trade from 1825 through 1843,
American destinations figured in only 7 percent of
the total. Although machinery values are unknown,
it may be surmised that a similar fraction, or even
less, of smuggled British machinery exports was
bound for America.

The influence of general business conditions in
the receptor economy presents another lesson to be
drawn about the circumstances for international
technology transfer. For the United States in the
period, noneconomic assets (a measure of political
stability; language and social and legal institutions
deriving from the originating country; a Puritan

257

258

work ethic; and a relatively high degree of social mobility) and economic ones (abundant raw materials and water power, a reservoir of skilled traditional craftsmen, and the will to make profits) looked unpromising when set against high capital and labor costs and constricted product markets. But combined with secure and expanding markets, or expectations of such, they enabled American capitalists to surmount their factor problems with the aid of technological solutions frequently imported by skilled industrial immigrants.

Prospects of greater political stability and the emergence of an economic infrastructure, together with the propagandist activities of a pro-industry pressure group, lent high expectations to new industrial ventures in the early 1790s. In this context, Arkwright cotton spinning technology was successfully transmitted to America. When European war indirectly choked such prospects and British textile imports flooded the American market, few early mills survived. Between 1807 and 1815 chances for American textile manufacturers improved again with the restraints on British imports and the demands of a wartime economy. Over these years cotton factory spindleage in the United States swiftly rose from under 50,000 to around 300,000 spindles. And in these expanding conditions, the diffusion of power loom technology occurred. Peace and the resumption of European trade in 1815 severely curbed American textile manufacturing; however, population growth, tariff protection, proliferating canal and road networks, and falling cotton prices greatly enhanced business prospects in the 1820s. New manufacturing investment in power loom weaving appeared profitable, and, later, calico

*Conclusion*

printing technology offered a likely way of maintaining unit profits. So the stage was set for the diffusion of calico printing technology.

Market conditions do not alone explain successful technological transfer. Even in the adverse business circumstances of the 1790s, immigrants like Slater and the Scholfields survived. Besides the backing of American capitalists and good measures of technical skill and business caution, they possessed the noneconomic assets of adaptable personalities, kin support, and integrity in their commercial dealings, all of which inspired confidence among the American community in them. The technologies they carried became more readily accepted. In contrast, dishonesty discouraged the employment of immigrants, delayed investment, and retarded technological transfer. Indeed a case might be made for the moral basis of the location of the American cotton industry; more scrupulous immigrants in the Philadelphia region might have secured the establishment of the cotton spinning industry in Pennsylvania and New Jersey ahead of Rhode Island. But this interpretation is taking one line too far. Reliability in the artisan was as necessary as skill or business prospects conducive to investment; technical transfer was the product of several, rather than one, economic and noneconomic predisposing conditions. At the same time market conditions unquestionably constituted a major variable in forming the contexts for technological transfer.

On the transatlantic movement of technology, this study confirms that rates of transfer could be surprisingly fast. Both spinning technologies, water frame and mule, appeared in the United States within twenty years of their emergence in Britain.

British power loom weaving technology took about
a dozen years to reach America, and within a decade of 1803, when Radcliffe's dresser opened the
way to mechanical weaving, the Boston Associates
were developing an alternative power loom
technology. At the level of individual mechanisms,
some transfer rates were faster still. Horrocks's
variable batten speed motion, patented in 1813, was
being built into Rhode Island power looms four
years later. On the other hand calico printing, the
most complex of the textile technologies under
review, took up to twenty-five years to reach
America, if the Thorp brothers' printworks at Bristol, Pennsylvania, marks the introductory stage. If
internal diffusion, as through the New England
firms, signaled successful transfer, then fifteen
years can be added to this rate. The timing of invention in Britain and changing business conditions in the United States largely explain these
differing westward transfer rates. In the opposite
direction, most major American innovations arrived in Britain between six months and twelve
years of their American patent dates. Exceptional
were the innovations of the Boston Manufacturing
Company and its offspring, the northern New England promotions of the Boston capitalists, which
designed, built, and consumed their own machinery and pursued a different course of technical innovation from that then prevailing in British cotton
manufacturing districts. For other American inventors, Britain possessed the best available workshops where new inventions might be perfected
and the largest market for equipment where they
might be sold. One wider observation is obvious:
the rapid movement of textile technology in both
directions across the Atlantic strengthens the im-

pression of close links between the two countries
and adds another dimension to the relationships of
the Atlantic community.

The second theme pursued in this technology-
diffusion study is the question of modifications to
imported technologies. Here a number of conclusions emerge. Foremost is the finding that a combination of economic and social mechanisms and
influences induced the reshaping of British
technologies, a finding that a company-level study
of the cotton industry leader in northern New England reveals with new clarity. In commodity markets, better-quality staples allowed Americans to
start power loom weaving with a technology cruder
(imposing greater physical forces on textile fibers
during processing) than the contemporary British
technology. Cruder technology was quicker to develop and get into production than more refined
mechanisms. Product markets favored innovations
suited to the mass production of coarse but durable
cotton fabrics, which led to the use of nonversatile
manufacturing equipment for long production runs
of standardized goods. When conditions in product
markets encouraged large-scale operations, the
Boston capitalists at Waltham met the problem of
rising overhead costs. Of these, labor seemed the
most easily reducible. At Waltham, therefore, when
mill expansion occurred, implying increased output
and falling prices and profit margins, managers
sought technical innovations designed to cut wage
bills and so restore profit margins. As the technology was modified to permit the substitution of
cheaper, unskilled, female labor for more expensive, skilled, male labor, so social circumstances in
New England prevented the gross exploitation of
young people in the cotton mills, in contrast to that

259

260

which disgraced England's cotton factories. Consequently the technology allowed a more humane treatment of the labor force than before. Competition in product markets led the Waltham capitalists to hedge their technology with patent protection, itself a further incentive to invent novel technical solutions. The technology selected as a result of all these pressures certainly lowered labor costs, but the higher throughput of materials that it attained surely conserved capital costs too. Rosenberg's insistence that the choice of technological alternatives conforms to the underlying mechanism of factor proportions and factor prices therefore seems to be borne out in this examination of the Waltham firm, with the proviso that one noneconomic influence also played some important part.[17]

For the labor-saving improvements contrived in the Rhode Island cotton manufacturing system and in American woolen manufacturing, commodity and product markets also diverted inventive efforts toward innovations appropriate to coarse, standardized products. In Rhode Island, however, versatility remained a technical goal, not least because of the persistent influence of British models, mediated by Slater and his disciples. And in the Rhode Island system, judging by comparative machinery costs, the smaller scale of cotton manufacturing led to efforts to save capital. Firm-level investigations should further illuminate the economic and social mechanisms behind the innovations of the Rhode Island system and of American woolen manufacturing.

This study also raises the possibility that immigration encouraged the employment of a more labor-saving technology in America. The high wages of immigrant mule spinners at Waltham

*Conclusion*

drew attention to labor costs, inducing the application of capital so that cheaper labor could replace more expensive labor. Supporting the direction of this tendency was the relative abundance of machine makers among British immigrants. Relatively low American machine makers' wages add substance to the impression, as do the comments of British visitors. A Glasgow cotton spinner, James Dunlap, who toured cotton factories between Massachusetts and Maryland before returning to Britain in 1822, claimed that most American machinery was built by immigrant Englishmen and Scots, among whom he especially noted "artificers in the metals, such as iron and steel turners" (the skills needed to make the key textile machine components of rollers and spindles).[18] If immigration was helping to lower the costs of capital goods manufacturers and machinery prices were consequently falling, then we have a further explanation for the greater application of capital in American textile manufacturing.

It does, however, remain to be seen whether the labor element in capital goods manufacturing was falling in the early nineteenth century. One detailed estimate of the cost of building a warping and dressing frame for an Exeter, New Hampshire, cotton factory in 1820 shows that capital charges amounted to nearly 63 percent and labor charges to 37 percent of the total cost of $136.50.[19] A number of other detailed cases are needed to discern a trend. And then, of course, an increase in native-born mechanics might have contributed as much as or more than immigrant machine makers to any fall in costs of making capital equipment. To settle these and other questions, a study on early American machine making that is far more thorough and systematic than any available will be needed.

The subject of capital and labor costs raises the
important question of regional variations. In the
Philadelphia region, labor seems to have been
much cheaper than in New England, for example.
Adult male cotton mule spinners on the Brandywine in 1811 earned only $1.50 a day, compared
to the Boston Manufacturing Company's mule
spinners, who averaged $20 a week in 1817.[20] But
neither the American nor the British, let alone the
transatlantic, labor markets behaved perfectly.
Variations between cotton mule spinners' earnings
in Manchester, for comparable work and hours,
differed between firms in the mid-1830s by as much
as 35 percent. More local studies are required to define the picture.[21]

One last theme to emerge from this investigation
concerns the impact of the new American textile
technology, particularly that created at Waltham:
the constant-batten-speed, cam power loom; the
high-speed dead spindle; the stop-motion warper;
the rapid-drying dresser fitted with warp measuring and shut-off device; the four-cone double
speeder with a low piecing requirement; the filling
frame with a cop winding motion; and the self-
acting loom temples. From management's viewpoint, these inventions unquestionably cut labor
costs. One dollar of labor in the Massachusetts cotton industry in 1831 processed 20 percent more
cotton, by weight, than the same amount of labor in
the Rhode Island cotton industry.

From the operatives' point of view, work in a
Waltham or Lowell cotton mill, or in most other
northern New England cotton mills, was free of the
exhaustion, oppression, and cruelty that marked
British cotton factories.[22] The very purpose of the
Waltham innovations with regard to operatives

was to admit educated daughters of respectable and
respected New England rural families to the cotton
mill. Not only would overseers treat their charges
with some caution, under a puritanical moral code
(which presumably applied as much to overseers as
to operatives), but the technology would have to
accommodate the social as well as economic characteristics of the new labor force. A high labor turn-
over pointed toward shorter operative learning
periods. Educated and active minds, which quickly
found machine minding very dull, required
equipment with maximum automaticity. And
younger, less physically strong and less well-
coordinated operatives needed less onerous and
hazardous machine-tending tasks. So the Waltham
innovations disposed of mule stretching and mule
spinning and their long learning periods; compensated for operative inattention with numerous stop
motions; reduced piecing in the roving frame and
in filling preparation with the filling frame; and
lowered operative interference with moving parts,
as with the self-acting temple. Only in dresser
tending did the work remain onerous, because it required prolonged and concentrated attention to the
sized warp yarns to check that none broke.

If the girls were made responsible for no more
equipment than they could comfortably manage,
the technology lowered the levels of skill, mental
alertness, and physical hazard involved in machine
minding. If overseers were tolerant, then boredom
or fear of offending the mill's narrow moral code
constituted the operatives' greatest afflictions. Waltham technology thus permitted a more humane regime at the purely physical (as opposed to the moral
or regulatory) level of operative work.

Whether the introduction of Waltham technology

261

262

to Britain—which hardly occurred in the period under review—would have ameliorated working conditions seems doubtful. Almost any technology can be applied either in a humanizing or a dehumanizing way; it all depends on the human priorities of those wielding the technology, which in turn are shaped by much wider social and economic considerations. The absence of enlightened and humane values in sufficient force in early-nineteenth-century Britain suggests that American innovations would have made little impact on British working conditions. Indeed even in New England the Waltham technology was operated with new severity from the 1840s, when the labor supply situation drastically eased.[23]

Some Lancashire manufacturers tried to follow the northern New England mills, but their efforts produced no more than poor imitations of American cotton goods.[24] Waltham technology, as a manufacturing system, was not implemented in Britain, in this period at least. Presumably the cost of quality cotton staples, higher in Britain than in the United States, limited Lancashire options. It is very likely therefore that in the Atlantic economy, New England manufacturers moved into a position that complemented rather than competed with British mills.[25]

In summary, this investigation finds that for a new technology still partially understood or not yet reduced to verbal or mathematical forms, the experienced practitioner must be the most efficient agent of international diffusion. Modifications to imported technologies conform to prevailing economic and social circumstances in the receptor country—in particular, conditions in commodity and product markets, relative factor prices, and so-

*Conclusion*

cial values. But technology diffusion is a complex
process. Although some generalizations can be
made about it, different technologies and different
regions demonstrate variations of some significance
on familiar themes. Using stage analysis concepts,
and trying to understand the interplay between the
technical and economic dimensions of the
technologies concerned, this study has sifted
technology diffusion in the cotton and woolen
manufacturing industries, which underpinned
America's industrial revolution. Clearly, without
technology diffusion, industrialization is long delayed or never takes place. For the United States,
transatlantic connections and circumstances facilitated technology transfer and so created the possibility of a transatlantic industrial revolution.

1. Hopkinson, *Account*.

2. Gallatin, "Report," p. 427 gives a maximum of 100,000 spindles in 1810–1811. For a U.S. cotton spindleage of 356,900 in 1820, see appendix D, an estimate 60 percent greater than the previously accepted figure of 220,000 spindles, cited by Rogers, *Transportation Revolution*, p. 339. Friends of Domestic Industry, *Report*, p. 16, gives a spindleage of 1,246,503 for 1831.

3. RIHS, Arnold Papers, box 1, Correspondence 1820–1855, W. A. Greene (Providence, Rhode Island) to Arnold (Great Falls, Somersworth, New Hampshire), October 22, 1826.

4. Friends of Domestic Industry, *Report*, p. 16. For the estimate of £20–£25 million for British fixed capital investment in cotton manufacturing in 1834, see Blaug's figures quoted in Chapman, *Cotton Industry*, p. 31.

5. Chaloner, "New Light," p. 41, quoting Arthur Redford et al., *Manchester Merchants and Foreign Trade, 1794–1858* (Manchester: Manchester University Press, 1934), p. 132. In November 1826 the Manchester merchants sent a memorial to the Treasury in which they opposed the free export of machinery for fear of competition in foreign markets, adding, "Such competition is no chimera; it has been felt in various markets, from the manufacturers of France, Switzerland Saxony and the United States of America. The race is begun, and we would not wantonly throw away any advantages." This American competition was publicly acknowledged in 1833. See PP (Commons), 1833 (690), 6:39, 171.

6. Josephson, *Golden Threads*, pp. 178–185.

7. Charles Dickens, *American Notes*, chap. 4.

8. For some recent discussion, see Gersuny, "Devil in Petticoats," and Horwitz, "Architecture and Culture." A good introduction to the Lowell girls is Wright, "Uncommon Mill Girls."

9. Dickens, *American Notes*, chap. 4.

10. Pennsylvania Senate, *Report*, p. 4 and passim. Wallace, *Rockdale*, pp. 73–239, has a vivid portrayal of the

336

development of an eastern Pennsylvania textile manu-facturing community within what was known as the Rhode Island system.

11: Examples were the Indian Head Factory at Nashua, New Hampshire, the Ware MC in Massachusetts, and the Matteawan MC in New York. Others can be traced in USNA, Census of Manufactures, 1820.

12. Baker MSS, Textron Inc. Papers, Indian Head Factory, Directors' Minutes 1825–1830, p. 53 (May 31, 1826).

13. Industrial exhibitions were held in Continental Europe and the United States well before 1850. Over two hundred European exhibitions before 1850 are listed in Carpenter, "European Industrial Exhibitions." Notice-ably, British participation tarried until after the repeal of the laws controlling machinery exports and the ac-ceptance of free trade. For late-nineteenth-century in-ternational exhibitions see Curti, "America at the World Fairs."

The transition from less random to more rational methods of technology diffusion deserves further study, as Hughes, "Comment," observed.

14. This, as one recent investigator readily acknowl-edges, is a major weakness in a purely case-study ap-proach to technology transfer. See Stapleton, "Transfer of Technology," p. 298.

15. Rosenberg, "Anglo-American Wage Differences," p. 224. His surprise seems to be warranted by subsequent work. Over a range of occupations it has been shown that the skilled-unskilled wage rate differential was greater in America than in Britain during the 1820s. See Adams, "Some Evidence."

16. Hamilton, *Espionage*, p. 75.

17. Rosenberg, *Technology*, p. 61.

18. PP (Commons), 1824 (51), 5:473.

19. Baker MSS, Nathaniel B. Gordon Papers, vol. 2, Memorandum Book 1816–1820, p. 8 (November 21, 1820).

20. EMHL, Acc. 500, Duplanty, McCall & Co. Papers,

General File, "Actual Expenses at Hagley Cotton Factory
for One Day" (February 18, 1811); compare this with the
figures in table 10.3 of this book. One Hagley spinner receiving these wages was David Our, a British immigrant
who described himself as a cotton manufacturer in 1812.
See USNA, RG 45, War of 1812 Papers, Marshal's Returns
of Enemy Aliens.

21. Ure, *Cotton Manufacture*, 2:444–445.

22. Gitelman's opinion that "the equipment adopted at
Waltham was of such design that children could not be
employed in its use" ("Waltham System," p. 231) is not
substantiated by a close study of Waltham innovations.
Quite the reverse: a greater use of inanimate power, the
integration of mechanical processes, increased mechanical work, and automatic shut-off devices permitted
a more extensive application of child labor. In Britain one
stop motion was invented in order to compensate for the
"inattention" of "the person who attends the frame (who
is for the most part a child)," but this did not positively
exclude child operatives; it simply reduced their number.
See Bradbury's GB Pat. 3990 (1816), printed specification,
p. 7. The diminishing employment of children in the
Waltham system surely reflected the influence of social
as much as or more than technical or economic factors.

23. Ware, *New England Cotton Manufacture*, chap. 10. In
the 1840s, wages in teaching left earnings in textiles well
behind, due to the influx of unskilled Irish workers. See
Gitelman, "Waltham System," p. 238.

24. Bagnall, *Textile Industries*, 2:2020, quoting Appleton's
January 17, 1828, speech in the Massachusetts House of
Representatives.

25. Potter, "Atlantic Economy," p. 279, raised this
possibility.

# [7]

## DIFFUSION OF KNOWLEDGE-BASED PRODUCTS: APPLICATIONS TO DEVELOPING ECONOMIES

M. L. BURSTEIN*

*Demand for a knowledge-based product often must be created by the product's innovator. But the process which diffuses knowledge, and so enhances demand, accelerates decay of the company's ability to appropriate benefits generated both by the product and by the company's complementary explanation of its qualities and uses.*

*Decay of the ability to appropriate profit arises from a pervasive free-rider problem. Rivals, as well as users, become educated by the innovator. As the market deepens and expands, entry will become feasible for firms capable of production but incapable of supplying effective educational packages.*

*At least two principal themes emerge: i) logically, innovative products will be sold bundled with education and other complementary services, at least in their early stages; ii) intensity of diffusion will be sensitive to the nature of governing property-rights regimes.*

*The analysis suggests that, because of free-rider and other problems, an innovator often finds initial product development plans unproductive and must resort to alternatives. Feasible diffusion often requires that the product space allocated to an innovator by grants of property rights be larger (contain more dimensions) than one confined to primary invention(s).*

### I. INTRODUCTION: KNOWLEDGE AND DEVELOPMENT

Typically, for a developing country to benefit from a new idea, it must learn to apply the knowledge imbedded in the idea and to control the flow of products and processes that ensue. Practical experience yields a corollary body of knowledge and more profound understanding of the core idea.

Markets for new products based on new ideas usually must be created, especially in developing economies where economically effective knowledge is relatively scarce. Effective use of knowledge-based products depends on diffusion of general knowledge as well as that which is needed to utilize the core product. To take an example which will be used again below, operators must be taught to use the software that makes hardware effective.

Markets for knowledge-based products often may be developed only through the intellectual and financial resources of the innovating companies. One implication of this fact is that hardware and software will often be sold as a bundle, especially in early stages of product development. Bundling demonstrates a characteristic of knowledge-based products and knowledge diffusion: complementarity.

*Professor of Economics, York University, Downsview, Ontario.

Successful diffusion of a knowledge-based product requires that marketers know their product's operational characteristics. Thus, vertical integration may become a characteristic mode of operation: innovative companies often are uniquely well-placed to explain how to use their ideas.

## A. *Free Riding and Property Rights*

Knowledge-based products typically cannot adequately be diffused, if indeed they can be created at all, without massive expenditures. The diffusion process must combat inertia: propagation of knowledge-based products, together with their modes of employment, may require that innovators incur huge initial losses. Often such losses cannot be recouped if innovators are denied property rights to their products and associated software; patents and copyrights are prominent devices protecting such property rights. Only recently has it become understood how sensitive economic actions are to existing regimes of property rights.

## B. *Marketing*

As a study of diffusion, this paper necessarily is concerned with marketing rather than initial invention. Its focus is on application rather than creation of fundamental new knowledge. Furthermore, less-developed economies naturally are more concerned with using knowledge developed in advanced economies than with incremental contributions to root knowledge they may make.

## C. *Decay*

In an important sense, decay complements diffusion. The proportion of social benefit appropriable by the innovating company may be small even under the most favorable (to the company) circumstances. A parallel may be found in the thermodynamic concept of *available heat energy*. When heat-fall is harnessed optimally, only a fraction of the falling heat-energy may be transformed into work.

The proportion of benefit from use of a knowledge-based product that can be appropriated by an innovator tends to decay over time. The decay factor may make it necessary for host economies to pay more in advance, or expand fields of protection, to induce entry of products requiring substantial market-development expenditure. Pakes and Schankerman (1978) make this point in an interesting way:

> The rate of decay in the revenues accruing to the producer of an innovation derives not from any decay in the productivity of knowledge but rather from . . . points regarding its market valuation, namely that it is difficult to maintain the ability to appropriate the the benefits from knowledge . . . Indeed, *the very use of new knowledge in any productive way will tend to spread and reveal it to other economic agents, as will the mobility of scientific personnel*. (1978, p. 3) (Emphasis supplied.)

"Decay" poses obstacles to hosts as well as innovators. The decay problem might inhibit diffusion effort, and simple patent protection might be an insufficient antidote. More complex product and marketing combinations might have to be devised. This possibility leads to applications and extensions of work by Burstein (1960a, 1960b) and Telser (1960, 1979) in the fields of tied sales, full-line forcing, resale price maintenance and monopolies of complements. The following colloquy supplies the flavor of the resulting analysis:

Q. If patent protection is given to *groups* of products, groups including products not particularly interesting to hosts, and perhaps available at lower prices from "pirates," is not monopoly being extended to products *B, C, . . .* from lead-product *A*?

A. It may be impossible for an innovating company to recover overhead (including product introduction and user education), let alone profits mandated by the cost of capital, unless a multi-product line is protected—especially if property rights to innovations from primary products are not created in favor of innovating companies. The field over which an innovating company finally exercises control may have to encompass broadly defined product groups.

Nor may a marketer find it worthwhile to offer rich packages of complementary services unless he freely can form bundles of complements designed for different sorts of users. Ideally, the lead product among a group of products easily is identified; in fact, it may be impossible. Such indeterminancy is endemic in "new-knowledge" situations. It is not clear *ab initio* what are the full implications of new discoveries. But when patents are respected in general, the profit motivations of innovating companies can assure that products with high-benefit potential are intensively enough marketed.

Pursuing property rights and optimality, if information were complete and the forces at work certain, innovating companies would be rewarded by lump-sum bounties. Real-world reward mechanisms must be politically feasible, and must take account of transactions costs and uncertainty. Thus, *ex post*, optimal feasible solutions will be second best, or *n*-th best.

D. *The Economic Development Context*

My analysis is not peculiar to less-developed economies. Indeed, advanced economies are notable for frequency of invention and rapidity of diffusion of sometimes radically new ideas. But the analysis may be crucial for less-developed economies: the modern theory of development emphasizes *knowledge and its diffusion*.

Economic development commonly has been perceived as an extensive undertaking. It has been hoped that as more and more capital, usually measured by constant-dollar expenditure on physical durables by producers, is accumulated per worker, output per head, net of capital cost, will increase. Such hopes commonly have been deceiving. Massive investment programs often have led to, or not prevented, *lower* per capita incomes: steel mills, chemical plants, state-owned airlines, etc. have absorbed more valuable resources than they have yielded. The extensive approach to economic development insufficiently considers, when it does not ignore, knowledge and investment in human agents.

Modern theory perceives development intensively: people learn to do different things in different ways, attracting profit-seeking capital in the process. Regarding intensively perceived development sequences, host-country resources are attracted to a nucleus based on imported knowledge. Then newly trained human capital may operate outside the original innovating companies, often becoming licensed to supply software, distribute hardware, etc. Still later, skills and experience accumulated in this way may lead new middle classes to import additional technology on their own. They may perceive possibilities which are not obvious to the innovating company's world headquarters; maps at world headquarters are on too large a scale. Host-country "second round" innovators thus may be able to tap local savings and evoke fresh savings attracted to newly created opportunities, since in simpler financial contexts savings are a function of investment opportunity. Derived development of this sort may lead to derived demand for the technology of highly developed economies through a feedback channel.

## II. THE MODERN THEORY OF THE DEVELOPMENT OF MARKETS FOR KNOWLEDGE-BASED PRODUCTS:   APPLICATIONS TO DEVELOPING ECONOMIES

Griliches (1957), in his landmark study of the diffusion of the hybrid-corn idea, introduced into economics the biometric *ogive* concept. (An ogive is the distribution curve of an accumulative frequency distribution.) Ogive-based processes such as marketing are strongly inertial,[1] but, once a critical mass is attained, rapid expansion becomes possible. Perhaps inertia may be overcome only by massive expenditures by the innovating company in the early stages of product development in new markets, markets which must be *created*. Initially at least the company is custodian of knowledge of what can be done with the product. The resulting financial plans, heavily front-loaded, may be financially justifiable only if innovating companies can foresee virtual protection for an extended period.

### A. *Bundling, Resale Price Maintenance (RPM), etc.*

Useful analogies to the problem at hand may be drawn from the theories of "bundling" and RPM, at least from that part of RPM theory concerned with assurance of substantial mark-ups for downstream promoters and distributors.

In modern economics, products are perceived as *bundles* of properties. Information and knowledge are stressed: economists now study the costs of acquiring and diffusing knowledge in the same way they long have studied the cost of extracting and distributing coal. [This and the following four paragraphs are based on Burstein (1982).]

If a layman bought a personal computer, the hardware would be but a toy for him until he learned to use the machine. Programs would have to be selected, or devised, for him. He would have to be taught to manipulate the materials. His progress would have to be monitored and his experience studied in order to improve programs assigned to him.

Contrast the last example with one based on the computer operations of a huge, technologically sophisticated company operating complex computer-guided processes. Such a company has vast software capability. It is likely to buy *unbundled* computer products. It will buy hardware, which it helps design; it will not pay computer companies for instructional materials.

The insurance industry supplies another example of the logic of "bundling." When an insurance product is new, commissions on sales dominate in agent compensation, encouraging agents to introduce and promote the product. As the market

---

1. See M. Trajtenberg (1982). Also see M. Trajtenberg and S. Yitzhaki (1982). Trajtenberg (1982, p. 3) writes that "there is no unique way to characterize diffusion processes; . . . the choice of parameters . . . always retains an element of arbitrariness."

matures, and if regulation permits, agent compensation will become dissociated from what had been a new product. The process will become unbundled. Conventional, annualized salaries will be paid for agents' services, which increasingly involve maintenance of the existing stock of policies.

Telser (1960) asks, "why should manufacturers want *fair trade?*" In the early stages of development of markets for knowledge-based products, it often is important for distributors to explain them to consumers. Explanation may be time-consuming and expensive. Distributors will want to do this work only if offered high mark-ups. Thus, in the early days of stereos, advanced cameras and similar products, manufacturers often *do* want fair trade (RPM).

Knowledge-based products may be very complex—for example, computers, together with associated software and operational techniques, or pharmaceuticals or agricultural chemicals. Then the most effective marketing (diffusion) mode may be vertical integration. The innovating manufacturer's unique command of the theory and praxis of the "product atom," comprised of a hardware nucleus orbited by software (praxis) electrons, may make vertical integration inevitable. Indeed, only recently, even in highly developed economies, have software industries truly become independent of original-equipment manufacturers (OEM). Originators may have to direct the diffusion of complex products if these are to become useful. The resulting huge "up front" expense is analagous to the large sums which have to be spent on new colleges before teaching may begin, and students may have to be trained for years before they can deal effectively with difficult problems.

### B. *Property Rights*

Here the *property-rights* idea intersects with two other basic concepts: *free riding* and *multi-product production and distribution.*

Efficiency of economic performance can depend significantly on how property rights are assigned, especially in the case of innovation and trade in the fruits of innovation.

The assignment to railroad companies of land rights along railroad rights-of-way in the nineteenth century in North America exemplifies the policy problem under study. External benefits from railway development were immense; whole regions became agriculturally viable. But the potential profit from hauling goods and people might not have sufficed to induce railroad entrepreneurs to make the huge, risky investments which were required. Assignment to the companies of rights to land, the value of which was enhanced by railroad-building, allowed the companies to internalize a substantial part of the benefit from their actions.

Property-right assignment may be important for more-or-less defensive reasons as well as for capture of external benefits. During early, loss-making phases of diffusion, the company generally will not have any rivals in the field. Once diffusion has begun, however, potential rivals are likely to *free ride.* The same factors which make the market potentially profitable are likely to make it impossible for the company to dominate the market once it has been developed. Indeed successful diffusion may accelerate *decay* of the company's appropriability coefficient: the diffusion process that develops the market also enhances understanding by the company's potential rivals. Unless the company can obtain sufficient property rights over affected products and processes for a sufficiently long period of time, it may decide against introduction of innovative products. And, surely, incomplete property-rights investiture will lead to socially suboptimal diffusion.[2] In more concrete terms, usually it is not feasible to market even breakthrough products except *en echelon* with what might be a large number of other products, most of which may be solid but unexceptional.

## C. *Transactions Costs*

Here transactions costs result from the political-economic setting itself. Transactions costs span the distance separating the real world from the friction-free economy of traditional pure theory, one in which information is perfect and search costless. Thus, it may not be feasible for an innovator to capture much of the external economies from his brilliantly engineered Product A. It would be no small feat for an alien railway company to be given vast tracts of land in a host country! The only feasible alternative may be to give the company privileged positions with regard to its otherwise unremarkable products *B*, *C*, etc. The outcome belongs to a family of results obtained from the author's analyses of full-line forcing and tied sales. [See Burstein (1960a, 1960b.)]

## D. *Free Riding*

An *honest free rider* "announces in advance his refusal to join a coalition; the coalition makes its optimal choice knowing this. A *dishonest free rider* promises to join the coalition that proposes an undertaking for all its members. Then, having promised to join the group, so that the group acts on the basis of this promise, the dishonest free rider breaks his promise." [Telser (1978), p.2.]

Dishonest free riding is beyond this study's purview, but honest free riding is central to it. Some remarks on honest free riding and patent strategies of less-developed economies can be usefully made now.

Denote an economy with a weak scientific base and perhaps a small home market as a *Beta economy*. Why should a Beta economy offer patent protection if it can buy patented products from pirate vendors at lower prices?

Responses may be made at two levels. The first response concerns quality control, delays in availability of new products in economies which do not offer patent-type protection, blocked development of products specialized to Beta conditions and

---

2. The following set of papers is concerned with *optimal diffusion*. Granted, the depletion analogy is not ideal for the study of inexhaustible knowledge. The papers include: Yoram Barzel (1968); Edmund Kitch (1977, 1980); and Douglas A. Smith and Donald G. McFetridge (1980); and an unpublished paper by S.N.S. Cheung.

Kitch (1977) reports that his ideas "crystallized in response to Barzel's essay." He also reports being influenced by Cheung's *development rights* concept. Kitch emphasizes how the property-rights *regime* affects the timing of diffusion:

> "[Barzel] points out that the exploitation of technological information has much in common with fisheries, public roads and oil and water pools—all resources not subject to exclusive control. If the rule of *first approporiation* controls, there will be an inefficiently rapid depletion of the resource. He suggests this problem can be solved if technological monopoly claims can be granted or auctioned off, giving their owner the exclusive right to develop the technological opportunity. What Barzel did not realize is that a patent system can be such a claim system and, indeed, that it is a more sensible system than an auction system would be." Kitch (1977, pp. 265-266)

Kitch uses the idea of a *prospect*, *i.e.* "a particular opportunity to develop a known technological possibility." (1977, p. 266) He is concerned with premature depletion along lines based on "over-fishing." It may seem that his interest is opposite to the author's: he is concerned about development being too fast; the author is concerned that it might be too slow. Such an impression is specious. In both cases, entrepreneurs may require pay-out periods too short for social optimization. In both cases, efficient exploitation of social possibilities requires grants of property rights to creators of new products. Both cases are affected by free riding.

insufficient appreciation within Beta economies of "clubbing possibilities." [See Burstein (1983).]

The second response is keyed to *diffusion*. It concerns both developed and less-developed economies, but the following remarks refer especially to the latter.

If host economies are to achieve optimal diffusion of new knowledge, they must not become obsessed with "hardware." The symbols of technological development should not be confused with its substance. The substance comprises profound understanding of the knowledge which permits productive use of hardware contrived in more-advanced economies.

It will become clear that, typically, it is in the interest of a Beta economy to make it easier for innovating guest companies to keep command of the hardware around which growth processes will cluster. Why? Successful diffusion is apt to require the sort of bundling of hardware and complementary goods and services (including the physical products consumed by the process and the intangible knowledge necessary for optimal use of the machine) that only can be provided, especially in early stages of diffusion in less-developed economies, by innovating companies. Innovating companies best are perceived as potential conduits for diffusion.

## III. A DYNAMIC MICROECONOMIC ANALYSIS OF MARKETING

### A. *Foundations: The Problem as Viewed by an Innovating Company*

In Section III.A.1 the simplifying assumption is made that the innovating company is diffusing a single product. In this way, it is possible to isolate problems of *timing* in considering the optimal path of prices, of promotional expenditures, *etc.* Partly because of demand inertia and partly because diffusion is enhanced by growing familiarity with products (so that today's sales significantly promote tomorrow's demand for new and strange products), large initial promotional expenditures are often required. It may also be appropriate, if innovating companies operate with long horizons, that prices will be discounted in the early stages of product diffusion.

The second subject of Section III.A.1 is the theory of vertical integration. It is difficult to coordinate the marketing efforts of independent distributors of knowledge-based products, especially in the early stages of diffusion. As a result, vertical integration can become a preferred scheme of organization. But a feasible vertical-integration strategy requires that the innovating company have exclusive rights to the production and distribution of the product, possibly for an extended period.

In Section III.A.2 the simplifying single-product assumption is dropped. Robust final results require postulation of real-world, $n$-product firms. The analysis of Section III.A.2, based on Burstein (1960b) and Telser (1979),[3] illuminates the principles of optimal pricing at a point in time; *inter alia*, it studies the pricing of complementary goods. As noted in Section III.A.1, today's output is significantly complementary with tomorrow's in that tomorrow's demand is influenced by knowledge gained from working with today's output. Section III.A.2 studies bundled (*vs.*

---

3. See also Burstein (1960a), Telser (1960), Warren-Boulton (1978) and R. G. D. Allen (1938, pp. 359-362). The bibliographies of Telser (1979) and Warren-Boulton (1978) are useful.

Allen (1938) is the first work to appear. Later authors, expanding on Allen's ideas, naturally obtained fuller development of the material.

unbundled) marketing in connection with provision of *education* by the innovating firm. Recovery of initial expenditures may require, along lines developed by Telser (1979), that a restricted set of bundles be offered, which may call for specific tie-ins.

Another theme of Section III.A.2 is based on Burstein (1960a). Feasible exploitation of profit possibilities may require that clients be forced to buy their requirements of certain goods other than *A* (a much-esteemed product controlled by the company) from the company if they are to have the right to buy *A*. *A* is the *tying* good, and *tied* goods include *B*, *C*, *D*, etc. In this analysis the tied goods may be unrelated to *A* in demand and production. It may not be possible for activity in the tying good to be profitable unless a degree of control can be established over the sale of tied goods. The costs of extracting the benefits from the use of *A* through a multipart tariff for *A* (*e.g.*, a user fee plus a charge per unit "consumed") may be prohibitive. The surcharges on tied goods comprise a system of excise taxes. [See Burstein (1975).]

### 1. *The Single-product Case*

Consider first a very simple two-period case: demand in the first period depends on prices (actual and expected) in both periods. Costs may be ignored. The problem is to maximize the present value of the sum of first- and second-period revenues.

Discount rates affect choices of first- and second-period prices and therefore the pattern of diffusion. The higher the first-period price, the more front-loaded will be the receipt profile. The higher the governing discount rate, the greater will be the incentive for front-loading. At very high discount rates, the company would "take the money and run." Thus, policies imposing short pay-out periods on technology-transferring companies may abort diffusion.

Recalling "ogives," consider a richer possibility. Diffusion proceeds slowly until a certain critical mass is attained, as measured by market penetration (*e.g.*, a "saturation" statistic) or the level of accumulated stock. Then the process accelerates for a time and eventually levels off.

Interpretation of this pattern may run as follows. Until the cumulation of diffusion reaches a critical level, it is not profitable for any but very heavily capitalized entrepreneurs to invest in ancillary industries. Thus, the free-rider problem reappears: when diffusion has reached its critical level, it may become highly profitable for outsiders to enter the market without having to invest much capital. The market no longer will need to be educated. But innovators may not make massive investments, needed to achieve critical levels of diffusion, unless they are assured of protection after critical levels have been attained.

To the extent that earlier sales comprise an input into the process-generating subsequent demand, current sales are profitable not only because current revenues exceed current costs, but also because current sales promote future profits. Indeed, innovating or technology-transferring companies with long horizons may make early-period sales *loss leaders*.[4] Early-period outputs then exceed levels suggested by models insensitive to diffusion. Thus, the value of telephone service to a subscriber depends on how many others subscribe.

In the remainder of this section, discounting is suppressed. It suffices to understand discounting qualitatively: with discounting, the optimal path of production

---

4. The *locus classicus* for this argument is Allen (1938, pp. 359-62). "For example a monopolist manufacturer of razors and blades may gain by marking down the price of razors (for which the demand is less elastic and strongly complemetary with that for razors)." (1938, p. 363). The author's counterpart to razors is "early period sales".

shifts as in a diminuendo effect: it pays to fell more trees *now* or to sell younger wine.
Where $q_1$ is first-period demand,

(1) $$q_1 = f(p_1, p_2)$$

(2) $$q_2 = g(p_2) .$$

Problem:

(3) $$\max p_1 f ( \cdot ) + p_2 g ( \cdot ) .$$

For simplicity it is assumed that second-period price is anticipated correctly in the first period, and that there is no forward market. To repeat, cost may be neglected.

Necessary conditions for a solution are that the partial derivatives of the revenue function with respect to $p_1$ and $p_2$ are zero. The resulting two equations are to be solved for $(\bar{p}_1, \bar{p}_2)$:

(4) $$f + p_1 (\partial f / \partial p_1) = 0$$

(5) $$p_1 (\partial f / \partial p_2) + g + p_2 g' (p_2) = 0 .$$

Equation (4) requires that the marginal profitability of a price change be nil.

A number of insights may be extracted from Equation (5). Consider a solution of Equation (5) in which $\partial f / \partial p_2 = 0$ so that the expectation of price in the second period does not affect demand in the first. But in the general case

(6) $$\partial f / \partial p_2 > 0$$

Thus, starting from a solution based on the assumption that second-period price has no effect on first-period demand, relaxation of that stipulation requires a higher second-period price, *i.e.*, a time-profile in which first-period price is relatively lower.

By way of proof, the marginal profitability of an increase in the second-period price will be positive—starting from a solution conditional on the "partial" of the first-period demand with respect to second-period price being nil: the sum of the second and third left-hand terms is zero; the first left-hand term now is positive.

Now consider a crude scheme in which the strength of second-period demand depends positively on first period sales:

(7) $$q_1 = f(p_1, p_2)$$

(8) $$q_2 = g(p_2, q_1) .$$

Conducting an analysis parallel to that of Equations (1)–(5), the problem is to maximize

(9) $$p_1 f (p_1, p_2) + p_2 [g (p_2, f ( \cdot ) ] ,$$

Setting partial derivatives equal to zero,

(10) $$f + p_1 (\partial f / \partial p_1) + p_2 (\partial g / \partial f)(\partial f / \partial p_1) = 0,$$

*i.e.*

(11) $$f + (\partial f / \partial p_1)[p_1 + p_2 (\partial g / \partial f)] = 0.$$

And

(12) $$p_1 (\partial f / \partial p_2) + g(\cdot) + p_2 (\partial g / \partial p_2) + (\partial g / \partial f)(\partial f / \partial p_2) = 0.$$

Equation (11) shows that the marginal profitability of an increase in first-period price is less because of the effect of first-period output on second-period demand. Equation (12) displays the advantage to the company of stimulating first-period demand by relatively discounting first-period price. Thus when the cumulation of diffusion stimulates demand (*e.g.*, as a result of a product becoming better understood) a long-run view influences rational calculations of innovating companies.[5] But such a long-run view is subverted by public policies, like short pay-off periods, encouraging innovators to use high discount rates.

As in Burstein (1960b), consider critical variables determining the demand for a good marketed through a vertically integrated process, which include:

a) retail prices at various outlets;
b) the number and location of such outlets;
c) the extent and kind of national advertising and other promotion;
d) the extent and kind of local advertising, promotion and service-provision.

The innovating company, controlling patents, *etc.*, will study manufacturing conditions and demand possibilities. Then it will decide upon an optimum mix of control-variable levels: the profit-maximizing combination of retail prices, number and location of outputs, national and local advertising budgets, special-service provision, *etc.*

It is difficult to see how the problem could be solved except within the organization. Oliver Williamson (1975) would say that the problem requires a hierarchical rather than a market solution. As suggested by Coase (1937), it might be impossible to establish a price system and communications network permitting coordination between the original producer and putatively independent downstream entities; or, even if feasible, a non-hierarchical solution might be unacceptably costly.

Earlier the author developed the problem at hand as follows:

> A familiar dilemma confronts manufacturers. High prices to distributors discourage them from pushing the item; low prices to distributors encourage promotion . . . but then profits simply might go to the distributors.

---

5. Thus the demand function at date $t$ can be written
$$q_t = f(p_t, p_{t-1}, \ldots, p_{t,T}, \int_0^t \psi(t)\, dt).$$

*The cumulant of past sales affects demand at any date $t$.* It is assumed that subsequent prices are known with certainty.

*Sans* vertical integration the solution runs as follows. Use the carrot and the stick. Set the wholesale price low enough to permit substantial mark-ups and encourage local promotion. Require wholesalers to purchase all their 'requirements' of other goods—over which the Company may have less market power and which otherwise might not be important for it—at a price substantially higher than the Company's cost. [Burstein (1960b, p. 85).]

Obviously, the multi-product analysis of Section III.A.2 has implications for the restricted analysis of Section III.A.1. Just as obviously, the implications for diffusion in markets of less-developed economies are clear: it is improper to treat in isolation one product, among many, in a company's line.

Potential *internecine strife* among distributors poses a troublesome difficulty. Uninhibited competition among distributors is inefficient. Thus, Distributor X does not value stimuli of demand for other distributors' products caused by his research and development, promotional and advertising outlays. For unintegrated market schemes to be optimal, complex coalitions of distributors must be formed; members must, in effect, agree to divide maximized joint profits. Resulting "supernormal" profits would have to be "taxed away" by innovating patent holders for their activities to be worth their while. *In reality, optimal development of property rights in knowledge-based products probably often requires that distributors simply be taken over, or never let go, by innovating firms—especially in early stages of product diffusion.* The analogy to bundled marketing is fairly obvious.

### 2. The Multi-product Case

Overhead, especially in the form of costs incurred in product introduction, user education, *etc.*, may not be recovered unless a broad product line can be established, especially in the absence of property rights to spin off. The field over which the innovating firm exercises control may have to be substantially enlarged. This problem is addressed immediately below and was studied by Burstein (1960a, 1960b) in related context. A more difficult line of analysis, based on Telser (1979), is addressed subsequently. It is based on the strong complementarity between knowledge-based lines; again, a commanding example is hardware/software. The analysis establishes that a marketer may not find it profitable to offer rich packages of complementary services unless he can freely form bundles of complements for different sorts of users. Stringent protection of intellectual property may be necessary to induce sufficient marketing effort.

The central problem can be stated in the following way. A host economy wishes to maximize consumer benefit from a set of products innovated by company $C$. The company's active participation in the diffusion of these knowledge-based products proves indispensable for the solution of the problem.

Assume it is politically and/or technically infeasible for the host economy to offer a lump-sum subsidy to the company to induce it to offer a socially optimal set of products with socially optimal promotion and prices. Indeed, in the real world, it is inherently infeasible to operate on deterministic lines. It is not possible to know what will be the fruits of new knowledge. The problems involved in the propagation of knowledge-based products and their associated technologies cannot be quantified *ex ante*, nor can a potential host economy know *ex ante* what products to covet.

The illustrative problem requires construction of optimal feasible standard operating procedures. In establishing any procedure, it must be recognized that, for an innovator to be willing to invest, he must expect to achieve at least some minimum rate of return. All the programs consistent with the company's rate-of-return constraint may require that protection be afforded to many products. The feasible program of greatest benefit to the host economy may require protection of multiple markets for the company's products.

The argument can be extended by exploring a modification of an example from Burstein (1960b). Assume that all consumers of a product the company can develop for $10 million are represented by the same demand curve. (See figure 1.) *If the company were constrained simply to establish a one-part tariff (a single price), it would charge $OA per unit and sell (n) (OB) units.* (There are n consumers.) Marginal revenue is nil at output OB. (Cost may be ignored.) *Total Benefit/ (n)  =  Area α; Total Revenue/(n)  =  Area β .*

Quite possibly, total revenue (here equal to profit before capital costs) cannot cover the $10 million investment. Yet social benefit may be massive. And it is easy to identify the cause of a socially sub-optimal outcome: the restriction imposed on the company's pricing policy makes it impossible for it to appropriate enough of the benefit to justify the $10 million investment.

It is useful to formulate first an interim problem. Once it is analysed, it may be transformed into what I call the *proper problem*, the problem that is truly of interest here: that concerning the conduct of the company and the government of the host economy.

Let *ED* represent the demand for coffee by the representative consumer faced by a coffee monopolist. It may be seen that the monopolist, if constrained to a single-part tariff, obtains $OBFA from sales of OB cups at $OA per cup. Maximum revenue in an unconstrained solution is $ODE.

## FIGURE 1

### The Representative Demand Curve Illustrated

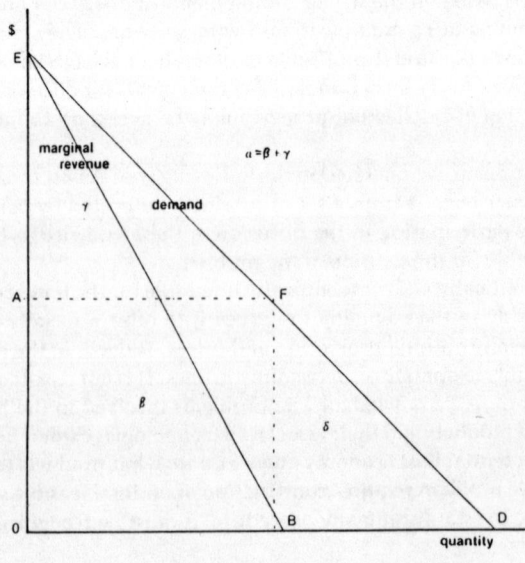

$ODE may be extracted from the market in the following way. Each consumer is told: to have the right to drink as much coffee as he wishes at zero price (equals marginal cost), he must pay an annual fee of slightly less than $ODE. In this way, benefit is maximized and then virtually totally expropriated. Profit is maximized and social efficiency achieved (in the sense that benefit is maximized).

Now assume that the company is released from the single-part tariff but barred from imposing user charges. A strategy that would preserve some, but not all, of the advantages of a user charge would be the following. Having been prevented from imposing lump-sum taxes, the company imposes a set of excise taxes. [See Burstein (1960a, 1975)]. The taxes are defined by the differences between the prices the company charges for *tied* goods and the competitive prices the consumer would pay if he quit the "coffee game." If he remains in the game, the consumer purchases his "requirements" of products $(X_2, X_3, \ldots, X_n)$ at prices $(p_2, p_3, \ldots, p_n)$. The array of excise taxes is $(T_2, T_3, \ldots, T_n)$. The arrangement comprises a *full-line force*. If the company could impose such a "force," the $10 million investment may become attractive to it.

Consider now the *proper problem*—the behavior of the company and the government. Abstracting from uncertainty, the government of the host economy sees that: i) the company cannot recover its investment if its property-rights position is confined to the power to establish a single-part tariff for $X_1$, the product prized by the government; ii) if the company's patents for products $(X_2, X_3, \ldots, X_n)$ were recognized by the government, and if the company were free to deploy profit-maximizing price strategies, it would want to operate in Beta.

The *proper problem* differs from the interim one in at least two significant respects:

(i) In the proper problem, the idea of recovering overhead expenditures, essentially by allocating or distributing them among a number of profitable products, instead of just one, is important. This idea is absent from the interim problem.

(ii) The two-person, non-zero-sum game played by the company and the government differs substantively from that of the interim problem in the following ways.

An agreement between the company and the government finds the latter abjuring opportunities to buy products $(X_2, \ldots, X_n)$ at prices $(\Pi_i)$ from "pirate" sellers. In return, the company makes a $10 million investment in a diffusion process that may generate substantial further gains to the Beta economy.[6] Not only will $X_1$ become available: it will be *explained*, as will be complements.

In the proper problem, Beta consumers do not face tie-ins or full-line forces. Instead the government agrees to submit to a quasi-"force," albeit only in the sense of recognizing valid patents and in exchange for the company's investment in the "A" diffusion process. (Once potential sunk costs have been incurred, the company's self-interest assures its follow-through, as indicated elsewhere in this paper.)

Perhaps the most interesting similarity between the interim and proper problems is that in both cases the company extracts revenue from a surrogate product field.

## C. *Decay of the Appropriability Potential [Pakes and Schankerman (1978)]*

Many political-economic policies, even if optimal, lead to *regret, pace* a critical necessary condition of dynamic programming. [See Sheffrin (1982) for a good summary.] *Patent policy* is an oft-cited example. Once patents are granted, and once

---

6. It is unlikely that "pirate" prices in fact would obey a free-competition norm. Instead, they are likely to float not much below prices set by patentees.

patentees have sunk costs, the state will regret that its longer-run interests dictate that it maintain its protection of the patentee. Thus, if no further invention from any source is anticipated (!), why not revoke protection and obtain the benefits of patented inventions more cheaply? In the real world, such regret is outweighed by continuing benefit from stimulation of further invention.

The rather rococo point just made is important for this paper. Once an innovating company has sunk substantial diffusion and other introductory costs, it normally will remain in the market even if the Pakes-Schankerman $\delta$ factor implies severe attrition of appropriable revenues: the *marginal* costs of continued operation in the Beta market are apt to be quite low once diffusion is extensive. The critical decision concerns sinking substantial diffusion cost in the first place.[7]

The Pakes-Schankerman problem may be schematically developed in the following way. The dynamic relationships between a variable standing proxy for the degree of diffusion, x, and the appropriability-of-benefit factor, $\delta$ are supplied by differential Equations (13) and (14)—*temporarily holding "y" fixed:*

(13) $$\dot{x} = \phi(x, \delta, y)$$

(14) $$\dot{\delta} = \psi(x, \delta, y)$$

Where,

(15) $$\partial \psi / \partial x < 0$$

(16) $$\partial \psi / \partial \delta < 0.$$

Potential appropriability falls as the market becomes more "saturated," where "saturation" refers to the progress of diffusion and the promotion of rivalrous skills in ways already described. [See Inequality (15) above.] The higher the level of the appropriability factor, the more attractive the prey. [See Inequality (16) above.]

An innovator, cognisant of Equations (13)–(16), will choose conditionally optimal paths such as $y = \phi(t)$ describing promotional expenditures as a function of time. The conditionally optimal path obviously will depend on alternative regimes of property rights. Choice of regime is likely to be sensitive to the properties of the economic structures just described.

To expand upon the last paragraph, the implicit optimal control problem has $(x, \delta)$ as state variables, and its control variables are expenditures $y$. Schematically, the company's problem is to maximize discounted profit as determined by

(17) $$\int_0^\infty e^{-rt} f[x(t), \delta(t), y(t)] dt$$

*subject to*

(18) $$\dot{x} = \dot{x}(x, \delta, y)$$

7. A chemical company may contemplate building a very expensive plant dedicated to a new process. Capital costs will be high. Operations costs (at least net of petroleum feed stock) will be low compared to a conventional plant. Once built, the plant will continue in operation even if the decision to build is rued.

$$(19) \qquad \dot{\delta} = \dot{\delta}(x, \delta, y)$$

*etc.*

## D. *Marketing and Diffusion of Complementary Goods*

### 1. *Telser's Theory*

Recall that R.G.D. Allen (1938) demonstrated that a monopolist of both razors and razor blades well might sell razors below cost in order to stimulate demand for razor blades. It has become clear that the success of such a strategy depends on the ability of a monopolist to appropriate blade revenues.

Also recall the intrinsic complementarity between knowledge-based goods and the techniques involved in associated services (often called "software" here). Let it be stressed that, especially in early stages of diffusion, current sales may be complementary with future ones. Increased familiarity with innovative products causes greater interest in, and comprehension of, them and increases future demand. Telser (1979) extends this argument.

Telser posits that there are $m$ components that can be combined to make up $n$ goods: a certain drug, together with the services of a doctor who has been trained by the innovating pharmaceutical company, comprise three components of one good, *i.e.*, treatment of patients in a certain way. The three components are the drug, the training, and the doctor's work.

Telser first shows that, if the "monopolist" were confined to selling components— if he were barred from selling *bundles* of components comprising goods—he would, in general, sell some of the components below cost.

Telser next shows that, if bundling is feasible, and if the $m$ components are complements,[8] it pays the company to sell particular bundles (goods) rather than component parts. He thus constructs a proper theory of bundled marketing. Telser writes:

> Assume that commodity 1 cannot be sold separately at a positive price. Perhaps it is information about another commodity 2 . . . Commodity 2 may be a new product sold by retailers some of whom furnish information about it to their potential customers. There is a free-rider problem if customers can obtain information . . . from one retailer who charges nothing for this service and then buy the product at a lower price from another retailer who furnishes little or no information . . . [The second] retailer can obtain a free ride . . . This situation cannot persist . . .
>
> The manufacturer wishes to have a method of distributing the product . . . so that no retailer can obtain a free ride at the expense of another. The optimal policy should tie the information and the physical commodity together so that the physical product cannot be sold by a retailer who does not provide the information . . . (1979, pp. 228–29)

My version of the free-rider problem occurs over time. Sales at Date $t$ contain, or generate, information about the perhaps physically identical product sold at Date

---

8. "For $m$ complementary commodities (components), a good is an appropriate combination of these $m$ constituents. The good is incomplete without all of these constituents. For $m$ substitute commodities, each of them contains the same good to some degree. Therefore, there is a set of prices, one for each of the $m$ substitute commodities, such that one dollar's worth of each of these $m$ commodities is essentially the same good." Telser (1979, p. 226).

$t + h$. However, the logical structure of the dynamic free-rider problem is isomorphic with that of the static one.

## 2. *Telser's Theory Extended*

In Telser's problem, consumers have all the information, together with the ability optimally to deploy information necessary for integrating the components of final products. Bundling is but an appurtenance of monopoly power. No consumer needs a bundle instead of a set of components. A learned, dextrous public comes to mind — indifferent in its choice between acquiring a fabricated mock-Georgian table and a kit of table components; or indifferent between buying stereo sets and their components; or, in the past, indifferent between buying completed model airplanes and kits containing balsa-wood parts, rubber bands, etc., together with arcane manuals. This public, at least in early stages of diffusion, cannot understand the manuals, so kits are worthless to it. Initially markets would exist only for bundles, *e.g.*, for completed model planes. The market would evolve towards Telser's (1960) model as information and comprehension deepened and expanded in the course of diffusion.

At the outset of many diffusion processes, only innovating companies may be able to assemble bundles. This likelihood will be less in more advanced economies where generalized conceptual and technical skills are widespread, perhaps enabling consumers or independent software houses quickly to develop assembly processes and indeed sometimes to devise advanced uses which may not have been foreseen by the company. Moving from early stages of diffusion, innovating companies often would want to compel bundling for reasons remote from those impelling Telser's monopolists. The companies would want to promote *enhancements* of simple programs imbedded in bundles initially supplied to the market.[9] As in the razor/razor blade parable of Allen 1938, companies might find it in their interest to provide free educational services (and even pay potential buyers of enhanced programs to take courses at spas) in order to avert stagnation of the market. And companies may take the strategic offensive to facilitate market development.

The logic of a compulsory bundling strategy resembles that of compulsory public education. It is expected that the students—or by analogy entrepreneurs and households—will absorb knowledge and acquire skills they cannot be expected to grasp before they attend school and in some cases even while they attend. In retrospect, most may be glad they were forced to go to school.

True, being forced to buy a package including the *right* to be educated is not the equivalent of marching to school under the baleful eye of a truant officer. The company could offer instruction to anyone on demand; but this would be an unfiltered process. Purchasers of the bundles are identified as high-potential prospects for costly training. Expenditures for educating non-purchasers mostly would be unprofitable, especially if hardware were available from others!

Although bundling of educational services is but quasi-compulsory, it suffices here because it increases, perhaps very substantially, the probability that susceptible users will become trained in the theory and praxis of the knowledge-based product.

9. Computers once were for the most part batch-processing units if not glorified accounting machines. A mass market for computers as agents of conceptual thinking took years to develop; this market now is accelerating.

### 3. *Spin-Off*

Diffusion provides spin-off opportunites in at least two directions. One concerns the equivalent of energy waves radiating towards other activities in the Beta economy (for example). The other concerns the company. Demand for enhancements of the programs *cum* hardware originally sold, together with that for new products, some not initially envisaged by the company, will increase. This result confirms Telser's conjecture:

> . . . the seller could then choose $p_i$ [*here the price of the hardware*] to maximize profits; and this would give him the maximum amount the customers [*here after being educated*] would be willing to pay. Observe that in effect every commodity is given away 'free' except for the first commodity. (1979, p. 227)

Ultimately the company will abandon bundling. Diffusion of knowledge about the products and their uses will become so complete that the market will be able to fabricate bundles more complex and varied than the company can hope to create. Indeed, in the end, the company may have no choice. Competition, intensified by expiration of patents, may preclude the company's having a choice of packaging. A profitable and successful diffusion process will give way to a regime of competition.

Successful diffusion is synergetic. Feedback from *external technical economies* generated by diffused products and processes will enhance the latter's productivity. Thus, it may be observed again that, unless profit-recapture possibliities are extended to encompass processes external to prime innovative products, knowledge-based economic progress may be blocked.

Still, due to the complementarity dominating much of the analysis, spin-off also enhances demand for related company products. Such enhancement inevitably results from intensification of activity on feedback channels. Marketing effort in the direct channel leads to spin-off, activating demand on a feedback channel that will be profitable for the innovating company. The economic system controlling the decisions of innovating companies is to some extent a closed loop.

### IV.  A DYNAMIC MACROECONOMIC ANALYSIS OF EFFECTS OF DIFFUSION OF KNOWLEDGE-BASED PRODUCTS

In this section, the effects of diffusion of knowledge-based products on the level and growth of per capita income in Beta economies are examined, together with effects on the balance of payments, domestic savings, etc.

Sections IV.A and IV.B are concerned with a variety of specific macro effects. Section IV.C offers a reminder that technology transfer leads to *disembodied technical progress*, equivalent to a persisting gentle rain of progress that cannot strictly be identified.

It has been demonstrated how a feedback channel propagating economic energy from the market back to the company leads to changes in company activity. There also would be propagation in other directions not amenable to revenue appropriation by the company. This analysis is presented in a standard diffusion scenario.

### A.  *The Standard Diffusion Scenario*

Once the feedback channel is activated and other induced creative activity gets under way, initial purchasers of patented products and processes may themselves become innovators. They may become the first to explore newly created sets of possibilities. "Energy waves" emitted by primary innovators will stimulate users of their products to develop ideas of their own. These ideas, comprising new uses of products or new operating processes, will be relayed back to primary innovators who will benefit from this feedback. Demand for skilled labor will be generated, permitting Beta workers profitably to upgrade their skills. Important productivity gains might follow.

Second, third, and later waves of technology transfer may unfold as increasingly more sophisticated economic agents in the Beta economy discover uses for innovating companies' products and processes the companies themselves may not have foreseen.

The first wave of technology transfer will be provoked by impulses supplied by innovating companies. Succeeding technology transfers may result from research stimulated by experience in host countries. Such "after shocks" stimulate *working* of patents.[10]

The "standard diffusion scenario" has been enriched by Gustav Ranis (1983) in his analysis of contrasting growth experiences in Latin America and East Asia:

> The fact that the more typical Latin American type economy tended to rely on its relatively abundant natural resources, plus foreign commercial bankers, . . . to finance a much more protected and costly development path. East Asia . . . [relied] on mobilizing its *human resources* (emphasis supplied)—first unskilled, then skilled—plus technology, by means of a more market-oriented, outward-looking strategy . . . (Ranis, 1983)

### 1. *Extension: The Balance of Payments Dimension*

It has been seen that impulses transmitted along the feedback channel induce demand for additional innovated techniques. A burgeoning middle class, a concomitant of accelerated economic development, causes a demand for sophisticated consumer products. Development may be blocked by balance-of-payments crises that can be averted if demands for sophisticated consumer goods can be satisfied by domestic production. A secondary technology transfer, rapid enough to foster sufficient import substitution to avert balance-of-payments pressures may be necessary if a Beta development process is to proceed.

The analysis thus shifts to a truly macroeconomic focus. The secondary technology transfers necessary to foster import substitution are not within the domain of activity of the innovating companies who may have triggered the macro growth process. Not even an immense multi-national company has the economic power to assure the profitability of its promoting secondary technology transfer sufficient to permit substitution for sophisticated imports.

### 2. *Protection of Intellectual Property*

Accelerated Beta growth induces many applications for licenses to use "Alpha" products. It may trigger direct entry by Alpha producers as it appears profitable for them to establish factories and marketing organizations in a booming Beta economy. But such entry, even if encouraged by a Beta government, would be discouraged by possible problems repatriating revenues earned in Beta because of either a depreciation of the Beta currency or a collapse of Beta currency convertibility. The consequence could be short required pay-off periods so that innovating guest companies would plan for declining rather than increasing receipts.

---

10. The author discusses working of patents in Burstein (1983). A central conclusion is that often the best evidence of "working" is the degree to which *imports* are stimulated and not that to which local manufacturing is stimulated.

The Beta authorities confront the following logic. The accelerator principle of investment requires that large, relatively short-term capital expenditure be made in order to generate a flow of sophisticated consumer goods. Necessary machinery must be imported from Alpha sources, which would appear as a spike in a curve plotting imports against time. The resulting deterioration of the current account must be financed by capital inflows. Consequent increased debt held by foreigners must be serviced over the life of the machinery by *subsequent* improvement in the Beta current account. As Ranis (1983) points out, one source of current-account improvement is continued expansion of knowledge-based industry in Beta. Once past the initial capital import requirement, the accelerator-principle-based strain on the balance of payments will disappear. As Ranis also demonstrates, failure of a capital-inflow episode to trigger a development process stimulating the subsequent Beta current account can be, and has been, disastrous.

### 3. *"Beta-ization" Requirements: Further Strain on the Balance of Payments*

Beta governments often require innovating companies penetrating Beta markets to incorporate Beta partners and agree to an early surrender of their technology to Beta entrepreneurs, or the Beta government. It has been seen that, even if such strictures do not prevent technology transfer from occurring, resulting higher internal discount rates can prevent optimal diffusion.

Even if mandated transfers of control to Beta entities are made at fair-market values,[11] the Beta strategy is self-defeating. Strains on the balance of payments become exacerbated, while at the same time capital inflows are discouraged. Domestic capital is squandered. It simply replaces foreign capital instead of financing local investment based on an influx of foreign capital.

### B. *Coda to Section IV.A*

Two rather intricate problems remain to be addressed. The first concerns the common practice of imposing minimum investment programs on "guest" entrepreneurs as a condition of Beta cooperation. The second concerns an implication of modern macro-consumption theory, that the very prospect of accelerated Beta development can lead to a lower saving rate so that still more capital must be attracted from abroad. To the extent that prospects of higher permanent income encourage acquisition of larger stocks of sources of consumer services (*i.e.*, consumer durables), there will be an increase in consumer expenditure (to be distinguished from consumption) imposing pressure on the balance of payments.

### 1. *Minimum Investment Requirements*

If a guest company must invest more capital in a project than it deems optimal, the resulting diminished profit opportunity may lead to project cancellation. The aggregate effect of such requirements may be reduced capital imports into Beta, and what is imported will be used sub-optimally.

Indeed, the better the policy works, the more harm it may do! Resulting over-investment will cause permanent strain on Beta resources. Either more resources

---

11. Such fair-market values are determined relative to a sub-optimal Beta policy. Resulting values might be far less than going-concern values based on uninhibited opportunity to deploy innovative technology and complementary devices.

must be dedicated to service of foreign debt or the balance of payments must collapse, doubtless with a spasmodically constrictive effect on foreign trade.

### 2. *Negative Effects on Savings Ratios*

Expectation of higher measured income in the future, or an increase in net wealth in general, leads to reduced saving from current measured income. (*Loci classici* include Friedman, 1957, Ando and Modigliani, 1963, and Modigliani and Brumberg, 1954.) This proposition, when combined with one from investment theory, implies additional pressure on the Beta balance-of-payments in certain important circumstances. If Beta consumers become convinced that the discounted value of net future receipts is rising, they logically increase consumption (reduce saving) now in order to distribute consumption streams evenly over their lifetimes. Firms, anticipating increasingly buoyant demand, will expand their plant and equipment (accelerating capital expenditures) in order to be better able to satisfy demand when it increases. Both forces work towards a shortfall of Beta sources of funds relative to credit demand. Both forces imply increased capital-inflow.

### C. *Disembodied Technical Progress*

> We start with *disembodied technical progress* which applies equally and alike to all resources of men and machines in current use . . . Such technical progress represents technical know-how falling like manna from heaven . . . We turn next to technical change in some variety or other of *embodied technical progress*. This applies . . . to certain branches of capital equipment . . . Machines of one vintage are different in kind from those of another; because of embodied technical progress, new machines are more productive than older [ones]. (R.G.D. Allen, 1967, pp. 236-237)

Disembodied technical progress obviously pertains to *unexplained growth of output*: "Only 40 per cent of the growth of output [in the United States] is explained by input growth." (Branson, 1979, p. 529). The large residual must be explained by technical improvement of human and non-human capital. The microeconomic analysis concentrated on embodied technical progress, but a macroeconomic analysis naturally assigns a major role to disembodied progress.

Disembodied progress is pure spin-off not directly capturable by innovating companies. The narrow calculations of innovating Alpha companies do not account for the extent to which company actions contribute to what will be identified as disembodied technical progress. Thus, it is important that, in their calculations, Beta authorities do so; they logically will widen the range of concessions offered to innovating guest companies. These concessions predominantly are in the form of grants of property rights to products and ideas of proved originality, together with important ancillary grants. These concessions reflect the need to enlarge the firm's product space to overcome its difficulty in appropriating external benefits of its innovative investment.

## V. CONCLUDING REMARKS

The modern theory of economic development takes an intensive, not an extensive, view of economic growth. In the modern theory, economic development consists of doing different things in different ways because of the effects of growing knowledge.

It has been found repeatedly that the proper concern of public policy in the field of knowledge-based products and processes is not the net revenues innovating companies might recover from their research-and-development expenditures, but rather the revenues *not* appropriable by them. The dog that does not bark is more important than the one that does. This paper has found that there is more reason to be concerned about too paltry grants of property rights in knowledge-based products than with the magnitude of quasi-rents to innovation.

The most interesting focus of the analysis concerns the difficulty of direct appropriation of social gain from innovation. In football metaphor, the innovator-passer, looking downfield, often finds his primary receivers covered and must pass off to secondary ones. Feasible diffusion of knowledge-based products often requires that the product-space allocated to the innovator by the grant of property rights be larger (contain more dimensions), perhaps much larger, than one confined to primary invention.

## REFERENCES

Allen, R.G.D., *Mathematical Analysis for Economists*, London: Macmillan, 1938.

_____, *Macro-Economic Theory: A Mathematical Approach*, St. Martin's Press, New York, 1967.

Ando, A. and Modigliani, F., "The Life Cycle Hypothesis of Saving: Aggregate Implications and Tests," *American Economic Review*, March, 1963, 53, 55–84.

Barzel, Yoram, "Optimal Timing of Inventions," *Review of Economics and Statistics*, February, 1968, 50, 348–55.

Branson, William H., *Macroeconomic Theory and Policy*, 2nd ed., Harper & Row, New York, 1979.

Burstein, M. L., "The Economics of Tie-In Sales," *Review of Economics and Statistics*, February, 1960a, 42, 68–73.

_____, "A Theory of Full-Line Forcing," *Northwestern University Law Review*, March–April, 1960b, 55, 62–95.

_____, "Review of W. S. Bowman," *Patent and Antitrust Law, Journal of Economic Literature*, December, 1973, 11, 1403–5.

_____, "Testimony before the Small Business Committee, Subcommittee on the General Oversight, U.S. House of Representatives, Sept. 21, 1981," published by the U.S. Government Printing Office, 1982.

_____, "An Economic Analysis of Effects of Protection of Intellectual Property on Economic Development: Public Interest and Private Profit," unpublished, 1983.

Coase, Ronald, "The Nature of the Firm," *Economica*, November, 1937, 4, 386–405.

Friedman, Milton, *A Theory of the Consumption Function*, Princeton University Press, Princeton, 1957.

Griliches, Zvi, "Hybrid Corn: An Exploration in the Economics of Technological Change", *Econometrica*, October, 1957, 25, 501–22.

Kitch, Edmund W., "The Nature and Function of the Patent System," *Journal of Law and Economics*, October, 1977, 20, 265–90.

_____, "Patents, Prospects, and Economic Surplus: A Reply," *Journal of Law and Economics*, April 1980, 23, 205–7.

Modigliani, F. and Brumberg, R.E., "Utility Analysis and the Consumption Function" in K.K. Kurihara (ed.), *Post Keynesian Economics*, Rutgers University Press, New Brunswick, 1954.

Pakes, A. and Schankerman, M., "The Rate of Obsolescence of Knowledge, Research Gestation Lags and the Private Rate of Return to Research Resources", Discussion Paper No. 659, Harvard Institute of Economic Research, Cambridge, 1978.

Ranis, Gustav, "For Latin American Economies, Lessons in Asia," *Wall Street Journal*, Oct. 12, 1983, 202, 34.

Sheffrin, Steven, *Rational Expectations*, Cambridge University Press, New York, 1982.

Smith, Douglas A. and McFetridge, Donald G., "Patents, Prospects and Economic Surplus: A Comment," *Journal of Law and Economics*, April 1980, 23, 197–203.

Telser, Lester G., "Why Should Manufacturers Want Fair Trade," *Journal of Law and Economics*, October 1960, 3, 86–105.

_____, *Economic Theory and the Core*, The University of Chicago Press, Chicago, 1978.

_____, "A Theory of Monopoly of Complementary Goods," *Journal of Business*, April 1979, 52, 211–30.

Trajtenberg, M., "The Effects of Regulation on the Diffusion of CT Scanners," National Bureau of Economic Research, Cambridge, 1982.

Trajtenberg, M. and Yitzhaki, S., "The Diffusion of Innovations: A Methodological Reappraisal," Harvard University & The National Bureau of Economic Research, unpublished, 1982.

Warren-Boulton, Frederick, *Vertical Control of Markets*, Ballinger Publishing Co., Cambridge, 1978.

Williamson, Oliver E., *Markets and Hierarchies: Analysis and Antitrust Implications—A Study in the Economics of Internal Organization*, The Free Press, Macmillan, New York, 1975.

# [8]

Excerpt from R. Culpan (ed.), *Multinational Strategic Alliances*, 123–65.

## Chapter 6

# Technology-Based Cross-Border Alliances

### Philippe Gugler
### John H. Dunning

## *INTRODUCTION*

Firms, particularly transnational corporations (TNCs), operate through a global network of related enterprises. One of the main characteristics of these networks is the creation, diffusion, and commercialization of technological innovations. Although the transnationalization of research and development (R&D) activities by TNCs is a comparatively new phenomenon, it is fast becoming an important one. Japanese companies, for example, have sent thousands of engineers to be trained at European and American universities and have set up R&D units in Europe (e.g., Sharp, Sony, Canon, and Hitachi in Ireland)[1] and in the United States (e.g., NEC in Princeton; Kobe Steel and Fujitsu in Silicon Valley). European and American companies have also set up innovatory centers overseas. For example, Dow, Corning, IBM, TI, DuPont, and W. R. Grace are actively pursuing basic research, and Kodak, Dow Chemical Pfizer, and Digital Equipment are actively pursuing applied R&D activities in their Japanese laboratories.[2] Ciba Geigy has innovatory capacity in the United States and Japan, as well as in Europe (e.g., France, Italy, Germany, and the Netherlands) and in some developing countries (e.g., Brazil, Mexico, and South Africa). In 1990, a Canadian multinational enterprise (MNE)–Northern Telecon–transferred part of its domestic R&D facilities to the southern U.S.

---

Figures 6.1 and 6.2 are reprinted from *Long Range Planning,* Volume 25(1), Philippe Gugler, "Building Transnational Alliance to Create Competitive Advantages, pp. 90-99, Copyright 1992, with permission from Pergamon Press Ltd, Headington Hill Hall, Oxford OX3 OBW, UK.

124        MULTINATIONAL STRATEGIC ALLIANCES

Some of these international R&D operations are realized through cooperative agreements. Over the past decade, technology-intensive industrial sectors have been affected by a wave of joint ventures and strategic alliances. Many of these involve large TNCs working together on different continents. By cooperating with foreign partners, firms may benefit from the sharing of risks and the pooling of assets, based on their allies' competitive advantages, and on the resource capabilities and markets of the nations in which they operate.

Cross-border strategic alliances, which differ from traditional collaborative arrangements in that they are concluded to advance the global strategy of the participating firms,[3] have increased dramatically in the past few years. For example, cooperative agreements concluded between U.S. and foreign firms are believed to outnumber fully owned foreign subsidiaries by a factor of at least four to one.[4] In the 1980s, U.S. firms formed over 2000 agreements with European corporations. A recent McKinsey study reveals that the number of U.S. international joint ventures (i.e., the creation of a new entity) established annually increased sixfold from 1976 to 1987.[5] In the European Commission (EC) as well, the number of cross-border coalitions also increased in the 1980s. Further details are set out in Table 6.1.

Tables 6.2 and 6.3 reveal that cooperative agreements have tended to be concentrated in high-technology industries (micro-electronics, aeronautics, new materials, biotechnologies) and in industrial sectors that are using more and more sophisticated technologies (e.g., robotics in the automobile sector). Thus, for example, over half of the firms in the fields of biotechnology and machine vision are engaged in strategic alliances,[6] and most of these agreements are R&D-related.

The data bases suggest that most alliances are concluded by large corporations that are headquartered in the U.S., Western Europe, or Japan. These corporations operate through a large web of formal and informal coalitions, most of which are in the advanced industrialized countries. These developments induce the formation of oligopolistic networks, with the major world producers at the hub of these networks. In the words of an Organization for Economic Cooperation and Development (OECD) report:

The Functional Dimension of Global Business Alliances          125

TABLE 6.1. Evolution of Cooperative Agreements Established Annually (1974-1989)

| | (a) | (b) | (c) | (d) | (e) | (f) | (g) |
|---|---|---|---|---|---|---|---|
| 1974 | | | 37 | | | | 169[1] |
| 1975 | 3 | | 14 | | | | |
| 1976 | 7 | | 16 | | | 31 | |
| 1977 | 7 | | 15 | | | | |
| 1978 | 7 | 2 | 14 | | | | |
| 1979 | 13 | 1 | 27 | | | | 317[2] |
| 1980 | 22 | 4 | 34 | 85 | | 94 | |
| 1981 | 28 | 22 | 40 | 169 | | | |
| 1982 | 23 | 19 | 35 | 197 | | | |
| 1983 | 39 | 16 | | 292 | 46 | | |
| 1984 | 66 | 42 | | 346 | 69 | | 1504[3] |
| 1985 | | | | 487 | 82 | | |
| 1986 | | | | 438 | 81 | | |
| 1987 | | | | | 90 | 180 | |
| 1988 | | | | | 111 | | |
| 1989 | | | | | 129 | | 2629[4] |

Sources :
(a) Alexis Jacquemin, Marleen Lammerant and Bernard Spinoit, Compétition européenne et coopération entre entreprises en matière de recherche-développement, Commission des Communautés européennes, Document, Luxembourg, 1986 : data on 212 cooperative agreements formed between 1978 and 1984 by at least one EC's firm.
(b) C.S. Haklisch, Technical Alliances in the Semiconductor Industry, Mimeo, New York University, 1986 : Cooperative agreements formed by the 41 major world's semiconductors producers.
(c) Karen J. Hladik, International Joint Ventures, Lexington Books, Lexington Mass., 1985 : U.S. International Joint Ventures created in high income countries between 1974 and 1982.
(d) G.C. Cainarca, M.G. Colombo, S. Mariotti, C. Ciborra, G. De Michelis, M.G. Losano, Tecnologie Dell'Informazione E Accordi Tra Imprese, Fondazione Adriano Olivetti, Edizioni di Comunità, Milano, 1989 : Arpa data base on 2014 agreements formed between 1980 and 1986 in the Information technologies sectors (semiconductors, computers and telecommunications)
(e) Commission's Reports on the EC's Competition Policy. See for example : Commission des Communautés européennes, Dix-huitième Rapport sur la politique de concurrence, Bruxelles-Luxembourg, 1989.
(f) Karen J. Hladik et Lawrence H. Linden, "Is an International Joint Venture in R&D For You ?", Research Technology Management, Vol. 32, No. 4, July-August 1989, page 12 : McKinsey Studies on U.S. International JVs created in 1976, 1980 and 1987.
(g) John Hagedoorn et Jos Schakenraad, Leading Companies and the Structure of Strategic Alliances in Core Technologies, MERIT, University of Limburg, 1990 : Cati data base on 9000 agreements formed until July 1989.
[1]: before 1974; [2]: 1975-1979; [3]: 1980-1984, [4]: 1985-1989.

TABLE 6.2. Distribution per sectors of U.S. International Joint Ventures created in 1980 and in 1987 (per cent)

| Sector | 1980 | 1987 | R&D JVs 1987 |
| --- | --- | --- | --- |
| Transport equipments | 10 | 12 | 14 |
| Computers | 3 | 8 | 14 |
| Semiconductor & optical disks | 8 | 8 | 18 |
| Other electric & electron. equip. | 4 | 7 | 4 |
| Pharmaceutical | 3 | 7 | 21 |
| Other chemicals | 19 | 21 | 14 |
| Instruments | 7 | 7 | 11 |
| Others | 46 | 30 | 4 |

Source :  Laren J. Hladik and Lawrence H. Linden, "Is an International Joint Venture in R&D For You ?", Research and Technology Management, Vol.32, No.4, July-August 1989, page 12.

126

TABLE 6.3. Distribution per sectors of JVs in the EC* (1984-1989)

| Sector | Number | Per cent |
|---|---|---|
| Food & beverages | 30 | 6,10 |
| Chemicals, rubber, glass and ceramics | 101 | 20,50 |
| Electric and electronic equipments | 102 | 20,70 |
| Instruments, machines | 60 | 12,15 |
| Information technologies equipments | 20 | 4,05 |
| Metal products and manufacturing | 48 | 9,75 |
| Transport equipments | 27 | 5,45 |
| Wood, furnitures, paper | 34 | 6,90 |
| Extractive industries | 15 | 3,05 |
| Textile, leather, shoes | 7 | 1,40 |
| Construction materials | 27 | 5,45 |
| Others | 22 | 4,45 |
| Total | 493 | 100,00 |

* JVs among firms : (i) from the same Member State; (ii) from more than one Member State; (iii) from Member States and outside States but with a significant effect on the EC market.

Source : Commission des Communautés européennes, **Dix-neuvième rapport sur la politique de concurrence**, Bruxelles-Luxembourg, 1990, page 332.

in the case of agreements between large firms in "world market" or global industries, the arrangements must be set in the
context of the mutual recognition and interdependence of decisions on the part of firms, which characterize concentrated
industries in which oligopoly prevails.[7]

Nevertheless, smaller companies are becoming more involved in
strategic alliances, particularly with their larger customers. According to the Arpa data base in Italy, asymmetrical agreements between
large and small enterprises are growing more rapidly than those
between large firms.[8] For example, in the computer sector, IBM is
collaborating with Microsoft to exploit its growing expertise in
software for desk-top computers. In the biotechnology sector, the
Swiss TNC Hoffmann La Roche has links with smaller firms such
as Genentech, Immunex/Ajimoto, Centocor, Cal Bio, Amgen, Ce-
tus, and Synergen.[9] Alliances among small- and medium-sized enterprises are also quite frequent, particularly in the EC, where these
operations are promoted by the Commission.

The causes, mechanisms, and consequences of these new organizational forms are still uncertain, since most cases of strategic
alliances are very recent. In this chapter, attention will be given to
these topics in order to evaluate the importance of R&D-related
alliances in the general context of the internationalization of technological activities. The discussion is divided into four sections. The
first examines the determinants of R&D alliances; the second deals
with the management of strategic alliances; the third studies the
mechanisms of R&D cooperative agreements; and the fourth contains some reflections on the potential impact of R&D alliances.

## THE ANALYTICAL FRAMEWORK
## OF THE DETERMINANTS OF STRATEGIC ALLIANCES
## AND SOME EMPIRICAL EVIDENCE

According to Bruce Kogut[10] the motivations for strategic alliances can be classified under three main headings: the enhancement of market power, the evasion of small number bargaining, and
the transference of organizational knowledge. A cost/benefit analysis of cooperative agreements may help to understand when, where,

why, and which types of firms cooperate. According to the main studies on strategic alliances, interfirm coalitions are intended to achieve the following benefits:

- Sharing of the cost of large investments needed for specific activities, such as R&D.
- Access to complementary resources (e.g., synergistic technologies).
- Acceleration of return on investments through a more rapid diffusion of the firm's assets.
- Spreading of risks.
- Efficiency creation through economies of scale, specialization, and/or rationalization.
- Co-opting competition.

Nevertheless, strategic alliances may also involve costs to one or the other of the participating partners. Foremost among these is the possibility that the other partner (who is also often a major competitor) may gain a *relative* strategic advantage.

A no less pertinent question is why other organizational forms may fail to achieve the same results as alliances at the same, or lower, net costs. Contractor and Lorange[11] present a comparison in terms of costs and benefits to answer this question. But one of the problems that still remains unsolved is why, and under what conditions, strategic alliances are likely to provide a more efficient form of operation than other modes (such as a complete integration or a market transaction).

The theories of games, oligopoly, the firm, international production, competitive strategies, marketing, and numerous theoretical developments may also help us to understand why firms are cooperating instead of pursuing a go-it-alone strategy or choosing a spot-market relation.

The theory of non-cooperative games may help to explain under what conditions cooperation may emerge and survive in a world of egoists without central authority (imposing the cooperative solution). Analysis of the prisoner's dilemma[12] and specific developments of this dilemma, such as the study of the tit-for-tat strategy,[13] shows that a non-cooperative agreement is possible when: (1) the game is repeated indefinitely; (2) the interests of the players are not in total conflict; and (3) the future is sufficiently important, relative to the present. Non-cooperative games theory also assumes that the partners to any coalition have similar needs and a willingness to share the risks of a particular venture and that reciprocity and the retaliation play an important strategic role.[14]

Cooperative games theory studies the negotiation tactics among the partners who have to decide the partition of the cooperative payoff. Under the development of model negotiation for two-person games[15] and n-person games,[16] it is possible to understand the rival interests among allies, particularly in the formation of interfirm coalitions. The interface between transaction costs (including negotiation costs) and the cooperative payoff varies according to: the relative power of the partners (conditioned by the relative position) when the firms don't cooperate; the maximum cooperative payoff available to each partner;[17] and the likely costs of the agreement relative, for example, to its duration.[18] The risk aversion of each partner to support a non-cooperative solution[19] determines all of these factors and under which conditions a potential partner will cooperate. Other developments, including studies of the characteristic function, the Shapley value,[20] and the bargaining set[21] are also useful in understanding the emergence of interfirm networks and the intra- and inter-network rival and cooperative interests. Thanks to these kinds of theoretical models, it is possible to evaluate complex situations and several kinds of joint-venture agreements, which strengthen all partners against outsiders while it weakens some partners vis-à-vis the others.

According to their main hypotheses, non-cooperative games (noncommunicability and unenforceable agreements) and cooperative games (communicability, enforceable agreements) are unable to fully cover the complex problems of strategic alliances whenever these occur in a market where rival behaviors coexist with cooperative behavior. In other words, where companies join forces in some areas, while pursuing go-it-alone strategies in others. New developments such as the almost non-cooperative games[22] and the almost cooperative games[23] simultaneously integrate rivalistic and cooperative behaviors into strategic models. Such models help (1) to evaluate strategic alliances in the context of the competitive incentives among allies and the competitive rivalry within industrial sectors[24] and (2) to provide a useful taxonomy of the transaction costs of the cooperative benefits related to the adhesion to specific coalitions.[25] Among other things, this approach also shows that a cooperative agreement is a constantly evolving bargain among the partners and that a long-term view is necessary for cooperation to emerge and survive in a non-zero-sum game situation.

The transaction cost problems, applied specifically to the firm, are studies in the theory of the firm and of the internalization. According to Ronald Coase, a firm will tend to expand until the cost of organizing an extra transaction becomes equal to the costs of acquiring this transaction on the open market.[26] This dichotomy between the market and the hierarchies, as organizational instruments, has been criticized in that it fails to acknowledge the fact that firms sometimes cooperate rather than compete with each other.[27] For example, G. B. Richardson suggests that complex networks of cooperation and association exist because of the need to coordinate closely complementary but dissimilar activities.[28] Likewise, David Teece argues that cooperative agreements may be explained by the need of firms to combine the use of different specific and co-specific assets to commercialize an innovation.[29] Thus, the comparison of transaction costs and governance costs is a possible way to understand why firms wish to cooperate with each other. They choose strategic alliances to minimize the costs of transaction and control, and to reduce risks. For example, Kogut writes:

> In summary, the critical dimension of a joint venture is its resolution of high levels of uncertainty over the behavior of the contracting parties when the assets of one or both parties are specialized to the transaction and the hazards of joint cooperation are outweighed by the higher production or acquisition costs of 100 percent ownership.[30]

Thus, R&D alliances may be concluded in order to avoid the control costs of an in-house development, a merger, or an acquisition (such as rigid structures, which prevent switching research capability, strategy, development orientation, etc.). The alliances may also be concluded to avoid high capital costs, investments, and risks as well as pure market transaction costs (such as costs induced by moral hazard and adverse selection).[31] Finally, the emergence and growth of generic technologies, while breaking down the traditional boundaries between industries, is creating additional pressures on firms to cooperate with their rivals in oligopolistic sectors. Indeed—and we are indebted to Francois Chesnais of the OECD for this point—future global competition may be as much between a collection or cluster of technology-related industrial or service firms as between firms according to the end products they produce.[32]

The above analysis, however, does not explain why firms choose to cooperate with *foreign*-based partners. The theory of international production argues that to invest abroad, firms must find it more profitable, or strategically worthwhile, to engage in foreign rather than domestic production. Similarly, cross-border strategic alliances may be presumed to offer *additional* advantages to the partnering firms than their intra-country counterparts. Comparative models developed, for example, by Hirsch,[33] Rugman,[34] and Buckley and Davies[35] are designed to compare the costs associated with the different organizational modes considered. The theory of transaction costs, applied to the internationalization of value-added activities, gives a good framework to: (1) compare the costs and benefits of foreign direct investment (FDI) and non-equity agreements;[36,37] (2) analyze "scale" and "linked" equity joint ventures;[38] (3) study firms' horizontal and collaborative joint ventures;[39] and (4) to explain vertical integration and collaborative agreements.[40]

However, it is questionable whether purely economic theories can adequately explain why firms cooperate and why they collaborate on an international level. Peter Buckley, for example, argues

> despite listing of these costs and classification (information costs, bargaining costs, enforcement costs, governance costs), nowhere do we find estimates of such costs. How significant are they in relation to transport costs, production costs, marketing distribution costs? Casual empiricism suggest that they are very high, and there are some wild estimates of the proportion of transaction costs in GDP. However, estimates are es-

sential if we are to move beyond heuristic models to concrete predictions about market configuration.[41]

Thus, many costs associated with the organizational forms compared are underestimated, as shown in a study on the transaction cost approach to the make-or-buy decisions in the automobile sector.[42] Furthermore, as noted by Farok Contractor, transaction cost minimization does not necessarily result in strategic optimization in the long run, or in profit maximization in the short run.[43] The static orientation of most of the transaction costs theoretical developments may explain this analytical weakness. Thus, as noted by Teece, transaction cost economics must be married to organizational decision theory if the dynamics of channel selection are to be properly appreciated and understood.[44]

The comparative models of choice of an organizational mode and the theory of transaction costs may be completed by John Dunning's eclectic paradigm of international production. This paradigm embraces the main vehicles of foreign involvement by enterprises (including several forms of strategic alliances) and suggests which route of exploitation is likely to be preferred.[45] The paradigm, also known as the OLI paradigm, suggests that it is the interaction between the level and pattern of a firm's ownership ("O") (or competitive advantages) and a country's location ("L") (or competitive advantages)–together with the strategy the firm chooses to adopt to organizing its cross-border activities in light of these advantages and those that arise from hierarchical control ("I" advantages)–that will determine its propensity to engage in FDI or some other form of international economic involvement.[46] For example, Dunning explains the more pronounced propensity by Japanese firms to engage in cross-border joint ventures relative to U.S. firms by observing that

> . . . their ownership advantages are of a kind that makes full internalization an inappropriate way to organize foreign manufacturing. They need the intangible assets of indigenous firms (e.g., better access, to local inputs, knowledge of product markets, law and customs, and how to negotiate with governments), while the technology they transfer is less idiosyncratic, more mature, and more readily marketable than that of US MNEs.[47]

The dynamic application of the eclectic paradigm of international production,[48] which takes into account the contribution of the literature on global strategic management,[49] offers a powerful framework for analyzing the different types of cross-border value-added activities.[50] Figure 6.1 presents a simple model, showing the contribution of the OLI parameters as explanatory tools of strategic alliances, particularly in the R&D field. The figure sets out the major factors that help to explain the emergence and growth of alliances in the 1980s and 1990s, according to the OLI configuration of the participating firms. In fact, new developments such as globalization of business, the development of cross-border organizational networks, and the emergence of core technologies (or technological systems) affect the way in which firms might create, maintain, and improve the "O" advantages and, consequently, the "L" and "I" parameters.

The eclectic paradigm may also be widened to take into account some new theoretical contributions such as the study of Charles Hill, Peter Hwang, and Chan Kim, which considers the major categories of variables influencing the entry mode choice of a firm.[51]

The distinction between structural and transactional market failures, identified by the eclectic paradigm, is particularly useful in analyzing the determinants of cooperation in specific sectors. Weijian Shan considers cooperative agreements to be the product of a double failure of the market. The first is the transactional difficulty in transferring the services of specialized assets. The second embraces the transactional problem of internalizing the functions of both production and transaction.[52]

At the same time, Shan and others also emphasize the role played by structural market failures such as entry barriers, government intervention, and recent changes in the world economy (such as globalization and unification of markets, increased investment risk and uncertainty, technological breakthroughs, convergence of system products, and faster rate of development [and obsolescence] of new products and processes). An overview of the main factors influencing the emergence of R&D alliances—technological innovation, convergence of technologies, and globalization of markets—seems relevant in understanding this issue. These three factors are related, but for analytical purposes, they are presented separately.

*The Functional Dimension of Global Business Alliances* 135

FIGURE 6.1. The Value of Strategic Alliances

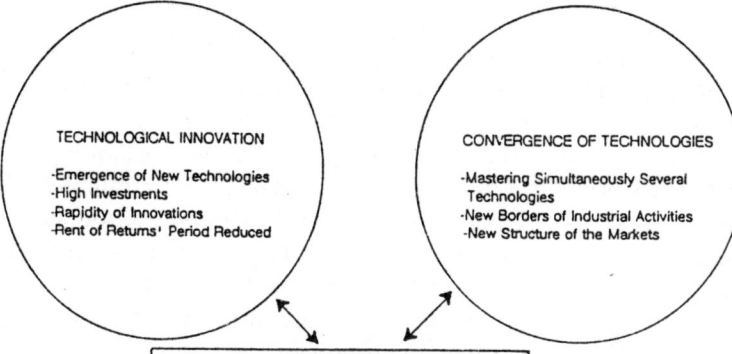

TECHNOLOGICAL INNOVATION

-Emergence of New Technologies
-High Investments
-Rapidity of Innovations
-Rent of Returns' Period Reduced

CONVERGENCE OF TECHNOLOGIES

-Mastering Simultaneously Several
 Technologies
-New Borders of Industrial Activities
-New Structure of the Markets

ADVANTAGES

OWNERSHIP

-Rapidity of New O-Advantages' Development.
-Rapidity of Existing O-Advantages' Exploitation.
-Higher Flexibility.
-O-Advantages Based on the Combination of
 Complementary but non Similar Assets.
-O-Advantages Based on the Supply of a Complete
 Range of Systemic and Compatible Products.
-O-Advantages Based on Products with a Dominant
 Standard.

LOCATION

-Access to Complementary Assets Based on the
 Nations' Competitive Advantages. Originated in
 the Partners' Home-countries.
-Access to the Main Worlds' Markets for the
 Inputs and Outputs when a Go-it-alone Solution
 is not Possible Because of the High Capacities
 Needed to Exploit them Alone.

INTERNALIZATION

-Sharing of the Costs and Spreading of the
 Risks in High Uncertainty Situations.
-Transaction Costs Less Important Because of
 the Technological Diffusion Rapidity.
-Benefits from Scales Economies.
-The Launching of Projects with High Sunk Costs
-New Oligopolistic Reactions to Replace Traditional
 Oligopolistics Strategies which are Inadequate
 Because of the Concentration, Unstability
 and Asymmetry of Oligolies.

GLOBALIZATION

-Concentrated, Asymmetrical and
 Unstable Oligopoly
-World's Products Adapted to
 Local Demand
-Systems Product
-Products Based on World's
 Accepted Standards

## Technological Innovation

The development and emergence of new technologies (defined as an industry paradigm shift) can trigger changes in the existing market structures.[53] A study conducted some years ago by International Management revealed that keeping pace with new technologies was the biggest problem affecting the performance of companies from 20 countries.[54] As earlier underlined by Raymond Vernon, competitive threat and/or profit opportunities are the main forces behind oligopolistic actions and reactions.[55] New technologies induce both a response by competitors and offer new profit opportunities. Consequently, they seem to be important in the choice of competitive strategies.

The commercialization of emerging technologies (such as microelectronics, biotechnologies, new materials) is characterized by more intensive competition in innovatory activities (first-mover advantage). Once an innovation is made, the innovator needs to commercialize it quickly, because of the erosion of its monopoly position (shorter product life cycles, diffusion of technology, new innovations, etc.) and the need to promote standards. In this context, strategic alliances may help to accelerate the marketing process (first-mover advantage) and provide the innovator with the maximum rent on his innovation.[56]

Furthermore, as uncertainty about the success of the commercialization of innovations is high, the development and production costs of technology are huge—for example, £2 billion for a new global car model and £500 million for a new mainframe computer.[57] Designing and manufacturing the new four-megabit dynamic random access memory (DRAM) cost $2 billion, while the current world market for all types of DRAMs is worth less than $10 billion in annual sales.[58] Thus, firms have to increase their investments, particularly in R&D.

In the electronics sector, the top-100 publicly held U.S. companies spent $18.2 billion on R&D in 1988, a 15.7% increase from the $15.7 billion logged in 1987.[59] Even large firms cannot muster the necessary financial resources and take the risks inherent in such huge R&D investments.[60] For example, a *Financial Times* survey in 1990 indicated that even with the growing concentration of European producers, the three largest chip-makers, Philips, Siemens, and SGS-Thomson Group, are still far from generating the return needed to fund the larger R&D investments necessary to remain competitive against their U.S. or Japanese rivals.[61] Thus, strategic alliances may be a relevant response to these challenges. For example, in the aeronautics sector, General Electric is seeking to spread the $1.2 billion to $2 billion cost of developing the world's largest commercial jet engine by negotiating risk-sharing partnerships with European and Japanese engine producers.[62]

## Convergence of Technologies

Innovation increasingly depends on combining incremental advances across a wide range of disciplines. The convergence of technologies, particularly of new technologies, is one of the major new trends in technological developments since the beginning of the 1980s.[63] This convergence may be observed within and between technological clusters. Thus, different technologies have to be mastered at the same time and, consequently, R&D expenditures for developing the latest generation of core products are higher.

Moreover, new core technologies tend to be more generic (i.e., more widespread in their use) than those that they replaced. This fact provides new opportunities for strategic alliances to benefit from synergistic technologies.[64] For example, between 1984 and 1989, Olivetti (computers and office automation) and AT&T (telecommunications) formed an agreement that was intended to exploit the maximum synergies from their complementary assets. AT&T is also collaborating with IBM for the development of network systems. The increasing use of optoelectronics in telecommunications has made it necessary for the major producing firms to master know-how and technologies in the field of optical fibers. Thus, Corning Glass, a leading firm in this field, is now at the center of an international alliance network.[65] Ciba Geigy, a leading Swiss chemical company, is cooperating with the U.S. firm Olin for the development of special chemicals used in the production of high-performance semi-conductors, and with the Chinon Corporation of California on research into human immune deficiency.

The robotics industry is also based on the integration of a multitude of sectors (e.g., computers, machine tools, software, etc.).

Firms have to acquire various knowledge and, thus, make large investments in R&D in order to maintain their global competitive positions. As noted by the Economic Commission for Europe/ United Nations Center on Transnational Corporations (ECE/ UNCTC) Joint Unit on TNCs, the main robotic producers have been driven to achieve economies of scale through collaborative agreements.[66]

## *Globalization*

One of the main forces behind strategic alliances is the need to compete internationally. As underlined by Fortune,

> for many major companies, going global is a matter of surviv- al, and it means radically changing the way they work.[67]

Globalization changes both the motivation for and the pattern of foreign direct investments,[68] and it creates a need for more flexible production and marketing systems and new forms of organization.

Firms trying to position themselves as global players face several problems, such as the cost of building a simultaneous presence in several product areas and foreign markets. For example, the limited size of the U.K. market, and the problems Apricot encountered in selling hardware abroad, convinced the company that it was neces- sary to have a partner who could offer both market access through new marketing channels and access to technology that Apricot felt unable to develop on its own.[69]

Globalization of markets mainly takes place in an oligopolistic competitive structure, characterized by high concentration and in- stability, and sometimes of asymmetry as well (e.g., in the computer sector, with IBM as the major producer). These oligopolistic fea- tures induce new kinds of actions and reactions. For example, in the case of market concentration, strategic alliances may be thought of as a new kind of "follow the leader" strategy, when the costs of hierarchical entry into a new market are too high.[70] Asymmetry may involve collaborative strategies such as technology leveraging and the protection of market positions.[71] For example, WordPerfect and Lotus signed a technological alliance with each other, in part because they were worried about Microsoft, which is cooperating with IBM.[72] Oligopolistic instability may also induce strategic alliances to meet the challenge of new competition; the major collaborative programs launched in the EC and U.S. are also aimed at competing against Asian producers.

Global markets may also be characterized by the convergence of consumer needs and preferences.[73] As noted by Kenichi Ohmae,

> for a firm with a good grasp of the shared needs of 630 million people and the courage to launch a product in the Triad market, it is essential to have networks that can deliver a newly developed product nearly simultaneously to scores of different points on the globe. . . . Presently, the most pragmatic and productive method of expanding a product's market is the formation of a consortia alliance.[74]

The emergence of product systems is another feature of a global market that may also increase the attraction of cooperative operations.[75] For example, in 1991, Silicon Graphic signed two R&D agreements—one with Microsoft and the other with the computer producer Compaq. These deals have helped create a powerful technological core in the computer industry, because they have linked together three leaders: one in software, the second in exploitation systems, and the third in computers.

The promotion of worldwide standards is another important factor to be considered. For example, European firms like Philips are cooperating for the promotion of a European standard for European high-definition television (HDTV). In May 1990, Philips and Thomson-CSF signed an agreement that involved joint R&D on integrated circuits, flat screens, liquid crystal displays, and broadcasting equipment. (This agreement aims to improve new standards.[76]) In 1991, AT&T and IBM signed an agreement to make their competing systems compatible.[77]

The completion of the European Single Market by the end of 1992 will also aid the path toward globalization of value-added activities. Europe 1992—involving a market of 340 million people—will modify the OLI parameters of many MNEs or prospective MNEs. For example, it will no longer be necessary for a firm to be located in every EC country to avoid non-tariff barriers. As a result, some kinds of intra-EC defensive investment may well be replaced by increased intra-EC exports. At the same time, non-EC firms will increasingly seek to establish a presence in the EC to counteract the possible emergence of "Fortress Europe"; harmonization of regimentations and the introduction of new legal structures (e.g., antitrust policy) are likely to influence the form of that presence. Confronted with the new conditions of the Single Market, firms are currently reconsidering their corporate strategies to create or sustain their "O" specific advantages. In this respect, strategic alliances are a means by which the exploitation of the new opportunities and challenges—including those of penetrating world markets—follow from the removal of intra-EC barriers.[78]

R&D strategic alliances have, then, to be studied in the context of the technology strategies that are concerned with exploiting, developing, and maintaining the sum of the company's knowledge and abilities.[79,80] The regionalization of market globalization frequently creates a need for the reorganization and restructuring of value-added activities, which involve interfirm links. The rising costs and uncertainty of R&D call for the sharing and pooling of risks. Convergence of technologies drives firms to seek quick access to know-how and capabilities they do not possess. All of these factors pose a threat to hierarchical growth and control. They are new parameters that need to be added to the explanations of a firm's expansion limits that have been suggested by several scholars, including Penrose,[81] Robinson,[82] and Chandler.[83] They require a new approach to the internationalization of production, since it may be that only by strategic alliances can firms properly create, exploit, and sustain their particular competitive advantages.


## MANAGING STRATEGIC ALLIANCES


Strategic alliances may provide unique opportunities to share the assets and capabilities of a wide web of partners, including customers, suppliers, competitors, distributors, universities, etc. But cooperative agreements are not risk-free. Howard Perlmutter and David Heenan, for example, found that in a number of U.S.-Japanese alliances,

Japanese colleagues took advantage of valuable US technology and marketing know-how only to discard their American
partners.[84]

Robert Reich and Eric Mankin assert that the Japanese partners
of U.S. firms keep the higher-paying, higher value-added jobs in
Japan and gain the project-engineering and production-process
skills that underlie competitive positions, whereas their American
partners are losing their competitive strength.[85] History has shown
that cultural differences; lack of agreement about the objectives of
the partnership, and/or of the right managerial handling of it; poor
communications; and partner opportunism have been among the
frequent causes of failure of collaborative arrangements. By cooperating, a firm may have access to some information without paying
the market price for it; yet it may use this knowledge to out-com-
pete its partner.

Such risks are likely to be especially high in R&D alliances. For
example, Daimler, which is cooperating with General Electric, concluded an alliance–through its MTU subsidiary–in 1990 with UTC's
Pratt and Whitney division. Immediately after the announcement of
this agreement, General Electric accused Daimler of fraud and mis-
appropriation of trade secrets by cooperating with UTC, one of
GE's competitors in the aero-engine market. GE alleged that this
alliance broke its agreement with MTU to cooperate in developing
the next generation of high-thrust engines. It also asserted that it had
provided MTU with comprehensive business and technological information, which would help a competitor develop an alternative to
the GE90 engine. Pratt and Whitney argued that the agreement with
MTU to develop a rival to the GE90 was a logical extension of the
close collaboration that had developed between the two firms over
many years.[86]

This kind of conflict is not unusual, and it seems to be becoming
increasingly frequent with the development of inter-organizational
networks. Firms often cooperate with other firms in one network,
while competing with them in other networks. For example, in
1974, Snecma (France) signed an agreement with GE for the production of a new type of engine, even though UTC's Pratt and
Whitney division held 10% of the Snecma stock. Aeritalia (Boe-

ing's partner) and Aerospatiale (Airbus industries' member) are cooperating in the development, production, and marketing of the commercial aircraft ATR. Boeing's Japanese partners (for the production of the 767) are now considering an alliance with McDonnell Douglas.

The case of GE-Daimler is an example of the kind of conflict that may occur between partners of a technologically-based strategic alliance. In his study of a group of collaborative ventures involving U.S. and non-U.S. firms, David Mowery concluded that the way in which technological development was managed by the partner firms was critical to their success or failure.[87]

In order to minimize the risk of one company using its partner's unique technological advantages to that partner's own disadvantage, firms have adopted various protective devices, five of which we will now briefly describe.

## Restrictive and Exclusivity Clauses

According to Jordan Lewis,

> one useful tactic, when law permits, is to agree to limit undesired market entry by a partner, an alliance, or product of the alliance.[88]

In 1986, Texas Instruments concluded an agreement with Hitachi to develop a 16-megabit dynamic RAM chip. The agreement precisely catalogued the intellectual property that belonged to each company. It was also agreed that whatever technological advances stemmed from the alliance would be jointly owned, but would not be shared with outsiders without the express permission of each partner.[89] To give another example, in January 1990, Siemens signed an agreement with IBM to jointly develop a chip capable of storing 64-million bits of information. One clause of this agreement prohibited either company from cooperating with any Japanese company in memory-chip development.[90]

In some cases, of course, such restrictions run counter to national or regional (EC) anti-trust provisions. Lewis quotes the interesting example of General Electric and Rolls Royce, which in 1989 formed an alliance to produce a pair of each other's jet engines.

Each firm also obtained access to the other's markets. The deal came apart when Rolls modified one of its engines to compete with the GE engine they shared. Beyond that point, further cooperation became impossible, since Rolls would have had access to sensitive GE sales information. While there was some feeling within GE that Rolls was obliged *not* to compete with GE, there was no denying that Rolls had an expected sales opportunity that created a strong incentive to move on its own. Anti-trust constraints had precluded a formal agreement on this.[91]

### Limitations Placed on Technology Transfer

Information is always pooled when firms cooperate. Yet, as noted by Hamel, Doz, and Prahalad, companies must carefully select the skills and technologies they pass on to their partners; they should also develop safeguards against unintended informal transfers of information.[92] Firms may also be tempted to minimize the transfer of technology to avoid a disclosure of their core know-how. This may be done, for example, by sharing only the *results* of applying the product or process technology. In the case of the consortia International Aero Engines (which includes Pratt and Whitney, Rolls Royce, Fiat, MTU, Ishigawajima-Harima Heavy Industries, Kawasaki Heavy Industries, and Mitsubishi Heavy Industries), Pratt and Whitney and Rolls Royce minimized the contribution of their own cutting-edge technology by designing the engine in modular form and by assigning the development of different modules to different partners.[93]

### The Division of Control and Responsibilities in Collaborative Ventures

The division of control is also an important issue that needs to be resolved prior to the commencement of any alliances. The partners must also find an organizational solution to minimize the uncertainties in their relationship. Some legal forms may ensure more policy control than others. Thus, collaborating firms may choose legal forms through which (1) the partners are more able to exercise significant control over the other's relevant activities and (2) "free riding" behaviors are avoided and the alliance is stabilized through the creation of mutual hostage positions, whereby all the parties stand to lose if the agreement is breached. Each kind of organizational mode implies a different level or kind of governance, which has to be optimized, depending on the resource commitments and the way in which risks are shared.

## Efficient Alliance Planning

Firms entering into alliances will wish to be clear of the terms of the agreement and to be able to monitor its consequences. In planning an alliance, each partner needs to form a precise picture of the agreement's implications (in terms of rights and duties) for each partner and to use this as a guide for partner choice and alliance design. The clarity of goals is vital. According to Robert Lynch, ambiguous goals, fuzzy directions, and uncoordinated activities are the primary causes of failure in cooperative ventures.[94] No issue of concern to either party should be taken on trust when it can be reasonably formalized. At the same time, trust and forbearance are critical ingredients of any successful alliance.

However, while it may be easier to cooperate when partner's non-competitive interests and partners are far apart, this does not mean that collaboration among competitors is necessarily an undesirable strategy. Alliances between rivals may be profitable for each partner, but the conditions to succeed (as identified by games theory, for example) have to be respected.

## Selection of Alliances Based on Trust, Commitment, and Compatibility

This selection could be a successful way to avoid insurmountable conflicts of interest among the partners. As one Chief Executive Officer has noted:

> you've got to be sure you're working with earnest and ethical people who aren't trying to undermine your company. Usually, a partner will have access to your trade secrets. He might attempt to complete a few projects, learn what you do, then exclude you from a future deal.[95]

Firms may also be more cautious of their potential partners' stability. This issue is also important to bolster trust and commitment. For example, the president of a Japanese company, complaining about its American partner (which had gone through three ownership changes over the preceding decade) remarked that "we never know who we are dealing with."[96] The purchase of MTU by Daimler may have also contributed to the destabilization of MTU's collaborative ventures. As suggested by the games theory, successful cooperation develops only with efforts over time.

## THE MECHANISMS OF R&D ALLIANCES: A NETWORK APPROACH

### *The Various Kinds of Alliances*

The literature identifies several different kinds of cooperative ventures. These include formal and informal alliances; vertical, horizontal, and conglomerative links; equity and non-equity agreements; production R&D, marketing, supply, or multiple agreements; and national, regional, or transnational alliances.

R&D alliances may be classified under four main headings. These are:

- University-located R&D strategic alliances that involve more than one industrial firm: the Semi-conductor Research Corporation (SEMTEC) in the U.S. falls into this category.
- Private strategic alliances negotiated and organized without the intervention of government.
- Interfirm agreements organized through governmental agreements (e.g., European Spatial Agency, and Airbus).
- National or international collaborative programs such as ESPRIT in the EC, Eureka in Europe, ICOT in Japan, Alvey in the U.K., etc.

As seen previously, R&D strategic alliances may involve the exchange and pooling of existing technology and/or the development of new technologies. For example, Hitachi is selling Texas

Instruments (TI) the secrets of how to stack semi-conductors on a single silicon chip, in exchange for TI's expertise in software.[97] In 1990, Siemens and IBM concluded a joint venture to develop a new generation of chips,[98] while AT&T and Tandem Computers agreed to jointly develop and market computer systems that combine Tandem's "fault tolerant" designs and AT&T's Unix operating system software.[99] Under the terms of another alliance between AT&T and NEC (concluded in 1990), the American company has the right to market, design, and produce chips licensed by NEC. In return, the Japanese firm will receive computer-aided design tools that AT&T has developed.[100] In the computer software sector, WordPerfect and Lotus have been working together since 1988 to develop a common interface for the next generation of their products.[101] In the aeronautics sector, Rolls-Royce (U.K.) and Snecma (France) signed a new cooperation agreement in 1989 to jointly develop a new-generation supersonic engine.[102]

However, in the majority of R&D-related agreements, technology is transferred in exchange for "something else," such as access to new markets or to financial resources. Coalitions involving the use of corporate-venture (CV) capital are frequent in some industrial sectors, such as biotechnologies. As noted by S. Mariotti and E. Ricotta,

> for a large company, CVC is an additional investment and in some cases, it is particularly well suited to take advantage of the continuous flow of new technologies produced by small and medium-sized companies. Through the CVC, a company can appraise its interest in a business in formation in "real time." It is the flexible financial activities of the small innovative units, which remain completely autonomous.[103]

In other cases, each partner contributes a different kind of value-added activity. For example, in 1990, AT&T and Mitsubishi signed an agreement covering technology sharing, worldwide marketing, and manufacturing of static random-access memory (SRAM) chips. Through this agreement, AT&T will have the right to manufacture and market SRAM chips designed by its Japanese partner—including current and future generations of products.[104] In April 1991, Toshiba and General Electric announced a wide-ranging alliance for the joint development and marketing of home appliances and the establishment of two joint ventures. In participating in this agreement, Toshiba is hoping to prevent, or bypass, any trade restrictions resulting from a possible surge in protectionism and anti-Japanese sentiments.[105]

Some R&D agreements, which are initially formed to pool technological assets and capabilities, are later extended to the manufacturing and/or marketing stages of production. For example, in 1990, British Aerospace and Aerospatiale (France) agreed to undertake a five-year feasibility study on a supersonic commercial airplane to replace the Concorde. The decision on whether to launch a $10 billion production program is to be made in 1995, based on the results of the study.[106] A questionnaire related to the strategies of 750 U.S. electronics firms showed that interfirm manufacturing and technology agreements will probably grow in the first half of 1990, while marketing alliances may decline. Some details are set out in Table 6.4

R&D agreements have also been formed as part of governmental collaborative programs. For example, more than 300 companies are involved in the ESPRIT program launched in 1984 by the EC. Around 1600 firms (99% of them are European) participate in the Eureka Projects (around 300 projects). Previously closed to "foreign" firms, national and inter-governmental collaborative programs are becoming more and more open to them. For example, AT&T and IBM participate in some EC programs through their affiliates established in Europe. In Japan, foreign companies have been invited to participate in a $195 million governmental project to develop jet engines for supersonic passenger aircraft.[107]

Small and medium-sized companies are also involved in several collaborative programs. For example, firms employing up to 500 people participated in 49% of the first round of the BRITE program and in 60% of the second round.[108] Nevertheless, EC-related programs represent only 4% of the private and public R&D investments in Europe.[109]

## The Dynamics of Networks

Firms are often involved in both private and public cooperative ventures within complex networks, which, like consortia, may have vast potential for affecting entire industries.

TABLE 6.4. Major types of Agreements formed by 750 U.S. Electronics firms in 1990 and expected for 1990-1995 (percentage of total respondent)

| Types of agreements | 1990<br>(A) | 1990-1995<br>(B) | (A)-(B) |
|---|---|---|---|
| Marketing agreements | 67 | 61 | - 9 |
| Providing technology licensing | 45 | 54 | + 20 |
| Research contracts | 38 | 43 | + 13 |
| Receiving technology licensing | 36 | 43 | + 19 |
| Receiving equity | 34 | 42 | + 24 |
| Manufacturing agreements | 33 | 44 | + 33 |
| Providing equity | 14 | 28 | +100 |

Source: Electronic Business, March 1990, page 58.

In these networks, R&D links involve complex relations based on the coexistence of cooperative and competitive interests. Such networks have to be studied within the general context of (1) inter-firm networks, which are both stable and changing and (2) the business transaction among firms that generally takes place within the framework of established relationships.[110] While these networks portray the interaction of the participants, they also provide the framework within which the interactions take place. Thus, any analysis of them must include the social environment as well as the intra- and inter-organizational interactions.[111]

According to K. S. Cook and R. M. Emerson,[112] an alliance may be defined as a "set of two or more connected exchange relations." Interfirm networks are articulated by a time structure, a power structure, an interest structure, and a capacity structure. The capacity structure determines the synergistic surplus that results from the web of connections.[113] As shown in games theory, the payoff for cooperation by any one firm in the network depends on several factors, including the "status quo" position (which means the firm's payoff when it doesn't cooperate). In most cases, the relationship between the assets controlled by the participating firms sets the basis for, and interaction within, a network.[114]

Industrial markets are, in fact, complex systems of formal and informal relations between economic, political, and social agents. Alliance networks are a particular kind of inter-organizational network. As observed by Lars Engwall and Jan Johanson,[115] in every exchange relationship within a network, there is a potential conflict between the participating firms over both the distribution of economic rent earned by the network and the course of its future development. At the same time, there are powerful forces making for cooperation and mutual forbearance.

Several kinds of industrial networks may be identified. These include intrafirm networks, non-cooperative interfirm networks, and cooperative interfirm networks. The relative importance of each kind of network depends on the country, industry, and firm-specific characteristics. For example, more than half of Corning Glass's profits are from its 23 joint ventures.[116] In 1988, 60% of Aerospatiale's sales were derived from operations realized in cooperation.[117] According to a recent survey, Japanese firms obtain 5%

of their supplies through non-cooperative transactions, 40% from
their internal networks, and about 55% from cooperative networks.
In contrast, only 6% of supplies of U.S. firms come from alliance
networks in the United States.[118]

In many industrial sectors (aeronautic, semiconductor, telecommunications, computers, semi-conductors, etc.), it is possible to
identify hierarchical galaxies, in which a pivotal group of firms are
surrounded by satellite partners.[119] For example, as illustrated in
Figure 6.2, the major world producers of semi-conductors have a
focal position in a complex web of alliances. At the same time, new
coalitions may induce new market structures. For example, Figure
6.3 reveals that in the international telecommunications sector, the
market is dominated by a limited number of interfirm clusters, each
of which comprises a network of affiliated or allied firms. These
companies are joined around systemic technological clusters and/or
technological trajectories. The result is the emergence of a network
of international oligopolists who perceive that their individual interests are best served by some degree of technological collaboration
with each other.

Alliance networks originate from a complex market structure in
which it is more and more difficult to discern rivalry from cooperation. By participating in several programs, one firm can be active in
various forms of strategic alliances, each involved with its own
cluster of partners, which often overlap.[120] Thus, a firm may cooperate with some partners in a specific network, and compete with
them in another.

## Networks and Innovation

As has been noted by David Teece, innovation demands complex
interactions and de facto integration among a multiplicity of organizational units.[121] One of the major effects of the emergence of
transnational alliance networks is their role in technological innovation. Hakan Hakansson has argued that an innovation should be
seen not as the product of only one actor but as the result of an
interplay between two or more actors; in other words, as a product
of a "network" of actors.[122] For example, in the semi-conductor
sector, new generations of chips have been developed through al-

FIGURE 6.2. Alliances Networks in the Semiconductor Industry (1990)

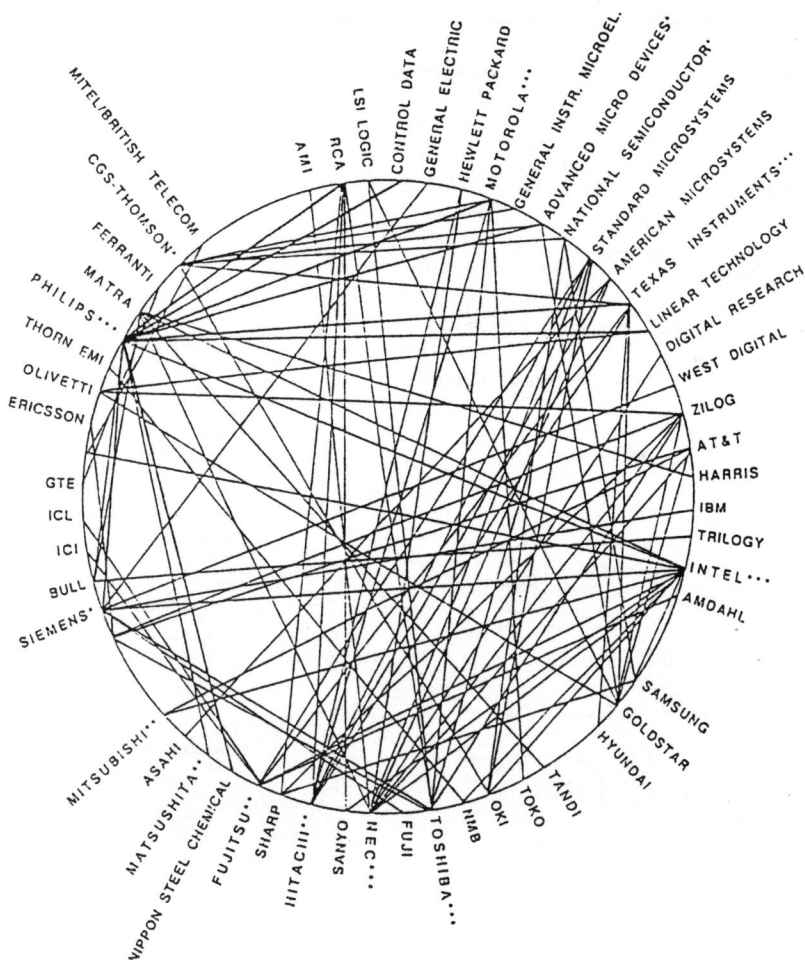

FIGURE 6.3. Interfirms Clusters and Networks in the Telecommunication Industry

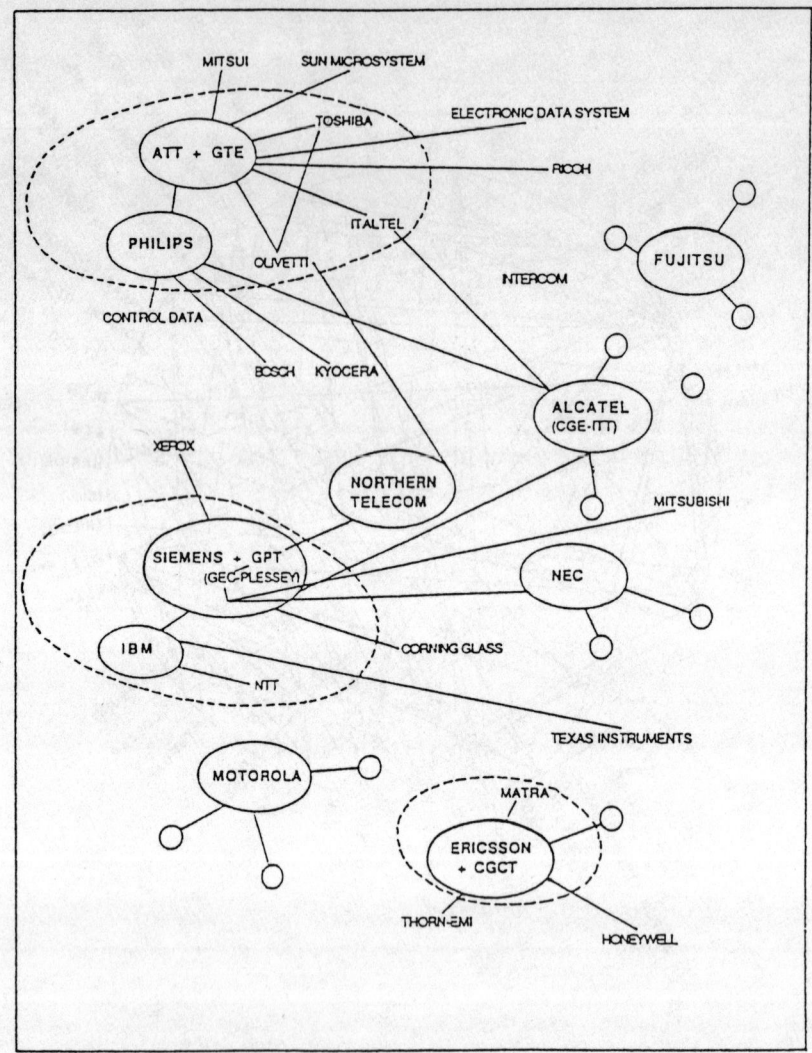

Source: Philippe Gugler, *Les alliances stratégiques transnationales*, Editions Universitaries Fribourg Suisse, Fribourg, 1991.

liances between Siemens and IBM, Siemens and Philips, Motorola and Toshiba, and Hitachi and Texas Instruments.

Formal and informal alliance networks may not only lead to more innovations, they may also help to diffuse existing technologies, know-how, and capabilities. In particular, linkages between firms belonging to different national collaborative programs–such as the European Eureka, Esprit, Race, or Jessi programs; the U.S. Sematech programs; and the Japanese VLSI and ICOT programs– may help promote an inter-network technological exchange and induce a greater synergistic surplus.

Nevertheless, interfirm agreements do not always induce new innovations or major technological transfers. In fact, there are some risks of this type of collaboration, the purpose of which is to delay or frustrate new innovations or to slow down the rate of technology diffusion.[123] Furthermore, knowledge sharing among partners may be limited because the firms do not wish to reveal too much of their cutting-edge technologies to potential competitors. For example, in the aeronautics sector, in several cooperative agreements, transfers of technology are minimized by assigning the development of different modules to different participants.[124] As a result, it took several years to negotiate how technology would be transferred between General Dynamics and Mitsubishi on the FSX fighter plane, in order to minimize technology transfer from the U.S. to Japan.[125]

In Japan, companies taking part in national R&D programs have been known to keep information about their own technologies in sealed envelopes, which are to be opened only in case of disagreement over the allocation of the results.[126] In some cases, alliance clauses prevent technological diffusion from the partners to other firms. For example, Siemens and IBM, who are cooperating in the development of a new generation of chips, have signed a clause that prohibits either company from cooperating with any Japanese firm in memory-chip development.[127] Thus, competition among interfirm networks may help preserve innovative vitality and prevent collusive restraint of technological innovation and diffusion. Nevertheless, networks' potential anti-competitive effects may be significant and need to be closely maintained by the authorities.

## SOME REFLECTIONS ON THE EFFICIENCY
## OF R&D ALLIANCES

Clearly, technological alliances have implications for the competitiveness of both firms and nations. Unfortunately, it is difficult to measure the impact of alliances in terms, for example, of the input and output related to R&D joint activities. What is the share of R&D jointly realized, as compared with all R&D activities? What is the importance of international technological alliances in the globalization of technological activities? What is the role of technical cooperation in new innovations? To what extent can one attribute any success of the commercialization of a particular innovation to its particular organizational structure? What is the contribution of each partner in terms of assets exchanged and/or shared? What share do foreign firms have in the local innovation realized through strategic alliances? In what location are R&D joint activities undertaken and what is their impact on the local economy?

According to the available data, it seems difficult to find an answer to these questions. Nevertheless, as we already mentioned, it is possible to glean some hints from specific industrial cases. For example, new generations of semi-conductors have been developed through alliances among the major world producers. The European HDTV standards may compete with the Japanese standard, thanks to collaborative ventures, particularly in the Eureka framework. Without the Airbus project—which involves intra-European technical, production, and marketing cooperation—it would have been difficult for the countries involved to have maintained an independent commercial aircraft industry.

Some years ago, Robert Hawkins observed that technological cooperation may be positive or negative for global industrial productivity, or it may be both good and bad from the national point of view. He went on to suggest that much will depend on the conditions of the individual case and upon the perspective from which goodness or badness is judged.[128]

The rising popularity of strategic alliances has not occurred without a measure of skepticism. Some scholars refer to the disappointing history of cooperative ventures in the past and point out the organizational problems incurred because of the incompatibility of goals and differing "corporate cultures" of the partners.[129] Further-more, some consortia, particularly in the U.S. (e.g., Sematech, Semi-conductor Research Corp. [SRC], and Micro-electronics and Computer Technology Corp. [MCC]), have been a major disap-pointment to their supporters. In encouraging the collaboration of firms that are strong competitors, some R&D consortia seem to have failed because of a lack of confidence and trust among and between partners.[130] In the U.S., opponents of government involve-ment assert that federal programs, far from encouraging innovation, may have stifled entrepreneurial efforts.[131] In Europe, some critics argue that the EC programs are spread among too many projects and/or too many firms to be efficient.

London's *The Economist* has been particularly scathing about government support for commercial R&D consortia. In an article published in January 1990, it argued that such consortia would

> neither improve education nor reduce the interest-rate-boost-ing federal budget deficit. Instead of encouraging the hearten-ing trends in R&D, consortia refill the troughs of big-company lobbyists.[132]

Yet there is some evidence to suggest that some cooperative projects do seem to have succeeded. One example is the VLST program in Japan, which has played an important role in helping Japanese companies to overtake U.S. competitors in semi-conduc-tor memory technology and production.[133] Another is the Eureka Project on HDTV for the development of an international standard, which may well help European producers to head off the threat of Japanese domination.[134] In some cases, government projects may also stimulate R&D; according to the EC, the European information technology industry doubled in the past four years, due in part to collaborative programs.[135]

Furthermore, it seems that the major criticisms of national con-sortia are more directed to their political management and organiza-tion than to the concept of R&D collaboration per se. The possible pitfalls of some government projects could also be attributed to the firms themselves. Some experts argue that certain companies are collaborating only on technologies that are not crucial to the com-petitiveness of the ultimate product. Sometimes, firms are participating in a given project only to keep an eye on competitors. There-fore, the cooperative goals are insufficiently determined, and the partners fail to achieve a successful commercialization of the technologies jointly developed.

Other critics point to the anti-competitive effects of cooperative agreements, and they advise governments to restrict some kinds of alliances, such as horizontal agreements.[136] On the other hand, strategic alliances may well fill a real need to share costs that are beyond the scope of many firms. As noted by Davidson,

> anti-trust policies that were appropriate in a domestic econo-my with few linkages to the international system can be self-destructive in an internationally competitive environment.[137]

Confronting this debate, governments adopt "reason rules" in their anti-trust regulations. Cooperative agreements are generally autho-rized whenever they appear to offer enough economic benefits to compensate for their possible anti-competitive effects. In fact, as we have seen, R&D alliances are rarely initiated to restrict competition but rather to counteract endemic market failure. Viewed from a national perspective, R&D collaboration may also induce higher productivity, avoid a duplication of research investment, reduce structural overcapacity, and improve the competitiveness of their national champions.

The socially beneficial effects of R&D strategic alliances depend upon the industry, market, and firm-specific characteristics. For example, as noted by Alex Jacquemin, R&D alliances are most likely to have positive effects in markets where there are strong externalities or spill-overs in the absence of cooperation, and in markets where a high rate of R&D sharing or between-member spill-over is feasible.[138]

Technological alliances seem to involve higher positive effects in the basic research field. In fact, it is usually hard to make a profit from basic research. For example, neither AT&T's introduction of the transistor nor IBM's Nobel Prize-winning discovery of higher-temperature superconductors seems to have revitalized their com-petitive positions.[139] Thus, private and public collaborative projects in basic research may promote innovations that otherwise wouldn't have been developed. For example, AT&T has joined with its rival

IBM to support a long-term research project into super-conductivity, at the Massachusetts Institute of Technology.[140]

Still, there are circumstances in which R&D agreements do involve high economic or strategic social costs. Where, for example, collusive behavior stifles future innovation or restricts the diffusion or dissemination of technology, or creates additional barriers to the entry of competitors, it might reduce the social product. It may also result in a strategically unacceptable technology drain from the home country. For example, in the information technology sectors, Sun, Hewlett-Packard, and MIPS conceded licenses to its Japanese partners Fujitsu, NEC, Sony, Toshiba, and Hitachi–all of whom are now able to build RISC-based workstations.[141] Some U.S. experts are opposed to these kinds of alliances. In the Boeing-Japanese firm links, experts argue that by cooperating, Boeing will lose its technological edge by helping its Japanese partners to acquire the know-how to develop and produce airplanes.[142] In the U.S., the FSX Fighter case (already mentioned above) also underlines the risk of technological drain to the benefit of the partners. The source of the debate was the project of the co-production with Japan (Mitsubishi was the prime contractor) of a new tactical fighter plane, based on the American F-16 produced by General Dynamics. Critics of this alliance feared it would lead to an erosion of American technology and add to the ability of Japan to launch a civilian aerospace industry that would directly compete with the American firms.[143]

The probable impact of technological strategic alliances on both the firm's and the nation's competitiveness is indeed complex. Clearly, much depends on the opportunity cost of such alliances and whether or not the alternative alliances, which might have been concluded, would have operated to the (net) disbenefit of the non-participating partner.[144] However, it can safely be concluded that technological cooperative agreements play an important role in both the ownership and location of innovatory activities. The potential impact of R&D joint activities may be appreciated from a private and public point of view. It constitutes a new factor to consider in the bargaining game between firms and governments, and particularly so in cases where the private interests related to strategic alliances oppose the social, economic, and political goals of governments.

## CONCLUSIONS

Within a web of strategic alliances, companies perceive they gain valuable technological synergism and risk-reducing benefits. This chapter has made an attempt to provide a broad overview of major developments related to strategic alliances, particularly in innovatory activities. It has also tried to indicate the probable impact of technological cooperative agreements from a private and a social point of view. Obviously, such an overview cannot be conclusive and comprehensive–if for no other reason than the reality it tries to analyze is complex, dynamic, and inadequately documented. Despite this caveat, we believe this chapter has demonstrated four main things:

1. International inflows and outflows of technologies organized through strategic alliances vary between industries.
2. Strategic alliances have to be understood not only as a substitute but also as a complementary organizational form necessary to create, maintain, and enhance the technological advantages of the firms and the location of their innovatory activities.
3. R&D cooperative ventures are part of complex intra- and inter-organizational networks in which technologies are created, disseminated and, improved upon on an international basis.
4. It is possible to incorporate strategic alliances into the mainstream economic and business theories, which have attempted to explain the internationalization of value-added activities, and to the level and pattern of both intra/interfirm exchanges and intra/inter-industrial exchanges.

## ENDNOTES

1. ECE/UNCTC Joint Unit on Transnational corporations, *Les sociétés transnationales japonaises en Europe: structures, stratégies et nouvelles tendances*, Geneva and New York, 1991.

2. Susan Moffat, "Picking Japan's Research Brains," *Fortune*, March 1991, pp. 54-59

3. Christopher Clarke and Koeron Brennan, "Allied Forces," *Management Today*, November 1988, p. 128.

4. Farok J. Contractor and Peter Lorange (eds), *Cooperative Strategies in International Business*, Lexington Books, New York, 1988, p. 4.

5. Karen J. Hladik and Lawrence H. Linden, "Is an International Joint Venture in R&D For You?," *Research Technology Management*, Vol. 32, No. 4, July-August 1989, p. 12.

6. Frank Hull and Gene Slowinski, *Strategic Partnerships Between Small and Large Firms in High Technology: A Theoretical Framework for Analysis*, Mimeo, Rutgers University, 1987, p. 1.

7. *Technical Co-operation Agreements Between Firms: Some Initial Data and Analysis*, DSTI/SPR/86.20, Part I, Paris, 1986, p. 3.

8. Gian Carlo Cainarca, Massimo G. Colombo, and Sergio Mariotti, *Cooperative Agreements in the Information and Communication Industrial System*, Milan, 1988, p. 5.

9. *Business Week*, February 19, 1990, p. 38.

10. For the strategic context, see Bruce Kogut, "Joint Ventures: Theoretical and Empirical Perspectives," *Strategic Management Journal*, Vol. 9, 1988, pp. 319-332.

11. Farok J. Contractor and Peter Lorange, op. cit.

12. Martin Shubik, "Game theory, behavior, and the paradox of the Prisoner's Dilemma: Three solutions," *Journal of Conflict Resolution*, Vol. XIV, No. 2, 1970, pp. 181-931.

13. Robert Axelrod, *The Evolution of Cooperation*, Basic Books, New York, 1984.

14. See, for example, Jordan D. Lewis, *Partnerships For Profit: Structuring and Managing Strategic Alliances*, The Free Press, 1990, p. 1.

15. John F. Nash, "The Bargaining Problem," *Econometrica*, Vol. 18, No. 2, April 1950, p. 152-162. John F. Nash, "Two-Person Cooperative Games," *Econometrica*, Vol 21, No. 1, January 1 953, pp. 128-140.

16. John Von Neumann and Oskar Morgenstern, Theory of Games and Economic Behavior, Princeton University Press, Princeton, 1947.

17. Ehud Kalai and Meir Smorodinsky, "Other Solutions to Nash's Bargaining Problem," *Econometrica*, Vol. 43, No. 3, May 1975, pp. 513-518.

18. Ariel Rubinstein, "Perfect Equilibrium in a Bargaining Model," *Econometrica*, Vol. 50, No. 1, January 1982, pp. 97-109. Ariel Rubinstein, "A Bargaining Model with Incomplete Information about Time Preferences, *Econometrica*, Vol. 53, No. 5, September 1985, pp. 1151-1172.

19. Frederik Zeuthen, *Problems of Monopoly and Economic Warfare*, Routledge and Kegan Paul Ltd., London, 1930.

20. L. S. Shapley, "A Value for n-person Games," *Annals of Mathematic Studies*, Vol. 28, 1953, pp. 307-371.

21. Robert J. Aumann and Michael Maschler, "The Bargaining Set for Cooperative Games," *Annals of Mathematic Studies*, Vol. 52, 1964, pp. 443-476.

22. John C. Harsanyi, *Rational behavior and bargaining equilibrium in games and social situations*, Cambridge University Press, Cambridge, Mass., 1977, pp. 273-288.

23. Philippe Gugler, *Les alliances stratégiques transnationales*, Editions Universitaires Friborg Suisse, Friborg, 1991.

24. Bruce Kogut, op. cit.

25. See also Peter J. Buckley and Mark Casson, "A Theory of Cooperation in International Business," in Farok J. Contractor and Peter Lorange (eds), *Cooperative Strategies in International Business*, Lexington Books, 1988, pp. 31-53.

26. "The Nature of the Firm," *Economica*, New Series, Vol. IV, November 1937, pp. 386-405.

27. M. A. Adelman, "The Large Firm and its Suppliers," *The Review of Economics and Statistics*, Vol. 31, No. 2, May 1949, p. 113-118. P. W. S. Andrews, "Industrial Economics as a Specialist Subject," *Journal of Industrial Economics*, Vol. 1, No. 1, November 1952, p. 72-79. Jacques Houssiaux, "Le concept de 'quasi-intégration' et le rôle des sous-traitants dans l'industrie," *Revue Economique*, No. 2, March 1957, p. 221-247 and "Quasi-intégration, croissance des firmes et structures industrielles," *Revue Economique*, No. 3, May 1957, pp. 385-411.

28. G. B. Richardson, "The Organization of Industry," *Economic Journal*, Vol. 82, No. 327, September 1972, p. 282.

29. David J. Teece, "Profiting from technological innovation: Implications for integration, collaboration, licensing and public policy," *Research Policy*, Vol. 15, No. 1, February 1986, pp. 288-290.

30. Bruce Kogut, op. cit., 1988, p. 321.

31. Alexis Jacquemin, "Cooperative Agreements in R&D and European Antitrust Policy," *European Economic Review*, Vol. 32, 1988, p. 552.

32. As is particularly the case in the strategic networking of firms in different branches of the telecommunications and biotechnology sectors. The notion of strategic business groups is also relevant for understanding the changing characteristics of oligopolistic competition.

33. Seev Hirsch, "An International and Trade Investment Theory of the Firm," *Oxford Economic Papers (New Series)*, Vol. 28, No. 2, 1976, pp. 258-270.

34. Alan Rugman, *Inside the Multinationals*, Columbia University Press, New York, 1981, pp. 54-60.

35. Peter J. Buckley and Howard Davies, *The Place of Licensing in the Theory and Practice of Foreign Operations*, University of Reading Discussion Papers in International Investment and Business Studies No. 47, November 1979.

36. David J. Teece, "Transaction Cost Economics and the Multinational Enterprise: An assessment," *Journal of Economic Behavior and Organization*, Vol. 7, 1986, p. 21-45, and "The Multinational Enterprise: Market failure and Market Power Considerations," *Sloan Management Review*, Vol. 22, No. 3, 1981, pp. 7-10.

37. Charles W. L. Hill and W. Chan Kim, "Searching for a Dynamic Theory of the Multinational Enterprise: A Transaction Cost Model," *Strategic Management Journal*, Vol. 9, 1988, pp. 93-104.

38. J. F. Hennart, "A Transaction Cost Theory of Equity Joint Ventures," *Strategic Management Journal*, Vol. 9, July-August 1988, pp. 361-374.

39. Paul W. Beamish and John C. Banks, "Equity Joint Ventures and the Theory of the Multinational Enterprise," *Journal of International Business Studies*, Vol. 19, Summer 1987, pp. 1-16.

40. Erin Anderson and Hubert Gatignon, "Modes of Foreign Entry: A Transaction Cost Analysis and Propositions," *Journal of International Business Studies*, Vol. 17, Fall 1986, pp. 1-26.

41. Peter J. Buckley, "The Limits of Explanation Testing the Internalization Theory of the Multinational Enterprise," *Journal of International Business Studies*, Vol. 19, No. 2, Summer 1988, p. 184.

42. G. Walker and D. Weber, "A Transaction Cost Approach to Make-or-Buy Decisions," *Administrative Science Quarterly*, Vol. 29, No. 3, 1984, pp. 373-391.

43. Farok J. Contractor, *Contractual and Cooperative Modes of International Business: Towards A Unified Theory of Model Choice*, Graduate School of Management Working Paper, No. 89-15, Rutgers University, August 1989, pp. 13-14.

44. David J. Teece, "Multinational Enterprise, Internal Governance, and Industrial Organization," American Economic Review, Vol. 75, No. 2, May 1985, p. 237.

45. John H. Dunning, *Explaining International Production*, Unwyn Hyman, London, 1988.

46. For a discussion of some strategies toward the internationalization of production see David Lei, "Strategies for Global Competition," *Long Range Planning*, Vol. 22, No. 1, 1989, p. 102.

47. John H. Dunning, "Non Equity Forms of Foreign Economic Involvement and the Theory of International Production," in R. W. Moxon, T. W. Roehl, J. F. Truitt (eds), *Research in International Business and Finance*, Vol. 4, JAI Press Inc., 1984, p. 34.

48. John H. Dunning, *Explaining International Production*, London, Unwin Hyman, 1988, Chapters 1 and 2, and *Multinational Enterprises and the Global Economy*, Reading, Mass., Addison Wesley, 1991, Chapters 3 and 4.

49. John H. Dunning, *Global Strategy and the Theory of Industrial Production: An Exploratory Note*, Reading and Rutgers University, 1990.

50. John Cantwell, *Technological Advantage As a Determinant of the International Economic Activity of Firms*, University of Reading Discussion Papers in International Investment and Business Studies, No. 105, October 1987, p. 3.

51. Charles W. L. Hill, Peter Hwang and W. Chan Kim, "An Eclectic Theory of the Choice of International Entry Mode," *Strategic Management Journal*, Vol. 11, 1990, p. 117-118.

52. Weijian Shan, "An Empirical Analysis of Organizational Strategies by Entrepreneurial High-Technology Firms," *Strategic Management Journal*, Vol. 11, pp. 129-131.

53. Alice M. Sapienza, "R&D collaboration as global competitive tactic–Biotechnology and the ethical pharmaceutical industry." *R&D Management*, Vol. 19, No. 4, October 1989, p. 285.

54. *International Management*, December 1984, p. 58.

55. Raymond Vernon, "International Investment and International Trade in the Product Cycle," *Quarterly Journal of Economics*, Vol. LXXX, 1966, p. 200.

56. Weijian Shan, op. cit., p. 131.

57. Christopher Clarke and Kieron Brennan, op. cit., p. 128.

58. *The Economist*, February 3, 1990, p. 66.

59. *Electronic Business*, August 1989, p. 50.

60. ECE/UNCTC Joint Unit on Transnational Corporations, *Recent Developments in Operations and Behavior of Transnational Corporations: Towards New Structures and Strategies of Transnational Corporations*, ECE/UNCTC Joint Unit Publications Series 7, Geneva, 1987, p. 19.

61. *Financial Times*, March 11, 1991, p. 1.

62. *Financial Times*, January 17, 1990, p. 26.

63. Rob van Tulder and Gerd Junne, *European Multinationals in Core Technologies*, John Wiley/IRM, 1988.

64. Margaret Sharp, "Europe: Collaboration in the High Technology Sectors," *Oxford Review of Economic Policy*, Vol. 3, No. 1, Spring 1987, p. 60.

65. *Fortune*, March 27, 1989, p. 58.

66. ECE/UNCTC Joint Unit on Transnational Corporations, op. cit., 1987, p. 20.

67. *Fortune*, August 28, 1989, p. 70.

68. Gaston Gaudard, "Transnationalization and Global Financial Equilibrium," in Zuhayr Mikdashi, *Bankers' and public authorities management of risks*, Macmillan Press Ltd., Londres, 1990, p. 234.

69. *Financial Times*, January 10, 1990, p. 23.

70. Frederick T. Knickerbocker, *Oligopolistic Reaction and Multinational Enterprise*, Harvard University Press, Cambridge, Mass., 1973. Knickerbocker study concerns national market concentration, but his contribution may easily be applied to an international market concentration.

71. ECE/UNCTC Joint Unit on Transnational Corporations, op. cit., 1987, pp. 22-23.

72. *The New York Times*, December 3, 1989, p. 14.

73. Kenichi Ohmae, "The Global Logic of Strategic Alliances," *Harvard Business Review*, March-April 1989, p. 144.

74. Kenichi Ohmae, *Beyond National Borders: Reflections on Japan and the World*, Dow Jones-Irwin, Homewood, Illinois, 1989, pp. 86-87.

75. OCE/UNCTC Joint Unit on Transnational Corporations, op. cit., 1987, p. 19.

76. *The New York Times*, May 17, 1990, p. D5.

77. *Financial Times*, March 27, 1991, p. 25.

78. Pierre Buigues et Alexis Jacquemin, "Strategies of Firms and Structural Environments in the Large Internal Market," *Journal of Common Market Studies*, Vol. XXVIII. No. 1, September 1989, p. 65.

79. *Business Week*, July 4, 1988, p. 109.

80. For an analysis of technology strategy, see David Ford, "Develop Your Technology Strategy," *Long Range Planning*, Vol. 21, No. 5, 1988, pp. 85-95.

81. Edith T. Penrose, *The Theory of the Growth of the Firm*, M. E. Sharpe Inc., New York, 1980 (reedition), pp. 43-63.

82. Austin Robinson, "The Problem of Management and the Size of Firms," *The Economic Journal*, Vol. XLIV, June 1934, pp. 242-257.

83. Alfred D. Chandler, *Strategy and Structure: Chapters in the History of the American Industrial Enterprise*, IT Press, 1962, pp. 7-17.

84. Howard V. Perlmuter and David A. Heenan, "Cooperate to compete globally," *Harvard Business Review,* March-April 1986, p. 142.

85. Robert B. Reich and Eric D. Mankin, "Joint Ventures with Japan give away our future," *Harvard Business Review*, March-April 1986, pp. 78-86. See also "Joint Ventures may damage your health," *Financial Times*, September 9, 1987, p. 8; "Are Foreign Partners good for U.S. Companies?," *Business Week*, May 28, 1984, pp. 48-52; "Corporate odd couples: Beware the Wrong Partner," *Business Week*, July 21, 1986, pp. 98-103.

86. *Financial Times*, April 7-9, 1990, p. 10, and *The New York Times*, March 28, 1990, p. D1 and D9.

87. David C. Mowery, "Collaborative ventures between U.S. and foreign manufacturing firms," *Research Policy*, Vol. 18, 1989, p. 26.

88. Jordan D. Lewis, op. cit. 1990 p. 60.

89. Louis Kraar, "Your rivals can be your allies," *Fortune*, March 27, 1989, p. 58.

90. *The New York Times*, February 2, 1990, p. D1 and D9.

91. Jordan D. Lewis, op. cit., 1990, p. 60.

92. Gary Hamel, Yves L. Doz, and C. K. Prahalad, "Collaborate with your competitors–and win," *Harvard Business Review*, January-February 1989, p. 136.

93. David C. Mowery, 1989, p. 23.

94. Robert Porter Lynch, "Building Alliances to Penetrate European Markets" *The Journal of Business Strategy*, March-April 1990, p. 8.

95. Cited in J. Michael Geringer, "Partner Selection Criteria For Developed Country Joint Ventures," *Business Quarterly*, Vol. 53, No. 1, Summer 1988, p. 61.

96. Cited in Jordan D. Lewis, op. cit., p. 250.

97. *The Economist*, May 20, 1989, p. 104.

98. *The Economist*, February 3, 1990, p. 72.

99. *The New York Times*, March 13, 1990, p. D4.

100. *The New York Times*, March 8, 1990, p. D9.

101. *The New York Times,* December 3, 1989, p. F14.

102. *Financial Times*, December 19, 1989, p. 14.

103. S. Mariotti and E. Ricotta, "Diversification, Agreements Between Firms and Innovative Behavior," paper presented at the Conference on Innovation Diffusion, Venice, 1986, p. 39.

104. *Financial Times*, February 16, 1990, p. 6.

105. *Financial Times*, April 14, 1991, p. 10.

106. *Financial Times*, May 10, 1990, p. 3.

107. *The New York Times*, December 1, 1989, p. D5.

108. *Financial Times*, January 1, 1990, p. 13.

109. Peter Kuentz, "Expériences réalisées en matière de coopération technologique européenne," *Vie Economique*, 6/1990, p. 13.

110. Jan Johanson and Lars-Runnar Mattsson, "Interorganizational Relations in Industrial Systems: A Network Approach Compared with the Transaction-Cost Approach," *International Studies of Management and Organization*, Vol. XVII, No. 1, 1987, p. 35.

111. Michael E. Porter and Mark B. Fuller, "Coalitions and Global Strategy," in Michael E. Porter, *Competition in Global Industries*, Harvard Business School Press, Boston, Mass., 1986, p. 316.

112. K. S. Cook and R. M. Emerson, "Power, Equity, and Commitment in Exchange Networks," *American Sociological Review*, Vol. 43, 1978, p. 725.

113. Dirk-Jan F. Kamman and Dirk Strijker, "Concept of Dynamic Networking in Economic and Geographical Space and their Application," in GREMI, *Innovative Milieux and Transnational Firm Networks: Towards a New Theory of Spatial Development*, International Workshop, Barcelone, March 1989.

114. Björn Aexlsson, "Supplier Management and Technological Development," in Hakan Hakansson, *Industrial Technological Development: A Network Approach*, Croom Helm, 1987, pp. 128-176.

115. Lars Engwall and Jan Johanson, *Banks in Industrial Networks*, Working Paper 1989/2, Uppsala University, 1989, pp. 4-5.

116. *Fortune*, March 27, 1989, p. 56.

117. Aerospatiale, *Le Groupe*, Paris, 1989, p. 25.

118. Frank Hull, Gene Slowinski, Robert Wharton, and Toya Azumi, "Strategic Partnerships between Technological Entrepreneurs in the United States and Large Corporations in Japan and the United States," in Farok J. Contractor and Peter Lorange (eds), op. cit., p. 451.

119. John H. Dunning, *Multinationals, Technology and Competitiveness*, Unwin Hyman, London, 1988, p. 177.

120. George Ferné, R&D "Programmes for Information Technology," *The OECD Observer*, August-September 1989, p. 10.

121. David J. Teece, "Inter-organizational Requirements of the Innovation Process," *Managerial and Decision Economics*, Special Issue, Spring 1989, p. 41.

122. Hakan Hakansson, *Industrial Technological Development: A Network Approach*, Croom Helm, 1987, p. 3.

123. Barry E. Hawk, "La recherche-dévelopment en droit communautaire et en droit anti-trust américain," in Jacquemin Alexis and Rémiche Bernard (eds), *Coopération entre enterprises: Enterprises conjointes, stratégies industrielles et pouvoirs publics*, De Boeck/Editions Universitaires, Brussels, 1988, pp. 230-231.

124. David C. Mowery, "Collaborative ventures between U.S. and foreign manufacturing firms," *Research Policy*, Vol. 18, 1989, p. 23.

125. *The Wall Street Journal*, March 21, 1990, p. A3.

126. George Ferné, op. cit., p. 10.

127. *The New York Times*, February 6, 1990, pp. D1 and D9.

128. Robert G. Hawkins, "Technical Cooperation and Industrial Growth: A Survey of the Economic Issues," in Herbert I. Fusfeld and Carmela S. Haklisch, *In-*

dustrial Productivity and International Technical Cooperation, Pergamon Press, 1982, p. 17.

129. Peter F. Drucker, "From Dangerous Liaisons to Alliances for Progress," The Wall Street Journal, September 8, 1989, p. A14.

130. Electronic Business, January 22, 1990, pp. 46-52.

131. Business Week, January 8, 1990, p. 25.

132. The Economist, February 3, 1990, p. 66.

133. Financial Times, December 3, 1990, p. V.

134. International Management, April 1990, p. 49.

135. Financial Times, March 11, 1991, p. VI.

136. Michael E. Porter, "The Competitive Advantage of Nations," Harvard Business Review, New York, The Free Press, 1990.

137. William H. Davidson, "Ecostructures and International Competitiveness," in Negandhi Anant R. and Savara Arun, International Strategic Management, Lexington Books, 1988, p. 20.

138. Alexis Jacquemin, op. cit. pp. 553-554.

139. The Economist, February 3, 1990, p. 66.

140. The Economist, February 3, 1990, p. 66.

141. Business Week, October 23, 1989, p. 110.

142. The New York Times, November 3, 1989, p. D1 and D5.

143. B. R. Inman and Daniel F. Burton, "Technology and Competitiveness: The New Policy Frontier," Foreign Affairs, Spring 1990, pp. 122-125.

144. For a general discussion of the social costs and benefits of the export of technology through joint ventures and foreign direct investment, see John H. Dunning, Multinationals, Technology and Competitiveness, Unwin Hyman, London, 1988.

# Part II
# Agencies

# A
# Artisans and Professionals

# [9]

## SKILLS, COAL AND BRITISH INDUSTRY IN THE EIGHTEENTH CENTURY

J. R. HARRIS

*University of Birmingham*

During the Industrial Revolution technical change undoubtedly extinguished some traditional skills, and annihilated some crafts altogether. Nevertheless the more complex industrial world which resulted demanded a whole range of new trades, subdivided old ones, and created new types of skilled worker, some of which became a permanent feature of the industrial scene.[1] Many of the new processes of the industrial revolution were developed by men steeped in craft skills, and were taught and learned in their turn in ways very similar to those by which older methods had formerly been transmitted. Craft skills were vital in the industrial use of coal and an essential part of the associated technology, and the purpose of this article is to show how important these skills were. Once they are properly understood, it is easier to see why in Britain, the fount of industrial invention and innovation, there was surprisingly little technological literature; why such literature as existed was of small use in the process of technological transfer and diffusion either at home or abroad, and (more speculatively) why English writers of scholarly and scientific pretensions largely ignored the craft-based advances which were to have an astonishing impact on the economy by the last years of the eighteenth century.[2] In some cases in which the technologies essential to industrial progress were contributed by the leading workmen rather than by managers and entrepreneurs, there may well be questions to be considered by the social as well as the economic historian.

Elsewhere I have made a number of suggestions[3] about the development of coal-fuel technology, in my view an important influence in pointing British industry in the direction it was necessary to take if an industrial revolution was to be achieved. These suggestions involved some comparison with France, which in the eighteenth century was our close rival in the industrial leadership of Europe, but which made little progress in acquiring the new type of technology. This was in part due to the fact that the success of coal-using processes was largely governed by the skills possessed by the

[1] Hence some illustrative examples, e.g. of cast steel making, are taken from later evidence, but of processes whose essential nature derived from eighteenth-century invention.

[2] In France a number of influences, the readiness of the State to regulate industry and even intervene in the details of production, the importance of France in the manufacture of articles of luxury and taste, the recruitment of scientists by the State to help solve technical problems, and the wish of the men of the Enlightenment to demystify craft secrets, all worked to produce publications in which craft processes were examined and illustrated, particularly the 'Descriptions des Arts et Métiers' and the 'Arts et Métiers' sections of the *Encyclopédie*. However praiseworthy, the technological gain from such publications seems to have been limited, probably because the most significant changes of method were not taking place in French industry anyway, and because of the difficulties of imparting craft skills by literary and graphic means, even with a high standard of exposition.

[3] J. R. Harris, *Industry and Technology in the Eighteenth Century: Britain and France*, an inaugural lecture published by the University of Birmingham (1972).

167

168            SKILLS, COAL AND BRITISH INDUSTRY

key workers involved, and the consequent difficulty of their transference. Indeed in an age when the spread of ideas through Europe was quite remarkably free, it is important to enquire why technological ideas (particularly some of great significance which had been conceived in Britain), did not get passed on with the same ease as other sorts of ideas. France, I argued, lacked the long industrial apprenticeship that Britain had undergone in learning how to evaluate, adapt and employ coal fuels. While agreeing emphatically with J. U. Nef on the length of the period during which the industrial use of coal had been gaining ground, I disagreed with his view that the post-Revolution period was technologically less revolutionary than the pre-Civil War period, and argued that it was more so. In developing her coal fuel technology Britain built up a broad band of techniques in furnace design, building and ventilation, in the choice and use of refractories, in boiler making, in the production of large iron plates and castings, in the provision of cokes, and in the important but now obscure craft techniques of furnace management. There was already by the eighteenth century a number of important coal-using processes which could not be used with other fuels. Because the methods of one industry after another could be converted to coal, British technology was given a long-term programme or target, technological optimism was generated and, as a spin-off, a battery of related techniques was distributed around British industry. As many of the innovated techniques were very much craft matters I suggested that this successful experience might have been one reason why there was so much British innovation at shop-floor level. Equally, the inter-relatedness of techniques made it difficult to remove them piecemeal to a rival but imitative economy. In the case of France, this reinforced the difficulties caused by fewer and poorer mines and the lack of a long tradition of familiarity with the arts of coal use.

Some of these points can now be taken further as we examine the relation between coal fuel technology and skills. The first contention is that the importance of coal fuel technology in British industrial development before the industrial revolution has been underrated because the craft element in the industrial processes using the fuel has meant that these processes are not well represented in contemporary technological literature.

The problems relating to the contemporary printed evidence of the period leading up to, and including the early decades of, the industrial revolution are little explored and some of these should be mentioned in order to view our particular problem in perspective. There was a considerable contemporary technological literature in European languages, including one dead one, in the eighteenth century.[4] French texts are probably the most celebrated, especially the *Encyclopédie* and the *Encyclopédie Méthodique,* though sets of the latter's many volumes are understandably hard to find. There are also a number of very useful works in French on mining and metallurgy.[5] The literature in English is not so extensive,

---

[4] Swedenborg, for instance, was still writing in Latin as late as 1734 though the part of the *Opera Philosophica et Mineralia* relating to iron was translated into French in 1762.

[5] For instance, R. A. F. Réaumur, *L'Art de Convertir le Fer Forgé en Acier* (Paris, 1722); Dietrich, Baron de, *Description des Gîtes de Minerai* 6 Vols (Paris 1786–99); J. F. C. Morand, *L'Art d'Exploiter Les Mines de Charbon de Terre*, 3 Vols. (Paris 1768–79).

## J. R. HARRIS 169

though by no means negligible. The concept of putting technological information into the ready form of a dictionary or encyclopaedia was early accepted in Britain, and the prospectus of the *Encyclopédie* acknowledged its imitation of the works of Harris, Chambers and Dyche,[6] though there is no doubt that in its items on the *arts et métiers* it transcended earlier works. There were of course later British encyclopaedic works and before the end of the century the *Encyclopaedia Britannica* was well established. Although I cannot pretend to have more than a very superficial knowledge of this vast literature, I am sure that we are in need of some broad surveys of this material, examining its evidence industry by industry, evaluating the accuracy of the various contributors and the up-to-dateness of their information as compared with contemporary practice, and discovering whether their accounts are heavily plagiaristic and dependent on writers of an earlier period or different country. For instance, it might be worth discovering by how many hands material from Neri's *Art of Glassmaking*[7] was translated and transmuted.

We ought also to consider whether possession of a considerable national technological literature necessarily means that the particular economy is ahead in the branches of technology treated. In the case of France, some of her main publications on mines and metals were produced by propagandists wishing to bring France up to the standards prevalent abroad, and some of the works were wholly or largely translations of foreign authors.[8] Again, many writers of the eighteenth century were part of the intelligentsia, savants to whom the esteem of the learned world was the first concern. It was more important to them to follow scholarly authorities and to cite the well-known works of the past—since failure to do this would reveal academic ignorance—than to make a simple enquiry of someone engaged in the day-to-day conduct of an industry, and in consequence it is now dangerous to take some of the literature at face value. A further difficulty is that there are considerable areas of technology which were of low interest to many of those compiling the technological works of the time, just because such people were often eminent 'philosophers' or members of the national scientific societies, and naturally took most interest in those industrial subjects which best reflected the new science of their times. These last points accentuate the general problems of contemporary technical literature and show why there was a particular lack of interest in the craft-based elements in the getting and using of coal. This is compounded, I think, by the tendency of modern historians of technology to be scientists by origin and particularly interested in those aspects of technological change which have been science-influenced. There seems to be a sort of 'pecking-order' by which developments which are clearly science-based, like scientific instrument-making or horology, take pride of place in their writings over

[6] J. Harris, *Lexicon Technicum* (1704; second volume 1710; Supplement 1744): E. Chambers, *Cyclopaedia* (1728): T. Dyche, *A New General English Dictionary* (1740). See Robert Collison, *Encyclopaedias: Their History Throughout the Ages* (N.Y. 1964), p. 121.
[7] Neri, A. *L'Arte Vetraria distinta* (Florence, 1612), English translation with additions by Merret 1662. Also a basis for German and French monographs and much drawn on by authors of encyclopaedic works.
[8] For instance, Monnet, *Traité de l'Exploitation des Mines* (Paris 1773) and Hellot, *De la Fonte des Mines, des Fonderies, etc.* (Paris 1764).

170          SKILLS, COAL AND BRITISH INDUSTRY

more empirical techniques which had a massive importance for industry and society.

In Great Britain the coal industry itself was remarkably little written about by contemporaries. Astonishingly in a country whose production at the end of the seventeenth century was greater (at about 2½ million tons) than her nearest rival's in the second decade of the nineteenth century, and which had achieved probably ten million tons—around a ton per head of population—by 1800, the native published technological literature was minute. There are only two works of value, one virtually anonymous by 'J. C.' in 1707, *The Compleat Collier* and another, John Curr's *The Coal Viewer* of 1797, much of which is involved with the related subject of steam engine construction.[9] Both, however, are valuable, the first because of its date and its earthy practicality, Curr's because of its author's own valuable innovations in winding and colliery transport. Valuable though these works are they are small, the one being a pamphlet and the other no more than a slim volume, but as far as I am aware, this remarkable poverty in the technological literature has passed without comment. There are of course references to the industry in topographical writing and there are manuscript materials in estate and other business records, but we ought to ask why no extensive and authoritative works on the technology of coal mining appeared in the country which pioneered it.

One reason, which I think affects most aspects of coal technology in the period, was the fact that the mining of coal was long familiar, and that continual improvements had come to be expected. This helps to explain the little notice taken by contemporaries of at least 1,000 steam engines entering the industry before 1800[10] and the too great readiness of modern historians to regard the eighteenth-century coal industry as without revolutionary technical change. Where aspects of the industry did attract the attention of scientific contemporaries it was often in connection with the phenomena rather than the technology—the emanation of coal gas, especially where permanently burning, explosions, deaths from damp and so on. The very able managers with their on-the-job training carried on with their work largely unregarded outside their own coalfields, though sometimes minor deities within them. Interestingly, they were often mathematically able, and had early in the century carried the surveying of coal below ground to considerable lengths—their 'dialling' was the *géométrie souterrain* whose lack in France was so deeply bewailed by her technocrats.[11] Nevertheless John Curr, whose own work is substantially filled by

[9] The *Compleat Collier* has recently been re-issued by Frank Graham (Newcastle 1968); for Curr see T. S. Ashton and J. Sykes, *The Coal Industry of the Eighteenth Century* (Manchester 1929) pp. 26, 61, 64–8.

[10] I would now regard my estimate of a total of about 1300 steam engines in Britain by 1800 (*History*, lii 1967) as much too low and that a figure in excess of 2000 should be substituted, taking into account the additions for Scotland of B. F. Duckham (*A History of the Scottish Coal Industry* Vol. 1, Newton Abbot (1970) p. 363) and information received from Mr. J. S. Allen and Dr. J. A. Robey on their continuing investigations in this field. It would be conservative to suggest that half the steam power employed in Britain before 1800 was at coal mines.

[11] Jeremiah Dixon, the American surveyor immortalized in the Mason-Dixon line, came from the Durham coalfield. Questioned about the mathematical education he had received he is said to have replied that it was gained 'in a pit cabin upon Cockfield Fell': *Sunday Times*, March 24, 1974.

## J. R. HARRIS                                      171

mathematical tables, clearly felt his little book to be of a different species from that intended for the learned world:

> It seldom falls to the lot of literary men, to be engaged in works of this sort, and therefore professing myself to be merely a mechanic, it can scarcely be expected that I should convey my ideas in all the elegance of expression, of which our language is capable; besides, it must occur to every reader, that such a work as this will not admit of any great choice of words, when it is considered that for the greater part it consists, of the various synonymous technical terms used in different parts of the kingdom, and of which, in order to convey a clear idea to every class of readers, there are unavoidably frequent repetitions, and explanations, and if I have the happiness to make myself understood on this head, I hope that it is all that will be required of me.[12]

It is interesting to see that in this country a celebrated coal viewer, of high managerial and technical talent, ascribes to himself a craftsman's role—merely a mechanic. It was otherwise in France where the three massive volumes of Morand[13] still impress the historian, though they do not seem to have had many contemporary British readers, if the difficulty in finding copies in our libraries today is any guide. Morand's main personal experience was of the Liége district, his English evidence partly second hand and much of his work compilatory though of a high order. There is also a considerable passage on English coalmining in Gabriel Jars' book,[14] and further passages (particularly some general panegyrics on coal) in Faujas de St. Fond.[15] There is more to be learned about coalmining (even British coalmining) from French than from English books. But the considerable skills in coalmining, which existed between the highly numerate expertise of the viewer at one end and the many rather mindless drudgeries above and below ground at the other, were not conveyed by literary means. A skilled coal-face worker was a very valuable man; in good times his pay was high though, as in a number of trades, peak earnings were achieved for only part of his working life, since a close combination of skill and strength was required. Good men were able to get out more sizeable, saleable coal by skilful under-and side-cutting, sometimes avoiding the waste of slack and dust by cunning wedging and use of natural roof pressures to break up the coal. Consequently when a new technique like longwall working was transmitted, it was done by recruiting colliers who knew how to do it.

> TWENTY COLLIERS, *industrious sober men* will be immediately employed by applying to JOHN MACKEY, ESQ., at PARR COLLIERY on the *Sankey Navigation near Warrington* those who will undertake on the long work way, will find great advantage in a small vein about three foot thick. Good lodgings will be immediately provided for those who

[12] John Curr, *The Coal Viewer and Engine Builder's Practical Companion* p. 6. Interestingly, Charles Hatchett described Curr 'as a very hospitable and civil man and appears to be an able mechanic': *The Hatchett Diary. A tour through the counties of England and Scotland in 1726 visiting their mines and manufactories* (ed.) A. Raistrick (Truro, 1967) p. 69.
[13] J. F. C. Morand, *L'Art d'Exploiter les Mines de Charbon de Terre* (1768–1779).
[14] G. Jars. *Voyages Métallurgiques*, Vol. 1 (Lyon 1774) pp. 238 *seq.*
[15] B. Faujas de Saint Fond, (trans. A. Geikie) *A Journey Through England and Scotland to the Hebrides in 1784* (Glasgow 1907) pp. 110 *seq.*

172          SKILLS, COAL AND BRITISH INDUSTRY

cannot return to their home at night; and those who chuse to occupy houses will be provided for in a decent manner, with gardens and other advantages, at MAY next.[16]

Boring and sinking similarly demanded skill and judgement. Borers were often specialists, working by contract and staying together and moving from one pit to another. Sinkers, though a little less peripatetic, were considerable specialists, hired by contract as a team, and in large collieries were prepared to endure periods of idleness between sinkings rather than hew or draw coal, though they might be forced to so demean themselves in small collieries. In both skills there was that common concomitant of a craft, a custom of hereditary succession.[17]

The craft element in coal-fuel technology may in a sense be said to have begun at the coal face, but it can be better instanced in some of the coal-using industries. One essential element was the building of suitable furnaces, together with ventilation systems and refractories, another was their firing and the technique of the furnacemen. There is very little in print (or manuscript) before the late eighteenth century about that major English breakthrough, the coke-iron furnace; again some of the most celebrated accounts are French, and I have suggested elsewhere[18] that interlinked craft elements were one of the problems in its transfer to France. Even more neglected, but I believe of very great importance, was the reverberatory furnace as used in England. Most regrettably there exists only one historical paper on the subject, though written by that founding Newcomenite and doyen of technological historians, Rhys Jenkins, in 1934—'The Reverberatory Furnace with Coal Fuel 1612–1712.[19] Not of course originally an English invention, attempts were made by a number of men to use the furnace with coal instead of wood in the seventeenth century, but understandably they did not have much success when they had the target of primary iron production in mind. In the early days the furnaces were appropriately called 'Furnaces of division, wherein the material or metal to be melted or wrought is kept divided from the Fewell',[20] indicating their main value in endeavouring to employ coal with its greater range and quantity of pollutants in place of wood fuel. Though at first they were sometimes used with bellows, the English developments were much involved with the improvement of natural draught by means of suitable chimneys, cones and below-ground air intakes. References to 'air' furnaces in the eighteenth century usually imply the reverberatory type. By the 1680s this sort of furnace was the key to successful English breakthroughs in smelting the non-ferrous metals, lead, copper and tin, and about 1700 there was increasing success with brass.[21] Initially coke was sometimes used, if only in part, as a fuel.

[16] *Liverpool Chronicle* 3 November 1768; for other examples of similar transferences by which the method was disseminated from its original home in Shropshire see T. S. Ashton & J. Sykes, *op. cit.*, pp. 30, 31 and B. F. Duckham, *op. cit.*, pp. 63–65.

[17] Ashton & Sykes, *op. cit.*, 14–16.

[18] *Industry and Technology in the Eighteenth Century* p. 8 *seq.*

[19] *Transactions of the Newcomen Society*, Vol. 14–15 (1933–35) pp. 67 *seq.*

[20] Rhys Jenkins, *op. cit.*, p. 70.

[21] See Joan Day's recent *Bristol Brass: The History of the Industry* (Newton Abbot 1973) pp. 32 *seq*; J. R. Harris, *The Copper King* (Liverpool 1964) p. 3 *seq.*

## J. R. HARRIS 173

The precedence of the non-ferrous metals over iron in coal smelting is underemphasized or unnoticed in texts on technological and economic history alike. Once again there seems little of significance on the reverberatory furnace in the contemporary English technological literature, beyond mentions of its older uses for laboratory and assaying purposes; the main published accounts and drawings seem to be those of Schulter[22] who described the English coal-using version. His book was translated into French by Hellot, and further plates or descriptions of the furnace were published in works by Jars and Swedenborg. Almost unremarked upon in Britain, the development of this furnace was a continual source of interest to foreigners. Talking of its use for reheating of iron for casting, Jars says 'le fourneau, dont on se sert pour cet usage est le fourneau à vent, que nous nommons en France fourneau Anglais'.[23] He mentions Swedish efforts to use their excellent iron for steelmaking, but noted that they were using the reverberatory furnace in the English fashion, even importing coal to do so. The De Givry-Wendel report in the 1780s also remarks on the proliferation of this furnace in British industry.[24] Nevertheless the astonishing thing is that despite Britain's leading position in the eighteenth-century European copper industry we have largely to rely for the use of the reverberatory furnace in copper smelting on the accounts of foreigners, Swedenborg, De Givry and especially the extremely rare Lentin account published in Leipzig in 1800.[25] My conclusion is that the diffusion of the coal-using reverberatory furnace through Britain was simply done by the open and honest, or the underhand and dishonest, enticement of skilled workmen from one area to another—for instance when the copper industry was set up at a Lancashire site my ancestor George Harris, 'Furniss-Builder', as he is described in the parish register, moved with his father, the works manager, from Kingswood near Bristol.[26] In these industrial operations, as we shall see later in steelmaking, my impression is that much of the technological knowledge remained with the workmen, especially the senior skilled workmen, not with the entrepreneur. When it came to the building of coal-fired bottle furnaces in France, in partial imitation of the English system, the basis of constructional skill was very narrow, being confined to one man, Jean Mol, of whom it was said in 1745 that he had built all the furnaces for coal-fired glass works in France, and had instructed the furnacemen in most of the works.[27]

From the furnaces it is a short move to the clays for refractory linings and for crucibles and other heat-resisting containers. Long before the end of the seventeenth century the introduction of coal saw special clays being moved for long distances for furnaces and particularly for crucibles. Stourbridge clay was early recognized as of exceptional quality,[28] but much of the art of the crucible maker was in the mixture of local and imported clays

---

[22] See above, footnote 8, *sub* Hellot.

[23] G. Jars, *op. cit.*, Vol. 1 p. 213.

[24] Archives Nationales, Paris, T.591⁴.

[25] A. G. L. Lentin, *Briefe über die Insel Anglesea* (Leipzig 1800) p. 79 *seq.*

[26] Lancashire Parish Register Society Publications Vol. 107, p. 101, entry of 1782; John Wesley, *Journal*, ed. N. Curnock, 1909–16, VI, p. 348, entry for 13 April 1782.

[27] Archives of the Company of Saint-Gobain. Report of D. Oury, 1745. I am most grateful to the Company, and particularly to M. Dominic Perrin, for access to their records.

[28] Robert Plot, *Natural History of Staffordshire* (Oxford 1686) p. 121.

174            SKILLS, COAL AND BRITISH INDUSTRY

to achieve economy with serviceability and in the addition of selected
ingredients, such as ground-down old crucible and coke dust, in order to
obtain the desired qualities. Crucible manufacture for each industry seems
to have been almost a separate craft. It was very important because of the
damage and loss a broken crucible could cause. At a furnace in the
Sheffield steel industry in the 1840s a crucible of Stourbridge clay could
last six meltings against a maximum of three for one from any other known
clay; as many as 108 crucibles were needed to supply a ten-furnace
crucible-cast works for a week. Huntsman allowed the Swedish observer,
Robsahm, to see round his works in 1761 but would not let him into the
secret of making crucibles 'even for £50'.[29] Some contemporary authorities
said that in the French wood-fired plate glass industry in the early eighteenth century the fracture of a crucible could cause a loss of £250.[30] Like a
number of the key craft processes crucible-making demanded great physical effort as well as skill. As late as 1891 an American writer noted with
some surprise that Sheffield crucibles were sometimes still made by treading clay with the feet for hours and cutting and turning it with a spade
rather than using a pugging mill because it was believed that human toil
produced a better result.[31] Even then, the practice was far from extinct. At
the beginning of the century the French savant and government inspector,
Hassenfratz, had pointed out that while one French works made crucible
cast steel for its own consumption, it did not produce any for commerce, as
they could not 'obtain crucibles capable of resisting the hot metal', and he
regretted the lack of research into crucible production: materials were not
always well chosen, the thickness was not right, and it was difficult to get
crucibles which would take the first effects of furnace heat or stand

[29] K. Barraclough (trans.) F. Le Play, *A report on the manufacture of steel in Yorkshire
and a comparison with the principal groups of steelworks in Europe* (Part II) n.p. This was
kindly supplied by Mr. Barraclough in advance of its publication in the *Bulletin of the Historical Metallurgy Group*, together with additional information. W. O. Henderson, *Industrial
Britain under the Regency, 1814–18*, p. 153 gives a translation of a report by the Prussian
factory inspector J. G. May on Sheffield cast steel works in 1814; 'I was only allowed to see
the furnaces and the running of the molten metal. The smelting crucibles were not shown but
were kept secret.'

[30] J. Beckmann, *A History of Inventions and Discoveries*, English translation Vol. IV (1817)
p. 206. *Encyclopaedia Britannica* (1797) Vol. VII p. 772. Both are clearly based on an earlier
source. Deslandes, the Director of the Saint-Gobain glassworks was greatly concerned with
the question of refractory clays, and was critical of contemporary science for its failure to
make progress in their chemistry. 'It is true that Chemistry has been pushed to a high point of
perfection throughout Europe. But it must also be admitted that our Chemists have not
acquired much knowledge in their little laboratories of the Arts of Glassmaking: it must also
be said that they have not discovered a good and reliable method of identifying clays'
(Saint-Gobain Archives: 'Historique de la Verrerie'). He says at another point that 'the
greatest service that could be rendered to plate glass manufacture and to glassworks in
general' would be the search for good clay. 'When I left Saint-Gobain [on retirement] in 1789
my plan was to travel throughout France in order to search out clays. I still possessed the
bodily strength and the enthusiasm to make such travels and to do so with greater ease I had
accepted the title of Inspector of the glassworks of France which the old Government had
offered. Our new one which for so long only steered us to barbarism, vandalism and death, did
not allow me to do it and caused me to lose precious years which I think could have been
employed with some usefulness.'

[31] H. M. Howe, *The Metallurgy of Steel* (Philadelphia 1891) p. 300. Mr. W. K. V. Gale tells
me that treading was used at Sheffield crucible steel works at least as late as 1931 and on the
Stourbridge clayfield in the 1950s.

J. R. HARRIS 175

being reheated after pouring, before being used again in the furnace.[32]

The same report remarks that crucibles could be lost during manufacture by 'the smallest carelessness and a fire forced on too rapidly or unequally will often be enough to break them'. This takes us to our next point, the importance of the craft skills of the furnacemen. Controlling the pace at which coal was fed to the furnace and its placing on the hearth, he had to cope with variations in the quality of the fuel and adjust his stoking accordingly, and sometimes add coal of various sizes and grades–large coal, small coal, slack, dust—at different points in the operation. Though there were sometimes flue and damper adjustments to be made, the state of the hearth itself (e.g. the amount of clinkering, the compression of the fuel over the bars) was frequently the main means by which ventilation and temperature were governed. All this was a matter of judgement, but in many instances this judgement governed the efficiency, or even practicability, of the process. This sort of judgement was not the kind of thing one learned from books. Of cementation steel it was said 'the competency of the workman shows above all in the regulating of the fire in such a way as to maintain continually a bright red heat which lends itself to cementation, without exceeding it and without ever allowing the grate to become empty . . . In the most usual furnaces in Yorkshire there is no damper to vary the draught so that the workman can only control the fire by the care which he gives to the firing of the grate'.[33]

Such skills in Britain are usually only referred to by foreign observers because they were taken for granted at home—as, by the early eighteenth century, were their continuous modification, adaptation and development. When there was an attempt to transfer coal-using processes abroad, there was, in addition to the problems of furnace building, furnace design, and ventilation systems, the overlooked problem of actually teaching people how to stoke a coal furnace. I have recently described the problems with which Oury and Deslandes, two outstanding French industrial experts, wrestled as they endeavoured with only partial success to introduce coal firing into plate glass production. The former struggled with the problems of the blowing process, the latter with those of the superior casting process.[34] So far as the fuelling techniques were concerned Oury tried to learn how to use coal by examining the coal-fired bottle works at Sèvres. He describes the importance and difficulty of acquiring the knowledge. . . . 'The method of stoking with coal seemed to Sr. Oury the most tricky part of the operation' . . . and he expended wine and bribes, spent much time with the workmen and even entered the *caves* under the furnaces, referring feelingly to 'observation of the different manual operations which he had the endurance to bear (not without discomfort)'. The chief furnacemen stayed on duty for great lengths of time throughout the refining period, 'they are very well paid and labour very hard', and a single furnaceman served throughout the whole period of sixteen or eighteen hours when the glass was worked. 'There is no question that the stoking is

[32] Conservatoire des Arts et Métiers, 'Rapport sur la fabrication d'acier fondu du Sr. Lenormant, par le Sr. Hassenfratz, inspecteur des mines'. Year 8. D103.
[33] Barraclough, *op. cit.,* Part I, *Bulletin of the Historical Metallurgy Group.* Vol. 7, No. 1 (1973) pp. 18–19.
[34] 'Saint-Gobain and Ravenhead', in B. Ratcliffe (ed.) *Great Britain and Her World 1750–1914* (Manchester, 1975).

176          SKILLS, COAL AND BRITISH INDUSTRY

the hardest and toughest operation in coal-fired glass-works, and it is
equally certain that this kind of worker is the hardest to find, the most
skilled and the dearest.' He goes into great detail as to how the fire should
be maintained by putting many compressed layers of small coal on the *grate*
during the period when the glass was being worked, when a mass of flame
and smoke was to be avoided. He described the special iron tools required
to deal with clinker and the air flow through the hearth. He feared that,
however handled, the furnace during the glass-working period would produce excess ebullition for 'a material so delicate and which so much needs
to be worked in a state of repose as glass for plate'. There would be 'une
fumée épaise et noire et de laquelle il aprehende beaucoup de vice pour la
verre des glaces'. Experts had to be found for this hard and responsible
work and the firm should not expect to have furnacemen for coal at the
same price as they paid those who worked with wood; they might have to
cast a wide net to get them, even as far as England.

Though coal *was* adopted, the concern had to use wood during the
refining period down to the end of the century; an Englishman was brought
in to advise on coal use within a few years of the introduction of the process
in 1745 and a secondary wood furnace seems to have been used in conjunction with the coal furnace to help burn out unconsumed coal particles
which would have harmed the glass.[35] A ten year period of continuous,
thorough and painstaking experiment at the Compagnie des Glaces' casting plant at their main Saint-Gobain works under an outstanding technical
director, Delaunay Deslandes, eventually came to nothing and coal use was
abandoned. Deslandes' detailed 'Essay on the means of applying to Glass
Furnaces the Fire of Coal' was submitted to the Académie Royale des
Sciences in 1784.

Perhaps the most striking part of this 'Essay' is the long section, closely
parallel to sections of the Oury document already cited, which deals with
the question of how a coal furnace is to be fuelled. It puts great emphasis
on the importance of the skill and technique of the furnaceman, and the
completely different method of working from that required with wood. The
workmen would normally use the new fuel by stoking vigorously at one end
of the hearth and piling up the coal, then doing the same at the other end
and waiting until it had burnt well down before refuelling. This was very
damaging, however. Initially, one got masses of heavy, black smoke from a
choked hearth. 'When the pots are not covered the thick smoke passes over
the surface of the glass and gives it a yellow and disagreeable colour.'
Eventually the fire got going and burned so furiously as to damage pots and
hearth, then died down considerably. 'This method though general is very
bad, with a very heavy fuel consumption and a very uneven heat.'
Deslandes then gives 'a method I worked out and which succeeded perfectly for the many years during which I made plate glass with coal'. It
resembled normal practice with wood in that there was a rapid alternation
between one end of the hearth and the other, a little fuel being added each
time. The Director had designed a special small shovel in sheet iron with
raised edges which would hold a proper quantity of coal. He then divided
the grille area into 16 separate sections and by the aid of a plan indicated

---

[35] Saint-Gobain Archives (*sub* Saint-Quirin Archives). Desrousseaux, 'Observations sur les
Glaceries de Tourlaville et St. Gobain etc.' Fructidor, Year 12.

how every shovelful should be placed on the hearth. 'One must accustom (the furnaceman) to regulate his speed in such a way that when he puts his sixteenth shovelful at point sixteen his first shovelful at point one is entirely burnt away.' There follows other advice about how to clear clinker from the grate with special iron tools, and how to reduce the heat of the fire by beating down layers of coal over the hearth, sometimes adding clinker. Here in this French work one has clearly indicated the effort needed to inculcate the skills so common in Britain as to be unremarked and unrecorded.[36]

The reverberatory furnace of course gathered round it a range of techniques as it was applied to different industries. The glass furnaces of Britain may be regarded as one type of reverberatory furnace with a highly specialized updraught system and air access passages. In the copper works the 'Welsh method' was perfected, though this may have emerged gradually from the method originally introduced in Bristol in the 1680s. The reverberatory furnace was first applied to the iron industry at the beginning of the eighteenth century for melting down pig for castings. It was gradually recognized that in terms of primary iron production it was better suited to refining than smelting iron. Eventually success was achieved, but not overnight, with the Cort process of puddling and rolling, the reverberatory furnace being of course the basis of the puddling part.

From that day to this—and *this* could be the last—for there is now only one puddler working occasionally in Britain and the trade will die with him—the craft has continued, demanding, like some other key trades in Britain's old coal-fuel technology, a balanced combination of physical strength and almost artistic judgement. Fortunately before it vanished two writers, one English and one French, have described the trade from its technical and its human side.

Keith Gale in his excellent paper 'Wrought Iron: A Valediction' has given a close and detailed account of how wrought iron was made from Cort to the present. He displays his remarkable ability to make the technical detail of the ferrous metals industries comprehensible to the layman and he shows how the operation of the puddling furnace and the mills of the forge train and rolling mill called for a sophisticated and varied series of manual operations. He sets out the full rotation of activities (with something different to be done every few minutes for a period of one-and-a-half to two hours) which constituted puddling. He explains that the essential point was the puddler's skill; the men seem to have been unsupervised and stood or fell by the quality of their iron once it reached the helve or (later) the steam hammer. Gale's remarks on the irrelevance of book-learning we shall mention subsequently, but his emphasis on the craft nature of the wrought iron trade concerns us here. 'Puddlers, underhands, shinglers, rollers and all the other trades . . . were tough, rough; uneducated and not at all out of the top drawer. But if they were uneducated . . . they were not ignorant . . . A new man in the trade started to learn in earnest, the hard way, by doing, not talking and he developed a taciturnity which lasted all his life. If he did not learn, of course, he was soon thrown out, for employers were tough, too. When he had learnt his trade a man was truly a

[36] Versions of this essay are contained in Deslandes' 'Historique' in the Saint Gobain archives and the Archives Nationales (26 AQ1).

178          SKILLS, COAL AND BRITISH INDUSTRY

craftsman, and he did his work well. His monetary reward was fairly small, but his satisfaction in his craft . . . was great.'[37]

A second powerful and even emotive 'valediction' is that of Jean-Paul Courtheoux in his paper 'Privilèges et Misères d'un Métier Sidérurgique au XIXe Siècle: Le Puddleur',[38] pregnant equally with Gallic verve and sociological insights. Unfortunately we cannot follow through this fascinating piece in its entirety, but a few quotations may show its relevance. Courtheoux opens by deploring the split between the evolution of the physical sciences, increasingly relied upon by producers to the exclusion of empiricism and tradition, and the evolution of the social sciences, considered only relevant to economic and 'human' problems. 'Knowledge of the process, ignorance of the craft, thus we may sum up what we know of puddling.'[39] He quotes with approval the view that the puddler's work was cardinal to the whole business of iron-making. 'Le puddleur constituait . . . le travailleur central dont l'art et la peine étaient les éléments autour desquels gravitaient les autres fabrications.'[40] Pudding, he suggests, is a technique and a craft belonging to a 'transitional economy' which was already industrialized but had not as yet reached its mature state. Puddlers, like some other key workers, became part of an industrial aristocracy or elite; the high quality of their workmanship inspired them with pride and self-esteem, but it also gave them status in the eyes of the whole industrial society they lived in. 'Their superiority was written into their work and was recognized by their bosses and their mates', and was reflected in the great pay differential they enjoyed. This superiority was, however, both economically and psychologically tied to the period between the full acquisition of the craft and the time when the strain of the great physical effort obliged the worker to bow his head and opt for some less arduous task, accepting at once a rank of economic, psychological, and social inferiority.

Another French writer, talking of the not dissimilar situation of workers in the rolling mills of the Ardennes emphasises the cultivated physical skill, the furnace environment, and true craft element involved. 'This aristocracy of the men of the forge train was above all built on the exceptionally hard and very specialized work at the fire. A craft which demands a long physiological apprenticeship is necessarily exercised by an elite. At a period when so much guts and manual strength were demanded this group of workers who sacrificed themselves so much, who worked hard and possessed a true craft, necessarily constituted a privileged group.'[41]

Once again, though Gale had no knowledge of Courthéoux's paper, the matter and tone of their conclusions are in notable harmony. 'Without regretting the passing of the excessive toil of transitional iron-making, one may hope that the present generation will not forget that to work well, that is to say work as becomes a man, is at least as important as to raise the standard of living. A generation which fails to grasp this point will not get

[37] K. W. V. Gale 'Wrought Iron: A Valediction', *Trans. Newcomen Society* XXXVI (1963–64) pp. 8–9.
[38] *Revue d'Histoire Economique et Sociale*, XXXVII, 1959, p. 161 *seq.*
[39] *Ibid.*, p. 162.
[40] *Ibid.*, p. 167, quoting R. Delavignette's impressionistic essay 'Le Marianneux'.
[41] *Ibid.*, p. 176, quoting M. Verry, *Les Laminoirs Ardennais* (Paris, 1955) p. 59.

to grips with the most profound problems of the human soul. It will not be a great generation.'[42]

If we can admit from such examples the existence of an important craft element within a coal-fuel technology which was one of the main bases of British industrialization, this may help to explain several things. Firstly, the assimilation and transmission of the technology by the watching and doing methods inseparable from learning a craft largely account for the silence of much contemporary technological writing on these processes, and its inadequacy and ineffectiveness from the point of view of technological transmission abroad. Secretly sketching machinery or plying the foreman with drink was not much help. For the foreigner attempting to master British technology there was a communication problem. This problem was particularly acute in the eighteenth century partly because it was a skilled worker or 'agent', not the entrepreneur, who had to be approached as the repository of knowledge. The British artisan was already familiar with using coal and largely unfamiliar (outside primary iron production) with wood fuels and consequently would find it hard to know what needed explaining. The gulf existed because so much knowledge was breathed in by the workman with the sooty atmosphere in which he lived rather than ever consciously learnt, or formulated into words.

Gabriel Jars, the greatest of contemporary technological observers, had language problems in interrogating craftsmen, even when he made great attempts to learn the tongues of all the countries he visited. 'Connaissance d'autant plus nécessaire qu'il avait principalement à traiter avec des gens qui n'entendoient la leur, au plutôt leur espèce de jargon plus difficle à entendre que la langue même.'[43] We have already noted John Curr's views on the separation of language between the 'mechanic' technologist and the educated reader. Over 40 years later a French inspector of industry, Le Play, visiting the Sheffield region where Curr had written, produced a remarkable passage, recently translated by Mr. Barraclough:

> The knowledge which the workers daily put to profit in the practice of their art and which alone could serve as a complete basis for the working of steel cannot at this time be considered within the domain of science. Among the obstacles which the savant has to conquer in this field of study I would particularly point out the following: The artists of all classes and in all countries are, in general, little disposed to communicate to others the results of their experience: among the bulk of the industrialists of Yorkshire this attitude is the established one. Moreover, such of the manufacturers who do show liberal intentions in this respect are themselves rarely in a position to give enlightenment on the operations which they only direct for commercial profit and of which they leave the technical direction to simple workmen. These latter are truly the metallurgists of Yorkshire and it is among them that one can gather the elements of steelmaking. But there, as elsewhere, there is barely a common language between the workman and the savant; it is, for example, extremely difficult to determine in many cases what qualities a workman means when he says that an iron has 'body', is 'sound', 'strong', 'tough', etc.; all of these, however, are expressions which have a very precise meaning and which distinguish properties which are perfectly

[42] *Ibid.*, p. 182.
[43] G. Jars, *Voyages Métallurgiques.* Vol. I (1774), xxiii. This passage comes from the introduction to Jars' posthumous work, edited by his brother who had accompanied him on some of his expeditions.

## 180        SKILLS, COAL AND BRITISH INDUSTRY

clear to the workman handling the iron. What increases the difficulty in this sort of study is that the expressions do not always have the same significance to two different branches of the steel trade.

If the works compiled by the savants say one thing and the workmen another, whom should one believe? Le Play tended to believe the worker:

> Many workers have assured me that the complex cementation mixtures indicated in most of the works which deal with steel manufacture and also in English technologies have never been generally employed. In their opinion, the bizarre recipes given on this subject, as well as the alleged fluxes necessary for the melting of steel have often been given with the aim of putting one off the scent of the real difficulties of steelmaking. The only reagent which I have seen charged with the iron in the cementation chests is crushed wood charcoal.[44]

Jars for some time wrongly doubted the information given to him by English workers, file makers in the North-East. They had told him that to harden their files they plunged them into water containing beer-lees and salt, and that the salt was the important thing. He was sceptical and tasted the water and thought that the taste came from the steel 'for the water appeared very clear to us'. Would he have believed them if they had told him that success depended on the angle at which the files entered the solution? Yet the man who was the last hand-forger of files in Lancashire and possibly in Britain said that it mattered a great deal.[45]

The gap between what could be conveyed by writing and what could be only learnt as a craft was particularly wide in puddling. As Mr. Gale confessed, 'while there are plenty of text books on the subject [of wrought iron making] puddling is an empirical trade; it cannot be learnt from a book'; however well he was able to expound the processes he knew that 'I . . . will fail to convey the "feel" of puddling. And in truth I could not earn my living as a puddler, though I have spent many hours at the furnaces and have handled all the tools. For proficiency I should have to spend years at the job . . .'[46]

I do not wish to suggest that the craft element was overwhelming in all branches of technology which were strongly dependent on the use of coal as fuel before (say) the mid-nineteenth century. Nor would I claim that the craft element was higher in the industries of a coal-using economy than in those which used wood fuels. The craft element did not remain constant and there were, of course, efforts to mechanize coal-using methods so as to abolish the need for skill or indeed to improve on human skills, witness attempts by the early nineteenth century to achieve automatic stoking. I would not deny that the coal-based technology of the industrial revolution replaced or reduced in importance a number of existing craft skills, even if it introduced others.

It is nevertheless the fact that there was a very large craft element in many of the newly-pioneered and progressively-developed technical processes by which industry was converted to the use of coal in Britain, a conversion which we could roughly regard as complete by 1790, but which had been growing in momentum for well over two centuries. The craft

[44] Barraclough, *op. cit.*, pp. 14–15, 17.
[45] G. Jars, *op. cit.*, Vol. I, p. 233: Information from Mr. F. Byrom of Prescot.
[46] W. K. V. Gale, *op. cit.*, p. 2.

element helps to explain several things. It helps explain why the great amount of industrial change involved in the shift to the new fuel occurs almost silently, so that the historian has to search hard for the evidence of it. It helps explain why contemporary technological literature is so limited in its coverage of developments which were making the British industrial situation remarkable and even unique. It helps explain why investigation by intelligent and learned foreign observers so rarely produced enough know-how to transfer British coal-using methods abroad. The multiplication of specialized crafts associated with coal explains why the emigration of the worker possessed of the 'central' craft in an industry was not enough. In the case of France I suspect that British workmen and managers emigrated there in more complete teams under the Bourbon Restoration and Louis Philippe than they had done under the *Ancien Régime* and that this concentration of crafts was critical to success.

The difficulty which contemporaries experienced in fully comprehending the significance of some of the coal-using crafts, and the near impossibility they found in explaining them in print, means that some bias has entered into the evidence available to those interpreting British technological progress before the Industrial Revolution, and even in its early decades. Contemporaries found it easier to convey in print the nature of mechanical invention, as in the textile industry, than they did the operation of furnaces, the nuances of empirical metallurgy, the making of refractories or even the cutting of coal. The distance between the British economies and other economies began to widen in the late eighteenth century at the point when increased mechanization began to gain ground. The French thus sometimes concentrated on bringing over machinery whose value in the division or abridgement of labour seemed, theoretically, so obvious; their subsequent pains and struggles in assimilating and domesticating the machinery were largely due to the lack or inefficiency of the supporting technologies, often with a craft element, which had a development period long anteceding the introduction of textile machinery or large machine tools.

Lynn White described history as 'a bag of tricks which the dead have played upon historians. The most remarkable of these illusions is the belief that the surviving written records provide us with a reasonably accurate facsimile of past human activity'.[47] One readily takes the point in respect of the middle ages, but I would suggest that the degree to which this applies down to the early decades of the nineteenth century, and into an era where there exists an apparently extensive and comprehensive literature on technology, is not appreciated.[48] In the coal-fuel technology built up in Britain

[47] *Mediaeval Technology and Cultural Change* (Oxford 1962) p. vii. H. Kellenbenz's chapter in the Fontana *Economic History of Europe: the Sixteenth and Seventeenth Centuries*, is an example of technological history heavily dependent on contemporary technological publications.

[48] There is an interesting relationship between the evidence in this paper (and the writer's earlier paper previously cited) and recent work by American historians. An examination of the problems of transferring British technologies to American industry is being made which parallels the author's interest in the problems of Anglo-French transference. The understanding of this subject has been greatly enhanced by Alfred Chandler's 'Anthracite Coal and the Beginnings of the Industrial Revolution in the United States' (*Business History Review* Vol. XLVI No. 2 1972). The attractiveness of the United States in terms of political liberty, economic opportunity and social mobility together with its common language recommended it

182            SKILLS, COAL AND BRITISH INDUSTRY

craft elements were embedded, and the essence of a craft is its dependence on a precarious combination of manipulative skill embodying a physical training and a judgement requiring both experience and intelligence. The resulting almost unanalysable pieces of expertise constituted the 'knacks' of a trade, and the essence of a 'knack' is its difficulty of communication. Chaptal, whose own work on the application of science to industry initiated an entire *genre* in the literature of technology, seems to be admitting this in an interesting passage on the transference of techniques.[49]

> We have seen several kinds of industry establish themselves and prosper in England, and render other nations subservient to their products over many years, we have made every effort we could to obtain for ourselves such manufactures, but in importing the machines, in putting dependence on various transmitted techniques, did we really believe we had naturalised these crafts in every particular? Can we believe we possess all the immensity of detail, all those *knacks*, those habits which are the soul of industry?[50]

to the emigrating British artisan who possessed coal fuel expertise. But a sufficient supply of cheap indigenous mineral fuel had to be identified and its technical characteristics understood before the coal-based technologies of Britain could be copied in large scale operations. W. Ross Yates in his 'Discovery of the Process for Making Anthracite Iron' (*The Pennsylvania Magazine of History and Biography,* April 1974) shows the important place of the Englishman George Crane and his expert employee, the Welshman, David Thomas, in the successful introduction of anthracite iron-making by the hot blast process, emphasising its foundation on experience and intuition. He points out that White and Hazard, expert mechanics themselves, decided to rely on British rather than American technologists as 'they were bargaining for the mechanical skill to make the process work as well as the process itself'. Again he refers to the need to buy much of the engineering equipment for the ironworks in Britain because of the lack of supporting technologies at a sufficiently high level in America; the firebrick came from Stourbridge, the blowing engines from Soho, Birmingham, and a Welshman with hard coal experience put the furnace into blast. The superior construction of the Thomas furnace he believes is the key to the success of this enterprise when other anthracite–iron furnaces failed.

    Mr. D. J. Jeremy, writing about the experience of the Philadelphia region in British technology transmission to the U.S. (*Business History Review* Vol. XLVII No. 1, Spring 1973), stresses the non-verbal, i.e. non-literary or non-graphic, nature of technological transference in the period 1770 to 1820, remarking that machines and models unaccompanied by a skilled mechanic were of very limited utility at this non-verbal stage.

[49] J. A. C. Chaptal, *De L'Industrie Française* (Paris 1819), Vol. II, p. 430.

[50] I would like to express my thanks to the Social Science Research Council for financial help with much of the research on which this paper is based. A number of friends and colleagues have helped me with advice and material, particularly Mr. W. K. V. Gale, Mr. K. Barraclough and Mr. D. Anderson; Miss Kathryn Ashworth has given valuable assistance with the research and Mrs. C. McCartney and Miss Celia Charlesworth with the preparation of the paper. A revised version of a lecture read at the University of Delaware in their *Technology and Work* series of 1974, this paper owes much to the stimulation of discussion with Eugene Ferguson, the most helpful and modest of scholars.

# [10]

*Textile History*, 13 (2), 225–248, 1982

# Immigrant Textile Machine Makers along the Brandywine, 1810–1820

David J. Jeremy

## Introduction

In the transfer of early industrial textile technology from Britain to the USA, a crucial phase in America's industrial development, the migrant artisan was the pre-eminent technology carrier. Between 1790 and the 1820s, a relatively small number — dozens rather than hundreds of British workers — conveyed information about the new manufacturing equipment and techniques across the Atlantic. If the movement is conceptualized in stages, it is clear that immigrant artisans were more important at the introductory stage (bringing information to America) in cotton spinning than at the introductory stage in cotton powerloom weaving, which Americans shared with British inventors in developing. At the prototype factory stage immigrants predominated where lengthy learning periods were required, as in mule spinning and cotton mill management. And at the stage of internal diffusion, immigrants' kin and trade networks, like those centred on Samuel Slater, played a major role before large capital goods firms were set up in the 1820s.[1] This broad picture does, however, need to be filled in, particularly with respect to the role of machine makers. How many immigrant machine makers were active in the respective mill districts in this early period? Were they introducing new equipment from Britain (acting as agents at the introductory stage of transfer)? Did they set up pilot plants (either pilot plants for the region or for the whole of the USA) or were they assuming more importance by loosening the supply of new but familiar equipment and by imparting their skills to native-born Americans? Only a series of local studies can answer these questions and this article is intended as one such study.

The focus of this paper is the Brandywine valley in Delaware, part of the much larger Philadelphia region which was then undergoing significant economic change. The whole commercial region comprised over 40 counties in Pennsylvania, Delaware and southern New Jersey. In Philadelphia and its hinterland, the conditions for industrial textile manufacturing ripened between 1800 and 1830. The city of Philadelphia, whose population doubled up to 80,000 in the three decades (and throughout the period was the second largest city in the USA), offered manufacturers a growing market, shielded from foreign competition by embargo and war between 1807 and 1815 and by some tariff protection in the 1820s. Though exports beyond the whole commercial region did not grow until after the 1820s, a water-borne freight trade was well organized along the Delaware River, while the western hinterland, stretching over 150 miles west of Philadelphia, was penetrated by turnpikes before 1810 and by canals in the 1820s.[2]

The three counties closest to Philadelphia and on the western bank of the Delaware River (Fig. 1) offered numerous waterpower sites for manufacturing. The most developed stream

*Immigrant Textile Machine Makers along the Brandywine*

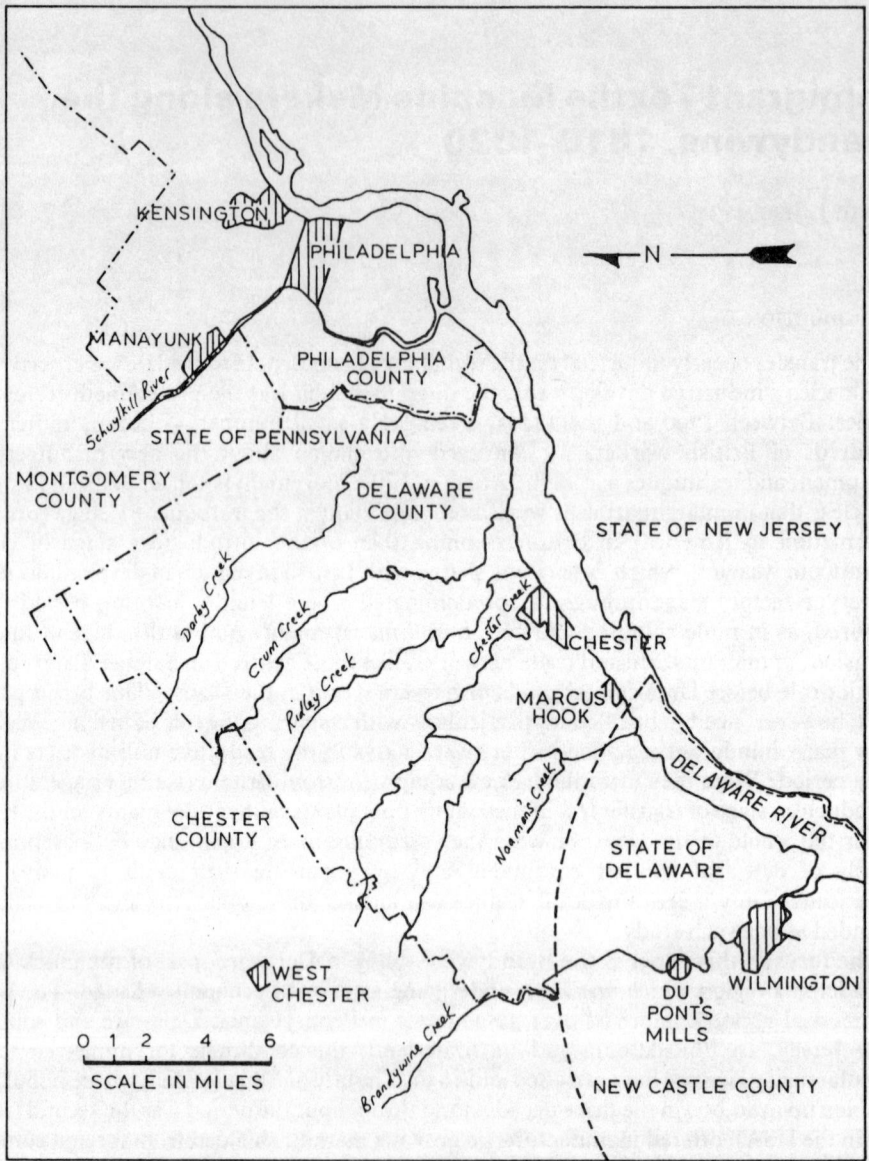

FIG. 1.   The Philadelphia–Wilmington area. Adapted from Anthony F. C. Wallace, *Rockdale, The Growth of an American Village in the Early Industrial Revolution* (New York: Alfred A. Knopf, 1978), 2.

DAVID J. JEREMY

TABLE 1.   COTTON SPINDLES IN THE VICINITY OF PHILADELPHIA, 1810–1830

|  | 1810[1] | 1820[2] | 1826[3] | 1831[4] |
|---|---|---|---|---|
| Pennsylvania |  |  |  |  |
| Philadelphia County |  | 14,038 |  |  |
| Philadelphia City |  | 7[a] |  |  |
| Delaware County |  | 1,680 | 15,116 |  |
| Total |  | 15,725 |  |  |
| Total Pennsylvania | 8,849 | 23,853 |  | 120,810 |
| Delaware |  |  |  |  |
| Newcastle County |  | 14,204 |  |  |
| Total Delaware | 1,822 | 14,204 |  | 24,806 |
| Pennsylvania and Delaware | 10,671 | 38,057 |  | 145,616 |
| Total USA | 122,647 | 342,266 (356,900)[b] |  | 1,246,503 |

*Sources*

[1] Tench Coxe, *A Statement of the Arts of the United States of America for the Year 1810* (Philadelphia: A. Cornman, Jr., 1814).
[2] Jeremy, *TIR*, Appendix D.
[3] George C. Leiper and William Martin, *Report* (1826).
[4] Friends of Domestic Industry, *Report*, 16.
[a] Spinning wheels probably.
[b] Highest figure from extrapolation.

was the lower Brandywine, where 14 merchant flour mills, pivoted between the farmlands of Pennsylvania and consumers on both sides of the Atlantic, ground half a million bushels of wheat in 1816.[3] These mills, in which Oliver Evans invented his automatic flour processing system, and other mills and industrial works in Philadelphia, created a reservoir of millwrighting and mechanical experience that would help to set up the new factories. By 1810 Pennsylvania reportedly had 7.2% of the USA's 120,000 cotton spindles and Delaware 1.48% (*see* Table 1). A decade later, Pennsylvania's share of the country's 350,000 spindles had not significantly changed while Delaware's had more than doubled up to about 4.2%. In absolute terms, however, while Pennsylvania's share went up 2.7 times, from 8,849 to 23,853 spindles, Delaware's shot up 7.8 times, from 1,822 to 14,204 spindles.[4] Obviously Delaware's cotton industry in the decade 1810–1820 deserves closer attention. In the 1820s the growth rate of Pennsylvania's cotton industry, measured in spindles, climbed more steeply than Delaware's: the latter's grew by 1.7 times and Pennsylvania's by 5 times. By 1831 Delaware had 24,806 cotton spindles compared to its giant neighbour Pennsylvania with 120,000 spindles.[5] Consideration of the Pennsylvania counties in the 1820s is planned for another article.

## MACHINE MAKING AND IMMIGRANT MACHINE MAKERS

### Machine making and immigrants

In the establishment of a spinning industry, throughout the USA as much as in the Philadelphia region, immigrant British machine makers played a key role and for several

*Immigrant Textile Machine Makers along the Brandywine*

broad reasons. Firstly, of course, Britain was the fountainhead of the new technology associated with factory organization and machine production. At the non-verbal state of that technology, before it was reduced to written or graphic forms and operating rules in the decades 1812–1832, the new knowledge and skill both in building and operating machine spinning systems resided in experienced artisans. While American craftsmen — carpenters, cabinet makers, wheelwrights, ironmasters, blacksmiths, wiredrawers and clockmakers — could obtain and to some extent shape the basic raw materials of wood, iron, steel and brass, only British sources could supply machine designs, constructional information for new components and assembly information for those components. And Britain's prohibitory laws, probably acting more effectively against the export of machinery than the emigration of artisans, increased reliance on skilled workers.[6]

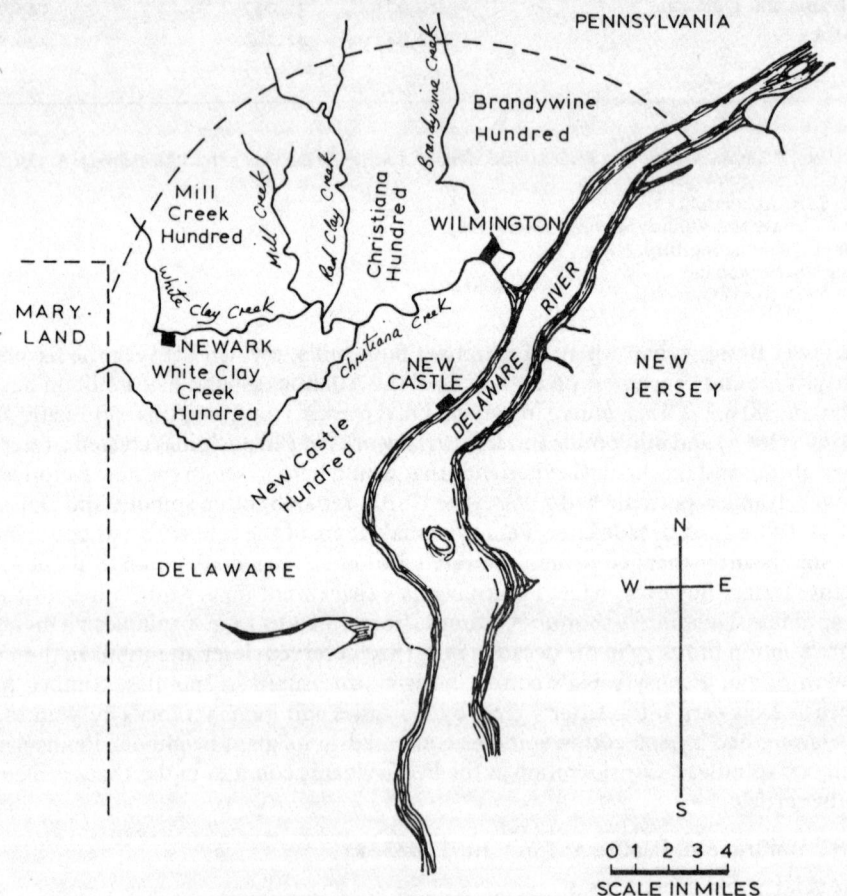

FIG. 2.   Northern Hundreds in New Castle County, Delaware.
Source: Daniel G. Beer, *Atlas of The State of Delaware* (Philadelphia, 1868).

228

DAVID J. JEREMY

By the War of 1812 the nature of textile machine making was changing fast. The first American machines, built in the early 1790s, were of wood, incorporating few metal parts (by quantity); thus two New York cabinet makers could advertise in 1791 that besides household furniture they would make spinning jennies and carding engines.[7] By 1814 iron was replacing wood in the cotton machines of the Philadelphia region mills. And whereas some of these early machines were suitable for home use, the later equipment was intended

TABLE 2.   IMMIGRANT MACHINE MAKERS IN THE VICINITY OF PHILADELPHIA DURING THE WAR OF 1812

| | Phila. Co. | Phila. City | Del. Co. | Total Pa. | Wilmington, Brandywine Hundred & Christiana H. | Mill Creek and White Clay Creek Hs. | Total De. | Total USA |
|---|---|---|---|---|---|---|---|---|
| *General* | | | | | | | | |
| Millwright | 3 | I | | 4 | I | I | 2 | 36 |
| Engineer | | I | | I | | | | 2 |
| Machinist | | | | | 6 | 2 | 8 | 10 |
| Machine maker | 3 | I | | 4 | 10 | | 10 | 45 |
| Steam engine builder | | I | | I | | | | |
| Tinplate workers | | I | | 2 | | | | 14 |
| Total | 6 | 5 | | 12 | 17 | 3 | 20 | 125[a] |
| *Textile* | | | | | | | | |
| Cotton machine maker | | I | | I | | | | 9 |
| Wire drawer (surrogate for card makers) | II | I | | 12 | | | | 15 |
| Spinning wheel maker | | | | I | | | | I |
| Turner (surrogate for spindle makers) | | 6 | | 6 | | | | 20 |
| Spindle maker | | | | | I | | I | I |
| Reed maker | | I | | I | | | | 4 |
| Shuttle maker | I | | | I | | | | 2 |
| Total | 12 | 9 | | 22 | I | | I | 60[a] |
| Grand totals | 18 | 14 | | 34 | 18 | 3 | 21 | 185 |

*Sources*

US National Archives, RG 45, 59, War of 1812 Papers, Marshals' *Returns of Enemy Aliens.*
[a] Including other workers and trades, *see* Jeremy, *TIR*, Appendix C.

## Immigrant Textile Machine Makers along the Brandywine

for factory production, requiring increased size, greater power, and a longer machine life. In addition, the new machines were part of a total interlocking production system, unlike the isolated attic jenny. A balance between the outputs of the equipment at each stage of processing (carding, drawing, roving and spinning) was therefore essential if mill managers were to avoid bottlenecks. While managers and overseers could regulate production line output to some degree, the parameters were settled by the original design of the machines ordered from the machine maker, who submitted suggestions for the mill manager to consider before the order was placed. A flow production system also intensified the need for a much higher level of mechanical reliability. This requirement was in turn reinforced by inputs of relatively lower operative skill with the Arkwright system of spinning, compared to the mule system. As in England, the changing nature of textile machines altered the character of machine making. Cabinet makers, clockmakers, and blacksmiths were replaced by specialist machine makers, numbers of whom came from Britain where a much larger market and industry supported a much greater division of labour.

This is one feature made clear in Table 2. Together, Pennsylvania and Delaware attracted nearly 30% of all recorded immigrant machine makers in the USA during the War of 1812 — compared to their combined share of 24% of all textile immigrants (operatives, managers, machine makers).[8] Within the two states there was a strong tendency for general and specifically textile machine makers to cluster in Philadelphia City and County and, more strongly, for general machine makers to concentrate on the Brandywine Valley, bounded to the east by Brandywine Hundred and to the west by Christiana Hundred, with the four thousand-strong city of Wilmington at the river's lowest bridging point (Fig. 2).

With these generalizations in mind we can now look at the work of several immigrant machine makers and try to discern the sort of part they played in the mechanization of textile manufacturing in the Philadelphia region during the formative period 1810–1820.

### Daniel Large

One of the first new specialist machine makers in the region was Daniel Large who emigrated from England in 1806 at the age of 25.[9] In 1808 he built and installed a 10 horsepower low pressure engine to grind paint at the white-lead works of Wetherill and Brother in Philadelphia. At the instigation of Benjamin Henry Latrobe, engineer of the Philadelphia waterworks and also an Englishman, Large formed a partnership with James Smallman, another immigrant and a former employee of Boulton & Watt. Smallman ran the rolling mill appended to the waterworks where he started building rather poor steam engines in 1804. They came together because Smallman needed Large's improvements, including the Watt parallel motion, and Large needed a loan of $10,000. But the partners fell out and in 1811 Smallman sued to recover his money, forcing Large, not for the last time, into bankruptcy. Large nevertheless continued to build steam engines at his foundry in Otter Street below the German Road and Latrobe regarded him, with Smallman and Oliver Evans, as the leading engine manufacturers of the region.[10] When Large registered as an enemy alien in August 1812 he gave 'Steam engine maker' as his occupation; at this point he lived on North Front Street with a wife and three children.[11] That his early machine building work was not confined to steam engines is clear from a later New England patent case in which one of the witnesses testified that Large had built four-cone double speeders in Philadelphia 14 years earlier, in 1807. Models of this important cotton roving

DAVID J. JEREMY

machine, patented in Britain in 1797, must have been among the first of their type built in the USA.[12]

It was therefore to Large that some of the region's cotton manufacturers turned during the Embargo and War of 1812, years when severed commercial and diplomatic relations between the USA and Britain lent indigenous American manufacturers widened prospects in expanding home markets. One of these manufacturers was Duplanty, McCall & Company, a five-man partnership, formed early in 1813, which set up a cotton factory on the Henry Clay mill site on the Brandywine with a modest capital of $30,000. Eventually the firm had a four-storey building measuring 73 × 44 feet containing 1800 spindles. The partners, tempted into the manufacture of relatively fine yarns (Nos 40–50 and later Nos 100–150), were eventually caught out by the inundations of imported English and Scottish yarn, thread and cloth in 1815–1816. Army contracts kept the firm going a few more years but on 12 August 1819 operations ceased with the two McCalls and the two du Ponts estimating their collective losses at around $50,000 (though the company's tangled financial affairs were not wound up until 1839).[13]

While the partners had previously associated in gunpowder and woollen manufacturing, none of them had any practical experience of cotton manufacture. Archibald McCall (1767–1843), the firm's general agent and a leading merchant in Philadelphia, sought out Large because he had 'made the machinery for Craig's Cotton Factory'[15] — located in the Globe Mill just north of the Philadelphia city boundary.[15] 'Large is unquestionably the best mechanician in this country, and his knowledge on all those subjects far before Coxan's', McCall told Raphael Duplanty, manager of the factory to be set up in the Brandywine mill.[16] McCall repeated the opinion of Samuel Wetherill (1736–1816), the Quaker chemical manufacturer who had earlier bought a steam engine from Large.[17] Wetherill rated Large as 'a very modest man', but thought that Coxan, an alternative British steam engine and machine maker in Philadelphia, was a 'visionary and scheming and always promises more than he can do, from not perfectly understanding the principles of his business'. Resting on Wetherill's opinion, for he regarded Wetherill as 'a man of sound judgement and one who would not deceive one', McCall urged Duplanty to employ Large. Duplanty had already made contracts, which will be examined later, with local machine makers for the mill's first set of cotton spinning machinery.[18] Fortunately, as it turned out, Large could only be hired on the fitting up side of the factory.

A week later Large, evidently preparing estimates for the mill's power transmission equipment, was asking for details of the projected spinning machinery:

Whether the works were to be calculated for mules or trussels [throstles]? If of each kind, what to be the proportion of mules to the trussels? How many spindles you wish to each trussel? If the trussels are to drive in the middle or at the ends? Whether with rims or without? If the mules are to be double or single? He says there is a material difference in the expence and therefore it is necessary for him to know which you wish.[19]

This was no salesman's bluff: much more power was needed to drive throstle spindles (about 100 per horsepower) than to drive mule spindles (700–1000 per horsepower).[20] Consequently the size and cost of gear wheels and shafts varied with the combination of spinning machines chosen — though no one had yet devised a means of accurately measuring an individual machine's power consumption. McCall, in discussions about these

231

## Immigrant Textile Machine Makers along the Brandywine

matters with Large, was especially heartened to learn that he 'has been a cotton spinner himself in England and appears to know a good deal of the business'.[21]

Duplanty and McCall decided to employ Large to build 'the pier head, water wheel and gearing' if his prices were reasonable. In this way McCall forecast that Large would also have to impart

all the information he possesses relative to a plan for the Factory. But, having once obtained his ideas, I mistake if you and Mr. Dupont are not much more capable of drafting a proper plan for a Factory than he is. However, of this you will be better able to judge when You converse further with him.[22]

McCall advised Duplanty to contract at fixed prices for items Large himself would make with a commission on the rest at a rate which depended on the total amount of the bills.[23]

The war years brought numerous demands for Large's services, and McCall could not get hold of him for days at a time. Eventually, on 15 April 1813, three weeks after the factory's spinning machinery had been ordered, Large was on his way to Wilmington and the Brandywine. Soon after, he signed a contract to build the Duplanty McCall & Co. factory's water wheel, gearing and shafting, and its heating system.[24]

In the months, and years, that followed, McCall and Duplanty's plans to use and exploit Daniel Large turned out far differently than they expected. Large rarely and reluctantly visited the Brandywine. He himself made only the more profitable lighter castings and did not closely superintend the heavy work which he subcontracted to the Weymouth Ironworks in New Jersey. A fly wheel and a crown wheel were mislaid in transit to the factory site.[25] And not until mid-January 1815, almost two years after the contract was settled, were the water wheel and its power transmission system ready to run. By this time McCall was exhausted and exasperated by what he regarded as Large's incompetence. He had discovered that Large dealt similarly with a Philadelphia merchant, Francis Makoe, when installing a steam engine in a saw mill. McCall concluded

I believe the fellow is an imposter who has undertaken things he knows nothing about and endeavours to pick up his knowledge as he proceeds and to learn his business at the expence of his employers. Such a man deserves to be punished.[26]

Harsh words! Were they really deserved?

From Large's correspondence it seems that his water wheel design was not unsound. Indeed its iron construction and its diameter are reminiscent of the wheels in the Belper North Mill, Derbyshire, belonging to the Strutts, which were well known and described in Rees' *Cyclopaedia* in November 1812.[27] Whether Large was personally familiar with Belper is unknown but not unlikely. Large specified a breast or undershot (which is unclear in the record) wheel 18 feet in diameter to sit in a head and tail race 10 feet wide and 4 feet deep. The shaft of the water wheel was to turn 11 feet above the bottom of the race or 9 feet over the usual water level. Enclosed within the factory (to reduce the hazard of icing in winter), the water wheel was located at one end of the building so that machinery in an extension could be easily powered.[28]

If his design was sound, Large was surely at fault in his choice of construction materials. Some of his shafts were of cast iron but some, including the main shaft on which the cast-iron water wheel turned, were of wood, unlike the Belper wheels which were cast in sections together with their axles. With a weight of 17,569 pounds of cast iron resting on a 25 foot long shaft, Large's arrangement proved to be a serious mistake. Over the years the

DAVID J. JEREMY

shaft wore and warped and produced desultory attempts by local millwrights to tighten and repair it and an unnecessary irritant for the mill managers.[29] The company bitterly quarrelled with Large over his water wheel installation and delayed settlement of his bills. Eventually Large brought an action against Duplanty which came to a head in 1819. In a deposition, James Blackie, Large's pattern maker, testified that his employer 'was very particular in giving directions about the said models [wooden patterns for castings]' when they were being made in 1814. Further, that Duplanty and McCall visited Large's workshop to monitor the progress of these patterns. The defendants countered with testimony from other millwrights, William Murphy and Englishman Thomas Oakes, arguing that Large 'to increase his percentage, [made] all the Casting . . . three or four times as Heavy as it should have been' so that the wooden members of the water wheel were overstrained.[30] Whether Large designedly or unwittingly erred remains a matter of doubt.

Some months before the final fracture and his total condemnation of Large, McCall attributed technical difficulties to Large's ignorance of America's natural resources:

I do not think Mr Large has been long enough in this country to be well acquainted with the qualities of our Timber. I have not seen him, but without knowing his opinion I can inform you [he told Duplanty], that Spanish Oak is not proper for a Mill shaft.[31]

Ship carpenters would not use this sort of oak if it was to be exposed alternately to wet and dry conditions; they preferred white oak or, if this was unobtainable, white pine. Several years after Duplanty, McCall & Co. stopped manufacturing, the surviving partners scrapped Large's cast-iron water wheel and replaced it in 1823 with a wooden one, signalling a reversion to a traditional but better known technology.

In fire-proof construction Large's proposals were progressive in concept, if not in execution, and were derived from British best-practice. The upper factory floors were to be supported by iron pillars, and piped water raised by a forcing pump was to be used both to flush water closets (his suggestion for the form of 'the necessaries') and to 'overflow the different floors of the factory in case of fire'. Duplanty feared that temperature variations would cause the metal pillars to expand and buckle the floors but Large assured him that they had been used in Manchester for about eight years with no such ill-effects.[32] In fact iron pillars had been employed in factory architecture as early as 1793–1795, at Belper, a feature also reported in the Rees' *Cyclopaedia* article of November 1812.[33] Duplanty evidently modified Large's ideas in some particulars. A staircase to the upper floors was placed outside the factory building instead of inside. And Alexander McCall offered another alteration: 'if the Necessaries could be over the water, I should suppose it would be very desirable & prevent much of the offensive smell arising.'[34] Whatever the extent of the modifications made by the mill managers and their Philadelphia agent, it is fairly clear that Large was not giving wholly bad advice. Sound in design principles, it was defective and damaging because it was poorly executed. And it was poorly executed chiefly because Large was too busy with other orders to go down to the Brandywine or to chase up subcontractors and supervise his share of the work in person and in detail.

Little is mentioned in the correspondence received by Duplanty about the heating system ordered from Large. In fact, Duplanty turned for advice on this matter to a New England manufacturer who recommended using Russian stoves.[35] In spring and summer 1814 the Brandywine manager was considering two rival plans for conveying heat from

15                                                                                      233

*Immigrant Textile Machine Makers along the Brandywine*

stoves around the factory — one using horizontal hot air pipes, the other vertical hot air pipes, both embedded in the mill walls — but Large was associated with neither.[36]

Though he made no machinery for the Brandywine company, Large certainly supplied advice about the equipment ordered and installed by Duplanty, McCall & Co. On the difficult but crucial question of balancing machine capacities, he told McCall that

to correspond with the 6 finishers, 24 In. in the wire & the main Cylinder 36 In. diameter, it will require 2 drawing frames of four heads each & 3 roving frames of 12 Cans each.[37]

Whether the recommendations, expressed with the precision associated with accuracy, were sound we cannot tell for the data concealed design details. But since he had the contract for the factory transmission system and other machine makers the contract for the manufacturing equipment, Large would have had to be extraordinarily devious to give deliberately bad advice at this point. Although he did not get the Duplanty, McCall & Co. machinery contract (because it was settled before he appeared on the scene), Large was evidently a competent machine maker as well as steam engine manufacturer. He built cotton machinery during the War of 1812, if not after 1815 when capital goods' demand suddenly fell. In spring 1818, when mills and machine makers were going out of business, Archibald McCall was searching for bargains in second-hand equipment. After failing to snap up Joseph Bamford's stock — Bamford, an English machine maker operated in Philadelphia from 1805 until he failed (and went to jail) in 1818 — McCall found a line of Large's machines at Jenkin Town, 12 miles north of Philadelphia. It consisted of a picker, 2 cards, a drawing frame (4 heads), a roving frame (6 heads), 2 throstles (108 and 132 spindles), a grinding cylinder, 3 bobbin winders, a warping mill and 3 handlooms. The equipment, costing $3000, had run for a few months before the peace (December 1814). Now, three years later, it was nearly as good as new. The vendor, who seemed fairly impartial, rated the throstles as excellent. They had cast iron frames and brass stands. The cards were good, though their clothing was damaged. The picker was small and no longer the latest design while the roving frame, like most, was 'apt to get out of order.' The lot was available for $1200 in 1818 and McCall, short of cash and credit, was negotiating to pay for them with a small house belonging to his mother.[38] This appraisal of Large's machines in the absence of their builder sufficiently demonstrates that Large was technically competent as a machine maker and that his problems arose less from his technical failings and more from his commercial ineptitude, his failure to organize his business activities efficiently. Essentially he was over-stretched and apparently lacked competent partners who, by their injection of skill and capital, could help his business to expand and meet growing wartime demands. Apparently Large went bankrupt again, sometime between 1815 and 1820, if John Price Crozer's memory was good.[39] What happened to him thereafter is unclear. His name appears in the Philadelphia directories until 1840 but in 1821 he was reported at Eastport, Maine — serving a customer or conveniently on the Canadian border far from his Philadelphia creditors?[40]

*The Hodgson Brothers*

Another firm of immigrant British machine makers active in the Philadelphia region during the formative period in cotton factory spinning, the Hodgson family, presented various contrasts to Large in their migration and activities on the Brandywine. These five brothers,

DAVID J. JEREMY

sons of John Hodgson, a turner and roller maker of Balloon Street, Manchester, emigrated to the USA between summer 1811 and the beginning of 1813 and settled in Christiana Hundred (Fig. 2). All were aged between 17 and 29 and arrived without family — as was typical of most textile immigrants from Britain at this time.[41] Family tradition has it that they evaded Britain's prohibitions on artisan emigration by posing as farmers and sending their tools separately, marked as fruit trees.[42] Be that as it may, they set up on the Brandywine in the vicinity of flour, gunpowder and woollen mills and at least one cotton spinning operation. We know little about their capital resources except that in August 1813, George, Thomas and Isaac Hodgson, soon after the fifth brother Henry arrived, were secured in business by several Brandywine manufacturers and Philadelphia merchants, at a time when the demand for textile machinery was rapidly outstripping supply. Several local capitalists (including Caleb Kirk, who had an older cotton and wool mill, and E. I. du Pont who operated gunpowder works, a woollen mill and was just venturing into cotton with Duplanty, McCall & Co.) formed the Brandywine Mill Seats Company to speculate in mill property. The Hodgsons contracted to purchase from this group a property on the Brandywine at Rockford previously belonging to Job Harvey, 'below the old dam', for the sum of $5000, repayable in nine annual instalments of $555.55 commencing on 1 May 1814.[43] They remained at Rockford until 1818; George Hodgson moved to another site near Rokeby also on the Brandywine before 1822, by which date the brothers had split up, one going West.[44] George stayed on the Brandywine until his death in 1849 though his machine works continued under his name until 1854 at least.[45]

As a kin-knit group of machine makers, they were in a much stronger position than Daniel Large, or any of the other eight or nine machine shops on or near the Brandywine,[46] to exploit opportunities for the specialization of labour presented in 1813–1814 by the fast-expanding regional capital goods market. Their work for Duplanty, McCall & Co. reveals the situation in which they found themselves and the ways in which they contributed to the mechanization of the region's cotton industry: primarily by adding to the supply of machine makers and lowering capital goods prices; by allowing their machine making skills to pass to native Americans; and by introducing the latest British equipment to the region.

At first the Duplanty, McCall & Co. partners approached other manufacturers for recommendations to machine makers: E. I. du Pont contacted the McKims, leading cotton manufacturers in Baltimore, while Archibald McCall made enquiries as we have seen, in Philadelphia.[47] The McKims' machine makers (John Duff and W. H. Richardson) had orders that would keep them busy until the end of 1813, so the Brandywine partners looked nearer home[48] — but now armed with a schedule of Baltimore machinery prices.[49] Within ten days of hearing from the Baltimore machine makers, Duplanty had three sets of quotations in front of him (Table 3). The Baltimore one could be ruled out because Richardson & Duff could not start work on a new order until the end of the year, nine months hence, and in any case their equipment imposed heavier transportation costs. The quotation from Alrichs & Dixon, a firm established in Wilmington in 1808,[50] specified prices which were fractionally higher than Richardson & Duff's; like the Baltimore machine makers', theirs did not include accessories (change gears, bobbins, cans, leather covering on rollers, chiefly); and once the makers 'deliver them to the waggons at the shop door, we consider [them] out of our charge.'[51] The Hodgson brothers, in contrast, offered

*Immigrant Textile Machine Makers along the Brandywine*

TABLE 3. MACHINE MAKERS' PRICES QUOTED TO DUPLANTY, McCALL & CO., BRANDYWINE VALLEY, NR. WILMINGTON, DELAWARE, 1813–1814 (PRICES IN DOLLARS)

| Machine | Richardson & Duff Baltimore, 12 March 1813[1] | Alrichs & Dixon Wilmington, 22 March 1813[2] | George Hodgson & Brothers | | | McFee & Newton, Wilmington, 1814[6] |
| --- | --- | --- | --- | --- | --- | --- |
| | | | Brandywine | | Contract, 22 March 1813[5] | |
| | | | Undated but March 1813[3] | Undated but (March?) 1813[4] | | |
| Blowing machine | 300 | 300–400 | 150 | 150 | 150 (returnable if not approved) | 400 |
| Carding machine (without card clothing) | 200 | 200 | 190 (white pine covering)* 195 (walnut)* 205 (mahogany)* | 200 | 200 (mahogany covering) | |
| Drawing frame (per head) | 35 | 25–40 | 38 | | 38 | |
| Roving frame (per head) | 30 | 25–40 | 31 | 31 | 31 | |
| Stretcher (variant of mule for roving stage) (per spindle) | | | | (108 sp.) | | 6.25 (114 spindle stretcher) |
| Mule (per spindle) | 2.50 (240 sp. mule) | 2.60 (240 sp. mule) | 3 (216 sp. mule) | 3 (216 or 240 sp. mule) | | |
| Throstle (per spindle) | 5.00 (120 sp. throstle) | 5.00–5.50 (without change gears, creel bobbins or covering on top rollers) | 5.50 (with all accessories) (120 sp. throstle) | 5.50 | 5.50 | |

* Each with iron frame and spur gears.

*Sources:* [1] EMHL, Longwood MSS, Group 6, Box 1, Richardson & Duff to E. I. du Pont, 12 March 1813. [2] Ibid., Group 6, Box 2, Alrichs & Dixon to R. Duplanty, 22 March 1813; ibid., Alrichs & Dixon quote, 1813. [3] Ibid., undated quote from Hodgsons: 'List of Machinery for Messrs Du Pond & Co.' [4] Ibid., another sheet listing Hodgsons' prices and written by one of the Duplanty, McCall & Co. partners. [5] Ibid., agreement dated 22 March 1813. [6] EMHL, Acc. 500, Duplanty, McCall & Co. Papers, accounts, 18 July 1814–25 January 1815.

DAVID J. JEREMY

prices the same or marginally less than their competitors', except for the blowing machine — half the price of their rivals' products — and mule spindles — twice the going price. In addition the Hodgsons offered their drawing, roving and spinning equipment 'compleate in Every thing',[52] except for the card clothing on their iron-framed carding engines. On 22 March 1813, the very day he received Alrichs & Dixon's quotation, Duplanty signed a contract with the Hodgsons in which he extracted even more favourable terms from them. The immigrant machine makers agreed:

The whole of the said machinery to be put up by ourselves, and in case some alteration is wanted, the same to be made at our own expense, it being well understood that the machinery herein stated shall be considered as delivered To & received by R. Duplanty & Co. at the time only when it shall have been Tried and have proved to work To satisfaction.'[53]

Duplanty was clearly suspicious of the low priced blowing machine and stipulated in the contract 'one Blowing machine, to be returned after Trial if R. Duplanty & Co. do not like its performance.'[54] The Hodgsons matched their keenly competitive prices and installation arrangements with the promise of speedy delivery dates: the blowing machine, one breaker and two finisher cards and the drawing frame would be set up by 15 October 1813; two breaker and three finisher cards, one roving frame and one throstle spinning frame a month later; and the rest, one breaker, one finisher, one roving frame and one throstle, by 15 December 1813, equipment totalling $4360. If the Hodgsons failed to keep to these dates, delivery was overdue by only a few months: the ten cards were recorded in the company's ledger as completed on 30 April 1814.[55] Perhaps to test the quality of the Hodgsons' work and certainly to avoid their high mule spindle prices, Duplanty ordered at least one more throstle frame of 120 spindles and a pair of mules (totalling 480 spindles) from Alrichs & Dixon.[56]

From their viewpoint the Hodgsons were not wholly exploited. Some of the prices they quoted were comparable with (i.e. slightly higher than) those prevailing in Manchester whence they had just come.[57] Above all the contract helped to secure their foothold in the Philadelphia region market. It no doubt led to the purchase of a millseat and workshop, as already noted, five months later.

The Duplanty, McCall & Co. factory ledger, opened in April 1814, shows that over the working life of the factory, until it closed in 1819, the Hodgsons supplied 19.4% of the factory's machinery, as measured in dollar costs.[58] They supplied much more than this of the original production line. A total estimate of the Hodgsons' contribution to this particular Brandywine mill is impossible to make because of the appearance in the 1814–1819 ledgers of items (like the 10 cards) which really belonged to the 1813 accounts.

What is especially interesting is that when machinery prices shot up in 1814, as Robert McCall remembered,[59] the firm bought few complete machines. Their purchases, as their account with the Hodgsons shows, were increasingly of spare or key parts. This account (Table 4) contains a number of features which throw further light on the Hodgsons' role as machine makers. In most cases their prices were standardized according to weight, except for gears, priced according to the number of teeth; and throstle spindles, flyers and bobbins, individually priced. The basis of the cast-iron price presumably was the cost of the iron plus labour costs for making the wooden patterns and pouring castings, plus profit. For steel spindles, flyers and rollers, requiring machining, the price reflected the wages of machinists as well as the cost of steel which was not easy to get in wartime.

*Immigrant Textile Machine Makers along the Brandywine*

TABLE 4.    STATEMENT OF ITEMS SUPPLIED BY THE HODGSONS TO DUPLANTY, McCALL & CO.,
1814–1815

|      | Date      | Item                                                            | Amount ($) |
|------|-----------|-----------------------------------------------------------------|------------|
| 1814 | 27 July   | Castings, 139 lb @ 12½ cts                                      | 17.37½     |
|      | 1 Aug     | Castings,  29 lb @ 12½ cts                                      | 3.62½      |
|      | 6 Aug     | Castings, 126 lb @ 12½ cts                                      | 15.75      |
|      | 6 Aug     | Castings,  39 lb @ 12½ cts                                      | 4.87½      |
|      | 9 Aug     | Lignumvitae, 62 lb @ 6 cts                                      | 3.72       |
|      | 9 Aug     | 14 roving cans @ $1.75                                          | 24.50      |
|      | 9 Aug     | 4 brass guides, 1½ lb @ 75 cts                                 | 1.12½      |
|      | 9 Aug     | 1 brass wheel, 37 teeth @ 4 cts                                | 1.48       |
|      | 15 Aug    | Castings, 12 lb @ 12½ cts                                      | 1.50       |
|      | 24 Aug    | 17 front and 34 back mule rollers, 136½ lb @ 5 cts            | 68.25      |
|      | 3 Sept    | Castings, 7 lb @ 12½ cts                                       | 0.87½      |
|      | 11 Sept   | 17 boxes of mule rollers, 137½ lb @ 5 cts                     | 68.75      |
|      | 21 Sept   | A Casting, 73 lb @ 12½ cts                                     | 9.12½      |
|      | 31 Oct    | Castings, 3 lb @ 12½ cts                                       | 0.37½      |
|      | 31 Oct    | 42 lb of cast wheels @ 15 cts                                  | 6.30       |
|      | 31 Oct    | 2 weights, 16 lb @ 10 cts                                      | 1.60       |
|      | 12 Nov    | 1 set of mule rollers, 139 lb @ 5 cts                         | 69.50      |
|      | 12 Nov    | 10 dozen throstle spindles and flyers @ $9.50                 | 95.00      |
|      | 17 Nov    | Mule weights, 619 lb @ 10 cts                                  | 61.90      |
|      | 21 Nov    | Castings, 20½ lb @ 12½ cts                                    | 2.56       |
|      | 21 Nov    | Castings,  4½ lb @ 12½ cts                                    | 0.56       |
|      | 28 Nov    | Castings, 51½ lb @ 12½ cts                                    | 6.43½      |
|      | 28 Nov    | Castings,  7½ lb @ 12½ cts                                    | 0.90       |
|      | 31 Dec    | 10 carding engines, $200 each                                  | 2000.00    |
|      | 31 Dec    | 1 set of mule rollers, 138 lb @ 50 cts                        | 69.00      |
|      | 31 Dec    | 2 sets of mule rollers, 276 lb @ 50 cts                       | 138.00     |
|      | 31 Dec    | Weights, 14 lb @ 10 cts                                        | 1.40       |
| 1815 | 17 Feb    | 10 dozen throstle bobbins                                      | 4.16       |
|      | 24 March  | 17 boxes of mule rollers, 145 lb @ 50 cts                     | 72.50      |
|      | 3 May     | Worm wheel, 30 teeth @ 5 cts                                   | 1.50       |
|      | 5 May     | 2 pair of card rollers, 73½ lb @ 40 cts                       | 29.40      |
|      | 12 May    | 1 set of mule rollers, 142 lb @ 40 cts                        | 56.80      |
|      | 16 May    | Castings, 65 lb @ 12½ cts                                     | 8.12½      |
|      | 8 June    | 6 carriage wheels, 36 lb @ 12½ cts                            | 4.50       |
|      | 8 June    | Turning wheels                                                  | 3.00       |
|      | 8 June    | 2 pinions, 19 teeth: 38 teeth @ 2 cts                         | 0.76       |
|      | 17 July   | 2 large end rollers, 10½ lb @ 40 cts                          | 4.20       |
|      | 23 July   | 12 large weights, 184 lb @ 10 cts                             | 18.40      |
|      | 24 July   | Quantity of plates, 159 lb @ 10 cts                           | 15.90      |
|      | 26 July   | 5 cast supports, 13½ lb @ 12 cts                             | 1.68½      |
|      | 9 Aug     | 1 cast wheel, 5½ lb @ 12½ cts                                | 0.68½      |
|      | 11 Aug    | Castings, 7½ lb @ 12½ cts                                     | 0.94       |
|      | 11 Sept   | Castings, 5½ lb @ 12½ cts                                     | 0.69       |
|      | 27 Sept   | 2 hearts [heart wheels], 4 lb @ 12½ cts                       | 0.50       |
|      | 27 Sept   | Brass worm wheel, 16 teeth @ 5 cts                            | 0.80       |
|      | 12 Oct    | 8 twisting rollers, 38 lb @ 40 cts                            | 15.20      |
|      | 17 Oct    | 18 boxes of throstle rollers, 173 lb @ 40 cts                | 69.20      |

238

DAVID J. JEREMY

| Date | Item | Amount ($) |
|------|------|------------|
| 17 Oct | Blower arms, cast, 37 lb @ 12½ cts | 3.62½ |
| 26 Oct | 8 twisting rollers, 35 lb @ 40 cts | 14.00 |
| 26 Oct | 2 drawing heads @ $35 | 70.00 |
| 11 Dec | 5 rollers, 21 lb @ 40 cts | 8.00 |
|  | Total | 3079.45 |

*Sources*

Duplanty, McCall & Co. Papers, In-file 1816 (128). Co. to George Hodgson Brothers. Eleutherian Mills Historical Library, Wilmington, Del.
Another version of this account is Longwood MSS, Group 6, Box 2. Also in the EMHL.

Some of the crucial elements in the spinning equipment, namely rollers and spindles, seemed moderately priced compared to the cost of a complete machine. Mule and throstle rollers, sold by weight, were four to five times the price of dead weight cast iron and averaged $1.34 each. Spindles and flyers were 79 cents a set (Table 4). A complete 120-spindle spinning frame (throstle) sold at $660; spindles and flyers at $94.80, were therefore 14.4% of the price of the machine and the rollers (assuming one set of three rollers served six spindles as in the throstle illustrated in Rees' *Cyclopaedia*), at about $80.40, 12.2% of the price of the machine — a little more than the proportion of 20% for rollers and spindles claimed for British machines a decade later.[60]

The relatively low prices of these castings and forgings apparently became an incentive to manufacturers to obtain only metal components from specialized machine makers, especially, perhaps, after the war when prices began to move downwards with rollers falling from 50 to 40 cents a pound. Having learned how the new machines were assembled and operated, the Duplanty, McCall & Co. managers used their millwrights or carpenters to build additions to their production line. William Boyd was one man employed in this way. Using wood, he made 'the end, roller beams and braces below' for one throstle and the ends for another. Other hands in the mill evidently completed assembly. William Murphy, a millwright, had one of his men make some of the reels.[61] In short, the new technology imported with immigrants like the Hodgsons was partially transferred to American manufacturers as two of the steps in machine making — machine design and the establishment of static and dynamic interrelationships between a machine's crucial parts — passed into the hands of American millwrights and carpenters.

The other vital stage, the manufacture of crucial metallic components like spindles and rollers, was also readily transferred to native Americans by the Hodgsons. Among their early pupils were Franklin and Titian Peale, sons of Charles Willson Peale (1741–1827), the Philadelphia portrait painter and naturalist. George Hodgson took the boys, under an initial arrangement for a six years' apprenticeship, in November 1813. Their father reported that the Hodgsons then had a foundry, 14 blacksmiths and 15 water-powered turning lathes at work.[62] Within a year, the two Peales had apparently almost finished learning how to make machinery, and, by spring 1815, were building equipment on their own account.[63] Whether they had in fact mastered all the skills of machine making in less

*Immigrant Textile Machine Makers along the Brandywine*

than two years is open to question. The important point is that the Hodgsons displayed none of the secretiveness frequently found among manufacturers and machine makers, immigrant and native-born alike, in this early industrial period.[64]

Besides helping to lower Philadelphia region textile machinery prices and sharing their new industrial skills with native Americans, the Hodgsons made a major contribution to mechanization in regional textile manufacturing by building a proportion of the region's stock of best practice equipment. Since they came from Manchester in 1811–1812 at the end of a period of innovation in cotton spinning,[65] theirs was knowledge of the latest British machine designs and techniques. What little is known about the design and the metal castings of the equipment they built on the Brandywine confirms their innovatory role. A blower, completing the separation of cotton fibres (a process begun with picking apart the impacted masses of cotton from the bale), was one of the opening machines reported in Rees' *Cyclopaedia* in November 1812 as 'used in some of the most improved mills' in England.[66] The Hodgsons' blower, one of these, apparently experienced some teething troubles for it took 12 months or more to come into operation.[67] Similarly progressive was their introduction of breaker and finisher cards of metal construction. Specified in their preliminary quotation, but not in the quotations of their rivals,[68] the two types of carding engine were well known in Britain and seem to have been introduced to America by Slater in the 1790s. The Hodgsons' design followed a radical improvement on the first generation of American cards because it used cast-iron framing and metal arches instead of wooden frames which, subject to warping, were liable to cause difficulty in operation and increase maintenance costs.[69] Details about the other equipment supplied by the Hodgsons are not known. Daniel Large, we have seen, was making throstles with cast-iron frames, again because this recent British practice minimized distortions in tolerances between key moving parts.[70] Presumably, the Hodgsons also supplied metal frames, if required, for their drawing, roving and throstle frames. This best practice equipment made by the Hodgsons was supplied to mills not only on the Brandywine but also to firms like Charles Willson Peale's cotton mill at Belfield in Philadelphia.[71]

Besides Large and the Hodgson brothers there were a number of other immigrant machine makers from Britain in the Philadelphia area. Among those used by Duplanty, McCall & Co. was William McKenzie who arrived in 1806 and by 1813 was a settled millwright in Wilmington, Delaware. In 1815 he installed a drive shaft for what proved to be inefficient waterlooms. The job took him 39 days and he charged $1.60–1.70 a day.[72] Another, James Wagstaff, a spindle maker who arrived in Wilmington in 1813, turned more than 800 mule spindles for the company in 1815, at an average of 20 cents each and later the same year, another 204 spindles at 30.5 cents each — thereby lowering the machinery prices charged by the Hodgsons.[73] William Hartley, reedmaker, who emigrated in 1811 and settled in Philadelphia, supplied the company with reeds between 1815 and 1817.[74]

*Thomas Siddall*

If British immigrants represented a substantial section of the Philadelphia region's pool of metal machine making skills, and if they also helped to introduce the latest technology from Britain, were they likewise at the frontier in developing new technologies in their new American situations? Duplanty, McCall & Co.'s efforts to acquire a powerloom reveal the

DAVID J. JEREMY

prominence of British immigrants among the inventors, builders and developers of this the 'sunrise' technology of the day. The one the firm dealt with most extensively was Thomas Siddall who introduced, and developed, an early powerloom from Britain.

Before 1815, several looms in the region were the subject of patents: those of Robert Lloyd (1809), Thomas Mussey (1811), Silas Shepard (1812), Seth Craige (1814), all of Philadelphia; and Thomas Siddall (1814) of Philadelphia County, Pennsylvania.[75] Some of these came to the attention of Alexander McCall. Two, belonging to individuals named Jarvis and Miller, he rejected on the advice of Daniel Large, as no better than existing unimproved waterlooms.[76] About the same time, early 1814, Thomas Cooper (1759–1839), the respected English-born chemistry professor and Manchester radical, told Duplanty, McCall & Co. about one in Germantown that worked 'well, quickly and neatly, and in that respect was like the looms in use in England and in Boston.'[77] News came too of Mr Levering's looms in Baltimore. McCall went to see the Germantown loom built by Thomas Siddall, a Lancashire immigrant who, with other Siddalls, arrived in Philadelphia in the War of 1812 period. The loom's salient improvement, claimed by Siddall as his own, was a warp protector stop motion — to halt the loom before the batten smashed a jammed shuttle through the warp shed.[78] Siddall planned to patent the device, and presumably it was in his patent of 9 July 1814. But he could not claim to be the first to design a warp protector stop motion. Large, the English millwright, accurately corrected Cooper's view that Siddall's warp protector stop motion was unique, by pointing out that it was already in use in Boston and had been employed for a long time in England — in fact since Cartwright's 1787 patent.[79] He could have added that it was in Thorp's American patent of 1812. Siddall in fact was pursuing an English type of powerloom technology, which as perfected by Horrocks was more complex than that currently being developed by Moody and Lowell.[80] Siddall's contrivance comprised an iron catch which held back the lathe or batten until the shuttle returned to one of its boxes. For his waterlooms, 34 inches wide and built by a machine maker under his direction, Siddall charged $50. He claimed they could each weave one yard in 20 minutes and four could be run by only one weaver.[81] If we suppose that there were 60 picks per inch, a not unusual weft spacing, then Siddall's loom ran at 108 picks per minute, about twice the rate of a handloom weaver. Since one weaver ran four of Siddall's looms, the productivity increase looked like a ten-fold one at least, and in summer 1815 a ten-fold production increase was indeed anticipated by McCall.[82] But there were problems. McCall found that Siddall's waterwheel was 'lop-sided' and conveyed irregular movements to the loom, and its inventor 'seemed shy of working it long at a time.'[83] Siddall, not unnaturally, rated his loom above others available, like that of Seth Craige in the Globe Mill, Philadelphia.[84] McCall wanted to examine powerlooms in Baltimore before deciding to purchase the Siddall loom.[85] But Siddall's machine in April–May 1814 looked good. Siddall had applied for a patent to guard and confine its use; he was installing 18 looms in a new weaving shop adjoining his factory; he made the daring bet that four of his looms could, under his supervision, weave 100 yards of No 30 yarn in 12 hours;[86] and, he claimed 'the finest Muslin's wch are made in Scotland & [are] wove in these looms.'[87]

What diverted Duplanty, McCall & Co. into powerloom weaving — most likely expectations of higher unit profits on fine cotton goods made from their mule-spun yarn — was never made wholly explicit in the correspondence from the company's Philadelphia agent to its Brandywine manager. What is certain is that, at the end of March 1814, nearly

*Immigrant Textile Machine Makers along the Brandywine*

months after the search for a powerloom began, McCall was warning that the American trade embargo and even the war would soon come to an end, the consequence of which would be 'that both yarns and cotton manufactures of every kind will fall considerably in price.'[88] Faced with this prospect, McCall suggested that, as a hedge against falling prices, the company take on a government contract. Duplanty's fear, that the government would continue the war and be unable to honour its contracts, was brushed aside by McCall who argued that in this case the contract could be cancelled (and the cotton goods sold to private customers).[89] Predicting peace, and the government's ability to pay contractors, McCall signed a contract with the Commissary General of Purchases on 13 April 1814 to supply.

Fifty Thousand yards of white twill'd cotton Drilling, 27 inches wide, weighing from 6½ to 7 ounces per yard, free from broken seeds and streams, fit for uniform clothing for the army with a regular & sufficient proportion of filling to chain for good & handsome work & substance, of a good & regular color, and well driven in the loom.[90]

Based on a standard sample and subject to inspection, the drilling would be bought by the US government at 40 cents a yard when delivered over the succeeding 12 months. McCall expected the company to spend up to $2000 on new powerlooms, a sum, he reckoned, amply covered by profit from the government contract, since the weaving could be done with powerlooms at 6 cents a yard. And, if the contract were cancelled, any unwoven yarn could be sold just prior to peace at 72 cents a pound.[91] With these calculations in mind, the company bought a powerloom from Siddall. Glimpses of its teething troubles are caught in McCall's letters.

Siddall would not let fellow Englishman Daniel Large alter the loom so that it would weave the twill agreed in the government contract.[92] Whether Siddall made the change in the shedding mechanism himself is unknown. The loom at last arrived, late in June 1814.[93] In transit from Philadelphia it was damaged — a shuttle box broke off, as did a cam and part of the warp protector stop motion. Without Siddall's warping and dressing machines, the company had to wait on Siddall for warps. And, to add to these frustrations, Siddall's bill increased to $101 — $75 for the loom (he claimed the one he sent was broader than his models), $25 for the patent rights and $1 for the packing box.[94] In March 1815 the Brandywine factory was still trying to get the Siddall loom into production, and was searching for a weaver who would make relatively fine goods on it. McCall located a man named Patterson who was running several of Miller's waterlooms and he gave advice from time to time until spring 1816.[95] In summer 1815 Siddall, now in partnership with James Phelps, was building new powerlooms at Frankford just north of Philadelphia. At his request the Brandywine mill sent some of its fine yarn warps up to Frankford so that Siddall could make adjustments on his machines that might be made in the powerloom in the Duplanty, McCall & Co. factory.[96]

At this point the company's experiment in stocking loom weaving was failing because its French hosiers were moonlighting.[97] It was therefore imperative that the company find a way to manufacture its fine yarns, which Philadelphia weavers would not handle, and so survive against importations without altering its fixed capital investment in fine spinning equipment. A Philadelphia weaver, William Huston, was deputed by McCall to try Siddall's loom.[98] Meantime McCall urged Duplanty, the mill manager, to expand his weaving operations and to order eight or ten more powerlooms. Duplanty could not see the

DAVID J. JEREMY

increasing importations but McCall's words had a note of urgency: 'you can never expect to do any good in the small way in which you now move. You must go upon a larger scale or you will be in a worse situation every day.'[99] Mechanization and economies of scale (i.e. higher turnover on lower profit margins) were the solution to falling cotton goods prices and the shortage of fine handloom weavers. If only a viable powerloom could be found and replicated. By early December 1815 the race to find such a powerloom, before the bottom dropped out of the cotton goods market, had been lost.[100] By spring 1816 one of the Miller looms turned up in the store of a Wilmington grocer. Patterson, one of the factory's weavers, insisted that it was the best available, but by now the company was relying on handloom weavers working coarser yarns for western and government customers.[101] In 1820, after the factory closed, 'one power loom on Siddall's plan' was inventoried without comment among the spinning equipment on the fourth storey.[102] Duplanty, McCall & Co. evidently did not buy one of Siddall's warping machines (patented 27 March 1815), possibly because it did not appear in time, possibly because it too could not handle fine warps.[103]

The failure of Siddall's loom on the Brandywine was obviously not wholly due to Siddall himself. Poor timing; a small investment in the new technology; separation, in ownership and distance, of research and development from the factory plant, were reasons implied in McCall's letters to Duplanty.[104] But the prime reason why the Siddall powerloom had eventually to be consigned to oblivion was the one which the Brandywine managers failed to realise until importations flooded the market. Committed to fine spinning, the Brandywine partners wanted their loom 'to work rather finer than common.'[105] Of course, the first powerlooms, with harsh, jerky mechanical movements, would strain and break the more fragile fine cotton warps. That the Brandywine firm inappropriately applied the available technology was suggested by E. I. du Pont's strong approval of the weaving in Siddall's factory in May 1815 and by Siddall's own successful production of 20 yards of bed tick per day per powerloom in 1816, using No 10 yarn, a very coarse and therefore strong yarn able to withstand the beating and shedding actions of a loom without Horrocks' variable batten speed motion (patented in England in 1813).[106] Thus it seems that if Siddall was trying to copy British examples, to adapt his powerloom to finer yarns, he singularly failed. His powerloom was no more suited to this purpose than the one developed at Waltham. Even so, Siddall's activities confirm that individual British immigrants were to the forefront in early attempts to mechanize weaving in the Philadelphia region.

CONCLUSION

On the Brandywine, one of the two centres of machine making in the Philadelphia region during the War of 1812 period, it is clear that immigrant machine makers played several roles. They were indeed introducers of a few specific pieces of new British equipment, like the four cone roving frame and the blowing machine. At the prototype stage, they were clearly active in setting up powerloom pilot plants in the Philadelphia region. But since this was done at a point when British manufacturers were also pioneering powerloom technology, they had the opportunity to pursue new directions. This, if Siddall was typical, they failed to do. Instead, immigrant machine makers in the Philadelphia region followed the lines indicated by British models: a more complex technology suited to the production of

*Immigrant Textile Machine Makers along the Brandywine*

higher quality cotton goods. But in missing Horrocks's variable batten speed motion, Siddall fell short of this goal. The concern for higher quality production was no doubt in part dictated by the established patterns of handloom manufacture in and around Philadelphia. Whether a Boston-type investment in a thoroughly American technology in the Philadelphia region would have yielded a different technology we cannot tell. It does seem, however, that the prominence of British machine makers as equipment suppliers for the new Brandywine factories helped to tie the first mills to the pursuit of a British style of powerloom weaving technology.

Equally importantly, the immigrant machine makers, by their numbers and competition for contracts, lowered the prices of machines and of machine parts. They also spread the use of cast iron in machine frames, like cards and throstles. They further accelerated the internal diffusion, within the region, of their Lancashire machine-making methods by training young Americans and by selling key metal components (rollers and their stands, spindles and flyers, gears, guides, plates, wheels, levers — all mostly cast iron but some steel and brass) which local millwrights and carpenters, copying immigrants' models, could assemble into woodframed machines. The complex process of industrialization in America, Professor Thomas Cochran has recently reminded us, was confined neither to New England nor to textiles. This paper certainly affirms the spread of advanced industrial technology through the activity of immigrant machine makers in the Philadelphia region, albeit in textile manufacturing. [107]

ACKNOWLEDGEMENTS

The preparation of this article would not have been possible without the use of collections in the Eleutherian Mills Historical Library at Greenville, near Wilmington, Delaware, USA. For this access I am most grateful to the Eleutherian Mills–Hagley Foundation Research Committee who awarded me a grant-in-aid to work there in summer 1969 and to the Academic Advisory Board of the Regional Economic History Research Centre (at the EMHL) who awarded me a Fellowship to spend a further month there in August 1979. This second visit was supported by a travel grant from the British Academy, to whom I am also grateful. I am further grateful to Ken Ponting for the invitation to present the first version of this article to the SSRC/Pasold Conference on Textile History held at St John's College, Oxford, on 26 September 1981. Finally, my thanks to Helen Jeffries and Alexandra Kidner, secretaries with the *Dictionary of Business Biography* project at the Business History Unit, who patiently and obligingly reduced my scrawl to legible typescript.

REFERENCES AND NOTES

[1] David J. Jeremy, *Transatlantic Industrial Revolution: the Diffusion of Textile Technologies between Britain and America, 1790–1830s* (Cambridge, Mass: MIT Press, 1981). This is hereafter cited as *TIR*.

[2] Diane Lindstrom, *Economic Development in the Philadelphia Region, 1810–1850* (New York: Columbia University Press, 1978), 66, 95, 100–101.

[3] Peter C. Welsh, 'The Brandywine Mills: a Chronicle of an Industry, 1762–1816', *Delaware History*, 7 (1956), 17–36.

[4] For the 1810 spindleages *see* Tench Coxe, *A Statement of the Arts and Manufactures of the United States of America in the Year 1810* (Philadelphia: A. Cornman, Jr., 1814) and for 1820 spindles, *TIR*, 277.

[5] The 1820 figures are from ibid. while the 1831 figures are from Friends of Domestic Industry, New York Convention, *Report on the Production and Manufacture of Cotton* (Boston: J. T. and E. Buckingham, 1832), 16.

[6] *TIR* chapters 1 and 4 for background and detailed references.

[7] *New York Daily Gazette*, 29 April 1791 cited in Rita S. Gottesman, *The Arts and Crafts in New York, 1777–1799: Advertisements and News Items from New York City Newspapers* (NY: New York Historical Society, 1954), item 369.

[8] *TIR*, 158.

[9] National Archives, Washington DC, RG 59, War of 1812 Papers, US Marshals' Returns of Enemy Aliens. Hereafter cited as DNA RG 59 Returns of Enemy Aliens.

[10] Carroll W. Pursell, Jr., *Early Stationary Steam Engines in America. A Study in the Migration of a Technology* (Washington, DC: Smithsonian Institution Press, 1969), 35, 43–44. For Smallman *see* ibid., 29–30, 32, 34, 35, 42. John P. Crozer, later to become a leading cotton manufacturer in Delaware County, years later recalled that Large was the leading steam engine manufacturer in Philadelphia. *See* John Price Crozer, *Biographical Sketch of John P. Crozer Written by Himself* (Philadelphia: Grant, Faires and Rodgers, ca 1861), 11–12.

[11] DNA RG 59, Returns of Enemy Aliens; *Paxton's Philadelphia Directory* 1813.

[12] *TIR*, 58–59, 90.

[13] For Duplanty, McCall & Co.'s history *see* Roy M. Boatman, 'The Brandywine Cotton Industry, 1795–1865' (Hagley Research Report, 1957), 20–49. The partners were two pairs of brothers, Archibald and Robert McCall and Victor and Eleuthère Irénée du Pont and Raphael Duplanty. Duplanty, who, like the du Ponts, fled from the French revolution, had worked in the USA for E. I. du Pont as an accountant and then was involved in the design of the du Pont woollen factory.

[14] Eleutherian Mills Historical Library, Wilmington, Delaware, (hereafter EMHL), Accession 500, Duplanty, McCall & Co. Papers (hereafter DMC), Archibald McCall to Raphael Duplanty, 31 March 1813.

[15] See Samuel H. Needles, 'The Governor's Mill and the Globe Mills, Philadelphia', *Pennsylvania Magazine of History and Biography*, 8 (1884), 379–382.

[16] DMC to Duplanty, 31 March 1813. Coxen was another steam engine maker and English immigrant; *see* Pursell, *Early Stationary Steam Engines*, 33.

[17] For Wetherill *see DAB* and for his steam engine, Pursell, *Early Stationary Steam Engines*, 43.

[18] All quotes from DMC, McCall to Duplanty, 31 March 1813.

[19] Ibid., same to same, 5 April 1813.

[20] *See TIR*, 91 and sources.

[21] DMC, McCall to Duplanty, 5 April 1813.

[22] Ibid., same to same, 14 April 1813.

[23] Ibid.

[24] Ibid., same to same, 14, 24 April 1813.

[25] Ibid., same to same, 26 January 1814, 18, 28 October 1814, 16 January 1815.

[26] Ibid., same to same, 21 January 1815.

[27] For the Belper water wheels *see* Robert S. Fitton and Alfred P. Wadsworth, *The Strutts and the Arkwrights, 1758–1830: a Study of the Early Factory System* (Manchester: Manchester University Press, 1958), 205–217.

[28] DMC, Daniel Large to Duplanty, 3 August, 4 September, 1 November 1813.

[29] Boatman, 'Brandywine Cotton Industry', 26. Dimensions and weights given by Robertson Buchanan: *Essays on the Shafts of Mills*, (London, 1914), 33 ff, show that Large's specification was well within the limits of the wooden water wheel (a wooden wheel of diameter 16 ft and a width of 9 ft was expected to weigh 10½ tons). This not only confirms that Large was confronted with a materials problem, but also suggests that his mill wrighting was competent to a degree. I am very grateful to Professor Louis C. Hunter for supplying this reference as well as comments and other sources on weights of water wheels.

[30] DMC, deposition of James Blackie in Horace Binney (the firm's lawyer) to Duplanty, 16 November 1819; ibid., Duplanty to Binney, 18 November 1819; ibid., joint deposition of Thomas Oakes and William Murphy, undated, in general file. For Oakes's English origins see *PMHB*, 37 (1913), 471–479.

*Immigrant Textile Machine Makers along the Brandywine*

[31] DMC, McCall to Duplanty, 17 March 1814. As early as August 1814 Duplanty was refusing to pay Large. *See* ibid., McCall to Duplanty, 29 August, 5, 9 September 1814.

[32] Ibid., Large to Duplanty, 23 April 1813; McCall to Duplanty, 14 June 1813.

[33] Fitton and Wadsworth, *Strutts and the Arkwrights*, 198–203 and *passim*.

[34] DMC, McCall to Duplanty, 1 September 1813. The external staircase at Waltham has been hailed as a novel feature but perhaps only the tower element was novel. *See* Bryant F. Tolles, Jr., 'Textile Mill Architecture in East Central New England: an Analysis of Pre-Civil War Design', *Essex Institute Historical Collection*, 107 (1971), 236.

[35] DMC, William B. Leonard (later manager of the famous Matteawan works) to Duplanty, 15, 27 September, 8 November 1813.

[36] Ibid., McCall to Duplanty, 5, 27 May, 14 August 1814.

[37] Ibid., same to same, 2 June 1814.

[38] Ibid., same to same, 21, 25 April 1818. Joseph Bamford emigrated to the USA in spring 1805. By August 1812, when he was aged 39 and living in Filbert Street, Philadelphia, with a wife and 5 children, he was working as a machine maker. *See* DNA RG 59, Returns of Enemy Aliens; also *Archives of Useful Knowledge* 3 (1813), 191.

[39] Crozer, *Biographical Sketch*, 10–12.

[40] US Federal Record Centre, Waltham, Massachusetts, First (Massachusetts) Circuit Court, Case files of *Paul Moody v. Medway Cotton Factory*, October 1821, affidavit of Samuel Stevens. My thanks to Ms. Betty-Bright P. Low of the EMHL for checking the Philadelphia directories.

[41] DNA RG 45, Returns of Enemy Aliens. John Hodgson was listed as a turner of 24 Balloon Street, Manchester, in 1804, and as a roller maker in 1808–1809. *See Dean's & Co Directory*. I am very grateful to Dr R. S. Fitton for checking the early 19th century Manchester directories for the immigrant machine makers in the Philadelphia region. See *TIR*, 144–159 for profiles of textile immigration during the War of 1812 period.

[42] Norman B. Wilkinson, 'Brandywine Borrowings from European Technology', *Technology and Culture*, 4 (1963), 3.

[43] EMHL, Longwood MSS, Group 6, Box 3, Articles of Agreement between Caleb Kirk (of Christiana Hundred), James Jefferies (of Chester Co, Pennsylvania), E. I. du Pont, John Warner (of Christiana Hundred), William Warner (Philadelphia merchant) and John Torbert of one part, and George, Thomas and Isaac Hodgson of the other part, 23 August 1813. For Caleb Kirk and the Brandywine Mill Seats Company, *see* Boatman, 'Brandywine Cotton Industry', 12–13, 79–80. For Kirk *see also* George H. Gibson, 'Fullers, Carders and Manufacturers of Woollen Goods in Delaware', *Delaware History*, 12 (1966), 34.

[44] Boatman, 'Brandywine Cotton Industry', 94–96. *TIR*, 231.

[45] Boatman, 'Brandywine Cotton Industry', 94–96.

[46] Ibid.

[47] EMHL, Longwood MSS, Group 6, Box 1. Alexander McKim to E. I. du Pont, 22, 24 February 1813; Robert McKim to E. I. du Pont, 24 February, 12 March 1813.

[48] Ibid.

[49] Ibid., Richardson and Duff to E. I. du Pont, 12 March 1813.

[50] Boatman, 'Brandywine Cotton Industry', 96.

[51] EMHL Longwood MSS, Group 6, Box 2. Alrichs & Dixon to R. Duplanty, 22 March 1813.

[52] Ibid., 'List of Machinery for Messrs Du Pont & Co', undated.

[53] Ibid., agreement for spinning machinery made by G. Hodgson & Brothers, 22 March 1813.

[54] Ibid.

[55] EMHL Acc. 500, Ledger 195, Duplanty, McCall & Co, Factory Ledger 1813–1825, f. 81.

[56] EMHL, Longwood MSS, Group 6, Box 2, Alrichs & Dixon to Duplanty, 29 March 1813.

[57] *TIR*, 188.

[58] EMHL, Acc 500, Ledger 195, 73, 81–82, 132–135, 164. The total sum expended in the machinery account 20 May 1814–30 April 1819, was $16,566.95 of which $3,212.74 was paid to the Hodgsons.

[59] DMC, Robert McCall to Archibald McCall, 18 January 1825. In 1814 'not one piece of machinery stated in the account [William Boyd's?] was done in that year, when the price of work was nearly double of what it was afterward'.

[60] *TIR*, 26.

[61] DMC, Robert McCall to Archibald McCall, 18 January 1825. *See also* ibid., 'Copy of Memo by R. Duplanty of work done by William Boyd'.

[62] American Philosophical Society, Charles Willson Peale Letterbooks, Peale to John de Peyster, 12 November 1813, (information from Roy M. Boatman's research notes in the EMHL).

[63] Ibid., Peale to Rembrandt Peale, 31 July 1814; Peale to John de Peyster, 17 November 1814; Peale to Angelica Peale, 14 February 1815; Peale to Rembrandt Peale, 19 February, 12 March 1815; Peale to Angelica Peale, 24 March 1815.

[64] *TIR*, 36–40, 322 n. 19 and *passim* (*see* index).

[65] Ibid.

[66] Rees, *Cyclopaedia* sv. 'Manufacture of Cotton'.

[67] DMC, Robert McCall to Archibald McCall, 18 January 1825. A blowing machine was known also to Richardson & Duff of Baltimore (EMHL, Longwood MSS, Group 6, Box 1, Richardson & Duff to E. I. du Pont, 12 March 1813) and to Alrichs & Dixon (EMHL, Longwood MSS, Group 6, Box 2, undated sheet headed 'Alrichs and Dixon').

[68] Ibid., and EMHL Longwood MSS, Group 6, Box 2, George Hodgson & Brothers 'list of Machinery for Messrs Du Pont & Co'.

[69] *TIR* 80, for illustration of Slater's card built of wood. The Hodgsons' preliminary quote specified 'Cards, with Iron frames' (source: previous note).

[70] *See* above, note 32 and for English practice which started around 1805–1810, *TIR*, 306 n. 41.

[71] Boatman, 'Brandywine Cotton Industry', 95.

[72] DNA, RG 45, Returns of Enemy Aliens; DMC, William McKenzie's bill, 8 December 1815.

[73] DNA, RG 45, Returns of Enemy Aliens; EMHL, Acc 500, Ledger 195, 144.

[74] DNA RG 59, Returns of Enemy Aliens; EMHL, Longwood MSS, Group 6, Box 20, William Hartley to Duplanty, 3 March, 25 May 1815.

[75] US Patent Office, *A List of Patents Granted by the United States from April 10, 1790, to December 31, 1836* (Washington, DC: Commissioner of Patents, 1872), *passim*.

[76] DMC, McCall to Duplanty, 17, 23 February, 13 April 1814.

[77] Ibid. *See DAB* for Cooper.

[78] DMC, McCall to Duplanty, 19 April 1814.

[79] Ibid., same to same, 26 April 1814; *Emporium of Arts and Sciences*, new series, 3 (1814), 459.

[80] *TIR*, 92–103.

[81] DMC, McCall to Duplanty, 19 April 1814. *See also* EMHL, Longwood MSS, Group 6, Box 2, Thomas Siddall to Duplanty, 29 November 1814 and 6 August 1815.

[82] DMC, McCall to Duplanty, 19 April 1814 and 11 July 1815.

[83] Ibid., same to same, 26 April 1814.

[84] Ibid., same to same, 26 April 1814. Craige was one of the few Philadelphia County manufacturers with powerlooms in May 1814. *See* ibid. McCall to Duplanty, 30 May 1814.

[85] Ibid., same to same, 26 April 1814, McCall wanted to run a Baltimore loom alongside Siddall's.

[86] Ibid., same to same, 20 May 1814.

[87] Ibid., same to same, 20 June 1814.

[88] Ibid., same to same, 31 March 1814.

[89] Ibid., same to same, 4 April 1814.

[90] National Archives, Washington DC, RG 92, Textual records of the Office of Quartermaster General, Special Files: Consolidated Correspondence, 1794–1890, Box 273, Duplanty, McCall & Co. contract with Callender Irvine, Commissary General of Purchases, 13 April 1814.

[91] DMC, McCall to Duplanty, 4 April 1814.

[92] Ibid., same to same, 27 May 1814.

[93] Ibid., same to same, 20 June 1814.

[94] Ibid., Duplanty to Siddall, 19 November 1814, draft on verso of Siddall to Duplanty, McCall & Co., 10 November 1814.

[95] Ibid., McCall to Duplanty, 31 March 1815. Also ibid., same to same, 23, 25 May 1816.

[96] EMHL, Longwood MSS, Group 6, Box 2, Siddall to Duplanty, McCall & Co., 13 August 1815.

[97] DMC, McCall to Duplanty, 18 June 1815.

[98] Ibid., same to same, 3 August 1815.

*Immigrant Textile Machine Makers along the Brandywine*

[99] Ibid., same to same, 6 October 1815.

[100] Ibid., same to same, 7 December 1815.

[101] Ibid., same to same, 6 April, 7, 21, 23, 25 May, 1816; Boatman, 'Brandywine Cotton Industry', 34–35, 40.

[102] DMC, Inventories, 1820.

[103] *TIR*, 99.

[104] For example, McCall expressed strong hopes that Thomas Siddall would move to the Brandywine, DMC, McCall to Duplanty, 16 May 1815.

[105] EMHL, Longwood MSS, Group 6, Box 2, Siddall to Duplanty, McCall & Co., 6 August 1815.

[106] DMC, McCall to Duplanty, 24 May 1815; ibid., same to same, 14 February 1816.

[107] Thomas C. Cochran, *Frontiers of Change: Early Industrialization in America* (New York: Oxford University Press, 1981). *See* especially chapter 4, where he emphasizes the relative importance of industries associated with the cotton gin, power saw, flour mill, steamboat and iron rolling mill. The 'backbone of production', he asserts, ran from Albany and Troy down the Hudson valley to New York then went from Paterson to Trenton, New Jersey, and then down the Delaware River through Philadelphia and beyond Wilmington into Maryland; in the west, heavy industry for local needs followed the Ohio from Pittsburgh to Cincinnati (ibid., 90).

# [11]

Excerpt from Kristine Bruland (ed.), *Technology Transfer and Scandinavian Industrialisation*, 229–67.

## 10

# The Norwegian Mechanical Engineering Industry and the Transfer of Technology, 1800–1900

*Kristine Bruland*

## 1  Introduction

From the early nineteenth century Norwegian engineers were able to establish a dynamic and solidly based mechanical engineering industry, which grew to become a core industry of Norwegian industrialisation. But how did they do it? What kinds of activities lay behind the construction of this new industry? In this chapter we argue that technology transfer – that is, the acquisition and adaptation of foreign technologies and engineering skills – was a central element in the development of this Norwegian industry. Norwegian engineers and entrepreneurs were alert to the extraordinary development of foreign engineering technologies in the early nineteenth century, and to the opportunities these technologies provided. With great energy they learned the key techniques and skills being pioneered abroad and successfully implanted them in Norway. What was the main source for this import of technology? As we shall see below, Britain was overwhelmingly important as a technological exemplar for Norway. Moreover it remained so to the end of the nineteenth century, despite a German challenge which seems to have emerged only slowly. This technology transfer from Britain involved a wide range of activities: foreign travel, purchases of foreign equipment, foreign education, contacts with foreign firms, agency relationships and so on. The technology transfer process was thus a complex one. But it was, and remained, a core element in the success of Norwegian mechanical engineering as a whole.

229

*Kristine Bruland*

## 2   Technology Transfer and Industrial Development

The modern industrial era began in Britain during the later part of the eighteenth century, with the emergence of a mechanised manufacturing industry based on continuous technological change and continuous productivity growth. Since then the world economy has been characterised by the existence of technological 'leader' and 'follower' economies, and by the transfer or diffusion of technologies from the former to the latter.[1] Technological leadership has, of course, changed hands several times during the past 200 years, often with momentous economic and political results: by the early twentieth century, British dominance had been supplanted by that of Germany and the USA, which in turn have been challenged by the economic rise of Japan and the Far East. For follower economies, especially small economies such as Norway, the identity of the leader at any particular time is perhaps less important than the follower's ability to imitate, adapt and utilise the technological innovations and advances on which leadership is based. Small countries can never hope to match the levels of investment, scientific activity and applied research and development of the major economies. Whether or not they become industrialised and remain at the forefront of advanced industrial performance, therefore, depends in large part on whether they can develop the skills to use technologies developed abroad. This is not just a matter of buying machines. It involves the development of 'technological capability', which Fransman has defined as follows:

> By a technological capability [we] mean the ability, embodied in people, to select the appropriate technology; to implement it; to operate the production facilities so implemented; to adapt and improve them, and possibly to create new processes and products. A technology may be purchased, but a technological capacity is operated only through the build-up of human capital.[2]

1. A. Maddison, *Phases of Capitalist Development*, Oxford, 1982, chap. 1.
2. M. Fransman, Introduction, in M. Fransman (ed.), *Machinery and Economic Development*, London 1986, p.xiv. Myra Wilkins presents a similar point of view, but refers instead to 'absorption' rather than 'capability': 'It is worth considering the difference between mere transfer and the absorption of technology within the host country. A company can export capital goods. In one country the machines installed might be allowed to break down and eventually fall into disrepair; in another country, the same machines might be used efficiently in modern industry, copied, adapted or produced locally' (M. Wilkins, 'The Role of Private Business in the International Diffusion of Technology', *Journal of Economic History*, (vol.39, 1974, p.171).

230

## Norwegian Mechanical Engineering, 1800–1900

In this chapter we treat 'technology transfer' or 'technological diffusion' as the ability to use such technological capability to deploy and develop technologies developed abroad. In fact relatively few countries have proved able to develop advanced technological capabilities or capacities in the sense used by Fransman, which in turn suggests that the development of such capabilities is a complicated and difficult process. The apparently simple process of technological diffusion, like that of technological development itself, appears to involve a particularly complex interweaving of cultural, educational, legal, economic and political factors, which is perhaps why few if any really adequate histories of technological change and development in the West have yet been written.[3] This chapter concentrates on one particular aspect of this process, namely the acquisition of skills in mechanical engineering, and the role played by the mechanical engineering industry in developing what we might call 'diffusion capability' within a small economy.

Mechanical engineering is a particularly important industry for industrial, technological and economic development as a whole. One key reason for this is that *product* innovations by mechanical engineering firms, for example in new machines and equipment, become *process* innovations when they are put to work by user firms; this can have a wide impact on technological change and productivity growth within the economy. Competition among mechanical engineering firms in terms of products thus embeds process innovation within the wider economy. The immense historical significance of this was emphasised strongly some years ago by Nathan Rosenberg:

> In both the US and the UK in the nineteenth century, technological change became institutionalised in a very special way – that is in the emergence of a group of specialised firms which were uniquely oriented toward the solution of certain kinds of technical problems. The rapid rate of technological change was completely inseparable from these capital goods firms. In fact I would regard the emergence of such firms as the fundamental institutional innovation of the nineteenth century from the point of view of the industrialization process.[4]

The development of specialised mechanical engineering firms

---

3. Thomas P. Hughes, 'History of Technology as Modern History', *ISIS: Journal of the History of Science Society* (forthcoming).

4. N. Rosenberg, 'Economic Development and the Transfer of Technology: Some Historical Perspectives', in idem, *Perspectives on Technology*, Cambridge, 1977, p.152.

*Kristine Bruland*

was clearly an important component of accelerated industrialisation in Britain during the early nineteenth century. But it had an international impact as well, for British capital goods enterprises increasingly sought markets abroad. Particularly after 1843, when prohibitions on some types of machinery exports were removed, this market-seeking by British engineering firms opened up possibilities for 'follower' economies to develop a range of industries using British techniques. In Norway, for example, a mechanised textile industry rapidly developed from the mid-1840s on the basis of 'packages' of technology – comprising technical information, equipment, skilled labour and managerial expertise – assembled by British textile engineering firms.[5] This was but one component of a general diffusion of technology from Britain at that time. Another important part of the spread of industrialisation was the spread of mechanical engineering industries themselves into those countries which succeeded in industrialising in the middle and late nineteenth century. Since mechanical engineering industries 'undertake technological change and adaptation as a matter of routine',[6] as Rosenberg puts it, this spread was of central significance in European industrialisation. But how did this happen, and how did the mechanical engineering industry facilitate the further spread of industrial technologies into growing economies during the nineteenth century? These are the fundamental questions which are asked in this paper.

In discussing the mechanical engineering industry from the point of view of technology transfer, we need to distinguish between two quite distinct phenomena. The first concerns the historical origins of the industry, while the second concerns its wider economic impact. These technology transfer issues are:

1. What role did technology transfer play in actually establishing the mechanical engineering industry in Norway? For example, how did embryo European (in this case Norwegian) engineers acquire both the general technical knowledge and the specific technical skills associated with the new technology developed in Britain, how did they acquire the competence to run specific techniques, what particular types of equipment were imported from abroad, and what types of labour, skill and managerial input

5. K. Bruland, *British Technology and European Industrialisation: The Norwegian Textile Industry in the Mid-nineteenth Century*, Cambridge, 1989.
6. N. Rosenberg, 'The Direction of Technological Change: Inducement Mechanisms and Focussing Devices', in idem, *Perspectives on Technology*, Cambridge, 1977, p.99.

*Norwegian Mechanical Engineering, 1800–1900*

were imported from abroad? In examining this question below, we shall be principally concerned with Norwegian absorption of technologies from Britain; later sections will deal with Germany.

2. What were the implications of the new mechanical engineering industry for the wider technological development of Norwegian industry? Specifically, how did the mechanical engineering industry shape the diffusion of foreign technological practice into other industries internally, by acting as an 'entry point' for foreign technologies?

The principal objective of later sections of this chapter will be to answer these questions empirically. Before doing so, however, we move to a brief description of the development of mechanical engineering in Britain, since it was Britain, as world technological leader, to which early Norwegian engineers looked first, and it was the British industry which was the source of virtually all of the early industrial technology deployed in Norway.

## 3   The Development and International Activities of Mechanical Engineering in Britain

The development of a substantial mechanical engineering industry in Britain in the early nineteenth century was perhaps the most significant long-term effect of the industrial revolution of the late eighteenth century. Early British industrialisation rested on a very narrow industrial base, with three 'staple industries' of coal, textiles and metal manufacture being the first to develop. Outside these core industries, output and productivity growth was definite but slow, and both investment and the use of new techniques lagged significantly. But within the core industries, growth was spectacular. In textiles, where expanded input supplies and falling input prices combined with new spinning techniques to produce high output growth, the industry grew from 1770 at a long-run average compound rate of approximately 3.5 per cent. This rate is sufficient to double output every twenty years. But in the first half of the nineteenth century textile growth accelerated: between 1810 and 1842 the growth rate averaged almost 5.5 per cent per year, and between those years annual output increased by over 500 per cent.[7] Over the same period, the capital equipment of the industry

7. These growth rates are calculated from P. Deane and W. A. Coles's estimates, *British Economic Growth 1866–1959*, 2nd edn, Cambridge, 1978, p.191.

*Kristine Bruland*

increased at an even faster rate: both the number of cotton spindles
and the total capital stock increased at approximately eight per cent
per year.[8]

These growth rates were unprecedented, and were associated
with a massively increased demand for machinery, power equipment and tools. This demand played a crucial role in the development of the fourth 'staple industry', namely engineering. 'In the
long run', as Landes has pointed out, 'the diffusion of mechanized
manufacture called forth major improvements in tool design.'[9]
During these years a combination of advances in iron manufacture,
new techniques in precision engineering pioneered by a small but
influential group of engineers[10] and a flow of labour from such
occupations as instrument and clockmaking, led to the establishment of a large number of engineering firms. Since the textile
industry was by far the largest market, a large part of the industry
was concentrated in Lancashire: by 1841 Lancashire had 115 engineering enterprises, employing over 17,000 workers.[11] By 1851,
according to the population census of that year, over 63,000 workers
were employed in making machinery or boilers. Employers' returns in the same census indicated over 800 'engine or machine'
enterprises; most were very small, but 155 firms employed over
twenty men each, and 34 firms employed more than one hundred
men each. The major firms in the industry were very large indeed:
'By the early 1840s, Fairbairn's were employing a total of between
1,000 and 2,000 men in their Manchester and Millwall works; in the
Atlas works of Sharp, Roberts and Co there were nearly 1,000 by
the early 1850s, while Nasmyth, Wilson and Company were by
then employing 1,500 in their Bridgewater foundry.'[12]

Naturally these enterprises looked not only to domestic but also
to foreign markets. A major constraint on machinery exports,
however, was the existence of laws prohibiting the export of a wide
range of tools and equipment, especially for textile and iron and
steel manufacture. Such prohibitions had existed, in various forms,

8. M. Blaug, 'The Productivity of Capital in the Lancashire Cotton Industry
during the Nineteenth Century', *Economic History Review*, 2nd Series, vol.13, 1961,
p.32; and *Select Committee on the Exportation of Machinery*, vol.7, 1841, Appendix 2,
p.230.

9. D. Landes *The Unbound Prometheus. Technological Change 1750 to the Present*,
Cambridge, 1969, p.105.

10. K. R. Gilbert, 'Machine Tools', in Ch. Singer et al., *A History of Technology*,
1958, vol.4, p.418.

11. K. Bruland, *British Technology*, p.33.

12. A. E. Musson, 'The Engineering Industry', in R. Church *The Dynamics of
Victorian Business*, London, 1980, p.95.

## Norwegian Mechanical Engineering, 1800–1900

**Table 10.1**   British exports of machinery to 1856 (£000)

|                  | 1814–16 | 1844–6 | 1854–6 |
|------------------|---------|--------|--------|
| Steam-engines    | –       | 319    | 753    |
| Other machinery  | 28      | 614    | 1537   |
| Total            | 28      | 933    | 2690   |

*Source*: R. Davis, *The Industrial Revolution and British Overseas Trade*, Leicester, 1979, p.27.

for many centuries, but were significantly strengthened in the 1780s. The regulations did not, in practice, stop the flow of machines and equipment to Continental Europe and North America: smuggling, espionage, emigration and so on led to a steady flow out of Britain.[13] None the less, the regulation constrained the sales of British engineering firms, and this became an increasing problem after the first quarter of the nineteenth century, as the engineering industry ran up against the constraints of the domestic market. A major political and economic debate ensued, as engineers and machine makers sought to repeal the prohibitions: as Mathias has put it, 'Once specialised as an industry in its own right, its leaders claimed the right to export markets of their own. Engineer after engineer argued thus before the Parliamentary Committees on the Export of Machinery in 1824 and 1843.'[14]

The prohibitions were repealed in 1843, and from that time British machinery exporters actively sought foreign markets. The growth of the trade can be seen in Table 10.1. These exports formed only a tiny part of British commodity trade; indeed they made up less than 10 per cent of exports in the 'metal and metal-ware' category. None the less they were of very great importance in non-British industrialisation, for these exports were a primary source of equipment for continental enterprises. In this sense, Britain genuinely was 'the workshop of the world'. But the trade in machinery ought to be seen, in my view, in two ways. On the one hand, there was the direct supply of equipment, and associated specific 'know-how' which enabled particular plants or production processes to be established. On the other, however, there was a less tangible effect, as foreigners acquired general technological

13. D. Jeremy 'Damming the Flood: British Government Efforts to Check the Outflow of Technicians and Machinery 1780–1843', *Business History Review*, vol.51, 1977, pp.1–34.
14. P. Mathias, *The First Industrial Nation*, 2nd edn, London, 1983, p.110.

*Kristine Bruland*

capabilities as a result of trade and other relationships with British engineering firms. It was these capabilities which underlay the spread of mechanical engineering itself, and this is the focus of subsequent sections of this chapter. We turn now to the international trade background to the development of mechanical engineering in Norway; later sections will examine the technological flow between British and Norwegian engineering directly.

## 4  Norwegian Industrialisation, Technology Imports and the Growth of Mechanical Engineering

Even before industrialisation began during the 1830s and 1840s, Norway was in some ways an unusual economy by virtue of its openness to foreign trade: fishing, timber and shipping were major export industries.[15] Two of these, shipping and timber (including especially sawmills), were also industries which, as the nineteenth century progressed, increasingly required the application of industrial techniques and thus engineering products. Even before Norway began the industrialisation phase of the mid-nineteenth century, therefore, there was significant scope for the activities of the mechanical engineering industry.

But fishing, timber and shipping also produced substantial foreign earnings, and Norway was therefore able to import a very wide range of manufactured goods, especially from Britain. Certainly this trade fluctuated according to economic circumstances, but by the mid-1830s Norway was importing large quantities of such products as cotton manufactures and yarn; woollen goods; canvas; coke and coal; pottery; soap; shoe leather; oil; paints; lead; ammunition and explosives; anchors and chains; iron and iron plates, castings and manufactures; steel and steel goods and wire, and so on.[16] Growth of demand in many of these product groups in turn opened up the possibility of import-substituting domestic manufactures, and this probably lay behind the growth of a number of important new industries in Norway. Growth here was interwoven with that of the traditional export industries.[17]

Naturally the industrial sector was, in absolute terms, small, but

15. See E. Hovland, H. V. Nordvik and S. Tveite, 'Proto-Industrialization in Norway, 1750–1850: Fact or Fiction?', *Scandinavian Economic History Review*, vol. 30, no.1, pp.45–56, for the structure of the pre-industrial economy and foreign trade.

16. A. Schweigaard, *Norges Statistik*, Christiania, 1840, pp.164–5.

17. F. Sejersted, *En Teori om den Økonomiske Utvikling i Norge i det 19 Århundre*, Oslo, 1973, p.37.

Norwegian Mechanical Engineering, 1800–1900

**Table 10.2**  Development of Norwegian industries: number of firms

|                    | 1865 | 1870 | 1875 | 1879 |
|--------------------|------|------|------|------|
| Metal industry     | 25   | 22   | 56   | 65   |
| Chemical industry  | 47   | 60   | 141  | 136  |
| Textile industry   | 80   | 107  | 122  | 145  |

Source: Statistik over Norges Fabrikanlæg (Kristiania, 1881), p.vii

by 1850 a significant part of the industrial workforce was employed in such industries as textiles, chemicals, pottery, iron foundries and engineering.[18] During the 1860s and 1870s some older industries began to decline in terms of the number of enterprises within them: quarrying, paper, timber etc. all contracted.[19] But in the newer industries growth in the total number of firms continued. As Norwegian industry grew from the 1840s, so in consequence did its demand for machines and equipment. At first much of this was imported. From the early 1840s, imports of machinery, in particular from Britain, accelerated, as Tables 10.4 and 10.5 indicate. Table 10.3 should probably be interpreted as a process of steady growth for the falls in exports in the late 1840s were the result, firstly, of financial crisis within Norway, and secondly, of recession and the profound political upheaval which swept Europe in 1848 and 1849. As Europe recovered from this crisis, from 1850 trade grew strongly, and this too is reflected in Norwegian imports of machinery. From 1850 imports of machinery increased sharply,

**Table 10.3**  Machinery and millwork exports from Britain to Norway, 1843–50 (£)

| 1843 | 1,392  |
|------|--------|
| 1844 | 2,483  |
| 1845 | 9,449  |
| 1846 | 15,518 |
| 1847 | 5,270  |
| 1848 | 5,727  |
| 1849 | 4,187  |
| 1850 | 12,175 |

Source: British Parliamentary Papers, 1854–5, vol.52, p.226

18. NOS, Historisk Statistikk, 1978, Oslo, 1978, table 42. See Table 10.2
19. See Statistikk over Norges Fabrikanlæg, Christiania, 1881, p.vii.

237

*Technology Transfer and Business Enterprise*

*Kristine Bruland*

**Table 10.4**  Imports of machinery by Norway, 1841–64 (thousand kr.) (from 1851, 3-year moving averages)

| | |
|---|---|
| 1841 | 28 |
| 1844 | 72 |
| 1847 | 93 |
| 1850 | 322 |
| 1851 | 142 |
| 1852 | 171 |
| 1853 | 363 |
| 1854 | 503 |
| 1855 | 549 |
| 1856 | 499 |
| 1857 | 408 |
| 1858 | 218 |
| 1859 | 409 |
| 1860 | 497 |
| 1861 | 467 |
| 1862 | 318 |
| 1863 | 280 |
| 1864 | 328 |

*Source*: Statistisk Sentralbyrå, *Historisk Statistikk 1978*, table 159, pp.276ff.

in value terms, until the rate of growth slowed noticeably towards the end of the decade.

These two series cannot be compared directly, but with an exchange rate of approximately 16 kr=£1 they are consistent with a British share of Norwegian imports of machinery varying between 60 and 90 per cent during the 1840s. The important trend here, however, is stabilisation and then decline of machinery imports from around 1860; since the Norwegian industrial sector continued to grow at this time, Table 10.4 implies that engineering needs were increasingly being met from domestic sources, that is that domestic engineering firms were successfully competing with foreign producers.

This growth of Norwegian mechanical engineering can be seen in Table 10.5, which traces employment from 1850 through to the end of the century. In response to general industrial growth and hence increasing demand for machinery, the Norwegian mechanical engineering industry expanded rapidly from the mid-1840s, and continued to do so until the turn of the century. But this occurred through a sharp upturn during the 1860s, which was subsequently maintained, with fluctuations in employment reflect-

## Norwegian Mechanical Engineering, 1800–1900

**Table 10.5**  Growth of employment in Norwegian mechanical engineering, 1850–1900

|  | Total engineering employees | Eng. as % of industrial employment |
|---|---|---|
| 1850 | 1,368 | 11.1 |
| 1860 | 1,608 | 7.8 |
| 1865 | 4,999 | 17.6 |
| 1870 | 7,161 | 20.6 |
| 1875 | 10,927 | 22.6 |
| 1879 | 7,929 | 18.3 |
| 1885 | 9,570 | 20.1 |
| 1890 | 13,663 | 21.4 |
| 1895 | 12,626 | 20.4 |
| 1900 | 16,790 | 21.1 |

Source: NOS, *Historisk Statistikk* 1978 (Oslo 1978), table 42, p.79

**Table 10.6**  Structure of the Norwegian mechanical engineering industry (number of enterprises)

|  | 1865 | 1870 | 1875 | 1879 |
|---|---|---|---|---|
| Mechanical workshops | 35 | 23 | 28 | 44 |
| Coach manufactories | 4 | 5 | 8 | 7 |
| Railway carriage Manufactories |  |  | 1 | 1 |
| Shipyards | 72 | 197 | 200 | 112 |
| Rifle manufactories |  |  | 1 | 1 |
| Musical instruments | 3 | 6 | 8 | 5 |

Source: *Statistik over Norges Fabrikanlæg* (Kristiania, 1881), p.viii

ing cyclical factors. Employment in the industry rose sharply from 1860 in absolute terms but also as a percentage of the industrial workforce, reflecting the increasingly important place of engineering in the industrial structure of the country.

In terms of the number of enterprises, this engineering industry remained strongly oriented, during the mid-nineteenth century, to the shipbuilding and repair industry which was so important to Norway's foreign earnings. The structure of the engineering industry during the 1860s and 1870s is given in Table 10.6. The

*Kristine Bruland*

material presented in Table 10.6 suggests that the growth of Norwegian industry was closely connected with foreign trade, in
particular machinery imports. But domestic mechanical engineering came to play an increasingly important role. This suggests
questions about whether the decreasing reliance on machinery
imports and the growth of domestic engineering, also have a
foreign dimension. How did this growing industry establish its
technological basis? In particular, what role did foreign influences
play as Norwegian entrepreneurs, managers and engineers acquired
the skills of the modern industrial era? We turn now to an empirical
examination of these questions.

## 5   Transfer of Technology into the Norwegian Mechanical Engineering Industry

In this and following sections we examine the role of the transfer of
foreign technology, primarily British, in the establishment of Norway's mechanical engineering industry. 'Technology' should not
be confused simply with equipment, nor 'technological diffusion'
with the purchase of machines (although of course it usually does
involve machine acquisition). Rather, technology and diffusion
capability consist of a complex combination of:

1. Information, which is in turn a complex phenomenon including, for example, (a) general information on the scope, range
   and structure of available technologies and on the main lines of
   technological advance at any particular time, (b) specific information on available techniques, (c) knowledge relevant to the
   construction, setting-up and operating of equipment.
2. Skills, both labour and managerial, in the construction, operation, supervision, maintenance and management of equipment.
   Training in all of these areas is thus an important component of
   technological diffusion.
3. Equipment acquisition, operation, adaptation and development.

In examining the details of technological diffusion within the
Norwegian mechanical engineering industry we shall be looking
closely at the practical ways in which information, skills and
equipment were acquired. In particular, the following sections will
investigate:

1. The role of foreign travel and foreign training in the develop-

## Norwegian Mechanical Engineering, 1800–1900

ment of Norwegian engineers and engineering firms
2. The nature of contacts between Norwegian and foreign engineering enterprises
3. The acquisition, adaptation and development of foreign equipment
4. The role of foreign workers in the development of the Norwegian engineering industry

During the nineteenth century the development of the mechanical engineering industry was extensive in Norway, with vigorous
creation of firms. The following examination will range fairly
widely over this often dispersed industry, but concentrate in the
main on the technological histories of eleven firms. They are:

Mesna Works (founded 1814)
Thune Mechanical Workshop (founded 1815)
O. Mustad (founded 1832)
Aker's Mechanical Workshop (founded 1841)
Trondhjem's Mechanical Workshop (founded 1843)
O. Jakobson's Machine Workshop (founded 1845)
Myren's Mechanical Workshop (founded 1848)
Christiania Nail Manufactory (founded 1853)
Kværner Works (founded 1853)
Nyland's Works (founded 1854)
Kampen Mechanical Workshop (founded 1865)

These firms had a wide range of primary and secondary activities,
but in general we can say that they cover the whole spectrum of
activities of mechanical engineering in Norway.

## 6  Technology Transfer (1): Foreign Travel and the Development of Norwegian Engineering

Britain's industrial development, from the inception of its industrialisation in the late eighteenth century, was of very great interest to
foreigners. The interested parties included those with a scientific or
technical concern over what was happening in Britain, those who
desired to become entrepreneurs on the British model, and –
perhaps most importantly – European governments who were
worried about the economic, political and military implications of
Britain's emerging industrial dominance. Thus by the middle of
the nineteenth century visits to Britain's industrial areas were a

*Kristine Bruland*

standard part of European entrepreneurial practice.[20] Like many European governments, Norwegian legislators were anxious for close contact with British industrial development, and accordingly the Norwegian parliament discussed, in 1836 and again in 1854, the desirability of promoting visits to Britain by Norwegian business-men and engineers. It was decided to subsidise such visits with official stipends, which became extensively used to support visits to Britain. Since these stipends were intended to support visits which would not otherwise have been made, there were of course many visits other than those which were officially supported. But we shall begin with a description of travel stipend applications as a way of getting some idea of the numbers and interests of Norwegian engineers who were visiting Britain.

Appendix 10.1 describes principal stipend-funded visits by Nor-wegian engineers against the background of overall travel stipend applications for the forty-five years from 1850. It can be seen that in any particular year engineers formed a significant proportion of those whose applications were granted. A total of 163 engineering stipends were approved during these years. Of the recipients, 101 (or 61.9 per cent) went to Britain; most of the remainder went to Germany, with a small proportion going to Denmark or Sweden.

A typical example of early stipend-backed travel might be that of the mechanic and machinist P. Nørbech of the Nylands Works. He had worked for a year in various mechanical engineering establish-ments in Britain, and spent five months in New York. In 1853 he received a travel stipend of 150 specie daler to visit several big workshops in Britain and France. He applied for a further 50 specie daler in 1855, and was clearly very familiar with British technical practice. The important engineer O. Jakobson received a similar amount for visits to English and Scottish workshops and the London Exhibition, on which he was expected to write a report for the Department of Domestic Affairs (Indredepartementet).

Many engineers travelled, however, without relying on stipends. Some of these visits are listed in Appendix 10.2. The visits covered a wide variety of activities. A. Jensen, of Myren Mechanical Work-shop, for example, who specialised in turbine construction, ac-quired his expertise in Germany, where he had 'studied turbines with first-class German experts'.[21] His brother, J. Jensen, was

20. See e.g. E. Robinson, 'The Transference of British Technology to Russia, 1760–1820', in B. M. Ratcliffe (ed.), *Great Britain and Her World 1750–1914*, Man-chester, 1975, p.3. For a description of visits undertaken by Norwegian textile entrepreneurs see Bruland, *British Technology*, chap. 5.
21. K. Anker Olsen, *Kværner Brug 1853–1953*, Oslo, 1953, p.71.

## Norwegian Mechanical Engineering, 1800–1900

rejected for a travel stipend in 1842, but apparently then used the resources of the University of Oslo Library for his informational needs; he subsequently travelled to England and Germany in 1851, 1857 (to study steam-driven saws in anticipation of the repeal of the 'sawmill privilege' which occurred in 1860) and 1860.[22] Subsequently Paul Holmsen, the manager of Myren's Fredrikstad filial, travelled in 1870 in Britain and Germany.[23] For Nyland's Works, the works manager Bang travelled to England in 1888 and 1890 'to study the progress made in England's best-known workshops'.[24] Later, in 1893, A. L. Thune in his correspondence with Babcock & Wilcox about business promotion in connection with the Chicago exhibition referred to at least fifty Norwegian technicians planning to visit the exhibition, and remarked that 'more will go than just those with stipends'.[25] Where firms themselves funded travel, conditions were sometimes attached: when the directors of the Trondhjem works gave their English works manager John Trenery 200 specie daler to visit the London Exhibition in 1862, he had to agree not to leave their employment for two years after the trip.[26] The firm subsequently engaged A. N. Olsen as works manager 'by telegraphing to England', and subsidised three visits to Britain for other employees.[27]

As in the textile industry,[28] foreign travel sometimes appears to have provided the initial impetus for enterprise formation itself: Steenstrup, of Aker's Mechanical Workshop, 'must have got the idea to establish the mechanical engineering firm from his stays in England and Sweden', where he had studied both mechanical engineering and shipbuilding.[29] Similarly Halvor Thune worked in Scotland in 1838, in a Glasgow workshop, and began boiler production in Norway on his return.[30] Steenstrup first visited England in 1834–5, then twice in the 1840s (during which time he visited the Maudsley works, perhaps the most important engineering shop in England at that time); later Aker documents record visits to

22. C. Gierløff, *Et Bruk ved Akerselven: Myrens Verksteds Hundre Års Minne*, Oslo, 1948, pp.69, 133.

23. Ibid., pp.158–9.

24. Nyland archives, Maritime Museum, Oslo, Styreprotokoll 1B; 11/11–88 and 8/8–90.

25. Thune archives, Norwegian Technical Museum, Oslo, Brevkladdebok; 24/4–93.

26. O. Henmo, *Trondhjems Mekaniske Verksted 1843–1918*, Trondheim, 1919, p.16.

27. Ibid., pp.24, 64.

28. See Bruland, *British Technology*, chap. 4.

29. L. Egge and H. Sandsbråten, *Gamle Akers Verksted*, Oslo, 1982, p.2.

30. Y. Hauge, *Boken om Thune*, Oslo, 1965, pp.16–17.

*Kristine Bruland*

England by Steenstrup to visit the Exhibition at the Crystal Palace in 1851,[31] then again in 1856 to order machinery, with reports of the visit and accounts of expenses. He visited London again in 1861, and visits by directors and foremen at Aker occurred in 1860 and 1882.[32] In fact, Aker exhibited at the London Exhibition in 1862, and won a prize.[33]

We note the beginnings of a reorientation among the Norwegian mechanics and engineers travelling abroad; the number of different destinations increased and were situated further afield as more countries, in particular Germany and the USA, developed into advanced industrial economies. In the 1870s H. G. Stub, for example, a director of Kværner, visited the United States, and the Kværner foundry-manager, W. Bergh, visited the USA around 1895: 'The trip resulted in several technical improvements in the production process.'[34] However, based on available material, Britain remained the dominating destination. It should also be noted that Norwegian firms often received visits from foreign engineering firms, which no doubt facilitated technology flow: Nylands, for example had visits from the English companies Foxwell & Co. in 1883, William Reid & Co. in 1884, Kenyon & Co. in the same year, and Turton & Sons in 1885.[35]

## 7   Technology Transfer (2): Foreign Education and Training of Norwegian Engineers

Foreign training was a frequent element in the development of early Norwegian mechanical engineering firms. This could involve either practical workshop training or formal academic education or both. Some patterns of formal education abroad are summarised in Table 10.7.

Foreign training was widespread through the firms studied here. For example, Mustad sent 'a man to learn in England' about 1847.[36] A director of Kværner trained at the Gewerbe Akademie in Berlin, and another – the technical director H. M. Smith – after

31. Aker archives, Norwegian Technical Museum, Oslo, Forhandlingsprotokoll, 1843–52, pp.49–52.
32. Aker archives, Norwegian Technical Museum, Oslo, Styreprotokoll, 1854–71, pp.69, 74; Styreprotokoll, 1871–1900, 27/4–82.
33. H. P. Lødrup, *A/S Akers Mekaniske Verksted 1841–1951*, Oslo, 1951, p.72.
34. Olsen, *Kværner Brug*, pp.163, 227.
35. Nyland archives, Maritime Museum, Oslo, correspondence in IIE 2316, 14/4–83; 2319, 13/5–84; 2320, 15/8–84; 2322, 28/7–85.
36. O. Wicken, *Mustad gjennom 150 År, 1832–1982*, Oslo, 1982, p.32.

*Norwegian Mechanical Engineering, 1800–1900*

**Table 10.7** Education and work experience from abroad

| Firm | Person | Visit |
|------|--------|-------|
| Thune Mechanical Works | H. Thune | Employed at workshop in Glasgow, 1838 |
| | S. Thune | Technical education in Germany, c. 1900 |
| Aker Mechanical Works | Steenstrup | Studied engineering and shipbuilding in England and Sweden; 1834, 1835, 1840, 1841 |
| | Bronn | Technical education from England |
| | G. Swensen | Educated in England 1872–74 |
| | J. G. L. Lie | Technical education and visits to England, Scotland, USA |
| | H. G. Stub | Educated in Berlin |
| | S. A. Weidemann | Visits to England and Scotland |
| Trondhjem's Mechanical Works | H. B. Holmsen | Educated in Germany and England |
| | A. Nørbecholsen | Educated in Sweden, and three years in England |
| | H. J. Olsen | Educated in Denmark and Germany |
| | W. H. C. Swenssen | Educated in the USA |
| | Trenery | Born and grew up in Britain |
| | S. A. Weidmann | Stays in England and Scotland |
| | H. G. Jürgens | Visits to France |
| Jakobsons Machine Works | H. Jakobson | Educated in USA |
| Myrens Mechanical Works | A. Jensen | Stays in Germany |
| Kværner Works | G. Onsum | Technical education from France and Germany |
| | H. G. Stub | Educated in Berlin, and visits to USA |
| | H. M. Smith | Worked in England for nine years, 1862–71 |
| Nylands Works | Morterud | Studied in England, $2\frac{1}{2}$ years |
| | L. Rode | Stays in Sweden in the 1860s |

*Sources*:  secondary literature referred to in this chapter; firms' records

245

*Kristine Bruland*

visiting Britain funded by a stipend in 1862, stayed to work as a draughtsman and foreman at Camell & Co. of Sheffield and New-ton Iron Works of Hull. He returned to Norway in 1871.[37] Another important Kværner director, G. Onsum, was educated in France and Germany, visited German workshops in the 1870s and sent back (with permission) detailed technical drawings of a travelling crane to the firm.[38]

Within the Aker firm, technical training from England was regarded as a 'family tradition', and nineteenth- and early twentieth-century directors (such as Bronn, G. Swenson and J. G. Lie) received engineering education in Britain and the USA.[39] Sverre Thune, of the Thune enterprise, received his technical education in Germany in the early twentieth century.[40]

The foreign influence was not simply a matter of initial training, but also of further education in mid-career. Some firms were very active in this regard. The Trondhjem's works, for example, apart from actively subsidising travel for its employees, specifically funded education. In 1878 works manager Helseth, engineer Nyhus and foreman Nielsen were all given 400 kroner each to study abroad. In 1872 Trondhjem's employee H. J. Olsen was given money for travel 'if he accepts to stay on . . . and if he tries to make the trip as technically useful as possible and uses this knowl-edge to our advantage'; in 1881 he was given more money for further education in Germany, Britain and Sweden, and in the same year assistant Falck was to spend two years abroad for further education.[41] Falck was granted two years abroad on full pay on condition that he remained with the firm for five years after his return. In 1878 the foreman, J. F. Nielsen, spent at least three months abroad on a state stipend and with full wages. He travelled again to Britain the following year.[42] This process continued: in the late 1880s, W. H. C. Svenssen visited the USA; in the mid-1890s H. B. Holmsen spent two years abroad, in Berlin, then studying shipbuilding techniques in Britain and Germany; in 1900 H. G. Jürgens (who had been educated in France) went abroad to study new machinery and work methods, and bought a substantial quan-tity of machinery for the firm while away.

37. Olsen, *Kværner Brug*, pp.62, 227.
38. Ibid., p.120.
39. Lødrup, *Akers Mekaniske Verksted*, p.35.
40. Hauge, *Boken om Thune*, pp.16–17.
41. Henmo, *Trondhjems Mekaniske Verksted*, pp.66–70, and Thune Archives, Norwegian Technical Museum, Oslo, Forhandlingsprotokoll, 6/7–72.
42. Henmo, *Trondhjems Mekaniske Verksted*, p.64.

## Norwegian Mechanical Engineering, 1800–1900

Although, as noted above, a significant number of Norwegian engineers were trained abroad, others were trained by foreign workers within Norway. Thus Andreas Jensen, the father of J. and A. Jensen of Myren, was taught in Norway by the Scottish mill builder John Wilson.[43] Similarly, one of Andreas's sons, J. Jensen, was trained by the textile engineer Gellertsen, who had himself trained abroad (at the Nordberg engineering firm in Copenhagen).[44] Others, such as Andreas Thune and O. Jakobson, studied English and German in order to be able to read the technical literature of the industry.[45] Jakobsons's son was trained as an engineer in the United States.[46]

Sometimes foreign training had a significant impact on the subsequent technological development of a firm. For example, Jens Jacob Jensen went on to be trained at a technical school in Zurich, 'the first of many young men from Myren', and this appeared to be an important factor in the switch by Myren from making turbines of the British type to new Continental turbine types.[47] Much later, foreign training extended away from purely technical questions into the area of engineering management: thus a Kværner manager, H. P. B. Lund, visited the Berwick Co. of Brooklyn, New York, in order to study the 'scientific management' techniques developed by F. W. Taylor.[48]

For most of the nineteenth century Britain was the dominant foreign source for technical education and experience. It remained important, but as the process of industrialisation spread, and in particular as German industry grew strongly, the choice of destination for further education and training open to Norwegian mechanics expanded. Furthermore, German industry was backed by a system of education – particularly within technical subjects – which at the time was unrivalled.[49] The material above, which indicates a gradual decline of Britain's dominance, does not reflect the dramatic changes which took place within the formal education of engineers. During the latter half of the nineteenth century aspiring Norwegian engineers increasingly sought technical education in Germany. Mainly based on source material from the educational

43. Gierløff, *Et Bruk ved Akerselven*, p.61.
44. Ibid., p.63.
45. Hauge, *Boken om Thune*, p.24.
46. *Jubileumsbok*, Oslo, 1946, p.32.
47. Gierløff, *Et Bruk*, p.164.
48. Olsen, *Kværner Brug*, pp.271–2.
49. Landes, *The Unbound Prometheus*, p.340.

*Kristine Bruland*

institutions, this has been demonstrated by, among others, Håkon
With Andersen and Fritz Hodne.[50]

## 8   Technology Transfer (3): Contacts With Foreign Engineering Enterprises

The central institutions initiating and receiving technology transfer
are business firms.[51] A key indicator of the existence of international transfer is therefore a high level of contact among firms
internationally. Much depends, of course, on the quality of these
contacts, but we should remember that even superficial relationships between firms may involve important information transmission. In tracing the international contacts of Norwegian
engineering firms we are, inevitably, limited by the available
archive material, especially correspondence and invoice files. Unfortunately this material does not survive for all the firms studied
here, and we are therefore limited to examining Thune, Jakobson,
Aker, Mustad, the Christiania Nail Factory, Kværner, Kampen,
Nyland and Trondhjem's mechanical workshop. The traceable
foreign contacts of these firms are listed in appendix 10.3, which
gives details of the names, locations and main functions of foreign
contacts, transactions which took place, as well as dates of contact
and sources.

The evidence suggests a very high level of contact: at the very
least, several hundred foreign engineering firms had some form of
contact with one or other of the Norwegian firms. During the
period 1830–1900 the Norwegian engineering enterprises were in
contact with 342 foreign firms, although it should be emphasised
that not all of these were mechanical engineering firms. Of these
firms 194 were British, thirty-four German, two from the USA
and twelve from various other countries. The level of contact
varied considerably between the firms; Thune, for example, had
some kind of contact with fifty-six foreign firms, virtually all of
which appear to be within the engineering industry. Over half of
these were in Britain; next most numerous were contacts with
German firms, with some contacts in the USA. A similar pattern
can be found with other Norwegian firms. Jakobson had contacts

50. H. With Andersen, 'Germany and the Education of Norwegian Engineers',
in *Berichte über das 2. deutsch-norwegische Historikertreffen in Bonn*, May 1987, NAVF,
1987, p.100; F. Hodne *Norsk Økonomisk Historie 1815–1970*, Oslo, 1981, p.247.
   51. See Wilkins, 'Role of Private Business'.

## Norwegian Mechanical Engineering, 1800–1900

**Table 10.8**  Foreign contacts of Norwegian firms

|  | 1841–50 | 1851–60 | 1861–70 | 1871–80 | 1881–90 | 1891–1900 |
|---|---|---|---|---|---|---|
| Thune mechanical works |  |  |  | 35 | 18 | 13 |
| O. Mustad | 1 |  | 5 | 1 | 1 | 9 |
| Aker mechanical works | 1 | 10 |  |  | 19 |  |
| Jakobsons machine works |  |  | 15 | 42 | 5 | 7 |
| Kværner Works |  |  |  | 1 |  |  |
| Christiania Nail Works |  |  |  |  | 5 | 8 |
| Nylands works |  | 15 | 51 | 10 | 68 | 1 |
| Kampen mechanical works |  |  |  | 1 |  | 17 |
| Total | 2 | 25 | 71 | 90 | 116 | 55 |

*Source*: appendix 3

with at least sixty foreign firms, Nylands with 119, Aker with thirty-one, Mustad with sixteen, the Nail manufactory fourteen, Kampen eighteen and Kværner one. This suggests a very high average number of contacts with strong variations around the median, but it must be emphasised that the numbers given above are lower limits. For some of the firms (especially Kværner) the available source material is very poor, so that the figures probably underestimate the real number of contacts – in Kværner's case by a large margin. As one might expect, general engineering firms, such as Aker, Nylands, Jakobson and Thune had a wider range of contact than firms producing a more limited product range (such as Mustad and the Christiania Nail Manufactory).

In Table 10.8 we have disaggregated the foreign contacts according to contact with Norwegian firms and time periods. We can trace the number of foreign firms active in the Norwegian market within each ten-year period. Because some firms were active during several ten-year periods, the table gives a higher total number, namely 359, than the total number of foreign firms operating on the Norwegian market during the whole period. The table shows clearly marked growth in establishing foreign contacts after the

249

*Kristine Bruland*

1840s. It culminates in the 1880s, then to decrease sharply in the 1890s.

How are we to interpret this decrease? It may be connected with a problem of source material. In particular the Nyland sources seem to be better for the 1880s than the 1890s. Nyland had contacts with a substantial number of foreign firms in the 1880s, and our ability to establish the real number of contacts for the 1890s might be seriously impaired by a possibly incomplete set of data for this period. On the other hand the sources may be equally incomplete for all the ten-year periods. We do not know this, but if it is so, the table probably reflects a real decrease in number of contacts. Since the 1890s was not a period of depression in Norway (actually it was the period when Norway definitely overcame the crises of the 1870s and 1880s and of industrial expansion and reconstruction), we cannot explain the decline in the number of foreign contacts by a general economic recession. What we see, therefore, may be a decline in Norwegian mechanical workshops' technological dependence on foreign supply, and a growth in technological independence of Norwegian enterprises. We note that the Norwegian firms' number of contacts as well as the profile of contacts over time vary considerably, but even so the 1890s show a sharp decrease in contacts for the firms – Thune, Jakobson and Nyland – which had definitely the largest share of contacts in the preceeding periods.

In the 1890s the geographical distribution of the foreign contacts also began to change. This is shown in Table 10.9. British contacts still made up the largest share of the total decreasing contacts, but the German share grew rapidly, from 11 per cent in the 1880s to 36 per cent in the 1890s.

Much depends, however, on the *content* of these relationships: what exactly did they mean for the process of technology transfer? Broadly speaking, we can distinguish from the available records four types of transaction between foreign and Norwegian firms:

1. Purchase of 'basic' inputs such as iron bars, iron and steel plate, glass, coal, oil etc. Such items were very frequently obtained abroad; where they were purchased domestically, it was often through importers such as Henry Hutchinson of Drammen.
2. Purchase of relatively simple but 'engineered' inputs, such as screws, washers, pins, tubes, springs, knives etc.
3. Purchase of machinery, machine tools and general engineering capital goods. Such equipment might include, for example, lathes, rolling machines, milling machines, boring machines,

250

## Norwegian Mechanical Engineering, 1800–1900

**Table 10.9** Geographical distribution of foreign firms active in the Norwegian market

|         | 1831–40 | 1841–50 | 1851–60 | 1861–70 | 1871–80 | 1881–90 | 1891–1900 |
|---------|---------|---------|---------|---------|---------|---------|-----------|
| Britain | 1       | 2       | 15      | 60      | 42      | 86      | 31        |
| Germany |         |         | 3       | 1       | 12      | 12      | 20        |
| USA     |         |         |         |         | 2       | 1       |           |
| Others  |         |         |         |         | 4       | 4       | 4         |
| Total   | 1       | 2       | 18      | 61      | 60      | 103     | 55        |

*Source:* appendix 3

turbines, pumps, jacks and lifting equipment, planing machines and steam-engines. From available records it is difficult to be precise about the number of foreign firms supplying such equipment, but for Aker, for example, out of thirty-one foreign firms supplying an identifiable product, seventeen were supplying equipment in this category. Jakobson and Myren appear to be roughly comparable.

4. Supply of technical or economic information. A number of these contacts involved, for example, information on prices and availability of equipment, or the supply of drawings, or catalogues of equipment, or information regarding patents and licensing.

Clearly it is the latter two categories which are important for the development of technological capability in Norway. Information flows have an obvious role in acquisition of technological competence. But we should remember that the supply of equipment (which will be discussed in more detail in the next section) is firstly usually associated with technological information flows, and secondly is normally associated with competence building through 'learning by doing'. The considerable extent of transactions in these areas, therefore, is prima-facie evidence of significant technology transfer.

## 9 Technology Transfer (4): Acquisition and Adaptation of Foreign Technologies.

Although, as I have argued above, technology transfer can never be reduced simply to the purchase of machines and equipment, nevertheless it normally does involve equipment acquisition. Machine

251

*Kristine Bruland*

acquisition frequently involved the acquisition or development of new skills, and, in the case of the mechanical engineering industry, was central to the development of new capabilities and new products. Our ability to trace such acquisition is of course limited by the availability of source material, but extant invoice and other documentary material enables us to give at least an outline of the overall process of machine acquisition by the firms studied here. Those purchases which can be definitely documented by invoice or shipment records, principally for Aker, Nyland and Trondhjem's Mechanical Workshop, are described in Table 10.10.

This Table almost certainly understates, because of limited sources, the extent of acquisition: there are very many other references – in correspondence or secondary literature – to other machines. But it can be seen that these firms maintained a consistent programme of machinery acquisition from abroad from the early 1840s. About eighty major items were purchased, mostly machine tools of various types. The principal source was Britain, and the trade involved some of the major engineering firms of the time, such as Whitworths; it is only towards the end of the period that purchases from Germany begin to appear.

In fact most of the firms studied here engaged in more or less extensive machine purchase, especially in their early years, and this was presumably an important element in the industrial learning process in Norwegian engineering. Thus Myren's new workshop in 1855 contained a large English boremachine, 'distinct by its complete arrangement and self-acting motion';[52] Mustad purchased all its early nail machines from William Thompson & Co. of Birmingham, and subsequently purchased lathes, a stick machine, a slotting machine and a planing machine.[53] At approximately the same time the Nail Factory purchased six nail-making machines, as well as patent rights, apparently for the Coates machine, for Sweden and Norway.[54] From the early 1840s, Aker acquired lathes, saws and machine tools (of various types) from England, as well as a steam hammer from the British engineer James Nasmyth.[55] Customs duty on the largest machine acquired, a lathe costing 802 specie daler, was refunded by the state.[56] Another round of pur-

52. Gierløff, *Et Bruk*, pp.113–4.
53. Wicken, *Mustad*, pp.13–14, 28.
54. Christiania Nail Works, archives, Norwegian Technical Museum, Oslo, Forhandlingsprotokoll, 1854, p.1; Olsen, *Kværner Brug*, pp.20, 79.
55. Akers archives, Norwegian Technical Museum, Oslo, Forhandlingsprotokoll, 1843– , pp.4, 17, 51, 52; Lødrup, *Akers Mekaniske Verksted*, p.39.
56. Lødrup, *Akers Mekaniske Verksted*, p.39; Nasmyth archives, England.

## Norwegian Mechanical Engineering, 1800–1900

**Table 10.10** Documented machine and equipment acquisitions

| Year | Firm | Purchase | Exporting country | Producer |
|------|------|----------|-------------------|----------|
| 1842 | Aker | 1 lathe | | |
| 1843 | Aker | 1 lathe | | |
| | Trondhjem's | 1 steam engine | | |
| | | 1 blowing machine | | |
| | | 1 lathe | | |
| 1847 | Mustad | 1 nail machine | | Thompson & Co. |
| 1850 | Aker | 1 bolt-boring machine | | |
| | | 1 expanding iron chuck | | Matthew Young |
| 1851 | Aker | 1 self-acting lathe | | |
| | | 1 shaping and planing machine | | Whithworth & Co. |
| 1854 | Aker | 1 steam hammer | | Nasmyth |
| | | 1 planing machine | | |
| | | 1 lathe | | |
| | | 1 chuck | | Parr, Curtis & Madeley |
| | Chra. Nail Works | 6 nail machines | | |
| 1855 | Myren | 1 bore machine | | |
| 1856 | Aker | [unspecified] | | |
| | Nyland | 1 planing machine | | |
| | | 1 circular saw | | Worsam & Co. |
| 1858 | Nyland | 1 shaping machine | | Whithworth & Co. |
| 1859 | Nyland | 1 plate-bending machine | | J. Buchton & Co. |
| 1860 | Aker | 1 lathe | | |
| 1861 | Nyland | 1 steam hammer | | R. Morrison & Co. |
| 1862 | Nyland | 1 steam riveting machine | | W. and I. Garforth |
| | | 1 steam governor | | W. Sergant |
| 1865 | Nyland | 1 [indecipherable] | | |
| | | 1 slotting machine | | |
| | | equipment for lathe | | |

*continued on page 254*

*Kristine Bruland*

**Table 10.10**   *continued*

| Year | Firm | Purchase | Exporting country | Producer |
|------|------|----------|-------------------|----------|
| 1866 | Nyland | 1 punching machine | | Collier & Co. |
| | | 1 slide lathe bed | | J. Hulse |
| | | 1 slotting machine | | |
| | | 1 drilling machine | | |
| | | 1 wheel-cutting machine | | Whithworth & Co. |
| | | 1 edge-planing machine | | |
| | | 1 plate-bending machine | | Smith, Peacock & Tanneth |
| 1870 | Trondhjem's | 1 slotting machine | | Hutton & MacDonald |
| 1871 | Mustad | 1 lathe | | |
| | | 1 stick machine | | |
| 1872 | Mustad | 1 lathe | | |
| | | 1 planing machine | | |
| | | 1 slotting machine | | |
| | Trondhjem's | 1 planing machine | | |
| 1873 | Trondhjem's | 1 riveting machine | | |
| 1882 | Nyland | manometer control app. | Germany | A. Barber & Co. |
| | | 1 blower & duplex engine | | Vulcan Iron works |
| | | 1 steam winch | | Clarke, Chapman & Guerney |
| | | 1 cylinder | | Hawkes, Crawshaw & Sons |
| 1883 | Nyland | [unspecified] | | |
| 1884 | Nyland | horizontal punching, beam-bending and angle-cutting machine | | Campbells & Hunter |
| | | 1 steam winch | | |
| | | 1 plate bending machine | | Clarke, Chapman & Co. |
| | | [unspecified] | | Scriven & Co |
| | | 1 mandrel (spindle) | | Foxwell & Son |

254

## Norwegian Mechanical Engineering, 1800–1900

| Year | Firm | Purchase | Exporting country | Producer |
|------|------|----------|-------------------|----------|
| | | [unspecified] | | Appleby Bros. |
| 1886 | Mustad | 1 wire-drawing machine | | |
| 1892 | Nyland | 1 steam hammer | | J. Rennie & Son |
| 1893 | Kværner | 1 slotting machine | Chemnitz | |
| 1894 | Nyland | 1 punching and cutting machine | | |
| 1898 | Chra. Nail Works | 13 nail machines | Germany | |
| | Kampen | 1 transmission steel-wire pliers | | |
| | | [unspecified] Worthington pump | Germany | Naxos Union Worthington by Fischer & Son |
| | | 1 lathe | Germany | E. Sonnenthal, by L. Ewald |

*Note*: unless stated otherwise machines and equipment are from Britain.
*Sources*: Aker: accounts, 14 December 1842; Protocol (Forhandlingsprotokoll); Nasmyth archives, Britain; Trondhjem's Mechanical Works: archives, box 1; Directors' Protocol (direksjonsprotokoll), pp.12, 20, 23; O. Mustad: O. Wicken, *Mustad gjennom 150 År 1832–1982*, Oslo, 1982, pp.13, 28; Christiania Nail Works: K. Anker Olsen, *Kværner Brug 1853–1953*, Oslo, 1953, p.20; Protocol (forhandlingsprotokoll), p.1; T. Parmer, *'Spigeren' som temmet jern og stål: Produksjon og arbeidsforhold ved Christiania Spigerverk 1860–1960*, Oslo, 1982, p.143; Myren's Works: C. Gierløff, *Et Brug ved Akerselven: Myrens Verksteds Hundre Års Minne*, Oslo, 1948, pp.113–14; Nyland Works: accounts (hovedbøker); correspondence; Directors' Protocols (styreprotokoller); Kværner Works: Olsen, *Kværner Brug*, p.162; Kampen Mechanical Works: correspondence.


chases – of similar but more modern equipment – was made in England in the early 1850s as Aker moved to Holmen, a harbour site, and formed a new company.[57] Purchases would sometimes be associated with a new product line: when Aker was considering whether to begin making coke ovens in 1851, Steenstrup went to England to acquire information and models, and to purchase appropriate

57. Egge and Sandsbråten, *Gamle Akers Verksted*, p.4.

255

*Kristine Bruland*

equipment.[58] From the late 1850s, Nyland appears to have purchased a very large part of its machine stocks in Britain: at least a
dozen major machine tools of various types.

Similar purchases continued to be made into the late nineteenth
and early twentieth centuries. In 1898, 1901 and 1912, for example,
the Nail Factory purchased equipment in Sweden and Britain (the
Swedish purchase including men to operate the equipment).[59] Both
Myren and Aker continued to buy rolling and lifting equipment
from the English firm Smith Bros. in the last years of the nineteenth century.

Where it is possible to trace the machine and equipment stocks of
firms, we find a large proportion of equipment to be of foreign
origin. For example, the Jakobson copy book records, in 1903, the
movement of a number of machines and lists their values: by far the
largest part are foreign, in particular boring and grinding machines
by Barnes and by the American firm Pratt & Whitney. The general
picture which emerges here is that machine tool acquisition in
particular was an important element in the development of technical capability by Norwegian engineering firms. The fact that these
transactions decreased in number in the 1890s suggests that such
capability actually had been developed.

## 10  Technology Transfer (5): Use of Foreign
## Technical Information and Imitation
## of Foreign Techniques.

The direct imitation or adaptation of foreign techniques, on the
basis of foreign information, was an essential element of early Norwegian engineering. Norway's first Scotch turbine, for example, was
constructed by the Myren firm in 1849, on the basis of British
technical drawings.[60] Likewise, Myren's first planing machine,
built in the early 1860s, appears to have been an imitation and perhaps
development of a British machine imported to Frederikstad in 1860.
Later, in the 1890s, Myren changed to an American model.[61] In
1870, Myren made a paper machine for Bentse Brug which was in
fact of copy of a machine delivered twelve years earlier by the

58. Lødrup, *Akers Mekaniske Verksted*, p.52.
59. T. Parmer, '*Spigeren' som temmet jern og stål: Produksjon og arbeidsforhold ved
Christiania Spigerverk, 1860–1960*, Oslo, 1982, pp.131, 143, 145.
60. Olsen, *Kværner Brug*, p.70.
61. Gierløff, *Et Bruk*, pp.138–9, 146.

## Norwegian Mechanical Engineering, 1800–1900

Edinburgh engineers James Bertram & Sons.[62] Within Myren 'English books of instruction were the most used', and the firm was often in receipt of technical information, such as the technical drawings of spinning machinery sent to Myren by Bertram and Sons in 1879.[63] In the mid-1840s, the engineer O. Jakobson visited the William Thompson engineering firm in Birmingham, collecting a nail-making machine on behalf of Mustad. He subsequently built a copy of it for Mustad, which was installed early in 1849; his agreement with Mustad stipulated that he would not build a similar machine for others. This pattern was repeated thirty years later: in 1876 one of Mustad's employees, one Topp, studied the American horseshoe machine in Christiania after the world exhibition, and subsequently – in 1881 – made a copy.[64]

Nylands were frequently in receipt of detailed technical information from British engineering firms. For example, in 1882, Clarke, Chapman & Gurney sent tracings of plans of a furnace which they had supplied in England. In the same year, Palmers Shipbuilding Co. of Jarrow sent similar tracings of a furnace, and ten days later sent photographs.[65] They also received patent information from UK patentees, one of whom suggested that Nylands take out a Norwegian patent for one of his inventions.[66] Nylands, like other Norwegian firms, were often also in receipt of circulars, catalogues, prospectuses and so on from potential suppliers: these were in themselves important sources of technical information.

The process of imitation became more sophisticated over the years. In Norway, as in other nineteenth-century industrialisers, railway construction was a key development. The Norwegian Railways had acquired most of the equipment abroad, in particular through the British firm of Beyer, Peacock: between 1866 and 1883, Beyer, Peacock sold fifty-seven locomotives to the Norwegian market.[67] But the railway also promoted domestic production. Thus, in 1901, Thune's first railway locomotive was successfully constructed. But like his later early locomotives, this was constructed using British drawings which had been acquired for him by the railway.[68] Another and important form of imitation was the

62. Hauge, *Boken om Thune*, p.129.
63. Gierløff, *Et Bruk*, p.164. These drawings are kept at the Norwegian Technical Museum in Oslo.
64. Wicken, *Mustad*, pp.14, 39.
65. Nyland archives, Maritime Museum, Oslo, correspondence in IIE 2315, 6/11–82.
66. Ibid., IIE 2321, 12/2–85.
67. Beyer, Peacock archives, Glasgow University Library.
68. Hauge, *Boken om Thune*, p.103.

*Kristine Bruland*

ability to use foreign technologies under licence. Aker, for
example, in 1912, licensed the marine diesel motor design of the
Danish firm Burmeister & Wain, receiving detailed drawings and
construction plans.[69]

## 11   Technology Transfer (6): Foreign Engineers and Workers in Norwegian Engineering Development.

Expertise or skill is a central component of any technology, and
equipment generally cannot be operated without it. One key way
in which technical skills can be acquired is through the import of
foreign workers, who may operate important equipment, transmit
their skills by teaching, or manage and supervise local workers.
The extent to which we can trace such workers in Norwegian
engineering is, as with other aspects of technology transfer, dependent on surviving records. These are unfortunately limited in this
area. But where records do survive, we find an important presence
of foreign workers, usually British. Sometimes these were involved in the establishment of firms in the first place: thus John
Trenery, of Trondhjem, was one of the founders of the firm.
Unlike the other founders, he had no financial liability for the firm,
which suggests that his role was primarily technological. Indeed
one of his earliest tasks was to travel to Newcastle to purchase a
steam engine and a wide range of other equipment which formed
the basic fixed capital stock of the firm. Some played important
roles in introducing major new technologies: William Stephenson
(himself a son of George Stephenson, builder of early steam locomotives and a figure of great importance in world industrial history) constructed Norway's first locomotive. Others played roles
less noticeable but perhaps no less important in the long run. Those
workers who can be traced are described in Table 10.11.

Unfortunately, the sources from which Table 10.11 is drawn do
not permit us accurately to describe either the lengths of stay or the
technological functions of British workers. It also probably understates numbers of workers, in particular omitting short-stay workers associated with the setting up of new equipment. We can see
however that all of the major Norwegian engineering firms at some
time employed foreign workers, usually British. Although it is
difficult to ascertain lengths of stay, we know that a significant
number of workers stayed for periods of several years: of the

69. Lødrup, *Akers Mekaniske Verksted*, p.106.

258

## Norwegian Mechanical Engineering, 1800–1900

**Table 10.11**  Foreign workers in the Norwegian engineering industry

| Name | Period | Function |
|---|---|---|
| *O. Mustad* | | |
| Hurst | 1887– | steel drawing |
| Henry Haynes | 1887–97 | filer |
| Holloway | 1880s | steel worker |
| John Croft | 1880s | fish hook worker |
| Tom English | 1880s | steel worker |
| Wm. Masters | 1880s | polisher |
| 'six English' | 1891–93 | start up needle production, machine purchase, management |
| 'English women' | 1899 | fishing fly makers |
| | | |
| *Aker Mechanical Works* | | |
| Asmundsen (Danish) | 1843 | |
| Charles Morris | 1846–48 | iron worker |
| Mellwright Spickles | 1855–56 | |
| Pickles | 1857 | |
| J. J. B. (Bing?) | 1884 | |
| | | |
| *Trondhjem's Mechanical Works* | | |
| J. Trenery | 1843–64 | founder |
| Wm. Trenery | c.1849–60 | |
| Wm. Stevenson | c.1850–68 | foreman |
| B. Cook | 1890s | |
| Stephenson | 1860s | |
| (Danish) | 1872 | formers |
| | | |
| *Myren Works* | | |
| Rollowy (?) | 1861 | foreman |
| Anton Harris | 1868–70 | drawer |
| Wilh Wettergren (Swedish) | 1869 | mechanical worker |
| | | |
| *Christiania Nail Works* | | |
| G. Hudson | c.1895 | |
| J. Hurst | 1854–1865 | |
| | | |
| *Nyland Works* | | |
| F. Ratcliffe | 1859 | turner |
| T. Jowsey | 1861 | foreman |

*continued on page 260*

*Kristine Bruland*

**Table 10.11**   *continued*

| Name | Period | Function |
|---|---|---|
| T. Ratcliffe | 1861 | mechanic |
| Samuel James | 1861–72 | foreman |
| James Rippon | 1862–67 | foreman |
| James | 1882 | |

*Note*: the workers are of British origin, unless otherwise stated.
*Sources*:   firms' archives, taxation records, secondary literature referred to in the footnotes

approximately sixty workers involved, about twenty-five appear to have been in Norway for periods of longer than two years. In other Norwegian industries, in particular textiles, I have been able to show that foreign workers performed a number of key technological functions. In equipment acquisition, setting up and operation of new techniques, information flows, training, production management and so on, they played a role out of all proportion to the small percentage of the workforce which they represented.[70] This was possible, however, only through access to detailed correspondence records of a type which does not appear to have survived in the engineering industry. None the less we can presume that similar functions were carried out in mechanical engineering, especially since foreign workers normally commanded wages higher than those being paid to Norwegian workers at that time.

Norwegian firms actively sought British labour: Nyland, for example, asked for help in this regard from the Norwegian firm of Bodin & Co., who were operating in Glasgow. But they also received applications for employment from workers in Britain. Often, interestingly, these were Norwegians who had trained or worked in Britain. Nyland received a number of detailed letters of application from such workers from the early 1880s, sometimes from previous employees: 'Having been in England for nearly two years now, and having gained experience from a variety of shops and drawing offices I take the liberty of writing, in accordance with Nylands Works' kind suggestion before I left for this country, to ask if there is any chance of a position at the works in the near future.'[71]

Klouman had worked for six months at the North Eastern

70. Bruland, *British Technology*, chap. 8.
71. Nyland archives, Maritime Museum, Oslo, correspondence in IIE 2318.

## Norwegian Mechanical Engineering, 1800–1900

Marine Engine Works, Wallsend on Tyne, in order 'to better acquaint myself with the English practical approach'. Subsequently Klouman worked as a draughtsman with the firm Cheesbrough & Roysden Engineers and Patent Agents, and spent nine months as an employee of Albion Foundry, Liverpool. Klouman in fact wrote three separate letters of application and forwarded references; the reference from Albion Foundry described Klouman as a 'good draughtsman, accurate in his work, punctual, steady and very obliging'.[72] Nylands received at least six separate similar applications during the early 1880s; a number had worked both in Britain and Germany, and for other Norwegian firms, as engineers, draughtsmen and so on. This kind of labour flow may well have been an important element, therefore, in the diffusion of technical skills. One such applicant, P. C. Pettersen, had in fact had his own firm in Christiania in the 1850s, but had sold up and gone to the USA, where he worked for an engineering firm in New York.[73]

Nylands actively sought British help in the management of labour. One of their letters, now lost, to the very important British firm of Whitworth & Co. apparently asked for such help, for it drew the reply that 'we shall be pleased to send out someone for the purpose you name – but in the meantime we shall be glad to know what class of work you are principally engaged with, and whether there is, or is likely to be any opposition on the part of your workmen to any reforms that may be suggested from time to time.'[74]

## 12  Technology Transfer (7): Norwegian Engineers and the Spread of Foreign Techniques in Norway.

One of the most important functions of Norwegian engineering firms was to operate as conduits for the flow of foreign techniques to other Norwegian firms. Myren, for example, in the mid-1850s offered to order machines from abroad for factories in Norway,[75] and in the late 1860s wrote to the textile entrepreneur Halvor Schou about his requirements for water pipes, steam engines and boilers,

72. Ibid., IIE 2318; 2319.
73. Ibid.
74. Ibid., IIE 2321 17/11–84.
75. M. E. Nord in *Skillingsmagasinet*, 1856, cited in Lars Thue, 'Framveksten av et Industriborgerskap i Kristiania 1840–1875', thesis, University of Oslo, 1977, p.32.

*Kristine Bruland*

about which 'we have written to several of our acquaintances in Manchester'.[76] In the early 1880s they were again advertising that they could use their connections abroad for the purchase of foreign machinery for Norwegian customers.[77] The engineer O. Jakobson obtained information from England for Mustad in 1851; they also received technical information on English hardening and plating techniques from the Bergen merchant Wallendahl in the late 1870s.[78]

The role of Norwegian engineering firms in the diffusion of foreign technologies is intimately connected with their activities in licensing foreign technologies and using foreign patents; some examples of this have been given in section 9 above. Even more important was the transition through which Norwegian engineers became agents of foreign engineering enterprises, constructing, selling and therefore spreading foreign techniques. The best-documented example of this is the firm of Thune, which acted as agent in Norway for the power engineering firm Babcock & Wilcox. Thune purchased his first Babcock & Wilcox boiler in 1890, and within four years had supplied seventeen such boilers to Norwegian customers. A formal agency relationship was not in place until 1895, but this led to the construction of sixty-three steam-engines using Babcock boilers within the following two years. Thune played a continuing and important role in marketing, skill transfer and the general diffusion of information relevant to power technologies in Norway.[79] Such activity was widespread and of considerable importance. Perhaps the most important technology involved here was steam power. For example, the English firm Clayton & Shuttleworth sold eighty-eight steam engines to Norway between 1857 and 1896; eighty-one of these were sold through Jakobson, and a further three were sold through Myren.[80] In the very late nineteenth century Thune sold a substantial number of high-pressure boilers for power purposes, from the firm of Babcock & Wilcox.

76. Hjula archives, Norwegian Technical Museum, Oslo; Correspondence in 7/3–68.
77. Gierløff, *Et Bruk*, p.170.
78. Wicken, *Mustad*, p.38.
79. E. Lange (ed.), *Teknologi i Virksomhet, Verkstedindustri i Norge etter 1840*, Oslo, 1989, pp.62–72.
80. Clayton & Shuttleworth archives, Reading University Library.

*Norwegian Mechanical Engineering, 1800–1900*

**Table 10.12**  Stipends for foreign travel; Germany and Britain as % of
all destinations

|  | Germany only | Britain only | Both Germany and Britain |
|---|---|---|---|
| 1851–1860 | 14.2 | 40.4 | 9.5 |
| 1861–1870 | 3.5 | 67.8 | 5.3 |
| 1871–1880 | 10.0 | 45.0 | 10.0 |
| 1881–1890 | 21.0 | 31.5 | 31.5 |
| 1891–1900 | 33.3 | 16.6 | 16.6 |

*Source:* calculated from appendix 1.

## 13  The Rise of Competing Leaders

From the mid-nineteenth century, the spread of industrialisation
began to produce challenges to Britain's economic and technological leadership; the most significant long-term development was the
rise of Germany as a machinery-producing and exporting nation. It
is important to note, however, that this was a gradual development, and only slowly became reflected in the international relations of Norway's engineering industry. But a changing focus on
the part of Norwegian engineers can be traced through most of the
technology transfer mechanisms which have been studied above. If
we consider foreign travel funded by state stipends, for example,
we can see a sharp increase in German destinations as a percentage
of all destinations from 1880. Although the sample numbers are
small, and there are wide fluctuations, Table 10.12 shows a clear
trend increase in German destinations, a trend decline in UK-only
destinations, and a steady increase in those visiting both Germany
and Britain.

Similar changes can be noted in terms of machine acquisition,
and in the general commercial contacts between Norwegian and
foreign engineering firms. But the slowness of the transition ought
to be emphasised: note, for example, that Table 10.10 above,
tracing machine acquisition, shows the beginnings of German
impact only towards the end of the nineteenth century. Table 10.13
shows ninety-eight larger machine and equipment purchases. The
British dominance is obvious.

These figures do not, however, reveal that German influence did
not begin until towards the end of the century, as we can see in
Table 10.14. However, machine acquisition, as I have suggested, is

263

*Kristine Bruland*

**Table 10.13**   Transactions (1831–1900)

|  | Britain | Germany | Other | Total |
|---|---|---|---|---|
| Machine parts | 31 | 6 | 2 | 39 |
| Machines | 30 | 7 | 2 | 39 |
| Machine tools | 18 | 2 | 0 | 20 |

*Source*: appendix 3, firms' records.

**Table 10.14**   Foreign contacts by Norwegian firms

|  | 1841–50 | 1851–60 | 1861–70 | 1871–80 | 1881–90 | 1891–1900 |
|---|---|---|---|---|---|---|
| *Thune Mechnical Workshop* | | | | | | |
| Britain | | 19 | 15 | 7 | | |
| Germany | | | 12 | 1 | 6 | |
| USA | | | 2 | 2 | | |
| Other | | | 2 | | | |
| *O. Mustad* | | | | | | |
| Britain | 5 | 1 | 1 | 9 | | |
| Germany | | | | | | |
| USA | | | | | | |
| Other | | | | | | |
| *Aker Mechanical Workshop* | | | | | | |
| Britain | 7 | | | 19 | | |
| Germany | | 3 | | | | |
| USA | | | | | | |
| Other | | | | | 1 | |
| *Jakobson's Machine Works* | | | | | | |
| Britain | 15 | 31 | 4 | 4 | | |
| Germany | | | 8 | | 3 | |
| USA | | | 2 | 1 | | |
| Other | | | 1 | | | |
| *Christiania Nail Works* | | | | | | |
| Britain | | | 5 | 6 | | |
| Germany | | | | | 1 | |
| USA | | | | | | |
| Other | | | | | 1 | |

Norwegian Mechanical Engineering, 1800–1900

**Table 10.14**  *continued*

|                          | 1841–50 | 1851–60 | 1861–70 | 1871–80 | 1881–90 | 1891–1900 |
|--------------------------|---------|---------|---------|---------|---------|-----------|
| *Nyland Works*           |         |         |         |         |         |           |
| Britain                  | 15      | 50      | 9       | 54      | 1       |           |
| Germany                  |         |         | 1       | 1       | 10      |           |
| USA                      |         |         |         |         | 1       |           |
| Other                    |         |         |         |         | 3       |           |
| *Kampen Mechanical Workshop* |     |         |         |         |         |           |
| Britain                  |         |         | 1       |         | 5       |           |
| Germany                  |         |         |         |         |         | 10        |
| USA                      |         |         |         |         |         |           |
| Other                    |         |         |         |         |         | 2         |

*Source*:  appendix 3

less important than other forms of business relationship. More significant are general business contacts. Table 10.14 is a development of Tables 10.10 and 10.11 above; it shows the countries of origin of the foreign contacts of seven Norwegian mechanical engineering firms during the period 1830–1900.

Within a general framework of sharply decreasing total number of contacts, Table 10.14 shows a clear general reorientation towards German contacts towards the end of the century. Even here, however, Britain remained overwhelmingly important to the end of the nineteenth century. Kampen Mechanical Workshop was strongly oriented towards German contacts, while Thune, Aker and Nyland appear to be shifting focus with regard to their foreign contacts.

None the less by the turn of the century German techniques were making an increasing impact. The German firm of Steinmüller were providing substantial competition to Thune's Babcock agency, and Jakobson would 'make all kinds of new machines of German construction'.[81] The decline of British engineering dominance clearly lies outside the time period of this study, though its beginning can be clearly seen.

81. Thune archives, Norwegian Technical Museum, Oslo, copy book, 1901–3, p.453.

Kristine Bruland

## 14   Conclusion

When, in 1892, the directors of Nylands Mechanical Workshop decided that they required a new hydraulic riveting machine, they considered purchasing it from England. But they found that their own workshop 'could make it for 8,000 kr., whereas the price from England was 12,000'.[82] This is an index of the fact that during the nineteenth century Norway had developed a mechanical engineering industry which ultimately became of world standard. In the opening section of this chapter I argued that this effort, for any small country, must involve the emulation of larger technological leader economies: the import of the technological capabilities developed in the dynamic industrial leaders. In the nineteenth century this meant Britain and then Germany. Norway made the effort, but the technology transfer process was a complex one: we have seen that it included foreign travel, training and education; machinery acquisition; the use of skilled labour from abroad; agency relationships; and a wide range of contacts with foreign engineering enterprises. It is useful to contrast this technology transfer process in the engineering industry with that which occurred in another important emerging industry in nineteenth-century Norway, namely textiles. In the latter, Norwegian cotton and wool entrepreneurs acquired machinery, expertise, information and labour from abroad in 'packages' which were put together by British textile engineering firms. The entrepreneurs themselves required commercial and marketing skills: they could, and did, remain relatively lacking in technical expertise.[83] In the engineering industry, by contrast, skill development and competence building were central: this is because engineering is not so concerned with the production of standardised products, but is much more a matter of technical problem solving in which competence is of critical importance. For that reason the role of technology transfer in the development of Norwegian engineering is much more a matter of training and education, of access to information about foreign technical developments, and possession of the ability to use that information. Norwegian engineers clearly had certain advantages in this process: the economy was a particularly open one, and Norway was – in cultural and political terms – outward-looking. Norwegian engineering entrepreneurs made the most of these

82. Nyland archives, Maritime Museum, Oslo, Styreprotokoll, (directors' protocol) IB 2113.
83. See Bruland, *British Technology*.

## Norwegian Mechanical Engineering, 1800–1900

advantages, deploying and developing foreign techniques to construct an industry which was central in Norway's transition from peripheral isolation to one of the richest economies of the advanced world.

# [12]

# The Transfer of Electrical Technology to Finland, 1870–1930

## TIMO MYLLYNTAUS

In the last thirty years of the 19th century, Germany and the United States emerged as the leaders in the development of electrical technology, and the rest of the industrialized world attempted to follow. The rapid spread of electrotechnical know-how resulted from intensifying international communication between technologists and engineering firms, increasing world trade, and the mushrooming growth of multinational firms specializing in electrical technology such as Siemens & Halske, Allgemeine Elektricitäts-Gesellschaft (AEG), Edison's companies (later General Electric), and Westinghouse.[1] Less developed countries became dependent on the American and German technology that they tried to purchase, imitate, and apply. Almost all countries considered electrical technology necessary for economic development. Individuals, private firms, and governments experimented with various methods for effecting the international transfer of electrical technology.[2]

Technology transfer is not just a matter of moving some piece of hardware from one place to another. Rather, it is a complex process of

Dr. Myllyntaus is a research fellow at the University of Helsinki. He has lectured on the history of technology at the Tampere University of Technology and is a docent at the University of Jyväskylä. In 1989 he received a Ph.D. in economic history from the London School of Economics. An earlier version of this article was presented at a 1988 symposium on technology transfer in historical perspective in Stavanger, Norway. The author thanks Leslie Hannah for reading and improving a preliminary version, and Risto Keskinen, Gunnar Nerheim, and other symposium participants for their comments. He is deeply grateful to Jane Morley for editing and critiquing subsequent drafts, and he appreciates the comments of the *Technology and Culture* referees.

[1]Thomas P. Hughes, "British Electrical Industry Lag: 1882–1888," *Technology and Culture* 3 (Winter 1962): 29–30; Malcolm MacLaren, *The Rise of the Electrical Industry during the Nineteenth Century* (Princeton, 1943), pp. 170–98; A. J. Körner, "Den elektriska industriens historia," in *Uppfinningarnas bok*, ed. Sam Lindstedt (Stockholm, 1927), 3:870–71.

[2]The term "international transfer of technology" refers essentially to the process whereby knowledge relating to the transformation of inputs into outputs is acquired by entities within a country (e.g., firms, research institutes) from sources outside that country (Martin Fransman, *Technology and Economic Development* [Brighton, 1986], p. 7).

294     Timo Myllyntaus

transformation that necessarily takes place in a specific economic, social, and cultural context. If certain contextual conditions are not fulfilled, the same technology that has been successfully applied elsewhere may completely fail in a new setting. In less developed countries, technology transfer is often impeded by a lack of the necessary societal infrastructure.[3] The recipient economy must have adequate prerequisites to receive, establish, and maintain a new kind of technology such as an electrometallurgical factory. A material infrastructure, such as water and electric power supply, as well as transportation and telecommunications facilities, is not enough. There must also be a sufficient nonmaterial infrastructure with components such as a stable, functional political system and a pool of skilled labor.

A new technology can be transferred from one country to another in many forms and through various channels. Contextual factors in the recipient country have considerable influence on the specific options.[4] Hence, when a new technology is transferred into another country, it must be "filtered" through an economic, sociopolitical, and cultural "sieve" (see fig. 1). This contextual filter affects the choice of technology as well as the channel through which it is transferred. The choice may essentially influence the pattern of adoption and its success.

At least in theory, the assortment of potential channels for technology transfer is roughly the same for all countries, but the contextual filter is always unique and nationally defined. This filter is very polymorphous. It can selectively close one channel completely, obstruct the functions of another, give the third a free hand, and overload the fourth. It does not generally remain stable over time but

[3] This infrastructure is also called "the recipient country's technological support network" (John M. Staudenmaier, Technology's Storytellers: Reweaving the Human Fabric [Cambridge, Mass., 1985], p. 123).

[4] Thomas Hughes has stated that "the differences found in the evolving regional (electricity supply) systems—the essence of style—stemmed mostly from the nontechnological factors of the cultural context" (Thomas P. Hughes, Networks of Power: Electrification in Western Society, 1880–1930 [Baltimore, 1983], p. 405). See also David J. Jeremy, Transatlantic Industrial Revolution: The Diffusion of Textile Technologies between Britain and America, 1790–1830s (Cambridge, Mass., 1981); Hans-Joachim Braun, "Technologietransfer im Maschinenbau in die USA 1870–1939," Technikgeschichte 50 (1980): 238–52, and "The National Association of German-American Technologists and Technology Transfer between Germany and the United States, 1884–1930," History of Technology 8 (1983): 15–35; Erich Pauer, "Technologietransfer und industrielle Revolution in Japan 1850–1920," Technikgeschichte 51 (1984): 34–54, and "Japanischer Geist—westliche Technik: Zur Rezeption westliche Technologie in Japan," Saeculum, Jahrbuch für Universalgeschichte 38 (1987): 19–51; and Hans-Heinrich Nolte, "Technologietransfer in Russland vor 1914, Möglichkeiten und Grenzen nachholender Industrialisierung," Technikgeschichte 51 (1984): 319–34.

*The Transfer of Electrical Technology to Finland, 1870–1930*     295

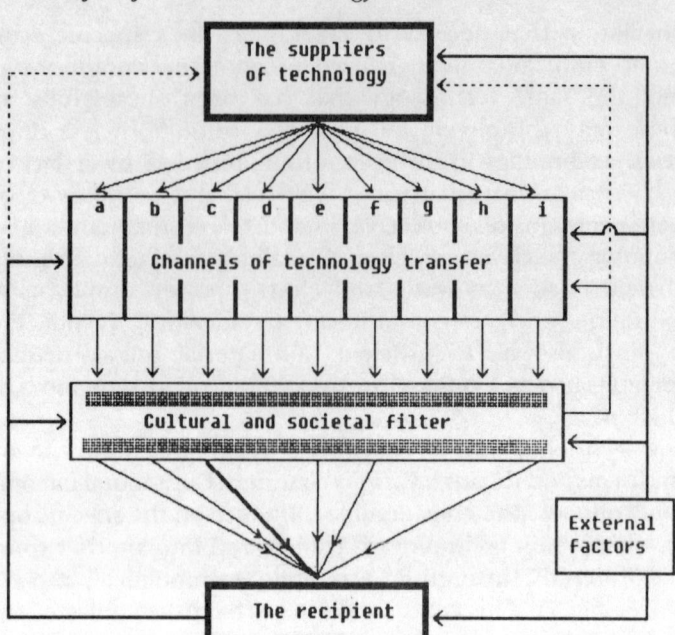

FIG. 1.—The mechanism of technology transfer

keeps shifting as a result of changes in government, trade relations, and the economic situation. The cultural and societal filter of technology transfer is an ambiguous combination of various elements from unconscious popular attitudes to strict laws; it is the autonomous mechanism that regulates the transfer and application of technology. The introduction of an innovation is successful only if the social environment of the recipient country is supportive of it. Therefore, technology transfer is not only a technical operation but also a societal procedure. The crucial factor is whether or not there are effective and socially accepted methods of carrying out the process. The technology itself and its transfer channels should be coordinated with the existing cultural and societal environment.

The function of a certain transfer model can be radically changed by some external factor such as war, economic crisis, or ecological calamity. External factors may have a considerable impact on the economies of the supplier or the recipient or both; they may also affect the modification of some channels or the entire transfer mechanism. The cultural and societal filter is very sensitive to alterations in external circumstances; it may react even if an external shock had practically no impact on the recipient economy or the channel apparatus. A crisis in the opposite hemisphere may change societal values, behavioral patterns, or government policy and thus remodel the filter.

In this article, my first aim is to analyze a set of transfer channels that Finland applied in the adoption of foreign electrical technology. Second, I attempt to apply my filter model to evaluate the effects of the contextual environment on the channels and to assess the results of the Finnish method of transferring electrical technology.

### The Arrival of Electricity in Finland

In the 1870s Finland, an autonomous grand duchy under the rule of the Russian tsar, was still a predominantly agrarian economy with limited mechanization.[5] Finland was an industrial latecomer, much like various eastern and southern European countries. Quite a few innovations such as the reciprocating steam engine, the steamship, and the steam locomotive had been introduced earlier in Russia, which is generally regarded as an extreme example of European backwardness.[6] One might expect that the transfer of heavy electrical engineering from industrialized countries to such a peripheral country as Finland—isolated from the rest of the European continent by a remote geographical location, icy waters, political restrictions, and language barriers—would have required external initiative and would have been delayed for many years until large multinational manufacturers of electrical equipment had started to market their products and to set up subsidiaries in the country. This was not the case. Heavy electrical engineering arrived in Finland through other channels—and comparatively quickly.

---

[5]The political history of Finland is characterized by three periods: from the 12th century to 1809 it was a province of Sweden, between 1809 and 1917 it was a grand duchy annexed to the Russian Empire, and since 1917 it has been an independent republic. During the 800 years of rule by foreign powers, the great majority of the Finns preserved their own distinctive language, culture, and personal freedoms without such restrictions as serfdom. The national identity, however, was formed only during the Russian period as a result of an active nationalist movement. See Eino Jutikkala, with Kauko Pirinen, *A History of Finland* (New York, 1974); and Lennart Jörberg, "The Industrial Revolution in Nordic Countries," in *Fontana Economic History of Europe*, vol. 4, ed. Carlo M. Cipolla (Glasgow, 1973), pt. 2: 375–402.

[6]Another example is a telegraph line that the German company Siemens & Halske built from the center of the Russian telegraph network, St. Petersburg, to Helsinki by order of the tsarist government during the Crimean War (1853–56) (Einar Risberg, *Suomen lennätinlaitoksen historia 1855–1955* [Helsinki, 1959], pp. 57–131; Peter Gatrell, *The Tsarist Economy 1850–1917* [London, 1986], pp. 29–47).

*The Transfer of Electrical Technology to Finland, 1870–1930*          297

Electricity as a source of light captured the attention of the Finns at an early date because in winter natural light was not sufficient to illuminate workshops, factories, offices, or homes even during the daytime. The existing sources of artificial lighting, such as wooden splints, candles, oil lamps, and gas lamps, were all fire hazards. In the late 19th century, the great majority of Finnish buildings were made of wood, which greatly increased that risk. (Almost all Finnish paper mills were destroyed by fire at least once during the 19th century.[7]) Additionally, the Finns were interested in electric light because of its quality and cleanness.

At the time when Paris installed arc lamps for regular use and electric lights were being demonstrated in numerous other cities— London, Berlin, Milan, St. Petersburg—the Finnish State Railways sponsored experiments with a Gramme dynamo and a Serrin arc lamp in its engineering works at Pasila in Helsinki on December 10, 1877. This first demonstration of electric light in Finland was carried out by the acting professor of physics at the University of Helsinki, Karl Selim Lemström, and the university engineer, Martin Wetzer. Lemström had imported the machinery from France, and Wetzer assisted him in installing and running it experimentally for some weeks. Lemström was a Finn who had studied in Stockholm, Paris, and St. Petersburg, whereas Wetzer was a German who had worked in Russia and Finland for many years.[8] The following summer Carl Kämp leased Lemström's machinery to light the garden in front of his restaurant in Kaivopuisto, a seaside park in Helsinki. With electric lighting and musical entertainment, he attracted customers to his establishment for half the summer season, from August to late September.[9]

In Finland, electrical equipment was first offered for sale by an official of the Telegraph Board, Daniel Johannes Wadén (1850–1930). Wadén had earned a master's degree in philosophy at the University of Helsinki and an engineering degree in St. Petersburg. In 1876, he set up an electrical installation firm that soon concen-

---

[7]Lars Gabriel von Bonsdorff, *Nokia Aktiebolag 1865–1965* (Helsinki, 1965), p. 123; Timo Myllyntaus, "Suomen graafisen teollisuuden kasvu 1860–1905," Research Report no. 12 (University of Helsinki, Institute of Economic and Social History, Helsinki, 1981), pp. 158–73.

[8]*Helsingfors Dagblad* (December 10, 1877, December 11, 1877, January 19, 1878, January 31, 1878); Tor Carpelan and L. O. Th. Tudeer, *Helsingin Yliopisto, Opettajat ja virkamiehet vuodesta 1928* (Helsinki, 1925), 1:534 ff.; Hj. Tallqvist, "Karl Selim Lemström," *Finsk Tidskrift* 57, no. 2 (1904): 292–98.

[9]*Helsingfors Dagblad* (August 3, August 4, August 9, August 16, 1878).

298    *Timo Myllyntaus*

trated on the telephone business, but as a secondary occupation he also sold and installed electric lighting systems in the southern parts of the country.[10] In Helsinki, he opened Finland's first electricity-supply utility in 1884, the same year that its earliest counterparts were commissioned in Germany (Berlin) and Sweden (Gothenburg).[11]

## The Channels of Technology Transfer

In studying the transfer of electrical technology to late-19th-century Finland, we can distinguish at least seven different channels through which foreign know-how was adopted: (1) the low-cost diffusion of easily accessible technology, (2) study by Finns in foreign technical institutions, (3) other educational visits abroad by business-men and engineers, (4) the recruitment of foreign specialists, (5) the purchase of foreign licenses and patents, (6) the importation of electrical equipment, and (7) direct foreign investments.[12] The utilization of these channels was not explicitly planned or organized by the government of the grand duchy or by any other institution. On the contrary, the Finns' own needs and initiative started technology transfer spontaneously, while the societal characteristics of the grand duchy defined the general framework of the process.

*The low-cost diffusion of easily accessible technology.*—Although it was a poor, peripheral country, Finland had a strikingly active press in the last quarter of the 19th century. By means of telegraph, the press obtained news from all over the world. Some capital-city newspapers, such as *Helsingfors Dagblad*, were eager to publish news about technology because their readers—the educated, urban elite—had a positive and optimistic attitude toward technology and were hungry for information about innovations. A number of people also read

---

[10]Wadén did not possess inherited wealth, but marriage to a baroness probably improved his chances of raising the necessary start-up capital (*Helsingfors Dagblad* [December 27, 1877]; Georg Christiernin, "Elektroteknikens pioniärer i Finland," *Tekniska Föreningen i Finland, Förhandlingar*, no. 2 [1930]: 53; Erik von Schantz, "Dan. Joh. Wadén, Minnesteckning," *Svenska tekniska vetenskapsakademien i Finland, Förhandlingar*, no. 6, pt. 4 [Helsinki, 1932], pp. 35 ff.).

[11]Wadén's undertaking as well as German Edison Company's Berlin utility and Edv. Bildt's central station in Gothenburg supplied electricity for indoor lighting to the subscribers in the city center. Wadén's was a rather small, steam-powered block station that delivered electricity initially to four subscribers and in February 1885 to 18 subscribers. See *Helsingfors Dagblad* (November 26, 1884, December 6, 1884, February 4, 1886); Johan Åkerman, *Ett elektriskt halvsekel, Översikt över ASEAs utveckling 1883–1933* (Västerås, 1933), p. 42; and Hughes, *Networks* (n. 4 above), pp. 72–77.

[12]Compare with Ole Börnsen, Hans H. Glismann, and Ernst-Jürgen Horn, "Der Technologietransfer zwischen den USA und der Bundesrepublik," Kieler Studien 192 (Institut für Weltwirtschaft an der Universität Kiel, Tübingen, 1985), pp. 32–70.

*The Transfer of Electrical Technology to Finland, 1870–1930*     299

foreign newspapers and journals; for example, in 1877 the Finnish pioneer of telephone technology, Daniel Wadén, read about the American Alexander Graham Bell's invention in the Swiss *Journal Télégraphique*.[13]

Information obtained through the press constitutes part of what I have called "low-cost diffusion of easily accessible technology," a process that also includes scientific and trade literature, business correspondence, and exhibitions.[14] Through this channel, the Finns received preliminary information on advancements in technology about which they could then start to obtain more detailed knowledge through other channels. In the last third of the 19th century and in the early 20th century, neither the Russian nor Finnish government attempted to censor or restrict the acquisition of foreign scientific, technical, or commercial publications. By gradually expanding higher education and supporting scientific libraries, the Finns promoted the diffusion of foreign know-how. In addition to the deliberate search for knowledge, natural curiosity was an important motivating factor in the absorption of up-to-date technological knowledge from abroad.

*Studies abroad.*—Instruction in electrical technology began at the Polytechnic of Helsinki in the 1880s. Until the transformation of the school into a modern university of technology, however, only supplementary courses for students of engineering were provided.[15] Up to the 1910s, it was customary for Finnish students interested in electrical engineering to travel abroad to study it. Knowledge about the first generation of Finnish electrical engineers and electricians is scarce. Although the membership records of the Association of Electrical Engineers in Finland (AEEF, established in 1926) give detailed information on engineers of subsequent generations who became members, unfortunately these records do not cover all of the early electrical engineers who practiced their profession in the grand duchy and in the young republic.

If we consider the age group of the AEEF members who could have earned an engineering degree before World War I, the clear majority (59 percent) had studied abroad at least one year. At the time, it was common to earn the engineer's degree in Finland first and then to

[13]*Helsingfors Dagblad* (December 27, 1877); Schantz (n. 10 above), pp. 39–46.

[14]Stanislaw Gomulka calls this "free exchange of scientific and technical information" "the natural diffusion" of technology in his *Inventive Activity, Diffusion, and the Stages of Economic Growth* (Århus, 1971), pp. 11–13.

[15]B. Wuolle, *Suomen teknillinen korkeakouluopetus 1849–1949* (Helsinki, 1949), pp. 127–28; Tauno Pyökäri, "Sähkömiehet opintiellä," in *Sähköinsinööriliitto 1926–1976*, ed. Jaarli Jauhiainen, Sakari Maaniemi, Tauno Pyökäri, and Nils-Holger Ståhle (Tampere, 1976), p. 79.

300    *Timo Myllyntaus*

TABLE 1
EDUCATION OF EARLY FINNISH ELECTRICAL ENGINEERS

|  | Number Born between 1858 and 1889 | Number Born between 1890 and 1899 | Total Number |
|---|---|---|---|
| Technical studies without degree: | | | |
| In Finland................... | 1 | 2 | 3 (1.5) |
| Abroad...................... | 1 | 2 | 3 (1.5) |
| Lower Finnish degree of engineer ..... | 15 | 14 | 29 (13) |
| Finnish degree of graduate engineer ... | 18 | 100 | 118 (53) |
| Lower Finnish degree of engineer plus supplementary studies abroad ....... | 22 | 1 | 23 (10) |
| Finnish degree of graduate engineer plus postgraduate studies abroad ..... | 4 | 6 | 10 (4) |
| Engineer's degree from abroad .......................... | 21 | 16 | 37 (17) |
| Total ......................... | 82 | 141 | 223 (100) |
| Studies by country: | | | |
| Only in Finland................. | 34 (41) | 116 (82) | 150 (67) |
| At least partly abroad ....................... | 48 (59) | 25 (18) | 73 (33) |
| Total ......................... | 82 (100) | 141 (100) | 223 (100) |

SOURCE.—Ole Fraser, ed., *Elektroingenjörer i Finland, Matrikel 1926–1951* (Helsinki, 1951).
NOTE.—Numbers in parentheses are percentages.

continue one's studies abroad. About 32 percent of these AEEF members followed that practice. A quarter (26 percent) of them went abroad after their secondary-school years and graduated from a foreign university, as shown in table 1.

Over the course of time, the number of Finnish electrical engineers increased but study abroad declined. In the case of those engineers born in the 1890s or later, three circumstances reduced the likelihood of their studying in foreign institutions. First, undertaking studies for a graduate engineer's degree in electrical technology became possible in Finland in 1911.[16] Second, the opportunities and level of education at home improved and the Finnish government cut grants for studies abroad. A modern electrical laboratory was opened at the Helsinki University of Technology (HUT) in 1926. Third, the outbreak of World War I greatly limited Finns' opportunities to study abroad. In

[16]Wuolle, pp. 330–31.

*The Transfer of Electrical Technology to Finland, 1870–1930*     301

TABLE 2

EARLY FINNISH ELECTRICAL ENGINEERS' STUDIES ABROAD BY COUNTRY

| Country of Studies | Number Born between 1858 and 1889 | Number Born between 1890 and 1899 | Total | |
|---|---|---|---|---|
| Germany | 36 | 18 | 54 | (62) |
| Sweden | 5 | 7 | 12 | (14) |
| France | 3 | 2 | 5 | (6) |
| Switzerland | 3 | 1 | 4 | (5) |
| Austria (Vienna) | 3 | 0 | 3 | (3) |
| United States | 1 | 5 | 6 | (7) |
| Russia (St. Petersburg) | 1 | 2 | 3 | (3) |
| Total | 52 | 35 | 87 | (100) |

SOURCE.—Ole Fraser, ed., *Elektroingenjörer i Finland, Matrikel 1926–1951* (Helsinki, 1951).
NOTE.—Numbers in parentheses are percentages.

addition, the economic and political instability of Germany discouraged Finns from going there to study after the war. Consequently, the percentage of people who undertook at least some foreign study fell to 18 percent among the group of electrical engineers born in the 1890s. The most conspicuous decline took place in postgraduate studies abroad. In 1930, only thirteen Finns were studying at the German universities of technology, while the number of students at HUT totaled 686.[17]

The German universities of technology had attracted nearly two-thirds of the early Finnish students of electrotechnology who left to study abroad. The most popular schools were the universities of technology in Karlsruhe and Charlottenburg-Berlin and the Ingenieure Technicum Mittweida. Nearly half the Finnish students in German technical schools had chosen one of these three institutions. As table 2 indicates, technical schools in other countries played a minor role. Many of those who had studied in Sweden were actually Swedes who immigrated to Finland after their university graduation. In the United States, Finns pursued rather short study courses; none of the AEEF members had earned a degree in electrical technology there.

Technical studies abroad were quite often supplemented by training in foreign companies in the field. The most fascinating training trip was made by Carl von Nottbeck (1848–1904) from Tampere. He

[17]*Deutsche Hochschulstatistik*, Band 5, Sommerhalbjahr 1930, ed. Hochschulverwaltungen (Berlin, 1930), p. 112; *Suomen teknillinen korkeakoulu, Vuosikertomus 1929–1930* (Helsinki, 1930).

302     *Timo Myllyntaus*

first studied at the University of Tartu (Dorpat in German, a town in the territory of modern Estonia, USSR), then graduated from the Zurich University of Technology (ETH) in Switzerland.[18] He was so inspired by the latest developments in electrical technology that he traveled to the United States in the late 1870s. There he hired himself out to Thomas Alva Edison in order to observe how the great inventor developed the incandescent filament lamp. Soon after Edison had completed his lighting system, Nottbeck ordered two of Edison's dynamos and a few hundred lamps and hurried back to Tampere to install them in a cotton factory managed by his father.[19] With the assistance of another former Edison employee, the Hungarian Istvan von Fodor, Nottbeck completed the installation. When electric lighting was introduced at the Finlayson cotton mill on March 15, 1882, Nottbeck and Fodor sent Edison a telegram: "Lighting installed at 61 degrees northern latitude. Complete success."[20]

The installation at Finlayson was the first permanent electric lighting plant in Finland. It was also the first lighting plant in the Nordic countries to use Edison's equipment and only the fifth permanent plant of its kind in Europe.[21] Nottbeck and Fodor, who had met in the United States while working for Edison, were genuine pioneers in transferring Edison's lighting technology to the Continent; they subsequently became Edison's first general agents for eastern Europe.

[18]Carl Nottbeck is an example of the first generation of Finnish electrical engineers. He was not included in the rolls of the AEEF, having died abroad before the association was founded. See Yrjö Raevuori, *Tampereen kaupungin sähkölaitos ja sähkön varhaisvaiheet Suomessa* (Tampere, 1938), pp. 17–25.

[19]These dynamos were among the first that Edison produced for sale. One of them had a serial number 3. After six months in operation, they were replaced by two more efficient dynamos with serial numbers 24 and 25 (ibid., pp. 22–23).

[20]Nottbeck's native language was German, but he knew Finnish and quite likely Estonian, too. The linguistic and ethnic affinity of Finns, Estonians, and Hungarians might have had an effect on the friendship between Nottbeck and Fodor. See *Bulletin, The Edison Electric Light Company*, no. 6 (1882); Esko Sarasmo, "Lisävalaistusta Finlaysonin sähkövalojen alkuvaiheisiin," *Yhdyslanka*, no. 2 (1957), pp. 5–6; and Hughes, *Networks* (n. 4 above), pp. 18–46.

[21]In the early 1880s, Nottbeck became the general agent for Edison electric equipment in both Russia and Finland, but he did not succeed as a businessman. He moved to Paris, where he died in 1904. In contrast, Fodor built a great career. He became an Edison agent in eastern Europe, then the managing director of the electricity supply utility in Budapest, and later the royal counselor of the court in Hungary. See Raevuori (n. 18 above), pp. 19 ff.; and Ilmari Killinen, "Sähkön käyttö ja sen kehitys Suomessa," *Voima ja valo* 2, no. 5–6 (1929): 158–59; see also Hughes, *Networks* (n. 4 above), pp. 47–78.

*The Transfer of Electrical Technology to Finland, 1870–1930*      303

Clearly, studies abroad and the contacts students developed there formed a very important channel of technology transfer, particularly before World War I. The first generation of Finnish electrical engineers, who entered the field in the 1870s and 1880s, men like K. S. Lemström, D. J. Wadén, C. S. von Nottbeck, F. Wilén, C. Wahl, and A. G. Strömberg, shared several attributes. First, all of them had a good command of foreign languages. In addition to their mother tongue (Swedish or Finnish), they knew from two to four other languages: German, French, Russian, or English.[22] Second, nearly all of them had studied technology or science at some foreign university or polytechnic. Not only the first generation of Finnish electrical engineers but also the majority of the second generation, men who reached adulthood at the turn of the century, still went to Germany or to other industrialized countries to study and to receive practical training. In the grand duchy, Finnish students' studies abroad were a widely accepted method of acquiring modern know-how. From the leaders of the early labor movement to big businessmen and senators, trendsetters recommended this channel, because they regarded it as the most efficient way to educate skilled labor and expert staff for the economy. Foreign studies by Finns were primarily considered an alternative to recruiting alien experts to work in Finnish industry. At the turn of the century, foreign studies were a must for Finnish engineering students specializing in electrical technology.

*Educational visits.*—Besides students, Finnish engineers, businessmen, factory owners, and government officials also traveled abroad. Some attended exhibitions; some visited factories; others collected information on some pertinent technical or organizational issue. These journeys used to be called educational visits (*opintomatkat*) even though many of them were actually business trips. According to membership records, nine out of ten AEEF members born in the 19th century reported having paid at least one educational visit abroad, although the data on these visits are not very accurate. Three-fourths of all these engineers visited Germany, and nearly three-fifths made at least one trip to Sweden. A third went to Denmark and a quarter to

---

[22]During about 700 hundred years of Swedish rule, a Swedish-speaking minority developed on the western and southern coasts of Finland as a result of immigration and the assimilation of local Finnish speakers to Swedish immigrants. In the 19th century, one-seventh of all Finnish citizens spoke Swedish as their mother tongue. It was the vernacular of sailors, fishermen, and farmers in the mentioned areas, while it was also the language of the upper class all over the country. See *Annuaire statistique de Finlande 1929* (Helsinki, 1929).

304　　*Timo Myllyntaus*

TABLE 3

EDUCATIONAL VISITS OF EARLY FINNISH ELECTRICAL ENGINEERS BY COUNTRY OF DESTINATION

| Destination | Number Born between 1858 and 1889 (N = 82) | | Number Born between 1890 and 1899 (N = 141) | | Total (N = 223) | |
|---|---|---|---|---|---|---|
| Germany | 69 | (90) | 95 | (84) | 164 | (86) |
| Sweden | 46 | (60) | 81 | (72) | 127 | (67) |
| Switzerland | 28 | (36) | 23 | (20) | 51 | (27) |
| Denmark | 22 | (29) | 45 | (40) | 67 | (35) |
| Great Britain | 18 | (23) | 32 | (28) | 50 | (26) |
| Norway | 16 | (21) | 27 | (24) | 43 | (23) |
| Russia/USSR | 16 | (21) | 11 | (10) | 27 | (14) |
| France | 15 | (19) | 32 | (28) | 47 | (25) |
| United States | 12 | (16) | 14 | (12) | 26 | (14) |
| Total number of engineers who paid at least one educational visit abroad.... | 77 | (100) | 113 | (100) | 190 | (100) |

SOURCE.—Ole Fraser, ed., *Elektroingenjörer i Finland, Matrikel 1926–1951* (Helsinki, 1951).
NOTE.—Numbers in parentheses are percentages.

Switzerland, Britain, France, or Norway. The main destinations of educational visits are listed by country in table 3.[23]

Educational visits had status among Finnish engineers. They were regarded highly by the general public and often subsidized by the government. These visits were significant from the standpoint of technology transfer because they promoted the flow of information, provided incentives and models, and helped establish personal contacts with foreign companies and experts.

*Foreign specialists.*—In Finland, native-born electrical engineers were not the only experts in the field. Foreign engineers came to work in importing and wholesale companies, factories, and engineering offices. In addition, alien instructors were hired by Finnish educational institutions. The number of these foreigners was rather small, but they held key posts, and their role was not unimportant.[24] Foreign

[23]Table 3 contains only a small number of countries visited. Nevertheless, the journeys of Finnish engineers primarily took them to the Continent. Educational visits were most frequently made to Scandinavia and the rest of developed western Europe, east-central Europe, and countries around the Baltic Sea. See Ole Fraser, ed., *Elektroingenjörer i Finland, Matrikel 1926–1951* (Helsinki, 1951).

[24]S. B. Saul considers the spread of technical and entrepreneurial experience through personal connections and immigrated foreign specialists to have been very significant for the growth of engineering in Switzerland and various other countries (S. B. Saul,

consultants and planning engineers probably contributed the most to the transfer of electrical technology to Finland. Most of them were Swedes or Germans. Oskar Faith-Ell (1873–1936) is an example: he was a Swedish engineer who came to Finland in 1899 to work for the subsidiary of Siemens & Halske AG. Between 1912 and 1936, he operated his own electrotechnical consulting office in Helsinki.[25] Another Swede, Bror Sjögren, was the managing director and main owner of a fairly large engineering office, Consulting Oy, from 1920 to 1943. His company designed many hydroelectric power plants in the interwar period. As an active proponent of hydropower, Sjögren also participated in the public debate on Finnish energy policy.[26]

The attitude of the authorities and the general public toward foreign experts and workers was reserved. The Finnish government did not encourage or recruit aliens with special skills to immigrate, as Russian regimes had from time to time since Peter the Great. Foreign technicians working in the grand duchy had either been drafted by the directors of factories in desperate need of technical expertise or moved to the country on their own initiative.

*The purchase of licenses and patents.*—In the transfer of technology between industrialized countries, the exchange of patents and licenses is very important. In less developed countries where a technological support network is weak—as evidenced by poor educational facilities, insufficient production capacity, and a lack of financial resources—this transfer channel plays a smaller role. In the grand duchy of Finland, the domestic manufacture of electrotechnical products started fairly early, despite the modest development of Finnish engineering and the accelerating importation of foreign electrical equipment. In 1880, only ten years after Gramme's invention of the dynamo, the first Finnish dynamo was built by a seventeen-year-old schoolboy, Axel Gottfrid Strömberg, assisted by his younger sister. Later, Strömberg became the first qualified teacher of electrical technology at the Polytechnic of Helsinki as well as the owner and manager of an engineering works.[27] The commercial production of dynamos was begun in 1887 in a small workshop belonging to Carl

"The Nature and Diffusion of Technology," in *Economic Development in the Long Run,* ed. A. J. Youngson [London, 1972], pp. 51–52).

[25]Fraser (n. 23 above), p. 68.

[26]*Consulting 1920–1930* (Helsinki, 1930); Bror Sjögren, *Water Power Development in Finland* (Helsinki, 1936).

[27]*Aktiebolaget Gottfr. Strömberg Osakeyhtiö, Juhlajulkaisu 1889–1919* (Helsinki, 1919), p. 5; V. J. Sukselainen, *Oy Strömberg Ab 1889–1939* (Helsinki, 1940), pp. 9–11.

306     *Timo Myllyntaus*

Wahl, who had been a student along with Gottfrid Strömberg at the Hanover University of Technology in 1886–87.[28] Besides manufacturing, Carl Wahl and his employees were interested in research and development; they took out patents for a dozen inventions. Meanwhile, Wahl also acquired foreign licenses. In 1889 he purchased a license from the English Electrical Power Storage Company in London that comprehended all of the most important battery patents at the time and the sole right to produce and sell "E.P.S." (electrical power storage) batteries in the Finnish, Danish, Norwegian, Russian, and Swedish markets. Because of Wahl's exclusive right, many manufacturers in those countries—for example, Wadén in Finland—had to give up making batteries.[29]

With the E.P.S. license, Wahl's engineering company attained a strong position in the Russian battery market, at least until the patents expired at the turn of the century. Although the purchase of foreign licenses undoubtedly had an impact on the success of Wahl's company, such licenses did not generally constitute a crucial channel for the adoption of electrical technology in Finland.[30] The modest utilization of this channel for technology transfer was not due to political or cultural factors—that is, the societal filter did not repel the acquisition of foreign licenses and patents. It was the modest economic and technological prerequisites of the Finnish engineering industry that constituted the bottleneck and prevented benefits deriving from the purchase and use of licenses and patents.

*The importation of electrical equipment.*—During the 1880s nearly all electrical equipment installed in Finland was foreign made. The most significant suppliers of generators were German, Swedish, American, and British manufacturers. At first, incandescent bulbs and arc lamps were imported from the United States and Britain, but German firms later became the main suppliers. In 1913, Germany provided nearly three-fourths and Sweden one-fifth of the electrical equipment imported to Finland. Because of the import of batteries, Denmark then ranked third, ahead of Russia and France.[31]

The relationship of domestic production to imports is shown in table 4. While the net supply of electrical equipment increased by a factor of 34 between 1890 and 1930, domestic production could not

[28]Christiernin (n. 10 above), p. 61.
[29]François Bertini, *Suomen teollisuus-sanakirja 1889* (Tampere, 1890), pp. D3–D4; Christiernin (n. 10 above), pp. 61–70.
[30]Christiernin, pp. 61–62, 82–83.
[31]*Official Statistics of Finland IA:33, Navigation et commerce extérieur 1913* (Helsinki, 1915).

## The Transfer of Electrical Technology to Finland, 1870–1930          307

TABLE 4

SUPPLY OF ELECTRICAL EQUIPMENT IN FINLAND, 1890–1930, AT CONSTANT 1913 PRICES

|  | Production (1,000 FIM) | Export (1,000 FIM) | Import (1,000 FIM) | Net Supply (Production – Export + Import) (1,000 FIM) | Index of Net Supply |
|---|---|---|---|---|---|
| 1890 . . . . . | 4.0 | .8 | 6.0 | 9.2 | 100 |
| 1900 . . . . . | 14.9 | 1.3 | 18.5 | 32.1 | 349 |
| 1913 . . . . . | 19.1 | .7 | 67.5 | 85.9 | 934 |
| 1920 . . . . . | 22.5 | .1 | 95.2 | 117.6 | 1,278 |
| 1930 . . . . . | 87.0 | .1 | 230.2 | 317.1 | 3,447 |

SOURCE.—*Official Statistics of Finland IA:10-50 Navigation et commerce extérieur* (Helsinki, 1893–1931).

keep up with imports. In 1895, the value of domestic production and imports was equal, but soon after Finnish producers started to lose their foothold in both the domestic and Russian markets. In 1930, domestic production was only 38 percent of the imports. From time to time, Finnish manufacturers of electrical equipment demanded higher protective tariffs against imports. The government, however, considered it vital to keep import tariffs at a moderate level for securing a sufficient supply of electricity for the export industries and for the general electrification of the country.

*Direct foreign investments.*—In the initial phase of the electrical age, foreign manufacturers sold their electrical equipment in Finland through their local agents; in the 1890s the largest of them set up their own Finnish subsidiaries. The Swedish Allmänna Svenska Elektriska Aktiebolag (ASEA) came first. In 1893 Gustaf Zitting's engineering office became ASEA's general agent in Finland and in St. Petersburg. Four years later in Helsinki, ASEA and Zitting founded a joint venture known as Finska Elektriska Aktiebolag (FEAB), the Swedish company's first subsidiary abroad. Finska Elektriska Aktiebolag was not only an import firm; it also began to manufacture electrical machinery for the Finnish and Russian markets as well. When competition intensified in these markets, ASEA had to close FEAB in 1904 and again rely on Zitting's engineering office as its agent. In 1913, the parent company opened a new sales subsidiary, Allmänna Elektriska Aktiebolag i Finland, which fared much better than its predecessor.[32]

[32] *Mercator* (January 3, 1914), pp. vi, 12; Åkerman (n. 11 above), p. 69; Jan Glete, *ASEA under hundra år 1883–1983* (Västerås, 1983), p. 38.

308    *Timo Myllyntaus*

In the late 1890s, German firms started penetrating countries with a weak domestic electrical engineering industry, such as Britain and Russia, and also Finland.[33] In 1898 Siemens & Halske AG founded an import and installation firm in Helsinki. It soon gained a foothold throughout the grand duchy, and between 1901 and 1905 Siemens & Halske AG became the leading supplier of generators and electric motors in Finland. Its biggest delivery was two Zoelly turbogenerators, each rated at 500 horsepower, installed in the Klingendahl textile mill in Tampere.[34] In 1900, the other German electrical giant, AEG, opened its first subsidiary in Finland and within a few years gained the position of market leader. Between 1899 and 1914, AEG delivered equipment for nearly half the installations of urban electricity-supply utilities, and it also became a prominent supplier of generators and electric motors for industry after the recession of 1901–3.[35]

The subsidiaries of foreign companies had a notable impact on the electrification of Finland and on the domestic electrical-engineering industry. First, the choice of electrical equipment available was expanded. In contrast to domestic firms, multinational companies marketed more modern technology, such as Nernst and Tantal lamps, and larger machines, such as Zoelly turbogenerators. Second, competition between manufacturers increased and the price of electrical machinery fell, which made electrification more attractive to consumers of energy. Domestic manufacturers, however, confronted difficulties. Wahl's engineering company continued to lose its market share in both Finland and Russia to the German firms until 1910, when it was taken over by AEG.[36] For a time after that, Gottfrid Strömberg's company was virtually the only Finnish manufacturer of generators, electric motors, and transformers. Third, not only did foreign electrical companies introduce new technology, they also brought capital and foreign technical personnel into the country and provided training opportunities for many Finnish engineers and electricians. When AEG purchased the electricity-supply utilities of Viipuri and Turku, it contracted for their expansion and also financed the building of tramway networks.[37]

Some direct foreign investments (DFIs), namely the sales and installation subsidiaries of multinational companies, undoubtedly

[33]I.C. R. Byatt, *The British Electrical Industry, 1875–1914* (Oxford, 1979), pp. 71–72.
[34]*Teknikern* (March 14, 1906), no. 455, p. 67; [Eino S. Repo], *Siemens 60 vuotta Suomessa 1898–1958* (Helsinki, 1958), p. 31.
[35]*Suomen kauppa, meriliike ja teollisuus, Helsinki I* (Helsinki, 1907–1915), pp. 57–61.
[36]*Kontrahti Viipurin kaupungin ja Berliinissä olevan yhtiön Allgemeine Elektricitäts-Gesellschaftin välillä (11.11.1910)* (Viipuri, 1932).
[37]Ibid.; [Oskar Schultz], *Sähkölaitos ja sähköraitiotiet Turussa* (Turku, 1908).

affected the supply of electrical equipment. They did not, however, help the nascent domestic industry; only ASEA attempted to manufacture in Finland, and it failed. Very few of the DFIs were directed toward the supply, distribution, or consumption of electricity, such as supply utilities, urban transport, and electrochemical and electrometallurgical industries. The few electrochemical factories set up by foreign capital proved unprofitable and were soon closed. In this respect, there was a marked difference between Finland and Norway.[38] Direct foreign investments might have influenced technology transfer to Finland much more if they had not persistently been opposed by the country's government and public opinion. This was a clear politico-cultural filter reaction; foreign involvement, especially in the utilization of Finnish hydropower resources, was considered a serious threat to the country's economic independence. As a result, direct foreign investment remained a transfer channel that had only a limited role in the electrification of Finland.

The main channels for the transfer of electrical technology to Finland were students' studies abroad, foreign visits by engineers and businessmen, and the importation of equipment.[39] The other four complemented the contributions of the main channels, but their potential was not efficiently utilized in the Finnish case. The unfavorable geopolitical location and linguistic barriers impeded extensive diffusion of easily accessible know-how; inadequate technological and manufacturing preconditions discouraged the exploitation of license and patent imports; and the significance of direct foreign investments and foreign expertise was much smaller in Finland than in many other peripheral countries.

## The Electrification of Finland

Finland made a promising start in electrification in the 1880s. During the following three decades, however, it could not keep up with the rest of Scandinavia and west-central Europe. The main reasons for this were economic. The grand duchy was less modern and had a lower standard of living. Purchasing power was weak, while electricity was expensive. Another important factor was that the low

[38]"The transfer of technology through direct international investment . . . was undoubtedly very much more extensive than is sometimes recognized, especially in the newer electrical and chemical industries. The electrical engineering industry in Italy in 1914, dominated by three foreign companies—Thomson-Houston, Siemens and Brown Boveri—is not an untypical example to quote" (Saul [n. 24 above], p. 55).

[39]Compare Braun, "Technologietransfer im Maschinenbau" (n. 4 above); Pauer, "Technologietransfer und industrielle Revolution" (n. 4 above); and Pauer, "Japanischer Geist" (n. 4 above).

population density in both urban centers and rural areas could not make large-scale electrification economical for utilities using the technology of the time.

Various applications of electrical technology quickly appeared in the manufacturing industries of Switzerland, Norway, and Sweden.[40] The characteristic "style" of electrification in those countries was the large-scale generation of hydroelectricity and the rapid emergence of electricity-intensive industries, such as electrochemical and electro-metallurgical production.[41] Finland's performance was modest in these respects, although in the electrification of installed power in manufacturing, Finland was not so far behind the hydropower countries or major industrialized economies.[42] As indicated in table 5, the countries with the highest per capita output and use of electricity were mostly hydro countries (Norway, Canada, Switzerland, Sweden, and New Zealand). It seems quite likely that during the early 20th century a high per capita electricity output can generally be attributed to two concomitant factors: abundant hydropower resources and high gross national product (GNP) per capita.

From the turn of the century, the electric motors of manufacturing industries consumed the most electricity in Finland. They constituted the crucial factor in the country's electrification because, before 1930, the manufacturing industries used 70–90 percent of all generated electricity. Nonindustrial consumption remained limited, although after 1914 there was at least one small electricity-supply utility in all 38 towns, and rural electrification had been in progress since the turn of the century.[43]

World War I caused the first energy crisis in Finland. The importation of coal slumped radically, and the supply of lamp oil (paraffin)

[40]James E. Brittain has stated that at the turn of the century Switzerland played an innovative role in electrical engineering disproportionate to its size (Brittain, "The International Diffusion of Electrical Power Technology, 1870–1920," *Journal of Economic History* 34 [March 1974]: 114–15; see also Frederick Bathurst, "Switzerland as the Present Electrical Center of Europe," *Electrical World* 23 [1894]: 731 ff.).

[41]On the style of electric light and power systems see Hughes, *Networks* (n. 4 above), pp. 404–60.

[42]Ingvar Svennilson divided countries into hydro and thermal. The former denotes economies that "produce more than 75 per cent of their electricity by water power." According to this definition, Finland was a thermal country up to the early 1930s, when it joined the club of hydro countries. See Ingvar Svennilson, *Growth and Stagnation in the European Economy* (Geneva, 1954), pp. 111–18.

[43]G. M. Nordensvan, "Suomen voima-ja sähköistysoloista," *Teknillinen aikakauslehti* 13, no. 10 (1923): 340; Timo Myllyntaus, "The Role of Industry in the Electrification of Finland," Discussion Paper no. 333 (The Research Institute of the Finnish Economy, Helsinki, 1990), pp. 16–18.

## The Transfer of Electrical Technology to Finland, 1870–1930        311

TABLE 5
ELECTRICITY OUTPUT PER CAPITA, 1900–1930 (Kilowatt-Hours)

| Country | 1900 | 1913 | 1920 | 1930 |
|---|---|---|---|---|
| Norway | 24* | 900 | 1,631 | 2,694 |
| Canada | ... | ... | 735 | 1,770 |
| Switzerland | 61 | 414 | 722 | 1,276 |
| United States | 75† | 260‡ | 531 | 931 |
| Sweden | 21 | 258 | 443 | 835 |
| Belgium | ... | 172 | 163 | 532 |
| New Zealand | ... | 17§ | 94 | 490 |
| Germany | 18 | 119 | 246 | 447 |
| France | 9* | 45 | 149 | 405 |
| United Kingdom | 5 | 91‡ | 195 | 386 |
| Austria | ... | ... | 274 | 374 |
| Australia | ... | 39§ | ... | 360 |
| Finland | 6‖•* | 60‖•* | 92‖•* | 350‖ |
| Netherlands | ... | ... | 104 | 313 |
| Italy | 5 | 63 | 125 | 261 |
| Japan | ... | 42** | 68 | 246 |
| Czechoslovakia | ... | 73‖ | 101‖ | 204‖ |
| Denmark | ... | 34* | 81 | 164 |
| Spain | ... | 25 | 45 | 111 |
| Iceland | ... | ... | ... | 92 |
| Poland | ... | ... | ... | 91‖ |
| Hungary | ... | ... | ... | 83 |
| Yugoslavia | ... | ... | ... | 56 |
| Russia/USSR | ... | 14‖ | 4†† | 52†† |
| Ireland | ... | ... | ... | 47 |
| Portugal | ... | ... | 16 | 38 |
| Greece | ... | ... | 16 | 35 |
| Romania | ... | ... | ... | 30 |
| Bulgaria | ... | ... | ... | 17 |

SOURCES.—Archive of the Board of Industry, National Archives of Finland, and Archive of the Central Statistical Office of Finland; "Statistics on Electricity Output, 1912–1925," Archive of the Central Statistical Office of Sweden; *Historical Statistics of the United States*, part 2 (Washington, D.C., 1975); G. T. Bloomfield, *New Zealand: A Handbook of Historical Statistics* (Boston, 1984), pp. 41, 206; J. Darmstadter with P. D. Teitelbaum and J. G. Polach, *Energy in the World Economy* (Baltimore, 1971), Table XI; T. Liesner, *Economic Statistics 1900–1983* (London, 1985); A. Maddison, *Phases of Capitalist Development* (Oxford, 1986); R. Minami, "The Introduction of Electric Power," in *Japanese Industrialization and Its Social Consequences*, ed. H. Patrick (Berkeley, 1976); B. R. Mitchell, *European Historical Statistics, 1750–1970* (London, 1978); *OEEC, Statistical Bulletins, Industrial Statistics, 1900–1959* (Paris, 1960); *Økonomisk'utsyn 1900–1950* (Oslo, 1955); W. Wyssling, *Die Entwicklung der schweizerischen Elektrizitätswerke und ihrer Bestandteile in den ersten 50 Jahren* (Zurich, 1946), p. 500.
*1901.
†1902.
‡1912.
§1910.
‖1937 borders.
*Estimate.
**1915.
††Post-1945 borders.

312    *Timo Myllyntaus*

gradually dried up.[44] The scarcity of foodstuffs, combined with price regulation and rationing, precipitated the rise of a black market. Rumors about exhausted supplies of food and other necessities increased anxiety in the country. People were afraid that the supply of candles and matches would be depleted.[45] The shortage of fuels and wartime inflation turned electricity into an unbeatable source of light. It did not vanish from the legal market like many other necessities because of hoarding and speculation; electricity could be generated continuously by means of indigenous energy sources (hydropower, firewood, and wood wastes). Owing to price regulation, electricity rates were not affected by the soaring inflation. As a result, in the areas that were wired, electricity became the cheapest and most reliable source of light within a few years. During World I and the Finnish Civil War (1918), remote rural areas were without paraffin for periods of several months.[46] Villagers had a strong incentive to jointly set up their own electricity-supply utilities. In a few years, the number of these rural undertakings rose from 100 to over 400. Especially in the western part of the country, the rural population proved to have a marked entrepreneurial flair, but rapid development also included unsound aspects. In the so-called electrification frenzy (*sähköistys-vimma*), people neglected to consider technological expertise and economic profitability. The lack of equipment made the situation still worse. Distribution networks were sometimes built so hastily that barbed wire was used for overhead lines and bottlenecks for insulators. Generating plants often failed to work. As a result, consumers were supplied with only flickering light, while transmission losses sometimes rose to over 50 percent of the distributed energy.[47]

During the interwar period, the supply of electricity was improved and extended. Many of the local village undertakings were merged into regional utilities. The municipalization of urban distribution utilities, which transferred private companies to municipal ownership, was nearly completed. New power plants and high-tension

[44]The total importation of coal and coke, which were supplied mainly by Britain, fell from 587,000 tons to 7,500 tons within the period 1913–15. The prewar figure was reattained only in 1924. Oil was imported from Russia until 1917, after which a great collapse in foreign trade took place. The imports of oil and oil products dropped from 39,100 tons to 2,300 tons between 1913 and 1918. Oil imports from the West rose to 39,100 tons by 1923. See *Official Statistics of Finland, Series IA Foreign Trade 1913–1924* (Helsinki, 1915–1926).

[45]*Mercator* (November 10, 1916), p. 778; *Kauppalehti* (September 18, 1918), p. 453.

[46]*Revue sociale* (1919), p. 148.

[47]Killinen (n. 21 above), p. 171; [Antti W. Lehtinen], *Vilppulan sähkö-osakeyhtiö 1918–1943* (Tampere, 1943), pp. 3–9.

*The Transfer of Electrical Technology to Finland, 1870–1930*     313

transmission lines were built. The largest of the new plants was the government-owned hydroelectric plant at Imatra, the first two phases of which, comprising a capacity of 75 megawatts, were completed in 1929 and 1930.[48]

Although the electrification of Finland did not proceed at the same rate as in the major industrialized countries during the years 1890–1919, it was not slow compared with that of eastern and southern European countries on a similar level of economic development. In the 1920s and 1930s the country considerably remedied the lag in the electrification of urban and rural households. The growth rate (on the average 14 percent per year) of Finnish electricity generation was the highest in interwar Europe except for the Soviet Union. By 1930, Finland had risen to thirteenth in the world in per capita output, as indicated in table 5. Finland had nearly caught up with some leading industrialized countries; it was only 9–22 percent behind Germany, France, and Britain in 1930.[49] Meanwhile, its output/population ratio had risen to anywhere from 250 to 2,000 percent ahead of the Iberian and Balkan countries.

By the end of 1930, Finnish industry had become highly electricity intensive, consuming just over 6,000 kilowatt-hours per worker, or about 80 percent of the corresponding figure for Swedish industry, which was ranked among the top countries in the world.[50]

## Conclusion

Measured by the GNP per capita, late-19th-century Finland was at about the same level of development as southern and eastern Europe. In terms of other cultural indicators, it was behind those areas. Consequently, how Finland managed to electrify its economy faster than the other late-to-industrialize countries with a similar standard of living is a puzzle.

The relatively rapid electrification of Finland was not based on any significant indigenous inventions or on a high level of scientific expertise. None of the great electrical inventors was from Finland, although surprisingly many of them were brought up in Scandinavia

---

[48]Jaakko Auer and Niilo Teerimäki, *Puoli vuosisataa Imatran Voimaa, Imatra Voima Oy:n synty ja kehitys 1980-luvulle* (Helsinki, 1982), pp. 48–58.

[49]By 1938, Finland surpassed these three countries and some others and was ranked seventh in the world (Joel Darmstadter with Perry D. Teitelbaum and Jaroslav G. Polach, *Energy in the World Economy* [Baltimore, 1971], Table XI).

[50]Before the outbreak of World War II, Finland caught up with Sweden in the industrial consumption of electricity per worker (*Official Statistics of Finland 18: Statistique des industries de Finlande 1930* and *1938* [Helsinki, 1932 and 1942]; *Sveriges Officiella Statistik, Industri 1930* and *1938* [Stockholm, 1932 and 1942]).

or eastern or southern Europe.[51] Hans-Christian Ørsted and Søren Hjorth were from Denmark, and Jonas Wenström, Gustaf de Laval, Birger and Fredrik Ljungström, and Oscar Kjellberg were Swedish.[52] Among the famous Russian inventors were Pavel Iablochkov, Nikolai Benardos, and Mihail Dolivo-Dobrovols'kii. Among the Hungarian pioneers were the inventors of the parallel connection for transformers, Otto Blathy, Max Déri, and Karl Zipernowsky. Nikola Tesla, the inventor of the alternating-current motor, was a Croatian. Corresponding lists of southern European inventors could be presented.[53] The peripheral location of those countries could not inhibit inventive activity, but it did delay electrification. In Finland, however, economic, cultural, and geopolitical factors hampered invention much more than they did electrification.

The early start and development of Finnish electrification cannot be explained in terms of technical education. Although the rate of literacy was very high in the country, teaching of and research in technology took place on a very small scale. Only after 1911 was it possible to earn the degree of graduate engineer in electrical technology at the new Helsinki University of Technology. In many other European countries, technical education was much better organized. In 1910, Germany had eleven universities of technology, while Austria-Hungary and Russia each had seven. In Sweden, it was possible to receive the electrical engineer's degree at two different institutions.[54]

Furthermore, Finland completely lacked coal and oil deposits and its hydropower resources were mediocre, whereas Norway and Sweden have benefited from the extraordinary gifts of nature in terms of hydropower resources. The generating capacity of Finland's techni-

[51]Eric M. C. Tigerstedt (1886–1925), "Finland's Edison," was the best-known Finnish inventor in the field of electrical technology and the one who came the closest to gaining international recognition. He received an engineer's degree in Germany and worked in the Nordic countries before immigrating to the United States. Financial and health problems hampered his inventive activity, and he managed to yield only minor income from his over 400 patented inventions before his early death in New York. See A. M. P. Kuusela, *E. M. C. Tigerstedt, "Suomen Edison"* (Helsinki, 1918).

[52]T. K. Derry and Trevor I. Willims, *A Short History of Technology* (Oxford, 1982), pp. 608–36; Klaus Schulz-Hanssen, *Die Stellung der Elektroindustrie im Industrialisierungsprozess* (Berlin, 1970), pp. 161–73.

[53]Harold I. Sharlin, "Electrical Generation and Transmission," in *Technology in Western Civilization*, ed. Melvin Kranzberg and Carroll W. Pursell, Jr. (New York, 1967), 1:582; Hughes, *Networks* (n. 4 above), pp. 94–105.

[54]Timo Myllyntaus, "Education in the Making of Modern Finland," in *Education and Economic Development since the Industrial Revolution*, ed. Gábriel Tortella (Valencia, 1990), pp. 153–72; Karl Remme, *Die Hochschulen Deutschlands* (Berlin, 1926), p. 12.

*The Transfer of Electrical Technology to Finland, 1870–1930*    315

cally exploitable hydropower resources was estimated in the mid-1960s to be only a third of Italy's, Yugoslavia's, or Spain's potential and approximately the same as the resources of Romania or Greece. In per capita terms, however, Finland was in a somewhat better position than those countries.[55]

Direct foreign investments in power plants, transmission lines, distribution utilities, and energy-intensive industry have considerably promoted the development of electricity output in many countries. Although DFIs were also made in Finland, they were far less extensive and significant than the foreign involvement in Norway or eastern Europe.

What, then, caused the early and relatively rapid electrification of Finland? First, although the country had a fairly narrow choice of indigenous natural resources, some of them, namely hydropower and timber, were appropriate for power generation. During its initial phase, Finnish electrification was not based on the country's "white coal." Hydropower became significant for electricity generation only after the turn of the century: the steep rise of output in interwar Finland was based primarily on the extensive utilization of hydropower. In the early 20th century, contemporaries had already observed that countries with considerable hydropower resources tended to have more electricity to consume than other countries.[56]

Second, the annual growth of industrial output averaged 5.3 percent in the period 1890–1913 and 7.9 percent between 1920 and 1938; the respective figures for gross domestic production were 3.0 and 4.7, which indicates that Finland was one of the most dynamic European countries at the time.[57] Since the rate of industrialization was fairly swift, it induced extensive investments in production capacity. The building of new factories presented opportunities to introduce up-to-date technology, and the most rapidly growing industries, such as wood processing, were energy intensive. Timber resources that were the second largest in Europe provided an ample supply of raw materials for the wood-processing industries. In addi-

[55]*The Hydro-electric Potential of Europe's Water Resources*, vol. 1, *Methods of Analysis and Their Application* (New York, 1968), p. 39.

[56]Walther Windel and Carl Th. Kromer, *Aufbau und Entwicklungsmöglichkeiten der europäischen Elektrizitätswirtschaft* (Berlin, 1928), pp. 255–56, 294–96; Walter H. Voskuil, *The Economics of Water Power Development* (Philadelphia, 1928), pp. 198 ff.; Adolf Ludin, *Die nordischen Wasserkräfte, Ausbau und wirtschaftliche Ausnutzung* (Berlin, 1930), pp. 482–90.

[57]Sakari Heikkinen, Riitta Hjerppe, Kai Hoffman, Timo Myllyntaus, and Birger Rabb, *Industry and Industrial Handicraft in Finland 1860–1913* (Helsinki, 1986), pp. 120–22; Riitta Hjerppe, *Finnish Economy 1860–1985* (Helsinki, 1989), pp. 192–96.

316    *Timo Myllyntaus*

tion, firewood and wood wastes from the forest industries alleviated the shortage of indigenous fuels. These factors provided electrification with a favorable economic basis.

Third, because the industrialization of Finland started in earnest only in the last third of the 19th century, electricity faced very few competing technologies. In Finland, there were only three urban gas utilities, in Helsinki, Turku, and Viipuri. In the early 1920s, their number was 17 in Norway, 37 in Sweden, and over 70 in Denmark.[58] Electricity was thus granted a lot of unoccupied technological Lebensraum.[59]

Fourth, a critical contextual factor was late-19th-century Finnish society's rather positive attitude toward industrialization and the adoption of innovations. The Finns were psychologically prepared to apply new technology that they viewed fairly unanimously as supporting their national aspirations. Consequently, electrification was backed by sociopsychological and political circumstances.[60]

Fifth, owing to reasonably high basic educational standards and a mental preparedness to adopt innovations, the Finns developed into eager, entrepreneurial transferers of electrical technology. The pronounced role of Finnish nationals in the successful transfer process is an interesting feature of Finnish electrification. This feature seems to conflict with the conventional wisdom about the unfortunate fate of a

[58]Timo Herranen, *Kaasulaitostoimintaa Helsingissä 1860–1985* (Espoo, 1985), pp. 32–33; Firtz Hodne, *The Norwegian Economy 1920–1980* (London, 1983), p. 47; Arne Kaijser, *Stadens ljus, Etableringen av de första svenska gasverken* (Malmö, 1986), p. 178; *Danmarks statistik, Statistiske Meddelelser* 4 række, 50 bind, 7 hæfte (Copenhagen, 1917), p. 15; *Danmarks statistik, Statistisk tableværk* 5 række, Litra A, nr 18 (Copenhagen, 1929), p. 124.

[59]Another advantage of relative backwardness was that Finland had an opportunity to adopt the most up-to-date, well-tested technical standards of the time. Therefore, it benefited from mistakes and debates in the trailblazing countries. For example, the battle of the systems was a minor episode in Finland. In its major electricity-supply utilities, alternating current was introduced in the 1890s. In addition, Germany and Switzerland had made their decisions in the 1910s about what technical standards they would apply in the unification of their electricity-supply systems, and Finland copied them just at the beginning of the great electrification boom. Thus, all Finnish utilities adopted the frequency of 50 hertz very early, and in a rather short period most of them managed to change over to the distribution tensions of 380/220 volts.

[60]Western countries, especially Germany and the United States, were regarded as the forerunners of scientific and economic development as well as the sources of new technology. It is illustrative that, if we consider the number and tone of the newspaper articles published in the 1880s and 1890s, the Finnish press rated the inventor Thomas Alva Edison as the most popular American at the time. Between 1898 and 1946, Finnish publishing houses issued four biographies of Edison. The first was a translation: W. K. L. Dickson and A. Dickson, *Edison, hänen elämänsä ja keksintönsä*, trans. Juhani Aho (Kuopio, 1898).

*The Transfer of Electrical Technology to Finland, 1870–1930*      317

latecomer to industrialization on the poor, agrarian periphery that relies mainly on its own efforts in technology transfer.[61]

Finally, there was societal demand for electrical technology in Finland. This need was fulfilled by the relatively efficient transfer of foreign technology and its rational application to local conditions. Meanwhile, the rapid growth of the Finnish economy and the very energy-intensive structure of the manufacturing industry constituted favorable economic preconditions for the transfer of electrical technology to the country. Although the contextual filter was very selective and idiosyncratic, it coordinated with and even enhanced the transfer channels society had accepted. The success of the electrification of Finland cannot be explained by the creation of indigenous technology but rather by the entrepreneurial and determined social engineering of technology transfer.

---

[61]Finns valued highly the benefits of the learning process in the design, construction, and maintenance of the electricity-supply system. They were aware that "reliance on borrowed technology perpetuates a posture of dependence and passivity." See Nathan Rosenberg, "Economic Development and the Transfer of Technology: Some Historical Perspectives," *Technology and Culture* 11 (October 1970): 568.

# [13]

Excerpt from David J. Jeremy (ed.), *The Transfer of International Technology: Europe, Japan and the USA in the Twentieth Century*, 90–117.

## 5. Aspects of the Anglo-American Transfer of Computer Technology: The Formative Years, 1930s–1960s

### Geoffrey Tweedale

---

## INTRODUCTION

Modern computer history begins after the Second World War with the development of the electronic stored-program digital computer. In another sense, though, the history of computers commences with the invention of the first calculating machines.[1] The humble abacus, which probably originated in Babylonia (now Iraq) five thousand years ago, was perhaps the first device to embody a momentous idea – the notion of using a machine to perform intellectual work. Through the ages numerous philosophers, mathematicians and inventors, amongst them the English mathematician Charles Babbage (1791–1871), who embarked upon the construction of his ill-fated Difference Engine in the 1820s, devised increasingly sophisticated machines. In the late nineteenth century, as calculating problems grew, an American, Herman Hollerith (1860–1929), built punched-card machinery to mechanize the tabulation of the 1890 US Census, so establishing the data-processing industry.

By the early twentieth century, mechanical calculation, either in the office or the mathematical laboratory, was well established. Punched-card machinery, analogue devices for solving complicated equations and manual desk calculators were commercially available for 'computing', though it should be noted that prior to 1940 this term meant only one thing – a clerk equipped with a hand-calculating machine, who could 'compute' the standard calculations required for wages, actuary tables and ballistics. During the period 1935–45 the application of better engineering techniques, especially in electronics, began to speed up calculations for specific problems. In the 1930s, firstly in America and then in England, the first differential analysers were built for the solution of differential equations. During the Second World War new challenges spurred two major developments: the COLOS-

SUS, a code-breaking machine constructed for the British Government at Bletchley Park; and the ENIAC (Electronic Numerical Integrator and Computer), which was commissioned late in 1945 for the US Army Ordnance Department for ballistics calculations.

Though these machines were an important step along the way, none were computers in the modern sense of the word. Some, such as the differential analyser, were analogue machines: these expressed numerical quantities by analogue instead of digital representation and were usually built for specific problems, such as simulating aircraft behaviour. None were *universal*: that is, machines capable of solving any problem that could be solved by mathematical means, once an appropriate program had been inserted. Above all, they lacked a key feature of modern computers – an internal store (or memory), which could hold both instructions (the program) and data, and whose contents could be selectively altered automatically during computation. Soon after the end of the Second World War both the conceptual and engineering problems involved in constructing and utilizing such a memory were solved, so ushering in the era of the stored-program digital computer. By the end of the 1960s, the computer revolution was well under way: digital computers were commercially available in increasing numbers, the market for punched-card machinery was rapidly declining, analogue machines were being phased out, and the silicon integrated circuit (or 'chip') had arrived.

During these formative years, leaving aside the importance of the military in virtually all aspects of computer development, a notable feature was the remarkably close relationship between Britain and the USA. It is a commonplace that during the heyday of the 'Atlantic Economy' in the nineteenth century, America drew heavily on British technologies in founding its own industries. Some aspects of this transfer have been well studied by historians: textile technologies, mechanical engineering, the pottery industry, silk manufacture and special steels have all been the subject of detailed case-studies.[2] Generally these accounts have emphasized how successfully and swiftly America absorbed foreign technologies (though some have highlighted US backwardness in certain areas). By concentrating on the nineteenth century they have also implied that technological transfer – at least from Britain to America – was largely over by the early twentieth century. More studies will be needed, however, before any final conclusions can be reached. What can be stated with certainty is that in computing technology the Anglo-American connection was far from moribund, even after 1945, and was of great importance for the emerging computer industry. The nature of this linkage has never been systematically explored: indeed in many recent accounts, which have tended to focus on American advances, it has been underplayed. This account aims to rectify this deficiency.

# DEVELOPMENTS BEFORE THE SECOND WORLD WAR

Charles Babbage is usually regarded as a decisive figure in the early history of computing, yet there is little evidence that his ideas were very widely disseminated. Many of the twentieth-century pioneers were admirers of Babbage, but this was often a retrospective interest and there is little evidence that Babbage influenced the design of the modern computer. One American who it is usually said was inspired by Babbage's ideas was Howard Aiken (1900–73), who began building the Harvard Mark I in 1937, and who often linked himself with Babbage and liked to present himself as his heir. According to one account, Aiken 'felt Babbage was addressing him personally from the past'. But Aiken's machines show no similarity to Babbage's designs, which were in any case unavailable to Aiken.[3] Inventors usually discovered Babbage's ideas afresh. Babbage may have influenced the American inventor and businessman, Herman Hollerith, but it seems more likely that his punched cards owed more to the Jacquard loom than to the Englishman.[4] It was with Hollerith, however, that transatlantic transfers of computer technology began.

By the 1900s, Hollerith's punched-card machines were being produced commercially by his Tabulating Machine Co. in New York. In 1904, Hollerith began exploiting his inventions in Britain, where he established a subsidiary which had the right to market and manufacture the American machines. The transfer of technology to the British company – which became the British Tabulating Machine Co. Ltd (BTM) in 1907 – was effected by C.A. Everard Greene, the first general manager of the British company, who visited the USA during 1902–4. He later recalled: 'My training consisted of getting an insight into the manufacture of parts, the assembly and wiring of machines, the making, planning and drawing up of cards for jobs, the investigation and organising necessary for installing and operating machines on the job.'[5] BTM purchased exclusive rights to manufacture and sell the machines, but the company was not without competition. In the early 1900s, rivalry over the US Census punched-card machine business developed between Hollerith and another inventor, James Powers, with the latter winning an important share of government business. This rivalry was transferred to Britain when, in 1915, an American-owned subsidiary of the Powers company was formed, the Accounting & Tabulating Corporation of Great Britain Ltd – often known as the 'Acc & Tab'.

In the heyday of the punched-card business in the inter-war period both BTM and the Acc & Tab prospered. Before the 1920s these two companies had essentially marketed and maintained machines which were largely American made. But gradually British manufacture and design prevailed: by 1920 Acc & Tab had opened a factory at Croydon and two years later BTM had begun manufacturing operations at Letchworth. R & D activity in these companies began shortly afterwards, though there was some patent activity before this time. Probably the most significant invention in the development of punched-card machinery, and unquestionably the most important British contribution, was the alphabetical printing unit, the work of Acc & Tab engineer, Charles C. Foster. This invention transformed punched-card accounting, since prior to alphabetical machines only figures could be printed, whereas henceforth names, addresses and descriptive material could be produced. Foster's unit was patented in 1916, but the First World War hindered its development, so that the prototype was not demonstrated until 1921. This led to a swift (and rare) reverse transfer of technology to the USA, since the American Powers company was eager to exploit it.

Documentation on other technical developments is scanty or non-existent at this time. But the broad picture is clear: British firms built upon the base of American technology, for which they sometimes paid dearly (BTM had to pay the patent holders – by that time IBM – 25 per cent after tax of its net revenues). American developments usually soon found their way to Britain. A good example was the introduction in the USA in 1928 of IBM's 80-column punched card with slotted holes in place of the standard 45-column card with round holes. The greater capacity of the new card was further extended by Remington-Rand, which offered a 90-column card soon after. These new-sized cards were adopted by BTM and Acc & Tab within a year or two of their appearance on the American market. However, on their own terms the British firms were able to make important contributions to punched-card design. One of the notable British achievements during the inter-war period was a smaller punched card, which after its introduction in 1932 by the British Powers Co. (by then known as Powers–Samas) enabled a range of low-cost machines to be marketed for small and medium-sized businesses. Another important development was the 'rolling total' mechanism developed at BTM by H.H. ('Doc') Keen. This device enabled values to be 'rolled' from one counter to another, so boosting the calculating possibilities of the machine. The BTM rolling total tabulator launched in 1936 was functionally superior to its IBM counterpart.

These machines were used for commercial applications. The first use of Hollerith tabulating equipment for large-scale scientific computations was pioneered in England by L.J. Comrie (1893–1950), who headed HM Nautical Almanac Office in London. In 1923–5, he had taught at two American universities, where he pioneered the introduction of computation as part of the curriculum. Comrie devised a method for calculating Fourier series on standard punched-card machines, and used this to calculate the motions of the moon for the years 1935 to 2000. These calculations were based on Brown's *Tables of the Moon* and, when E.W. Brown visited England in 1928, he was shown the computations in progress. Brown returned to the USA with the technique and discussed it with his friend Wallace J. Eckert, a mathematician and astronomer. Eckert was later able, in 1930, to persuade the president of IBM, Thomas J. Watson, to fund an Astronomical Computing Bureau at Columbia University, so that work could be done along similar lines to that of Comrie in London. It was not long before standard punched-card accounting equipment was modified to further facilitate large-scale calculation. In 1932, Comrie managed to persuade the Hollerith firm to modify their equipment so that the contents of one register, which previously had been limited to taking part in addition operations, could now be transferred to other mechanical registers in the machine. IBM equipment was modified in the same way about ten years later.[6]

The appearance of punched-card machinery in Britain in the early twentieth century was a visible symbol of the Anglo-American transfer of technology. Less apparent, but perhaps of greater importance for the subsequent history of computing, were linkages concerning analogue machines for mathematical calculations. Among the most important analytical tools in science and engineering are differential equations: a branch of calculus, these equations enable the prediction of the behaviour of moving objects by relating them to certain variables. They are very difficult to solve, but Lord Kelvin (1824–1907) argued in a remarkable paper published in 1876 that a mechanical 'differential analyser' capable of solving complicated differential equations was theoretically possible. His brother, Professor James Thomson, had first thought of the idea. Basically the machine used an 'integrator' – a wheel and disc arrangement with attached drive shafts – to effect the mathematical process of integration. But the technology of the day was incapable of realizing Kelvin's ideas and so his machine was never built.

In about 1930, however, Vannevar Bush (1890–1974), a professor of engineering at the Massachusetts Institute of Technology, returned to the problem and constructed a working differential analyser. Bush overcame the technical problems that had defeated Kelvin. In particular he was able to incorporate into his machine a 'torque amplifier', an indispensable device which 'stepped up' the smallest forces of the numerous shafts and gears. How much this design was based on Kelvin's previous published work is difficult to establish. According to Bush's memoirs, his project was original and he did not read Kelvin's paper until after he had built his machine.[7] Be that as it may, Bush's device was soon imitated.

In the summer of 1933, Bush received a visit from Professor Douglas R. Hartree (1897–1958), who at that time held the Chair of Applied Mathematics at Manchester University, and whose career highlighted not only the transition from analogue to digital techniques but also the importance of the Anglo-American connection. Hartree had realized that Bush's machine would much lighten the enormous calculations involved in his own work on quantum theory regarding self-consistent fields and he therefore decided to build such a machine for himself. The plans were generously given to him by Bush, who hoped that Manchester would produce an even better machine. With the help of a Manchester undergraduate, Arthur Porter, Hartree had constructed his first machine from *Meccano* (a child's construction set) by January 1934. Hartree and Porter then began building a full-scale machine, which was completed in 1935, having been engineered by Metropolitan Vickers. For a time it was the largest, most sophisticated and widely used differential analyser outside America.[8]

In 1937, Arthur Porter was awarded a Commonwealth Fund Fellowship, which enabled him to study at MIT under the supervision of Vannevar Bush. By now Bush had a new machine under way, the Rockefeller Differential Analyser, which made increasing use of electronics and high-precision engineering. Porter was to examine the problem of loading information into the machine using punched tape.[9] The first demonstration of the incomplete analyser was made in December 1941 and in the following year it was used for important war work on the calculation of firing tables and the profiles of radar antennas. By then, however, computing was swiftly developing in other directions.

## THE SECOND WORLD WAR AND THE ADVENT OF THE STORED-PROGRAM DIGITAL COMPUTER

After 1940, military needs were the overriding influence on the development of computers, as the war threw up a host of computational problems. Computing devices were needed for gunnery control, bomb aiming, flight simulation, radar signal processing, atomic weapon calculations and code-breaking. In Britain a number of centres, such as the Telecommunications Research Establishment (TRE) and the Post Office Research Station, provided new ideas and, above all, the personnel, who were to direct post-war British efforts in this area.

Of particular interest is the special-purpose 'computer' named COLOSSUS, produced in great secrecy by the Post Office Research Station for the Government Code and Cipher School, Bletchley Park, Buckinghamshire. This machine, which was operational by the end of 1943, was built to crack German coded messages, which it did with great success. Absurdly, these operations are still classified, though the secrecy surrounding the project has relaxed sufficiently for the broad details to be known. Apart from its impact on the allied war effort, two aspects are particularly important. Firstly, although it is clear that the COLOSSUS was in no sense a modern computer, the final version of the machine contained all the necessary elements except an internal program store. Since it used almost 1 500 thermionic valves, it was also a remarkable demonstration that large numbers of electronic circuits could be made to do reliable calculations at speed. Secondly, the project brought together mathematicians and engineers in a fruitful collaboration, which alerted them to the possibilities inherent in computing. As one of them later commented: 'The value of the work ... was that we acquired a new understanding of and familiarity with logical switching and processing because of the enhanced possibilities brought about by electronic technologies which we ourselves developed. Thus when stored program computers became known to us we were able to go right ahead with their development.'[10] Amongst those involved with the building of the COLOSSUS were the mathematicians M.H.A. (Max) Newman (1897–1984) and Dr Alan M. Turing (1912–54), both of whom were to play an important role in subsequent British computing developments.

COLOSSUS was independent of American efforts. However the US armed forces had established contacts with the British cryptanalysts well before the USA came into the war, and there were regular missions for the exchange of information. Certain members of the COLOSSUS team, such as Alan Turing, were already familiar with the American scene. A Cambridge-trained mathematician, Turing had first visited the USA during 1936–8, when he held a visiting fellowship at the Institute of Advanced Study at Princeton University (IAS). This brought him into contact with that other major figure in twentieth-century computing, John von Neumann (1903–57). The visit coincided with the publication of a famous mathematical paper by Turing, 'On Computable Numbers', in which he outlined his idea for a theoretical universal automaton. The impact of this abstract and, to some, incomprehensible paper, is difficult to assess, though it now takes its place as a significant milestone in computer history. Personal contact between Turing and von Neumann appears to have been slight, though apparently by 1939 von Neumann knew and admired Turing's ideas and, certainly, Turing's paper was known at Bletchley Park.[11] At the end of 1942, Turing was in America again, charged with liaising with the Americans in connection with his work on COLOSSUS. This trip included a visit to Washington, followed by a tour of the Bell Laboratories, where he immersed himself in the electronic technology of speech encipherment and discussed matters relating to his work with prominent engineers, such as Claude Shannon. This was probably the only journey Turing made to America during the war, though the legend grew up that he and von Neumann met during the conflict, and that this was somehow an important event in the evolution of the modern computer.[12] There is no evidence for this, however: the modern computer evolved from the mainstream of developments in twentieth-century mathematics and engineering and did not need a meeting between Turing and von Neumann to bring it about.

Nevertheless, John von Neumann's involvement with the development of the computer was crucial. It stemmed from his introduction to the work on the ENIAC at the Moore School of Electrical Engineering at the University of Pennsylvania, which was a major source of technical and computational support for the US Army's Ballistics Research Laboratory. The ENIAC contained 18 000 valves and 1 500 relays, weighed over 30 tons and was built by a team of 200 people between 1943 and 1945. Significant less for its achievements – it did not become operational until some months after the war – than for what it promised, ENIAC gave scientists and the public their first dramatic glimpse of the computer age. It was a development in which British scientists played a part. Douglas Hartree had visited the USA again in 1945, as soon as the European war was over, and had seen the ENIAC, which was still incomplete. In the following year he was invited for a second visit so that he could advise on its use. According to Herman Goldstine, the representative of the Ballistics Laboratory, Hartree 'was of great help to the ENIAC group coming when he did, because he helped keep up the morale and intellectual tone of the ENIAC operating staff'.[13] Hartree was able himself to use the ENIAC for a problem in laminar boundary flow, though a program error was later found to have vitiated the results. Nevertheless he had shown that ENIAC had a broad range of applicability. On his return to England, where he took up a chair in mathematical physics at Cambridge University, Hartree was able to publicize American activities in *Nature* and in his inaugural lecture. His support for US computing continued in 1948, when he spent three months as acting director of the Institute for Numerical Analysis, newly established by the Bureau of Standards on the UCLA campus. A series of lectures he gave at the University of Illinois on this trip was repeated in Cambridge on his return and also published.[14] Hartree continued this championing both of American advances and computers throughout his life.

The development of the ENIAC and the personalities involved – notably, the driving force behind the project, J. Presper Eckert and John Mauchly – have been well described elsewhere.[15] Here attention will be confined to its impact on Anglo-American technology, which was profound. The ENIAC team were to be responsible for the sudden flowering of the stored-program concept, when they began work on their next machine at the Moore School, the EDVAC (Electronic Discrete Variable Automatic Calculator). Before then it had always seemed logical to store data outside the machine, but this became increasingly difficult once machines such as the ENIAC could process thousands of instructions per second. The ENIAC engineers took the crucial step of considering storing data *within* the machine. Building on Eckert's and Mauchly's work (though unfortunately not acknowledging them), von Neumann elucidated the concept of the stored-program digital computer –

one in which both data and instructions would be held in a memory – in his document, 'First Draft of a Report on the EDVAC' (1945).[16] Available in a limited number of copies, this report began percolating into the scientific community at about the same time as a special summer course was run at the Moore School in 1946. The course was entitled, 'Theory and Techniques for the Design of Electronic Digital Computers', which attracted 28 people from both sides of the Atlantic.[17] Together with von Neumann's report, these seminars helped trigger the post-war development of the stored-program computer both in Britain and the USA. By the late 1940s, the race was on to overcome the engineering problems involved in building such a computer. England, with its wartime experience with electronics and code-breaking machinery, was well placed to take up the challenge with a number of major initiatives. These projects and especially their relationship to American developments will now be considered.

## MANCHESTER UNIVERSITY MARK I

After the war the key problem facing engineers was the construction of a satisfactory computer memory. Clearly whoever solved this problem stood a good chance of being the one to build the world's first operational stored-program computer. In the event, the prize was to be claimed by a team at Manchester University, led by Professor (Sir) F.C. Williams (1911–77) and (Professor) Tom Kilburn. Both men had acquired a wartime grounding in radar electronics at the Telecommunications Research Establishment at Malvern. In 1946, Williams accepted a chair at Manchester University and with the help of Kilburn (who was seconded from the TRE) began work on a novel form of computer storage using the cathode ray tube (crt). This was at a time when a number of mathematicians and engineers with expertise in the fields of electronics and mathematics had contrived to secure jobs at Manchester University. These eventually included Max Newman (who had secured a Royal Society grant to set up a 'Computer Laboratory'), Alan Turing and I.J. Good, all former COLOSSUS men. The development of the Manchester University project, and its relationship to American activities, is known in some detail because of later patent litigation.[18]

During the war Williams had visited the USA in connection with his radar work. He was one of the editors of a Radiation Laboratories multi-volume reference work (known as the 'five-foot shelf' because of its bulk) and to discharge his responsibilities in this direction he visited the Laboratories in Boston in November 1945, where he observed that there was much discussion of crt storage. In June 1946, Hartree had secured an invitation for him to view the ENIAC, which in Williams's words was the first time he 'personally came across computation in terms of numbers'. Methods of storage using crt's were being explored at this time in the USA, particularly at the Moore School, but without any success. On his return to England in July 1946, however, Williams, whose own background and inventive genius had given him a unique insight into the technical difficulties involved, began work on crt storage at the TRE. The result of his efforts, which by 1947 had been transferred to Manchester University, was the Williams tube – the first electrostatic random-access memory for a computer – in which Williams discovered a relatively simple method to regenerate, and thus store, electronic pulses. The device became the basis for a prototype machine (the forerunner of the Manchester Mark I) which was built at Manchester University and on 21 June 1948 became the world's first stored-program computer to operate.

As for the logical design – the stored-program concept – that had been explained to Williams by Newman in 'all of half an hour'. It is not known if this reflected American influence, though Newman was certainly aware of US progress after the war. The Royal Society grant enabled him to send a member of the Manchester group, David Rees, to the famous Moore School lectures and, in 1946, Newman himself visited the Moore School and saw ENIAC, later spending a term at Princeton where he discussed computers with von Neumann. On the other hand, Newman professed ignorance of American ideas before that date: after all, he was able to draw on the theories of his student and fellow COLOSSUS worker, Alan Turing. Williams and other members of his team always emphasized that their work was an independent growth and owed nothing to US precedents: as Williams was wont to say, 'it was all in Babbage'.

A provisional patent on the Williams tube was filed on 11 December 1946 and by the end of the following year Kilburn had produced an unpublished report, 'A Storage System for Use with Binary Digital Computing Machines', which summarized their findings. This was duplicated in considerable numbers and taken by Hartree to the USA, where Williams refused permission for it to be copied too widely because it would pre-empt publication in the *Journal of the Institution of Electrical Engineers*. 'Nonetheless,' as Williams stated, 'there is no doubt that the information spread through the United States.' This had two results. Firstly, in 1950, Williams found himself defending his crt patent against American inventors led by Eckert, who refused to acknowledge Williams's work and implied that previous US efforts along those lines were responsible for the regenerative technique. With the help of the National Research Development Corporation (NRDC), which had taken the Manchester work under its wing on behalf of national interests, Williams successfully refuted this claim.[19] Secondly, since for a brief period the Williams tube represented the best available storage device, it was used under licence in some of the first US computers.

In 1949, recalled Williams, he was invited to IBM with his wife and son 'and given the VIP treatment. They went in for c.r.t. storage in a big way and, of course, offered me a lucrative job, which I refused, much to Mr. Watson's surprise.' Here Williams scandalized IBM workers with a jibe at their president's THINK motto, when at a lecture someone asked him:

> Can you explain to me how it comes about that you with one scientific colleague and two technical assistants have managed to build a computing machine, and we with all our resources have not succeeded? I said yes, it's very simple: we pressed on regardless, without stopping to THINK too much.

In July 1949, negotiations took place between Lord Halsbury of the NRDC and IBM, which licensed the use of the Williams tube in early IBM computers. In 1952, IBM introduced the 701 design, the first large scientific calculator made in any quantity, which incorporated 144 Williams tubes. By 1955, IBM had taken over six of Manchester's US patents and this was eventually to earn the NRDC £125 712 in royalties. IBM became involved in research to improve the technical capabilities of the crt store, but magnetic core storage soon made the Manchester invention redundant.[20]

The Williams tube was also used in the machine built at the IAS in Princeton, which was the result of an initiative by John von Neumann in 1946. Finding an adequate storage device had proved a headache, since American ideas had run into technical problems. At this point the Princeton team heard about the work at Manchester University and Julian Bigelow, the project leader, travelled to Manchester in 1948. Here he met Williams – according to Bigelow 'a true example of the British "string and sealing wax" inventive genius'[21] – and became acquainted with the Williams tube. Eventually the IAS machine incorporated a 40-tube Williams memory, arranged in something like a 'V-40' engine configuration. Thus it appears in a famous photograph, with von Neumann standing in front of the computer. The Manchester store was also used in the American ILLIAC, SWAC and SEAC and was initially the preferred memory in the Whirlwind.[22]

Meanwhile at Manchester University, Williams and Kilburn developed their own machine – the Manchester Mark I – which became the basis, at the government's prompting, for a series of collaborative commercial ventures with the local electronics firm of Ferranti Ltd. This company's interest in computers had begun in earnest in the summer of 1948, when it dispatched one of its technical experts, Dr D.G. Prinz, to America to report on computer activities. Prinz toured all the major US computer centres, discussed technical matters with scientists and engineers, collected printed material and assessed the possibility of an alliance with a US firm. Reported Prinz:

A few words may be said about the reaction of the various people to the F.C. Williams tube of which I had taken a description along with me. Patterson and Snyder (EDVAC, Moore School) were greatly impressed; they pointed out that they had worked on similar lines some years ago, but gave it up in favour of the mercury delay line, and they appreciated that Williams had discovered some trick that they themselves had not found. Similarly, Eckert and Mauchly were very much interested. Rajchman thought the Williams tube 'the second-best'. Klemperer (formerly Raytheon, now MIT) was impressed because he knows the difficulties inherent in secondary-emission storage tubes such as those developed by Haeff and himself (the charge tends to spread out on the screen). Only Aiken dismissed the idea.[23]

An oft-repeated story, that on his visit Prinz was told to his surprise that the world's most advanced computer project was on his doorstep, was perhaps apocryphal: probably Prinz learned about Williams's work at an IEE (Institution of Electrical Engineers) lecture. Nevertheless the burden of his report was that Ferranti should utilize the work of Manchester University in building its own computers. The first result of this collaboration was the Ferranti Mark I, a commercial version of the University machine, and this was followed by a number of influential computers.

The Williams–Kilburn partnership (which continued until about 1952, when Kilburn assumed direction of computer projects, as Williams turned to other research) was responsible for a number of pioneering innovations in computer design – to say nothing of the work of those such as Gordon Black, who pioneered the *use* of computers for such tasks as optical ray tracing. Of particular importance was the Manchester invention of index registers (or B-lines, as they were known by the designers), a feature now seen on every modern computer, and the early combination of a small, but fast, random-access store backed by a slower (but larger-capacity) sequential store, which was to produce significant Manchester innovations in the ATLAS computer (considered below). These ideas had their impact on American design, though documenting individual transfers of technology is not always possible. The Anglo-American connection remained close in the 1950s. Kilburn visited America in 1952 and admitted to being 'fairly familiar' with US projects, though his designs followed a very independent line. The US influence on Manchester was perhaps strongest in terms of hardware: by the early 1950s, Williams was requesting samples of the first American transistors to arrive in the UK and, by the end of 1953, Kilburn had built the world's first transistorized computer (an experimental model).[24] American interest in Manchester University and Ferranti computers remained strong and Kilburn remembers regular visits by IBM staff. In the mid-1950s, when he was engaged in work on the Ferranti Mercury computer, 'the fact that we were well on the way with this floating-point [arithmetic] startled IBM at one time when they visited us'.

Manchester innovations also reached the USA via Ferranti's activities in Canada. In September 1952, Ferranti sold a copy of the Mark I (named FERUT) to the University of Toronto, where it helped with design calculations for the St Lawrence Seaway. Lord Bowden (1910–89), Ferranti's first computer salesman, attested to the impact the success of this venture had on IBM. At Ferranti Electric Ltd (later Ferranti–Packard) in Toronto, engineers under Arthur Porter (Hartree's co-worker) were responsible for important pioneering designs in the mid-1950s. They built a naval data acquisition and target tracking system that included the first computer network in which three physically separate processing systems were interconnected through radio channels and operated as a single system. This was copied by the Americans. Ferranti–Packard also built the FP6000, the world's first time-sharing, multi-tasking machine, which in the UK in the mid-1960s was used as the basis for International Computers Ltd's 1900 series.[25]

## THE CAMBRIDGE UNIVERSITY EDSAC (ELECTRONIC DELAY STORAGE AUTOMATIC CALCULATOR)

The EDSAC was perhaps the most influential of the post-war British computer projects. Although Manchester University had run the first stored program, it was the Mathematical Laboratory at Cambridge University that was the first centre in the world to provide a useful and reliable computing service. The EDSAC is particularly interesting, since it was inspired directly by American example.

The EDSAC team was headed by Professor Maurice V. Wilkes, whose experience in mechanical calculating had been provided by the Cambridge differential analyser and the friendship of Douglas Hartree. In May 1946, L.J. Comrie gave him a copy of von Neumann's report on the EDVAC and Wilkes 'recognised this at once as the real thing, and from that time on never had any doubt as to the way computer development would go'.[26] Armed with introductions from Hartree, Wilkes next attended the Moore School meetings and even whilst in Philadelphia began formulating ideas for the EDSAC. The keynote was pragmatism: on his return Wilkes intended to build a usable machine, rather than one which attempted to experiment with the most advanced techniques. For the memory he followed the EDVAC idea for a store using mercury delay lines, even though a working version had not yet been built even in the USA. EDSAC ran its first calculation at the Mathematical Laboratory, Cambridge, in May 1949, almost a year behind the Manchester team, but still well ahead of the Americans. However, the Cambridge machine was more than simply a testbed: by the end of the year it was able to offer a useful computer service.

A significant feature of the Cambridge approach was the attention paid to user convenience and programming. No sooner had EDSAC been completed than work began at once on the development of a programming system and a library of subroutines. Here the expertise of the Cambridge team's programmers, such as (Professor) Stanley Gill and particularly (Professor) David J. Wheeler, became evident. Besides providing a well-stocked library of subroutines, a report on the preparations of programmes and subroutines was also prepared and sent by Wilkes to interested users throughout the world. This was especially useful because, until then, only theoretical studies on programming had been available and these had not been based on practical experience. This report aroused a good deal of interest, particularly in the USA. At about this time, Wilkes received a visit from Zdenek Kopal, then at MIT and later a professor of astronomy at Manchester University, who took the report away with him and brought it to the attention of Addison-Wesley, then a small-scale publishing operation in Cambridge, Massachusetts. In 1951, Addison-Wesley published the report under the authorship of Wilkes, Wheeler and Gill as *The Preparation of Programs for an Electronic Digital Computer: With Special Reference to the EDSAC and the Use of the Library of Subroutines*. The economy and elegance of the EDSAC programming, largely the work of Wheeler, was much in advance of any US or British group and the book (usually referred to as Wilkes, Wheeler and Gill, or simply as WWG) was very influential.

Wheeler's work, which included the use of a symbolic notation for programming, the so-called 'initial orders', subroutines, the subroutine library, and debugging techniques, was recognized in America as the 'leading example of programming virtuosity'.[27] In the USA the programming system developed for the MIT Whirlwind by Charles W. Adams was strongly influenced by the Cambridge group. In 1951–3, Wheeler joined the University of Illinois at Urbana as an assistant professor, at a time when the University was constructing the ILLIAC computer (and its 'twin' the ORDVAC, which was later transferred to Maryland) for the Ballistics Research Laboratories. Wheeler, as the only member of the team with any practical experience of stored-program computers, influenced the programming system of the machine, which was based on the EDSAC. He also influenced aspects of the design, such as extending the address space, though the conservatism of the chief engineer of the project, Ralph Meagher, meant that Wheeler's suggestion that they double the power of the machine by installing index registers was not utilized.[28] Nevertheless the ILLIAC influenced several other computers in the USA, besides those in other countries, such as Israel (the WEIZAC) and Australia (the SILLIAC). After Wheeler's return, the Cambridge connection at the University of Illinois was maintained by Stanley Gill and (Professor) A.S. Douglas. The latter was closely involved in studying the library routines in order to advise on the design of ILLIAC II. Douglas and a US programmer, David Muller, produced a report outlining their conclusions regarding word-length and the number of registers in the mill, as well as their cross-connections. Of particular importance was the idea that all registers on the mill should be interconnected in such a way that numbers could be moved from one to the other directly, via shift paths or through adders under micro-program control – a scheme that was later implemented on the MANIAC III, constructed at the University of Chicago in the late 1950s. Douglas concludes, however, that British influence was strongest on the programming side:

> David, Stan and I taught successive classes of 50 using the techniques we had developed (cf. Wilkes, Wheeler & Gill). My instructors (i.e. assistants) were Gene Golub (now a senior professor at Stanford), Werner Frank (sometime vice-president of a major software house, Informatics Inc.), and my students went all over the US – one turned up in an LP team at IBM some ten years later – whilst we certainly influenced many whom we talked to on our travels.[29]

Wheeler also highlights the importance of this 'grapevine' aspect in the spread of computer technology. It was fostered by the integration of the ILLIAC and the EDSAC into teaching courses and the operation of these computers in an 'open shop' manner. Users were expected to programme their own calculations, assisted where necessary by staff, and this facilitated the flow of information. While users waited their turn at the machine, problems and new procedures could be discussed. Lectures were given to newcomers and the first summer school for training outsiders was held in Cambridge in 1950. Notes Wheeler: 'The use of the computer spread rapidly as successful users infected their friends.'[30]

Naturally these men, and also Wilkes, who remained a regular visitor to the USA at this time and was present when the electrostatic core memory was run on the Whirlwind in August 1953,[31] returned from the USA with fresh ideas, which proved useful in the design of EDSAC II. The ILLIAC II ideas were also directly reflected in the English Electric machine, the KDF9.

Cambridge University, too, developed links with industry. This was in the unlikely form of a close relationship with the catering firm of J. Lyons & Co., whose forward-looking management had become interested in the latest developments in computing as a way to free the firm from the cost and tedium of its enormous accounting operations. The link was spurred by American developments. In May 1947, representatives of Lyons visited Princeton, where they met Herman Goldstine who, besides explaining projects such as the EDVAC and the new IAS machine, told them of Wilkes's computer at Cambridge. Following a communication from Goldstine to Hartree, the latter then wrote to Lyons inviting their representatives to Cambridge. After their visit in the summer of 1947, Lyons agreed to support the Cambridge project and began building the LEO (Lyons Electronic Office). It was the beginning of an involvement in computing which was eventually to see the catering firm, somewhat incongruously, enter the field as computer manufacturers themselves.[32]

## THE NPL PILOT ACE (AUTOMATIC COMPUTING ENGINE)

The projects at Manchester and Cambridge were by far the most successful in the early history of British computing; as such, their relationship with America is relatively easy to assess. This is less true of another post-war effort centred at the National Physical Laboratory at Teddington.

In 1944, a Mathematics Division was established in the NPL, in order to coordinate facilities and techniques related to machine-aided computation. Douglas Hartree and L.J. Comrie were father figures in the foundation of the Division, but, as regards the building of a stored-program computer, the NPL received most of its impetus from the disbanded COLOSSUS team from Bletchley Park, particularly Alan Turing.

The NPL project is of particular interest, since at the end of the war it became the centre of what it was hoped would be a national attempt to build a computer. The superintendent of the Mathematics Division, J.R. Womersley, began canvassing support and recruiting staff in 1945 and was responsible for Turing's appointment. Events in the USA had undoubtedly inspired Womersley's actions. Early in 1945, he had visited America and had been one of the first foreigners to see the ENIAC and be given a copy of the EDVAC report. The threat of America moving ahead in the computer race provided the impetus for his plans, though it should be noted that a strong, independent line was already evident. Womersley had read Turing's paper 'On Computable Numbers' before the war and was well acquainted with the idea of 'Turing machines'. Hence his desire to have Turing join the NPL.

The fragmentation of British technical personnel and resources after the war, however, meant that Womersley's plans for a national machine were never realized, but an NPL computer did emerge, based on a draft report submitted by Turing to the executive committee of the NPL in 1946. Turing's report detailed the design and operation of an electronic universal machine, christened the ACE (Automatic Computing Engine). From this blueprint emerged the Pilot ACE, so named because the prototype became the completed version of the machine. In recent years this draft report has become the subject of renewed interest, mainly because of Turing's far-sighted recognition of the possibilities of the stored-program computer. Among the topics discussed in Turing's report, but not found in that of von Neumann, are address mapping, instruction address register and instruction register, microcode, hierarchical architecture, floating point arithmetic, hardware bootstrap loader, subroutine stack, modular programming, subroutine library, link editor, symbolic addresses and the ability to treat programs as data. Turing's vision was also apparent at a London Mathematical Society lecture he gave in February 1947. Before the first stored-program computer had even been demonstrated, he predicted the end of the human computer, the use of terminals connected over telephone lines, analysts, programmers, operators, on-line curve followers and – perceptively – the protective mystique and gibberish affected by systems programmers.[33]

The relationship of Turing's ideas to US developments is obscure. On the one hand, Turing made no great claims to the originality of his ideas, recommending that his proposal should be read 'in conjunction' with von Neumann's EDVAC report. The hardware for the memory of the ACE, the mercury delay line, was also an American idea. Indeed, at Hartree's suggestion, a member of the ENIAC team, Harry D. Huskey, had joined the ACE project to introduce expertise on the hardware side. On the other hand, a biography of Alan Turing has argued persuasively that his ideas were original and owed most to his idea of a universal machine outlined in his paper 'On Computable Numbers'.[34] Significantly Turing resisted Womersley's suggestion that he cooperate with Wilkes on an EDVAC-style machine, scathingly arguing that Wilkes's plans were 'in the American tradition of solving one's difficulties by means of much equipment rather than by thought'. When Turing himself visited the USA at the end of 1946 to attend a Symposium on Large Scale Digital Calculating Machinery held at Harvard and also met the ENIAC and von Neumann groups, he later commented that he had not 'brought any very new technical information to light'. He thought the numerous American projects were dissipating their energies over too wide an area and concluded: 'We ought to be able to do much better if we concentrate all our effort on the one machine, thereby providing a greater drive than they can afford on any single one.' Jim Wilkinson (1919–86) and Donald Davies, members of the ACE team who brought the project to a successful conclusion after Turing prematurely left the NPL (he appears to have become disillusioned with bureaucratic indecision there), also emphasized the independence of the NPL work.[35] Wilkinson highlighted the fact that, although they had the EDVAC report to hand, they made little use of it in the design for the ACE, which had an idiosyncratic design with user convenience, never one of Turing's priorities, sacrificed for speed. Moreover he believed that Huskey, with whom Turing violently disagreed, had brought little to the project, apart from basic circuitry.

Whatever the assessment of American influence on the ACE may be, there is little doubt that Turing's work represented a considerable achievement: when completed the 800-tube Pilot ACE had a computational speed several times better than the 3000-tube EDSAC. Hartree took a copy of the ACE report to the ENIAC group in 1946 and Turing's work was also known at Princeton, which he himself visited in January 1947. But the ACE design, as was often the case in England, was too far outside the mainstream of computer development to have any widespread impact, whatever admiration Turing's ideas may arouse today. Nevertheless it is worth recording that, when Huskey returned to the USA to work on the SWAC (taking with him, as it happened, the technology of the Williams tube, which he recommended for use in that machine), he utilized the ideas of the ACE. During a sabbatical year at Wayne State University in 1953, he designed a computer for the Bendix Corporation that was strongly influenced by Turing's ideas. This computer, the Bendix G-15, was produced successfully – over 400 were eventually sold – and its speed made it a favourite for certain classes of engineering design.[36]

## A.D. BOOTH AND THE ARC (AUTOMATIC RELAY COMPUTER)

In the early days of stored-program computing there was still room for the old-style 'tinkerer', working with very limited resources. Such an individual was Dr Andrew D. Booth, who had first become involved in automatic calculators during the Second World War, while working on the determination of crystal structures using X-ray diffraction data. The computations involved were extremely tedious and there was ample incentive for automating the process. Booth was employed as a mathematical physicist in the X-ray team at the British Rubber Producers' Association (BRPRA), Welwyn Garden City, Hertfordshire, from August 1943 to September 1945. Subsequently he moved to Birkbeck College, London University, though he was still retained as a consultant by BRPRA. This consultancy later proved fortuitous in respect of workshop facilities for his Automatic Relay Computer (ARC), which he designed during 1947–9.

Booth's interest in universal automatic digital computers was fostered by contact with the indefatigable Douglas Hartree, who looked him up after reading some of Booth's mathematical papers. Booth was already working on a digital calculator, but Hartree gave him his first knowledge about what was going on in the wider world of computing, especially in the USA. Booth was to be heavily influenced by the logical design of American machines. In the summer of 1945, he crossed the Atlantic, funded by the Rockefeller Foundation, London University and the proceeds of a lecture tour on which Booth was after-dinner speaker to 'earnest old ladies'. He visited a number of groups, including that of von Neumann and Goldstine at Princeton, J.W. Forrester at MIT, Howard Aiken at Harvard, Eckert and Mauchly in Philadelphia and Morris Rubinoff at the Moore School. After this visit, which had enabled Booth to evaluate the respective merits of the various projects, the Rockefeller Foundation offered him a fellowship to work at an American institution of his own choice. Booth had by now read the EDVAC report, courtesy of Hartree, and he had no hesitation in choosing the IAS at Princeton, 'not because I had seen anything of the hardware of their project, which I had not, but because the rigorous knowledge and precise thought of von Neumann attracted me: it had none of the airy imprecision of other workers'.[37] This visit to the IAS, during March to September 1947, firmly launched Booth on the design of a stored-program computer.

At Princeton, besides exploring the joys of von Neumann's approach, Booth was also able to refine some of his own earlier ideas on electronic storage, which he regarded as 'the whole key' to computing technology. He believed that magnetic processes were the only ones that had any potential for large-scale, long-term storage and while in America sketched out, but took no steps to construct, various magnetic storage devices.

On his return to the UK, Booth and his assistant, Miss Kathleen Britten (later Mrs K.V.H. Booth), began to build a machine. The project was tiny: apart from his future wife, Booth never had more than one engineer and he was dependent for facilities on the BRPRA, since no laboratories were available at Birkbeck College. Booth's first experiments concerned a 'floppy-disc'-type memory, using oxide-coated paper discs from the US firm which manufactured 'Mail-a-Voice' recording machines. Though the principle was very similar to that used for modern disc drives, Booth's device proved unstable and he had no resources to overcome the technical problems. So, with the help of his father, who was a fine mechanical engineer, he returned to his original idea, the magnetic drum. This was more successful and by 1948 he had installed a working magnetic drum in the experimental ARC. Booth and his father continued to experiment and by 1952 they were producing large drum memories for others. Some of these drums were exported to the USA, where they performed reliably for as long as ten years. Closer to home, Booth's attempt to design low-cost computers that would be useful to smaller scientific organizations attracted the interest of BTM. Eventually, the firm used Booth's designs to build the HEC (Hollerith Electronic Computer), which was later marketed as the BTM 1200 series.

By 1962, Booth's work in England was complete: in that year he took his expertise to Canada, where he continued his computing career.

## THE ANGLO-AMERICAN COMPUTER RACE

Within a decade of the pioneering projects outlined so far, the frame of reference had changed drastically. By the end of the 1950s, the race for computing supremacy was no longer being run in technical terms in university laboratories, but in the market-place where British and American electronics firms competed for the rapidly growing world business in computers.

In the late 1940s and early 1950s, Britain rivalled American firms in innovation. The first operational stored-program computer, the first computer service, the first transistorized computer and the first commercially available computer – these were a string of remarkable technical achievements, especially considering the acute shortages of manpower and resources in post-war Britain. The way in which several companies – Ferranti, Elliott Bros, Lyons, English Electric, GEC and Marconi – soon became involved in production shows also that the commercial potential of the computer was not overlooked. However, the vulnerability of these firms to the large US corporations was soon apparent. By 1964, it was estimated that there were almost 22 000 computers installed in the USA, including approximately 1 767 in civil government, 2 000 in the Department of Defense, and another 2 000 or so used by government contractors at the state's expense. In contrast, less than 1 000 computers had been installed in the UK by the same date, of which only 56 were in civil government departments.[38] Thus the American companies set the agenda for the commercial development of the computer in the 1960s. Yet British scientists and firms continued to make important contributions to computer design in this period and Anglo-American transfer of computing technology therefore continued unabated.

American competition operated across the board, but the threat was perceived initially in terms of large computers for scientific use. By the mid-1950s, it had become clear that Britain was rapidly falling behind in this field and that a concerted effort would be needed if the country was to stay in the computer race. This fear was inspired by an awareness of several major computer projects to build high-speed computers, such as the IBM STRETCH. In the summer of 1956 a team of British experts – A.S. Douglas from Cambridge, D.W. Davies and J.H. Wilkinson from the NPL and Jack Howlett from Harwell – had visited the USA and had confirmed America's rapid progress for themselves. It was felt that Britain needed its own fast computer project, responsibility for which was assumed by the NRDC, which had become closely identified with the progress of the British computer industry. But NRDC attempts to nurture such a project amongst computer firms and government establishments had only limited success and resulted in costly delays and only a half-hearted commitment to a cooperative project. Eventually a close approximation to a national computer emerged in the form of the

Ferranti–Manchester University ATLAS, under the direction of Tom Kilburn. This machine was to be a commercial failure and only three ATLAS computers were ever installed after deliveries began in 1963. Whether this was because of the greater resources of the Americans or of bureaucratic indecisiveness remains debatable.[39] What is certain is that ATLAS was a technical achievement of great importance. A number of innovations in its architecture were to be very influential in later computer developments around the world. Amongst these were: multiprogramming, job scheduling, spooling, extracodes, interrupts, pipelining, interleaved storage, autonomous transfer units, virtual storage and paging. Virtual memory and the ATLAS operating system – by which blocks of information could be swapped between different stores to increase the speed and capacity of the machine – were especially influential. ATLAS proved that it was possible to build a multiprogramming machine with a paged memory and sophisticated operating system (there was even a pioneering provision for time-sharing, which had to be discarded owing to budget limitations). These ideas eventually featured in US machines.

Meanwhile the early 1960s also saw a 'technology gap' and a rapid escalation of US competition in the commercial sector. In the 1950s, American competition in the European market was almost non-existent, but by 1967 the leading US computer manufacturers such as IBM, NCR, UNIVAC, Burroughs and Honeywell had nearly a 70 per cent share of the British market. The dominant company was IBM, which had 40 per cent of the market in that year. In the electronic data-processing market British firms had fallen well behind. By 1960, virtually all US computers had second-generation transistorized processors; their software and programming languages (such as FORTRAN and COBOL) were in advance of Europe; and their magnetic tape and disc storage technology were vastly superior. Not surprisingly, British firms, such as ICT (formed in 1959 as a result of a merger of BTM and Powers–Samas), had to go to America for their technology. In 1961, ICT signed an agreement with RCA, which gave the British firm a licence to use RCA computer technology. Another arrangement allowed ICT to import the RCA model 301 computers, to be resold as the ICT 1500. (English Electric was another firm which had a long-standing technology-sharing agreement with RCA: in 1961, it introduced a version of the RCA 501 which was known as the KDP10 and also adopted the RCA Spectra 70 architecture for its System 4 in the late 1960s.) In the following year, ICT made a similar deal with UNIVAC to import and resell their model 1004 calculating tabulator.

American dominance was underlined in 1964, when IBM launched its highly successful System/360 range of computers. This event, which is regarded as a watershed in the history of commercial computing, saw the introduction of the first compatible family of third-generation computers. The range encompassed six distinct processors and 40 peripherals, which were intended to replace all of IBM's current mid-range computers. The concept of compatibility and the several-fold increase in price/performance offered by the System/360 range sent shock waves through the industry. In IBM the brilliance and success of the range took on a symbolic significance. Yet, even here, transatlantic influences were at work. An important feature of System/360 computers was the use of control stores. This was a key hardware feature that allowed computers with very different hardware implementations to appear identical to the user, except for different costs and speeds. Such stores were based on the idea of 'microprogramming', first presented in 1951 by Maurice Wilkes at Cambridge. An important part of the development work for the control store was directed by John W. Fairclough, a graduate of Manchester University who had worked on STRETCH at IBM's Hursley Laboratory in the UK. This was the first time an important aspect of machine development had been undertaken by one of IBM's overseas subsidiaries. Thus 'the existence of control-store technology and system design experience in the Hursley Laboratory facilitated an important achievement in IBM System/360: first use of read-only control stores in a series of commercial computers'.[40] Several serious shortcomings in the System/360, particularly a weakness in virtual memory address translation, later resulted in the model 67. Although this model was not entirely compatible with the /360 line, it did introduce an ATLAS-like paging system with the hardware address translation and other facilities that were necessary for efficient operation of a machine via time-shared terminals.

Almost immediately after the IBM announcement of the System/360, the other large computer companies followed with their own 'families' of compatible machines. The two larger British firms responded with the ICT 1900 series and the English Electric–Leo–Marconi System 4. The story of these computers, and that of the eventual merger of the British industry into ICL in 1968, has been recently documented.[41]

## CONCLUSION

Herman Goldstine remarked that, in 1945, 'a number of British visitors came to the Moore School, and from these visits stemmed the computerization of Great Britain'.[42] Although this statement contains an element of the truth, the discussion above has shown that it is also a gross simplification. In fact it might be more accurate to argue that the development of the computer in the UK (and in the USA) would have proceeded independently, whatever was occurring across the Atlantic. As it happened, US events did exert a powerful influence on British computerization; but then so, too, did UK technology on its US counterpart. A more complete picture would aim to show that the development of the computer in these two countries was a product of a complex interaction. This has been attempted here, though it should be emphasized that it is an impressionistic picture. The international diffusion of computer technology is a huge field and more research needs to be done to delineate fully the web of Anglo-American links.[43] Nevertheless, a number of significant trends and conclusions have emerged.

Perhaps the overriding impression is of the speed of the Atlantic transfer of computing technology. This was due to a number of factors. Firstly, a common military cause focused both British and American minds on crucial areas of electronics technology, providing the seedbed for the development of the stored-program computer. This in itself spurred transatlantic exchanges of technology. British and American electronics engineers not only spoke a common everyday language they also spoke a common scientific language – one that shared an involvement with radar work, valve and crt storage technology, and weapons development. The military, both during and after the Second World War, also underwrote all the major projects. To this extent, the transfer of computer technology was not, as it was for nineteenth-century technologies, a product of 'free enterprise'.

Secondly, information could be passed relatively swiftly and freely between interested parties, especially university and research institutions. Although many computer projects were funded by the military, once the war was over there were surprisingly few restrictions on the dissemination of technical information. It is interesting to note, for example, the ease with which Dr D.G. Prinz on his post-war tour of American computer establishments was granted unrestricted access virtually everywhere, despite his German nationality. The latest ideas and news of projects could be transferred by articles in technical journals, newsletters (notably those of the Office of Naval Research and the National Bureau of Standards), patents, lecture courses (especially at the Moore School), the activities of overseas subsidiaries and, above all, by personal contacts. Often the lag between British and American developments was no more than the time it took for a technical paper to be posted across the Atlantic or for a scientist to arrive by sea (or, towards the end of this period, by air). Gradually, as computer science developed, the opportunity for contacts increased (in fact, an interesting feature of the growth of information technologies is that their very success facilitates the further spread of the technology). Soon major international scientific meetings on computing technology were a regular event. One of many such meetings was the Joint AIEE–IRE Computer Conference at Philadelphia in December 1951, where Maurice Wilkes and Fred Williams reviewed progress in their respective projects.

Thirdly, as in the nineteenth century, it was the individual technologist who was paramount in the spread of transatlantic information in the early stages. The computing fraternity was a very small one in the late 1940s: in England at that time the leading scientists could all meet comfortably in one room, such as at the Cambridge Mathematical Laboratory's regular colloquia. Maintaining contacts was relatively easy, especially before the mid-1950s, so that in reviewing this period one is impressed by the closeness of the transatlantic scientific community. As Sandy Douglas has remarked:

> It would be wrong to underestimate the importance of personal contacts at that time, since we were not a large group world-wide and many ideas 'floated' from place to place with whoever moved around. Ideas from EDSAC I, Manchester Mark I and ACE were a 'common heritage' in the UK. Whilst I was in the US, I visited the Whirlwind at MIT, SEAC at the National Bureau of Standards, SWAC in UCLA, ORACLE at Oak Ridge, JOHNNIAC in Princeton, the MANIAC II in Los Alamos and so on, where we exchanged ideas freely – and I know David Wheeler and Stan Gill did the same. Monty Phister, who was at Hughes and later became chief engineer of Max Palevski's group at Xerox, was an old Cambridge alumnus and had several ex-Ferranti people working for him, like Ted Braunholtz. Another alumnus was Tony Oettinger from Harvard and ideas even went via Australia, where John Bennett, also ex-Cambridge and Manchester, built SILLIAC (a copy of ILLIAC I), so that his engineer, Barry Swire, was at Illinois when I was, picking up technology. Whilst no one intentionally plagiarised, we tended to treat knowledge as internationally available and not necessarily attributable – sometimes one didn't know who had a particular idea first anyway. Also, of course, we were far from being the only transatlantic travellers. Douglas Hartree, Maurice Wilkes and, after 1954, Jim Wilkinson visited frequently, as did Freddie Williams. Only Tom Kilburn seemed to expect everyone to come to him![44]

A truly comprehensive study would examine more fully the transfer of personnel, which has continued to the present day. It was particularly intense during the 'brain-drain' era of the 1960s, when the dominance of American firms in both semi-conductor and computer technology had reached new heights, and when many British graduates and computer engineers were attracted to the USA by better pay and job prospects. There they seem to have revelled in the more open managerial and scientific ambience of American firms, occasionally returning to the UK with fresh ideas.[45]

The speed with which knowledge on computing is disseminated, the openness of the Western research community, the fact that individual engineers hold much of the most important technological information, and that patents have little protection for a firm's innovations (often patent cases are economically moot by the time they are settled) – all these ensure both the necessity of continuous heavy investment in new technology and the inevitability of the internationalization of the industry.[46] This study has highlighted

114          *Twentieth-century Transfers Originating in the West*

aspects of the international nature of computing technology – in particular, the continuing importance of the Anglo-American connection in the twentieth century. It shows that the development of a technology can rarely be considered in isolation, by viewing that technology solely within a national boundary. If this is true of the development of the computer, which in the popular mind is strongly associated with a single country (that is, the USA), then how many other industries might repay closer attention from an Anglo-American perspective?

## NOTES

This chapter draws on a wide range of secondary literature. In addition, I have been able to use the documentary material I have been collecting and cataloguing on behalf of the National Archive for the History of Computing, Manchester University. During the course of this work I have also benefited from discussions with many key individuals, whose comments have informed this study. I am particularly grateful to: Professor Gordon Black, the late Lord Bowden, Professor Tony Brooker, Professor Sandy Douglas, Professor Dai Edwards, Professor W.S. Elliott, Professor Tom Kilburn, Dr D.G. Prinz, Professor Bernard Richards, Professor David Wheeler and Professor Maurice Wilkes. Dr Martin Campbell-Kelly kindly read my initial draft and made several useful suggestions.

1.  For general background, see Stan Augarten, *Bit by Bit: An Illustrated History of Computers* (New York, 1984); Michael R. Williams, *A History of Computing Technology* (Englewood Cliffs, NJ, 1985). *Annals of the History of Computing* (Arlington, Va., 1979) also contains much that is relevant to Anglo-American technology transfer. Simon H. Lavington, *Early British Computers* (Manchester, 1980), provides the essential story of UK developments, with an expert appreciation of the technology. G. Tweedale, *Calculating Machines and Computers* (Aylesbury, 1990) is a brief, illustrated account.
2.  For details of these case-studies, see my annotations in Lewis Hanke (ed.), *Guide to the Study of United States History outside the US, 1945–80*, 5 vols. (New York, 1985); and my *Sheffield Steel and America: A Century of Commercial and Technological Interdependence, 1830–1930* (Cambridge, 1987).
3.  I. Bernard Cohen, 'Babbage and Aiken', *Annals of the History of Computing*, **10**, (1988), pp. 171–93. The American philosopher Charles Sanders Peirce was also aware of Babbage's work and saw that his concept could be realized electromechanically, with relays. But Peirce's ideas had no impact on computing technology. See Alice and Arthur Burks, *The First Electronic Computer: The Atanasoff Story* (Ann Arbor, Mich., 1988), p. 260 and *passim*.
4.  Geoffrey D. Austrian, *Herman Hollerith: Forgotten Giant of Information Processing* (New York, 1982), pp. 16–17; Anthony Hyman, *Charles Babbage: Pioneer of the Computer* (Oxford, 1984), pp. 254–5.
5.  C.A. Everard Greene, *The Beginnings – Reminiscences* (1958), p. 3. For the subsequent account, I have relied on Martin Campbell-Kelly, *ICL: A Business and Technical History* (Oxford, 1990).
6.  Williams, *Computing Technology*, p. 254. See also Mary Croarken, *Early Scientific Computing in Britain* (Oxford, 1990).
7.  V. Bush, *Pieces of the Action* (New York, 1970). However Kelvin's work was common currency amongst engineers at this time. See Larry Owens, 'Vannevar Bush and the Differential Analyzer: The Text and Context of an Early Computer', *Technology and Culture*, **27**, (1986), pp. 63–95; and 'Straight Thinking: Vannevar Bush and the Culture of American Engineering', Princeton University PhD thesis, 1987. For useful surveys

on the pre-digital era, see Paul Ceruzzi, *The Prehistory of the Digital Computer, from Relays to the Stored-Program Concept, 1935– 45* (Westport, Conn., 1983); and James Small, 'Analogue Computers: Technical Change and Designer History', Manchester University MSc, 1988. Small is presently working on the definitive study of analogue computers, a neglected area in the literature.

8.   Douglas R. Hartree and Arthur Porter, 'The Construction and Operation of a Model Differential Analyser', *Memoirs of the Manchester Literary and Philosophical Society*, **79**, (1934–5), pp. 51–73.

9.   Arthur Porter, recorded by Dr Christopher Evans for the 'Pioneers of Computing' (Science Museum, London, 1976), tape no. 20.

10.  T.H. Flowers quoted in Brian Randell, 'The COLOSSUS', in Nick Metropolis, Jack Howlett and Gian-Carlo Rota (eds), *A History of Computing in the Twentieth Century* (New York, 1980), p. 87. Another member of the team, W. Gordon Welchman, later went to work in the USA, joining the Whirlwind project at MIT and then various US and British companies.

11.  According to S. Frankel, who worked on the atomic bomb at Los Alamos and was one of the first to use the ENIAC, von Neumann 'firmly emphasised to me, and to others I am sure, that the fundamental conception [of the computer] is owing to Turing – insofar as not anticipated by Babbage, Lovelace and others'. See Randell, 'COLOSSUS', p. 79.

12.  See Lord Halsbury, 'Ten Years of Computer Development', *Computer Journal*, **1**, (1959), p. 154, who refers to the cross-fertilization of Turing's and von Neumann's ideas when they met during the war.

13.  H.H. Goldstine, *The Computer from Pascal to von Neumann* (Princeton, NJ, 1972), p. 246. Sir Charles G. Darwin wrote in Hartree's obituary in *Biographical Memoirs of Fellows of the Royal Society*, **4**, (1958), p. 109, 'I do not think it would be an exaggeration to say that it was he who taught [the Americans] the way in which advantage could be taken of [the ENIAC's] extreme rapidity of action.'

14.  D.R. Hartree, 'The ENIAC, an Electronic Computing Machine', *Nature*, **158**, (1946), pp. 500–6; also *Calculating Machines: Recent and Prospective Developments and Their Impact on Mathematical Physics* (Cambridge, 1947); *Calculating Instruments and Machines* (Urbana, Ill., 1949, and Cambridge, 1950): these two books have been reprinted, with an introduction by M.V. Wilkes (Cambridge, Mass., 1984).

15.  Nancy Stern, *From ENIAC to UNIVAC: An Appraisal of the Eckert-Mauchly Computers* (Bedford, Mass., 1981).

16.  The Report is partly reprinted in Brian Randell (ed.), *The Origins of Digital Computers* (Berlin and New York, 1975), pp. 355–64. See also W. Aspray, *John von Neumann and the Origins of Modern Computing* (Cambridge, Mass., 1990).

17.  Martin Campbell-Kelly and Michael R. Williams (eds), *The Moore School Lectures* (Cambridge, Mass., 1985).

18.  For this account I have relied upon Simon H. Lavington, *A History of Manchester Computers* (Manchester, 1975); the Williams and Kilburn 'Pioneers of Computing' tapes (n. 9), nos 5 and 7; and the papers of Manchester University Department of Computer Science, deposited in the National Archive for the History of Computing (NAHC/MUC/ Series 1–3). Particularly useful are documents relating to the Williams tube interference suit, 1946–52 (NAHC/MUC/Series 1. C. 1a).

19.  Eckert, however, continued to imply that his crt store (which he called the 'iconoscope') was responsible for the electrostatic storage tube. In 1976 he wrote: 'I worked on a storage-tube device at the University of Pennsylvania modeled on the ideas of the iconoscope. I showed this work to F.C. Williams, who came over from England when the Moore School Lectures on Computer Design were given in 1946. Williams went back to Manchester and applied for patents on iconoscope ideas, first in England and then in the US.' See *Metrpolitan History of Computing*, p. 534.

20.  Charles J. Bashe *et al.*, *IBM's Early Computers* (Cambridge, Mass., 1986); Bryon E. Phelps, 'Early Electronic Computer Developments at IBM', *Annals of the History of Computing*, **2**, (1980), pp. 253–67; Emerson W. Pugh, *Memories that Shaped an Industry: Decisions Leading to IBM System/360* (Cambridge, Mass., 1984).

21.  Julian Bigelow, 'Computer Development at the Institute for Advanced Study', in *Metropolitan History of Computing*, p. 304.

22.  The Whirlwind was operational by 1953 at MIT and pioneered the magnetic core store of Jay W. Forrester, which was the next major advance in computer memories. Though the machine was planned with crt storage, Forrester later wrote that it 'inherently lacked the high signal levels, the high signal-to-noise ratio, the ability to give good signals from the noise, that we would require for our high-reliability application ... [so] ... we did not stay with the Williams tube idea for very long'. See Kent C. Redmond and Thomas M. Smith, *Project Whirlwind: The History of a Pioneer Computer* (Bedford, Mass., 1980), p. 181.

23.  Papers of Dr D.G. Prinz. NAHC/PRI/C.1a. Report of a visit made by Prinz to US, September 1948, p. 3.

24.  On 24 January 1952, F.C. Williams wrote to E. Cooke-Yarborough at the Atomic Energy Research Establishment, Harwell: 'I saw Sir John Cockcroft the other day in London and he said that you were the proud possessor of about a dozen American-made transistors, and that you had high hopes of more to come. I probed him on the possibility of your sparing us a representative sample from relatively speaking your abundant wealth.' NAHC/MUC/Series 1. B.1d.

25.  Author's interview with Lord Bowden, 1 November 1988. See also Beverley J. Bleakley and Jean LaPrairie, *Entering the Computer Age: The Computer Industry in Canada: The First Thirty Years* (Agincourt, 1982), pp. 10–12, 51–2; NAHC/FER/C.32. Records of Ferranti–Packard Electric Ltd, c. 1962–8.

26.  M.V. Wilkes, *Memoirs of a Computer Pioneer* (Cambridge, Mass., 1985), p. 109.

27.  Charles W. Adams, quoted in Wilkes, Wheeler and Gill, *The Preparation of Programs etc.* (Cambridge, Mass., 1982; reprint of 1951 edition). Introduction by Martin Campbell-Kelly, p. xviii.

28.  Author's interview with David Wheeler, 28 February 1989. See also James E. Robertson, 'The ORDVAC and the ILLIAC', in *Metropolitan History of Computing*, pp. 347-64.

29.  Letter to author, 12 October 1988.

30.  D.J. Wheeler, 'Programmed Computing at the Universities of Cambridge and Illinois in the Early Fifties', in Stephen G. Nash (ed.), *A History of Scientific Computing* (Reading, Mass., 1990), pp. 269–79.

31.  Wilkes, *Memoirs*, admits to having had a 'love-affair' with the USA ever since he met Americans in Germany during the Second World War. In 1980, he became senior consulting engineer at Digital Equipment Corporation in Massachusetts and an Adjunct Professor at MIT.

32.  John Hendry, 'The Teashop Computer Manufacturer: J. Lyons, LEO and the Potential and Limits of High-Tech Diversification', *Business History*, **29**, (1987), pp. 73–101. John Pinkerton, who helped develop the LEO, remembers a visit by an IBM vice-president: 'The gossip that got around was that they thought, well, if a tea company in Britain can do it, well surely we ought to be able to do it. The impression was that it was rather disgraceful for a company which was already the dominant company in office machinery at that time, not to be able to supply what people were going to ask for.' Quoted in Peter Pagnamenta and Richard Overy, *All Our Working Lives* (London, 1984), pp. 254–5.

33.  B.E. Carpenter and R.W. Doran (eds), *A.M. Turing's ACE Report of 1946* (Cambridge, Mass., 1985).

34.  Andrew Hodges, *Alan Turing: The Enigma of Intelligence* (London, 1983). Uncited quotations in this paragraph are from this source. See also the comments by Mike Woodger, 'The Foundations of Computer Engineering', *Radio and Electronic Engineer*, **45**, (1975), pp. 598–602.

35.  Davies and Wilkinson, 'Pioneers of Computing', nos 1 and 10. See also 'The Birth of a Computer: An Interview with James H. Wilkinson on the Building of a Computer Designed by Alan Turing', *Byte*, (February 1985), pp. 177–94.

36.  Martin Campbell-Kelly, 'Programming the Pilot ACE: Early Programming Activity at the National Physical Laboratory', *Annals of the History of Computing*, **3**, (1981), pp.

133–62; Huskey, 'Pioneers of Computing', no. 13; 'The SWAC: The National Bureau of Standards Western Automatic Computer', in Metropolis *et al.*, *A History of Computing*, pp. 419–31.

37.  Metropolis *et al.*, *A History of Computing*, p. 553. Other material on Booth is derived from 'Pioneers of Computing', no. 9, and reports and working drafts in the Papers of Andrew D. Booth, NAHC/BOO.

38.  Barry White, 'State Intervention in Technology in the Post-War Years: Case Studies in Technology Policy', Aston PhD thesis, 1985, p. 124.

39.  John Hendry, 'Prolonged Negotiations: The British Fast Computer Project and the Early History of the British Computer Industry', *Business History*, **26**, (1984), pp. 280–306; *Innovating for Failure: Government Policy and the Early British Computer Industry* (Cambridge, Mass., 1989). See also Paul Drath, 'The Relationship between Science and Technology and the Computer Industry, 1945–1962', Manchester PhD thesis, 1973. An unpublished history of Ferranti computers provides an interesting comment on the relative size of UK and US computer projects: 'One of the remarkable features of the ATLAS project was the small number of staff employed compared with the large numbers used by IBM on STRETCH and by the Bull Company on the Gamma 60. We have seen that, probably, the former had some 300 graduates and the latter about 200 programmers. ATLAS never had more than ten programmers on the supervisor and about 15 working on compilers. The total number of engineers on one computer must have been of the same order.' See Bernard Swann, 'The Ferranti Computer Department' (1975). Copy in NAHC/FER/C.30.

40.  Pugh, *IBM System/360*, p. 202. See also Emerson W. Pugh *et al.*, *IBM's 360 and Early 370 Systems* (Cambridge, Mass., 1991). IBM (UK) Ltd was founded in 1951 and has since become the largest British computer manufacturer.

41.  Campbell-Kelly, *A Business and Technical History*.

42.  Goldstine, *From Pascal to von Neumann*, p. 217.

43.  William Aspray, 'International Diffusion of Computer Technology, 1945-1955', *Annals of the History of Computing*, **8**, (1986), pp. 351–60.

44.  Douglas (n. 29).

45.  Pagnamenta and Overy, *Working Lives*, pp. 256–7. See generally P. Stoneman, *Technological Diffusion and the Computer Revolution: The UK Experience* (Cambridge, 1976).

46.  Kenneth Flamm, *Creating the Computer: Government, Industry and High Technology* (Washington, DC, 1988), pp. 203–34, has a particularly useful chapter on technology transfer and international competition. Rapidity of technological transfer also typified the semi-conductor industry. See John E. Tilson, *International Diffusion of Technology: The Case of Semiconductors* (Washington, DC, 1971).

# B
## The Role of the State

# [14]

By David J. Jeremy

CECIL JONES HIGH SCHOOL

# Damming the Flood: British Government Efforts to Check the Outflow of Technicians and Machinery, 1780–1843*

*In the waning decades of the age of mercantilism, Great Britain intensified her efforts to keep at home the new industrial technology—technicians and machines—that the inventiveness of her people had produced. From his researches in the records of the Board of Trade, which played a major role in coordinating these efforts, Mr. Jeremy shows that as time wore on the policy of prohibiting emigration and exports became more and more internally contradictory and incapable of enforcement despite great ingenuity on the part of those responsible.*

Once British entrepreneurs had demonstrated the superiority of machinery in the manufacture of textiles, in the decades following the inventions of Hargreaves, Arkwright, Crompton, and Cartwright, traditional efforts to contain British technology within the kingdom were intensified. Checks against the outflow of Britain's early industrial technology were applied both by private businessmen and the government. The attitudes of suspicion and secrecy displayed by numerous early–nineteenth–century manufacturing firms are well known. Even more familiar are numerous aspects of the protectionist system

*Business History Review*, Vol. LI, No. 1 (Spring, 1977). Copyright © The President and Fellows of Harvard College.
* I am grateful to the Pasold Research Fund for a grant to work on Board of Trade records at the Public Record Office in summer 1974. I am also indebted to Dr. Charlotte Erickson, Reader in Economic History at the London School of Economics, to Miss Julia de L. Mann, and to my wife Theresa for their keen criticisms of earlier and longer drafts of this article. This research is part of a doctoral thesis being written under Dr. Erickson's supervision, on the transmission of cotton and woolen manufacturing technologies from Britain to the United States, 1790–1840.
The cover illustration, entitled "Wharf Shed of the Trafalgar Dock, Liverpool," shows one of the scenes where Customs officers attempted to prevent the illegal export of machinery before 1843 — Trafalgar Dock being opened in 1836. This print, drawn by Hair and engraved by Jackson, was published in Charles Knight (publisher), *The Land We Live In: a Pictorial and Literary Sketch Book of the British Empire* (4 vols., London, ca. 1847–1850). The author is indebted to H.M. Customs & Excise, in the person of Mr. P.J.T. Machin, Deputy Librarian, for providing this print for reproduction, and to Mr. J. Smith of the Liverpool Record Office for identifying its source.

constructed by seventeenth– and eighteenth–century governments. But relatively little has been written on the working of British legislation aimed at the retention of technology.[1] With what consistency and effectiveness were the prohibitive laws applied? When modification came in 1824–1825, what sort of logic freed artisan emigration but continued governmental control over machinery exports? What policies did the Board of Trade follow in licensing machinery for export in the period 1825–1843? How effective were the continuing prohibitions? These are some of the questions that surround the importance of governmental checks on the international movement of technology, especially textile technology, in this critical period.

## I. THE PERIOD OF MAXIMUM LEGISLATIVE PROHIBITION, 1780s–1824

In the early 1780s no skilled artisan or manufacturer was legally free to leave Britain or Ireland and enter any foreign country outside the Crown's dominions for the purpose of carrying on his trade. Textile printing workers were even forbidden to leave the British Isles, the implication being that other workers could at least travel within British possessions. It was an offence, moreover, to entice artificers or manufacturers to emigrate to foreign parts. It became illegal to export or to prepare to export to any place outside Britain and Ireland any pre-industrial or industrial textile, metal-working, clock-making, leather-working, paper-making or glass manufacturing equipment. But whereas the laws relating to textile implements and machinery were thoroughly comprehensive, encompassing existing equipment and any invented in the future, those dealing with metal-working tools and utensils were not exhaustive. As we would expect, the penalties for removing the living instruments of trade, the artisans, were much higher than for exporting the machines they were capable of reproducing with their own hands. Emigrants lost their nationality and property and recruiting agents were fined £500 per migrant enticed and received twelve months' imprisonment. In contrast, a £200 fine, forfeiture of equipment and twelve months' imprisonment (or a £500 fine and forfeiture in the case of textile printing tools and machinery) was laid down for the export or attempted export of the hardware. These massive penalties threatened a middle class manufacturer more than an unpropertied artisan.[2]

---

[1] The major exception is A. E. Musson, "The 'Manchester School' and Exportation of Machinery," *Business History*, XIV (January, 1972), 17-50, which looks at the question from a significant local vantage point.

[2] The textile-related prohibitory statutes were 7 & 8 Wm. 3, c. 20 (1695); 5 Geo. 1, c. 27 (1718); 23 Geo. 2, c. 13 (1749); 14 Geo. 3, c. 71 (1774); 21 Geo. 3, c. 37 (1781); 22 Geo. 3, c. 60 (1782); Ireland 25 Geo. 3, c. 17 (1785); 26 Geo. 3, c. 76 (1786), which repealed clauses in the 1774 and 1781 statutes regarding cheap wood cards.

At least six government departments were involved in enforcing the prohibitive laws. The Foreign Office collected consular reports of artisans domiciled in foreign countries or of machinery illegally exported. The actual policing of the ports and coasts was the province of the Customs Commissioners. The Home Office, to whom the local magistrates reported, intermittently received information about foreign recruiters operating in the British manufacturing districts or artisans who were making arrangements to emigrate. Permissions to emigrate or licenses to export machinery were issued by both the Privy Council and the Treasury, while the former was also capable of issuing orders that could clarify the statutes. The key administrative department was the Board of Trade, as reconstructed in 1786 by William Pitt. It advised the Treasury on Customs policy and on commercial relations with other nations,[3] and it is clear from the Board's Minutes between 1780 and 1843 that it was usually informed or consulted by all the other departments except the Privy Council about infractions, interpretations, or modifications of the prohibitive laws. Thus, records of the Privy Council and the Board of Trade provide a good picture of central government policies, the Privy Council being concerned with these at the highest level, particularly in wartime, and the Board of Trade, with their commercial implications.[4]

After the 1780s, the loss of the American colonies, the prolonged European war with France (1793–1815), and expanding industrialization at home made the problem of emigrant workers critical. Foreign demand for the advanced technology created by Britain's pioneering industrialization grew rapidly, while a wide variety of incentives made hundreds of individual British industrial workers willing to risk the penalties of the prohibitory laws to meet, east and west, the rising demand for their skills.[5] From 1783 to 1812, some

---

The metal trades were covered by 25 Geo. 3, c. 67 (1785) and 26 Geo. 3, c. 89 (1786) as well as the 1718 Act. The act 26 Geo. 3, c. 89 included the prohibitions on the export of paper-making, glass blowing and cutting, pottery and leather-working equipment. The act 39 Geo. 3, c. 56 (1799) placed the seducing of colliers to emigrate on the same footing as the seducing of manufacturers.

[3] Lucy Brown, *The Board of Trade and the Free Trade Movement, 1830–1842* (Oxford, 1958), 20–21.

[4] This article is based on the following collections in the Public Record Office: P.C.1 (Privy Council papers, unbound); P.C.2 (Privy Council Minutes), vols. 128–226 (1782–1844); B.T.1 (Board of Trade In-Letters), letters from the period 1791–1830s; B.T.5 (Board of Trade Minutes), vols. 1–45 (1784–1838); B.T.6 (Board of Trade Miscellanea), vols. 151 and 152 (Machinery Books, 1825–1843). I should like to thank the staff of the North Room and the Rolls Room at the P.R.O. for their cooperation in locating and producing documents.

[5] On the emigration of industrial workers to Europe, see W. O. Henderson, *Britain and Industrial Europe, 1750–1870* (3rd ed., Leicester, 1972). On the emigration of industrial workers to the United States, see Mildred Campbell, "English Emigration on the Eve of the American Revolution," *American Historical Review,* LXI (October, 1955), 1–20; Herbert Heaton, "The Industrial Immigrant in the United States, 1783–1812," *Proceedings of the American Philosophical Society,* CV (1951), 519–527; David J. Jeremy, "British Textile Technology Transmission to the United States: the Philadelphia Region Experience, 1770–

EMIGRATION OF BRITISH TECHNOLOGY     3

100,000 persons left Ulster for the United States, an average of 3,000 a year. Between 1785 and 1800, however, immigration from Ulster was at a rate of 5,000 a year.[6] Incomplete Returns of Enemy Aliens made during the War of 1812 show about 7,500 male Britons resident in the United States in 1812–1814; of this number some 1,300 were workers in the textile or textile-related trades.[7] How aware was the central government of this widespread defiance of its prohibitory laws and what actions did it take to reduce it?

Relatively few consular reports about British artisans working abroad were received by the Board of Trade during the period. This is somewhat surprising because we know that at least one consul, Phineas Bond at Philadelphia, bombarded the Foreign Office with missives on the subject between 1787 and 1805.[8] The matter of emigration was first brought to the Board's attention by a letter dated October 5, 1791 from Sir John Temple, Consul General at New York. It complained of an influx from Ireland. A few months later another communication to the Governor of Lower Canada, suggesting that the Irish migration to Philadelphia, estimated at 1,500 persons a year, might be deflected to Canada, brought no response from the Board of Trade, which, according to the minutes of its meetings, feared that action might "have the appearance of encouraging still further Emigration." Such passivity disappeared when war against France broke out in 1793.[9]

The period of the negotiations leading up to the Peace of Amiens in March 1802 found the subject again under consular consideration. This time there were reports of English artificers in Germany, detained on the Continent by fear of criminal prosecution should they return home. Apart from information on English artificers in Austrian factories in 1805 and a general comment on West of England workers in the United States in 1812, the consular service provided the Board of Trade with very limited indications of the leakage of skill.[10] Even so, the Board was hardly ignorant of the scale of the leakage, since a variety of informal means of communication, such

1820," *Business History Review*, XLVII (Spring, 1973), 24–52; Charlotte Erickson, *Invisible Immigrants. The Adaptation of English and Scottish Immigrants in Nineteenth Century America* (Coral Gables, Fla., 1972), 229–389. Anthony F. C. Wallace and David J. Jeremy, "William Pollard and the Arkwright Patents, ' *William and Mary Quarterly*, forthcoming.

6 Maldwyn A. Jones, "Ulster Emigration, 1783–1815," in E. R. R. Green, ed., *Essays in Scotch-Irish History* (London, 1969), 46–68.

7 My analysis of U.S., Marshals' Returns of Enemy Aliens, War of 1812 Papers, Record Groups 45 and 59, National Archives, Washington, D.C.

8 Joanne L. Neel, *Phineas Bond. A Study in Anglo-American Relations. 1786–1812* (Philadelphia, 1968), 75–76, 129, 149, 151.

9 B.T. 5/7, p. 310; 8, pp. 102–104.

10 B.T. 5/13, pp. 62–65; 15, p. 292; 21, p. 117.

as manufacturers, customs officials, and informers, could bring news from foreign sources.

Internal reports of specific enticements or migrations were first recorded in 1807 with advance news of the visit of an industrial spy from Russia, Charles Baird (an agent of Sir Charles Gascoyne), who was on his way to obtain information and machinery from the Scottish and Manchester manufacturing districts. Notification came from Joseph Seddon, Chairman of the Society for the General Protection of Trade, a Manchester organization with correspondents in Glasgow. In the years just before and after the depression of 1819, moreover, manufactures in the Midlands metal trades made complaints of workers emigrating.[11] Customs reported goldsmiths leaving for the United States in 1809, and in 1811 they relayed an anonymous report about a Bristol mechanic taking machine-making tools to Falmouth with the intention of embarking for the United States. The same year Customs submitted reports on the emigration of Bristol glass workers and Lancashire cotton workers, and the following year they relayed an anonymous letter about workers from the Birmingham arms manufacturers taking their tools with them to America.[12]

There appear to have been no financial inducements to inform against skilled workers, as there were in the case of emigrant seamen after 1793, and informers sometimes risked intense odium. One tradesman found that he had been named as a prosecution witness after he patriotically informed in 1809 against a mechanic who openly bragged of plans to emigrate. Fearing for his reputation and his business, the tradesman had to make a desperate appeal to the Board of Trade for his discharge from appearing at the Stafford Assizes.[13]

### EFFORTS TO IMPROVE ENFORCEMENT

Cognizant of this extensive emigration during a period of rapid industrial change and European war, the government was obliged to consider improving the enforcement of the laws. Logically, questions of definitions and interpretation came first. Could any exceptions to the laws governing the emigration of artisans be allowed and, if so, who should legally be permitted to go abroad? A number of limited exceptions were made, although the arguments behind the permissions are rarely recorded. Presumably they were thrashed out

---

[11] B.T. 5/17, p. 277; 26, p. 419; 27, p. 337; 30, p. 406; 31, p. 436. For this Manchester society, precursor of the Chamber of Commerce, see Arthur Redford, ed., *Manchester Merchants and Foreign Trade, 1794–1858* (Manchester, 1934), 67–68.

[12] B.T. 5/19, p. 64; 21, pp. 70, 364. B.T. 1/76, ff. 29–38.

[13] P.C. 2/137, pp. 468–470. B.T. 1/44, f. 176; 45, f. 292.

orally. For at least one category of artisan — steam engine erectors — the government's hand was forced when an anomalous situation arose. Whereas the prohibition against the emigration of artisans was technically universal, that against the export of machinery was selective. For reasons that are not clear, neither Savery's, Newcomen's nor Watt's steam engine ever came into the machinery clause of the prohibitory statutes. Obviously, if the government permitted a foreign customer to buy a British steam engine, it could hardly deny him the services of British millwrights and mechanics to set it up. The Privy Council, which apparently controlled the licensing of emigrant artisans during the French Wars, allowed two erectors to set up three steam engines at St. Petersburg in 1806. Later that summer Samuel Davies was licensed to set up a steam engine at New Orleans, while three other men were permitted to sail to Sweden for the same purpose. Between 1815 and 1824 the Board of Trade approved only three applications for steam engine erectors to emigrate. Presumably most of those who went abroad did so without government licenses.[14]

After it was decided in 1797 to allow the export of mint machinery, and especially after 1799 when Matthew Boulton obtained an act of Parliament to send both coining machinery and millwrights to Russia, millwrights were often released to complete similar foreign orders. Assuming that the same principle could cover the printing of bank notes, the Swedish government obtained Privy Council approval in 1806 for a person to go to Sweden and set up a paper mill for bank-note-paper making.[15] During the French Wars, the Privy Council received applications from the Swedish government for the release of four sinkers to help open up some new coal mines, and two manufacturers of farming implements. All were allowed abroad. It seems likely that political considerations played a part, but evidence is lacking.[16]

No artisans at all were licensed by the Privy Council to emigrate between 1814 and 1824. And in the period 1780–1824, apart from the three permits to steam engine erectors, the Board of Trade allowed only one skilled worker, Richard Smith, to go abroad. Halted at the Liverpool Customs in 1817, he admitted that he had once been a master spinner, was presently assistant to a Staffordshire land surveyor, and was going to Philadelphia to recover unspecified property taken thence by James Slater, late of Cheadle.[17] Only one application (this to the Board of Trade) for permission for artisans to emi-

14 P.C. 2/169, p. 629; 170, pp. 429, 674. B.T. 5/25, p. 337; 26, p. 140.
15 B.T. 5/10, pp. 435–437. P.C. 2/160, p. 384; 170, pp. 543–544.
16 P.C. 2/164, pp. 128, 177; 175, p. 67.
17 P.C. 2/205 pp. 420–421. B.T. 1/119, ff. 8–10. B.T. 5/26, p. 140.

grate was rejected before 1824. The Turkish government was emphatically refused (in 1802) permission to hire British artisans. But these were cannon founders, and Britain was still at war against France.[18]

One other important question arose on the interpretation of the laws regarding artisans. Did the definition of artificer and manufacturer cover the non-manufacturing trades? In wartime the old prohibitive laws might be interpreted more rigorously than in peacetime, and only the positively unskilled be let through Customs. Perhaps for this reason the issue did not reach the central government until 1817. That year the Liverpool Customs officers squabbled with the Liverpool magistrates over the interpretation of the law on this point. The Liverpool Collector wrote to the Customs Commissioners asking, "whether common workmen, such as shipwrights, coopers, ropers, joiners or house carpenters, brick makers &c. [are covered by the laws]." The Board of Trade wrote to the Customs Secretary asking what the established practice had been, and to the Attorney General and Solicitor General for their opinion of the law, but Board of Trade minutes record no settlement.[19]

The greatest loophole in the prohibitory laws related to the problem of detection. The Chairman of the Customs readily acknowledged in 1824 that his officers had no way of identifying a skilled worker.[20] Even if the embarked passengers were mustered, as was possible under the 1795 proclamation or the 1803 Passenger Act, determined artisans would willingly perjure themselves to escape abroad.[21] Some joined ships after they had been cleared from English ports. In 1807, it was claimed, "emigrant artists & manufacturers" took small boats out to ocean-going vessels cleared from Liverpool "on their passage along the Cheshire shore." Revenue cutters or Royal Navy vessels might occasionally intercept these emigrants, but the problem of identification remained.[22]

Despite the odds against success, the Privy Council determined to

---

18 B.T. 5/13, pp. 114–115.
19 B.T. 5/26, p. 261. B.T. 1/120, ff. 90–95.
20 G.B. *Parl. Papers (Commons)*, 1824 (51), V, "Six Reports from the Select Committee on Artizans and Machinery," 50.
21 Few cases of artisans who perjured themselves are documented. One of them, Samuel Mearbeck, later recalled his questioning by Boston Customs officers who wondered how he had managed to escape from England with his cutlery tools: "They asked how did I to get away from England if I did not take a false oath. I told them what Butler says: 'He that imposes an oath makes it, not he who for convenience takes it.' They laughed at me and said that would do and desired me to tell it them again and they wrote it down."
Samuel Mearbeck to his mother, from Boston, Mass., January 1, 1817, Mearbeck Letters, typescript copies made from those in Sheffield Central Reference Library and kindly brought to my notice by Dr. Charlotte Erickson. The quotation is from Samuel Butler's seventeenth-century satirical poem, *Hudibras*.
22 B.T. 1/35, ff. 192–193.

apply the laws more effectively in the early years of the wars against France. An Order in Council of April 8, 1795 forbade foreign vessels to sail without their captains first submitting to the port officer a list of passengers, specifying names, ages, occupations, and nationalities. Any British artificers, manufacturers, or seamen discovered were to be arrested. This new measure against the emigration of artisans apparently owed much to the indefatigable reporting of Phineas Bond, Consul General at Philadelphia. Doubts arose over its legality, but Grenville and the President of the Board of Trade were satisfied and made the order.[23]

Although the Customs officers implemented this Order so well that at first few foreign vessels were cleared at all and the Privy Council began receiving complaints from masters, the problem of accurate identification defeated them. The obvious badge of the artisan was his bag of tools, but this he generally concealed carefully, or sent by a separate ship, or never took at all. Otherwise the indiscreet possession of some kind of documentation might announce the artisan. It was often only by accident that true identities were revealed, as with David Wooding, a Manchester spinner who, in March 1811, boarded the ship *Union* at Liverpool, bound for New York. During the Customs mustering of passengers, his marriage certificate was found in his sea chest and letters of introduction to a Philadelphia inhabitant in his pocket. Wooding, haled before the magistrates, betrayed his father-in-law, who was also on the *Union*, and the pair were committed for trial at the next Lancaster Assizes.

James Kipping, the Surveyor on Special Service, suggested, after his detection of Wooding, another way of solving the identification dilemma. Wooding's marriage certificate and the testimony of Manchester magistrates had given him away. Why not require all aspiring emigrants to produce a certificate, attested by their local magistrate or parish officer, "stating that in [his or their] opinion the Parties were not prohibited emigrants." Essentially this meant extending and reinvigorating the old Settlement and Removal Laws to cover the skilled as well as the poor, emulating the Prussian (1731) and French (1803) workers' passport laws. The Board of Trade received this proposal in May 1813, some two years after the Liverpool officers wrote to their superiors in London, and silently passed the papers to the Privy Council, which was expected to advise the Treasury. No more was heard of the plan. The bureaucratic procrastination, countenanced perhaps because of the War of 1812 (which tem-

23 P.C. 2/137, pp. 468–470; 142, pp. 361–362. P.C. 1/25, A48 and A49. Neel, *Phineas Bond*, 75, 149.

porarily closed emigration to the United States), suggests an unwillingness to pursue protectionist policies to their logical conclusion.[24]

While improving the means of detection, the authorities also tried the tactics of deterrence. Prosecutions followed arrests so consistently for English artificers on the Continent, as late as the 1820s, that they feared returning home to meet fines and imprisonment. Fear of loss of citizenship was the major legal deterrent. If the country was at war, moreover, some emigrants would be open to the charge of treason.[25] But another type of legislation supplemented the prohibitory laws: the Passenger Act of 1803, patterned on Phineas Bond's proposals. The purpose of the first Act is disputed, but recent evidence seems to favor a Machiavellian interpretation. In limiting the number of passengers that British vessels could carry to one for every two tons of a ship's unladen capacity, the Act not only relieved the conditions of transatlantic travel but also struck a blow at emigrant traffic in general, and thereby also at emigrant artisans. According to Castlereagh (President of the Board of Control in 1803) in an admission to John Quincy Adams, this was the real purpose of the Act.[26]

In attempting to hold back the artisan from emigration, the central government adopted no wide-sweeping measures. Owing perhaps to contemporary ideas on the free circulation of labor, it opposed increasing legislative restrictions (like the Settlement laws) on labor migration. Instead, it limited itself to sniping actions against individuals. When informed, the Board of Trade could alert Customs of impending depredations of foreign recruiters, like those of Baird in 1807 and Edward Thomas in 1819 (the latter was returning to collect artificers to set up a steam engine factory near Elberfeldt). If recruiters were already active in manufacturing districts, or artisans were known to be preparing to emigrate, the Home Office could send out Bow Street runners (then the country's only police force) to bring miscreants before magistrates, as in the case of James Patterson in 1809. But the scale on which those procedures took place was insignificant, considering the size of the problem.[27]

[24] For passport systems, see Arthur Redford, *Labour Migration in England, 1800–1850* (2nd ed. Manchester, 1964), 87–88. B.T. 1/76, ff. 33–40. B.T. 5/22, p. 276.

[25] See P.C. 2/159, p. 254; 180, pp. 539–540 and B.T. 5/19, pp. 55–56 for prosecutions. G.B., *Parl. Papers (Commons)*, 1824 (51), V, 569.

[26] Oliver MacDonagh, *A Pattern of Government Growth. The Passenger Acts and Their Enforcement, 1800–1860* (London, 1961), 57–58. This may be compared with Jones, "Ulster Emigration," 57–58 and Neel, *Phineas Bond*, 87, 149–150.

[27] Redford, *Labour Migration* pp. 81–84. B.T. 5/17, p. 277; 26, p. 419; 27, p. 329. B.T. 1/44. f. 176.

## Efforts to Reverse the Flow of Emigrant Artisans

Since the root of the matter was a complex of changing economic and social conditions, it was hardly possible for an enlightened and constitutional government to take adequate preventive steps against emigration. Indicative of the very limited possibilities for collective measures was the Board of Trade's reaction to the unemployment and emigration of the Birmingham arms manufacturers who, in 1812, were reported to be taking their tools with them to America. It suggested that the Board of Ordnance be asked to find work contracts for the 200 unemployed manufacturers. This was the solitary occasion in the records examined, when the central government tackled the emigration of artisans with a sound preventive economic measure, impelled by wartime conditions when arms manufacturers would plainly claim much greater government attention than handloom weavers.[28]

Apart from improved detection, deterrence, or prevention, the only other avenue in implementing the laws was to attempt a reversal of the migrational flow. Several times this tactic was tried or considered during the period of the French Wars. But if the law penalized the emigrant artisan, how could he be expected to return willingly to Britain? Either some clauses of the prohibitory laws must be set aside or else even more drastic and effective penalties must be involved to dislodge the fugitive artisan from his foreign abode. The former possibility was clearly more practicable. In 1797 Phineas Bond recommended a compensatory payment to returning artisan James Douglas (sometimes named as John or William), a native of Dumfriesshire, for his invention of three labor-saving machines (a cloth shearing frame, a sailcloth powerloom, and a brick-making machine). Douglas had emigrated to the United States in 1792 or 1793. Ironically, after his return Douglas was recruited by Chaptal, Napoleon's Minister of the Interior and eminent chemist, and introduced the latest British carding and spinning machinery to the French woolen industry. The relevant point is that the Board of Trade acted on Douglas's memorial and Bond's covering letter with a surprising recommendation to the Treasury: that Douglas be rewarded not for his inventions, but "for his expences in returning to this country, by the recommendation of His Majesty's Minister and Consul at Philadelphia." [29]

---

28 B.T. 5/21, pp. 221, 275, 364.
29 Henderson, *Britain and Industrial Europe*, 30–31; G.B., *Parl. Papers (Commons)*, 1841

Twice again this kind of waiving of the law was considered by the Board of Trade. A fortnight before the Treaty of Amiens was signed on March 27, 1802, a number of letters from Continental sources indicating the activities of English mechanics in Germany and the Low Countries prompted the Board to recommend that a bill be proposed to Parliament offering amnesty to returning workers. Artisans availing themselves of the amnesty would escape the penalties of the prohibitory law 5 Geo. 1, c. 27 and the Traitorous Correspondence Act, 33 Geo. 3, c. 27, provided they came back within a fixed period and had not emigrated for treasonable or criminal purposes. At the same time the Board advised the Foreign Secretary to initiate the process of artisan re-emigration. British ministers and consuls, having located British manufacturers and artificers in their districts, were to be instructed "to warn such Artificers to return into this Kingdom within six months according to the provisions of the said Act of 5 Geo. 1 cap. 27." To facilitate the exiles' return, the Postmaster General was to instruct masters of Post Office packets to grant any artificers holding consular certificates (of evidence of identity) a free passage to Britain. Presumably this amnesty scheme derived from the impending reopening of the Continent under the Amiens Treaty. The proposed bill never reached the statute book. And though the instructions went to the Postmaster General, it is difficult to see how he could carry them through without the complementary act of Parliament. Still, the Board did not abandon the possibilities of attracting artisans back to Britain. In September 1805 it considered John Rae's report on the activities of British artificers in developing the cotton manufacturers of Austria and his plan to offer financial inducements to these machine builders and manufacturers to return home. But again, the proposals produced no response because two months later, after the battles of Ulm and Austerlitz, Napoleon for the third time in ten years overran Austria and removed her from the Third Coalition. Nevertheless in 1824 Sir Charles Stuart was reported to have sent many artisans back to England from the Continent, paying their fares for them.[30]

If re-emigration could not be induced by this method, the only alternative was to strengthen the sanctions against remaining in a foreign country. Loss of citizenship could be a serious matter, especially for those who migrated to Europe. As the 1793 proclamation

(201), VII, "First Report from the Select Committee Appointed to Inquire into the Operation of the Existing Laws Affecting the Export of Machinery," 64. B.T. 5/10, pp. 427-429. B.T. 1/15, ff. 120–126.

[30] B.T. 5/13, pp. 62–65; 15, pp. 292–293, 322. G.B., *Parl. Papers (Commons)* 1824 (51), V, 581.

reminded seamen, if they lost their British nationality and fell into the hands of Algerine corsairs or Turks, they would not be "reclaimed" as British subjects. But only if the migrant happened to be captured while crossing the Mediterranean or while in the Eastern Atlantic when the Straits of Gibraltar were open to Algerian cruisers, was this a remotely likely, and very unpleasant, fate. The fortunes of war (as the foundation of a royal woolen factory at Seville in 1781 by English prisoners of war illustrates) increased the possibility of this sort of eventuality. Rae believed that if refusal to return home was elevated to a felonious offence some emigrants would come back from Europe.[31]

Migration to the United States was not susceptible to this kind of penalty. British immigrants could become naturalized Americans even though this required five years' residence, after Jefferson secured repeal of the Federalists' Naturalization Act of 1798. About this the British government could do little, until war broke out in 1812 and erstwhile British citizens served on the American side. With its forces invading America, the British government was in a position to flush out Britons who had become naturalized Americans, especially since British law at this time generally rejected foreign naturalizations and recognized only "natural Allegiance" or citizenship endowed by birth. Thus the royal proclamation of July 23, 1814 gave British citizens serving in the American forces four months to withdraw or else be deemed guilty of high treason, which no doubt tested the motivations and adaptation of some immigrant workers.[32]

## PROHIBITING THE EXPORT OF MACHINES

The ease with which Continental and, despite the greater distance involved, American manufacturers could obtain British machinery is well known. Until the 1820s, during this early non-verbal period before the new skills and designs were reduced to writing and drawings, the role of the artisan in the diffusion of technology seems to have been more important than that of machines, models, or specifications. Even so there was a sizeable traffic of undetermined volume in the export of tools, machinery, parts, and plans. In 1794 the Wiltshire clothier, Henry Wansey, found all the most up-to-date British cotton and woolen equipment in operation in the United

---

[31] P.C. 2/137, pp. 468–470. H. G. Barnby, *The Prisoners of Algiers. An Account of the Forgotten American-Algerian War, 1785–1797* (London, 1966), 103–106. J. Clayburn La Force, "Technological Diffusion in the 18th Century: the Spanish Textile Industry," *Technology and Culture*, V (Summer, 1964), 330.
[32] P.C. 2/196, pp. 36–38.

States. Much was surely constructed in America, but some vital components, drawings, and specifications were clearly smuggled from Britain. Likewise the Continental industries obtained Britain's new technologies. At first, equipment was mostly accompanied by artisans. When overseas reservoirs of skill had developed, the foreign demand for British machinery could be sustained by three factors: (1) its cost competitiveness, for even with the charges involved in smuggling, British equipment could be cheaper than foreign-built machinery; (2) its superior construction, though the durable construction of British–built machines could retard innovation; and (3) in many cases, its improved or new design. The continuing movement of British machinery abroad was readily acknowledged when the Parliamentary Committee investigated the problem in 1824.[33]

The Board of Trade minutes show that the first report of unlawful machinery exports was made in 1799. Then the Home Secretary relayed to the Board a memorial from "The Committee of Merchants and Manufacturers of Manchester and the neighbourhood for the general protection of trade" complaining that its unrestricted export to Ireland allowed British machinery to be conveyed to foreign rivals. The problem was solved administratively, if not in reality, by the Act of Union the next year. In 1807 the same body of Manchester merchants and manufacturers notified the Board of the impending arrival of Charles Baird on a machine-collecting assignment for the Russian government. That year Customs reported its first seizure of illegally exported machinery, though significant quantities were escaping undetected through the Customs net. The central government seems to have been apprised of only a fraction of the leakage.[34]

Whereas before 1807 there was only one report of unlawful machinery exports, between 1807 and 1811 a handful of complaints came to the Board about workers emigrating and taking their tools with them to the United States — London goldbeaters, Birmingham arms makers, and a Bristol textile machinery maker. This was the period of the Embargo that hurt America's foreign trade but diverted capital and labor into the manufacturing sector of the economy. New industrial opportunities in America naturally offered fresh migrational "pulls." It was also the period of Napoleon's Continental system, which theoretically closed Europe to British trade and traffic.

[33] David J. Jeremy, ed., *Henry Wansey and His American Journal, 1794* (Philadelphia, 1970), *passim; idem,* "British Textile Technology Transmission to the United States," 24–52. Henderson, *Britain and Industrial Europe, passim.* G.B., *Parl. Papers (Commons),* 1824 (51), V, 1, 2, 3, 6, 9, 10, 13, 36, for example.
[34] B.T. 5/11 pp. 367–368; 17, pp. 277, 302–303.

EMIGRATION OF BRITISH TECHNOLOGY   13

After Waterloo the two or three complaints about unlawful export of machinery concerned Italy and France: a consignment of paper-making and cotton-stocking machinery reached Naples in 1816, and in 1819 English manufacturers were sending models of machinery to France.[35]

As with artisan emigration, reports on the export of machinery reached the Board of Trade through the Consular system, and domestically through informers, Customs, and manufacturers or masters. Anonymous tip-offs were an important source of "counter-intelligence." One informer, identifying himself as "Bill," in 1822 submitted a memorial that detailed a smuggling route to the Continent: "the means practiced for conveying off machinery is by concealing it in different Packages and sending them to Rye and other places from whence they are taken to the Continent by Fishermen." When, in 1824, leading Custom officers had the opportunity to summarize their experience of the problem, they observed that they had insufficient staff for searching; that wharf facilities for unpacking and inspecting items for export were usually lacking; that large cargoes could not be closely examined; and that, when parts of different machines were mixed up and packed together, it was impossible for Customs searchers to identify the original equipment. Manufacturers, on the other hand, when complaining to the Board of Trade, assumed that the leakage of hardware was capable of being stopped, and that an arrest could be made simply by notifying a magistrate of a suspected intention to export, but this the law did not allow. Trade associations at Manchester in 1783 and Birmingham in 1799 expressed strong feelings for the more effective enforcement of the prohibitions on machinery export. The 1806 Report of the Parliamentary Committee on the Woolen Manufacture, reflecting for the most part the opinions of the Yorkshire manufacturers, likewise expressed a protectionist verdict on the laws, as being "founded in sound policy."[36] But, in spite of manufacturers' and Customs' complaints, the Board shied away from tougher statutory measures. Instead it simply urged Customs officers to greater vigilance, and the Board and the Privy Council allowed a series of relaxations of enforcement, some serving the manufacturing interest.

35 B.T. 5/18, pp. 138, 159, 160, 205; 21, pp. 70–71, 364; 25, p. 153; 28, p. 136.
36 B.T. 1/169, bundle 26. G.B. *Parl. Papers (Commons)*, 1824 (51), V, 49–59. Redford, ed., *Manchester Merchants*, 7, 67–68. G.B., *Parl. Papers (Commons)*, 1806 (268), III, "Minutes of Evidence Taken before the Committee Appointed to Consider the State of the Woollen Manufacture," 4–5; Julia de L. Mann, *The Cloth Industry in the West of England from 1640 to 1880* (Oxford, 1971), 147.

### INTRODUCTIONS OF EXEMPTIONS

Simple pre-industrial "tools and implements of a plain and simple construction" were virtually removed from the terms of the prohibitory laws in 1785. Under this principle the Privy Council permitted the export of implements for the construction of canals to Sweden (1809) and, in 1811, allowed Messrs. Burley & Co. to despatch "a spinning wheel made of mahogany, ebony &c" to Malta, "the said wheel being intended to be sent to Egypt as a present to a female at Cairo." Similarly the Board of Trade allowed the export of farming utensils (including hemp implements and a threshing machine) to Quebec (1805) and permitted a wife in England to send cutlery tools to her husband in Canada (1819). Perhaps it was under the same principle that the export of hand calico printing blocks to Calcutta was permitted in 1824.[37]

The second protectionist exemption, established in 1797, was in the case of equipment used to process raw materials needed for the British cotton manufactures. The Privy Council pronounced that the prohibitory laws "[were] not intended to prevent the exportation of machines [in this case cotton packing presses], the object of which is to facilitate & increase the importation of the raw material into this Kingdom." Between 1797 and 1806 the Privy Council passed five applications to export cotton packing or bailing presses, to the West Indies, the United States, the East Indies, and India. The Board of Trade passed four more between 1814 and 1824, to Surinam and Egypt, among other places. Because of this economically prudent policy, mechanized cotton packaging opened the distribution bottleneck between Whitney's recently-patented American cotton gin (1794) at the harvesting level and the new manufacturing technology at the processing stages.

Under the 1797 decision the Privy Council made other exceptions, releasing a skin-packing press to Philadelphia in 1801, a cotton gin to New Orleans in 1806, and a wool press to Lisbon in 1810. The Board, most active after the French Wars, passed a wool press to Bilbao, a cotton gin to Valparaiso or Lima, and four almond presses to Bengal. Primary processing equipment could also support traditional handicraft manufactures in the colonies, and possibly this explains why cotton printing machinery went to India under license.[38]

[37] P.C. 2/130, p. 313; 184, p. 31; 218 (for Oct. 5, 1811). B.T. 5/15, p. 282; 28, p. 143; 32, pp. 152, 342.
[38] The sources for the preceding two paragraphs are P.C. 2/147, pp. 572–573; 148, pp.

Another exceptional case was created when the Board of Trade in 1797 advised the Treasury on the export of steel rollers for the U.S. Mint at Philadelphia. Since mint machinery would not compete with British manufactures, the Board urged the Treasury to allow its export. This led to further exports of mint machinery, mostly Matthew Boulton's, to Russia (confirmed by act of Parliament in 1799), Denmark (1804), and America again (1816).[39]

The last excepted type of equipment was the steam engine. In 1786 the Privy Council refused to prevent three "Fire Engines" from going to Spain, although allegedly "extremely injurious to the copper mines of this Kingdom," on the grounds that Parliament alone could prohibit machinery exports not covered by the law. Other permissions followed. Under Privy Council license Hugh and Robert Baird sent two to St. Petersburg in 1802; Robert Francis shipped one to New York in 1805; Baron Henry de Bode despatched three more to St. Petersburg in 1806; Messrs. Mann & Barnard exported a 14 h.p. one to New Orleans to drive a saw mill in 1806; and two more went to Lima in 1814 and 1816. Numbers were regularly exported by Boulton & Watt during the French Wars. Thereafter the Board of Trade licensed three for export to Germany and Austria.[40]

By the early 1820s imperial preference was the other ground on which central government allowed exemptions from the prohibitions. Whereas Customs were ordered to seize iron–working rollers destined for France, Alexander Galloway was licensed to export copper–working rollers to Egypt, and Peter Fary was even permitted "to carry machinery for spinning stocking worsted to Van Dieman's Land" (Tasmania).[41]

In the most significant area of machine designs, government found it quite impossible to apply the letter of the law. First laid under the prohibitions in 1781, machine plans became increasingly available to foreigners by three routes. Of private communications, we have only traces of foreign visitors and little idea of the extent to which technology transfer occurred through correspondence. Published drawings and sketchy specifications came with the appearance of Rees' *Cyclopaedia* (45 vols., London, 1802–1820) and operating rules with Montgomery's *Carding and Spinning Master's Assistant* (Glasgow, 1832).[42] Censorship could check the publication

432–433; 153, pp. 529, 539; 158, p. 112; 170, pp. 70–71; 189, p. 205. B.T. 5/11 pp. 210–212; 22, p. 449; 25, p. 378; 26, pp. 10, 260; 29, p. 80; 32, p. 15; 33, p. 43.
    [39] B.T. 5/10, pp. 435–438; 11, pp. 3–7; 14, p. 165; 25, p. 25. 39 Geo. 3, c. 96 (1799).
    [40] P.C. 2/131, p. 528; 161, p. 112; 167, p. 178; 169, p. 629; 170, pp. 70–71; 173, p. 233; 174, p. 252; 205, pp. 420–421. B.T. 5/24, p. 155; 25, pp. 24–25, 337.
    [41] B.T. 5/29, p. 304; 32, pp. 2, 15, 345–346; 33, pp. 24–25, 30.
    [42] Jeremy, "British Textile Technology Transmission to the United States," 37–42.

of technical information, but this the government left to the discretion of individual authors. Patents of invention, the third source of machine plans for foreigners, offered government its greatest opportunity to enforce the law on machine designs. But secret patents, not uncommon in contemporary France, were scarcely allowed in Britain. Of the 5,000 patents enrolled before 1824, only two were secret, i.e. placed under covers, sealed and lodged with the Lord Chancellor "in order to prevent the invention getting to foreign countries," a procedure involving statutory approval.[43] Ironically, both patents, for textile processes, turned out to be technological failures. Instead, British patents were loosely guarded from prying foreigners by a chaotic search system evolved over centuries of accretion in patent procedures. This was supplemented by the inadequacy of published patents, which were often abridged and haphazardly chosen for inclusion in trade journals, where only a quarter of the enrolled patents could be found. But all this would not prove insuperable to the persistent investigator, who, at the cost of time and money, could legally obtain accurate copies of any British patent, copies which might then go abroad.

Only twice, between 1785 and 1824, did the Board of Trade recommend any statutory alteration of the prohibitions on machinery exports. In 1785–1786 it heeded the pleas of wool card manufacturers from Bocking and Colchester in Essex and Southwark, London, for permission to export their products to the United States. They argued that Dutch cards rivalled English ones, that hogsheads of Irish cards were going to America, and that Americans had John Kay's card–clothing engines at Boston, leaving only middle and southern American markets to Britain. They further maintained that the inferior cards that the British exported, with teeth loosely set in thin leather, were suitable merely for coarse wool, and that the prohibition on exports was causing unemployment among poor families in England's textile districts. Following the Board's recommendation, Parliament passed an amending act allowing the export of coarse wool cards to the United States.[44] The other occasion when the Board expressed willingness to change the law came in 1798 over the export of primary processing machinery, as noted above.

Though the Board made limited efforts to enforce the prohibitory laws down to about 1807, after that date it simply recorded its in-

---

[43] For the operation of the early–nineteenth–century patent system, see G.B., *Parl. Papers (Commons)*, 1829 (332), III, "Reports from the Select Committee on the Law Relative to Patents for Invention." Censorship of technical publications and the role of the British patent system in the process of technology transfer will be considered in more detail in my forthcoming doctoral thesis.

[44] B.T. 5/2, pp. 290, 296, 300–305. P.C. 2/130, pp. 278, 309–314.

EMIGRATION OF BRITISH TECHNOLOGY 17

ability to halt the emigration of artisans and the export of machinery. Finally, in 1820 the Privy Council suspended its 1795 wartime Order-in-Council, which was cited by the Board as late as 1818 and used by Customs until early in 1821.[45] It seems to have been this administrative impracticability in applying the laws that spurred the changes of the 1820s.

## II. Liberalizing Amendments of 1824 and 1825

All restrictions on artisan emigration were lifted in 1824, and the following year the prohibitions on machinery exports were consolidated and placed under a licensing system that allowed some relaxations. These changes came in the period of economic liberalism associated with Lord Liverpool's enlightened Tory government of 1822–1827 and in the years of expanding trade, 1823–1825.

Initiatives for the political assault on the prohibitory laws originated with two prominent free traders, the Ricardian economist J. R. McCulloch, and the Radical M.P., Joseph Hume. In the *Edinburgh Review* early in 1824, McCulloch ended a long essay against the Combination Laws (which in the 1823 session of Parliament the Radicals were planning to repeal) with condemnation of the restraints on emigration: all the great jurists pronounced them unjust; in a prospering economy they were unnecessary; in declining trades they prevented the relief of unemployment; they were unenforceable; and they deterred only the "poor and ill-educated class of artificers" from flocking to the United States, rather than the most skilled artisans. Regarding machinery export restrictions, he contended that these too were unenforceable; that they undermined the principles behind Britain's industrial growth; and that they spurred the construction of foreign machine shops.[46]

Less than a month after McCulloch's article appeared, Hume moved the appointment of a House of Commons Select Committee to investigate artisan emigration, machinery exports, and the combination laws. William Huskisson, President of the Board of Trade, spoke in his support. Both M.P.s unreservedly urged the removal of all prohibitions on the emigration of workers. Additionally, Hume pointed out that Parliament itself had, in attempting to reduce unemployment and spiralling poor–law rates, promoted emigration

---

[45] B.T. 5/17, p. 277; 19, p. 64; 26, p. 419; 27, p. 337. P.C. 2/202, p. 3. English Customs Board, *General Orders or Letters*, vol. 2 (1814–1823), 991–992.

[46] *Edinburgh Review* 39 (1823–1824), no. 78 (January, 1824), 315–345. For McCulloch's authorship, see Graham Wallas, *The Life of Francis Place, 1771–1854* (4th ed., London, 1925), 208.

schemes to the Cape and Canada. On machinery export, both Hume and Huskisson were more cautious. The law was inoperable, however, and if smuggling was enabling foreign rivals to construct their own manufacturing equipment, the British machine-making industry ought to be impeded no longer. Even so, Hume felt that the industry needed the encouragement of free export because it was hobbled by a patent system that allowed the piracies of foreign rivals.[47]

These views may be compared with those of the machine makers, manufacturers, and Customs officials who gave evidence before Hume's Select Committee between February 17 and May 20, 1824. Over the question of artisan emigration, witnesses from the textile industry, in contrast to the solidly protectionist iron masters, were divided. Those opposed feared the loss of technical secrets, a shortage of skilled workers, or the development of foreign competition. In the opinion of Thomas Ashton (a spinner and manufacturer in the Manchester area), only "rambling mechanics; no steady men, but loose sort of characters," and inferior workmen would resort to emigration; while William Fairbairn, the Manchester millwright, thought that repeal might even bring some artisans back to Britain. These diverging attitudes enabled Hume's Select Committee to recommend repeal of the prohibition on artisan emigration, on the grounds that the law was both unjust and impracticable. Repeal came on June 21, 1824.[48]

A very different response emerged from the textile industry over the proposed repeal of the prohibitions on the export of machinery. Apart from the London machinists, the textile manufacturers and machine makers before the Select Committee all registered their strong opposition to free export, fearing loss of overseas markets in the long term and, more immediately, curbs on the current expansion of business activity. Such resistance prompted Hume's Committee to advise a further enquiry. It came in the following session of Parliament, when business activity was just beginning to slow down. The poor response to Hume's appeals for witnesses possibly reflected the deepening depression. Only the Birmingham Chamber of Commerce and the master cotton spinners of Renfrewshire sent in evidence. Consequently, Hume's Committee was forced to rely on a reappraisal of the evidence given in 1824 and on the testimony

[47] *Hansard* 2nd ser. 10, cols. 141–151 (February 12, 1824); Adam Smith, *An Inquiry into the Nature and Causes of the Wealth of Nations*, Edwin Cannan, ed. (New York, 1937), 121.

[48] G.B., *Parl. Papers (Commons)*, 1824 (51), V. Unless otherwise stated, this section is based on these six reports on the export of machinery, emigration of artisans and on trades unions; and on the 1825 report on the export of machinery, G.B., *Parl. Papers (Commons)*, 1825 (504). Musson, "'Manchester School'," p. 24. 5 Geo. 4, c. 97 (1824).

EMIGRATION OF BRITISH TECHNOLOGY   19

of four London engineers (John Martineau, Timothy Bramah, Henry Maudslay, and Alexander Galloway), together with an assessment of the pertinent statutes. When the Report on the Export of Machinery finally appeared on June 30, 1825, the Privy Council's continued discretionary control over the export of tools and machinery was recommended. Hume fought for his bill another eighteen months, but eventually gave way to the manufacturers' protectionist sentiments.[49]

As Joseph Hume argued in 1825, free emigration made nonsense of the restraints on machinery exports. How can we explain the incongruity arising from repealing the prohibitions on the live instruments of trade while preserving those covering the dead instruments? Obviously many more people were immediately affected by emigration restraints than by the machinery export prohibitions, and the former were more difficult to enforce. Public opinion had long regarded freedom to travel as one of the free born Englishman's rights, a feeling reinforced by Adam Smith, Thomas Paine, and the French Revolution.[50] All of these points help to account for the freeing of emigration in the liberal climate of 1824.

A change in the economic and political climate strengthened the protectionists' hand in 1825–1826. Depression led the Manchester machine makers to secede from their alliance with the manufacturers who used their machinery, since free machinery exports would allow them to accept £20,000 worth of foreign orders. But jobs for machine makers now meant unemployment for manufacturers, and politically the manufacturers held the whip hand: the government dared not risk alienating wealthy new industrialists or aggravating the threat to law and order posed by rising unemployment in their segment of the labor force.[51] Economic depression may account for the postponement of repeal of machinery export prohibitions, but it does not explain the manufacturers' long–term faith in the practicability of the machinery prohibitions despite, and concurrent with, free artisan emigration.

The incompatibility of the policies of free emigration on one hand and restriction of machinery exports on the other reflects differing assumptions about the locus of technology. Protectionists based their views on Manchester machine-making practices. Here the division of labor was reaching levels unknown before, particularly in the

    49 *Hansard* 2nd ser. 15, cols. 908–911 (May 5, 1826), 1118–1122 (May 11, 1826); 16, cols. 291-298 (December 6, 1826).
    50 *Hansard* 2nd ser. 12, col. 652 (February 24, 1825); Smith, *Wealth of Nations*, 625; E. P. Thompson, *The Making of the English Working Class* (London, 1963), Chap. 4.
    51 See sources in note 49.

handicraft arts of turning and filing. Consequently, specialization was diffusing traditional skills over a multitude of sub-trades, at least six of which, for example, were required to make a spinning frame. Now the international movement of machine–building initially required the emigration of teams of skilled artisans, in place of individual craftsmen. Since workers emigrated singly or in family groups, free artisan emigration seemed unlikely to represent a leakage of technology.

Hume and the free traders, drawing upon London's machine–building experience, arrived at the opposite conclusion. In the first two decades of the nineteenth century Henry Maudslay and his pupils had wrought a machine-making revolution in the London area. Their metal-cutting tools — screw cutters, planes, and lathes — permitted a precision in metal working that could hardly be matched by the most skilled hand craftsman. The new machines dispensed with the need for highly-developed skills and also facilitated accurate copying of machinery from models, specifications, or drawings. As Alexander Galloway testified in 1824, metal-framed cotton machines were as well built in Paris as in Manchester; the only difference was their price.[52]

Some Manchester machine makers, like Sharp, Roberts & Co., were operating the new machine tools in the 1820s, but northern manufacturers seem to have been unaware of this, assuming that the locus of technology still resided in teams of specialized craftsmen. In fact it had shifted from men to machines, and the manufacturers' imposition of regulated protection between 1825 and 1843, based upon the Board of Trade's licensing system, became a vain effort to shut the stable door after the horse had bolted.

### III. Period of Discretionary Limitation of Machinery Exports, 1825–1843

In February 1825, four months before the Select Committee published its Report on the Export of Machinery, Board President William Huskisson admitted to the Commons that he had assumed discretionary powers and was allowing some prohibited machinery, such as hydraulic presses, to be exported. After a Commons debate on the subject in mid–June, it was agreed that the Board of Trade

---

[52] Besides the evidence in the 1824 (p. 17 for Galloway's comment) and 1825 reports, see L. T. C. Rolt, *Tools for the Job. A Short History of Machine Tools* (London, 1965), 83–121, for a general survey of the machine tool revolution. A. E. Musson and Eric Robinson, *Science and Technology in the Industrial Revolution* (Manchester, 1969), 459–509, deal with Manchester developments. Also Thompson, *Making of the English Working Class*, 245–246.

should exercise this kind of control. No statutory power was ever conferred on the Board; instead, the Customs Regulation Act of 1825 consolidated and continued all the earlier prohibitions on the export of machinery. They were renewed in 1833 and finally repealed in 1843. During this period the Board advised the Treasury on applications for machinery export licenses "under the general power of the Lord High Treasurer," but Huskisson apparently never received an increased staff to cope with the flood of applications that he expected.[53]

Under the new system an exporter petitioned the Board of Trade, disclosing the nature and value of his machinery, its port of shipment, and destination. If the Board believed that the machinery came under the prohibitions but might be exported, it wrote to the Treasury recommending a license. The Treasury then informed Customs at the port of exit, and the exporter learned of the license by contacting Customs there. The license, which was valid for twelve months, cost two guineas. If the Board did not decide to permit export, it simply notified the exporter and the case was closed.

Short of staff, the Board developed several tactics in doubtful cases to transfer the burdens of investigation and even of decision to other parties. It placed the burden of proof upon the applicant that his machinery came under the 1825 Act, and the onus of detecting inaccurate machinery descriptions upon Customs officials, who also advised the Board on the mechanical aspects of identification, relying on technical textbooks or the opinions of professional engineers. On rare occasions the Board noted in its Machinery Books, "left to Treasury to decide."

On the wharf, Customs men inspected machinery to verify that it corresponded with its license description. They made a record of the shipment in the Customs House books and cleared the equipment or, in doubtful cases, detained it. In clearly fraudulent cases, officials seized the shipment and placed it in the King's warehouse. If the shipper was convicted in the Court of the Exchequer (Common Pleas), his impounded machinery was auctioned off and the proceeds paid to the Crown and the Custom officers involved. No other penalty was exacted.

Occasionally the Treasury issued export licenses without apparent

[53] *Hansard* 2nd ser. 12, col. 652 (February 24, 1825); 13, cols. 1136–1138 (June 14, 1825). G.B., *Parl. Papers (Commons)*, 1841 (201), VII, "First Report from Select Committee Appointed to Inquire into the Operation of the Existing Laws Affecting the Exportation of Machinery," 8. Unless otherwise stated, the material in the remainder of this paper derives from this report and from its companion, the Second Report on the Exportation of Machinery, *Parl. Papers (Commons)*, 1841 (400), VII. Precise citations from these major sources are provided only for verbatim quotations.

reference to the Board of Trade. Between 1825 and 1839, the Board in every year except 1826 dealt with more applications than the Treasury, but in 1826 the Treasury handled sixteen applications untouched by the Board and in 1836, two. Nevertheless, the evidence suggests that the Board of Trade was the key member of the administrative triumvirate (the Treasury, the Board of Trade, and the Customs Board) responsible for running the licensing system.

## THE BOARD OF TRADE'S LICENSING POLICY

Under the licensing system there were three broad categories of machinery: machines not specified in the 1825 Customs Act, and therefore freely exportable; machines listed in the Act but exportable under a Treasury license; and machines of which export was prohibited by the Act and by the Board's licensing principles. In a system that evolved slowly, these categories were not always clear-cut.

At the beginning of the licensing system, Huskisson tried to avoid charges of prejudice by laying down a principle to guide the Board in its licensing decisions. Addressing himself to a petition from the Manchester Chamber of Commerce, Huskisson explained during a debate at the end of 1826 on one of Hume's motions to repeal the prohibitory laws that "machinery of great bulk and much raw material" was exportable, but machinery of modern construction, depending on "the ingenuity and excellence of the mechanism, and where the raw material was trifling," was prohibited.[54] John Kennedy, a leading Manchester cotton spinner and machine maker, had first formulated this principle in his pamphlet *On the Exportation of Machinery* (dated April 6, 1824). It was hardly meaningful. According to A. E. Musson, Kennedy referred to steam engines and millwork when he wrote about heavy machinery.[55] But these had been exported for decades, as shown above. If, on the other hand, the principle applied to processing–machinery, it could only be confusing and naive. Spinning frames, for example, were both heavy and ingenious in design.

A different principle was formulated and relayed to Customs collectors by a General Order dated November 14, 1838. As Deacon Hume explained in 1841, licenses were withheld for all textile machinery used in the first, and all subsequent, processing stages in which the mass of raw material was divided up ready for spinning.

[54] *Hansard* 2nd ser. 16, col. 293 (December 6, 1826).
[55] Musson, " 'Manchester School'," 25–26.

EMIGRATION OF BRITISH TECHNOLOGY    23

The only exceptions were made for foreign heads of state.[56] But this principle was not borne out by the Board of Trade's licensing decisions, for textile machinery was not the only type denied export licenses in years 1825–1835. Of twelve non-textile applications denied licenses in this decade, six were for iron rolling equipment, three were for machine tools (turning lathes and teeth cutting engines), and one for wire weaving reeds. However the late 1820s saw iron rolling and metal-working equipment and machine tools passed more and more often.[57]

As for textile machinery, the first ten years of the licensing system were marked by numerous violations of the 1838 rule. Spinning wheels went to Greece in 1827; more spinning apparatus to missionaries in New Zealand in 1829; cotton spinning machinery to missionaries in Madagascar in 1825; flax spinning machinery to Guernsey in 1825 and to Hamburg in 1831. The American ambassador, Albert Gallatin, obtained permission in 1827 to send a silk tramming machine to the United States. Mehemet Ali continued to gain as Egypt received over 500 power looms from Alexander Galloway in 1826 and 1828, and four drawing frames, 320 slubbing spindles, 580 fly frame spindles and 4800 mule spindles in 1835. Some 200 power looms went to Calcutta in 1828, and handlooms the next year, to be followed by power loom equipment in 1833. Three jacquard card cutters, though no jacquard looms, were exported, as were weavers' mails in 1838 and, earlier, shuttles and pickers in 1830. Marc Isambard Brunel was allowed to export his knitting machine (patented in 1816) to France in 1828, and the following year two stocking frames went to Upper Canada. Brunel's "tricoteur" was released on the grounds that it was a French invention that had not proved popular in England.[58] All of this suggests that the Board spent over ten years working towards the seemingly clear-cut criterion enunciated in 1838. A close study of the Board's Minutes for the first decade of the licensing system reveals that a rather more flexible and larger principle guided decision-making about machinery export licensing.

Hume and Huskisson recognized this principle in the Commons in 1825, and Huskisson took the opportunity to reiterate it early in 1826. He explained to a M. Andelle, who pressed for permission to export fifty cases of cotton spinning machinery to France, that the Board of Trade's "discretion was to be limited to such articles of machinery as might *in the opinion of the manufacturers* of this country, be per-

[56] Customs Board, *General Orders*, 1838 vol., for October 24, 1838. G.B., *Parl. Papers (Commons)*, 1841 (201), VII, 4.
[57] B.T. 6/151, *passim*.
[58] *Ibid.* B.T. 5/34, pp. 48, 417. B.T. 1/254, no. 33.

mitted *to be exported without injury to their interests.*"⁵⁹ On other occasions, too, refusals were made on the grounds of "the sentiments of the manufacturers."⁶⁰

Manufacturing opinion came mostly from petitions sent spontaneously to the Board. In one case the Board did solicit opinion, asking the Customs, through their officers, to "ascertain among the woollen trade" whether the export of shearing machines would be "injurious to that branch of trade." Within a month the Leeds merchants and woolen manufacturers sent a memorial to the Board objecting to the export of cropping and finishing machinery, and John Jones of Leeds was denied a license.⁶¹

## A Coherent Licensing Policy

Thus, there is some consistency in the apparently conflicting permissions and denials of the first ten years of the system. Missionary societies operating in remote regions hardly posed a threat to Britain's manufactures. Likewise a solitary emigrant worker taking a few tools with him no longer represented a critical leakage of technology. Thus, A. McKnight of Tarbolton was allowed to take two stocking looms to Matilda township, Upper Canada, in 1829, and Benjamin Levi took engraving tools with him to the United States in 1835. In each case it was emphasized that the tools were necessary for their owner to carry on his trade. On the other hand, in the opinion of Nottingham manufacturers the single lace machinery maker represented a major leakage of technology, and the Board did all in its power to keep his tools and machinery from leaving the country.

The practice of making gifts of machinery to heads of state was followed only in accordance with the interests of British manufacturers. Egypt, like India, lacked the complex of factors conducive to economic growth found chiefly at this time in Europe and the United States and represented no threat to British manufactures. Nevertheless there were limits. The Turkish ambassador in 1836 was confined to choosing one model of each kind of absolutely prohibited machine up to a total of ten models. But if machinery from a major branch of the textile trade was intended for a potential manufacturing rival, the chances were that no amount of diplomatic intervention could pry a license from the Board. Both the Russian ambassador in 1827

⁵⁹ B.T. 5/35, p. 155. Italics added.
⁶⁰ B.T. 5/36, pp. 78–80.
⁶¹ B.T. 5/41, pp. 86, 109.

and the Swedish chargé d'affaires in 1830 failed to gain export licenses for cotton machinery.[62]

Not only the future destination and ownership served to indicate whether export would injure British manufacturing interests, but also the type of textile machinery concerned. Between 1825 and 1841 three classes of textile equipment seem to have had no difficulty in gaining export licenses. The first was obsolete equipment. Pre-industrial implements like spinning wheels or wool combs and their broaches were regularly exported, as were textile printing blocks, and also some old if not obsolete industrial machinery. In 1833 "some old defective machinery which had originally been applied to the spinning of worsted by hand" was passed for export to Rotterdam.[63]

Most equipment for preparing fibers for manufacture, such as cotton picking, baling, willowing, batting, scutching, blowing, and spreading machinery was allowed to leave the country, but not carding, drawing, roving, or spinning equipment. In the woolen industry, balers and pickers or devils and also rag or shoddy pickers (developed in the 1830s) and carding engines that did not divide up the woolen lap (i.e., that lacked Goulding's condenser or a piecing attachment) went out, but not billies, jennies, or mules or combing machinery. In the silk industry, preparatory equipment meant machines for reeling or winding the raw material and preparing waste silk, but not for throwing or twisting it. In the flax and hemp industries, preparatory machinery was understood to include breaking and heckling devices. But in the case of tow or coarse bast-fibre processing, for a number of years spinning machinery was allowed abroad, presumably on the grounds that Britain's fine flax spinners would not be threatened. Weaving and knitting machinery was rarely licensed, and lace-making equipment, never. Accessory textile equipment such as card grinders or card–clothing machines were not licensed for export, though card clothing frequently was. Weaving accessories were also sometimes allowed.[64]

The pre-1825 protectionist principle of allowing the export of machinery for initial processing in order to expedite the major manufacturing stages carried on in British mills was at the base of these policies. But two problems developed with this process-based criterion. New preparation machines were developed that revolutionized some branches of textile manufacturing in the 1830s. The rag picker, (on which the shoddy trade was based), waste silk machinery, and

62 B.T. 5/38, p. 179; 39, pp. 319–320; 42, pp. 189, 357; 43, p. 440.
63 B.T. 5/41, p. 489.
64 B.T. 6/151, *passim*.

cotton spreaders, for example, tended to reclaim previously lost raw materials and, by increasing productivity and raising yarn quality — notably in the cotton industry — threatened British manufacturers in overseas markets. Tow preparing and spinning machinery, moreover, was so similar to that used in fine flax spinning that it could easily be adapted for this purpose; after 1835, therefore, its export was consistently prohibited.[65]

In the third group of textile machine types generally licensed for export were various finishing machines. Engraved copper rollers, calico printing machines, and calenders for cotton and linen cloth; cloth washing, drying and pressing (but not fulling, raising, or shearing) equipment for woolen cloth working; and fabric printing pieces in the silk industry were most frequently released. One curious discrimination was made in the case of calico printing machines. Their export to the United States was prohibited for a while, probably because New England cotton manufacturers were turning to cotton printing in the second half of the 1820s, but by 1830 the Board recognized that printing machines were licensed for Continental ports and then re-exported to America.[66]

Over two-thirds of the 2098 applications for licenses to export machinery between 1825 and 1843 involved textile machinery, a proportion that rose from 50 per cent in the 1820s to over 70 per cent after 1838. In the vast majority of cases the recorded destinations were European ports. About 10 per cent of the applications were denied licenses, and over 90 per cent of these refusals covered textile machinery.*

## THE OUTFLOW OF MACHINE TOOLS

Since the Board of Trade developed its licensing policy primarily as a reaction to pressure from British manufacturers, the absence of such pressure from machine tool makers must have been responsible for inattention to the outflow of such items. Export of rollers, slitters, presses having screws over 1½ inches in diameter, cutting-out presses, and lathes for "plain, round and engine turning" was prohibited under the 1825 and 1833 acts. Yet between 1830 and 1840 over 1300 metal working rollers, for iron, copper, tinplate, and zinc; 41 lathes, including planing, boring, and screw-cutting lathes; and 20 cutting engines including gear cutters and punching presses, were licensed

[65] B.T. 5/43, pp. 103–104, 151.
[66] B.T. 5/39, p. 272.
* The author has detailed data on destinations, fibers, and processing stages for the machinery involved in these applications, which he will be glad to supply on request.

for export. Inasmuch as these tools would have incorporated screws larger than 1½ inches in diameter, the total number of machine tools exported must have been much larger. The swiftness with which Americans became self-sufficient in machine tools is reflected in destination data, which reveal that most of this British technology was leaking to continental European rivals.[67]

It was not merely the failure of manufacturers to understand the central role of machine tools in the new technology, however, that accounts for the approval of increasing numbers of applications to export metalworking equipment and machine tools. There is evidence that the Board, a haven for free trade intellectuals in the 1830s, biased machinery export licensing decisions towards free trade in these years. Petitions from the Birmingham Chamber of Commerce in 1827 and the Manchester Chamber of Commerce in 1828 and 1834 to strengthen the prohibitory laws met with the Board's refusal or silence. And in 1827 the Board recorded that it did "not think it would be advisable to introduce a system of more general prohibition."[68] The Board's free trade influence was also revealed in its relations with the Treasury. In the case of tow–spinning machinery Treasury overruled the Board's recommendation of free export. However, when Goole Customs officers confiscated bobbins of yarns exported by Manchester merchants and Treasury sided with Customs, the Board brusquely informed the Treasury that "there is no ground whatever for classing those little wooden rollers with machinery in the cotton manufacture."[69]

### INEFFECTIVENESS OF GOVERNMENT

The attempt to safeguard Britain's textile technology from foreign rivals was defied in many ways. The licensing system itself, for example, might be used to cover illegal exports. Sometimes would-be exporters applied for "precautionary" licenses, in case their machinery should after all be covered by the law. At first, makers had attempted to export illegally by means of misrepresentation in applications, but after 1835 the Board considered applications defective if their specifications inadequately described the machinery they purported to cover.[70]

A piece of machinery might be broken up and exported in parts in the hope that Customs would be unable to identify it. Of the 51

67 B.T. 6/151, *passim.*
68 B.T. 5/36, pp. 133–135; 37, p. 262; 41, p. 554. These petitions from Birmingham and Manchester were evidently inspired by Manchester Chamber of Commerce. See Musson, "'Manchester School'," 36–37.
69 B.T. 5/42, p. 306 to B.T. 5/43, p. 65 *passim.*
70 B.T. 5/41, pp. 320, 493, 498, 518; 42, p. 383.

seizures made by Customs at Liverpool 1830–1839, indeed, some nine were of unidentified machinery parts, but at London this proportion was halved to nine out of the 121 seizures.[71] Customs evidently followed the policy of detaining unidentifiable parts and placing upon their owners the burden of establishing their true nature. Related to this problem of machinery identification was the ambiguity implied in the use of discretionary powers to waive or enforce the statutory prohibition, a weakness to which Deacon Hume drew attention in 1841.

Smuggling, or totally by-passing the licensing system, was facilitated by the coming of steamboat service across the English Channel. Operating by published timetables, the cross-Channel steamers began loading at a fixed time and cleared two hours later. The smuggler's ruse was to bring cases of machinery down to the dock at the last moment in hopes of avoiding Customs inspection. Or small items of machinery might be taken aboard by passengers in their baggage, which Customs men never examined. Smugglers could send machinery to the Continent by means of the coastal trade, depending upon the inadequacy of the fleet of Revenue cruisers, of which there were in 1839 only 49, plus 21 tenders, manned by 6183 sailors, to monitor the 7000-mile coastline of England, Scotland, and Ireland.[72]

Whatever the success of efforts to keep machinery at home, drawings of them continued to offer a major loophole in the licensing system. When the Nottingham bobbin net manufacturers petitioned for tougher enforcement of the law, the Board observed that "the transmission of drawings cannot be prevented." Eight years later, in 1841, even verbal descriptions were reckoned sufficient for a machine maker to copy a design. The artisan, meanwhile, continued to threaten the retention of technology in two ways. Where technology was centered in the artisan, as in the Nottingham lace trade, manufacturers still pressed the Board to restrain workers from smuggling parts and plans abroad. The artisan or mechanic might also, as agent for a foreign firm or government, engage directly in industrial espionage: in 1826 the Polish government had an agent in Britain pursuing the clandestine export of plans and machinery, while foreign apprentices were often deliberately sent to serve part of their time in British machine shops.[73]

[71] Customs House Archives, Customs 37/56, 'Seizures, 1825–1856," ff. 131–152. I wish to thank the staff of H.M. Customs Library and Archives for facilitating my work with their records, which were inspected by kind permission of H.M. Commissioners of Customs & Excise.

[72] See A.J.D., "Historical Notes on the Coastguard Service" (London, H.M.S.O., 1907), 23–24.

[73] B.T. 5/35, p. 282; 41, pp. 82, 417; 42, p. 273.

EMIGRATION OF BRITISH TECHNOLOGY   29

What might be done to seal these leakages? Before 1841 all that the Board did was to tighten its application procedures, to urge Customs to greater vigilance, and to contemplate more intensive Customs inspection of machinery presented for export clearance. The Manchester Chamber of Commerce assisted Customs, apparently at Liverpool, London, and Hull, in the late 1820s in identifying and arresting illegal exports of machinery.[74] That Customs kept very busy trying to prevent illegal exports is clear from seizure and detentions data, compiled by Customs. The Second Report of the Select Committee on the Export of Machinery in 1841 contains Customs data on detentions and seizures for the period January 1, 1824 through May 15, 1840. A Customs volume of "Returns Relating to Seizures" covers only seizures for the period 1830–1839 inclusive, but, unlike the Committee Report, it identifies types of machinery.[75]

By its very nature, however, the total amount of machinery smuggled abroad is unknown, so we have no means of precisely estimating the efficiency of British Customs houses. From the 1830–1839 return, the London officers were most busy, making 121 seizures in the period while Hull and Liverpool made less than half this number, 59 and 51 respectively. Then came Dover with 23 seizures, and Goole with 12. No other Customs house managed more than three seizures over the decade. The published return of 1841 likewise showed London most active with 84 detentions and 93 seizures. Hull followed with 13 detentions and 76 seizures; then Liverpool with 9 detentions and 70 seizures; Dover with 35 seizures; and Goole with 7 detentions and 5 seizures. The distinction between detentions and seizures was not clarified in the two returns, but presumably the 1830–1839 report included detentions because, on this basis, the figures in the two returns would seem to agree. The seizures in the 1830–1839 return, moreover, included numerous cases of mill work, machine tools, and large rollers being returned to owners.

These returns also suggest that the risk of losing machinery through confiscation was no real deterrent. Both the evidence presented to the Select Committee in 1841 and the 1830–1839 return of seizures give the strong impression that the illegal export of machinery was boldly conducted at major ports, whose well-built wharves could accommodate heavy machinery, and not on remote moonlit beaches, the romanticized setting for smuggling. Unscrupulous exporters appear to have had a low regard for the efficiency of the Customs,

---

74 Musson, " 'Manchester School'," 36.
75 Second Report, 1841, p. 107 (see note 53); for Customs seizures see note 71.

which was shared by the companies that offered them insurance against detection!

## IV. THE ENDING OF THE PROHIBITORY LAWS

In the early 1840s pressures for repeal mounted. When the severity of the 1825–1826 depression was repeated in the textile industry in 1841–1842, the cotton manufacturers no longer presented a united front.[76] Manufacturers, who supported the Anti-Corn Law League in the interest of lower food prices, moved logically towards a concomitant policy of free machinery exports; and free traders rapidly advanced their cause in the late 1830s in Lancashire, taking over the Directorate of the Manchester Chamber of Commerce early in 1839.[77] By the spring of 1841, the manufacturers were at the end of their tether. In the opinion of a protectionist Manchester cotton spinner, Holland Hoole, "the general expression of feeling is, that the legislature may do whatever they please, they cannot make things worse." [78] The engineering industry in Manchester, moreover, had expanded so far that machine makers rivalled manufacturers as a major industrial pressure group. Furthermore, there is evidence that provincial men were becoming increasingly involved in the textile machinery export trade in the 1830s and 1840s. A quadrennial sampling of applications for textile machinery export licenses, 1826–1842, shows that the percentage of London applicants fell from 78 per cent in 1826 to 33 per cent in 1842.[79]

Mark Philips, one of the two Manchester M.P.s, successfully moved, on February 16, 1841, for a Commons Select Committee to investigate the export of machinery.[80] In a masterly synthesis of the evidence, the Committee's Report summarized the objections to the mixed system of law and license, and surveyed the major changes that had occurred since 1825. Among the former were the difficulty in applying the 1838 principle, the insufficient expertise of Customs officers in both law and mechanical engineering, confusion and uncertainty arising from the conflict between the statutory prohibitions and the Board of Trade's licensing principles, and the sheer impossibility of preventing smuggling.

[76] R. C. O. Matthews, *A Study in Trade Cycle History: Economic Fluctuations in Great Britain, 1833–1842* (Cambridge, 1954), 127–151.

[77] Norman McCord, *The Anti-Corn Law League, 1838–1846* (2nd ed. London, 1968), 39. See also Musson, " 'Manchester School'," for diminution of the activities of Manchester Chamber of Commerce in the 1830s.

[78] Second Report, 1841, p. 52 (see note 53).

[79] Musson, " 'Manchester School'," p. 41 and Table 1. Author's data based on London directories.

[80] *Hansard* 3rd ser. 56, cols. 670–692 (February 16, 1841).

A series of developments had produced a new situation since 1825: the free emigration of artisans; the almost free export of coal; the revolution in machine tools, also freely exportable; important improvements in flax, waste silk, and cotton preparatory machinery; the rise of machine-making industries on the Continent and the United States; the easy transmission of new inventions through models, drawings, letters, and oral descriptions; active efforts to acquire British technology on the part of foreign governments; and the discriminatory effects of the domestic patent system against British inventors. To these considerations, which substantiated the futility and injustice of protection, were added familiar free trade arguments.

The protectionists could only urge the repeal of the Corn Laws and reductions in foreign duties before, rather than after, machinery export was freed. To maintain efficient protection for machinery exports, they suggested the registration of machine makers, stamping makers' names on machinery, and the funnelling of machinery exports through five or six ports for inspection by competent Customs officers. Meanwhile, manufacturers and machine makers alike faced up to the irritating revelation by the Select Committee's First Report, that more types of machinery than were generally assumed lay outside the export prohibition. Evidently export agents and smugglers had exploited the ignorance generated by the licensing system to rack up large profits, mostly at the expense of foreign purchasers.[81]

In July 1841 the Board allowed free export of spinning machinery to the colonies, if bonded against re-export, and in August 1842 it made a more sweeping relaxation. Most types of cotton and woolen machinery were released to foreign markets through the three ports of London, Liverpool, and Hull where specially competent officers examined the machinery, as Sharp, Roberts & Co., leading Manchester machine makers, had urged in 1827.[82] Though the Select Committee voted eight to one in favor of repealing the prohibitory laws, Parliament tarried for another two years. Gladstone, as President of the Board of Trade, at last took up the question of repeal in 1843. The delay, he claimed, stemmed from the new Tory government's initial preoccupation with reform of the tariff schedule chiefly in 1841–1842.[83] In the summer of 1843 he acted swiftly. His Machinery Exportation bill was reported on July 10 and by August 22 was on

[81] See note 53.
[82] B.T. 6/151 and 152 *passim;* B.T. 5/36, p. 246.
[83] *Hansard* 3rd ser. 70, col. 830 (July 10, 1843).

the statute book. After nearly one hundred and fifty years, all types of British machinery were at last freely exportable.[84]

## V. Conclusion

The most important fact about laws to prevent export of technology is that, despite spasmodic yet determined efforts, they were administratively impracticable. In addition to communication difficulties between the half dozen government departments involved, intractable problems at the ports made it difficult to filter emigrants and exports. Discriminating between skilled and unskilled artisans raised practical questions about definition, detection, and deterrence, while hitting upon positive measures to reverse emigration proved impossible. In wartime, the government tried simply to discourage all emigration, but even wartime pressures could not induce it to override the public's faith in the rights of the freeborn Englishman and its fear of the Census apparatus that a passport system would have required.[85] The tide of emigration continued to rise, and it could only be a matter of time before law followed practice. Free emigration was enacted soon after the post-war repression was replaced by a more liberal political and economic climate.

Illegal machinery exports were also very difficult to contain. Not only were machines readily transmogrified into parts, models, plans, specifications, patents, and mental images, but also some machines were legally released because their export was thought beneficial to British manufactures, as with primary processing machinery, or because they apparently held little threat for British manufacturers, as with steam engines, mint machinery, and obsolete equipment. Such relaxations and those dictated by diplomacy helped to open the door to free export. That this movement was aborted for a decade and a half in 1824–1825, when free emigration began, resulted from the manufacturers' failure to perceive the shift in the locus of technology from men to machines.

Administratively the licensing system of 1825–1843 was likewise difficult to implement. The success of machinery smugglers can be surmised from the ease with which foreign rivals acquired British models, the availability of insurance cover for illegal exports, and the infrequency of Customs machinery seizures, averaging less than thirty a year in the 1830s. The principles guiding the award of export

[84] *Hansard* 3rd ser. 71, cols., 493–517 (August 10, 1843), cols. 545–547 (August 11, 1843); 6 & 7 Victoria, c. 84.
[85] See David Glass, *Numbering the People: the Eighteenth–Century Population Controversy and the Development of Census and Vital Statistics in Britain* (Farnborough, Hants, 1973), 19–20. I am obliged to Dr. Charlotte Erickson for this reference.

licenses were never closely and satisfactorily defined, while the avowed principle — the interests of the manufacturers — was too vague to be consistently applied to complex and changing industrial technologies, especially those of machine building and textile manufacturing.

While quantification is impossible, it is likely that laws prohibiting the diffusion of technology did have some effect. They surely discouraged some potential emigrants from setting out on the international tramp; they certainly kept many emigrants from returning home. But despite the laws, thousands of artisans did go abroad during the French Wars, taking their machine building, operating, and managing skills with them. The licensing system likewise reduced the volume of machinery going abroad, but by 1841 no knowledgeable contemporaries believed that new technology would long remain safe from the hands of enterprising foreigners, especially as drawings and machine tools made copying extremely easy. British manufacturers unwittingly made their rivals' depredations easier still. Not comprehending the shift in the locus of technology, they persuaded the Board of Trade to shape the whole licensing system into an engine directed against textile manufacturing equipment: processing machines were halted, but not the machines that made processing machines.

The dominance of textile machinery in applications for export licenses confirms the familiar impression of Britain's early lead in textile manufacturing. The official demand pattern, however, is not readily quantifiable, because machinery values are irregularly, and technical data never, recorded in the Board of Trade papers. The only firm conclusion to be drawn is that northern Europe showed the greatest interest in Britain's textile machinery, especially in 1834 and 1838; expectedly, the Board's licensing policies noticeably weighted official demand toward preparatory machinery.

Britain prohibitory laws thus failed signally to stem the flood of technological information spreading abroad, either via men or machines, in this early industrial period. Administering and policing the sort of protection envisaged by the laws required Draconian measures that public opinion would not tolerate and internal economic and social conditions could not support. After 1825 British manufacturers made an inadequate system of regulated protection more ineffectual still by biasing it away from machine tools and towards manufacturing machinery. In these circumstances the flood could hardly be dammed.

# [15]

# British Response to the American System: The Case of the Small-Arms Industry after 1850

RUSSELL I. FRIES

I

The Great Exhibition of 1851 marked the beginning of a substantial transition within the British small-arms industry. For the first time British ordnance officials and British manufacturers saw the fruits of the American System of production exhibited in large quantities.[1] As a result of several related events—the subsequent production of American guns in England, the investigation of American machinery by an English commission, and a detailed parliamentary inquiry into small-arms manufacture—military gun making changed greatly almost overnight.[2] On the other hand, the methods of production changed far less in the construction of shotguns for sale at home and abroad and muskets for shipment to Africa. Indeed, techniques changed so little in some cases that examples of early 19th-century craft techniques exist to the present day. An examination of this diverse British response to a changing technology can increase our understanding of the nature of British industry and the American System of manufactures.

The American System consisted of several components which were usually blended together but which are separated here for analytical purposes. In the first place it tended to involve the use of power-

DR. FRIES is assistant professor of history at the University of Maine. This article is based on research conducted at the Johns Hopkins University during 1967–70, and the author wishes to thank Prof. Alfred D. Chandler, Jr., of Harvard University for his assistance; he is also grateful to Paul Uselding and Edwin F. Battison for their comments on an earlier version of the paper.

[1] Guy Hubbard, "The Development of Machine Tools in New England," *American Machinist* 60 (1924): 129; Charles T. Haven and Frank A. Belden, *A History of the Colt Revolver* (New York, 1940), pp. 327–35.

[2] Samuel Colt started a gun factory in London in 1852 which lasted until 1857 (see Haven and Belden, pp. 86–89, 345–49; Great Britain, *Parliamentary Papers*, vol. 18 [1854], "Report from the Select Committee on the Cheapest, Most Expeditious and Most Efficient Means of Providing Small Arms for Her Majesty's Service" [hereafter P.P.], vol. 50 [1854–55], "Report of the Committee on the Machinery of the United States").

378     *Russell I. Fries*

driven machinery to replace hand power. This usually had the primary effect of increasing output but could have other advantages as well. Thus the substitution of the water- or steam-powered trip-hammer for the hand hammer of the barrel welder resulted in both increased output per worker and a more sturdy barrel because the trip-hammer blows were faster, harder, and more regular. A second aspect involved the transfer of skill from man to machine. Thus the tracer finger of a profiling machine duplicated the memory function of a skilled craft filer in following a complicated pattern. Yet a third aspect, and probably the most novel part of the system, was the use of techniques to produce interchangeable parts. This was also probably the least separable aspect of the system, since it was almost impossible to make interchangeable parts without extensive power-driven machinery embodying a great deal of transfer of skill. Combined, the various components of the American System allowed American producers to make very complicated products at low cost by assembling machine-made interchangeable parts.

Previous analyses of the English small-arms industry have tended to focus primarily on the adoption and use of the American System in *military* production. Such authors as H. J. Habakkuk, Edward Ames, Nathan Rosenberg, and Peter Temin have concentrated on shifts in technology within the military-arms manufacturing segment of the arms industry rather than on analyzing the overall changes within Great Britain.[3] This slights the diversity of the British response. In fact, there were substantial reasons for British failure to adopt the system before 1854, even if the manufacturers had a complete picture of the way it worked. The technology adopted after 1854 varied primarily according to the *type* of firearm produced. Military-arms manufacturers adopted American technology and the American System of interchangeable-parts manufacture in wholesale fashion; shotgun producers adopted some American machines but not the American System; African-musket makers appear to have changed their methods little, if at all.

Professors Ames and Rosenberg have recently suggested that Professor Habakkuk's statement that British consumers tended to prefer variety to quantity may help to explain nonadoption of the American System by the British producers prior to 1850. The American System, they say, could manufacture large quantities of guns, but only of a

[3]H. J. Habakkuk, *American and British Technology in the Nineteenth Century* (Cambridge, 1962); Peter Temin, "Labor Scarcity and the Problem of American Industrial Efficiency in the 1850's," *Journal of Economic History* 26 (September 1966): 277–98; Edward Ames and Nathan Rosenberg, "The Enfield Arsenal in Theory and History," *Economic Journal* 78 (1968): 827–42.

standard pattern. Thus the English arms maker's decision not to use the American System was completely logical since he faced a market requiring variety.[4]

British consumer preference undoubtedly played a significant part in the history of the *civilian* industry before and after 1850. However, the American System should have appealed to the British Ordnance Office precisely because it sought standardization, not variety, in military muskets. English ordnance officials early developed inspection standards and gauges in order to enforce relative uniformity on the producers.[5] Uniform standards should have encouraged producers to develop and use the American System in England prior to 1850. However, Ames and Rosenberg also suggest that British military purchasing was so erratic as to preclude mechanization, since producers were not assured employment for expensive machines.[6] Fluctuating demand naturally discouraged mechanization, but the procurement and production systems would have prevented producers from realizing the benefits of machine introduction, even if demand had ceased to fluctuate.

## II

Before describing the changes that took place in the British industry in response to the challenge set by the American System it is necessary to look at the structure and production methods in the English industry as of 1850. All observers were agreed that the industry was predominantly a handicraft system. The industry can be conveniently divided in two different ways that tell us something about its nature. In the first place it is possible to segregate the industry by the *product* that it turned out. In the main there were three product segments: the military-arms segment, the African trade, and the sporting-arms segment (see table 1). Segregating the industry by the *stage of production* reveals three types: those who made parts, those who put parts together, and those who did both (complete shops).

There were few producers who made only one type of product in the pre-1855 industry. For example, most firms turned readily from making sporting shotguns to making military muskets as the market conditions warranted.[7] On the other hand, many producers can be

[4]Ames and Rosenberg, p. 836.

[5]Howard L. Blackmore, *British Military Firearms, 1650–1850* (London, 1961), pp. 262–65, 269–71.

[6]Ames and Rosenberg, p. 838.

[7]A. Merwyn Carey, *English, Irish and Scottish Firearms Makers* (New York, 1954) lists many smiths who did government and civilian work. Barbara M. D. Smith, "The Galtons of Birmingham, Quaker Gun Merchants and Bankers, 1702–1831," *Business History* 9 (1967): 132–50, details the history of one such firm.

## TABLE 1
### BRITISH FIREARMS EXPORTS IN 1844

| COUNTRY OR REGION | NUMBER OF | | | VALUE IN £ | | | AVERAGE VALUE IN £ | | |
|---|---|---|---|---|---|---|---|---|---|
| | Muskets | Shotguns | Pistols | Muskets | Shotguns | Pistols | Muskets | Shotguns | Pistols |
| Australia | 332 | 57 | 2 | 306 | 271 | 11 | 0.92 | 4.7 | 5.5 |
| Brazil | 9,301 | 176 | 20 | 5,098 | 748 | 43 | 0.55 | 4.2 | 2.1 |
| British North America | 2,471 | 202 | 28 | 2,628 | 478 | 14 | 1.1 | 2.4 | 0.50 |
| France | 705 | 67 | 2 | 364 | 557 | 2 | 0.55 | 8.3 | 1.0 |
| Germany | 317 | 24 | 5 | 152 | 332 | 22 | 0.48 | 13.8 | 4.5 |
| India | 41,466 | 3,777 | 8,808 | 93,074 | 23,393 | 10,463 | 2.2 | 6.2 | 1.2 |
| Russia | ... | 15 | ... | ... | 85 | ... | ... | 5.7 | ... |
| United States | 8,894 | 1,560 | 41 | 10,709 | 2,968 | 60 | 1.2 | 1.9 | 1.5 |
| West Africa | 80,530 | 6 | 337 | 37,874 | 30 | 108 | 0.47 | 5.0 | 0.32 |
| Africa | 83,721 | 350 | 596 | 39,964 | 2,025 | 259 | 0.48 | 5.8 | 0.43 |
| North and South America | 39,233 | 2,268 | 518 | 31,396 | 5,019 | 517 | 0.80 | 2.2 | 1.0 |
| Asia | 44,888 | 3,914 | 9,218 | 95,617 | 24,333 | 10,752 | 2.1 | 6.2 | 1.2 |
| Europe | 3,210 | 367 | 82 | 1,824 | 2,856 | 227 | 0.57 | 7.8 | 2.8 |
| South America alone | 22,499 | 257 | 108 | 13,465 | 906 | 189 | 0.60 | 3.5 | 1.7 |
| All regions | 171,052 | 6,899 | 10,474 | 168,802 | 34,233 | 11,755 | 0.99 | 5.0 | 1.1 |

SOURCE.—Great Britain, Public Record Office, Customs 9, Exports by Article.

segregated by their place in the production process. The complete shop, where all the stages of production were carried on, was very unusual. It existed only in the luxury sporting-arms trade, where the maker depended on the reputation of his products for continued business. Thus he employed a highly skilled staff of craftsman within his shop or "factory" in order to control quality and finish carefully at all times.[8] Below the level of the top-quality producers, outwork was the rule, with the small shop of from two to ten workers the dominant unit. A survey of the industry in 1855 yielded thirty-two different specialties among the parts makers and sixteen special types of assemblers (setters up).[9]

Production took place primarily on the basis of orders received. In the case of a military order, the Ordnance Office placed contracts with the various parts makers, who then shipped the completed parts to the government inspectors at the Towers in London or Birmingham. The parts that passed inspection went on to the setting-up contractors, who then made subcontracts with the necessary assemblers. After each stage in the assembly process the guns went back to the Towers for inspection and were then shipped on to the next subcontractor.[10]

In the case of a civilian order, one man generally took over the function of the Ordnance Office, making all the subcontracts as necessary for parts and assembly. In executing both military and civilian orders the division of production into subcontracts with different shops required extensive shipping of parts and semifinished guns from place to place within the gun-making districts.[11]

The parts makers and assemblers used very little machinery. Perhaps the only trade which used large quantities of capital in production equipment was that of barrel making. The best sporting barrels were produced carefully and slowly by hand forging, but military barrels were made from iron sheets rolled and welded into tubes, then drawn out to the required diameter and length by machinery.[12] There is a suggestion that one high-quality gunlock maker began to use machinery during the 1850s and that the Enfield Arsenal had also

[8]Interview with Mr. Leyton Greener, July 1969. Mr. Greener headed the firm of W. W. Greener, Ltd., a manufacturer of fine guns for more than 100 years prior to its purchase by Webley & Scott, Engineers, Ltd.

[9]John Dent Goodman, "The Birmingham Gun Trade," in *Birmingham and the Midland Hardware District*, ed. Samuel Timmins (1866; reprint ed., London, 1967), pp. 388–412.

[10]Ibid., p. 411.

[11]*P.P.*, vol. 8 (1854), *Committee on Small Arms*, Testimony of John Anderson, p. 65; Smith, p. 137; G. C. Allen, *The Industrial Development of Birmingham and the Black Country* (1929; reprint ed., London, 1966), pp. 116–20.

[12]Timmins, pp. 388–91.

382    *Russell I. Fries*

begun to introduce some few machines in the setting-up process.[13] However, by and large it can be said that the English industry was unmechanized. The file and bow-and-breast drill were the chief tools of the average worker in the industry.

The subdivided contract system for the procurement of military weapons made it economically impossible for the *producers* to reap the fruits of mechanization. The American System was uneconomical under the British contract practice prior to 1855 because its very nature as a "system" precluded benefits to separate contractors. Savings were available only when it was adopted as a complete entity.

The American System involved several elements, not all of which promoted direct savings. As mentioned above, one element involved the transfer of skill from craftsman to machine, thus allowing the use of lower-cost labor and/or permitting greater output per worker. An example of such a machine is the Blanchard gunstock lathe, which permitted an unskilled worker to replace a skilled craftsman in forming the rough blank stock. The lathe used a pattern model of the stock to control the action of a cutter, which duplicated the pattern on a rough wood blank rotated side by side with the pattern. The pattern replaced the trained eye and hand coordination of the stock maker using files on the wooden blank. It was estimated that the saving resulting from the use of the Blanchard lathe was twenty-five cents per stock, notcounting the capital cost of the machine.[14] On a musket price of $13.00 in 1820 this was a relatively insignificant saving of only about 2 percent, given the constant volume of military production. The cost saving would have been somewhat greater in civilian industry since the machine eliminated the constraint imposed by an assumed skilled firearms labor supply curve which had a sharp upward slope in the United States.[15] Use of the Blanchard lathe also might have led to increased volume due to the price elasticity of demand, if production costs were lowered. The Blanchard lathe was perhaps the most revolutionary single machine introduced during the period. Thus the short-range savings attributable to a single machine should not be exaggerated.

English barrel-rolling machinery, another innovation introduced by the Springfield Armory managers during the 1850s, demonstrated another aspect of the American System. In this case the same workers continued to use the new machinery so that the skill level of the workers remained the same. On the other hand the workers accepted

[13]*P.P.*, 1854, 18:6–7, 72, 84; 1849, 9:39–40, 399.

[14]Felicia Johnson Deyrup, *Arms Making in the Connecticut Valley* (York, Pa., 1970), pp. 97–98. This is a reprint of vol. 33 of Smith College Studies in History, 1948.

[15]Ibid., pp. 152, 230, 246–48.

a two-thirds reduction in the piece rate per barrel welded, which both gave them an effective increase in wages and reduced the cost per barrel because production went up fivefold. The net saving to the government (again neglecting machinery cost) was about seven cents per barrel welded, or one-half of 1 percent on the total musket price of $13.00.[16] Even had all the cost benefits of increased output gone to the armory, the total saving would have been only about ten cents, or 0.7 percent. In other words, single machine innovations probably saved little in extant costs, though they affected future costs by shifting the supply and skill of labor required. The Blanchard lathe eliminated the necessity for the use of skilled workers in rough stock making. Barrel-welding machinery allowed the same number of barrel welders to turn out five times as many units as before. The government gained by reducing dependence on critical skilled labor or getting along with a smaller number of employees. It did not necessarily gain *substantial* direct dollar savings by bringing in machines from the extant-cost standpoint (the only analysis that was used at the time in assessing a machine's value).

Despite the relatively marginal cost savings stemming from individual machine introduction alone, the American System did reduce costs dramatically by its use as a system. An examination of data on the price of Springfield Armory muskets shows that the average cost of the musket remained constant over a long period of time, though there were considerable fluctuations from year to year as production increased or diminished, or as new models entered. The first high volume production occurred at the armory during the years 1808–12 and, according to a modern estimate, musket cost averaged $12.62. The average cost during 1848–52, after the full interchangeable system had been introduced, remained at $12.68.[17] Material prices had remained virtually constant, but the average wages of armory workers had nearly doubled.[18] Since labor was 38 percent of cost, the 91 percent increase in wages should have resulted in a 34 percent increase in the total cost. Since this change did not occur, labor had become a priori more efficient. Output per worker shows an upward trend, though not enough to explain alone the constant price of the musket. Dividing the average production from 1808 to 1812 by the average number of production workers yields an output per worker of 53.2 muskets. For 1848–52 the figure is 67.8 muskets per worker, indicating a 27.5 percent increase in productivity.[19] The additional

[16]Ibid., p. 230.
[17]Ibid., pp. 229–30.
[18]Ibid., pp. 232, 234–39, 241–43.
[19]Ibid., pp. 233, 245.

384     *Russell I. Fries*

savings in cost may be explained by the larger output in the later years and the consequent economies of scale. There was also some simplification of gun design with the change to the simpler percussion-ignition musket and hence a reduction in cost.[20]

Changes in weapons design or individual machine introduction did not significantly decrease the labor in musket manufacture; thus an explanation for the large 27.5 percent increase in productivity must be offered in place of these factors.[21] It is suggested that the increasing interchangeability of parts accounted for a great deal of the savings noted above. Samuel Colt's testimony and the history of his firearms company support this interpretation. Colt's Paterson Arms Manufacturing Company turned out revolvers in the middle 1830s through a combination of machine and hand manufacture. From all the available evidence it appears certain that his early guns were not interchangeable and required expensive hand assembly and fitting. The resulting revolving rifle was very expensive—$50.00 cost for the first production lot of 100 rifles.[22] Colt's first venture failed in the early 1840s, but he reentered the manufacturing business during the Mexican War. Using the system of interchangeability in his second factory at Hartford, Connecticut, in the early 1850s Colt managed to lower the direct cost of one of his revolving pistols to $6.00.[23] Colt's testimony in 1854 before the Parliamentary Commission indicated that he achieved the greatest proportion of his cost reduction through assembly of interchangeable parts.[24] If it is true that the interchangeable system produced most of its benefits in the *assembly* stage of production then there are strong reasons for English rejection of the American System prior to 1854. (Similarly one can explain the failure to adopt the principle of the comparable Brunel-block–manufacturing system.)

[20]Ibid., p. 233.
[21]Lacking exact figures, part can be attributed to greater volume, part to design simplification, and part to reductions in material and tool waste.
[22]Detailed records on the early Colt venture are found mainly in the Samuel Colt Manuscripts, Connecticut Historical Society. This is from Samuel Colt to Dudley Selden, November 18, 1837.
[23]Records of the Hartford operation are spread between the Connecticut Historical Society and the Connecticut State Library. This is from Contractor's Book for 1851, Samuel Colt Manuscripts, Connecticut State Library. While design simplification took place between 1838 and 1851, and rifles were more expensive than pistols, it is believed that the cost of a directly comparable rifle would have been only $8.00–$12.00, or a fourfold reduction over hand manufacture.
[24]*P.P.*, 1854, 18:130 (see also the testimony of John Anderson [p. 74] and Thomas Hastings [p. 100]; James Lovell testified on the same subject in 1849); 1849, 9:399; Nathan Rosenberg, ed., *The American System of Manufactures* (Edinburgh, 1969), pp. 56–57, stresses that he believed the assembly of interchangeable parts saved a great deal.

Basically there were two types of benefits that followed from the adoption of the interchangeable system of manufacture. The first, as mentioned in the case of Colt, was the reduction in cost, achieved mainly in assembly, that occurred in the production phase. This usually meant lower prices and better profits on increased volume. The second benefit was the possibility of quick repair and replacement of parts with little difficulty by even unskilled customers. This second aspect of the benefits of interchangeability occurred in the use, not production, of the firearm. The producer received no direct economic benefit from ease of repair except insofar as this factor encouraged sales of his firearms over those of other producers. The manufacturer might initiate the use of interchangeable parts on his own, but only if he stood to gain the direct economic benefits in the production stage. If the cost savings were not available to the producer because the production process was subdivided, then the *consumer* had to force the adoption of interchangeable parts to get the benefits of quick repair. Under the American Ordnance Department's policy of giving out contracts for complete guns the producer benefited from interchangeable manufacturing, since any cost savings in the production process accrued to him.[25] However, the English Ordnance Office issued separate contracts for parts and assembly.[26] Thus the parts-production contractors could not gain the economic benefits of low-cost assembly of interchangeable parts, and they had no economic reason to initiate the system of interchangeability. English contracting policy, rather than stimulating innovation, reinforced the traditional craft structure of the industry.

## III

This analysis of English and American arms-procurement policies, and the nature of the benefits from the American System, helps to explain the actual course of events in England after 1854. The American System was imposed on the English industry from above by the English Ordnance Office (the consumer of military weapons), rather than coming up from the contractors. The reasons for Ordnance Office sponsorship of the American System were a mixture of primary (lower-cost muskets) and secondary (easy repair in the field) benefits.[27] The system operated after 1854 primarily through contracts for complete guns, both at the Enfield Arsenal and in contracts

[25]Deyrup, p. 42–45.

[26]Blackmore, pp. 39–40, 275.

[27]In the Summary Report and Recommendations of the Committee only economic production is alluded to directly. However, testimony throughout also pointed to the noneconomic benefits (*P.P.*, 1854, 18:10).

386     *Russell I. Fries*

made with private producers.[28] These facts accord with the suggestion that the English private contractors could not adopt the American System under the pre-1854 procurement policy with any substantial benefit to themselves. Similarly, the complete change in the contracting process indicated that both the ordnance officers and the manufacturers sensed that economic benefits came only when the system was used in its entirety within a factory. The function of the setting-up contractor was virtually eliminated from the military industry.

The Parliamentary Small Arms Commission of 1854 recommended a small-scale reconstruction of the Enfield Arsenal along the lines of the American System.[29] This began the sequence of changes in the English arms industry. Probably the most sweeping changes occurred in the military-musket production segment, and the Enfield Arsenal led the way for the contractors. The Crimean War shortly afterward caused the Ordnance Office to abandon plans for a small-scale armory and substitute a much larger one.[30] The Enfield Arsenal purchased quantities of machines from America in order to equip the factory, and capital was poured in.[31] Between 1854 and 1856 the British government expended £106,526 on buildings and machinery at Enfield.[32] Under the supervision of the former superintendent of the Harper's Ferry Armory, James Burton, the Enfield Arsenal became a virtual duplicate of the American System. Production began in 1858 at 26,739 rifled muskets and reached 100,370 rifles in 1863.[33]

While Enfield began the process that changed the firearms industry in Great Britain, change came more slowly in the private sector. Only one factory, the London Armoury Company, was already using machine-production techniques as of 1860. Its output was accepted by the English government as being fully interchangeable with that of Enfield production.[34] The other private producers, primarily former contractors of the War Office, did not alter their production methods as rapidly. In part their hesitance was quite logical. The Enfield Arsenal remained an experiment up until 1858. War demand lasting through 1856 produced orders that Enfield could not supply, so that the Ordnance Office had to settle for mostly noninterchangeable

[28]After 1860 Birmingham Small Arms, London Small Arms, and others almost always received complete contracts, though sometimes they were issued separate parts from store such as rough barrels and stocks as a credit against price.

[29]*P.P.*, 1854, 18:10.

[30]Timmins, p. 402.

[31]See Rosenberg.

[32]*P.P.*, 1856, 40:437.

[33]Timmins, pp. 2–4.

[34]Christopher Roads, *The British Soldier's Firearm, 1850–1864* (London, 1964), pp. 83–85.

muskets from the hand contractors. The government in desperation placed an order with the American firm of Robbins & Lawrence, which was incomplete at the war's end.[35]

The end of the Crimean War and the beginning of Enfield Arsenal production changed the response of the civilian producers rapidly. Overall musket demand dropped off, and the British government soon indicated that it would not place any more orders for noninterchangeable arms. The gun makers were given to understand that any further contracts would depend on their building a plant capable of meeting interchangeable standards.[36] In 1860 fifteen of the largest Birmingham contractors, members of the former Birmingham Small Arms Trade Association and many of them setting-up contractors, formed the Birmingham Small Arms Company, Ltd. (B.S.A.).[37] This company was again a virtual duplication of the American System. The managers were given access to the Enfield Arsenal in drawing up the plans for B.S.A., and they also hired an experienced American, Cory McFarland, to supervise the factory.[38] The new firm purchased special-purpose stocking machinery in the United States from Ames and Company of Chicopee Falls, Massachusetts, though Greenwood & Batley of Leeds furnished most of the metal-cutting machinery.[39]

From the time of the formation of the Birmingham Small Arms Company in 1860 competition for British military contracts took place solely within the ranks of those companies that could produce interchangeable arms. The London Armoury Company disappeared after 1865 as a rival of the Birmingham firm. It produced a great many muskets for the South during the Civil War and ceased operations at the end of the conflict.[40] It was replaced in 1867 by a new firm, the London Small Arms Company, a reorganization of the London Armoury Company plant under new management.[41] The Birmingham company and the London company quickly formed a pooling agreement to regulate competition among those building interchangeable muskets.[42] However, this agreement did not prevent the 1872 formation of a third large firm, the National Arms and Ammu-

[35] Ibid., pp. 89–90; see also Hubbard, pp. 130–32.

[36] [Leslie B. Taylor], "The Birmingham Small Arms Company, Ltd." (unpublished manuscript in possession of the company [1924]), pp. 27–33 (hereafter "B.S.A.").

[37] Timmins, p. 403; Taylor, pp. 1–3.

[38] Taylor, pp. 6, 10–11, 21, 60.

[39] Ibid., pp. 4, 9–10.

[40] Great Britain, Public Record Office (P.R.O.), Board of Trade 41-378/2138 and 31/177, The London Armoury Company, Ltd.; William B. Edwards, *Civil War Guns* (Harrisburg, Pa., 1962), pp. 78, 85–88. A. W. F. Taylerson, "The London Armoury Co.," *Arms and Armour* 2, no. 3 (September 1956): 45–58.

[41] Great Britain, P.R.O., Board of Trade 31/30754, The London Small Arms Co., Ltd.

[42] Taylor, pp. 83–84, 92.

388    *Russell I. Fries*

nition Company, Ltd.[43] Relatively unsuccessful due partly to its failure to secure British military orders and partly to financial mismanagement, the National Arms Company failed in 1882.[44] The firm of
C. G. Bonehill of Belmont Row, Birmingham, bid for several contracts in 1878 and 1879 but was never a major competitor, probably
because it lacked capital and machines.[45]

The decision to buy only interchangeable muskets after 1858 severely circumscribed the business of the civilian makers who had formerly made military muskets in addition to sporting weapons. They
were no longer able to compete for British military contracts, because
the government required standards of interchangeability which they
could not meet. Only war demand, which led to a shift in military
buying preference from quality (i.e., interchangeable muskets which
could only be made in machine factories with a long lead time) to
quantity (noninterchangeable muskets which could be turned out
rapidly by almost all makers), allowed the noninterchangeable producers to sell to foreign buyers (see table 2 and fig. 1).[46] The fate of
this sector was sealed gradually by the steady adoption of
interchangeable-parts arms techniques in other countries. As these
countries built their own equivalents to the Enfield Arsenal even the
companies producing machine-made interchangeable guns
suffered.[47] The best evidence for this trend is the decline in British
rifle exports over the years. During the American Civil War (1862–64)
British rifle exports were 76 percent of the total value of arms
shipped abroad, while in the years 1880–81 rifles were only 30 percent. The actual value in current pounds of rifles exported declined
by more than two-thirds over the same period.[48] The remaining rifles
exported in the 1880s were probably mostly civilian models made by
sporting-arms makers.

The history of the sporting-arms segment of the industry is
perhaps the most interesting because the producers adopted relatively little of the interchangeable system. Prior to the 1850s the
sporting-arms producers used the same standard techniques of craft
production as in the military segment. Unlike the military contractors,
the sporting makers continued to use almost the same technology as

[43] Ibid., pp. 153–60.
[44] Ibid., "Appendix," pp. o.4–o.38; Leslie B. Taylor, *A History of the Westley Richards
Firm, 1812–1913* (Stratford-upon-Avon, 1913), pp. 38–39, 46–47.
[45] Taylor, "B.S.A.," pp. 301, 332–35.
[46] Ibid., pp. 123–25. Table 2 and fig. 1 show the large fluctuations in the business and
the steady fall off after the 1870s.
[47] Taylor, "B.S.A.," pp. 302–4.
[48] Figures tabulated from Great Britain, P.R.O., Customs 9, Exports by Article.
Deflation would account for only 26 percent of this drop.

## TABLE 2

### BRITISH EXPORTS OF RIFLES

| Year | N | Value in £ | % of Total | Year | N | Value in £ | % of Total |
|---|---|---|---|---|---|---|---|
| 1862 | 454,674 | 343,045 | | 1880 | 55,652 | 94,540 | |
| 1863 | 223,714 | 653,136 | 58.4* | 1881 | 59,493 | 110,020 | 30.3† |
| 1864 | 77,458 | 206,139 | | 1882 | 222,762 | 315,077‡ | |
| 1865 | 85,322 | 261,690 | | 1883 | 259,169 | 354,862 | |
| 1866 | 68,726 | 171,781 | | 1884 | 300,872 | 385,036 | |
| 1867 | 137,420 | 318,538 | 63.0 | 1885 | 250,411 | 374,074 | |
| 1868 | 258,443 | 627,971 | | 1886 | 159,729 | 298,702 | |
| 1869 | 76,212 | 188,400 | | 1887 | 121,742 | 213,551 | |
| 1870 | 195,435 | 655,059 | | 1888 | 142,676 | 230,006 | |
| 1871 | 153,585 | 638,625 | | 1889 | 199,498 | 281,189 | |
| 1872 | 59,130 | 191,632 | 59.2 | 1890 | 178,232 | 262,227 | |
| 1873 | 86,312 | 276,579 | | | | | |
| 1874 | 60,527 | 203,178 | | | | | |
| 1875 | 147,049 | 472,980 | | | | | |
| 1876 | 31,710 | 106,340 | | | | | |
| 1877 | 49,198 | 93,551 | 44.8 | | | | |
| 1878 | 62,752 | 94,895 | | | | | |
| 1879 | 75,431 | 74,624 | | | | | |

SOURCE.—Great Britain, Public Record Office, Customs 9, Exports by Article.
*Rifles not tabulated separately before 1862. Percentage given is total of rifle exports for 1862–64 divided by total arms exports for 1860–64. Rifles were 76.0 percent of total arms exports for 1862–64.
†Percentage is for years 1880–81 only, since rifles were not separately tabulated after that date.
‡All figures from 1882 on include muskets and fowling pieces as well as rifles. While the separate course of the types are thus hidden, an estimate can be made. It is probable that rifle exports remained constant or declined. Musket exports probably increased slightly to 1884, then joined the general decline. Fowling-piece exports probably continued to rise in importance relative to the other two categories, though declining in absolute numbers after 1884.

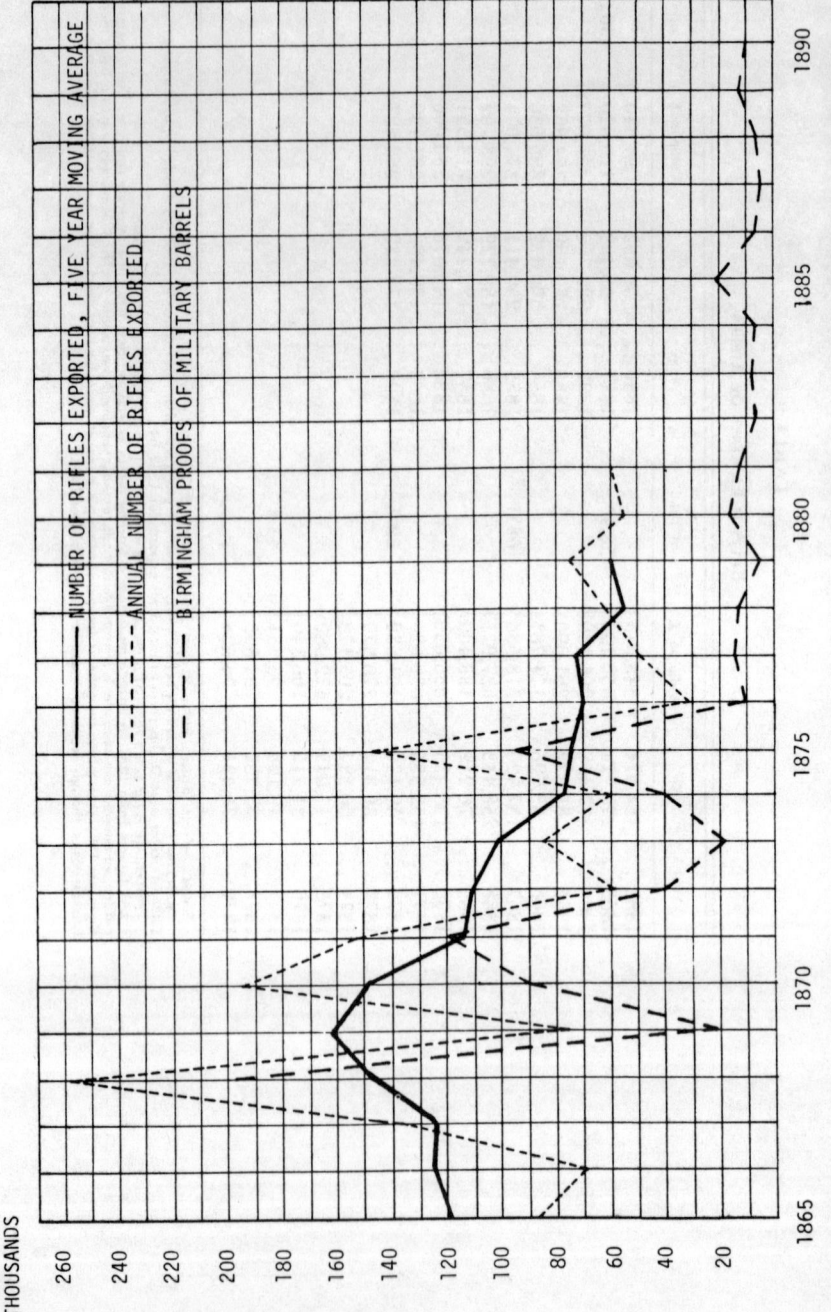

FIG. 1.—British proofs and exports of military-type arms, 1865–90. (Source: Great Britain, Public Record Office, Customs 9, Exports by Article, Birmingham Proof House Records.)

before, though there were distinct variations in machinery used within the segment. Generally the highest-quality sporting-arms makers adopted the greatest amount of the new technology. One of the premier Birmingham makers, William W. Greener, described the system of manufacture as of the early 1900s.[49] The adoption of steel for barrels had forced some mechanization on the gun makers in the form of drilling, boring, and grinding equipment. Yet this specialty of barrel making had been mostly mechanized in the craft system (witness the use of English barrel machinery in the United States), so that, rather than increasing their mechanization, barrel makers simply shifted to other machines. Lead lapping the inside diameter followed barrel boring, and this was usually done by machine after the introduction of the breech-loading shotgun because the finish could be achieved much faster than by hand lapping. The greatest change was the use of machinery in forming the breech action of the shotgun, that is the metal body of the gun which contained the lock work and closed the rear of the barrels of breech loaders.

The increasing use of machinery in the sporting-arms segment occurred for several reasons. In lead lapping the barrel, machines served primarily to increase the speed of the operation. In forming the barrels and breech action the higher cutting rate of power-driven machinery was but one of the reasons. The milling machine, one of the new American machines, shaped plane surfaces far more accurately, as well as faster, than the hand filer. The profiling machine and milling machine with shaped cutters cut much squarer edges and truer patterns than were obtainable through hand techniques, a transfer of skill to the machine. Accurate locating of plumb holes and of true radii for mating surfaces made subsequent fitting operations in which barrels and breech action were mated easier. There was, however, no desire to make the guns fully interchangeable through the use of machinery.[50] The production of such guns was considered to involve too much of a compromise with the fitting of the gun so necessary to the individual. The fact that the factory owners introduced more machinery than garret workers again verifies, in a circular fashion, the hypothesis that producers of complete guns benefited most from machine introduction.

The differing use for machinery in the large factory for interchangeable manufacture and the much smaller sporting factory implied different economic uses of tools. The interchangeable-arms factory developed enormously specialized machines, tools, and fixtures

[49] W. W. Greener, *The Gun and Its Development*, 9th ed. (1910; reprint ed., York), pp. 229–87; see also Timmins, pp. 396–400.

[50] Greener, pp. 284–87.

392     *Russell I. Fries*

in order to reduce the time necessary to perform each operation to a bare minimum and to eliminate the necessity for changing tools (tool changing and machine adjusting were skilled and time-consuming activities) as much as possible. Thus the Birmingham Small Arms Company had some 771 machines in place by 1871.[51] This policy paid because tooling and fixturing costs could be applied over a large production run. In contrast to this the sporting-arms makers used each individual machine for a variety of purposes. One lathe was used for chambering several different types of guns by changing tools, while simple chucks held the barrels for chambering, adjusted as necessary to hold the particular gun barrel under construction. The factory owner did not build special expensive fixtures to hold each individual type of gun because the cost of the fixture could not be amortized over a sufficient number of units, and because the varying sizes of the barrels and actions often made it impossible to build fixtures that could hold more than one particular gun properly. Where one dimension on several models could be standardized, such as the radius joining barrel-lug and action, then the milling machine came into its own.

In the other major segment of the arms industry, the African-musket builders, there were no extensive changes in technology as far as can be discerned. As recently as the 1950s African muskets were made by hand craftsmen in the Birmingham area for less than twenty dollars.[52] It is this low price which provides the most feasible explanation of why mechanization did not take place in this segment. By using the cheapest materials and the lowest grade of skilled labor, the makers held down the cost of the African-trade musket, probably close to the cost if made in volume by machine operators.[53] Production volume for any one contractor at this price level was probably insufficient to allow rapid amortization of tools, given easy entry, a highly competitive market, and consequent low profit margins. Lack of constant volume probably discouraged contractors from trying to break out of the subcontracted and subdivided production system. Quality of construction played little part in African purchases because the dealer who bought the guns sold in a monopoly market and had little interest in the satisfaction of his customers. It was better for the dealer if the gun fell apart rapidly so that he could sell another. This

[51]Taylor, "B.S.A.," p. 151

[52]Catalogs of the American Stoeger Arms Company (The Shooters Bible) list muskets of this type, produced by either English or Belgian workmen. Most musket manufacture had ceased by 1900, however. Clive Harris, *The History of the Birmingham Gun Barrel Proof House* (Birmingham [1946]), p. 75.

[53]*P.P.*, 1854, 18:43.

removed any incentive to use machines because they could make more accurate cuts than the hand filer. There was no technological change in the product itself, the flintlock musket, since that was virtually dictated by colonial laws which prohibited sales of rifles or breech-loaders to natives.[54] Last, the labor supply tended to remain adequate precisely because the workers displaced due to lack of military orders tended to drift into the musket segment. Thus models and techniques tended to remain fixed, orders fluctuated, and skilled labor was cheap and readily available. It would have been uneconomical to use machines so long as these unfavorable conditions prevailed.

## IV

Up to this point it has been suggested that the determination to adopt, or not to adopt, machine technology was based on the economic realities of the situation—the nature of the market. The makers of sporting arms, however, rejected the manufacture of interchangeable-parts shotguns by machines (i.e., the full American System) for reasons besides the actual nature of the market. They claimed that their customers demanded custom-fitted, individually tailored shotguns, and were thus precursors of Habakkuk's argument on variety.[55] This perception of the market helped to determine the conservative pace of change.

In fact, both the British domestic and export markets seem to have been adaptable to mass-produced sporting guns. The extent of the English domestic market has usually been overemphasized. By comparing actual English production with statistics of export and import, one can derive some indication of the extent of the home market. Based on the statistics available (Birmingham proofs and British exports; see fig. 2 and tables 3, 4), home consumption appears to have been at the maximum about 135,000 guns annually during the years 1855–64,[56] or between 30 and 40 percent of total production. This figure probably overestimates British home consumption. Even so, it seems that British producers thus concentrated on pleasing the tastes of British consumers at the expense of pleasing their foreign purchasers and sacrificed a part of a large market for a smaller, though

[54] Artifex and Opifex (pseuds. W. W. Greener and ?), *The Causes of Decay in a British Industry* (London, 1907), pp. 128–32.

[55] Greener, pp. 417, 284–85.

[56] The estimate has been drawn up by adding the total number of private proofs at London and Birmingham (Timmins, p. 415). From that the total number of British exports are subtracted; then British imports minus reexports are added to give an estimate of domestic consumption. Figure 2 and table 4 give total Birmingham Proofs; table 3, British exports in constant and current pounds.

PROOFS IN
THOUSANDS

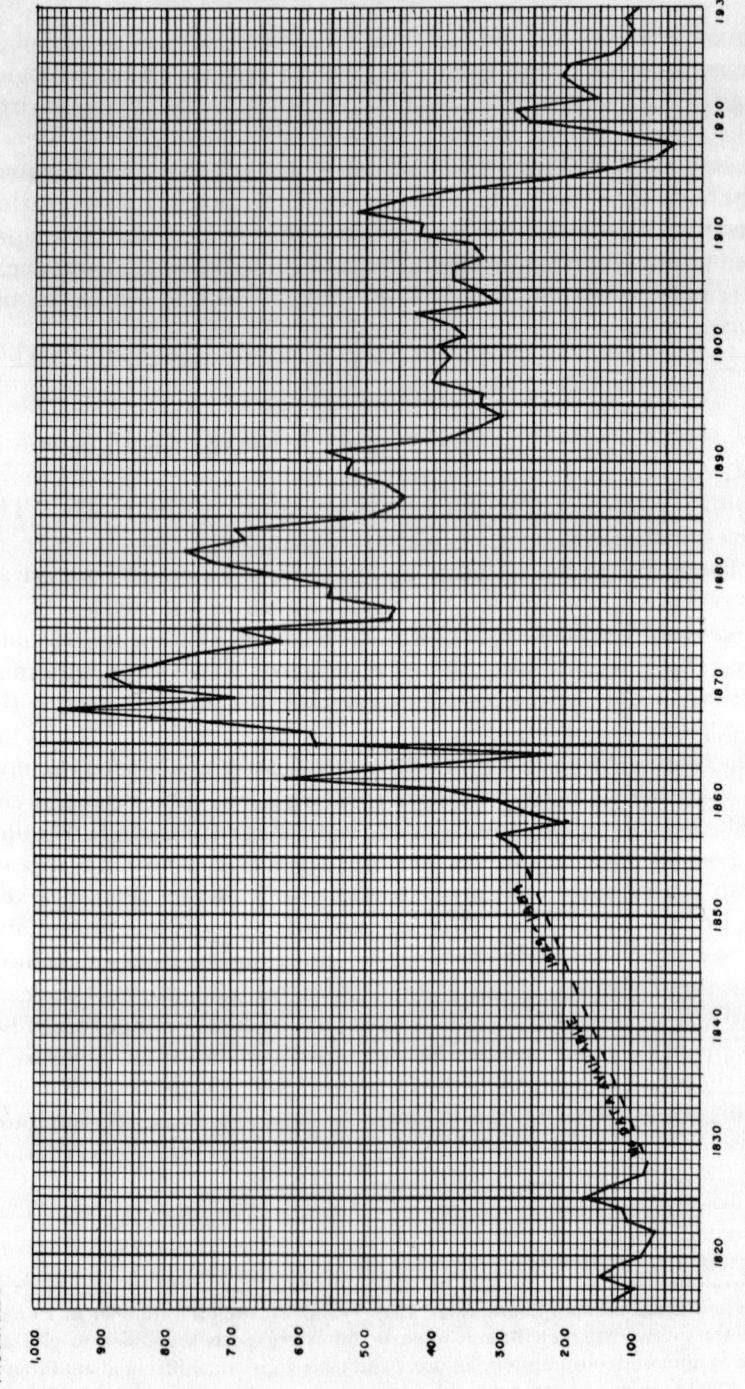

FIG. 2.—Birmingham trade proofs, 1816–1930. (Source: Clive Harris, ed., *History of the Birmingham Gun Barrel Proof House* [Birmingham (1946)], pp. 152–55.)

## TABLE 3

GREAT BRITAIN TOTAL ARMS EXPORTS

(**£**)

| Year | Current | Index | Constant | Five-Year Average |
|------|---------|-------|----------|-------------------|
| 1845 ....... | 312,438 | 99 | 315,600 ⎫ | |
| 1846 ....... | 270,696 | 99 | 273,400 | |
| 1847 ....... | 290,373 | 104 | 279,200 ⎬ | 320,000 |
| 1848 ....... | 375,135 | 92 | 407,800 | |
| 1849 ....... | 282,304 | 87 | 324,500 ⎭ | |
| 1850 ....... | 236,330 | 93 | 254,100 ⎫ | |
| 1851 ....... | 219,106 | 89 | 246,200 | |
| 1852 ....... | 184,700 | 93 | 198,600 ⎬ | 222,000 |
| 1853 ....... | 272,853 | 112 | 243,600 | |
| 1854 ....... | 211,374 | 126 | 167,800 ⎭ | |
| 1855 ....... | 163,284 | 112 | 133,800 ⎫ | |
| 1856 ....... | 288,677 | 120 | 240,600 | |
| 1857 ....... | 447,855 | 124 | 361,200 ⎬ | 234,000 |
| 1858 ....... | 325,543 | 112 | 290,700 | |
| 1859 ....... | 168,297 | 116 | 145,100 ⎭ | |
| 1860 ....... | 358,847 | 117 | 306,700 ⎫ | |
| 1861 ....... | 515,361 | 114 | 452,100 | |
| 1862 ....... | 1,597,477 | 124 | 1,288,300 ⎬ | 614,000 |
| 1863 ....... | 925,493 | 128 | 723,000 | |
| 1864 ....... | 373,527 | 125 | 298,800 ⎭ | |
| 1865 ....... | 429,571 | 118 | 364,000 ⎫ | |
| 1866 ....... | 361,382 | 118 | 306,300 | |
| 1867 ....... | 502,980 | 114 | 441,200 ⎬ | 444,000 |
| 1868 ....... | 820,387 | 112 | 732,500 | |
| 1869 ....... | 375,136 | 100 | 375,000 ⎭ | |
| 1870 ....... | 879,073 | 109 | 806,500 ⎫ | |
| 1871 ....... | 955,176 | 112 | 852,800 | |
| 1872 ....... | 511,230 | 127 | 402,500 ⎬ | 571,000 |
| 1873 ....... | 557,711 | 129 | 432,300 | |
| 1874 ....... | 416,643 | 115 | 362,300 ⎭ | |
| 1875 ....... | 709,522 | 110 | 645,000 ⎫ | |
| 1876 ....... | 291,856 | 107 | 272,800 | |
| 1877 ....... | 289,647 | 103 | 281,200 ⎬ | 370,100 |
| 1878 ....... | 318,985 | 92 | 346,700 | |
| 1879 ....... | 268,173 | 88 | 304,700 ⎭ | |
| 1880 ....... | 345,167 | 95 | 363,300 ⎫ | |
| 1881 ....... | 329,714 | 92 | 358,400 | |
| 1882 ....... | 342,893 | 95 | 360,900 ⎬ | 390,000 |
| 1883 ....... | 381,955 | 94 | 406,300 | |
| 1884 ....... | 409,518 | 89 | 460,100 ⎭ | |
| 1885 ....... | 397,849 | 85 | 468,058 ⎫ | |
| 1886 ....... | 316,802 | 79 | 401,000 | |
| 1887 ....... | 233,571 | 79 | 295,700 ⎬ | 364,000 |
| 1888 ....... | 244,657 | 82 | 298,400 | |
| 1889 ....... | 299,836 | 84 | 356,900 ⎭ | |
| 1890 ....... | 290,814 | 83 | 350,400 | |

SOURCE.—Great Britain, Public Record Office, Customs 9; index of metal products from B. R. Mitchell and Phillis Deane, *Abstract of British Historical Statistics* (Cambridge, 1962), pp. 471–72 (Rousseaux Price Indices).

396    *Russell I. Fries*

TABLE 4

BIRMINGHAM PROOF HOUSE: TOTAL PROOFS

| Year | N | Year | N |
|---|---|---|---|
| 1816 ........ | 127,431 | 1872 ........ | 815,863 |
| 1817 ........ | 96,673 | 1873 ........ | 756,056 |
| 1818 ........ | 144,778 | 1874 ........ | 626,478 |
| 1819 ........ | 120,617 | 1875 ........ | 695,554 |
| 1820 ........ | 84,568 | 1876 ........ | 466,748 |
| 1821 ........ | 74,579 | 1877 ........ | 458,656 |
| 1822 ........ | 65,954 | 1878 ........ | 559,815 |
| 1823 ........ | 108,447 | 1879 ........ | 552,152 |
| 1824 ........ | 113,324 | 1880 ........ | 638,070 |
| 1825 ........ | 171,295 | 1881 ........ | 730,634 |
| 1826 ........ | 137,170 | 1882 ........ | 771,597 |
| 1827 ........ | 81,955 | 1883 ........ | 681,439 |
| 1828 ........ | 76,977 | 1884 ........ | 694,035 |
| | | 1885 ........ | 501,634 |
| 1855 ........ | 264,477 | 1886 ........ | 459,052 |
| 1856 ........ | 275,468 | 1887 ........ | 440,334 |
| 1857 ........ | 302,670 | 1888 ........ | 460,211 |
| 1858 ........ | 198,238 | 1889 ........ | 529,082 |
| 1859 ........ | 250,922 | 1890 ........ | 520,949 |
| 1860 ........ | 301,021 | 1891 ........ | 561,631 |
| 1861 ........ | 380,781 | 1892 ........ | 379,086 |
| 1862 ........ | 622,372 | 1893 ........ | 335,271 |
| 1863 ........ | 460,140 | 1894 ........ | 299,273 |
| 1864 ........ | 221,726 | 1895 ........ | 328,791 |
| 1865 ........ | 576,884 | 1896 ........ | 324,898 |
| 1866 ........ | 582,127 | 1897 ........ | 402,115 |
| 1867 ........ | 766,893 | 1898 ........ | 392,939 |
| 1868 ........ | 961,459 | 1899 ........ | 375,513 |
| 1869 ........ | 693,572 | 1900 ........ | 390,268 |
| 1870 ........ | 852,079 | | |
| 1871 ........ | 891,228 | | |

SOURCE.—Clive Harris, ed., *The History of the Birmingham Gun-Barrel Proof House* (Birmingham [1946]), pp. 152–55.

closer and less divided, one.[57] For comparison, similar statistics in the American case (table 5) show that consumption at home ranged upward from 50 to 85 percent, and thus the need to cater to foreign tastes was less essential.[58]

[57]Since exports were not dutied there was less tendency to count them accurately. Hence British exports almost surely underestimate British arms actually shipped abroad. This would reduce the percentage of domestic consumption.

[58]United States, Bureau of Statistics, *Foreign Commerce and Navigation of the United States* (Washington, D.C., 1868–1900); United States, Bureau of the Census, *Census of Manufactures* for 1869, 1879, and 1889 (Washington, D.C., various). The same procedure was followed as in n. 56 above, except that value was used rather than number. The same bias applies to U.S. shipments abroad, since they were not taxed (table 5, U.S. arms production, exports, imports, and consumption).

## TABLE 5

### UNITED STATES ARMS INDUSTRY: PRODUCTION, EXPORTS, AND CONSUMPTION

| | 1849 | 1859 | 1869 | 1879 | 1889 | 1899 |
|---|---|---|---|---|---|---|
| a) Firearms production ($) | 1,173,014 | 2,368,931 | 5,582,258 | 4,736,936 | 2,922,514 | 5,444,659 |
| b) Firearms export ($) | 58,650* | 236,893* | 2,777,398 | 2,169,230 | 820,933 | 681,440 |
|   % of production | 5* | 10* | 49.7 | 37.8 | 28.1 | 12.5 |
| c) Home consumption of U.S. product (a − b) ($) | 1,114,364* | 2,132,038* | 2,804,860 | 3,567,706 | 2,101,581 | 4,763,219 |
|   % of production | 95* | 90* | 50.3 | 62.3 | 71.9 | 87.5 |
| d) Firearms imports ($) | 232,110 | 358,370 | 267,209 | 576,702 | 1,159,157 | 1,044,039 |
| e) Reexports of foreign arms ($) | 12,148 | 48,289 | 5,551 | 6,254 | 14,275 | 7,574 |
| f) Net imports ($) | 219,962 | 310,081 | 261,658 | 570,448 | 1,144,882 | 1,036,465 |
| g) Total firearms production (c + f) ($) | 1,334,326 | 2,442,119 | 3,066,518 | 4,138,154 | 3,246,463 | 5,799,684 |
| h) Imports as % of total consumption (f/g) | 16.5 | 12.7 | 8.5 | 13.8 | 35.3 | 17.9 |
| i) U.S. production as % of total consumption (c/g) | 83.5 | 87.3 | 91.5 | 86.2 | 64.7 | 82.1 |
| j) Production in constant dollars | 755,000 | 1,580,000 | 2,460,000 | 4,280,000 | 2,520,000 | 4,640,000 |
| k) Domestic consumption of U.S. product in constant dollars | 719,000 | 1,432,000 | 1,224,000 | 2,660,000 | 1,810,000 | 4,070,000 |
| l) Total U.S. consumption in constant dollars | 860,000 | 1,626,000 | 1,352,000 | 3,090,000 | 2,800,000 | 5,000,000 |

SOURCES.—For production: U.S. Bureau of the Census, *Thirteenth Census: 1909*, 8:486–87. For imports: U.S. Treasury Department, *Commerce and Navigation of the United States* (annual volumes, 1820–66), and U.S. Treasury Department, Bureau of Statistics, *Foreign Commerce and Navigation of the United States* (annual volumes, 1867–1900). For Exports: U.S. Treasury Department, Bureau of Statistics, *Foreign Commerce and Navigation of the United States* (annual volumes, 1867–1900). Figures on production and consumption corrected to constant dollars using wholesale price indices of metal products in U.S. Bureau of the Census, *Historical Statistics of the United States, Colonial Times to 1957*, series E-7 and E-20, pp. 115–17. All from Government Printing Office, Washington, D.C.
*Figures on exports for 1849 and 1859 are estimated at 5 percent and 10 percent, respectively, since no reliable export series for these years exist.

398     *Russell I. Fries*

Most people have accepted the luxury nature of the British market as a given. Yet there is evidence that many of the shotgun purchasers, even within England, were buying guns of relatively low price. Such purchasers were forced to buy a poor-quality shotgun because they could not afford one of high quality. They might well have preferred getting a solid, unfitted interchangeable gun at the same low price. This kind of purchaser should have been suited precisely by a standardized and inexpensive shotgun manufactured entirely by machinery, and this is indeed the characteristic type of production in England today. Testimony of the gun makers themselves during consideration of the 1870 Gun Licensing Act supported the existence of a large market for low-priced guns.[59]

Because of the nature and exclusions of the Gun Licensing Act it is impossible to use it as any evidence of the total size of the firearms market within Great Britain. However, it can be used to demonstrate the rough outlines of the aristocratic market. It is probably fair to assume that almost all those paying the license fee were "sportsmen" using relatively high-priced rifles or shotguns for either game or target shooting, not utilitarian purposes.

Examination of registrations under the subsequent Gun Licensing Act gives some indication that the luxury market absorbed relatively few of the guns actually produced in Great Britain. Between 1870 and 1880 there were roughly 60,000 new licenses issued.[60] Assuming that some 50 percent of the 93,679 registered gun owners in 1870 also purchased a new firearm during the intervening years would mean that the actual number of luxury guns sold between 1870 and 1880 was 106,000, or about 11,000 out of the estimated total British consumption of upward of 100,000 per annum. This meant that there were a substantial number of nonluxury sporting arms sold which could have been produced cheaper or better by machine.

The export market was equally well adapted to consumption of machine rather than custom-made firearms. After 1860 the United States was the single largest purchaser of British arms (see table 6). Between 1865 and 1889 it bought an increasingly heavy share of the total value of British exports, moving from a low of 6 percent in 1865–69 to a high of 29.7 percent in 1880–84. In the category of shotguns (fowling pieces) the United States was an even heavier purchaser, taking between 20 and 60 percent by value between 1866 and 1881 (see table 7). Most gun makers who spoke on the subject reported that the United States was their single most important

---

[59]Birmingham Proof House Records, Annual Report for 1870, p. 7; Resolution to Parliament dated April 3, 1870; *Times* (London) (April 23, 1870), pp. 9–10.

[60]*P.P.*, 1871, 37:209; 1872, 36:765–73; 1875, 42:689; and 1880, 40:573.

TABLE 6

GREAT BRITAIN, EXPORT OF FIREARMS: PERCENTAGE TAKEN BY SELECTED COUNTRIES

| Country | 1845–49 | 1850–54 | 1855–59 | 1860–64 | 1865–69 | 1870–74 | 1875–79 | 1880–84 | 1885–89 |
|---|---|---|---|---|---|---|---|---|---|
| United States | 6.1 | 25.3 | 11.8 | 41.5 | 6.6 | 11.6 | 16.3 | 29.7 | 21.2 |
| India | 57.3 | 16.7 | 40.6 | 2.0 | 3.8 | 3.3 | 5.1 | 5.2 | 8.6 |
| West Africa | 10.6 | 17.8 | 16.9 | 7.1 | 7.5 | 7.7 | 10.7 | 10.3 | 6.3 |
| Australia | 0.4 | 6.4 | 2.5 | 5.1 | 3.0 | 3.2 | 12.3 | 13.8 | 20.6 |
| Canada | 1.2 | 3.0 | 2.6 | 1.0 | 1.3 | 1.8 | 2.5 | 2.4 | 5.6 |
| Brazil | 2.2 | 2.6 | 0.1 | 0.5 | 1.3 | 0.3 | 0.6 | 4.1 | 3.0 |
| China | 0.1 | 0.3 | 0.3 | 0.7 | 8.0 | 0.6 | 0.1 | 2.0 | 1.3 |
| Egypt | 1.1 | 3.5 | 5.3 | 2.5 | 9.2 | 2.0 | 0.6 | 0.3 | 0.3 |
| Total | 79.0 | 75.6 | 80.1 | 60.4 | 40.7 | 30.5 | 48.2 | 67.8 | 66.9 |

SOURCE.—Great Britain, Public Record Office, Customs 9. The countries listed include most of the consistently heavy purchasers and representative countries from most regions outside Europe. European countries made occasional large purchases.

400     *Russell I. Fries*

TABLE 7

GREAT BRITAIN: FOWLING-PIECE EXPORTS TO UNITED STATES

| Year | N | Five-Year Average | Value | Five-Year Average | % of Total Exports |
|------|------|------|------|------|------|
| 1860 ......... | 7,558 | | 16,688 | | 13.5 |
| 1861 ......... | 2,805 | | 6,573 | | 6.0 |
| 1862 ......... | 119 | 2,268 | 346 | 5,240 | .8 |
| 1863 ......... | 527 | | 1,612 | | 3.6 |
| 1864 ......... | 334 | | 982 | | 2.2 |
| 1865 ......... | 973 | | 2,055 | | 6.2 |
| 1866 ......... | 15,448 | | 30,919 | | 45.5 |
| 1867 ......... | 11,844 | 9,997 | 29,342 | 20,932 | 40.3 |
| 1868 ......... | 4,836 | | 10,792 | | 19.3 |
| 1869 ......... | 16,882 | | 31,550 | | 38.0 |
| 1870 ......... | 27,216 | | 46,705 | | 49.5 |
| 1871 ......... | 31,304 | | 61,010 | | 58.2 |
| 1872 ......... | 36,023 | 25,514 | 87,316 | 58,038 | 58.0 |
| 1873 ......... | 22,107 | | 59,437 | | 45.5 |
| 1874 ......... | 10,922 | | 35,722 | | 34.9 |
| 1875 ......... | 13,910 | | 43,682 | | 37.8 |
| 1876 ....... | 7,218 | | 30,252 | | 27.6 |
| 1877 ......... | 6,594 | 11,301 | 25,673 | 41,227 | 25.7 |
| 1878 ......... | 14,196 | | 52,534 | | 40.4 |
| 1879 ......... | 14,585 | | 53,994 | | 44.7 |
| 1880 ......... | 19,340 | 21,066 | 71,005 | 73,249 | 53.3 |
| 1881 ......... | 22,791 | | 75,492 | | 52.0 |

SOURCE.—Great Britain, Public Record Office, Customs 9.

customer.[61] It would be difficult to pick a country which had demonstrated greater enthusiasm for mass-produced articles than the United States, though it was becoming more of a luxury-importing market than earlier. Australia was the next most important export buyer, and it seems likely that guns possessing the relatively high price and high utility value characteristic of American goods would have found a ready market there as well. Possibly luxury buyers in India, another important British market, would not have accepted the relatively plain and unfitted guns turned out by machine firearms factories, but after 1875 was less than half as important as the expanding Australian market (table 8).

British purchasers demonstrated no aversion to machine-produced revolvers. By the middle of the 1880s most of the handicraft revolver makers had been driven from the industry by the machine revolver producers, particularly Webley & Scott. Despite the fact that fitting was probably less important for revolvers than shotguns, the success

[61]Greener, p. 286.

TABLE 8

BRITISH EXPORT OF FIREARMS: AVERAGE VALUE OF FOWLING PIECES
(£)

| Year | Shipped to All Countries | Shipped to W. Africa | Shipped to United States | Shipped to Australia | Shipped to India |
|------|---------|---------|---------|---------|---------|
| 1862 .......... | 2.36 | 0.55 | 2.91 | 2.73 | 4.96 |
| 1863 .......... | 1.71 | 0.53 | 3.30 | 2.85 | 2.44 |
| 1864 .......... | 1.08 | 0.46 | 2.94 | 2.04 | 1.83 |
| 1865 .......... | 2.15 | 1.93 | 2.12 | 2.04 | 1.38 |
| 1866 .......... | 1.86 | ... | 2.00 | 1.97 | 1.25 |
| 1867 .......... | 1.56 | 0.42 | 2.47 | 2.72 | 1.20 |
| 1868 .......... | 1.35 | 0.44 | 2.23 | 3.09 | 1.69 |
| 1869 .......... | 1.50 | 0.44 | 1.86 | 1.85 | 1.46 |
| 1870 .......... | 1.43 | 0.43 | 1.72 | 2.20 | 2.06 |
| 1871 .......... | 1.59 | 0.45 | 1.95 | 1.79 | 1.95 |
| 1872 .......... | 1.79 | 0.48 | 2.42 | 1.99 | 1.90 |
| 1873 .......... | 1.78 | 0.48 | 2.62 | 2.02 | 2.08 |
| 1874 .......... | 1.90 | 0.47 | 3.27 | 1.80 | 3.04 |
| 1875 .......... | 2.06 | 0.47 | 3.14 | 1.89 | 2.46 |
| 1876 .......... | 2.06 | 0.43 | 4.19 | 2.11 | 2.48 |
| 1877 .......... | 2.01 | 0.41 | 3.90 | 2.63 | 3.09 |
| 1878 .......... | 2.22 | 0.39 | 3.70 | 2.60 | 5.00 |
| 1879 .......... | 2.39 | 0.39 | 3.70 | 2.79 | 7.14 |
| 1880 .......... | 2.38 | 0.37 | 3.67 | 2.67 | 13.89 |
| 1881 .......... | 2.31 | 0.37 | 3.31 | 2.43 | 10.61 |

SOURCE.—Calculated from Great Britain, Public Record Office, Customs 9.

of the large-scale revolver manufacturers suggests that machine-made products were readily accepted there.[62] American revolvers, such as the Colt and the Smith & Wesson, sold well in England.[63]

The factors of an existing lower- and middle-class domestic arms market, large overseas arms market suited to machine-produced arms, and insufficient verification of the hypothesis that machine-produced arms would not serve consumers, suggest that British producers were influenced more by tradition and prejudice than reality in rejecting the full technology of the American System. William Greener's use of the term "degenerate specimen"[64] to describe the

[62]The firm of Philip Webley (now Webley & Scott, Ltd.) was the only significant revolver producer in England by the mid 1880s. At the time of its incorporation in 1897 the firm had a plant valued at about £70,000 (Webley & Scott, Ltd., Annual Report for 1897) (see also William Chipchase Dowell, *The Webley Story* [Leeds, 1962]).

[63]The Colt Patent Fire Arms Manufacturing Company sold an average of £14,000 worth of revolvers through its London sales agency in 1854 and 1855. Correspondence and orders in the 1870s and 1880s suggest annual sales of $10,000–$20,000 (Foreign Letter Book, Samuel Colt Manuscripts, Connecticut State Library; "London," n.d. [ca. 1856], Samuel Colt Manuscripts, Connecticut Historical Society).

[64]Greener, p. 286.

machine-produced double shotgun illustrates the emotional nature of many arguments against mass-produced guns. In a similar fashion, American repeating and automatic shotguns are listed by Greener under the heading "miscellaneous," separate from other shotguns, though by 1900 they were a major part of American production.[65]

There are two pieces of factual evidence against this thesis of British entrepreneurial failure. Despite tariffs that ranged upward to more than 30 percent, English firms continued to sell shotguns in America, particularly high-value breech-loading double guns, after 1871, though in decreasing volume. Also, with the exception of the Colt and Smith & Wesson revolvers, American firms made relatively few sales in the British market, despite the fact that English arms makers were no longer protected by tariffs. The reason for this seeming British strength may be twofold. American purchasers were buying guns of a type which could not be manufactured readily in the United States, given the high cost of skilled labor and the very great percentage of skilled labor that went into making a fine double gun. Second, the high cost of labor and extensive capital employment in the United States made even the cheaper American guns high priced, compared with the least-expensive British models.[66]

This evidence does not affect the hypothetical benefits which the British *might* have derived from a switch to the American System of production for sporting arms, but it does help to explain why such a switch was not *necessary*. Since American direct competition could not be very effective in England, due to the higher American labor cost, the British were not forced to alter their production pattern for some time. So long as profits remained barely adequate in the present production mode, businessmen were certainly reluctant to shift to an entirely different method of manufacture, which in most cases made their old talents obsolete. Couple this lack of pressure for *immediate* change with the demonstrated emotional commitment to old products and manufacturing skills and you have the explanation for the conservative pattern in sporting-arms manufacture.

The evidence suggests that after 1855 the introduction of machine techniques into British firearms manufacture proceeded at a pace consonant with economic realities except among the sporting-arms makers. With the exception of the African segment of the industry, all the producers eventually adopted one or more facets of the American

[65] Ibid., pp. 503–6.

[66] Substantial British shotguns could be made and sold for as little as £2 or £3 wholesale ($10–$15), compared with £.45 ($2.20) for inferior grades. American shotgun manufacture did not begin much before 1870. Most double guns sold above $50, while even repeaters were $25 or $30.

System. Yet the fully interchangeable system of manufacture was adopted only by the government's Enfield Arsenal and the producers for the government. The sporting- and the African-arms makers refused to adopt the cost-saving system of fully interchangeable manufacture because they believed it offered them no benefits. Thus even so revolutionary a technological advance as the American System, which offered cost benefits of up to 25 percent or more in the United States, was not always adapted to market or resource conditions in the receiving country.

Similarly, the historical process of technological adoption is not always one that proceeds at what might be termed an economically optimal pace. The British sporting-arms makers generally ignored the complete American System since it did not fit their existing pattern of production and their mental image of the market that they thought they were serving. Of course, the argument in the last section cannot prove that the British firms would have been able to profit from the adoption of the new technology, since most never did adopt it. However, most of the old firms also suffered either substantial decline or total extinction, while most major existing British small-arms firms today do use American methods. Even the one segment which adopted the complete American System, the military-arms makers, did so mainly as a response to economic pressure from the War Office, rather than firm conviction of its benefits. Thus mental attitudes toward technology undoubtedly played an important role in the British response to American technology. It might prove interesting to observe whether the pattern was different in any industry close to the machine-producing centers of Manchester, Leeds, or Glasgow. Perhaps closer proximity to machine producers would have resulted in more ready acceptance of the output of machines.

While the focus of this study has been narrow and specific, it suggests that technological change is a complex process in which receptivity to change is an important variable. Economic analysis may help to explain why a particular process was or was not adopted, as in Section II. It cannot always show a sufficient reason for rejection of a seemingly economically advantageous process, as was the case in British sporting-arms makers' rejection of the full American System. The real reason for this rejection seems to lie in the fiercely held belief in advanced ingenious mechanisms which the gun makers thought could only be created by skilled craftsmen and the assumption that the small aristocratic percentage of the market necessarily dictated retention of a dying system of manufacture for the market as a whole. Fear of corporations and loss of individual initiative also acted as a brake on some makers.

# [16]

## Transferring Technology to a Peripheral Economy: The Case of Lower Danube Transport Development, 1856–1928

JOHN H. JENSEN AND GERHARD ROSEGGER

This paper deals with some of the technical, economic, and sociopolitical forces which shaped the development of a modern transportation system in the lower Danube basin.[1] Our focus will be on the persistent pattern of competition between the river, as the traditional mode, and the railroads, as the technological intruders into established commercial and political relationships.

During the first phase (1856–65) this competition was dominated by several factors: the respective performances of an international public enterprise (the European Danube Commission) and of a private concern (the Danube and Black Sea Railway and Kustendje Harbour Company, Ltd.) in transferring and adapting Western technology to a backward region;[2] the political and economic motives of the participants in trade and transport; and the emergence and subsequent decline of the region as a major grain supplier to the industrialized countries of western Europe.

For our story, it is important to note that, in this first phase, technological leadership came entirely from outside the area. The engineers employed by the European Danube Commission and the

---

DR. JENSEN is professor of history at the University of Waikato in Hamilton, New Zealand, and DR. ROSEGGER is professor of economics at Case Western Reserve University.

[1] For purposes of this paper, this term includes the areas drained by the Danube and its tributaries—the Sereth and Pruth—and delineated by the Balkan and Carpathian Mountains and the Iron Gate.

[2] Other aspects of development have been covered in our papers, "British Railway Builders along the Lower Danube, 1856–1869," *Slavonic and East European Review* 46 (1968): 105–28; and "The Danube–Black Sea Shortcut: A Problem in Technology Transfer" (paper presented at the Third Colloquium of the International Cooperation in the History of Technology, Jablonna, Poland, August 1973). Some early results of the present study were reported in "Rail-River Competition in the Lower Danube Basin, 1856–1865" (paper presented at the IIIᵉ Congrès International d'Etudes du Sud-Est Européen, Bucureşti, September 1974).

676 *John H. Jensen and Gerhard Rossegger*

young men, half entrepreneurs, half engineers, who directed the work on the railway were from Britain or from western Europe. Local Turkish officials had some considerable influence on the environment in which these foreigners acted, but the latter took all the initiatives. The bulk of the local people were passive onlookers or, at best, menial participants in the work forces of the development projects. Their technological knowledge proved irrelevant to the tasks at hand. They were unable to take an active part in the transformation of their own country until they had formed new social groupings to conform to Western models of labor-management relations, and until they had acquired working skills, habits, and outlooks which satisfied the requirements of modern technology.

During the second phase (1896–1928), a new set of forces dominated advancement in transportation—Romanian nationalism together with the professional ambitions of a growing class of native engineers and technicians, and the development of petroleum exports as a key element in the integration of the region's transport system into a more broadly conceived network. Certain of the projects undertaken in this period, albeit still with foreign technical assistance, were characteristic of a new national consciousness. Saligny's bridge across the Danube at Cernavodă joins the Budapest Danube bridge, the Yangtze bridge, and the Aswan High Dam as a useful symbol of that drive toward technological and cultural maturity so characteristic of a politically motivated, but as yet unproven, elite.

But the region's traditional activity, agriculture, managed to maintain its role in the midst of these modernizing trends. Foreign exchange was essential for the acquisition of capital goods and technology, and foreign exchange could still be best earned through the bulk export of foodstuffs. Indeed, the requirements of nonagricultural development increased the pressure on the agricultural sector well into the 20th century. This pressure helped to force additional rural areas into the market economy just as it exacerbated rural social relationships. The great peasant revolt of 1907 served to demonstrate the farming population's impotent fury at modernization as it struck out against the sharecropping arrangements by which surplus grain was extracted from the countryside to be used to maintain Romania's balance of payments in equilibrium. After 1918, agrarian reform made the collection of grain for export more difficult. However the efforts of economists and government officials alike were devoted to repairing this situation.[3] Agricultural exports remained the major

[3]On the specific pressures of the market economy on Romanian agriculture, see H. L. Roberts, *Rumania: Political Problems of an Agrarian State* (New Haven, Conn., 1951), pp. 10–17.

source of foreign exchange earnings in spite of the rapid growth of Romania's petroleum industry. On the surface, these interests might appear to have unified efforts in transport development, but competition between river and rail continued.

Economic pressures also worked upon the Romanian engineering profession. Those who were dependent on government employment resented the use of foreign engineers in better-paid consultancies or in specially created top posts. Those who had political connections tried to apply leverage in gaining the places of foreigners. And those who went into business for themselves (e.g., I. G. Cantacuzino with his Brăila cement works)[4] saw their success rooted in obtaining monopolies on government contracts. The slender economic base of these native engineer-entrepreneurs made their struggle all the more intense. However, until the 1920s they participated only marginally in transport development. By that time, liberal governments insisted that nationalistic policies extend not only to state-sponsored technological ventures but also to the utilization of native private enterprise in the execution of these undertakings. But this is to get well ahead of our story, and perhaps to trespass beyond its limits.

The thread which unites our account of both phases of development is the continuing interaction between foreign and Romanian interests. Western economic concern begins with the Treaty of Adrianople (1829) which abolished Turkish rights of preemption in the Danubian principalities, a monopsony which ensured tribute payments and provided livestock, grain, wine, and cooking oils for the bazaars of Constantinople. But the local landowners and merchants were equipped with neither the know-how nor the capital requirement to take advantage of the opportunities created by the terms of Adrianople. Instead, western European diplomats at the Porte gained rights to trade and Western businessmen quickly stepped into the breach.

Definite regional patterns of trade and transport soon developed. But these were more successful in getting grain from the interior to the maritime Danube ports of Brăila and Galaţi than in overcoming the handicaps of navigational conditions on the Lower Danube and in the delta. (The sketch map in fig. 1 shows the relevant geographic features discussed in this paper.) Between the Iron Gate, that major obstacle to shipping on its middle course, and Cernavodă, where it runs to within approximately 70 kilometers of the Black Sea, the river offered reasonably good conditions for small vessels. However, below

[4]On I. G. Cantacuzino and the Brăila cement factory, see Societatea Politecnică din România, *60 (Sase) de ani dela deschiderea primei căi ferate realizată de ingineri Români şi dela înfiinţarea Societăţii Politecnice* (Imprimeria CFR, Bucureşti, 1941), p. 45.

678    *John H. Jensen and Gerhard Rossegger*

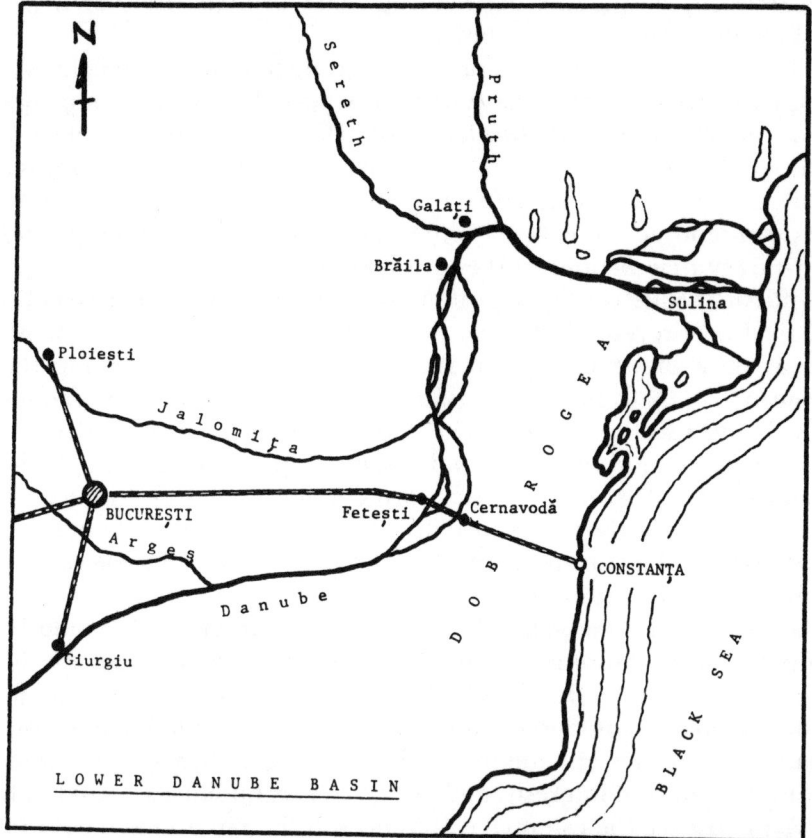

Fɪɢ. 1.—Sketch map of the Lower Danube basin

Cernavodă the Danube turns sharply north, meandering through marshlands, forming numerous secondary arms, and posing continuous hazards through the shifting of the shipping channel. At Brăila, which was accessible to oceangoing ships of the size then used in the Levant trade, the river forms a single course again and shortly thereafter flows eastward through the delta into the Black Sea about 110 kilometers away.

Here, natural as well as political conditions posed major obstacles to transport. Predominant among the former was silting. Before 1829, the Turks had maintained adequate channel depths by requiring all vessels to drag along iron rakes, thus stirring up the silt which would then be carried farther to sea by the current. Not only did cessation of this primitive remedy increase the problem, but the spread of extensive farming along the Danube's tributaries, without regard to soil conservation, meant that each spring flood brought another heavy load of silt into the delta channels.[5]

The major political factor inhibiting transport improvement was Russian control of the delta during this period. Any such improvement for the grain trade on the Lower Danube would have been in clear conflict with Russia's interest in her own exports out of Odessa. Thus, instead of assuring safe navigation, the authorities in charge of the delta ignored all technical work and winked at the pilferage and near-piracy that victimized Western merchant vessels.

Until the 1850s, conflicting commercial interests and intensive political rivalries prevented any real progress for transport development, just as they prevented any more general cooperation among the Western powers to put the Turkish empire on the road to economic growth and fiscal soundness. The Turks, in their turn, viewed Western efforts to find investment and trading opportunities and to bring modern technology into the backward country with great suspicion.[6] Russia's defeat in the Crimean War brought a radical change in these conditions, and it is at this point that our account of technological and commercial competition begins.

At the outset, some general observations are in order. It would be tempting, from the purely technological point of view, to equate the history of transport development in the region with the seemingly similar problems confronted in many areas of the United States.[7] But whatever analogies one might find by focusing on topography, competing commerical interests, and alternative technologies, would only serve to obscure the essential differences in the political and cultural settings. Nor are there any parallels between the forces driving the several developers of the Danubian transport system and, say, Friedrich List's vision of the railway as a symbol of German national unity.

Instead, we must look to the history of quasicolonial and postcolo-

[5]The progressive settlement in the Sereth and Pruth watersheds to the north can be studied in R. F. Kaindl, *Das Ansiedlungswesen in der Bukovina seit der Besitzergreifung durch Oesterreich. Mit besonderer Berücksichtigung der Ansiedlung der Deutschen* (Innsbruck, 1902). The process of agricultural expansion, with the related removal of forest cover, can be followed in S. Iana, "Regresia limitei pădurilor in timpuri istorice in raionul Oltenița," *Analele Universitații C. I. Parhon: Seria Știinţele Naturii (Geologie-Geografie)* no. 23 (1960): pp. 165–69.

[6]F. S. Rodkey, "Ottoman Concern about Western Economic Penetration in the Levant, 1848–1856," *Journal of Modern History* 30 (1958): 348–53.

[7]See, for example, J. Rubin, "Canal or Railroad?" *Transactions of the American Philosophical Society*, n.s. 51, vol. 7 (November 1961); and A. Fishlow, *American Railroads and the Transformation of the Ante-Bellum Economy* (Cambridge, Mass., 1965).

680    *John H. Jensen and Gerhard Rossegger*

nial infrastructure development for our more general models. There is the interest of advanced countries in the provision of exporting facilities to supply cheap foodstuffs, raw materials, and energy to the metropole; there are local landlords, merchants, and shippers who influence development either through their narrowly conceived economic interests or merely through their traditionalism; there is a new state's need to earn large amounts of foreign exchange so that a stable currency, banking facilities, and modern technology can be acquired; there are the professional men, bureaucrats and intellectuals, strongly xenophobic and yet profiting from foreign stimulated growth and thus ambivalent in attitudes which gradually emerge as a heightened brand of elite nationalism.

So far, we can follow the Romanian experience as a replication of many stories of emerging national consciousness in the political, economic, and technological spheres. But then we come to a stop as we look at the special characteristics of our case. Shifting and faltering economic interests of foreign investors and traders; competition between an emerging nation-state and one of the earliest international development institutions; and the continuing failure of that state to achieve anything like an optimal coordination of public and private investment in transport—these are some of the differentiating features that lend uniqueness to the story of the Lower Danube basin's hesitant progress.

There are no easy comparisons in this story, and we believe that it is best to look at the particularities of development in a region in full consciousness of the difficulties in drawing lessons for development out of a special environment and a special experience. Our belief is based on the view that the value of the historical approach to development lies in its acting as a spur toward the careful examination of dissimilar similarities, in the search for the factors accounting for the openness or laggardness to change which are so central in the transfer of technology and the concomitant economic growth.

### Foreign Engineers and Entrepreneurs: 1856–65

At the conclusion of the Crimean War in 1856, the Danubian principalities (the nucleus of present-day Romania) were placed under the supervision of the European great powers, while the Ottoman empire as a whole was formally included in the European family of nations. Now Western entrepreneurs found new vigor in the hope that commercial opportunities would be safeguarded by appropriate political arrangements, especially when it came to the negotiation of concessions for investment projects and protective agreements in Constantinople.

In this promising atmosphere, two major projects were initiated in order to realize the full potential of the principalities' grain production: (a) the European Danube Commission's work to open up the delta and thus prepare the way for the safe progress of merchant ships to the ports of Brăila and Galați; and (b) the private Danube and Black Sea Railway and Kustendje Harbour Company, Ltd.'s line between Cernavodă and Constanța (the shortest distance from the middle Danube to the sea), together with handling and storage facilities at both terminals, which founding British investors hoped would provide the equivalent of a direct link of river and deep-water port well away from the obstacles and hazards of the delta.

Both projects were designed to establish an outlet for the grain supplies produced in the region. But in every other respect they were fundamentally different, and it is these differences which add interest to their comparative roles in the Lower Danube basin's economic development. At the same time, they bring to the fore questions about the relative viability of public and private enterprise and of traditional versus "exotic" technologies in infrastructure improvement.

Before these questions can be properly examined, however, the experiences of the commission and of the company should be narrated briefly to provide the necessary background and to illustrate the difficulties of technology transfer to a primitive region.

*a*) In the Treaty of Paris (1856) Turkey was given political sovereignty over the maritime Danube in exchange for the Porte's acceptance of technical and navigational control by the European Danube Commission.[8] For three years thereafter, no real progress was made in improving the conditions for shipping. The Turkish government had too little money to pay for extensive works, and the lack of improvements prevented the commission from collecting any substantial amounts of passage fees, which were meant to be its major source of revenue. The commission's structure was truly European, with a Turkish chairman, French naval support, Austrian consular supervision, and a British chief engineer, Charles Hartley.

There was much bickering among the administrators and politicians about the best technical approach to navigational improvement and even about the proper delta channel to select for the work. The Sulina mouth was already the most used, but the commissioners initially favored the development of the larger St. George's channel. At this point, the chief engineer's determination to do something and his technical ingenuity managed to overcome the commission's dis-

---

[8]D. A. Sturdza, *Les Travaux de la Commission Européenne des Bouches du Danube, 1849 à 1911* (Wien, 1913), pp. 1–3. This is the essential source for the commission's work.

682    *John H. Jensen and Gerhard Rossegger*

agreements and penury. Hartley obtained permission to start work with the Sulina "in a provisional way"—after all, there was traffic to be accommodated there—and his first efforts were so successful that the alternatives were quickly forgotten.[9]

Hartley set out to build two inexpensive stone jetties out into the sea, thus creating a funnel through which the increased water pressure swept the worst of the silt into deep water where it dispersed. He spoke confidently of increases in channel depth of 2 feet per year through this device alone. A Turkish subsidy of £40,000 annually was sufficient to carry on with the project, and by 1861 the jetties extended 4,630 and 3,000 feet, respectively, eastward into the Black Sea. At this time, depth increases had exceeded the chief engineer's forecast. Occasionally, floods and strong east winds drove heavy loads of silt back into the channel, but he pushed on doggedly. The record of his achievement is quite remarkable, especially in view of the essential technical simplicity of this approach: a minimum depth of 17½ feet permitted all common vessels to proceed up the delta with safety.[10] Traffic responded quickly: tonnage moving through the Sulina increased from just over 300,000 in 1859 to nearly 600,000 in 1864. This initial success enabled the commission to generate sufficient fees income to proceed with the gradual extension of navigational improvement upriver.

In 1865 Hartley tackled the next big project—straightening the winding Sulina itself. In the course of the next decade, twenty-seven elbows of the channel were cut off. In addition, carefully prepared charts and markings enabled ship captains to reach the grain ports in safety, and the commission's efficient administration prevented all those costly abuses under which shippers had suffered during Russian control. Chief beneficiaries were clearly the British—the tonnage of their vessels increased from 68,000 in 1861 to 216,000 in 1874.[11]

*b*) During the years immediately after the Crimean War, however, no one would have predicted such a rapid and complete fulfillment of

[9]These projects are set out clearly in a bound manuscript volume in the possession of C. W. Hartley, CBE, entitled *European Danube Commission: Projects for the Improvement of the Lower Danube* (1857[?]). C. W. Hartley, of Amberly near Stroud, Gloucestershire, is the grandnephew of Sir Charles Hartley and has provided us with much biographical detail on the latter's career. We would like to express our gratitude here for C. W. Hartley's generous assistance with our research and wish him well in his projected biography of his great-uncle.

[10]Sturdza, pp. 9–11. See also Sir Charles A. Hartley, "Description of the Delta of the Danube, and of the Works Recently Executed at the Sulina Mouth," *Minutes of Proceedings, Institute of Civil Engineers* 21 (1861–62): 272–308.

[11]A. Beer, *Geschichte des Welthandels im 19. Jahrhundert* (Wien, 1884), p. 280, n. 11.

the commission's charge. Indeed, a group of British investors viewed the public enterprise's chances with enough pessimism to undertake a competitive scheme—the trans-Dobrogean railway from Cernavodă to Constanţa. They saw the answer to the agricultural export problem in combining the most advanced form of transport technology (which had just brought such tremendous growth in the development of the British Isles) with modern grain-cleaning, storage, and handling facilities such as could not be matched by the merchants and dealers in the traditional Danubian ports.[12]

Work on the line began in 1857–58, and the railway was opened for traffic in September 1860. Although its length was only about 70 kilometers and across relatively level terrain for most of the way (see fig. 1), the builders ran into numerous difficulties. All skilled labor and even the most rudimentary tools had to be brought out from Britain. Local workers and materials proved to be more expensive and less satisfactory than the promoters had anticipated. Swamplands near the Danube posed problems of drainage and track construction, as did the one steep grade from the Dobrogean plateau down to the port of Constanţa. The cost of providing breakwaters and docks came near to bankrupting the company.[13]

Part of the responsibility for these unforeseen difficulties must rest with the entrepreneurs, whose visions of unprecedented profits closed their eyes to technical details, and part with the works' engineering staff. The young Englishmen in charge (led by the Barkley brothers—J. Trevor, George, Robert, and Henry) were largely self-taught and relied mostly on some brief practical experience in mining and railway employment in Britain or in Turkey. Careful design and planning were not their strengths; they had to make up in determination, courage, and a sharp eye for the main chance what they lacked in formal engineering knowledge. They shone in "muddling through," in handling large numbers of recalcitrant employees, and in overcoming the barriers of language and culture between British and Balkan ways of doing things. Despite their failures in the narrowly technical aspects of executing the project, who is to say that, under the circumstances, their other skills did not stand them in better stead? They,

---

[12]T. Forester, *The Danube and the Black Sea: A Memoir on Their Junction by a Railway between Tchernavoda and a Free Port at Kustendje* (London, 1857), pp. 6–7; and see *Times* (London) (May 28, 1857), p. 10.

[13]The technical aspects of construction are covered in greater detail in our "British Railway Builders. . . ." An interesting account by one of the engineers in charge of the works is H. C. Barkley, *Between the Danube and the Black Sea, or Five Years in Bulgaria,* 2d ed., (London, 1877).

684    *John H. Jensen and Gerhard Rossegger*

like so many British technical "change agents" of the period, seem to have brought a unique combination of dilettantism and sophistication to their task.

By 1861, or just about at the time of the European Danube Commission's first major successes, the original share capital of the Danube and Black Sea Railway and Kustendje Harbour Company, Ltd., had been exhausted. The volume of grain traffic over the line never reached anything near the amounts predicted by the promoters, forcing the company to mortage its future through the issue of debentures at very costly terms. These injections of additional financing proved inadequate in the face of new disasters: heavy storms in 1862 and 1863 damaged the barely completed works in Constanţa port and flooded the low-lying tracks near the Danube. Poor harvests in 1865 and 1866 ended any real hopes of dividend payments to the original shareholders.[14]

The directors of the railway gave up the struggle and offered their entire port facilities to the Ottoman government in return for a fixed annuity which was expected to cover the deficit from railway operations. Meanwhile, navigational development in the delta was going from strength to strength, reinforcing the merchants' and shippers' natural inclination to adhere to established connections and routes. Round 1 in the competition between railway and river had clearly gone to the latter.

*c*) If we set to one side all exceptional factors in the struggle between public and private enterprise (such as Hartley's luck in showing quick success and the natural disasters which befell the railway project), we can concetrate on the more fundamental determinants of each scheme's fortunes. In the first place, the trading and financing patterns of the Lower Danube basin had developed with an orientation toward the delta ports before either project was begun. The commission simply took advantage of these patterns, and the merchants did not have to change their practices. In this connection, it is also important to recognize that the commission, despite the conflicts of its member countries in other spheres of activity, did not become a political football. Though the Romanians would later chafe under such "tutelage," from 1856 to 1914 the members worked together in quite reasonable harmony to manage such mundane technical affairs as dredging, the erection of signal lights, the establishment of quarantine and health-inspection facilities, port improvements, user-fee as-

---

[14]See C. Baicoianu, *Handelspolitische Bestrebungen Englands zur Erschliessung der unteren Donau* (München, 1913), pp. 61–65; and our "British Railway Builders . . . ," pp. 125–26.

sessment and collection, and the straightening and deepening of channels. We have here one of the first examples of a principle that has since then become commonplace: an international agency and its bureaucracy are most likely to function effectively when presented with a well-defined and primarily technical set of tasks.

The primitive character of land transport and the marginal navigational facilities on the Lower Danube's tributaries were a second major handicap for the railway. With respect to grain exports, the river and rail routes were not parallel outlets for otherwise unchanged transport arrangements. These arrangements could serve either the maritime ports or the railway. In this region (and age) of strong traditional ties between producers, brokers, and shippers, a shift to rail transport might have been difficult to accomplish even under more favorable conditions for the company.

The third long-range influence on the course of the two ventures derived from the relationship which existed in each case with the local authorities and the Ottoman central government. The commission had the direct support of the great powers and freed the Porte from annoying and expensive administrative obligations of one of the empire's most unstable frontiers. The fact that the commission's chairman was a Turk undoubtedly smoothed the path for its undertakings. Here, on the outskirts of Ottoman sovereignty, such a high official from Constantinople could wield considerable power over the local bureaucrats.[15] On the other hand, the private company could rely on no more help in its dealings with these same authorities than came from the influence of the British ambassador's remonstrations at the Porte. This influence proved to be minimal.

Finally, we cannot overemphasize the importance of the difference in the two projects' technical approaches: the construction of jetties and the digging of channels were entirely within the ken of the local labor force's experience and they required no exotic tools or materials. The more sophisticated tasks, such as the installation of lights and other navigational aids, could be handled by the commission's own staff. By contrast, construction and utilization of the railway im-

[15]See R. H. Davidson, *Reform in the Ottoman Empire, 1856–1876* (Princeton, N.J., 1963). The contrary experiences of the railway and the frustrations of the directors may be followed in the records of the British Embassy at Constantinople. These are preserved in the Ashridge House annex of the Public Record Office, with the relevant files in series F.O. 195, for the years 1861–77. Even during the railway's construction, management pleaded with the British ambassador to station a gunboat nearby for protection. This plea was ignored. (cf. J. T. Barkley to Sir Henry Bulwer, September 4, 1860; enclosure with Bulwer's no. 607 to London, MSS, F.O. 78/511, Public Record Office).

686    *John H. Jensen and Gerhard Rossegger*

plied a technological and commercial revolution. As an island of modern technology in an otherwise indifferent or hostile environment, the company could not effect this revolution.

### *Romanian Engineering and Capital Inputs: 1882–1913*

Our story does not end with the triumph of an international commission in bringing efficient transport to the Lower Danube basin. The railway's physical capital was still there when Romania acquired the Dobrogea in 1878 and when the country became an independent kingdom in 1881. Typically, Romanian nationalism grew as the indigenous educational system became capable of bringing along a new generation of administrators, politicians, and engineers. These men saw the outcome of the "War of Independence"—known to non-Romanians as the Russo-Turkish War of 1877–78—as their great opportunity. If the Dobrogea was a poor consolation prize for the young enthusiasts, and if only the European Danube Commission prevented a renewed Russian stranglehold on Romania's trade through the delta, the Dobrogea was still Romanian and something might be made of this underdeveloped "frontier."

We need to understand more about the experiences which shaped Romanian nationalism in the last quarter of the 19th century if we are to appreciate the reasons for a revival of the seemingly hopeless Cernavodă-Constanţa rail link. Romania's small railway system connected with Hungarian Temesvar at the southwestern end and with Austrian Cernowitz at the northeastern end. The Austrians demonstrated their commercial domination over the young country by strengthening their control over transport on the middle and Upper Danube and by imposing strict limits on Romanian livestock exports to Central Europe through the use of various nontariff barriers such as complicated veterinarian regulations.[16] Furthermore, bitter resentment of Magyar discrimination against the Romanian majority in Transylvania added venom to Romania's reaction against foreign economic mastery.[17]

While the European Danube Commission, which controlled Romania's only other outlet to the world markets, was not offensive in this same sense it nevertheless was not Romanian, and the newly

---

[16]Cf. C. Baicoianu, *Problema conventiunilor veterinare: politica veterinară şi comerţul nostru de vite de la 1860 şi până în present* (Bucureşti, 1903), with documents.

[17]While Romania was still maintaining a modest position of clientage vis-à-vis the Soviet Union, her historians were very critical of the passive attitude of conservative governments toward the Hapsburg monarchy (cf. I. Georghiu and I. Negoiu, "Politica Antinaţională a Claselor Conducatoare din Rominia faţă de Transilvania în Ultimele decenii ale secolului al XIX-lea," *Analele Universităţii C. I. Parhon: Seria Ştiinţe Sociale, Istorie* 9, no. 16 [1960]: 131–44).

educated elite chafed under its presence.[18] Besides, Romanian grain exports, the only hope for development, now faced recurrent problems in western Europe. Britain especially, which had played such an important role in the first phase of transport expansion, had turned to the more plentiful and more easily accessible grain harvests of North America. Clearly, some major venture was needed to boost Romanian self-confidence. The newly acquired Dobrogea offered an opportunity for such a venture.

The Cernavodă-Constanţa railway had deteriorated steadily while the territory was under Turkish control. Even the most rudimentary maintenance had been neglected by the disillusioned company. Further damage was inflicted upon the property during the 1877–78 war, and after it purchased the railway in 1882 the Romanian government had to spend a good deal of money to bring the line back to a reasonable standard of operation.[19] At the same time, the meager financial resources of the new state were also devoted to improving the rail links between Moldavia and Wallachia—the home provinces—and the ports of Brăila and Galaţi. In these provinces agriculture expanded rapidly, with the result that a large percentage of Romania's harvests was poured onto an already depressed world market.[20]

One important component of agricultural development was the colonization and intensifying cultivation of Bărăganul, the broad Danubian plain stretching from Bucureşti east toward the river. A railway line was pushed into this region to Feteşti, just across the Danube from Cernavodă.[21] The technical imagination of the new generation of Romanian engineers was challenged by the task of linking the Bucureşti-Feteşti line to the old railway from Cernavodă to Constanţa. For this purpose, a giant bridge had to be constructed

[18]See N. Dascovici, "Dunărea internaţională şi ţinutul Românesc al gurilor," *Analele Dobrogei* (1928), pp. 771–77; published separately as *Dobrogea: 1878–1928. Cincizeci de ani vieaţa Românească* (Bucuresti, 1928).

[19]Material on the state of the railway and harbor at the end of the period of Turkish control can be found in the Ottoman Empire Series, 78/3194-3195, Public Record Office, London, Foreign Office. For Romania's approach to the harbor problem after 1881, see I. G. Cantacuzino, "Memoir," in *Analele Ministerului Lucrărilor Publice* 4 (1897): 128–35.

[20]D. Chirot and C. Ragin, "The Market, Tradition, and Peasant Rebellion: The Case of Romania in 1907," *American Sociological Review* 40 (1975): 428–44. Mention should be made here also of Chirot's book, *Social Change in a Peripheral Society: The Creation of a Balkan Colony* (New York, 1976), which arrived too late to be assimilated properly into this article.

[21]The several stages in which this line was built are explained in detail in a typed manuscript held in the Centre for Documentation of the Ministry of Transport, Bucureşti, entitled, "Cercetare Documenteara: Istoricul Căilor Ferate Romine (cu referat de literatură)" (no author, no date, but apparently from the early 1970s; hereafter cited as "Cercetare Documentarea").

across the floodplain and the river—just the sort of project to fire national self-esteem. This first major technical achievement (see fig. 1) was completed under the leadership of the Romanian railway system's chief engineer, Anghel Saligny.[22]

One could argue that this bridge project marked the growing to maturity of the engineering profession in the young country. Having had to content themselves for years with relatively minor and unspectacular tasks, these bright young men now found a highly visible focus for their energy and pride. Their ambitions were no doubt furthered by the fact that they were largely excluded from any real political influence by the ring of interlocking families who formed the impenetrable "political nation." Be that as it may, the Romanian state railway (Căilor Ferate Românei—[CFR]) managed to obtain foreign credits but to retain technical control over construction in native hands. The Regele Carol I Bridge was completed in 1896, thus allowing for the integration of the old trans-Dobrogean railway into the national, and therefore also the European, network.[23]

One, seemingly simple, technical consideration in the design of the bridge served to seal the fortune of Cernavodă as well as to dictate the future direction of Constanţa's development, and thus the nature of competition between rail and river. This consideration was the height of the new structure. Two factors played a role in its determination: one was the level of the river in flood which required the elevation of the approaches to the bridge; and the other was the continued passage of vessels, including sailing ships, under the bridge itself. Since about 20 kilometers of floodplain had to be traversed, and since the shipping channel shifted both seasonally and over longer periods, any sort of movable structure was impractical. Therefore, it was decided that the main arm of the river had to be crossed by a permanent and fixed bridge, 37 meters above the highest watermark.[24]

On the other hand, the old railway line into Cernavodă had been built at river level all the way. In the 1850s, the British engineers did not consider the possibility of a bridge across the Danube; they were concerned with easy access to the grain transshipment facilities.

The new line from the bridge led directly to a promontory south of Cernavodă whose height obviated the need for any earthworks be-

[22]See N. Perciun, " 'Anghel Saligny' în Museul Căilor Ferate," *Revista Căilor Ferate Române* 3 (1973): 684–88, for a useful review of Saligny's career. There are many Romanian studies of Saligny and his work. A good place to begin is the special issue of the *Buletinul Societăţei Politecnice* (November–December 1925) dedicated to his career.

[23]G. C. Măinescu, "Evoluţia căilor ferate in Dobrogea dela 1877 până în zilele noastre, din punct de vedere constructiv," *Analele Dobrogei* (1928), pp. 431–54, contains appreciations of the bridge and of Saligny's genius by foreign observers.

[24]Planning details for the bridge are given in N. Ciorcîrlan, "Cum s-a ajuns la construcţia podului de la Cernavodă," *Revista Căilor Ferate* 16 (1968): 28–34.

tween the structure and firm land. But now the railway bypassed the town and port, even though a showpiece station (Cernavodă Pod) was built on the high ground well away from the river and its terminal. This station could only serve passenger traffic, and it was not until considerably later that the spur into Cernavodă was integrated with the main line.[25]

Once this break with the original economic justification for the railway had been accepted, much of the development of Constanţa also took a new direction. With the river port essentially bypassed, attention was concentrated on building the straightest possible line across the Dobrogea. The British engineers, taking their model from railway planning in the United Kingdom, had built their project from village to village; all of these interim stops had some economic justification at the time but derived little stimulus from the railway. The Romanians rejected these twists and turns: let the towns come to the railway if they wanted to enjoy its benefits! So the line was cut almost straight across the Dobrogea and at a higher level than the old one.

The new state's limited resources had permitted little development effort in the Dobrogea before 1896. The Danube bridge seemed to offer an opportunity for linking the region to both heartland and port. Clearly, it represented an almost untapped potential for increased grain production and thus for supporting Romania's thirst for foreign exchange. Even more important for the future role of the railway line was the growth of petroleum exploitation as the Black Sea area became one of the world's centers for this new industry. As a vital link between the Ploieşti oilfields and Constanţa port, the railway across the Dobrogea received continuing attention and improvement.[26]

Taken together, the technical decisions with the bridge design and these economic developments meant that the role of the railway as a connection between the Danube and the Black Sea declined to insignificance. And from being a reasonably important entrepôt, Cernavodă slipped into becoming a murky, deteriorating eastern village. Today, only a small port and the still-inhabited wrecks of pompous merchant villas testify to Cernavodă's glory years.

At Constanţa, another decision confronted the engineers: How should the new tracks be brought down from the highlands, over some sharp drop-offs, to sea level?[27] A foreign contractor (Hallier of Paris) had been brought in to enlarge and modernize the port

[25]Today, the spur seems to serve no more than the limited local requirements of Cernavodă.

[26]The Ploieşti-Slobozia line, which linked the oilfields with Constanţa, was completed in 1910. "Cercetare Documentarea," p. 21.

[27]Măinescu, pp. 437–38.

690    *John H. Jensen and Gerhard Rossegger*

facilities.[28] He provided what appeared to be the impetus toward an answer to the question posed above by building a short railway spur to the stone quarries at Canara some 12 kilometers from Constanţa.[29] Originally, all stones for the harbor works had been brought from sources near Cernavodă.[30] It is not clear whether the new source yielded better materials or whether the cost of hauling stones all the way from Cernavodă was considered too high. In any event, the CFR engineers, working on a line that was already being rebuilt in its entirety, saw in this short quarry line a suggestion for an entirely new approach to the port.

Their new trace cut across the old British railway near where the Canara spur joined it. And since other Dobrogean quarries were also being surveyed for use in the Constanţa port project in the 1890s, maintenance of satisfactory entry to the harbor became an imperative matter which had to be taken care of hand in hand with improvements on the main line.[31] The original British engineers had never really solved the problem: their main track ended high on a bluff above the port and from there a spur ran down into the dock area.[32] In the Romanian literature this section came to be christened "the zig-zag line" (*linia zig-zagului* catches the eye in Romanian). Its layout can sometimes be seen in open-pit mining operations (the first port and railway engineer, J. Trevor Barkley, learned his trade in coal-mining operations in County Durham, and later in Asia Minor),[33] but it hardly represents more than a makeshift solution to traffic on a trunk line. Since the bluff was too steep to take the rails down in a series of curves, a number of straight sections and switches were constructed, with trains heading into the end of the first section, then backing through a switch onto the next lower section, repeating this switching and backing until the port level had been reached.[34] It was clearly a clumsy and time-consuming business.

---

[28]Direcţiunea Serviciuli Porturilor Maritimu Constanţa, "Evolutia portului Constanţa," *Analele Dobrogei* (1928), pp. 457–58.

[29]Măinescu, p. 439.

[30]Barkley, pp. 192–95.

[31]On the role of the Dobrogean quarries, consult the interesting short articles by C. Alimanistianu in *Buletinul Societăţei Politecnice,* 12 (1896): 131–33 and 175–76. The contrast is made clear there between the high-quality materials from the Canara and Murfatlar areas and the poor stone available near Cernavodă.

[32]The difficulties of constructing this spur are described in Barkley, pp. 145–47.

[33]For a full obituary notice giving details of J. Trevor Barkley's career, see *Journal of the Iron and Steel Institute* 2 (1882): 651–53.

[34]V. V. Stoica, "Materiale de baza privind istoria C.F.R.", manuscript (Bucureşti, Centre for Documentation of the Ministry of Transport, Library and Archives, c. 1953), 1:225–26 (hereafter cited as "Materiale de Baza").

Even the British company regarded the zig-zag line as a very "temporary" solution, but it never had the funds or the volume of traffic to justify any improvement. According to Romanian accounts, all grain was unloaded at the town level and fed into elevators through chutes.[35] It was obvious that any expansion of the railway along the Romanian engineers' ambitious plans must have, as a key component, some better solution to the access track, especially since-grain haulage was no longer the sole purpose of the line.

Hallier did not push ahead his work with great energy or dispatch. The contract did not prove as lucrative as he had hoped, and close scrutiny of accounts as well as supervision of the work by Romanian officials and engineers were probably resented by the French firm. A quarrel promptly developed over the quality of materials supplied, and the contract was put into arbitration after a government commission found against Hallier.[36] A settlement was reached, and the firm withdrew from its half-finished projects in 1899. We may surmise that this conclusion was greeted with considerable relief by both sides and perhaps with a sense of triumph by the Romanian engineers.

With Hallier out of the way, the Constanţa harbor engineers now joined (and not always very happily) with the CFR technicians to finish the job. The access problem was solved by tunneling through the bluffs directly from the west—and another triumph recorded for Romanian know-how in the process!

One would like to know more, of course, about the influence of Hallier's experts on this solution. It is not clear whether the project was conceived before or after their departure, and Romanian accounts of this phase of construction are noticeably reticent on the point. However, the question of the origins of the tunnel idea can now be settled, at least partially, on the basis of new evidence which comes from the diaries of Charles Hartley, the European Danube Commission's chief engineer. He has always been credited by Romanian historians with one of the original plans for the port of Constanţa, plans which were subsequently modified by Romanian engineers like Gheorghe Duca, I. B. Cantacuzino, and Anghel Saligny. While the question of Hartley's contribution needs fuller investigation on the basis of Romanian archival materials which have not yet been made available to us, it is clear that he presented very detailed designs for the proposed harbor as early as November 1881 and that these designs had a shaping influence on all subsequent projects. One of his

[35]Direcţiunea Serviciuli Porturilor Maritimu Constanţa, n. 28 above, p. 456.
[36]See the article on the Hallier dispute in *Buletinul Societăţei Politecnice* 14 (1898): 4–6; and V. Cotovu, *Portul Constanta* (special issue, *Buletinul Societăţei Politecnice* 50, no. 4 [1936]), pp. 10–11.

692    *John H. Jensen and Gerhard Rossegger*

major preoccupations at the time was with the technology of railway tunnels and especially with their ventilation. Hartley was a thoroughly practical man; his sudden interest in this latter subject obviously had its roots in the drafting of the tunnel proposal. This deduction from his notes must yet be confirmed through an investigation of his design materials.[37] But there is little else in his diaries that would suggest a fascination with engineering ideas for their own sake—without a view toward his own immediate and practical concerns.

Both the spectacular Danube bridge and the equally remarkable tunnel project reflect another important aspect of indigenous intellectual development in a young country. The murkiness of the Romanian sources on the matter of responsibility for the tunnel idea only underlines this issue. Here we are thinking of the "technological vanity" of native engineers, who enjoy a monopoly of knowledge in their field and whose work—since they are small in number—tends to feed their personal reputations as well as the esteem in which their guild is held by the public. While there is no doubt some element of national pride in these accomplishments, there is as yet little evidence of the engineers' identification with the broader objectives of government or industry.

The engineers and their spokesmen did not seem to mind at all that much of the actual work was done by foreign firms, as long as they provided the plans. After all, their technological ambitions far outran the domestic economy's ability to provide the necessary materials and equipment. And if they commented on the government, it was mainly to deprecate the stinginess with which their projects were supported. On the other hand, they derived some satisfaction from having selected and hired the most prestigious foreign concerns to serve their ambitions, and they continuously emphasized that all work was done to Romanian plans and specifications and under Romanian technical supervision. As the names of firms in table 1 suggest, the list of contractors for bridge construction and railway improvement was indeed impressive.

But the advance of the new rail line and the development of Constanța illustrate two other characteristics of such undertakings in young countries: the rise of technical experts to the role of bureau-

[37]Sir Charles Hartley's diaries for 1881 (in possession of C. W. Hartley, see n. 9 above): entries for April 30 (Sinaia), May 10 (Sulina), May 24–25 (București), June 5–9 (București), June 14–21 (Constanța), page after entry for July 20 devoted to construction and ventilation details of Mont Cenis tunnel, July 21 (London), July 30–September 17 (London), October 19–November 1 (Constanța), November 2–December 5 (Sulina), entry for November 28 ("Read aloud account of St. Gotthard tunnel"), December 8–10 (București).

*Lower Danube Transport Development*    693

TABLE 1

WESTERN EUROPEAN FIRMS EMPLOYED FOR RAILWAY AND BRIDGE
WORK IN THE DOBROGEA SECTION

| Firm | Country | Work Assignment |
|------|---------|-----------------|
| Gaertner ............... | Germany ⎱ | Bridge across Borcea |
| Schneider-Creusot ........ | France ⎰ | branch |
| D. F. Ozinga ............. | Netherlands | Foundations, piles |
| D. Gratzoski ............. | Austria | Masonry work |
| John Cockerill .......... | Belgium | Bridge metalwork |
| Schwedler ............... | Germany | Metal superstructure |
| Rottenberg et Cie. ........ | France | Masonry work |
| Comp. Fives-Lille ......... | France | Bridge construction |

SOURCE.—G. C. Măinescu, "Evoluția căilor ferate in Dobrogea dela 1877 până în zilele noastre, din punct de vedere constructiv," *Analele Dobrogei* (1928), pp. 434–37.

crats and the irregular forward movement of projects dependent on the erratic benevolence of an essentially oligarchic political structure. Thus we see three or four generations of engineer-bureaucrats (generations in the sense of training and levels of service in the CFR and the Constanța Port Authority) working on the projects from about 1891 to 1912.

For the port, the work was originally managed (1888–95) by I. B. Cantacuzino, who was also chief engineer of the government's hydraulic service. It was taken in hand in 1895, by G. I. Duca, first as head of the Port Construction Authority and then, after the merger of this agency with the Port Authority proper, as project manager. After Duca's death (1899), he was succeeded by Anghel Saligny himself, who brought the harbor works to a conclusion in 1910.[38]

While the Danube bridge was under construction, Saligny also had charge of the entire railway; in 1897–98 Stefan Ghiorghiu became engineer-inspector-general and chief of division, assisted by V. Christescu and N. Teodorescu.[39] In 1898, yet another chief engineer, I. Pâsla, took over and pushed the railway work forward until lack of funds forced its suspension in 1900. At this point, Pâsla became head of the Port Authority, which obviously still had some financial resources, and completed the tunnel project. In 1904, railway construction was resumed under a new supervisor, P. Zahariade. From 1909, Engineer-in-Chief Victor Bruckner (a Siebenbürgen German) took care of the last stages of construction, doubling trackage throughout and bringing all of the line's new facilities into operation. We list all of these names only to show how, in a relatively rapid turnover of per-

[38]Direcțiunea Serviciuli Porturilor Maritimu Constanța, pp. 457–58; and Societatea Politecnică din România, *60 (Sase) de ani*, p. 18.
[39]Măinescu, p. 434.

694     *John H. Jensen and Gerhard Rossegger*

sonnel, a number of individuals managed to gain both technical and administrative experience. In fact, many of them later moved on to even higher positions in both the CFR and other government departments.[40] Thus, the two projects also served as unique training grounds for a cadre of future top bureaucrats, all of whom had risen out of purely technical functions.

The work was completed in its entirety in 1912, only a few months before the CFR moved to lay a petroleum pipeline along its right-of-way in order to bring the Cîmpina-Ploieşti oilfields into direct contact with the port of Constanţa. We may be allowed some impolite conjecture at this point: Was the pipeline, which must have been in the planning stage for several years, the result of internal competition among the CFR's top engineers, or was it simply a natural response to advances in transport technology which the Romanian engineers wanted to complete as further proof of their close contact with the edge of the technological frontier?[41]

Whatever the answer, the pipeline project proved to be difficult. Work was begun in 1913, after the disruptions caused by Romania's participation in the Second Balkan War, and was expected to be completed by mid-1915. However, the specialized materials required had been ordered from the United States, and Turkey's entry into the European war in 1914 ended shipments of this kind through the Dardanelles.[42] The pumps for the system also were not delivered until after the war, and it appears that reconstruction of the war-damaged Danube bridge and additional delays held up completion of the pipeline until about 1921–22. With this project, the heroic early years of Romanian transport engineering came to a close.

These generations of engineers and technical experts, with their contributions to the advancement of trans-Dobrogean transport and their future greatness in the bureaucratic galaxy, should be suggestive for students of economic development. The friendships and rivalries forged in the Dobrogea would continue as each generation inexorably moved up the ladder. Romania had too few trained engineers to ignore any of them, while she had a superabundance of lawyers, social science and philosophy dons, and journalists. So the engineers were

[40]See ibid., pp. 432–38, for detailed "job histories" of several of these leading engineers.

[41]D. Cioriceanu, *La Roumania économique et ses rapports avec l'étranger de 1860 à 1915* (Paris, 1928), and M. Pizany, *La Situation de la Roumania dans le commerce mondiale du Petrole* (Bucureşti, 1938), are suggestive along these lines.

[42]*Buletinul C.F.R.*, November 16, 1913, reviews the construction plan; on capcity, see Constantinescu, "Conductele de petrol în România," in *Societaţea Politecnică din România Semicentenarul 1881–1931: Istoricul Dezvoltării Tecnice în România* (Bucureşti, 1931), 3:253–72; "Materiale de Baza," 4:161–63 for delays in the construction.

carried forward not only by their national pride but also by the knowledge that they could share out all bureaucratic-technical posts early in their careers when they were at the height of their mental and physical productivity. Yet, does an "engineer-inspector-general-in-chief" have full opportunity to utilize his professional capabilities or does he get drawn quickly into a network of administrative and political involvements which lay waste to his original comparative advantage? The problem is a familiar one in the scientific and technological establishments of developing countries even today, and Romania at the turn of the century is a fine example of the dilemma.[43]

The other, economic, aspect of the projects is simpler to chronicle. Port and railway construction, relying heavily on foreign contractors and suppliers, could not proceed without foreign exchange earnings. Romania was unique among the Balkan states in this period in that her exports provided her with adequate resources. But the pressures which this disposal of her "grain surplus" imposed on rural social relationships became clear in the great peasant uprisings (and army massacres) of 1907.[44] At the same time, exactly because of her outpouring of wheat, livestock, and petroleum, Romania remained sharply exposed to the vagaries of the world markets for these commodities, which also may help to explain the erratic progress of her exchange-dependent projects. Technological advancement thus became a function of foreign trade; in the general economic downturn of 1900–1905, for example, projects came to a virtual standstill.[45]

While the harbor works were impressive by 1909, they were far from complete. Port activity continued to be hampered by inadequate storage and handling facilities, poor tugboat service, and meager repair works. Thus the actual capacity of the port always lagged behind its theoretical capacity, that is, the level of operations which would be achieved a bit later (*mai tirziu* is the famous Romanian phrase, similar to the Latin *mañana*) when all of the necessary equipment was acquired. But the fact is that a series of bottlenecks kept preventing the realization of the port's planned potential. The important question is:

[43]See n. 22 above for sources on Saligny's fate along these lines.

[44]See Chirot and Ragin, n. 20 above; also A. Otetea et al., *Marea Rascoală a Tăranilor din 1907* (Bucureşti, 1967); P. G. Eidelberg, *The Great Romanian Peasant Revolt of 1907* (Leiden, 1974); and J. Tucker, "The Rumanian Peasant Revolt of 1907: Three Regional Studies" (Ph.D. diss., University of Chicago, 1972).

[45]The pages of the *Buletinul Societăţei Politecnice* for this period are full of complaints by Romanian engineers about the employment of foreign engineers on the few projects still in progress during this general retrenchment in public works. See especially the number for November 1905, vol. 21, pp. 460–63, which reports the society's protest to the minister of public works about a contract let to a German firm for water and sewage reticulation work in and around Bucureşti itself.

696    *John H. Jensen and Gerhard Rossegger*

Why was not the work on its components pursued at a more modest pace and kept in reasonable balance ... why build a double-track railway when a fully operative single track would have been more than adequate in terms of other facilities ... why expend vast funds on both an inner and an outer port, when a fully developed single port could have handled all shipments for quite some time?[46]

There are several answers. One, of course, lies in the technical vanity of the project leaders who wanted their work to be the biggest and technologically most advanced. A second possible answer is impossible to document: it lies in the opportunities for special profits in the various arrangements with foreign contractors and domestic suppliers. Here we draw close to the mysteries of higher politics. For example, What was the role of the royal family and the inner ring of politicians who were in a position to help or hinder the engineers' schemes? We see the construction of a short, but expensive, railway spur from the main line north to Mamaia where the king and his court took their holidays. How did this commercially unjustified project enhance the bargaining position of engineers when it came to their more grandiose undertakings? How much in the way of limited funds did it divert from the immediate needs of the port and the main railway? Similar spurs to Eforia and Carmen Sylva in the south fall into the same category. Efforts by CFR apologists to justify these sidelines on economic grounds remain singularly unconvincing.[47] These suggestions ring familiar, even in the context of today, but their resolution must remain a matter of conjecture.

There remains one final and essential argument, however, which takes us back directly to the competition between rail and river, between Constanţa on the one hand and the ports of Brăila and Galaţi on the other, and between the Romanian government and the European Danube Commission. Although the Dobrogea was scorned by the Romanians in 1878 as poor compensation for the blood and treasure spilled in the war with Turkey and for the loss of Bessarabia to czarist Russia, this scorn evaporated as the region of Baraganul was colonized and the bridge completed.[48] The Bărăganul experience demonstrated what could also be done in the Dobrogea, and the bridge enabled Romania to bypass any influence of the European Danube Commission on political, if not on economic, grounds. As Romanian publicists and the engineers themselves expressed it, the Dobrogea was to be a "lung" through which the country could breathe

---

[46]Detailed data on all these projects can be found in Directinuea Serviciuli Porturilor Maritimu Constanţa.

[47]See Măinescu, pp. 450–51, and see also our earlier reference to I. G. Cantacuzino and his Brăila cement works (n. 4 above).

[48]"Cercetare Documentarea," pp. 20–21.

freely. The image of a foreign stranglehold on the country's economic windpipe (the maritime Danube) was very real in the minds of the nationalist elite.[49] Arguments about balanced development faded to insignificance in the face of the opportunities offered by a radical tracheotomy!

Thus the railway and the port of Constanţa were seen as Romania's hostages to the future. They must be built bigger than the requirements of the present; they must serve to validate the potential of foreign trade on a grand scale, and never mind the fact that present commerce came only to 20 or 30 percent of the port's design capacity. Independence from the delta was all-important.

When we descend from this high plane of aspiration to the realities of river-rail competition, a somewhat more sobering picture emerges. In the years between 1900 and 1914, when Romania's grain exports were pushed to close to 2 million tons annually and constituted over 80 percent of the total value of exports, the port of Constanţa was handling only between one-sixth and one-third of these cereal movements in spite of all the government's efforts to discourage shipment through the delta.[50]

These efforts were not well coordinated. As one keen observer of the state of grain shipments through Constanţa noted early in the century, much depended on the state's policy for fixing freight rates on the railways. So long as a flat kilometer rate was charged for grain shipments, these would be more likely to move through Brăila and Galaţi than through Constanţa. The reason was simple, and a glance at a map will confirm it: distances for shipment from all over the country's grain-producing regions were shorter to the Danube ports except for a few districts in the southeast. Competition among the ocean shippers minimized the freight differential between Brăila and Galaţi on the one hand, and Constanţa on the other. Furthermore, government improvements in shipping on the middle Danube meant that port taxes could be evaded by loading the grain straight from river barges onto the ships lying in the Brăila roadstead. Under these conditions, Constanţa fought a losing battle for grain shipments.[51] De-

---

[49]Mǎinescu, p. 437, writes of " . . . baloanele de oxigen cu care se hrania plǎmânul României" ( . . . oxygen supply for the nourishment of Romania's [economic] lungs); see also, for example, E. M. Brancovici, *Un raspuns Uniunei scriitorilor, savantilor şi artistilor Bulgari* (Bucureşti, 1920).

[50]I. Teodorescu, *Der Getreidehandel in Rumänien* (Tübingen, 1910). The importance of grain shipments to the Romanian railway system, and the care with which they were observed, can be seen in the regular and detailed reports and tables given in *Buletinul C.F.R.*, which give quarterly breakdowns of all grain tonnage moved through all ports and border points.

[51]These points are made in detail by G. Christodorescu in his *Portul Constanţa: Mişcara Comercialǎ şi maritimǎ în anul 1903* (Constanţa, 1905), pp. 239–44. Christodorescu

698    *John H. Jensen and Gerhard Rossegger*

spite the supposedly improved and superior technical facilities at Constanţa, grain traffic through the port actually declined between 1909 and 1913. Round 2 had once again gone to the river system. Only the growing shipments of petroleum rescued the port improvements from total irrelevance. The lack of bulk grain business rendered the railway an economic disaster despite its technological brilliance.

### Recovery and Renewal: 1919–28 and Beyond

Although Romania did not enter World War I until 1916, the Turkish blockade put an effective stop to her export trade and to imports of capital goods through the Black Sea at a time when rail and port facilities were only just in full operation and the pipeline had been barely begun.[52]

The end of the war brought a new situation for the Lower Danube and for the Romanian economy as a whole. Wartime destruction had to be repaired while the government was faced with new responsibilities and opportunities. Two large areas were added to Romania's territory: backward Bessarabia brought little further economic strength, but at least its acquisition removed Russian authority from the vicinity of the Danube delta; the more advanced regions of Transylvania, Banat, and Bucovina enhanced the kingdom's wealth, but tying them to the heartland would require considerable outlays on improved transport. Extensive rural landholding reforms made Romania's peasants into proprietors of their lands but reduced the rent burdens which had helped to squeeze out agricultural-export surplus before the war.[53]

The hostilities had also added tremendously to the world's appetite for petroleum products, while postwar protectionism weakened the market position of Romania's cereal exports. Finally, the liberal governments of the early and middle 1920s took a strongly nationalistic line in economic matters, emphasizing Romanian independence and self-development but at the same time impairing the country's access to international capital markets.[54]

How did the transport system of the Lower Danube basin fare

---

was secretary of the Constanţa Chamber of Commerce in the early years of this century and wrote a dozen important studies on the port, its hinterland, and its commerce. His works have provided the basis for much later Romanian scholarship on these questions though this is seldom acknowledged!

[52]Cioriceanu, pp. 352–54.

[53]H. L. Roberts, *Rumania: Political Problems of an Agrarian State* (New Haven, Conn., 1951). The position is also clearly summarized in Chirot and Ragin, p. 443.

[54]A good discussion of these problems, with valuable bibliographical suggestions, can be found in L. S. Stavrianos, *The Balkans since 1453* (New York, 1958), chap. 35.

under these new conditions? For Brăila, Galaţi, and the delta the story is simple—the bulk of Romania's shrinking grain exports continued to be shipped on the river route, and the Romanians continued to chafe under the European Danube Commission's control over the delta. Where they could, Romanian engineers carried out minor improvements along the Danube's middle section and also tried to build up rail links from Transylvania to the maritime ports. However, this well-fed province never produced export surpluses sufficient to generate any strong trade through the delta.[55] Thus, the pattern of commerce via the Danube was little changed from the prewar period.

Along the rail line, the Danube bridge complex had suffered considerably, and much of the network of spurs in the Dobrogea was destroyed. The years of wrecking by the retreating Romanians, advancing and then again retreating Germans, Austrians, and Bulgarians had taken their toll. The German command tore up one of the two tracks between Cernavodă and Constanţa and used the rails to improve north-south connections elsewhere in the Dobrogea. And, of course, the work on the petroleum pipeline, which had been barely begun, now had to be taken up again.[56]

Reconstruction of the bridge works began immediately at the conclusion of hostilities. Even after her victory, Romania let stand the long-term repair contracts and orders for pipeline equipment which had been let originally by the pro-German government before 1916. The rail line to Constanţa was not fully restored to traffic until 1922. By 1925, a steady upward trend in freight movement on the railway had begun and was interrupted only by the Great Depression.

The port of Constanţa suffered modest wartime damage. By 1928, it had been restored to its 1913 standard and capacity. This relatively slow recovery can be explained by a chronic shortage of funds, which also made any thoughts of further expansion utopian. In addition, as traffic picked up the loss of the double track was a clear disadvantage.[57]

The Romanian engineers' and planners' financial handicap is best documented by the fact that, whereas annual expenditures for port projects averaged 3,322,800 gold lei between 1896 and 1916, they

[55]M. Popă-Veres, *Comerţul nostru de cereale sub aspectul vieţii economici româneşti* (Bucureşti, 1928); see also S. D. Zagoroff, J. Végh, and A. D. Bilimovich, *The Agricultural Economies of the Danubian Countries, 1935–45* (Stanford, Calif., 1955), chaps. 2–4; and W. E. Moore, *Economic Demography of Eastern and Southern Europe* (Geneva, 1945).

[56]"Cercetare Documentarea," pp. 28–35.

[57]In his *Memoirs* (New York, 1961), Admiral K. Doenitz claimed that the line was double tracked again only in the course of World War II in order to provide adequate logistical support to the German campaign in the Crimea. However, Romanian sources indicate that this work was in fact done in the early 1930s, apparently with the help of a French loan (see "Cercetare Documentarea," pp. 35–43, 54).

700    *John H. Jensen and Gerhard Rossegger*

amounted, on average, to only 843,300 gold lei per annum from 1919 to 1927.[58] In general, the government attempted throughout the 1920s to limit investments to the surplus of port income after coverage of the authority's operating costs and of debt service. The results were what one might expect: unfinished equipment, inadequate capacity of service facilities relative to dockage space, and imbalances even among finished facilities. Only two of three major storage buildings were equipped for actual use; there was machinery sufficient to discharge 1,600 tons of grain per day into vessels but the elevators could receive only 1,000, and the capacity of the cleaning equipment was limited to 800 tons.

No wonder the grain exports through Constanţa stagnated. In the peak prewar year, 1911, they had reached 600,000 tons. Between 1920 and 1927 they never exceeded 350,000 tons and were well below that amount in most years.[59] No matter what the engineer-administrators' ambitions, it appears that the government had resigned itself to the status quo just as the British railway entrepreneurs had done fifty years before: grain traders and shippers preferred to move the bulk of all cereal exports through the delta; only crops that could easily be hauled by rail from Bărăganul would go through Constanţa.

Petroleum was another matter. The port's capacity here was steadily built up to meet a burgeoning world demand. By 1928, six branch pipelines carried crude oil, diesel fuel, heavy fuel, and gasoline from the tank yards and the Ploieşti pipeline to the docks, where five vessels could be loaded simultaneously. Between 1920 and 1927, petroleum and petroleum product shipments increased from 150,000 to 1,850,000 tons.[60] Nevertheless, the port continued to suffer from inadequate storage and pumping capacity relative to the capacities of other handling equipment. But for the time, Constanţa was considered a satisfactory outlet. Needless to say, there was general cargo traffic as well, but it played a minor role compared with the booming petroleum business.

It is significant that throughout this period the continuing work of the European Danube Commission went almost unnoticed. The Romanians clearly had no interest in publicizing the international body's accomplishments even though they were their chief beneficiaries, especially as long-distance haulage on the middle and Upper Danube had declined with the dismemberment of the Austro-Hungarian empire and assertions of autarchy by the successor

[58]"Evoluţia Portului Constanţa," *Analele Dobrogei* (1928), p. 475.
[59]Ibid., p. 480.
[60]Ibid.

states. But whatever political and economic troubles beset southeastern Europe after World War I, the Danube delta continued to be safe, and the maritime ports were accessible to the shipping of all nations. In fact, the record of the commission as one of the first, if not the first, successful international public development venture remained unblemished until the outbreak of World War II despite the disagreements and conflicts of its member governments in other areas.

## Conclusions

Several salient points emerge from our comparison of the efforts of three distinct institutions—an international commission, a private enterprise, and the administrative arm of a small state—to build up an efficient transport system in a slowly developing, primary-producing region.

1. Despite many initial expectations to the contrary, the international organization reached its objectives because it enjoyed three fundamental advantages—the support of the powers having the greatest political and economic interest in improved transport, the predisposition of traders to utilize traditional connections and outlets, and the adequacy of very conservative technological means for the task of providing safe and efficient navigation.

2. The venturesome project of private enterprise failed (at least on the investors' terms) for lack of any of the above conditions. We may suspect that even with more plentiful financial resources at their disposal the railway builders would have found it difficult to overcome both the problems of transferring a very modern technology into a backward region and the handicap of having to divert traffic from its established patterns. In terms of modern approaches to development, one might judge that railway investment in the 1850s was premature or "inappropriate." But of course, as in all cases of direct foreign investment, commercial failure did not prevent the physical capital (albeit deteriorated) from remaining in the host region.

3. When a small, young state took over the development of railway and port, considerations of economic feasibility were clearly overridden by technological and political ambition. To some extent, the resulting overcommitment of resources was validated later on by the role of petroleum in Romania's export trade. But as is so often the case with projects of this type in developing regions, technological ingenuity proved insufficient. What was missing was the projects' integration into the overall economic system through appropriate plans and policies. The most obvious example of this lack of a coordinated policy was the Romanian government's failure to institute a rate and fee structure that would have provided incentives for a redirection of

702    *John H. Jensen and Gerhard Rossegger*

traffic from the delta. However one wishes to judge mercantilist policies in a broader framework, when carried out badly and piecemeal they are bound to run afoul of the marketplace's adaptability.

4. It is probable that the political influence of the established trading community and of the European Danube Commission militated against the Romanians' effort to turn the trans-Dobrogean transport system into an economic, as well as technological, success. However, because of the government's public attitude toward the commission, this suggestion would be difficult to prove.

5. The hesitant and imbalanced way in which the petroleum export facilities were developed and improved is indicative of a young economy's difficulties in effectuating, on its own, adjustments to new opportunities in world trade. Again, the internal complexities of Romanian politics, of vested economic interests, and of financial arrangements would need to be examined in greater detail before a full picture of these factors could be drawn.

When all this has been said, however, there remains the basic fact that an alternative outlet to the Black Sea was in Romania's economic and political interest and that railway and port development provided an important training ground for the country's new technical and administrative elite. In this sense, at least, the proprietors of the Danube and Black Sea Railway and Kustendje Harbour Company, Ltd., built better than they knew when they laid their track across the Dobrogea. It is equally basic that its heirs fell short of realizing the full potential of this technological legacy even after one takes into account the delta's competitive role.

# [17]

## Technology Transfer as War Booty: The U.S. Technical Oil Mission to Europe, 1945

ARNOLD KRAMMER

By the last month of the fighting in Germany, as the Allied armies rolled across the Rhine, combat-weary GIs were used to seeing groups of intelligence officers moving about the war zone. They were no longer startled to see small groups of scholarly looking American officers drive up to bombed-out and newly captured factories and, apparently unmindful of the smoke and sometimes nearby gunfire, systematically investigate the plant. They watched as tons of records were hauled out into the open for eventual crating and shipment and as nervous and obsequious German scientists were questioned by these visitors who wore neither rank nor unit designation on their American uniforms. The onlookers would have been surprised to learn that these investigators were not really army officers at all but industrial scientists and government experts, and that the plants they investigated had one thing in common: All had produced strategic materials under the Third Reich. Ultimately, more than 3,000 separate teams—involving more than 10,000 investigators, industrialists, engineers, and technicians—visited thousands of enemy factories, scientific institutions, business premises, and other objectives in an effort to explore and exploit the fundamental industrial knowledge of the enemy.

The Second World War, unlike any war in the history of civilization, was a war of science. The brains and industrial techniques of Allied scientists and engineers were matched against those of the enemy to produce the most advanced and effective devices in pursuit of military supremacy. The forces of science and industry had been marshaled to invent and develop new devices, all of which were held under a tight cloak of military security. Each warring nation spent hundreds of millions of dollars on fundamental and applied research which, while primarily intended for the purpose of war, presented a

DR. KRAMMER is professor of history at Texas A&M University.

unique form of war booty as the conflict came to a close. Germany's substantial advances in numerous industries—especially rocketry, optics, plastics, industrial chemistry, pharmaceuticals, and synthetic fuel technology—would be of substantial value to both American industry and to the continuing war effort against the Japanese. Moreover, it was clear that there was another contender for the information: the Soviet Union, whose own teams of investigators were already crisscrossing Germany from the East. Speed, therefore, was as important as thoroughness. Of all the areas of concern, the most representative of the swift and sweeping transfer of industrial technology was synthetic petroleum and related materials. It was a story whose roots went back to Germany during the 1920s.

### Early Synthetic Fuel Research

Numbed by her defeat in the cataclysmic First World War, and casting about desperately for an explanation, Germany found herself in agreement with Lord Curzon's dictum that the winning side had floated to victory on a sea of oil. It was a bitter lesson which the Germans hoped to avoid in the future. Consequently, a number of German chemists, building on their earlier successful research in synthetic aniline dyes, ammonia, and nitrogen, now attacked the problem of converting Germany's vast resources of coal to the otherwise unavailable petroleum products. By the mid-1920s, they had developed two major processes. The first, developed by Friedrich Bergius (later to win the Nobel Prize for his work), forced a mixture of powdered coal, recycled oil, and a catalyst into a high-pressure vessel filled with hydrogen, where the coal was eventually liquefied. The resulting product was then separated into gasoline, middle oil, and heavy oil. The final products were gasoline and diesel fuel, and about 4 or 5 tons of coal were required to make a ton of gasoline, including the coal required to produce the power, steam, and hydrogen used in the process.

In the second process, discovered by two German chemists, Franz Fischer and Hans Tropsch, the powdered coal was broken up by superheated steam to produce a mixture of hydrogen and carbon monoxide gases. This gas mixture, after purification to remove all sulfur compounds, was passed over a metal catalyst under lower temperatures and pressure than required in the Bergius process, to produce a low-octane gasoline, a high-grade diesel oil, and wax which could be further processed into lubricating oils. Like the Bergius direct hydrogenation process, about 4 or 5 tons of coal were required to produce a ton of petroleum product. Whatever the advantages of

70     *Arnold Krammer*

either process, the Germans could now obtain gasoline from a coal mine.[1]

However, it could not be accomplished by just anyone. The production of synthetic fuel required acres of pipes, ovens, conveyor belts, and compressors capable of building from between 3,000 and 10,500 pounds per square inch. This was no operation for small, independent factories, since the capitalization could easily run into the tens of millions of Reichsmarks. Ruhrchemie bought the patent rights to the basic Fischer-Tropsch process, and I. G. Farbenindustrie, in turn, bought the Bergius patents and built the first hydrogenation plant at Leuna in 1927. By the outbreak of war in 1939, fourteen large hydrogenation and Fischer-Tropsch plants were in full operation, with six additional plants under construction. Their combined production of synthetic oil from brown coal of 1,200,000 metric tons (or some 19 million barrels) per year provided more than one-third of the total fuel capacity with which Nazi Germany went to war.[2]

In the United States during the same period, research on synthetic fuel had been nearly nonexistent—and with good reason. America rode the 1920s and 1930s on a sea of natural petroleum. The number of American petroleum refineries, for example, jumped from an

[1] For technical information on these processes, see Combined Intelligence Objectives Subcommittee (CIOS), "The Fischer-Tropsch Process," CIOS report, item 20, file VI-22, X-18 and 22, XV-5; British Intelligence Objectives Subcommittee (BIOS), "Medium Pressure Synthesis with Iron Fixed-Bed Catalyst: Interrogation of Dr. H. Kolbel," BIOS final report no. 1712, item 12; BIOS, "Synthetic Oil Production in Germany: Interrogation of Dr. Bütefisch," BIOS final report no. 1697, item 30; these CIOS and BIOS Reports at Modern Military Branch, National Archives, Washington, D.C.; H. H. Storch, N. Golumbic, and R. B. Anderson, *The Fischer-Tropsch and Related Syntheses* (New York, 1951); Howard R. Batchelder, "Synthetic Fuels," in *Advances in Petroleum Chemistry and Refining*, ed. John J. McKetta, Jr. (New York, 1962), 5:3–75; Neal P. Cochran, "Oil and Gas from Coal," *Scientific American* 234, no. 5 (May 1976): 24–29; Warren F. Faragher, "Germans Made High Aromatic Aviation Gasoline by Coal Hydrogenation," *Refinery Management and Petroleum Chemical Technology* 37, no. 45 (November 7, 1945): 851–55; Arnold Krammer, "Fueling the Third Reich," *Technology and Culture* 19, no. 3 ( July 1978): 394–422.

[2] Burton H. Klein, *Germany's Economic Preparations for War* (Cambridge, Mass., 1959), p. 40; "United States Strategic Bombing Survey, Oil Division Report," mimeographed (Washington, D.C.), p. 18; and C. C. Hall, "The Fischer-Tropsch Process," *Chemical Age* ( January 18, 1947), p. 115. For statistics on Germany's fuel production and consumption, as well as their available stocks of liquid fuels, rubber, and strategic chemicals, see CIOS evaluation report no. 19, C.I.C., 7S/12, May 23, 1945, Roll no. A-1007, 119.0412-19, Albert F. Simpson Historical Research Center, Maxwell Air Force Base, Ala. Lest the reader be overly impressed with the percentage of Germany's fuel requirements which was provided by synthetics, Batchelder notes that "this amounted to 100,000 barrels per day, which was only 2.2% of the United States demand in 1944, and about 1.2% of the United States current [1962] demand" (Batchelder, pp. 10–11).

already impressive 289 in 1919 to 427 in 1929. Production soared from 300 million barrels to 857 billion barrels a year during the same period. The development of tetraethyllead improved both the quality and quantity of the fuel.[3] Even so, the production of gasoline barely kept pace with the voracious demands of a public which had embraced the mass-produced automobile. In fact, thanks to Henry Ford, the increased use of crude oil in the United States rose so spectacularly that even the man on the street began to wonder if the nation would not soon run dry. This concern faded into insignificance, however, with the discovery of the vast new Oklahoma and west Texas fields. Production continued to increase throughout the 1930s, reaching 1 billion barrels in 1939 and nearly 1.5 billion by 1941. America went to war as the producer of a whopping 60 percent of the world's output.[4] Moreover, it was done without any real consideration of synthetics. In fact, with the exception of Standard Oil of New Jersey's acquisition of a number of German patent processes from the I. G. Farbenindustrie combine, American synthetic fuel development had not gone beyond the laboratory or small pilot-plant stage. There was obviously an abundant, if not unending, supply of natural crude petroleum.

## The Approach of World War II

It was the nature of the new war that brought petroleum into sharp focus. The blitzkrieg conquest of Poland, with its emphasis on fighter planes and hordes of tanks, foreshadowed the role that petroleum was to play in the conflict. Three points became evident: The conflict was to be a war of mechanical equipment, the internal combustion engine was the heart of the military machine, and the machine's lifeblood was oil. In World War I, for example, the U.S. military and all its Allies used less than 39,000 barrels (or 1,639,000 gallons) of gasoline daily; in World War II, daily consumption averaged 800,000 barrels (or 33.6 million gallons)! Indeed, during a period of seven weeks in June and July 1944, the U.S. Fifth Fleet alone used 630 million gallons of fuel oil; and during one month of attacks on Japanese shipping and installations in 1944, the Far Eastern Air Force burned 143,257,000 gallons of aviation fuel. Each sortie by the Ninth Air Force in its daily bombardment of Germany used an average of 634,000 gallons of 100-octane gasoline.[5]

[3]Harold F. Williamson et al., *The American Petroleum Industry: The Age of Energy, 1899–1959* (Evanston, Ill., 1963), p. 748; *Federal Register* 6 (June 1941): 2760.

[4]Williamson et al.

[5]U.S. Congress, Senate, *Investigation of Petroleum Resources, Senate Resolution 36, 78th*

72    *Arnold Krammer*

### Harold Ickes and the Oil Industry

To meet this challenge, President Roosevelt, girding for the war still seven months away, appointed Secretary of the Interior Harold L. Ickes to the additional position of Petroleum Coordinator for National Defense. His duties, as outlined in the president's letter of May 28, 1941, were to gather from various government agencies information on military and civilian requirements for petroleum and petroleum products and to take the necessary action to meet those needs. The awesome authority of the position suddenly made the crusty Ickes one of the most powerful men in the nation. The nation's powerful oil industry, having tangled with Ickes some years earlier over governmental conservation measures, had no doubts that he would use that power against them if necessary. The question of what the new petroleum coordinator intended to do, then, was uppermost in the minds of the more than 1,000 most prominent oil executives who were called to his first meeting on June 19, 1941. "It may have been my imagination," Ickes later recalled with a chuckle, "but I think that as I entered the room someone frisked me for concealed weapons."[6]

Their fears were quickly dispelled, however, as Ickes assured them the oil industry was "threatened not by our government but by totalitarian powers that have already stamped their ruthless way across the prostrate democracies of Continental Europe." They were heartened when he admitted that the Office of Petroleum Coordinator for National Defense had not been established by a specific legislative act and was therefore "not based on the theory of punitive action." They beamed when he announced that his deputy petroleum coordinator, vested with authority equal to his own, would be Ralph K. Davies, ranking vice-president of Standard Oil of California. And, finally, they applauded as he proposed that they form a government-industry team "in our effort to prepare and defend America."[7]

---

Cong., 1st sess., June 19–25, 1945, pp. 94, 59, 266–67, 282–83; J. Stanley Clark, *The Oil Century: From the Drake Well to the Conservation Era* (Norman, Okla., 1958), pp. 139–40.

[6]The Editors of *Look*, *Oil for Victory* (New York, 1946, p. 69; see also Harold L. Ickes, *The Secret Diary of Harold L. Ickes*, vol. 3, *The Lowering Clouds, 1939–1941* (New York, 1944), p. 529, and *The Autobiography of a Curmudgeon* (New York, 1943), pp. 229–305. Curiously, even President Roosevelt himself appears to have had little knowledge about Ickes's intentions. A reporter asked him during a press conference, "Mr. President, has the Deputy Coordinator of petroleum been selected yet?" The president responded, "No. Is there to be one?" "Yes, sir," assured the reporter. (See *Complete Presidential Press Conferences of Franklin Delano Roosevelt*, no. 746, June 3, 1941 [New York, 1972], 9:377–78.)

[7]*Oil for Victory*, pp. 69–70.

Still, in view of nearly eight years of hostility between Ickes and the oil company executives and the fact that the new petroleum coordinator had no real power to force industry's compliance, the partnership required a rather dramatic gesture on the government's part. Consequently, Ickes approached Attorney General Francis Biddle with a startling request: that the Justice Department suspend all outstanding antitrust suits against the oil industry during the defense emergency. Biddle was appalled but reluctantly acquiesced under White House pressure. The resulting bargain was that the Department of Justice agreed to temporarily suspend its antitrust suits against oil companies and to turn a blind eye toward the industry's future violations of the Sherman and Federal Trade Commission Acts. In return, the oil industry gallantly, though cautiously, linked arms with the Roosevelt administration for the war to come.[8]

Within sixty days, by mid-August 1941, the partnership proposed by Ickes had progressed into a fairly firm structure. The petroleum coordinator divided the country into five districts, roughly parallel to the areas serviced by the major oil companies. In each such district, industry representatives and government functionaries worked amiably and responsibly to supply sufficient fuel for the coming winter, while they provided petroleum for beleaguered England and stockpiled for America's own uncertain future. It was, in fact, a very successful and often unselfish relationship. During the week of July 30, 1941, for example, the oil companies on the East Coast (district I), at their own expense, began an intensive newspaper, radio, and letter campaign to inform the public about economizing gasoline consumption. In mid-August, the nation's oil companies voluntarily cut their deliveries to their own wholesale and retail outlets by 10 percent.[9] The oil industry made similar gestures in areas of swifter rail and ship transportation, increased crude oil production, and higher refining capacities. The gratified Office of Petroleum Coordinator for National Defense responded wherever possible, by cooperating on matters of materials allocation, bypassing bureaucratic obstacles, and defraying financial burdens. The Japanese attack on Pearl Harbor on

---

[8] Ickes had not sold out to the oil interests, as his liberal critics—particularly the editors of *The Nation*—had charged. He was, above all, a pragmatist. Nor was he always as successful in shaping policy, as evidenced by his repeated failure to initiate the construction of the so-called Big Inch pipeline, connecting the oil fields of east Texas with the petroleum-starved Atlantic coast, and his frustrating inability to halt American petroleum exports to the nation's near-foe, Japan. For an outstanding survey of the relationship between government and the petroleum industry, see Gerald D. Nash, *United States Oil Policy, 1890–1964* (Pittsburgh, 1968), pp. 157–79.

[9] *Natural Petroleum News* (September 3, 1941), p. 10.

## 74   *Arnold Krammer*

December 7 firmly welded the partnership between Ickes's government organization, now renamed the Petroleum Administration for
War (PAW), and the representatives of the oil industry, calling themselves the Petroleum Industry War Council (PIWC). By all accounts,
they did an admirable job—voluntarily pooling supplies, producing
their wells at an uneconomical rate, overrunning their refineries, and
straining their transportation network to the breaking point (see fig.
1).[10] Under their joint leadership, the petroleum industry produced
5.8 billion barrels of crude oil from January 1942 to August 1945, and
over 13 trillion cubic feet of natural gas. The production of gasoline
increased from 462,852,000 gallons in 1941 to an astonishing
4,832,722,800 gallons in 1945.[11]

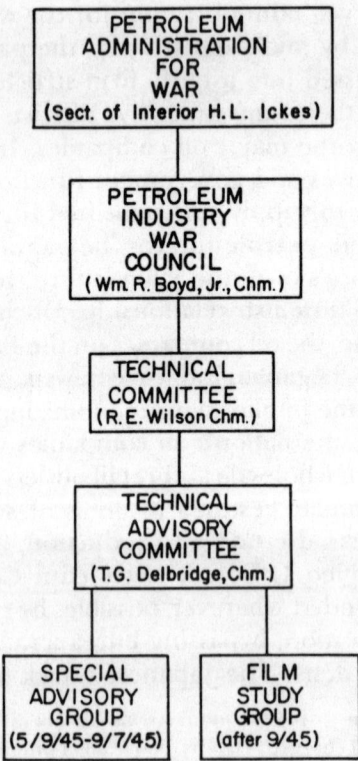

Fɪɢ. 1.—World War II government–petroleum industry cooperation

[10]*Oil for Victory*, pp. 71–72; John W. Frey and H. Chandler Ide, eds., *A History of the
Petroleum Administration for War, 1941–1945* (Washington, D.C., 1946).
[11]Frey and Ide, eds., pp. 228–43; also J. F. Barkley, Thomas C. Cheasley, and K. M.
Waddell, *The National Fuel Efficiency Program during the War Years, 1943–1945*, Bureau
of Mines Bulletin 469 (Washington, D.C., 1949).

Technology Transfer as War Booty     75

### The Exploitation of German Industrial Secrets

The idea of exploiting the industrial secrets of the Third Reich first arose in connection with German patents. During the First World War, the Trading with the Enemy Act of 1917 authorized the seizure of enemy property and records in the United States and suspended enemy patent restrictions for the duration of the war. Ultimately, later amendments to the act begrudgingly authorized the payment of up to $10,000 to compensate former enemy nations whose "money or other property" was held in trust by the U.S. Alien Property Custodian's Office.[12] Thus, when America went to war in December 1941, the precedent with regard to enemy industrial secrets was clearly established. The old Trading with the Enemy Act was resurrected, and the private and corporate records of suspected enemy agents were again seized by the Alien Property Custodian.[13] This time, the government not only suspended the restrictions on enemy patents but, democratically, made the patents available to the general public for the nominal fee of fifteen dollars each.[14] The attitude among American industrialists and scientists was that "when technical skill and industrial knowledge is used for evil ends, it is right that it should be canalized for beneficent purposes. If the original possessors of that knowledge cannot be trusted to use their skill in that way, then the rest of the world should be told how to do so."[15]

Thus, when an oil industry representative, in the autumn of 1943, first suggested to the government that both parties could only profit from a more thorough investigation of Germany's petroleum secrets, Ickes's PAW was impressed with the possibilities. Germany was well known to have made significant advances in the fields of synthetic fuel and shale oil production,[16] and such an investigation of its technical

[12]See Bess Glenn, "Private Records Seized by the United States in Wartime—Their Legal Status," *American Archivist* 25, no. 4 (October 1962): 399–405.

[13]For a brief overview of the varying opinions surrounding such captured records, see Robert W. Lovett, "Property Rights and Business Records," *American Archivist* 21, no. 3 (July 1958): 259–69.

[14]More than 45,000 patents were made available to American manufacturers by the Alien Property Custodian's Office at traveling industrial exhibitions; see, e.g., "Delve into Axis Tricks: Manufacturers Flock to Patent Flow at City Hall," *Kansas City Star* (April 18, 1944), p. 5.

[15]"The Reports on German Industry," *Chemical Age* 61, no. 1438 (January 18, 1947): 111–12.

[16]See, e.g., Richard Kenyon, "The German Chemical Industry, Past and Present," *Chemical and Engineering News* 25, no. 20 (May 19, 1947): 1437–39; B. Orchard Lisle, "Synthetic Fuels in Germany: 1. Introduction," *Petroleum* 9, no. 4 (April 1946): 74, 93, "Synthetic Fuels in Germany: 2. Hydrogenation," ibid., no. 5 (May 1946), pp. 102, 109, "Synthetic Fuels in Germany: 3(a). Fischer-Tropsch Process, Lubricating Oils and

76    *Arnold Krammer*

secrets would allow the United States to take up where Berlin had left off. It was the right of the victor to exploit the vanquished. On the other hand, however enthused the government was, the war was still far from over, and the U.S. Army could offer little early encouragement. The government, meanwhile, was embarking on a different venture which would ultimately make it as interested in the advances of German petroleum technology as was American industry: the creation of a domestic synthetic fuels program.

### The Synthetic Liquid Fuels Act of 1944

Washington was becoming increasingly concerned about the continued availability of petroleum. German submarines were taking a tremendous toll of Allied tanker traffic in the Atlantic; overland truck and rail transportation was insufficient to move oil products from California and the Gulf Coast region to the large cities in the Northeast; and the refusal of the War Production Board to allocate the steel and strategic materials for the construction of the so-called Big Inch and Little Inch pipelines, and for scarce oil exploration tools and machinery, led both civilian and military authorities to fear an imminent oil shortage by 1943. This concern led to a new interest in synthetic fuels.

In January 1943, Michael Straus, director of the War Resources Council, wrote a memo to his superior, Secretary Ickes, to tell him how tired he was of "piddling around" with the tiny appropriations provided by Congress for the synthetic fuels research that had been carried on by the Bureau of Mines since 1935. The bureau's scientists had been successful, after all, in producing gasoline, kerosene, and heating oil from American coals and were eager to move on to the pilot-demonstration-plant stage. Yet despite these steady though unheralded successes, the bureau's request for a modest $100,000 to design such a plant had been unceremoniously rejected by the defense-strapped Bureau of the Budget some months earlier in 1942. Ickes, attracted by Straus's seductive references to Ickes's "splendid record for public interest and foresight" and still deeply suspicious of the oil companies and their lack of interest in anything which did not immediately increase their short-run profits, instructed Straus to go ahead.[17]

---

Acetylene," ibid., no. 7 (July 1946), p. 158, "Synthetic Fuels in Germany: 3(2)," ibid., no. 8 (August 1946), p. 191, and "A Century of Oil-Shale Patents (1845–1945)," *Chemical and Engineering News* 24, no. 17 (September 10, 1946): 2342–43.

[17]Ickes to Straus, January 29, 1943, RG 48, Department of the Interior, central classified files, 1937–53, box 3762, file 11-34 (synthetic fuels), National Archives, Washington, D.C. For an outstanding analysis of these congressional machinations, see

*Technology Transfer as War Booty*     77

After taking pains to assure the equally suspicious oil industry that the government did not "intend to take up the natural petroleum trade," Straus and Ickes picked important congressmen from the states with the largest coal reserves[18]—and thus with the most to gain by the passage of a major coal-to-oil appropriations bill—to introduce a bill which would authorize the Bureau of Mines to undertake a five-year, $30 million program of research which would culminate in the construction of several demonstration plants capable of producing gasoline and oil from coal.[19] By midsummer of 1943, experts of all varieties enthusiastically endorsed the synthetic fuels program and soothed the ruffled feathers of isolated congressmen who saw the enactment of a bill to underwrite the development of an expanded research program as another example of government competing with private industry and thus a further erosion of American free enterprise. Congress approved the project, and the Synthetic Liquid Fuels Act became Public Law 290 on April 5, 1944.[20] In this unplanned way, the government now found itself with a vested interest in a thorough examination of Germany's synthetic fuel industry.

## The Creation of Technical Missions to Europe

Now that both the government and industry shared a desire to search out Germany's petroleum secrets, the only question concerned the method. They found that a small trickle of technical teams had already begun to investigate various areas of the enemy's military and

---

Richard H. K. Vietor, "The Synthetic Liquid Fuels Program: Energy Politics in the Truman Era" (working paper HBS 78-54, Cambridge, Mass., 1978).

[18]They were Sen. Joseph O'Mahoney of Wyoming (chairman of the Subcommittee on Public Lands and Surveys); Sen. C. Wayland Brooks of Illinois; Rep. Jennings Randolph of West Virginia (speaking not only for the state with the largest annual coal production but also as the chairman of the House Mining Committee); and, finally, Sen. James J. Davis from Pennsylvania.

[19]U.S. Congress, Senate, Subcommittee on Public Lands, *Hearings on Synthetic Liquid Fuels*, 78th Cong., 1st sess., August 1943.

[20]The act contained a major problem, the full extent of which would not become clear until the program's maturity. The project was required to be overseen by a Technical Advisory Group which met periodically to review and authorize the bureau's projects. The problem was not the scrutiny by an outside group but the fact that, of its eighteen members, no less than thirteen were from the nation's largest oil companies. Thus the government's single major effort to develop a synthetic liquid fuels program was watched over by the very industry which stood to lose the most if the program became too successful. Indeed, several of these representatives—C. C. Kemp of Texas Co., E. V. Murphree of Standard Oil Development Co., and Eugene Ayers of Gulf—later held important positions on the National Petroleum Council, which ultimately discredited the bureau program in 1952. In 1944, however, the impact of this built-in conflict was still years in the future.

78    *Arnold Krammer*

industrial secrets. From the end of 1942, for example, the army's Ordnance Department periodically dispatched special teams to combat zones to collect enemy equipment for later study at the Aberdeen Proving Ground in Maryland. A more general mission, staffed largely by experts from the Massachusetts Institute of Technology, slipped into Italy from December 1943 to March 1944 to investigate the alarming rumors of German research on nuclear fission. As it turned out, the MIT mission obtained very little information—that would be left to the later cooperative venture by the army, navy, and Office of Scientific Research and Development, known as the ALSOS mission[21]—but it did set the pattern for the scientific and industrial teams to follow. The best method, it was found, was for the mission to establish a primary headquarters in the field from which it would send small teams to investigate previously targeted enemy factories. Captured records, transcripts of interrogations, and initial reports were fed back to headquarters by the separate units, after which they were forwarded to Washington for analysis. The real key to future missions, however, lay in the uniquely high quality required of investigators:

> They had to be top-flight scientists of broad interest, able to comprehend wisely and quickly, able to speak the language of the enemy, men whom their German colleagues would respect for scientific attainment so that they might more quickly respond to interrogation. Moreover, these scientists could not be kept in a mere interrogating position because their judgment as to where the best clues were was likely to be better than that of the soldiers; at the same time soldierly guidance was needed in the matter of transport, communications, billeting, and safety, for the group would often be close on the heels of, or in the confused areas, even ahead of the troops.[22]

Following the Normandy invasion in June 1944, the few existing teams and missions were overwhelmed by the sheer volume of suddenly available enemy records. In July, Eisenhower's Supreme Headquarters attempted to meet the demand by creating special intelligence-gathering units called T-Forces. Within a month, however, it became clear that even these new technical investigators, with their large red T's emblazoned on their helmets, could not handle the deluge of information. Consequently, on August 21, 1944, the Allied Combined Chiefs of Staff established the broad umbrella agency that

[21]Samuel A. Goudsmit, *ALSOS* (New York, 1947), and "The Nazis Came Close— But," *Chemical and Engineering News* 24, no. 16 (August 25, 1946): 2176, 2263; and Leslie R. Groves, *Now It Can Be Told* (New York, 1962).

[22]John Burchard, *Q.E.D.—M.I.T. in World War II* (New York, 1948), p. 110.

## Technology Transfer as War Booty 79

would henceforth plan and administer the orderly exploitation of enemy secrets. Called the Combined Intelligence Objectives Subcommittee (CIOS), this new control center was responsible for compiling lists of "target plants," allocating the technical experts to investigate them, and processing and distributing the resulting reports. Its wide scope and cooperative purpose was reflected in the number and nationality of the agencies it represented: from the *United States*—the Department of State, War Department, Navy Department, Army Air Forces, Economic Warfare Division, Office of Scientific Research and Development, Office of Strategic Services, and Mission of Economic Affairs; from *Great Britain*—the Foreign Office, War Office, Admiralty, Air Ministry, Economic Advisory Branch of the Foreign Office, Department of Scientific Research, Ministry of Fuel and Power, and Control Commission for Germany.[23] High on the agency's priority blacklist of categories to be investigated were synthetic fuels and lubricants.

Thus, in August 1944, when the American petroleum industry, through its Petroleum Industry Council, again petitioned the government to investigate Germany's fuel secrets, it found Washington fully receptive. Within weeks, correspondence among Dr. Robert E. Wilson, chairman of the Technical Committee of the PIWC; Dr. T. L. Delbridge, chairman of the Technical Advisory Committee of the PIWC; and Bruce K. Brown, assistant deputy administrator of the PAW, had reached Ickes who, in turn, passed his strong recommendation for the mission to Admiral William Leahy of the Joint Chiefs of Staff. The Germans were unquestionably far advanced in the area of synthetic fuel operations, and speed was of the essence if the information was to be found intact and made useful against the Japanese. Admiral Leahy agreed with the urgency of the mission and assigned the arduous task of formulating the necessary plans to the Technical Advisory Committee (TAC) of the PIWC.

It was to be a complicated assignment. First, the members of the TAC had to determine the most important areas of German petrochemistry and offer their opinions of the best methods to conduct such an investigation. Recommendations were discussed in great detail at a special executive session of the TAC on August 23, 1944, in New York City. The result was a "Report on Oil Processing Information from Enemy Sources," in which the oil industry determined that the subjects of greatest interest, in decreasing order of priorities, were

[23]T. J. Betts, brigadier general, United States, chairman; and R. P. Linstead, British Ministry of Supply, deputy chairman, "The Intelligence Exploitation of Germany," Report of the Combined Intelligence Objectives Subcommittee, G-2 division. SHAEF (Rear) (Washington, D.C., September 15, 1945), p. 16.

80    *Arnold Krammer*

refining technology and engineering, catalysts, instrumentation, and analytical methods. The TAC concluded its report to the government by dramatically and patriotically offering to join the invading Allied armies to be among the first into the enemy plants.[24]

The following month, on September 25, at a routine meeting of the TAC in New York City, the normal course of business was interrupted when the government's representative, Dr. Malvin R. Mandelbaum, rose to announce some welcome news. The PAW had decided to act on the industry's recommendations and was now ready to consider volunteers to form a Technical Oil Mission (TOM) to Europe to accompany the front-line combat troops into Germany. The TAC was galvanized to action, and within two days, telegrams went out to all member companies requesting each to submit the names of three competent men, experts in refining engineering and foreign languages, for the immediate investigation of hydrocarbon technology in conquered countries. "Prompt action is imperative," noted the telegram, "or important information may be lost permanently,"[25] Within days, the industry responded with their candidates, twenty-six of the highest qualified synthetic fuel experts available, and waited anxiously for Washington to authorize the next step toward shipment overseas.[26] It was to be a long wait.

The project now bogged down in the bureaucratic mire. September passed, then October, and then November. The embryonic Technical Oil Mission to Europe was foundering for lack of direction. Moreover, the idea of investigating the hidden industrial processes of the collapsing enemy nations was spreading to other government agencies, and various missions were popping up like mushrooms. Indeed, it was the appearance of so many conflicting groups which finally forced the government into action. A directive came on November 15, 1944, from the Joint Chiefs of Staff, the final arbiter of

[24]Albert E. Miller, "The Story of the Technical Oil Mission" (paper presented to the Twenty-fifth Annual Meeting of the American Petroleum Institute, Chicago, November 14, 1945); also mimeographed (Morgantown, W. Va., 1945), p. 6.

[25]Ibid., pp. 3–4.

[26]Companies offering candidates for the Technical Oil Mission were: Atlantic Refining Co.; California Research Corp.; Gulf Oil Corp.; Houdry Process Corp. (Agency, Sun Oil Co.); Humble Oil and Refining Co.; M. W. Kellogg Co. (Agency, Atlantic Refining Co.); Koppers Co., Inc. (Agency, Gulf Oil Corp.); Phillips Petroleum Co.; Pure Oil Co.; Shell Oil Co., Inc.; Sinclair Refining Co.; Socony-Vacuum Oil Co.; Standard Oil Development Co.; Standard Oil Co. (Indiana); Texas Co.; and Universal Oil Products Co. (Agency, Atlantic). When this list of nominees was exhausted, the following companies offered a total of eleven more candidates: California Research Corp; Kenyon and Kenyon (Agency, TAC Panel); Shell Oil Co., Inc.; Sinclair Refining Co.; Standard Oil Co. (Indiana); Standard Oil Co. (Ohio); Union Oil Co.; Universal Oil Products Co.; and Worthington Pump Co. (Agency, Atlantic).

*Technology Transfer as War Booty* 81

military related matters, to create a Technical Industrial Intelligence Committee (TIIC) responsible for the coordination of all such missions. It was based on the policy developed in 1943 by the War Department in partnership with the British War Office which provided for the overseas collection of certain captured documents, their immediate military use overseas, and their transmission to newly established Military Intelligence Research Sections (MIRS) in London and in Washington for further military use.[27] More important, it designated the TOM as the single channel through which captured enemy data was to be funneled. Finally, the Joint Chiefs directed that in order to coordinate this American effort with a similar interest by British industry, the new TIIC hierarchy was made responsible to the Allied CIOS with headquarters in London (see fig. 2). Suddenly, the TOM to Europe was very much alive once again.

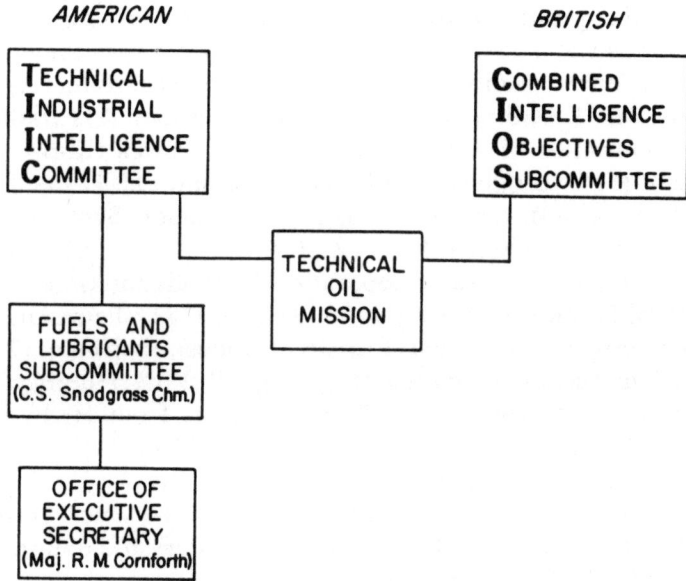

FIG. 2.—Organizational structure for Technical Oil Mission 1944–45

[27]"History and Operations of MIRS London and Washington Branches, 1 May 1943–4 July 1945," Agar-S document 1440, Modern Military Branch, National Archives, Washington, D.C. One of the groups which prompted the government to expedite the creation of the TIIC was Gen. H. H. Arnold's Army Air Force technical intelligence teams. Established to collect enemy air force equipment and records which might help maintain AAF superiority, General Arnold's teams combed Germany between December 15, 1944, and April 21, 1945, interrogating technicians and seizing substantial quantities of enemy data. High on their list of priorities were aviation and rocket fuels (see Colonel Donald L. Putt, "The AAF and Petroleum Research," *American Petroleum Institute: Addresses and Reports* 25:30–37.

82    *Arnold Krammer*

Now, with official authorization and a clear definition of responsibility, the TOM project found itself propelled forward at breakneck speed. The Allied armies were well across France; German collapse seemed apparent; and unless the American petroleum specialists could successfully prepare their team and quickly join the troops overseas, all might be lost.

Consequently, the various candidates for the mission—both those remaining from the original applicants and new replacements—were brought to Washington, where they underwent a series of briefing sessions to prepare them for the tasks ahead. Under the tutelage of C. S. Snodgrass from the PAW and Dr. W. C. Schroeder, the candidates were bombarded with information from representatives of the PAW, Army Service Forces, Navy Bureau of Ships, Ordnance, Military Intelligence Service, Office of Naval Intelligence, and the Department of State. By January 10, 1945, the candidates had divided themselves into specialist groups to digest more efficiently the vast amount of enemy industrial data as well as to dovetail better with the efforts of the various British teams.[28] These groups were: (1) Products and Test Methods, Paul K. Kuhne (Gulf Oil Corp.), chairman; (2) Classification and Carbonization of Solid Fuels, Alfred R. Powell (Koppers Co.), chairman; (3) Hydrogenation of Coal, Tar, and Oil, Lester L. Hirst (Bureau of Mines), chairman; (4) Hydrogenation Synthesis of Oil from Water Gas: Fischer-Tropsch, Irvin H. Jones (Koppers Co.), chairman; (5) Recovery and Separation Methods for Gases, Liquids and Solids, Horace M. Weir (H. W. Kellogg Co.), chairman; (6) Oil Shale Processes, W. W. Odell (Bureau of Mines), chairman; (7) Materials of Construction, including Metallurgy, E. Voss (Humble Oil and Refining Co.), chairman; and (8) By-Products from Hydrocarbons Produced in Synthetic Oil Processes, Vladimir Haensel (Universal Oil Products Co.), chairman.[29]

After two solid days of detailed study, the men became "temporary" colonels in the U.S. Army (to provide them a salary and to insure their safety under the Geneva Convention), with uniforms and credentials to follow. They were given a no-nonsense lecture about the urgency of their mission and the need for absolute secrecy; were cautioned that "every member of the team should assume that no German can be trusted in any degree"; and, finally, were ominously "urged to prepare their will and testament."[30] "It is important to emphasize,"

---

[28]For a brief but thorough summary of the BIOS, see L. Ivanovszky, "BIOS Trip to Germany, 1946," *Petroleum* 10, no. 2 (February 12, 1947): 34–48.

[29]For a complete list of team members and their specialties, see "Combined U.S.-British Technical Oil Mission for Investigation of German Oil Industry," secret, typewritten, February 12, 1945.

[30]Detailed Summary of Meetings held in New Interior Building, January 9–10, 1945.

recalls one senior member, "that most of the civilian investigators were far from youngsters and were thrown into a pretty rugged existence without benefit of a six-week basic training period. Indeed, unless they had served in World War I, most were 'babes in the woods' concerning even the most routine Army procedures."[31] Nonetheless, the experts of the TOM, brimming with optimism and driven by the urgency of the mission, anxiously looked forward to their departure overseas.

By mid-February, with the end of the Battle of the Bulge and with the Allied armies approaching the Rhine, Schroeder led the first half-dozen team members, technicians, and military escorts to London and into the war zone. Over the next two months, the remaining twenty TOM members—each an acknowledged expert in his field, well versed in German, familiar with many of their German colleagues, and thoroughly briefed on all military related matters—were dispatched to join behind the invading troops.

The first synthetic fuel plant to fall into Allied hands, a relatively small installation at Moers on the west bank of the Rhine, was a disappointment to the eager team members. The plant had been so badly damaged by Allied bombs that little could be gleaned from the smoking rubble and twisted steel. Earlier optimism gave way to the more realistic expectations of future finds. Still, several valuable lessons were learned. This first experience quickly taught oil mission members the best methods of interrogating the bewildered German technicians and plant personnel and gave them an opportunity to perfect their report-writing techniques.

These first few weeks provided the American investigators with a variety of other lessons in the realities of collecting industrial data in the country experiencing military defeat. Germany's final collapse was swift and resulted in immediate chaos—an atmosphere hardly conducive to the thorough, orderly, and systematic study of its technology. No such thing as government or law and order existed. German soldiers were surrendering in such numbers that the beleaguered forces under General Omar Bradley were instructed not to accept any more prisoners until more supplies became available. The situation was complicated by the additional burden of millions of homeless German civilians and newly liberated slave laborers—all of whom looked to the victorious American Army for food, fuel, transportation, and government. It was not an easy atmosphere in which to work.

---

confidential, typewritten, file no. 105987, Department of the Interior Library, Washington, D.C., p. 20.

[31] Walter J. Murphy, "The Job Is Still Unfinished," *Chemical and Engineering News* 23, no. 17 (September 10, 1945): 1529.

84    *Arnold Krammer*

In addition, the investigators soon learned that the much-vaunted cooperation and coordination between the American and British missions did not always exist as planned; moreover, there was a considerable duplication of effort. Not only was the whole operation unnecessarily top heavy with committees and the representatives of various intelligence services, but, in the opinion of one American investigator, "one vitally essential element was lacking—a top coordinating head or group with full and complete overriding authority to direct the activities of the several intelligence groups; to establish priorities and uniform practices; to decide and define the exact functions and objectives of each group; to enforce cooperation and avoid duplication; and to provide immediate information concerning the activities of each intelligence group."[32] Under these conditions, it is little wonder that CIOS was frequently facetiously referred to by the investigators as CHAOS. Nevertheless, the members of the TOM to Europe, sobered by their first exposure to the realities of the situation, remained buoyantly optimistic and continued deeper into Germany with the advancing Allied troops.

The next installation to fall into their hands was the substantial plant at Wessling, south of Cologne on the west bank of the Rhine, which had manufactured synthetic aviation fuel before it had been shut down by Allied bombs. While an interrogation of the German scientist in charge ultimately revealed a very important comprehensive report covering everything from plant design to valuable technical data, the physical installation was otherwise useless. "The examination was a tiring job of climbing over bricks, rubble, tanks, and destroyed stairways," recalls W. C. Schroeder. "It was necessary to be on guard at all times against brick work that might fall or floors and walls that might give way. Pipelines and other connections between various units of the plants seemed particularly susceptible to damage by concussion, and were down in all areas of the plants. We became so accustomed to these conditions that it actually felt strange . . . to be in an undamaged building."[33]

On March 23, the Allies captured a massive I. G. Farbenindustrie plant at Ludwigshafen, and the very next day no less than sixty-five technical experts—the largest team assembled for any single target installation—swarmed over the huge chemical plant which covered an area of many square miles. Within six weeks, the Ruhr valley was completely occupied and the various members of the oil mission were

[32]Ibid., p. 1528.
[33]Interview with Dr. Wilburn C. Schroeder, March 13, 1977; W. C. Schroeder, "Investigation by the U.S. Government Technical Oil Mission" (paper presented to the Twenty-fifth Annual Meeting of the American Petroleum Institute, Chicago, November 14, 1945), p. 2.

dispatched to the major synthetic fuel plants at Leipzig, Hamburg, Zeitz, Böhlen, Leuna, Lutzkendorf, Pölitz, Blechhammer, Heydebreck, Auschwitz, and Schaffgotsch—with particular emphasis on the last five, which were clearly to fall within the Russian zone.[34]

The mission's search for captured German documents—the key, it was hoped, to American synthetic fuel production on a national scale—ranged from one extreme to the other, with a variety of ludicrous experiences in between. For example, the industrial Ruhr valley, which held so much initial promise, proved to be nearly void of documents. The members of the mission became more and more frustrated as they found the great research facilities at Ruhrchemie in Sterkrade-Holten and the Kaiser Wilhelm Institute at Mulheim-Ruhr with scarcely a filing cabinet full of technical data. The location, or even existence, of this material was not discovered until investigators accidentally tracked down and interrogated a fleeing Dr. Martin and his assistant, Hagerman, and discovered that the tons of missing technical records had been evacuated to the so-called Bavarian Redoubt where fanatical Nazis were supposed to be preparing for a last stand. The investigators piled into a 3½-ton truck and began a wild and often harrowing journey across collapsing Germany in an effort to secure the documents before they could be destroyed or fall into the hands of the advancing Russian forces and, presumably, their scientific teams.[35] After a frustrating detour through tons of Reich financial records found in an iron mine belonging to the Hermann Göring Works near Salzgitter, the trail led the team to a picturesque 13th-century castle, complete with surrounding moat, near the hamlet of Reelkirchen. "There were two German old maids in the castle," Schoeder recalls, "who hovered around like frightened old hens protecting their chicks." Eventually, however, when all the documents were collected, they filled six tremendous rooms of the castle from floor to ceiling—"a fantastic cache of information on German synthetic-fuel operations and research."[36]

[34]See "The Operation of Two Fischer-Tropsch Plants," *Industrial Chemist* (May 1946), pp. 253–57; "The Western German Hydrogenation Plants," ibid. (November 1946), pp. 637–41; "Inspection of Hydrogenation and Fischer-Tropsch Plants in Western Germany during September 1945," BIOS final report no. 82, item 30; "Steinkohlen-Bergwerk Rheinpreussen Moers-Meerbeck," CIOS XXV-6; "Plant of Klocknerwercke A. G. Castrop-Rauxel," CIOS XXV-7; and Leonard C. Halpenny, "German Plants Extract Gasoline from Coal," *Petroleum Engineer* (August 1945), pp. 188–92.

[35]For a thorough examination of the Allied race against Soviet scientific teams, see the outstanding book by Clarence G. Lasby, *Project Paperclip: German Scientists and the Cold-War* (New York, 1971); or the less scholarly work by Michael Bar-Zohar, *The Hunt for German Scientists* (London, 1967).

[36]Ruth Sheldon, "The Hunt for Nazi Oil Secrets," *Saturday Evening Post* (October 6, 1945), p. 121. The most important finds concerned the production of higher alcohols, isoparaffins, and aromatics (see W. C. Schroeder, "Technical Oil Mission Studies Ger-

86    *Arnold Krammer*

On other occasions, such as that at the Wessling installation, the American petroleum experts found the documents somewhat closer to home. In this case, the investigators were convinced that the plant's records were nearby but could not move the German personnel to cooperate. Finally, as a last resort, the investigators tried a bluff. Acting as though they really knew where the data were located, the American scientists simply handed the ranking German scientist a spade and, displaying their most knowing looks, ordered him to dig them up. To their amazement, he led them to a nearby field, where he ruefully dug up the missing data. But, as a general rule, Schroeder today recalls, "the German scientists—caught as they were between us and the advancing Russians—were terribly helpful in organizing their records and making them available to us."[37]

Finally, the TOM investigators reached the mother lode: the main I.G. Farbenindustrie installation in Frankfurt. An enormous colossus of steel and buff-colored stone, the six seven-story buildings were virtually undamaged by the fighting, and the long rows of undisturbed filing cabinets were bulging with every imaginable technical document and international contract agreement. However, by the time the members of the mission arrived on the scene, on April 28, the buildings had become the temporary living quarters for over 11,000 displaced persons and liberated slave laborers. According to an eyewitness account, "the rooms, corridors and stairwells of the building had become snowbound in a blizzard of paper . . . as desperate people took care of the needs of warmth, sanitation and revenge in one mad assault on the neatly ordered rows of cabinets."[38] Just as the American scientists began to restore order and, with dozens of German PWs, swept the precious documents into an isolated wing of the building, the advance representatives of SHAEF Headquarters arrived to requisition the building as a future forward HQ for General Eisenhower and ordered the building cleared of all this "refuse." Army trucks began backing up to the entrances and hauling off tons of paper for burning. No sooner had the crisis been solved than a

man Petroleum Research Activities," *Oil and Gas Journal* 44, no. 29 [November 24, 1945]: 113).

[37]Schroeder interview. On the other hand, another senior American investigator sourly notes that "despite the general spirit of cooperation in evidence among most of the German scientific personnel, it is plain that the Germans made elaborate plans to conceal scientific data. Day after day caches were delivered and the contents brought to one of the several documentary centers established in various parts of the American and British occupied zones of occupation" (Murphy, p. 1530).

[38]James Stewart Martin, *All Honorable Men* (Boston, 1950), p. 58. For the first complete investigation of slave labor in this area, see Terry Hunt Tooley, "Nazi Technocracy and Coerced Labor: A Case Study of the Synthetic Fuel Industry," *Red River Valley Historical Journal* 3 no. 2 (Autumn 1978): 161–72.

battery of clerks from the American Third Army had to be convinced not to discard records in favor of salvaging badly needed file cabinets and folders—though not before "more than a hundred tons of records lay dispersed over an area larger than a city block."[39] Ultimately, the many hundreds of tons of Farben records were saved by moving them to the Reichsbank building next door for later detailed examination. "We did it in three days," recalls one investigator, "with two tractor-drawn vans, two hundred PW's and three hundred I. G. Farben employees."[40] So important were these industrial records that the Final Report of the CIOS program equated the capture of these I. G. Farben records and patent files with the many months of interrogation of the nation's industrial director, Albert Speer, Reichminister of Armaments and War Production, and his deputies.[41]

Of equal importance to the rescue and investigation of Germany's fuel records and experimental data was the Technical Oil Mission's success in locating the most influential and knowledgeable German scientists in the field of synthetic fuels. For example, at the Kaiser Wilhelm Institute, the largest government-financed research laboratory in Germany, the members of the mission found director Dr. Helmut Pilchler and his entire staff anxious to cooperate. At I. G. Farben, Dr. Matthias Pier, the aging but brilliant father of modern hydrogenation, not only responded to their skillful questioning but ultimately instructed his very sizable personal staff at the Leuna plant to help the Americans as well. Pier's colleague, Dr. Ernst Donath, himself a world-renowned petroleum chemist, recalls that it was not simply a case of having been ordered to cooperate with the TOM investigators, but, rather, it was "a continuation of the exchange of information that we had begun before the war. We generally knew one another and knew who was working on what specialty. Many of the investigators, in fact, had spent months or years with us at Ludwigshaven before the war as representatives of their companies to I. G. Consequently, although there was a rule against fraternization during our many meetings, as soon as we were alone in a room, the T.O.M. experts [generally Dr. Faragher of Houdry Process or Sir Paul White of Britain's Imperial Chemical Industries] always asked me about my wife and family, and I about theirs, so the rule was not taken too seriously."[42] The more than ninety other influential German fuel engineers who were interrogated at length by the team members—many of whom were old friends from prewar years—represent a "who's who" of German science: Drs. Ernst Mommsen,

---

[39] Josiah E. DuBois, Jr., *The Devil's Chemists* (Boston, 1952), p. 36.
[40] Martin, p. 60.
[41] Betts and Linstead (n. 23 above), p. 73.
[42] Interview with Dr. Ernst Donath, March 6, 1979, Virgin Islands.

Martin Cunradi-Muller, Fritz and Karl Winkler, Gebhard von Krupp, Erich Frese, Heinrich Bütefisch, Hermann Mayer, Ernst Donath, Leonard Alberts, Walter Oppelt, Ernst Graff, Otto Hubmann, Wilhelm Gunz, and Heinrich Tramm, to name but a few. On one occasion in Frankfurt, a tall, thin, gray-haired German named Karl Fischer walked into the office of the petroleum investigators and, in nearly perfect English, offered his services as a chemist. "We were just starting operations involving . . . the German chemical industry," recalled M. H. Bigelow, director of Technical Service, "so, making certain Fischer was not an active Nazi party member, we gave him the job of technical consultant. Not until several months had passed, because of the extreme modesty of the man, did we realize he was the Fischer who had discovered the reagent widely used in oil technology for moisture determination, the Karl Fischer reagent."[43]

The Technical Oil Mission (as well as the other teams from the United States, plus those from Great Britain) worked smoothly through the spring of 1945. Operating first by plane from London to Brussels or Paris, and from there by automobile to the German installations, individual team members—generally in concert with fellow British members—flew or drove directly to specific target plants. As each group of two or three investigators completed its work at an installation, the men returned to London to write up their report and pick up their next assignment. The makeup of the TOM itself was constantly shifting as companies recalled their experts or sent new representatives to London. Within the remarkably short period between the team's first arrival in London in mid-February 1945 and the phased withdrawal of the last TOM members in August, Schroeder's two dozen experts (functioning within an army of 1,876 other CIOS investigators whose interests ranged from midget U-boats to insecticides) had visited and analyzed the workings of no less than forty of Germany's largest synthetic fuel–producing installations. Among them were the giants listed in table 1.

As each plant came under investigation, the tons of industrial records were loaded aboard trains and shipped out of Germany to CIOS headquarters in London. The volume, understandably, was enormous; the quarters close; and the inefficiency and duplication of effort between the different agencies' teams was frustrating. Of these problems, the greatest, by far, was the massive and unwieldly mountain of papers and reports. As a result, the documents were run through another selection process to eliminate unimportant material, and the remainder was microfilmed—ultimately to become 306 reels—each containing about 1,000 pages of German data. These

[43]M. H. Bigelow, "New Fischer Techniques for Studying Oils and Waxes," *Chemical and Engineering News* 25, no. 33 (August 18, 1947): 2366.

TABLE 1

GERMANY'S LARGEST SYNTHETIC FUEL-PRODUCING INSTALLATIONS

| Name | Location | Products of Interest to Technical Oil Mission |
|---|---|---|
| Reichs-Marineamt (Admiralty) | Kiel | Diesel fuel |
| Betriebstoff Laboratorium | Wilhelmshaven | Fuel standards |
| Rhenanin-Ossag (Mineralowerke A.G.) | Hamburg | Petroleum products |
| Deutsche Vacuum Oel A.G. | Hamburg | Petroleum products |
| I. G. Farbenindustrie A.G. | Leuna | Synthetic fuel and by-products |
| I. G. Farbenindustrie A.B. | Ludwigshafen-Oppau | Synthetic fuel and by-products |
| Braunkohle Benzin A.G.: | | |
| Brabag I | Bohlen-Rotha | Synthetic liquid fuels |
| Brabag II | Magdeburg | Synthetic liquid fuels |
| Brabag IV | Troglitz-Zeitz | Synthetic liquid fuels |
| Gelsenberg Benzin A.G. | Gelsenkirchen | Hydrogenation |
| Hydrierwerke Scholven A.G. | Scholven-Buer | Hydrogenation |
| Union Rheinische Braunkohlen Kraftstoff A.G. | Wesseling | Hydrogenation |
| Ruhrol A.G. (Matthias Stinnes) | Bottrop-Welheim | Hydrogenation |
| Wintershall A.G. | Lützkendorf Mucheln | Gasification |
| Ruhrchemie und Ruhrbenzin A.G. | Sterkrade-Holden | Gas synthesis or Fischer-Tropsch process |
| Friedrich Krupp | Wanne Eickel | Synthetic liquid fuels |
| Klochner Werke A.G., Gewerkschaft "Victor" | Castro-Rauxel | Gas synthesis |
| Hoesch Benzin A.G. | Dortmund | Gas synthesis |
| Gewerkschaft Stein: | | |
| Kohlenbergwerk | Homberg | Gas synthesis |
| Rheinpreussen | Homberg | Gas synthesis |
| Chemische Werke, Essener Stein-Kohle A.G. | Kamen-Dortmund | Gas synthesis |
| Braunkohle-Benzin A.G.: | | |
| Brabag III | Ruhland-Schwarzheide | Synthetic liquid fuels |
| Schaffgotsche Benzin A.G. | Odertal (Deschowitz-Beuthen) | Synthetic liquid fuels |
| Kaiser Wilhelm Institute für Kohlenforschung | Mulheim-Ruhr | Research on gas-synthesis process |
| Studien und Verwertungs G.m.b.H. | Mulheim-Ruhr | Synthetic liquid fuels |

SOURCE.—W. C. Schroeder, "Investigation by the U.S. Government Technical Oil Mission" (paper presented to the Twenty-fifth Annual Meeting of the American Petroleum Institute, Chicago, November 14, 1945), pp. 3–4.

reels of microfilm, covered by a variety of "Secret" classifications, were then shipped back to Washington. There, the Technical Advisory Committee of the PIWC created yet another group, called the Special Advisory Group, whose task it was to review the incoming reports from the Technical Oil Mission and recommend to the government which of these should be released to the oil industry. As the Special Advisory Committee met to consider the importance of the first incoming reels of German documents, it was clear that if it was to help the American petroleum industry (which, after all, had been the central reason for the entire investigation), the information had to be swiftly declassified and then distributed effectively to each company. The solution to the classification problem came during a routine Special Advisory Group meeting in New York on July 23, 1945, when a representative from the PAW, P. J. Byrne, Jr., announced the government's welcome decision to reduce the classification level from "Secret" to the far more manageable "Restricted." Later, on October 15, 1945, even that classification was done away with.

As it became increasingly evident that the volume of captured data required more analysts than those available on the TAC's Special Advisory Group, and since the obstacles of secrecy were being rapidly dismantled, the PAW announced, on August 7, 1945, the creation of small, independent study groups whose task was to aid in the screening and translation of the material. This PAW-PIWC plan provided for the participation of additional industrial representatives who could review the microfilm away from Washington in their own offices or libraries.[44] Expenses for this work were to be borne by the individual companies or the PIWC, and their main task was the formidable one of indexing the thousands of frames of randomly photographed documents on each reel. That everyone concerned saw in this information the key to American synthetic fuel production is evident from the speed with which they completed the first segment of the project: the first index, covering reels 1–119, was completed and distributed to the 450 names on the TAC mailing list by October 1945![45]

With Germany's official surrender came a rapid dismantling of the majority of wartime agencies, including those involved in examining the enemy's technical, industrial, and scientific secrets. On July 13,

[44]Miller (n. 24 above), p. 14. Companies which set up Technical Advisory Committee Film Study groups include Atlantic Refining Co.; Bureau of Mines; California Research Corp.; Gulf Research and Development Co.; Humble Oil and Refining Co.; M. W. Kellogg Co.; Koppers Co., Inc.; Phillips Petroleum Co.; Pure Oil Co.; Shell Development Co.; Sinclair Refining Co.; Socony-Vacuum Oil Co., Inc.; Standard Oil Co. (Ohio); Texas Company; and Universal Oil Products Co.

[45]Ibid., p. 16.

1945, with the dissolution of SHAEF, the entire CIOS operation—of which the Technical Oil Mission was a part—was terminated, and its remaining work was taken over by the newly organized Consolidated Advance Field Teams or CAFTs.[46] Soon thereafter, President Truman wrote to Secretary of the Interior Julius A. Krug that, "in keeping with the Administration's policy . . . the Petroleum Administration for War, which has so successfully completed its wartime assignment . . . is terminated."[47] Finally, the petroleum industry received a joint letter of thanks from representatives of the army and navy for its contributions to the Allied victory.[48] The American investigators of the Technical Oil Mission began to make their way home, and by December 1945, all were back on their old jobs. They were to find, however, that their participation was not quite over.

### A New Oil Mission to Europe

In December 1945, the members of the former Technical Oil Mission as well as representatives of the petroleum industry, the PAW, and the Bureau of Mines gathered at a final conference in Washington to summarize the findings of Germany's oil secrets. One of their initial discoveries concerned the need for additional investigation. As good as the work of the Technical Oil Mission had been, the teams had clearly worked under adverse conditions, and, due to the pressure of competing Soviet investigators as well as the need to make each discovery available for the war against Japan, the time allotted for each investigation had been limited. Moreover, it was certain that many people with valuable information had not been located and that large caches of important documents had not been properly examined. Now that both Germany and Japan were defeated, a more thorough exploitation of their industrial secrets was "not only a right but also a duty of our Government, since a knowledge of the details of the commercial activities, and of the manufacturing processes of the conquered countries constitutes the only reparations we would receive or require."[49]

The result was the creation of a new organization, to be attached to the United States Military Government for Germany, called the Field Intelligence Agency (Technical), or simply FIAT. Originally designed to explore areas of fundamental scientific research, FIAT was soon expanded to serve as an umbrella agency for numerous new missions from American industry which hoped to learn about everything from

[46]Betts and Linstead, pp. 26–29.

[47]Frey and Ide, eds. (n. 10 above), p. 297; *Federal Register* 4 (May 3, 1946): 4965.

[48]*Oil for Victory* (n. 6 above), p. 73.

[49]W. F. Faragher, "Collecting German Industrial Information," *Chemical and Engineering News* 26, no. 52 (December 27, 1948): 3817.

92    *Arnold Krammer*

German pharmaceuticals and optics to synthetic rubber and leather tanning.[50] It was to this new agency, then, that many of the former TOM members were once again dispatched in May 1946, to reexplore the German synthetic petroleum industry. Operating first from the city of Höchst, and then Karlsruhe, the five to twelve American engineers bypassed the relatively unproductive investigation of bombed-out plants and factories and instead turned directly to the responsible German scientists and engineers to author reports on their individual specialties. Ultimately, FIAT employed 650 Germans and twenty Allied civilians who authored some 28,000 short reports and more than fifty comprehensive volumes from mid-1946 to mid-1947. The work on synthetic petroleum was directed by no less than the aging Franz Fischer himself, the co-inventor of the Fischer-Tropsch process. In addition, a staff of 175 German technicians was engaged to read, translate, and classify 30,000 pages of captured documents and blueprints daily for inclusion on the proliferating reels of microfilm.[51]

The extraction of information went beyond the documentary stage. American industry was not only interested in reports, documents, and blueprints but also in seeing the actual plants in operation, tracing the production line, learning the industrial techniques, interviewing the German engineers and foremen on the job, and bringing back samples and models for the development of major industry reports. Through the spring of 1947, a steady stream of selected industry scientists and engineers traveled to Europe under joint industry and government sponsorship. In the area of synthetic petroleum, these industrial teams brought back such unique and massive items as a high-pressure, forged, low-chrome steel liquid phase converter weighing 170 tons from Dortmund-Hörder Hüttenverein; a spe-

[50]See, e.g., E. C. Kleiderer, "The Pharmaceutical Industry of Germany," *Chemical and Engineering News* 27, no. 17 (April 25, 1949): 1206–8; Cortex F. Enloe, Jr., "The War and the German Drug Industry," ibid., 24, no. 22 (November 25, 1946): 3046–48; E. R. Weidlein, Jr., "Synthetic Rubber Research in Germany," ibid., no. 6 (March 25, 1946), pp. 771–74; Harry A. Kuhn, "Developments in the German Chemical Industry," ibid., 23, no. 17 (September 10, 1945): 1516–22; W. C. Goggin, "Advances in Plastic in the United States and Germany," ibid., 24, no. 3 (February 10, 1946): 339–43; and "New Tanning Agents Based on German Technology," ibid., 26, no. 27 (July 5, 1948): 1980–81, 2029.

[51]Ivanovzky (n. 28 above), p. 48. All of these activities were unexpectedly concluded following General Lucius Clay's decision to terminate the majority of FIAT's projects on June 30, 1947; he cited as the reason the burden on his own budget, which had just been reduced by Congress, and FIAT's apparent interference in his plans for the rehabilitation and reoperation of Germany's factories (see W. F. Faragher, "Collecting German Industrial Information," p. 3819; and Richard L. Kenyon, "The FIAT Review of German Science," *Chemical and Engineering News* 25, no. 14 [April 7, 1947]: 962–63).

cialized, high-pressure, 54-foot long, Wickel-type heat exchanger from Krupp; and a variety of high-pressure, hydraulic paste injection pumps, preheaters, lens rings, fittings, and special values and instruments. Of particular importance, especially to the later Bureau of Mines Synthetic Fuel pilot plant at Louisiana, Missouri, was a huge Linde-Fränkl Oxygen Plant, of 1-ton/hour capacity, at 98 percent purity, from Höchst.[52] While American industry did not feel obliged to purchase the equipment, "for we do not want to pay American dollars to our defeated enemies," it took pains to assure the public that neither "do we want to loot German establishments and take from them equipment for our benefit." "Accordingly," noted a government spokesman, "we have listed and are listing those pieces of physical equipment which our technical experts tell us deserve examination in this country, asking the Military Government to set them aside for us and the Allied Reparations Commission to allocate them to the United States as its share in war reparations. When this is successfully accomplished, the German Government assumes the cost of furnishing us the equipment and we in turn will make it available to all in this country with a legitimate interest."[53]

Indeed, the representatives of American industry could hardly wait to delve into the secrets of Germany's synthetic fuel production. Never in the history of the modern work has a sophisticated industrial nation had at its complete disposal the industrial secrets of another nation. Among the many areas of interest, the most valuable subjects

[52]Merritt L. Kastens, L. L. Hirst, and C. C. Chaffee, "Liquid Fuel from Coal," *Industrial and Engineering Chemistry* 41 (May 1949): 870–85; L. C. Skinner et al., "Thermal Efficiency of Coal Hydrogenation," ibid., 41 (January 1949): 87–96; and J. T. Donovan, B. H. Leonard, and J. A. Markovitz, "Design and Development of High Pressure Injection Pumps for Hydrogenation Service" (American Society of Mechanical Engineers [ASME], paper no. 51-A-71, New York, 1951).

[53]John C. Green, "Scientific Information from Enemy Sources and Government-sponsored Research," *Chemical and Engineering News* 24, no. 13 (July 10, 1946): 1798. The Department of Commerce as well as the War and Navy Departments did indeed arrange to make enemy industrial equipment available to the general public through numerous exhibitions. The British government held similar industrial exhibitions, known as "Germany under Control" shows, and the President of the Board of Trade, Sir Stafford Cripps, personally urged British industry "to make the fullest and speediest use of this new knowledge. All of it will become out-of-date, some sooner than others, so that there is no time to waste if we are to get the full advantage of all this effort that has been undertaken. I would particularly appeal to the smaller firms who have not their own research departments to allow BIOS to help them introduce the latest manufacturing methods and processes" ("German Industrial Information," *Chemical Age* [December 14, 1946], p. 741; and "Germany under Control Exhibition: Petroleum Production and Refineries in Germany," *Petroleum* 9, no. 10 [October 1946]: 228–29, 247).

94    *Arnold Krammer*

concerned the production of synthetic acetylene, alcohol, and ammonia; various catalysts; fatty acids; several processes for butadiene
and synthetic rubber; fractionation; gasification; high-pressure hydrogenation; hydroforming; various patents; and analytical and testing methods.[54] "Our discoveries in Germany were of immense value
in terms of national security," declared Edward B. Peck of Standard
Oil Development Company. "We shall not have to go down already
beaten paths in our own research and experimentation. We shall be
able to profit by German mistakes and we can eliminate years of work.
. . . [Moreover], in case of another war, development of an adequate
synthetic industry would obviate . . . the necessity of dependence on
foreign oil supplies."[55] Michael Straus, assistant secretary of the interior under Ickes, went so far as to declare that the German records,
together with Bureau of Mines research and industry's help, had
"brought a synthetic fuels industry within the nation's reach. This," he
continued, "will make the United States ready for a declaration of oil
independence at any time that our dwindling domestic natural petroleum reserves make it necessary to turn to synthetics. That time,"
he added prophetically, "may be sooner than we think."[56] The
Bureau of Mines, in fact, was already preparing a program to test the
German research discoveries at pilot and demonstration plants in

[54]W. C. Schroeder, "Capture of German Records Will Boost United States Synthetic
Fuel Development," *Oil and Gas Journal* 44 (June 30, 1945): 78, "Technical Oil Mission
Studies German Petroleum Research Activities," ibid., 44 (November 24, 1945): 113,
and "German Synthetic Liquid Fuels Manufacturing," *Gas Age* 96 (November 29,
1945): 21; Kenyon (n. 16 above), pp. 1437–39, and "Fischer-Tropsch Held One of Most
Valuable Enemy Processes," ibid., 25, no. 12 (March 24, 1947): 839.
[55]Sheldon (n. 36 above), p. 122.
[56]Schroeder, "Capture of German Records Will Boost U.S. Synthetic Fuel Development," pp. 78, 88; and "German Synthetic Liquid Fuels Manufacturing," p. 21. However, not all American engineers were as convinced as Straus or Schroeder about the
overall value of German technology. While no one could deny the fact that industries
had scored many scientific successes, the enemy was often found to be surprisingly far
behind Allied scientists in a variety of areas. The notable exception, however, was
synthetic fuels, and, as if to perk up any disheartened American petroleum engineers
among his readers, Walter J. Murphy, editor of *Chemical and Engineering News* and a
ranking member of the investigative mission to Germany, noted that "it must be remembered that in totalitarian Germany technologists mobilized for war early in the
thirties, while American men and women of science focused attention solely on ways
and means of raising the standard of living of all mankind—not its ultimate enslave-
ment" (Murphy [n. 31 above], p. 1530; "The Reports on German Industry," *Chemical
Age* [January 1947], p. 112; and "What Can We Learn from German Technology?"
ASME round table discussion, Battelle Memorial Institute [Columbus, Ohio, December
13, 1945]; also see "Columbus Section Hears Discussion on German Technology,"
*ASME News* [February 1946], p. 179).

Pittsburgh, Pennsylvania; Rifle, Colorado; Laramie, Wyoming; and the later controversial and successful plant at Louisiana, Missouri. On the occasion of the bureau's acquisition of the basic plant in Louisiana, Missouri, Secretary Ickes declared that "American advances, plus German developments of recent years . . . will make the bureau's synthetic fuels demonstration plant the most modern and efficient yet known."[57] Despite the soaring optimism of both government and industry, the end of the war period led to an understandable waning of interest.[58]

Middle East petroleum was again flowing steadily, and new American fields were producing inexpensive crude for the postwar market. That the Germans were able to produce in excess of 4 million metric tons of synthetic fuel yearly was no longer important. Indeed, a division of the government's Foreign Economic Administration, Enemy Branch, called the Technical Industrial Disarmament Committee, recommended the total elimination of Germany's synthetic industry, not only to prevent a resurgence of German militarism and competition with the Allied nations in oil exports to Western Europe but "because the global supply of petroleum for the foreseeable future will be one of abundance."[59] Despite all the wartime predictions that the secrets of German industry would jolt America's synthetic fuel development forward by several decades and perhaps even allow the United States to achieve the elusive goal of fuel self-sufficiency, such was not to be the case.

"It is estimated," noted the secretary of the interior's first postwar report to Congress on the status of American fuel, "that gasoline can be produced from . . . coal or oil shale for 7½–9½ cents per gallon.

[57]"Missouri Ordinance Works Becomes Bureau of Mines Demonstration Plant," *Chemical and Engineering News* 24, no. 1 (January 10, 1946): 39.

[58]Two of the very few exceptions to industry's lack of interest in developing synfuels along the German model were the Hydrocol Process of Carthage-Hydrocarbon, Inc., which erected a $15 million plant in Brownsville, Texas, to produce 300,000 tons per year of high-grade gasoline; and a facility constructed by Stanolind Oil and Gas Co. in Kansas (see P. C. Keither, "Gasoline from Natural Gas," *Oil and Gas Journal* [June 15, 1946], pp. 102–12).

[59]Bruce K. Brown, "The Petroleum Industry in 1945," *Chemical and Engineering News* 24, no. 3 (February 10, 1946): 330–33. For various discussions regarding the dismantling of Germany's postwar synfuel industry, see "Committee Recommends Elimination of Germany's Synthetic Industry," *World Petroleum* 17, no. 2 (February 1946): 25; "Engineers Recommend Permanent Abolishment of Nazi Synthetic-Oil Facilities for Peace," *Oil and Gas Journal* (October 7, 1944), p. 50; "No Revival for I. G. Farben," *Chemical Age* (March 30, 1946), pp. 327–28; "Control of German Industry," *Mechanical Engineering* 67, no. 11 (November 1945): 754–57; and Roger Adams, "The Future of German Industry," *Chemical and Engineering News* 24, no. 18 (September 25, 1946): 2486–89.

96     *Arnold Krammer*

These are . . . higher than the present estimated cost of gasoline from petroleum."[60] The following year, the secretary's next report went on to note that "Preliminary estimates indicate that to produce 2 million barrels of [synthetic] oil a day—less than 40 percent of our current daily consumption—would require about 16 million tons of steel and the expenditure of around 9 billion dollars. . . . It is doubtful," the report sadly concluded, "that private industry will find the investment sufficiently attractive to warrant starting plant construction without governmental assistance in some form"[61] Despite the report's urging that just such a synthetic oil program be undertaken, industry concluded, as predicted, that the German experience was not economically feasible for the United States, with its union issues, standard wage scales, competitively cheaper petroleum fuel, and rising postwar inflation.[62] The Department of Commerce, however, was quick to point out that "the inclination and ability of our major oil companies to commence production of synthetic fuel, should conditions require it, is well known."[63]

[60]"Synthetic Liquid Fuels Studies from Several Viewpoints," *Chemical and Engineering News* 26, no. 9 (March 1, 1948): 610–11.

[61]U.S. Department of the Interior, "Report of the Secretary of the Interior on the Synthetic Liquid Fuels Act from January 1, 1947 to December 31, 1947," mimeographed, file no. 30409 (Washington, D.C., 1948), p. iii.

[62]The United Kingdom, incidentally, found itself in similar economic straits, and as noted in the Ministry of Fuel and Power's so-called Gordon Report of 1947, also elected to ignore the mountains of German industrial data which its own CIOS and BIOS investigators had worked so hard to obtain (British Ministry of Fuel Power, *Report on the Petroleum and Synthetic Oil Industry of Germany*, BIOS overall report no. 1 [London, 1947], p. 2; and Kenneth Gordon, "Progress in Hydrogenation of Coal and Tar," *Chemical Age* [October 28, 1946], pp. 795–804).

[63]R. L. Trisko, "United States Petroleum Import Prospects," *Industrial Reference Service, Part II: Metals and Minerals* 5, no. 2 ( July, 1947): 11. Whether or not the Department of Commerce realized it, conditions already required the serious support of synthetic fuels by the oil industry. The expected abundance of natural petroleum had not yet materialized; postwar demand for gasoline and other petroleum products was rising at a level above all expectations, and, by 1947, there was still no leveling off in sight. In fact, the predicted U.S. consumption of oil products for 1948 was to exceed by 14.2 percent the wartime peak of 1945. The country was clearly facing an energy crisis in the not-too-distant future, while the petroleum industry continued to resist synthetics ("Report of the Secretary of the Interior on the Synthetic Liquid Fuels Act from January 1, 1947 to December 31, 1947," p. i). Curiously, the petroleum industry was still unconvinced of a shortage. According to an industry spokesman, "We should anticipate no long-term shortage of crude oil and we certainly should not see ourselves running out of liquid hydrocarbon products. Even with unprecedented demands for crude, our known reserves are larger than at any time in the past" ("No Long-Term Shortage of Crude Oil Anticipated," *Chemical and Engineering News* 26, no. 16 [April 19, 1948]: 1149).

Technology Transfer as War Booty 97

## Making the Captured Industrial Data Available

And what of the captured information which American petroleum engineers had worked so hard to locate, abstract, and ship home? With the end of the war, the few remaining activities of the Technical Oil Mission were transferred to the Bureau of Mines, still optimistic about the $30 million Synthetic Fuels Act mandated to them in a long-gone moment of wartime cooperation. The bureau also inherited the thousands of catalyst types, lubricating oils, greases, coal samples, and the many miscellaneous materials brought home by the TOM specialists and produced a number of significant, if unheralded, reports on these items.[64]

The estimated 175 *tons* of captured documents went a different route. The most immediately important data, refinery records and experimental data, microfilmed either on the spot in the captured factories or later in London, were made available as quickly as possible. As early as August 1945, Deputy Petroleum Administrator Ralph K. Davies was able to announce that the PAW was ready to release the first reels of microfilm through the offices of the Technical Advisory Committee of the Petroleum Industry War Council. However, "while it is the PAW's desire," Davies stated, "to make these technical data freely available to anyone interested as a matter of public information . . . much of the information is still classified as restricted material and cannot, therefore, be distributed except on a restricted basis."[65] If military security prevented the swift distribution of the captured enemy records, the lack of a central clearinghouse led to disputes over the ownership of particularly critical documents. In at least one case, the struggle over ownership became sufficiently public to warrant the following plea by the editor of the prestigious *Chemical and Engineering News:* "Currently it is being rumored in Washington that I.G. files are the subject of a tug-of-war between various American agencies. . . . If such is the case, the President should stop it im-

---

[64]See, e.g., H. Hollings, "Report on the Investigation by Fuels and Lubricants Teams at the Wintershall A. G., Lutzkendort, New Mucheln, Germany," *Bureau of Mines Information Circular 7369* (July 1946); R. Holroyd, "Report on the Investigation by Fuels and Lubricants Teams at the I. G. Farbenindustrie A. G. Leuna Works, Merseburg, Germany," *Bureau of Mines Information Circular 7370* (July 1946); John W. Buch, "Design and Operation of the Coal Planer, Ruhr District, Germany," *Bureau of Mines Information Circular 7377* (July 1946); and Thomas Fraser and M. G. Driessen, "Coal Preparation Practice in Western Germany," *Bureau of Mines Information Circular 7389* (September 1946). The bureau also translated several hundred mimeographed reports of special interest between 1947 and 1949 known as the "T," "K," "L," "S," and "SR" series.

[65]"German Records to Be Available," *Petroleum Engineer* (August 1945), p. 63.

mediately and appoint a special commission to supervise *all* German documentary material. Far too much is at stake. We must not lose valuable scientific knowledge, the fruits of the labor of thousands of highly qualified German scientists, as a result of interagency quarrels, jealousies and the petty ambitions of 'Empire Builders.' "[66]

In response to the need for this central clearinghouse in order to speed the declassification and distribution of the mountains of incoming captured documents to the American public, in autumn 1945, President Truman issued Executive Order 9568, which created the Office of Publication Board within the Department of Commerce.

Composed of representatives of all major governmental and military agencies, the Publication Board was, in essence, a massive administrative conduit through which the captured data was routed to the proper authority for declassification. As each report was cleared, it was indexed and then forwarded to the Government Printing Office for publication and general distribution.[67] It soon became apparent, however, that no matter how efficient, the Publication Board could not manage the enormous scope of incoming data unaided. Consequently, President Truman issued a second Executive Order, no. 9604, to substantially enlarge the operation. The Publication Board was rechristened the Office of Technical Service, and it spent the next three years preparing and making available some 28,000 reports per year on the secrets of German industry. Public response was startling, and even a year after the war, the Office of Technical Services was handling orders at the rate of 1,000 items a day.[68] Three years later, in 1948, the OTS could still announce that "more than 5 million microfilmed pages of technical documents, all in German, containing drawings, flow sheets, reports of chemical experiments and meetings of German technical societies are now being processed with the cooperation of [1,500 volunteer members from] 422 technical and scientific societies."[69] As to subject matter, however, reports dealing with rocketry, dyes and synthetic fibers, optics, and plastics, seemed to be of greatest interest to American (and Russian)[70] customers, with curiosity in synthetic fuel lagging far behind.

[66]Murphy (n. 31 above), p. 1531.

[67]John C. Green, "Scientific Information from Enemy Sources and Government-sponsored Research," *Chemical and Engineering News* 24, no. 13 (July 10, 1946): 1795–99.

[68]L. Lester Walker, "Secrets by the Thousands," *Harper's Magazine* (October 1946), p. 335.

[69]Al Leggin, "Potomac Postscripts," *Chemical and Engineering News* 26, no. 11 (March 15, 1948): 733.

[70]The most insatiable customer was Amtorg, the Soviet Union's foreign organization. One of its representatives reportedly walked into the Publication Board Office with the

Nor was there much interest in the numerous summary reports by the team of German technologists brought to the United States to continue their research under the direction of W. F. Faragher of the Houdry Process Company. From mid-1946, several hundred German scientists worked on a variety of projects for the military services, and although the Bureau of Mines also hoped to obtain its share of German technicians, it was ultimately decided to invite the seven most important German synthetic fuel experts. These scientists—Drs. Ernst Donath, Leonard Alberts, Ernst Graf, Kurt Bretschneider, Hans Schappert, Walter Oppelt, and Erich Frese—all eminent men who had cooperated closely with their American colleagues in the Technical Oil Mission, produced a significant amount of research on behalf of their hosts.[71] So successful was the utilization of German technicians that plans were announced by the government in April 1947 to bring over 1,000 more scientists for various purposes, and a cry of opposition arose in Congress. Recalling the lessons of the war which was only two years past, Congressman Lane (D-Massachusetts) introduced in the House of Representatives H.R. 2763, a bill "to prohibit the use by the United States of Nazi scientists . . . [i.e.,] any member of the German National Socialist Party or any alien who was employed by the German Government between 1933 and 1945. . . ."[72]

---

bibliography in hand and said, "I want copies of everything" (Walker, pp. 335–36; see also Valentin L. Sokolov, *Soviet Use of German Science and Technology, 1945–1946* [New York, 1955]).

[71] For a verbatim record of the frank discussions of the numbers and problems involved in bringing such German scientists to the United States, see "The Proceedings of the Technical Oil Mission, Held under the Auspices of the Petroleum Administration for War and the Bureau of Mines," mimeographed, file no. 10160 (Washington, D.C., December 13–14, 1945), pp. 89–98. In addition to the synthetic liquid fuels projects, German scientists were brought to the United States to participate in a variety of other ventures (see Joint Chiefs of Staff, "Statistical Report of the Aliens Brought to the United States under the Paperclip Program, December 1, 1945," mimeographed [Office of Technical Services, Department of Commerce, Washington, D.C.]; and Lasby [n. 35 above]). Relations with former enemy scientists quickly became so congenial that it became popular to regard the German scientists (or all scientists, for that matter) as neutrals in the previous war (see Gavin de Beer, *The Sciences Were Never at War* [London, 1960], and "Science Has No Nationality," *Science Illustrated* 2, no. 2 [February 1947]: 13).

[72] Al Leggin, "Potomic Postscripts," *Chemical and Engineering News* 25, no. 17 (April 28, 1947): 1165; ibid., no. 20 (May 19, 1947), p. 1415; and U.S. Congress, House, "Let Us Not Give Away Our Weapon Secrets to the News," *Congressional Record*, 80th Cong., 1st sess., March 25, 1947, 93, pt. 2:2540–41. The feelings of those most sharply against allowing German technicians in the United States are best summarized by an editorial in an industrial journal which stated, "It may be said that after a period of re-education when Nazi ideas have been eliminated from the German mind, the potential danger of

100    *Arnold Krammer*

Although the bill did not go on to become law, it would have made little difference to the few German scientists working for the Bureau of Mines: They never amounted to more than a handful of men; the majority went on to become American citizens; and most important, the public was hardly concerned with synthetic fuel.

Indeed, a substantial portion of the captured data was neither cataloged nor checked, and as late as 1948, the Office of Technical Services acknowledged that "at the present time there are stored in a warehouse over 2,500 trunks of German documents which have not been opened."[73] The OTS spokesmen publicly expressed their fear that "the great bulk of the information will be lost to business and industry, and that present plans call for selection of only the most important [documents] to be made available to industry and educational institutions."[74]

The most important documents in the field of synthetic fuels, as assembled by the TOM and the second team of petroleum technicians with FIAT, were generally microfilmed and forwarded by air, lest their secrets be delayed in the war effort against Japan, to the offices of the Petroleum Administration for War. These reels were then made available, on a security restricted basis, to the Petroleum Industry War Council through its Technical Advisory Committee. Ultimately 306 such reels, each containing some 1,000 frames, were completed, and are simply known as the Technical Oil Mission reels.[75] The overall responsibility for the efficient distribution of all of this information was placed in the hands of the industry's American Petroleum Institute (API). To do so, they, in turn, created an API Technical Oil Mission Study Group which continued the wartime work of the TAC in distributing and indexing the TOM data. Under the direction of Albert E. Miller of Standard Refining Company, the study group continued to evaluate and index the reels through 1950. All other records and correspondence dealing with the TOM, as well as those of the Petroleum Administration for War, the Petroleum Industry War Council, its Technical Advisory Committee, and,

---

this industry will also vanish. . . . Meanwhile, however, the teams of technicians, many of whom are tarred with the Nazi brush, must be allowed to die out" ("No Revival for I. G. Farben," *Chemical Age* [March 30, 1946], p. 328).

[73]"Potomac Postscripts," *Chemical and Engineering News* 26, no. 11 (March 15, 1948): 733.

[74]Ibid., no. 25 (June 21, 1948), p. 1823.

[75]An additional fifty reels of material pertaining to the Allied air offensive against the Axis oil industry were made at the Albert Simpson Historical Research Center at Maxwell Air Force Base, Montgomery, Ala.

## Technology Transfer as War Booty    101

finally, the American Petroleum Institute, were placed on deposit at the National Archives in Washington (see fig. 3).[76] The Bureau of Mines, since absorbed into the Department of Energy, maintains a substantial volume of records at its Research and Development Division Technical Library at Pittsburgh, including 225 reels of microfilm of captured German technological records and reports of the BIOS and TOM missions, the United States Naval Technical Mission to Japan, the ALSOS Mission, the United States Strategic Bombing Survey, and correspondence and technical studies of the Technical Industrial Intelligence Committee's Liquid Fuels and Lubricants Subcommittee, 1944–45.[77] The remaining documents were placed in

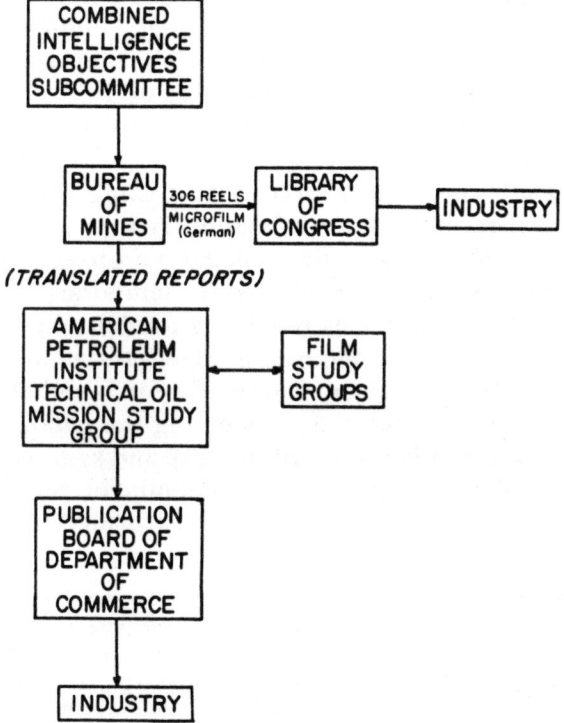

FIG. 3.—Flow of Technical Oil Mission information after 1945

[76]This substantial collection of documents comprises Record Group 330, which, according to a detailed examination by National Archives personnel, has "seldom, if ever, been consulted by the public" (Janet Hargett, assistant director of the General Archives Division, General Services Administration, to author, October 12, 1976).

[77]For a more detailed list of their holdings, see *Federal Records of World War II*, vol. 1, Civilian Agencies (Washington, D.C., 1950), pp. 816–17.

102    *Arnold Krammer*

dead storage at various government facilities across the country and largely forgotten.[78]

Also forgotten were the reasons for the entire effort to obtain information about the enemy's progress in synthetic fuels and the lessons locked within the hundreds of thousands of pages of technical data. Indeed, for the next twenty years, while the bulk of captured data languished in storage in the United States or in West Germany, the American oil industry grew more and more dependent on relatively inexpensive petroleum from the Middle East and Venezuela. Between 1953 and 1973, in fact, America's annual reliance on foreign oil rose from less than 1 percent (19 million barrels) to more than 29 percent (1.1 billion barrels). While the oil industry did invest heavily in the nation's great coalfields during the same period, it was a program prompted by a diversification of its huge holdings rather than with a view toward future synthetic fuel production. It was the Arab oil embargo and the subsequent price hikes of 1973–74 which suddenly caused the government's attention to be riveted on the question of synthetic liquid fuels.

Energy related legislation poured out of Congress. The Energy Supply and Environmental Coordination Act of ( June 22) 1974 (Public Law 93-319), for example, initiated a thorough investigation of the nation's fuel supplies and allowed the temporary suspension of certain air pollution restrictions regarding the burning of fuels. On October 8, Congress passed the Energy Reorganization Act of 1974 (Public Law 93-438) to create an Energy Research and Development Administration. On December 31, lamenting the nation's "past and present failure to formulate a comprehensive and aggressive research program" and calling on the nation's commitment "similar to those undertaken in the Manhattan and Apollo projects,"[79] Congress

---

[78]However, the documents were not forgotten by the German authorities. By 1949, German archivists and newspaper journalists were developing a public case for the return of their records to Germany. On June 21, 1950, the German Bundestag passed a resolution requesting the return of the archives that had been removed from Germany, and for the next three years, each NATO agreement and conference served to bring the matter one step closer to completion. Finally, in 1954, the Department of the Army prepared a master plan for the return of the records: declassification, microfilming of selected portions, and shipment to the Federal Republic of Germany (see Dagmar Horna Perman, "Microfilming of German Records in the National Archives," *American Archivist* 22, no. 4 [October 1959]: 433–43; and Seymour J. Pomerenze, "Policies and Procedures for the Protection, Use, and Return of Captured German Records," in *Captured German and Related Records: A National Archives Conference*, ed. Robert Wolfe [Athens, Ohio, 1974], pp. 5–32).

[79]U.S. Congress, *U.S. Statutes at Large*, 1974, 88, pt. 2: 1878–85.

passed the Federal Non-nuclear Energy Research and Development Act of 1974. By this act (Public Law 93-577), Congress allocated nearly half a billion dollars "to accelerate the commercial demonstration of technologies for producing syncrude and liquid petroleum products from coal."[80]

That little serious progress ensued may be deduced from President Carter's speech before a special joint session of Congress, on April 20, 1977, to announce a national energy program which would be "the moral equivalent of war." High on the nation's energy goals, the president declared, would be a significant increase in the use of coal—and especially "clean energy sources like liquefied and gasified coal . . . [on which there has been] . . . very little research and development."[81]

The history of the development of synthetic fuels in the United States, despite the transfer of wartime technology from Germany, had come full circle.

[80]Ibid., p. 1882.

[81]U.S. Congress, House, *Congressional Record*, 95th Cong., 1st sess., April 20, 1977, 123, no. 65:3330.

# [18]

## Technology Transfer and Foreign Trade in the Early Years of the Federal Republik of Germany[1]

Reinhard Neebe

### 1. Preliminary Remarks: Technology and Foreign Trade

In the mid-sixties slogans about "the sell-out of German technology"[2] became popular; there was talk of the decline of German research[3] and that West Germany was so behind in technological know-how that there was little chance of catching up. The discussion about the loss of technological competence was conducted not just by the West Germans. During this period, as the journalistic success of Jean-Jacques Servan-Schreiber's book "Die Amerikanische Herausforderung" – "The American Challenge"[4], published in 1967, showed, the public and also government circles throughout Western Europe were gripped by the fear that the Old World could be subjected to "technological colonisation" and reduced to being the 51st state of the USA. Massive American direct investments in Western Europe, the obvious superiority in the fields of "big science", space travel, aircraft construction and nuclear technology as well as the continual exodus of European scientists and technicians to the United States, the so-called "brain drain", were impressive evidence of Europe's economic and political loss of competence. In this situation the

---

[1] The following essay is an extended version of the lecture I held at the "Historikertag" in Bamberg (Section "Technologytransfer and international markets in the 19th and 20th Century") on 13.10.1988. The article is part of the research project "The Return to the World Market. German Business and Transatlantic Relations in the Adenauer Era" sponsered by the Volkswagenwerk Foundation and conducted at the Philipps University, Marburg, West Germany under the direction of Prof. Dr. Gerd Hardach. I thank the Volkswagenwerk Foundation for their generous support for this study.

[2] *Blauhorn*, Kurt, Ausverkauf in Germany? München 1966. Characteristic for the state of awareness at this time (among many other articles): *Der Spiegel* No. 9/1966, pp. 28–44. "Nach 30 Jahren Niedergang: Notprogramm für Deutschlands Forschung?"; ibid., pp. 52–56 "Wir müssen einen gewaltigen Rückschlag wettmachen." SPIEGEL-Interview with the Minister for Scientific Research Dr. Gerhard Stoltenberg about the promotion of research in West Germany.

[3] In the first "Bundesbericht Forschung" (1965) the Minister for Scientific Research had to admit that for the period 1958–1963 the expenditure for research and development, measured in % of Gross National Product, in the Federal Republic was not even half the expenditure levels of the USA and was even behind the levels of Great Britain and Sweden. Thus in 1958 the Federal Republic spent 0.9% of the GNP on research and development compared with 2.1% in Great Britain and 2.5% in the USA, in 1962 1.3% compared with 1.6% in Sweden or 3.0% in the USA. Bundesbericht Forschung I. Bericht der Bundesregierung über den Stand und Zusammenhang aller Maßnahmen des Bundes zur Förderung der wissenschaftlichen Forschung mit einer Vorausschau des Bedarfs an Mitteln des Bundes für 1966 bis 1968. Der Bundesminister für Wissenschaftliche Forschung, Bonn 1965, pp. 114f.

[4] *Servan-Schreiber*, Jean-Jaques, Die amerikanische Herausforderung, with a preface by Franz-Josef Strauß, Hamburg 1968. (The original edition was published in 1967 in Paris under the title "Le défi americain".)

Italian Foreign Minister, Admintore Fanfani, even suggested a "technological Marshall-Plan" and the OECD dedicated a series of detailed studies to the subject "Gaps in Technology".[5]

However, from West Germany's specific point of view it is surprising that the subject of technology was not a matter of intense public discussion long before the mid-sixties. From 1945 until well into the 1950s West Germany, as a conquered country, was faced with a number of practical problems and in particular was subject to special legal restrictions, which hampered any attempt to develop systematic technological and industrial policies. To mention just three points:

1. The complete loss of patents and trademarks. After the Agreement on the Treatment of German-Owned Patents was signed in London, 27th July 1946, the protection of all former German patents was suspended and these were made available to interested parties in finance and industry in all signatory countries free of charge and without restrictions.[6] On 1st October 1949 a "German Patent Office"[7] was in fact opened in the German Museum in Munich for the newly created Federal Republic. But only on the occasion of the Chancellor Adenauer's first visit to the United States in April 1953 did the USA declare the programme of dispossession of German assets and patents to be over once and for all. And it took until 1956 before the signing of the "Patent Interchange Agreement"[5] established a new legal basis for the trade in all licences including those involving sensitive technological knowledge.

2. The loss of know-how due to the war.[9] An example for this is the American operation "Overcast" from July, 1945 and especially the project "Paperclip".[10] In the course of this programme from 1946 to 1952 about 1000 top quality German scientists emigrated to the USA, the most famous being probably Wernher von Braun with almost the entire staff from the missile research institute Peenemünde. In 1947 the US Army estimated that at least 700 million dollars had been saved by the work of German missile researchers.[11] However

---

[5] Gaps in Technology. General Report, (OECD) Paris 1968. Gaps in Technology. Electronic Components, (OECD) Paris 1968. Gaps in Technology. Scientific Instruments, (OECD) Paris 1968. Gaps in Technology. Electronic Computers, (OECD) Paris 1969. Gaps in Technology. Plastics, (OECD) Paris 1969. Gaps in Technology. Non-ferrous metals, (OECD) Paris 1969. Gaps in Technology. Analytical Report. Comparisons between Member Countries..., (OECD) Paris 1970.

[6] *Stamm*, Thomas, Zwischen Staat und Selbstverwaltung. Die deutsche Forschung im Wiederaufbau 1945–1965, Cologne 1981, pp. 48f.

[7] *Hallmann*, Ulrich C. & *Ströbele*, Paul, Das Patentamt von 1877 bis 1977, in: Hundert Jahre Patentamt. Festschrift issued by the Deutsches Patentamt, Munich 1977, pp. 420ff, p. 424.

[8] *National Archives Washington* (NA) RG 59 762A.5–MSP/1–656 (Box 3563): Here is the complete text of the Agreement signed on 6.1.1956 and various accompanying texts.

[9] *Gimbel*, John, U.S. Policy and German Scientists: The Early Cold War, in: Science Quarterly, No. 3 (1986), pp. 433–451; *Kurowski*, Franz, Alliierte Jagd auf deutsche Wissenschaftler: Das Unternehmen Paperclip, Munich 1982; Lasby, Clarence G., Projekt Papierclip: German Scientists and the Cold War, New York 1971; *Bar-Zohar*, Michel, Die Jagd auf die deutschen Wissenschaftler (1944–1960), Frankfurt/Berlin 1966.

[10] Cf. most recently *Bower*, Tom, The Paperclip Conspiracy. The Battle for the Spoils and Secrets of Nazi Germany, London 1987.

[11] *Stamm*, Zwischen Staat und Selbstverwaltung, p. 45.

the actual brain drain only really began in the 1950s. Between 1949 and 1966 about 1800 scientists and 4200 technicians emigrated to the USA.[12]

3. Bans and restrictions on scientific research and production.[13] Thus the Allied High Commission (AHC) in law No. 22 dated 15.3.1950 forbade the construction of nuclear reactors and particle accelerators as well as the possession and purchase of the raw materials required for nuclear technology. AHC law No. 24 dated 30.3.1950 included production, import and application bans or precise maximum production limits for steel and rolling mill products, chemicals, ship construction, electrical engineering and machine tools.[14] These restrictions only ceased to apply with the Paris Agreements and the military integration of the Federal Republik in May 1955[15], although they had already been successively rescinded in certain fields, especially against the background of the Korean War[16] and the beginnings of the Western European economic integration (European Coal and Steel Company – ECSC).[17]

Typical for the situation at this time was the appeal of the President of the Chemical Industry Federation (Verband der Chemischen Industrie), W. A. Menne, addressed to Chancellor Adenauer, calling for the repealing of the production and research restrictions before the Treaty with the Allies defining the European Defence Community finally came into effect. Menne wrote:

> "In the course of time the ban and control system has developed so ruthlessly that it has latently had a paralysing effect on the fields of research and production. The registration system is exercised as strictly as possible and demands the divulgence of such exact details that the questions amount to a systematic investigation of new developments. This is often so off-putting

---

[12] *Servan-Schreiber,* Die amerikanische Herausforderung, Preface by Franz-Josef Strauß, p. 16. A comparison with Great Britain and Canada shows, however, that the Federal Republic was by no means worst hit by the exodus to the USA. For example, in 1956 1003 scientists and engineers emigrated from Canada, 433 from Great Britain and 339 from the Federal Republic. In 1966 Great Britain was at the top of the table with 1251 emigrees, followed by Canada with 1105 and the Federal Republic with 346. Figures from OECD: Gaps in Technology. Analytical Report, Paris 1970, p. 55.

[13] Cf. *Stamm,* Zwischen Staat und Selbstverwaltung, pp. 54ff.

[14] *Politisches Archiv des Auswärtigen Amtes,* Bonn (AA.PA)II–242–01, Vol. 1: 20.12.1950 Memorandum from the Chancellor to the AHC. In the field of electrical engineering, for example, the production of X-ray equipment without a licence was limited to 150 000 volts, although by 1950 instruments using 250 000 volts had become standard in American industrial production. Another example: the production of electron tubes above 250 megahertz was not allowed, in spite of these being needed for television manufacture.

[15] Bundesanzeiger No. 92, 13.5.1955.

[16] AA.PA II–242–01, Vol. 1. Memorandum by the Chancellor to the AHC from 20.12.1950.

[17] AA.PA II–242–02–12: extensive documents on the laws AHC No. 23d 24 and the negotations with the Allies about the rescinding of the controls. Cf. also: *Lüders,* Carsten, Die Bedeutung des Ruhrstatuts und seine Aufhebung für die außenpolitische und außenwirtschaftliche Emanzipation Westdeutschlands (1948–1952), in: *Knapp,* Manfred (Ed.), Von der Bizonengründung zur ökonomisch-politischen Westintegration. Studien zum Verhältnis zwischen Außenpolitik und Außenwirtschaftsbeziehungen in der Entstehungsphase der Bundesrepublik Deutschland (1947–1952), Frankfurt 1984. The restrictions for steel production were rescinded by law No. 78, 28.7.1952, after the Regulating Ordinance No. 6 to law No. 24, 26.7.1951 had already allowed the production limit of 11.1 million tonnes to be exceeded, "as far as mutual defence requires", loc cit: Die Neuordnung der Eisen- und Stahlindustrie im Gebiete der Bundesrepublik Deutschland. Ein Bericht der Stahltreuhändervereinigung, Munich & Berlin 1954, pp. 52f.

that many research projects are never begun and many applications for planned production are never even made."[18]

However, the question of technological competence was not a subject of public interest in the 1950s. There was no general awareness of these problems, which, viewed from today, would appear to be so serious. The fact that the Federal Republic was extremely successful on the export market in the 1950s was probably one reason for this state of affairs. German industry reconquered her old markets at a speed other countries often found frightening and at the same time opened new outlets for her products. The "economic miracle" was largely based on the "export miracle" of the 1950s.

The following exposition is only a first attempt to illuminate the questions related to "Technology Transfer and Foreign Trade in the early years of the Federal Republik of Germany". Although the subject of technology has been the subject of much attention in Anglo-Saxon historiography[19], the reappraisal of the history of technology and technology policies is only just beginning in West Germany.[20]

Three aspects will be handled in more detail in this article:
1. the regional structure of German foreign trade after 1945 and in the early years of the Federal Republic in connection with the subject of technology;
2. technology and technology transfer in German-American economic relationships;
3. definitions of the so-called "technology gap" and evaluation of the indicators.

The discussion will be conducted from two perspectives, firstly from the historical, i.e. from the point of view of the main actors in finance, science and politics at this time and secondly analytically on the basis of the available statistical material and the knowledge we have today of long-term trends and developments.

## 2. The regional structure of German foreign trade and the subject of technology in the 1950s

The reintegration of the West German economy into the world economy in the years after 1945 and after the founding of the Federal Republic took place at a speed which surprised even contemporaries. In 1950 the balance of trade still showed a debt balance of over 3000 million DM but in 1951 the volume of foreign trade exceeded prewar levels (1937) with a near balance of exports and imports.[21] And in 1953 there was already an export surplus of 2500 million DM, which increased to 5000 million DM annually by the end of the 1950s.

---

[18] AA.PA II–242–02–12: 26.11.1953 Menne to Adenauer.

[19] Cf.: *Gimbel*, John, U.S. Policy and German Scientists: The Early Cold War, in: Science Quarterly, No. 3 (1986), pp. 433–451. *Nau*, Henry R., Technology Transfer and U.S. Foreign Policy, New York 1976. *Yakemtchouk*, Romain, L'Europe face aux Etats-Unis. Relations politiques et strategies militaires, contentieux économiques, competition technologique, in: Studia Diplomatica (1986), pp. 331–632; ibid., Transfers de technologies sensibles entre l'Est et l'Ouest, in: Studia Diplomatica (1984), pp. 395–552.

[20] *Krieger*, Wolfgang, Technologiepolitik und Forschungsförderung in der Bundesrepublik, in: Vjh. f. Zeitgesch. 2 (1987), pp. 247–271.

[21] Bevölkerung und Wirtschaft 1872–1971, issued by Statistisches Bundesamt, Wiesbaden 1972, p. 191.

The general conditions for a successful return of the partitioned state of West Germany to the world market were favourable: the German industry, which was traditionally export orientated, was able to develop quickly within the new world economic system established by the USA after World War II on the basis of multilateralism and liberalization.

The Korean War accelerated this development by creating an increased demand for capital goods. German industry, whose capacities were not tied up in armament production[22], was particularly well placed to meet this demand. At the same time Allied production bans and restrictions for West German firms were speedily rescinded, reflecting the rapid gains in sovereignty as a consequence of the Cold War.

The special quality of the West German offensive on the world market in the early 1950s was that, at first, all efforts were concentrated on returning to the traditional prewar markets for German exports. Therefore, it is characteristic that in 1952 80% of the German export of patented goods carried brand names that had been registered trademarks at home and abroad before the war.[23]

**Figure 1:**    The development of foreign trade in the Federal Republic 1950 – 1959[24]

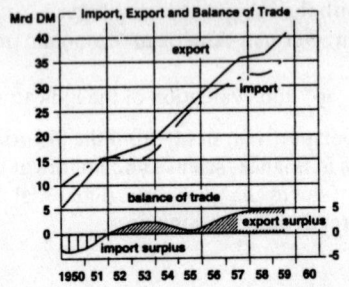 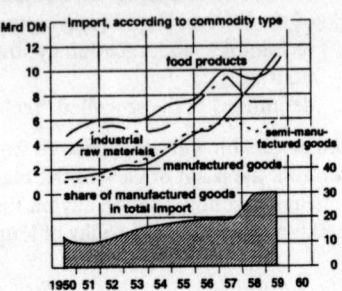

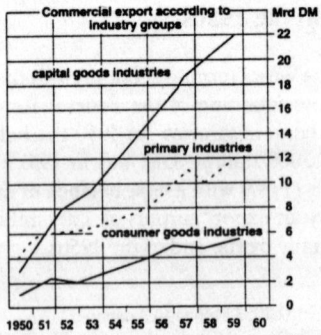

---

[22] *Private Archive Dietrich von Menges,* Essen Vol. 147: 21.11.1952 von Menges to Ferrostaal Overseas Corp., New York: Remarks by the American Minister of Economic Affairs Sawyer to West German industrialists during his visit to the Federal Republic in November 1952 saying that a reduction in the armament programme was expected and thus American production capacity for exports would be increased led initially to considerable disquiet in the circles of West German export business.

[23] *Erhard,* Ludwig, Deutschlands Rückkehr zum Weltmarkt, Düsseldorf 1953, p. 255.

[24] Monatsberichte der Deutschen Bundesbank, March 1960, p. 8.

How did West German exporters themselves view their sales prospects abroad? A special internal study by the IFO-Institute in August/September 1952, which was also made available to the Minister of Economic Affairs, Ludwig Erhard, provides informative data on this subject. The IFO-Institute asked export firms the following questions: 1. Which country is your main competitor on the foreign market? and 2. Where do you see your best sales chances (country or group of industries)?

First of all, it is interesting to note that, apart from the markets in Western Europe, the West German economy saw its best sales chance in Central and South America. Great importance was also attached to the Near East (Turkey, Iran, Levant). On the other hand the export industry rated the sales chances in North America (USA) as poor. Only certain textile products (knitting wool, jute thread and material, gloves) were seen as having very good sales chances in the USA, as well as cameras from the precision engineering/optical industry and household sewing machines from the mechanical engineering industry.

Above all, the exporters named USA and Great Britain as their main competitors in the various market segments. Great Britain was seen as the most important competitor, even more important than the USA, especially in the motor and textile industry. On the other hand Japan did not yet play a noteworthy role as a competitor in the eyes of German industry. However, it is striking that Japanese products were named as the main competition in those fields in which German firms named the USA as their main sales market. From today's point of view it is interesting to note that Great Britain in 1952 was still considered to be Germany's strongest rival in many areas of the foreign markets.

**Table 1:** Sales Prospects Abroad (Survey by the IFO-Institute August 1952)[25]

| Product | Main Competitor on Foreign Markets | Best Sales Chances exist in |
|---|---|---|
| *Textiles:* | | |
| Machine knitting wool | GB, France | North America: USA<br>Western Europe: Sweden |
| Worsted fabrics | GB, Italy Switzerland | Western Europe: esp. Scandinavia, France |
| *Chemicals:* | | |
| Paints and varnish (esp. oil colours) | GB, USA, Holland | Asia: Near East<br>Western Europe<br>Africa |
| *Electrical engineering:* | | |
| Electric Motors | GB, USA | Western Europe<br>Asia: Near East, Turkey<br>South America |
| Accumulators | GB, USA | South America<br>Africa<br>Asia |

---

[25] Selected and compiled from: *Bundesarchiv Koblenz* (BA) B 102 Vol. 57772. IFO-Institute for Economic Research to Minister Erhard. Special Study "Absatzchancen im Ausland", 25.9.1952.

Technology Transfer and Foreign Trade in the Early Years of the Federal Republic of Germany   139

**Table 1:**   *(continued from page 138)*

| Product | Main Competitor on Foreign Markets | Best Sales Chances exist in |
|---|---|---|
| Major Electrical Equipment | USA, GB Benelux | Western Europe South America: Uruguay |
| *Precision Engineering / Optics:* Cameras (incl. box cameras) | USA, GB, Japan | North America: USA Western Europe: Scandinavia Australia |
| *Motor industry:* Automobiles | GB, France, Italy, USA | Western Europe, esp. Scandinavia, Benelux, Switzerland South America South Africa |
| *Mechanical Engineering:* Land cultivation equipment | USA, GB, Czechoslovakia | South America South Africa Asia: Near East, esp. Turkey, Levant Western Europe: Benelux |
| Knitting machines | USA, GB, Belgium | South America Western Europe: Italy, France, Scandinavia |
| Household sewing machines | Japan, Italy, Switzerland | USA South America Western Europe: Holland Asia: Near East, esp. Turkey, Iran |

A systematic inspection of the regionalised structure of West Germany's foreign trade in the 1950s shows that imports and exports took place within a network of four submarkets, each with its own structure. These were: 1) the West European market and the member states of the EPU (European Payments Union), 2) the dollar area i.e. USA and Canada, 3) the "outside" markets in Central and Middle America, Africa, Asia and Australia and finally 4) the state-trading countries (East and Southeast Europe, China).

There are certain factors specific to each of these four areas:

1.   EPU countries. The hypothesis often presented in the relevant literature is that the economic integration of German industry into Western Europe opened up new markets which formed the base of the export boom. But this was not really the case. It is more correct to say that in the years after 1945 the percentage of exports going to Western European countries was much higher than before the war (76.8% in 1949 compared with 54.8% in 1936) because overseas business had been disrupted by the war and Germany had not yet been fully reintegrated into the world economy. After 1949 it became the declared aim of West German export policy to reduce this percentage to a more balanced level. As is shown by

the large proportion of exports to highly industrialised Western European
countries (especially Holland), the question of technological acceptance did not
pose a problem for German exports to Western European countries in the years
after 1945.

2. Dollar-Area/USA: Compared to the other economic areas the German-
   American trade and economic relationships in the 1950s are of course the most
   interesting with respect to the question of technology. Therefore, the dollar
   market will be explored separately and in more detail in the next section. For
   the moment it suffices to state the following:

   In the case of the dollar area the German (and also the Western European) trade
   balance and balance of payments at the beginning of the 1950s were
   characterised by an almost overwhelming asymmetry. This was mainly the result
   of large imports from USA and Canada, above all raw materials and food stuffs.
   The percentage of imports from the USA remained roughly constant until the
   middle of the 1960s and, on average, made up about 12-15% of the Federal
   Republic's total imports. Therefore, in international statistics the USA was the
   Federal Republic's most important importer. On the other hand, the West
   German exports to North America stagnated at around 6-7% of the total
   exports and were therefore at about the same level as before World War I (1913
   7.1%) or between the wars (1929 7.4%).[26]

3. Outside markets: The first target of the West German export offensive was,
   above all, the countries of Central and South America. Here West Germany was
   soon able to bring first Great Britain and then the USA under pressure. By 1951
   10.8% of West German exports were going to South America; this meant that
   the prewar level of 10.6% (1935/37) had already been met and surpassed two
   years after the founding of the Federal Republic.

   The conclusions that Dietrich Wilhelm von Menges, then board chairman of
   Ferrostaal AG and one of the pioneers in dealings with South America, drew
   after his first journey to all the important Central and South American countries
   from September to November 1949 contain interesting information about the
   background to the success of German industry on this market. Von Menges saw
   the following reasons for the very good chances of German business:

   While a lack of dollars in almost all of South America had led to a stagnation of
   American private business, the English influence was "declining everywhere".
   This was because of "their very long delivery times" and because the quality of
   goods delivered after the war by English industry had often been poor. In South
   America there was a strong general desire to do business with Germany again.
   There were three reasons for this:

   "a) the realization that Germany is a better sales outlet for many products from
   these countries than the United States,
   b) the quality of German products, which had stood the test especially during
   the war when there was no servicing or new deliveries,

---

[26] Cf. C. E. H. *Hames,* Sternstunde des deutschen Exports, in: Der Volkswirt, 13, 1959, pp. 2183–85.

c)   the ability of German companies to adapt commercially and technologically to the customer's individual requests."[27]

It was technological considerations, among other factors, which played an important role in the first major breakthrough in business with South America, the so-called Trolley-Bus-Deal of 1951. The town of Buenos Aires awarded the contract, worth 80 million DM, for the delivery of 800 busses to a consortium of German companies. The deal was shared by Daimler-Benz AG, Henschel und MAN under the overall control of Ferrostaal AG. Although at this time the USA practically controlled the motor vehicle market in South America, there was a "technological gap in the market" for trolley busses, which gave the Europeans an advantage. Because of special electronic steering equipment for such purposes, the "Kobold", the consortium offered a special incentive for the deal.[28]

4.   State-trading countries: Trade with Eastern Europe represented the opposite problem with respect to technology. COCOM (Coordinating Committee of East-West Trade Policy still active today), after being set up on 22.11.49 in Paris, began to draw up more and more comprehensive lists of strategic goods, which, after the Battle Act came into force in 1952, brought East-West trade to an almost total standstill.[29]

**Table 2:**   Percentage of international commodities included in the COCOM List[31]

| Period | Number of Items | Percentage of international Commodities |
|---|---|---|
| November, 1949 | 100 | 6.7% |
| 1951/52 | 400 | 26.7% |
| June 1953–15. Aug. 1954 | 450 | 30% |
| 16th August 1954 | 255 | 17% |
| Juli 1958 | 203 | 13.6% |
| August 1958 | 170 | 11.3% |
| 1964 | 150 | –10% |

---

[27] *Haniel-Archiv Oberhausen* (HA), Nl. Paul Reusch, No. 4001012022/20: Dietrich Wilhelm von Menges, Report on a journey to South America from 12th September to 26th November 1949, dated 5.12.1949. In: *Neebe*, Reinhard, Überseemärkte und Exportstrategien in der westdeutschen Wirtschaft. Aus den Reiseberichten von Dietrich Wilhelm von Menges, Zeitschrift für Unternehmensgeschichte, Beiheft 68, Stuttgart 1991, pp. 90–155.

[28] *von Menges,* Dietrich Wilhelm, Unternehmensentscheide, Düsseldorf/Vienna 1976, pp. 86f. The steering equipment had been developed by the company Kiepe in Düsseldorf.

[29] Essential for the presentation of the Western embargo policy: *Adler-Karlsson,* Gunnar, Western Economic Warfare 1947–1967. A Case Study in Foreign Economic Policy, Stockholm 1968. On the origins of the U.S. economic security policy cf.: *Pollard,* Robert A., Economic security and the origins of the Cold War 1945–1950, New York 1985. The diverging interests in trade with Eastern Europe between the Federal Republic and the USA are discussed in: *Jacobsen,* Hanns-Dieter, Die Ost-West-Wirtschaftsbeziehungen als deutsch-amerikanisches Problem, Baden-Baden 1986. This also contains an extensive bibliography on trade with Eastern Europe. For a more detailed discussion of the technology aspect in East-West trade from the American point of view see: *Bertsch,* Gary et al., East-West Technology Transfer and Export Controls, in: Osteuropa-Wirtschaft, 26, 1981, pp. 116–136.

**Figure 2:** Number of items on the COCOM lists 1948–1964[30]

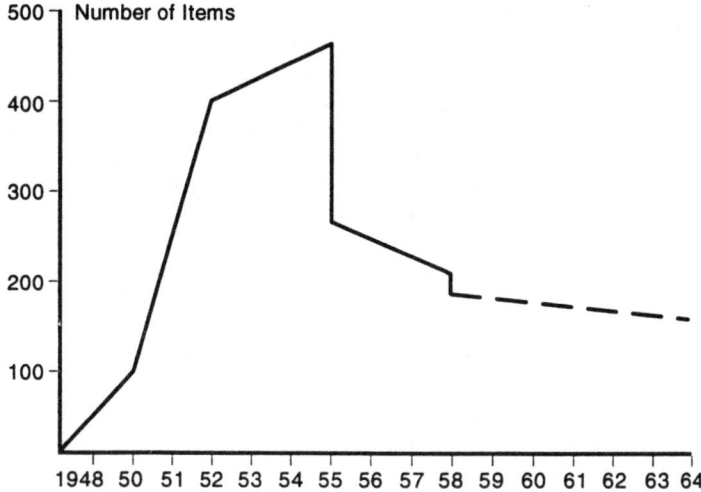

The memorandum of the US National Security Council (NSC 68), 14th April 1950, stated explicitly the internal relationship between the American Marshall Plan, the embargo policy against the Soviet Union and the politically desirable re-orientation of the West German and Western European export trade[32] to the dollar-area.[33] In May 1951 the delegation of the European Cooperation Administration (ECA) in West Germany informed the West German Minister responsible for the Marshall Plan that in future counterpart funds from the Special Funds of the European Recovery Program (ERP) could only be given to firms which did not take part in "illegal" trade with the Soviet Union and its statellite states.[34] In July, 1951 the

---

[30] Adapted from *Adler-Karlsson*, Western Economic Warfare, p. 146.

[31] Ibid.

[32] Documentation of the most important embargo regulations for West Germany between 1949 and 1955 in *Lambers*, Hans Jürgen, Das Ost-Embargo, Frankfurt/Berlin 1956.

[33] 14.4.1950 A Report to the National Security Council by the Executive Secretary (Lay) (NSC 68) in: Foreign Relations of the United States (FRUS) 1950, Vol. I, pp. 234–292. Detail on NSC 68 in *Gaddis,* John Lewis, Strategies of Containment, New York 1982, pp. 89ff. Cf. also *Pollard,* Economic Security, pp. 222ff.

[34] NA RG 469, Mission to Germany, Box 2, 10.11.1952. Alexander F. Kiefer, Confidential Security Information from HICOG Bonn: "Development of the Export Control Procedure in the Federal Republic of Germany during the Period 1945–52".

Marshall Plan authorities froze the allocation of ERP counterpart funds to a West German firm for the first time because of a violation of the embargo policy on the export of strategically relevant goods to the Eastern Bloc. By June 1952 the blacklist of West German firms accused by the USA of illegal practices in East-West trade totalled 87. Apart from freezing ERP counterpart funds the sanctions consisted firstly of refusing to deliver strategic goods from the USA (including licences) and secondly of crossing the firms involved off the Marshall Plan delivery programmes.[35]

It was in trade with Eastern Europe that the real structural changes are to be found in a comparison with the prewar period. Before the war about 15-17% of German exports went to Eastern European countries but in 1949/50 this percentage was only about 5% and even dropped to 2% in 1952. From a contemporary point of view, it was the breaking up after 1945 of the Central European trade links, which had developed through history, that posed the greatest problems for the West German export industry.

Apart from the major political constellation and the global opposition of the Soviet Union and the USA, another important condition at the beginning of the 1950s for the successful reintegration into the world economy lay in the extending of trade with East and Southeast Europe. West German manufactured and semi-manufactured goods in exchange for raw materials and foodstuffs meant a reduction in the high level of imports from the USA and an improvement of the dollar balance (dollar saving) and this seemed to the top organisation of German industry (BDI) and to the foreign trade department in the Ministry of Finance to be the right way of solving certain foreign trade problems.[36]

However, from 1950 until at least 1954, i.e. the height of the embargo policy in the Cold War era, this strategy could not be put into action for political reasons. When the obstacles in East-West trade started to fall at the beginning of the 1960s, it became evident that the prewar pattern of Central European trade currents had lost its attractiveness for West Germany and also for the other industrial countries in Western Europe.

The politically initiated redirecting of the trade currents at the end of the 1940s and the beginning of the 1950s towards the Western hemisphere and the outside markets put increased pressure on West German industry to modernise. The successful reintegration into the world economic system demanded an increasingly international division of labour. Therefore, the complementary division of labour, whereby raw materials were exchanged for manufactured goods, lost more and more of its relevance. This pattern was supplemented and superseded by a rapidly

---

[35] *Trautmann*, Walter, Osthandel Ja oder Nein?, Stuttgart 1954, pp. 116ff.

[36] For detail on trade with Eastern Europe together with extensive evidence: *Neebe*, Reinhard, Optionen westdeutscher Außenwirtschaftspolitik 1949–53, in: *Herbst*, Ludolf et al., (Ed.), Vom Marshallplan zur EWG: Die Eingliederung der Bundesrepublik Deutschland in die westliche Welt, Munich 1990, pp. 163–202. For direct evidence it suffices here to mention: HA, Nl. Hermann Reusch, No. 400101401/85: 3rd March 1950, confidential meeting of BDI in Cologne "Der Handel mit ost- und südeuropäischen Ländern" (minutes, 17 pages); BA B 102, Vol. 7204: "Aufzeichnung über den gegenwärtigen Stand und die weiteren Perspektiven der Handelsbeziehungen der BRD mit der sowjetischen Besatzungszone und den Ostblockstaaten" from 9.6.1952 ("Strictly confidential!"), circulated by Minister of Economic Affairs Erhard as a discussion paper for the cabinet on 16.6.1952 and approved by the cabinet on 24.6.

developing intra-industrial network within the Western world.[37] East-West trade, which had faced a political barrier in the postwar years, was now, that is at the end of the 1950s and beginning of the 1960s, impeded by structural handicaps, which became more and more obvious.

## 3. Technology and Technology Transfer in German-American Relations

### 3.1 West German Industry and Commerce Interest in the Dollar Market

In the case of the dollar area it can be seen that West German industry, apart from certain branches, did not make any considerable effort[38] to establish itself in this market in the early 1950s. The reservations about the US market were rooted not only in a possible deficit in advanced technology. It was rather that many branches of German export industry saw little opportunity to sell their products which were made to a high standard of quality on an American consumer market geared to the mass consumption of low-priced products and were not prepared to adapt accordingly.

A report by the Hamburg Chamber of Commerce on the New York industrial fair of April 1949 gives a very good insight into the differences in business mentality. This report is also interesting, because here in New York German export trade had its first opportunity to present itself after World War II and the experience gained here formed the guiding principles for future reactions to the US market. The Chamber's report stated:

> "One of the most interesting experiences ... is the fact that the American market prefers the cheap standardised mass product. If we in Germany believed that we could open up the largest market with the greatest purchasing power using durable quality products, this assumption has been proved wrong. For most consumer articles it is important that they are fashionable, have the appearance of quality rather than genuine quality and replace durability with low prices ... The branches of industry affected will have to consider whether they are in a position to make a qualitative adjustment and whether such an adaption is in fact desirable."[39]

---

[37] The proportion of raw materials in West German exports sank between 1950 and 1960 from 14.0% to 4.6%, while the proportion of manufactured goods during the same period rose from 64.8% to 82.4%. For imports the proportion of manufactured goods rose from 12.6% to 32.2%, while the proportion of raw materials sank from 29.6% to 21.7% and the proportion of food stuffs fell from 44.1% to 26.3% of the total imports. Bevölkerung und Wirtschaft 1872–1972, issued by Statistisches Bundesamt Wiesbaden, 1972, pp. 192ff.

[38] In earlier literature the development of the West German foreign trade offensive on the North American market after 1949 was portrayed too smoothly. Cf., for example, *Knapp*, Manfred, Politische und wirtschaftliche Interdependenzen im Verhältnis USA-(Bundesrepublik) Germany 1945–1975, in: *Knapp* et al. (Ed.), Die USA und Deutschland 1918–1975, Munich 1978, pp. 187ff.

[39] *Archiv der Handelskammer Hamburg*, V 30 70e/4 (XIII b): Eindrücke von der New Yorker Industrie-Ausstellung 9.–24. April 1949. Report by the Hamburg Commerce of Trade, Department for Foreign Trade (Dr. Stephan) from 12.5.1949. My thanks to Mr. Hanno Sowade (Institut für Zeitgeschichte) for drawing my attention to this document.

The characteristics of the American market were thus in clear contrast to the traditional philosophy of quality adhered to by German firms. In the self-assessment of the German export trade the quality but not the price was an important factor in the success of a product on the foreign markets. According to a survey by the IFO-Institute in 1958 on the "Causes of West German Export Successes in the Opinion of Industrialists"[40] the quality factor dominated at 85% in the production and capital goods industry. Even in the consumer goods industry the quality factor came first for 77% of the exporters, while value for money only came a poor second (41%). Value for money was only considered to be more important in the leather and shoeware industry. Ability to deliver and service facilities (especially in the capital goods industry) were also considered to be crucial criteria for the success of German firms.

In the first phase of the revival of trade with America mechanical and precision engineering and products from the optics industry played a larger role.[41] After 1953, immediately following the liquidation of IG Farben, the chemical industry became active in a major way in the USA with direct investments and thus was able to take up its prewar position once again.[42] The rapid rise of the motor industry to become the most important West German exporter on the US market however only took place in the second half of the 1950s. In 1952 only 19.4 million DM came from the export of vehicles to the USA but in 1959 the export of cars was worth over 1300 million DM and represented more than a third of all German exports to the USA (3830 million DM).[43] However, in spite of the success in certain sectors, the proportion of exports to the USA did not exceed 6-7% of the total exports until well into the 1960s and remained thus below the levels which had already been reached before World War I and between the wars.

But it would be wrong to judge the importance of the US market for German export trade in the 1950s only on the basis of the export percentages. For a number of major firms (e.g. Siemens & Halske AG)[44] an outlet in the USA was of great importance in order to be able to organize deals for large-scale industrial units in other countries, especially in South America, using their personal contacts to the large international consulting engineering firms. As the financing of such projects was usually crucial, cooperation with American banks was often imperative. For these firms it was of secondary importance that by being present in New York they could observe and prepare to supply the American market.

---

[40] IFO Schnelldienst. Articles on the economic situation. A 9 No.1/2, 9.1.1958 "Die Ursachen der westdeutschen Exporterfolge im Urteil der Unternehmer."

[41] In 1952 the mechanical engineering industry exported goods worth 142.0 million DM to the USA, the precision engineering and optical industry goods worth 121.0 million DM. In 1952 the total export to the USA was 1049.0 million DM. Stat. Jb. für die Bundesrepublik Deutschland, 1953, p. 317.

[42] BA B 102 Vol. 6928 No. 1: This volume of documents gives a very good insight into the forced expansion of the IG-Farben successors Farbwerke Hoechst AG und Farbenwerke Bayer AG beween 1953 und 1957 on the US market. Cf. also *Der Spiegel* No. 7, 7.2.1966, Deutsche Chemie. Weltmacht aus der Retorte, pp. 47ff.

[43] Stat. Jb. 1960, p. 307.

[44] E.g. Siemens, cf.: BA B 102, Vol. 6935, No. 1: 22.1.1954. Bavarian Ministery for Economic Affairs and Trade to the Federal Minister for Economic Affairs.

At the same time West German business and official institutions in the Federal Republic observed the technological development in the USA attentively. Regular reports by the German Consulate General in New York and by the Embassy in Washington (Dr. Krekeler) on new technology were sent to the federal ministeries, research institutes and business associations with the specific aim of introducing those technological innovations deemed necessary as quickly as possible to West Germany and in accordance with American high technology. Special emphasis was placed in these reports on recent technological developments in the USA, especially the peaceful use of nuclear energy, transistor technology, computer and television industry.[45]

Ambassador Krekeler commented on the fundamental value of this information in a report from 20.3.1952 titled "Recent technological developments in the USA":

"One of the main tasks of political-economic reporting from the USA is the foresighted evaluation of future technological development. On the one hand this is necessary because of the especially high level that research and technological development have recently reached in the USA. On the other hand it is particularly important for Germany because both scientific and industrial research in Germany were cut off for many years from other countries by the war and have found it particularly difficult to re-establish the contact with countries abroad in the first postwar years because of the very troubled economic situation. It is true that German research and technology did achieve considerable advances during the war, which were only revealed to other countries after the war. However, these achievements were made available with every last detail to other countries in the manner we all know, while German researchers and technicians were not given corresponding access to the results achieved abroad."[46]

---

[45] The reports by the German Consulate General in New York and by the Embassy in Washington listed below can be found in the archives: AA.PA IV-Ref. 414, Vol. 81, 86, 87, 88, 89 and: BA B 102, Vol. 6156, No. 2; Vol. 6158, No. 1; Vol. 6161 No. 1.
9.4.1951 Entwicklung und Exportbemühungen der amerikanischen Fernsehindustrie (4p) (Krekeler/Opfermann)
20.6.1951 Bemerkungen zur amerikanischen Automobilproduktion (Krekeler/Opfermann)
10.3.1952 Verwendung von Atomenergie für friedliche Zwecke im Lichte amerikanischer Veröffentlichungen (24p) (Krekeler/Opfermann)
20.3.1952 Neuere technische Entwicklung in den Vereinigten Staaten (9p) (Krekeler)
18.6.1952 Verwendung von Atomenergie zum Antrieb von Schiffen (Nautilus) (3p) (Dr. Meyer)
28.10.1952 Neuere Entwicklungen auf dem Gebiet der Erzeugung von Atomkraft (3p) (Krekeler)
28.1.1953 Verwendung von Atomenergie im Kriege (3p) (Krekeler)
24.3.1953 Entwicklung und Verwendungsmöglichkeiten des Transistors (5p) (Krekeler)
30.4.1953 Entwicklung einer neuen elektronischen Rechenmaschine durch die IBM (3p) (Opfermann)
21.5.1953 Exporterfolge europäischer Elektrounternehmen auf dem US-amerikanischen Markt (3p) (Opfermann)
17.6./14.7.1953 Ausfuhr elektromedizinischer Geräte nach den USA (2p) (Correspondence from Department of Trade Policy in the Foreign Office to the Embassy in Washington)
2.7.1953 Wirtschaftliche Erzeugung von Atomenergie (2p) (Krekeler)
7.4.1954 US-Sammelwerk über friedliche Verwendung von Atomenergie (2p) (Dr. Walter)
19.11.1954 Amerikanische Einfuhr von Kameras und Kamerateilen im Jahre 1953 und im ersten Halbjahr 1954 (4p) (H. Bodden)
13.1.1955 Wirtschaftlichkeit der Elektrizitätserzeugung durch Ausnutzung der Atomenergie (3p) (Krekeler)
[46] BA B 102 Vol. 6156, No. 1: Report Dr. Krekeler No. 330B.427/52 from 20th March 1952 "Neuere technische Entwicklung in den Vereinigten Staaten", p. 1.

In spring 1952 Krekeler saw the technological innovations which would decide the future as lying in two fields, firstly in the peaceful use of nuclear energy and secondly in the applications of the transistor. Krekeler wrote as follows on the relevance of nuclear energy to West Germany:

> "I would like ... to draw the conclusion that we should now seriously consider the fact that, with a probability greater than 50%, electric energy produced from nuclear power will be available at economically acceptable conditions with a period of time that one cannot afford to ignore in the forward planning of economic development. I think that this development deserves the most serious attention from the Federal Government, since it can be assumed that it will be completed within the next 25 years. The radical economic changes which will be triggered off when the conjectures or forecasts are fulfilled will be so great that they will be accompanied by massive social disruption if state economic policy is not adapted accordingly in good time, i.e. many years beforehand."[47]

The report has the following to say about the development of transistor technology and its future use, particularly for electronic calculating machines:

> "The development of the transistor may become ... so important that it deserves serious interest, even outside the specialist industries, in the whole German business world and in the Federal Government... There can be ... no doubt that this field of research, whose future possibilities exceed the most daring fantasies in their manifoldness, could be of central importance to the competitiveness of precisely those economic sectors which produce a large share of our exports. These developments cannot be disregarded as scientific games or as excessive tendencies towards a dangerous technological takeover of our lives, but must be subjected to a very serious, but rational and critical review."[48]

The report on the peaceful use of nuclear energy was received with particularly great interest in West Germany. The original Krekeler-Report "Recent technological developments in the United States" was sent to the Undersecretary in the Foreign Office, copies to the Ministery of Economic Affairs and to the Ministery for the Marshall Plan with the express request to bring its contents to the attention of Vicechancellor Blücher and the Minister of Economic Affairs Erhard. Another copy was sent to the Undersecretary in the Cabinet Office in order to inform the Chancellor as well. Further copies were sent to, among others, the Federation of German Industry (Bundesverband der Deutschen Industrie – BDI), the Emergency Association of German Science (Notgemeinschaft der Deutschen Wissenschaft, later the Deutsche Forschungsgemeinschaft) in Bad Godesberg, the Chemical Industry Federation, the Central Association of the Electrical Industry (Zentralverband der Elektrotechnischen Industrie), the Association of German Electric Power Companies (Vereinigung Deutscher Elektrizitätswerke – VDEW), the Association of Industrial Power Stations (Vereinigung Industrieller Kraftwirtschaft), the Association of German Precision Engineering and Optical Industry (Verband der Deutschen Feinmechanischen und Optischen Industrie), the Organisation of German Mechanical Engineering Institutions (Verein Deutscher Maschinenbau-Anstalten) and, last but not least, the Max-Planck-Society in Göttingen.[49] The history of the development of nuclear policy in West Germany

---

[47] Ibid., p. 3.

[48] Ibid., pp. 8–9.

[49] For the distribution list of the reports cf.: HA No. 40010146/667 10.4.1952 Executive Mangement of BDI to Hermann Reusch. Also AA.PA IV-Ref. 414, Vol. 88, Correspondence April to July, 1952.

cannot be described in detail here. Krekeler's report arrived at a time when a number of nuclear scientists, at their head Werner Heisenberg, began to pressurise the Federal Government to address the development of nuclear technology in all seriousness.[50] And the Max-Planck-Society stated to the Foreign Office in July 1952 explicitly "that these reports are of great value to us and that we are most interested in obtaining other such reports".[51]

The German representative in Washington was also directly involved in the concrete negotiations for the issuing of licences for American high technology to West German companies. Licensing by American firms, not least in the field of sensitive technology including transistor technology, was subject to strict security conditions laid down by the United States Government. Ambassador Krekeler said with respect to the security policy aspect of American licensing policy:

> "As it concerns a field of considerable military importance, the participation of German representatives in negotiations on issuing licences is subject to certain security precautions. The diplomatic corps is negotiating at the present time with the American authorities about the admission of German representatives. Here it is expressly welcomed that the German specialist industry is so keenly interested in this development."[52]

One year later in March 1953 Krekeler was then able to report that several German companies had been able to complete licensing contracts with the Bell Company.[53] It is characteristic that the licensing of German firms in the USA (and also the suggestions and information in the political-economic reports from the embassy) were not aimed at entering the US market directly. Instead the object was to use the technological innovations coming from the USA to be able to survive the international competition on the outside markets.

### 3.2 The Problem of the "Technology Gap" from the American Point of View

From the American point of view the problem of the "technology gap" in Western Europe presented a conflict of goals: on the one hand the USA systematically encouraged the German and European brain drain to the USA after 1945 and in the 1950s but on the other hand it saw that a technological decoupling of Western Europe could be become a problem for political reasons. As much as the United States was interested in expanding its own lead in high technology, there was also a vital interest in stabilizing the Western European economy permanently in the context of the containment strategy towards the Eastern Block.

Therefore, the USA developed a special "Technical Assistance Program" within the Marshall Plan.[54] However, it was characteristic that the ECA, and its

---

[50] *Radkau* Joachim, Aufstieg und Krise der deutschen Atomwirtschaft 1945–1975, Hamburg 1983, pp. 39ff. Most recently: *Eckert,* Michael, Die Anfänge der Atompolitik in der Bundesrepublik Deutschland, in: Vjh. f. Zeitgesch. 1(1989), pp. 115–143.

[51] AA.PA IV-Ref. 414, Vol. 58: 15.7.1952 Prof. Dr. K. Wirtz to the Foreign Office, Bonn.

[52] BA B 102 Vol. 6156, No. 1: Report Dr. Krekeler No. 330 B.427/52 from 20 March 1952 "Neuere technische Entwicklung in den Vereinigten Staaten"

[53] BA B 102, Vol. 6158, No. 1: from 24 March 1953 Report Dr. Krekeler No. 330–00 B.6067/53 "Entwicklung und Verwendungsmöglichkeiten des Transistors"

[54] Review in: Wiederaufbau im Zeichen des Marshallplanes 1948–1952, 12., Final report of the Federal Government presented by the Minister for the Marshallplan, Bonn 1953, pp. 34–40.

Technology Transfer and Foreign Trade in the Early Years of the Federal Republic of Germany   149

successor MSA (Mutual Security Agency), did not see the technology aspect itself as its main goal but the increasing of productivity in the Western European national economies.[55] This was to be achieved mainly by an intensive transfer of American management methods.[56]

The projects for German participants were prepared and coordinated by the newly introduced "Board for the Rationalization of the German Economy" (Rationalisierungskuratorium der deutschen Wirtschaft – RKW). This programme led above all to numerous journeys by German and European experts to study special issues in the USA. Of particular importance were also the "German-American manager talks", which were organised by the BDI and the NAM (National Association of Manufacturers) and conducted in different regions of the Federal Republic between 1951 and 1954.[57] In the course of its "education programme" the Rationalization Board also issued various publications on the subject of modernizing West German training of new staff and executives. In particular the texts by Heinz Hartmann had a pioneering influence.[58]

It is beyond the scope of this essay to go into the importance of American direct investments in Germany, which began to reach decisive levels at the end of the 1950s and at the beginning of the 1960s and were aimed at transfering American "business culture", modern production techniques, management methods or newly defined "human relations" in plant management. These interrelations have already been analysed in detail in contemporary surveys[59] and have been taken up again and expanded in modern business research.[60]

When the subject of technology became topical in the 1960s and the "American challenge" became one of the most widely discussed subjects in Western Europe, competent American experts emphatically drew attention once again to the European deficits in modern management and innovation capacity. According to them the most important problem was not the discovery or the scientific-technological breakthrough but the transposition of technology into a product

---

[55] From the American point of view cf. NA RG 59, 762A.5–MWP/1–2953: 29.1.1953 HICOG Berlin to Department of State, Washington, Desp. No.616 "Technical Assistance Program".

[56] The following areas are mentioned by name: methods of plant management, industrial production methods, personell relations, improvement of distribution of goods and encouragement of competition in business, increase in agricultural production, public administration and finance, and general transport questions. Loc cit: Wiederaufbau im Zeichen des Marshallplans, 1953, p. 35.

[57] Extensive material in HA No. 40010146/663: Cf. also 2.6.1951 Circular from the executive management of the BDI (Dr. Beutler/Stein). Still indispensible for the history of German-American businessmen relations: *Link,* Werner, Deutsche und amerikanische Gewerkschaften und Geschäftsleute 1945–1975, Düsseldorf 1978, pp. 130ff.

[58] *Hartmann,* Heinz, Unternehmerausbildung. Die Rolle der deutschen Hochschulen, Munich 1958 (Rationalisierungs-Kuratorium der Deutschen Wirtschaft. Auslandsdienst, No. 55); ibid., Der deutsche Unternehmer: Autorität und Organisation, Frankfurt 1968 (First published: Authority and Organization in German Management, Princeton N. J. 1959).

[59] Cf. esp. *Hartmann,* Heinz, Amerikanische Firmen in Deutschland. Beobachtungen über Kontakte und Kontraste zwischen Industriegesellschaften, Cologne/Opladen 1963.

[60] *Berghahn,* Volker R., Unternehmer und Politik in der Bundesrepublik, Frankfurt 1985, esp. pp. 249ff.

which was financially useful and could be sold at a profit. "In this particular point ... the USA obviously has a considerable lead over Europe at present."[61]

## 4. Definitions of the "Technological Gap" and Evaluation of the Indicators

As can be seen in the comparison of the American and German emphasis with respect to "technology", the term "technology deficit" was understood to mean something quite different depending on point of view and interests.

An attempt to present the phenomenon of technological competitiveness and the "technological gap" in more general terms and to find a clear and operable definition[62] of the same presents by no means negligible methodological difficulties. Naturally, the field of "big science" played a key role from a Western European point of view in the 1960s: in other words, space travel, military and nuclear technology or the so-called technological industries such as computers, electronics and aeronautical engineering. And, as we have seen, this was the sector to which the German consulate and the embassy in Washington paid particular attention.

On the other hand the OECD, in the report "Gaps in Technology" from 1968 mentioned above, emphasised a different point: according to the OECD study group three interrelated aspects lay behind the gaps in advanced technology. These were:

1. the differences in the ability to create the scientific and technical conditions necessary for technical and technological progress,
2. the diffences in the ability to apply new scientific and technical knowledge and use this in the production process, i.e. technological innovations,
3. the economic effects of 1. and 2. in connection with the influence of the international exchange of goods and technological knowledge.[63]

The similarity between the OECD definition and the American approach, which saw the heart of the Western European technology deficit less in the field of "big science" than in the lack of innovation, is striking. However, the criteria chosen by the OECD do not amount to a comprehensive and objective measurement of the technological standards of a country at an international level. A look at the West German balance of licence transactions and patents 1950-1963 illustrates how problematic the use of exact measureable quantities can be. The balance of revenue and expenditure on inventions, copyrights etc. in transactions with other countries is often used to determine more precisely a country's technological standard.[64]

---

[61] *Moline,* Edwin G., Das Problem der "technologischen Lücke" zwischen den Vereinigten Staaten und Europa. Ausmaß, Ursachen und Lösungsmöglichkeiten. Speech by the Envoy for Financial Affairs at the United States Embassy in the Federal Republik of Germany on 27th April 1967 at a meeting of the European Committee for Economic and Social Development (CEPES), in: Europa-Archiv 22. Jg., 1967, pp. 427–434.

[62] *Standke,* Klaus-Heinrich, Die "Technologische Lücke" zwischen den Vereinigten Staaten und Europa. Zu Definition des Problems, in: Europa-Archiv 22. Jg., 1967, pp. 593–600. *Majer,* Helge, Die "Technologische Lücke" zwischen der Bundesrepublik Deutschland und den Vereinigten Staaten von Amerika, Tübingen 1973.

[63] Gaps in Technology. General Report, (OECD) Paris 1968, p. 11.

[64] For fundamental link between patents and technological development cf.: *Grefermann,* Klaus, Patentwesen und technischer Fortschritt, in: Hundert Jahre Patentamt. Festschrift issued by the Deutsches Patentamt, Munich 1977, pp. 37–64.

Technology Transfer and Foreign Trade in the Early Years of the Federal Republic of Germany   151

**Table 3:**   Revenue and Expenditure for Inventions, Techniques, Copyrights etc. in Transactions with other Countries 1950–1963[65]

| Time | Millions DM | | | | Balance |
| | Revenue | | Expenditure | | |
| | total | Including: copyrights* | total | Including: copyrights* | |
|---|---|---|---|---|---|
| 1950 | 10 | * | 22 | * | – 12 |
| 1951 | 17 | * | 65 | * | – 48 |
| 1952 | 32 | * | 89 | * | – 57 |
| 1953 | 50 | * | 135 | * | – 85 |
| 1954 | 79 | * | 177 | * | – 98 |
| 1955 | 76 | * | 222 | * | –146 |
| 1956 | 80 | * | 262 | * | –182 |
| 1957 | 94 | * | 305 | * | –211 |
| 1958 | 116 | * | 364 | * | –248 |
| 1959 | 144 | * | 416 | * | –272 |
| 1960 | 155 | * | 510 | * | –355 |
| 1961 | 169 | * | 619 | * | –450 |
| 1962 | 186 | * | 631 | * | –445 |
| 1963 | 216 | 17 | 637 | 96 | –421 |

\*   Besides copyrights and design patents payments for trademarks also form a small part of these figures.

**Transactions in patents and licences with foreign countries**

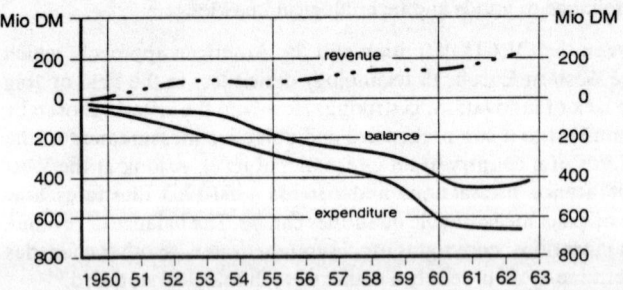

As the figures show, there was an increasing debt balance between 1950 and 1961. While the deficit from trade in patents and licences was only 12 million DM in 1950, it had already reached 146 million DM in 1955 and by 1961 450 million DM. Only in 1962 was there a drop of 5 million to 445 million DM. However, the conclusion so fondly drawn that the continual negative balance reflects West Germany's ever increasing technological deficit in the 1950s[66] is not tenable.

---

[65] Monatsberichte der Deutschen Bundesbank, 16. Jg., No. 4, 1964, p. 22.

[66] Cf., for example: *Der Spiegel* No. 9, 21.2.1966, Leading story: "Nach 30 Jahren Niedergang: Notprogramm für Deutschlands Forschung?", p. 33.

Firstly, one has to take into consideration that the revenue from German licence exports (in contrast to the expenditure) was de facto new business because of the confiscation of the old patent and licence rights after World War II. Secondly, the deficit from the patent and licence transactions is relatively small compared with the surplus in the overall balance of all service transactions with abroad (1963 4300 million DM surplus in the invisible balance and 421 million DM deficit from patents and licences). However, the data does reveal, as was also stated by the German Federal Bank, that on balance in the 1950s the Federal Republic "imported technical knowledge on a not inconsiderable scale" in all fields of applied technical research.[67]

But even here it is necessary to make a differentiated evaluation: the decision as to whether a company wants to manufacture the products developed through its research and to export these to other countries itself or wants to issue the licences in these countries depends very much on a country's relative cost advantages. Therefore, it can be assumed that the USA, which held a leading position in technological know-how in the 1950s but was at a disadvantage with respect to production costs (wages) in comparison with many countries, was more likely to issue patents and licences than West Germany, for example, which saw a better chance in exploiting technical knowledge for its own products, which were competitive from a cost point of view on the world market. Thus the German Federal Bank was correct to state in its first report on the patent balance (1964):

> "For these reasons there is without doubt an interrelation between the high level of German exports on the one hand and the relatively low German licence revenue on the other, which ... does not permit a direct assessment of the state of technical research solely on the basis of the development of the licence balance."[68]

In this context a look at the Japanese technology imports in the 1950s and 1960s reveals interesting information: from 1950 to 1959 the Federal Republic was in third place behind the USA and Switzerland with respect to the number of technology-import-agreements signed by Japan. The situation was even more favourable from 1960 to 1964: in this period the Federal Republic advanced to second place behind the United States in the ranking of licensers.

**Table 4:** Number of technology-import-agreements signed by Japan according to the countries of origin 1950–1959 and 1960–1964[69]

|  | 1950–1959 | | 1960–1964 | |
|---|---|---|---|---|
| 1. USA | 658 | 67.1% | 1228 | 62.7% |
| 2. Switzerland | 81 | 8.3% | 155 | 7.9% |
| 3. West Germany | 72 | 7.3% | 255 | 13.0% |

---

[67] Monatsberichte der Deutschen Bundesbank, 16. Jg., No. 4, 1964, pp. 23–24.
[68] Ibid., p. 23.
[69] OECD: Gaps in Technology. Analytical Report, Paris 1970, p. 204.

Technology Transfer and Foreign Trade in the Early Years of the Federal Republic of Germany   153

**Table 5: Number of patents issued abroad (1963)**[70]

| | | |
|---|---|---|
| 1. USA | 65300 | 38.5% |
| 2. West Germany | 29900 | 20.5% |
| 3. Great Britain | 15200 | 10.4% |
| 4. Switzerland | 9400 | 6.5% |
| 5. France | 9300 | 6.4% |

The Federal Republic also held a leading position with respect to the number of patents issued abroad. At 20.5% in 1963 West Germany was in second place behind the USA with 38.5%.

It is revealing that in the trade in patents and licences the Federal Republic had by far the highest surplus with Japan (1963 – 35.3 million DM), more than half of which came from the electrical industry (18.1 million DM). However, the chemical industry, with 30.1 million DM, was once again able to derive a large part of its revenue from transactions with the USA (39%). But with a debt balance totalling 165.5 million DM in the mutual trading of patents and licences, the USA remained the most important exporter of technical know-how to the Federal Republic.

A more precise understanding of the differentiation of technological innovations in specific branches can be gained from a comparison of the main inventions and their first industrial applications. It is not possible to discuss here in depth the development of nuclear technology, which became the main focus of West German technology policy in the second half of the 1950s. The West German nuclear policy has already been examined and written about in detail.[71] West German researchers and companies did not play a significant role in the development of the semi-conductor technology.[72] It can be assumed that the discontinuation of most military research in the early years of the Federal Republic had its effect on this field. By the end of the 1950s the USA had a de facto worldwide monopoly status in transistor technology.[73]

The situation was similar in the field of computer technology: the Federal Republic had lost its lead from before and during World War II in this field completely and only entered the market in 1959 with the transistor computer ER 56 from Standard-Elektrik-Lorenz AG (SEL), a daughter company of the American ITT.[74] However, the technological competitiveness of the West German chemical industry was incomparably better. The companies Hoechst and Bayer established an international lead[75], for example in the first economic use of the plastics polyethylene, polypropylene and polycarbonates in 1955-7. In the introduction of pharmaceutical products from 1950 to 1967 the Federal Republic was in third place worldwide (with 15 of 138 new medicines, 10.9%) behind the USA (67 from 138,

---

[70] Ibid., p. 205.
[71] Cf. esp.: *Radkau,* Aufstieg und Krise der deutschen Atomwirtschaft, 1983.
[72] See above, Report from the chargé d'affaires in Washington "Neuere technische Entwicklung in den Vereinigten Staaten".
[73] Gaps in Technology. Analytical Report, 1970, Tab. 6, p. 196.
[74] Ibid., Tab. 1, p. 190.
[75] Ibid., Tab. 3, p. 192.

48.6%) and Switzerland (20 from 138, 14.5%).[76] In the sphere of scientific instruments the company Siemens was able to take up its prewar innovation (1939) with an improved electron microscope in 1954 and in 1960 Telefunken AG brought the "radio pill" invented in West Germany in 1957 on to the market for the first time.[77] West German companies also held a leading position internationally in metallurgical technology and paper production, among other fields.[78]

All this shows that the question of the "technology gap" in the Federal Republic in the 1950s cannot be answered in one or two sentences. As an examination of the patents issued and of the revenue and expenditure from inventions, techniques and copyrights etc. in transactions with abroad proves, it can hardly be said that there was a general technology deficit in West Germany at this time. This, however, does not contradict the statement that there was, at the same time, a significant deficit in important sectors involving modern technology (e.g. semi-conductor techniques, computer technology).

## 5. Conclusion

West Germany's quick and successful return to the world market after 1945 took place against a background of favourable starting conditions: the economic and political framework of the "pax americana", multilateralism and liberalization in world trade offered good terms to the German economy, which was traditionally export-orientated, and especially to the capital goods industry. The allied controls on production and trade could be dismantled relatively quickly, particularly as a consequence of worsening international relations during the Korean War.

In this context it is also important that the structural continuity of the foreign trade relationships maintained a more or less unbroken continuity from prewartime, with the exception of trade with Eastern Europe. The return to the world market was in this respect also a return to former markets and the West German economy had few difficulties in reconquering its traditional position very quickly, especially in competition with Great Britain.

The balance of trade with the dollar area remained a problem for the German economy until about the middle of the 1970s – here again in remarkable continuity of the structural patterns prevalent between the wars. The particular importance of this market for the export economy was not just as an outlet for German products. It was especially relevant in that the USA, as the most important licenser for German firms, played a key role in transfering technological know-how to West Germany.

The sectorial deficits of the West German, and of the Western European, economy only became evident and developed to a subject of political concern in the 1960s. In the field of "big science" the technological deficit, compared to the USA, could no longer be ignored: The Western European shock at the "American challenge" was triggered off mainly by massive American direct investments in the 1960s, although in the 1950s there had still been complaints about the lack of American private capital.

---

[76] Ibid., Tab. 4, p. 193.
[77] Ibid., Tab. 5, p. 195.
[78] Ibid., Tab. 7, p. 197. Cf. also Tab. 2, p. 191.

There is no simple answer to the question as to whether developments had been missed[79]: it is definitely true that the 1950s, seen from the point of view of the 1960s, were indeed an "economic miracle" but not a "research miracle". But to what extent had the government's technology policy in the triangle formed by research, business and state possibly missed important chance? It cannot be denied that the Federal Republic, while hoarding foreign currency in the so-called Julius Tower[80], was at the bottom of the international table with respect to expenditure for science and research.[81]

However, in the face of the overcritical comments from the 1960s it is advisable to remain somewhat sceptical. The question to be asked here is whether the pressure of the obviously overdue modernization of society in the 1960s did not tempt an excessive tendency to paint the 1950s black in all respects. For example, a more thorough analysis of the patent and licence balance, especially with respect to transactions with Japan, shows that care is needed with overgeneralized judgements.

The most detailed examination to date on the technological gap between West Germany and the USA in the 1950s and 1960s has confirmed that one cannot speak of a general technological deficit between West Germany and the USA.[82] However, it is safe to assume that especially in the fields of management methods and openness for innovations there was a lot to be learned from America after 1945 and in the 1950s.[83] The transfer of modern industrial economic attitudes and methods from the United States, as is confirmed by the latest research in business history, only began to take effect in top West German management and in West Germany's "economic culture" during the 1960s and 1970s. This lengthy modernization process did not take the form of a "total Americanization" but represented instead an "industrial cultural synthesis".[84] Viewed from today, the successful synthesis of modern American industrial culture with traditional West German business mentality and business structure was perhaps an essential foundation for the long-term success of the West German economy and one of the main prerequisites for the fact that Europe's technological deficit with respect to the USA, which seemed so frightening in the 1960s, has now, for the most part, been overcome.

---

[79] Cf. also the cartoon in *Der Spiegel* No. 9, 21.2.1966, p. 49: While the economic miracle is celebrated with gusto on the upper floors of the West German house, down below the research foundations are breaking up.

[80] The Julius-Tower is a metaphor in Germany for accumulating public funds and refers to a tower in the fortress Spandau in Berlin, where Imperial Germany after 1870 had hoarded war treasure.

[81] See data on promotion of research in Note 3.

[82] *Majer*, Die "Technologische Lücke", p. 322.

[83] Cf. *Standke*, Die "technologische Lücke" zwischen den Vereinigten Staaten und Europa, p. 598.

[84] *Berghahn*, Unternehmer und Politik, pp. 228ff., p. 255.

# [19]

Excerpt from Jay Tuck, *High-Tech Espionage: How the KGB Smuggles NATO's Strategic Secrets to Moscow*, 180–206.

## THIRTEEN

## CoCom – The Toothless Watchdog: What Are We Doing?

The snowplough dredging its way through the drifts of a winding mountain road in the foothills of the French Alps came to a grinding halt. A car parked carelessly on the shoulder of the road was blocking its path; its door hung half-open. Someone might be in distress, the plough driver thought as he yanked up his handbrake. He heard music blaring from the car radio as he approached. But at first glance he knew that it was too late for help. A snowy corpse sat drooped over the wheel, blood from a gaping gunshot wound clinging to the frozen skin. It was an apparent suicide.

Homicide experts arriving at the grisly mountain scene a few hours later found a handgun about six feet from the body. Three spent cartridges were in the chamber. The man had been killed, the police discovered, by a single bullet that had penetrated the head, exited, and was nowhere to be found. An autopsy would reveal that the fatal shot had been fired at a distance of several yards. The 'suicide' was a cheap set-up. It was clearly an assassination – and a professional one.

The dead man, found on 25 February 1983, was Lieutenant-Colonel Bernard Nut, an undercover agent of the French counter-espionage organization DST (Direction de la Surveillance du Territoire). Monsieur Nut was no ordinary spy. He had succeeded in penetrating the innermost circles of the KGB's Directorate 'S', a top-secret arm of Soviet intelligence which infiltrates Soviet-born agents into Western societies using assumed identities. Soviet 'S' agents had scored major successes in the piracy of French technology. For a long time, no one had suspected that the Frenchman in their midst was a double agent. Even diplomats at the Paris

Embassy had trusted Bernard Nut as a loyal Soviet agent, speaking openly in his presence about KGB operations.

The intelligence funnelled out by their 47-year-old officer proved priceless to the French, leading to the exposure and arrest of several high-ranking KGB operatives. Among them was Patrick Guerrier, a 25-year-old archivist at the state-owned coal company Charbonnages. Guerrier was apprehended at a clandestine meeting in the suburb of Meaux as he passed documents to a Soviet Embassy attaché. Nut also uncovered major espionage operations against the French naval base at Toulon, where the nuclear hunter-killer submarine *Rubis* was stationed. His biggest coup, however, was the unmasking of the Bulgarian/KGB connection behind the attempted assassination of Pope John Paul II. It was a tip from Bernard Nut that led to the arrest on 12 February of Viktor Pronin (an Aeroflot official in Rome) in connection with the Vatican shooting. Three days after the arrest, Bernard Nut lay dead in an Alpine snowbank.

Contrary to many popular notions, spooks seldom kill each other. It is an unwritten rule of their game. Given the stakes in this case, however, the KGB hierarchy apparently thought it would be worth risking an exception. It was, as it turned out, an epic miscalculation.

The murder of Bernard Nut – a French military officer gunned down on French soil – touched off a wave of high-level resentment, even fury, within the Paris Government. It sent shock waves right up to the very top of the Elysée Palace. Unfortunately for Moscow, the incident coincided with several other developments that day and the slaying was the straw that snapped French patience, prompting a historic decision that François Mitterrand had been contemplating for several months.

In November 1982, the French President had ordered a DST investigation into the damage done by Soviet technology theft. Originally, concern had been aroused by the remarkable increase in Soviet staff at the Paris Embassy. Since 1973, the number of Soviet diplomats in the French capital had risen from 200 to 700, the total number of Soviet citizens in France from 1,000 to 2,400. While three representatives of the USSR had been sufficient to staff the offices of the United Nations in 1973, they now numbered forty. In the same period, the number of personnel at consulates in Paris and Marseille had risen from six to thirty-six.

The DST report, which hit Mitterrand's desk shortly before the news of Bernard Nut's murder, confirmed his worst fears: the new arrivals in Paris were hardly dedicating themselves to diplomatic

## 182 Consequences

affairs. They were using their assignments as a cover for espionage.

The sheer number of warm bodies, counter-intelligence officials warned, had turned their job into a nightmare. The French reckoned they needed a twenty-man team to observe a suspected spy for any extended period of time. There were already more suspected spies than they could handle, and the number was growing by the day. Line 'X' – the overseas arm of the KGB technology specialists at Directorate 'T' – had planted its agents at every available institution. They were posing as representatives of Aeroflot and Intourist, Tass and Soviet trade missions. Following the tightening of US embargo measures across the Atlantic, DST officials had noted a sharp increase in illicit activities. France had become a major hunting-ground for technology spies. 'We discovered their field workers were suddenly working overtime,' one expert reported, 'and those who normally did office paperwork were out on the streets looking for contacts.'

Moscow's interest in military projects came as no particular surprise. The French had a lot to offer. They were international leaders in the development of miniature neutron warheads; their computer-guided Exocet missile had demonstrated its lethal efficiency in the Falklands War; and the Mediterranean coastline was studded with several choice targets. There were intermediate-range nuclear missiles on the Albion Plateau and long-range strategic bombers in Orange. At a nearby airstrip, secret test flights for the Mirage-2000 fighter-interceptor were under way, and Marignane Arsenals were tinkering with an advanced French-German combat helicopter.

Most of all, however, the French were disturbed by blueprint bootlegging of civilian know-how. From computer electronics and glass-fibre cables to infra-red optics and navigational instruments, France was on the cutting edge of much advanced technology which would one day decide the long-term economic and military potency of Western Europe. All of these fields were targets of the experts from 'X'. Only half of all Eastern espionage operations, the DST report warned, were still aimed at traditional targets in the armed forces, in defence industries and in public politics. The Soviets wanted technology. And they were getting it. Nearly 30 per cent of all top-of-the-line French developments had already been plundered.

For President Mitterrand, who had his own ambitious plans for the French defence industry, this was not good news. He knew that the Rogers Plan for expanding conventional forces in Europe would mean billion-dollar contracts, and he wanted French companies to get their piece of that lucrative pie. Ever since assuming office, Mitterrand had carefully and quietly expanded ties with relevant West German and British industries. Widespread Soviet theft was now threatening his plan. With the Concorde copy TU-44 and the Airbus forgery IL-86 still fresh in French memory, the Paris government decided to act.

On 31 March 1983, thirty-three days after the slaying of Bernard Nut, preparations were complete. Prime Minister Pierre Mauroy summoned Soviet Ambassador Yuli Voronsov to the Foreign Ministry and announced the immediate expulsion of forty-seven Soviet citizens from France. The reason, Mauroy stated flatly, was espionage. Forty diplomats, five commercial officers and two journalists – among them KGB station chief Nikolai Chetverikov – were granted time only for a brief farewell party and some hurried shopping on the Champs-Elysées. On 5 April television cameras and newspaper photographers recorded the historic departure of the forty-seven at Charles DeGaulle International Airport. Laden with bulging shopping bags and clutching bouquets of flowers, the Soviets were driven in six grey mini-buses to a waiting Aeroflot IL-86 plane and jetted back to the USSR. It was the most sweeping house-cleaning of Kremlin spies since 1971, when London threw out 105.

Reaction from Moscow was surprisingly mild. Tass huffed and puffed about 'unfriendly acts' and 'hysteria', but no retaliation followed. The French had done their very best to discourage it. When Mauroy announced the purge at the Quai d'Orsay meeting, he presented Ambassador Voronsov with a second list of forty additional names – Soviets who would be asked to leave should their government 'overreact'. Mauroy added in no uncertain terms that any positions left vacant by *persona non grata* would not be open to future Soviet applicants.

'Retaliatory steps against French citizens living in the USSR,' Soviet leader Yuri Andropov explained in an interview with the German magazine *Der Spiegel* two weeks later, 'would have been a simple matter. In demonstrating restraint, we were guided by the overall interests of Franco-Soviet relations, which we appreciate and which have helped maintain détente in Europe for many years.'

When the expulsions were announced, Paris cited 'a systematic attempt by agents of several secret services of the USSR to acquire scientific, technical and technological intelligence'. The expulsions were directed primarily at the activities of Line 'X,' and this was new. For the first time, technological piracy figured in a major cam-

## 184  Consequences

paign of Western counter-intelligence. For years, the illicit deeds of this highly effective arm of the KGB had been registered with mounting concern in a number of Western capitals. The fact that it was François Mitterrand, a socialist leader, who took the first radical step to end Soviet snooping caught the world by surprise. *Le Monde*: 'The action washes Mitterrand, if that was necessary, of all suspicion that his freedom of action is limited by his coalition with the Communist Party.'

The real effect of the bold French stroke, however, was the global chain reaction it touched off. From Bonn and Brussels to Berne, Bangkok and Bangladesh, authorities were soon sending Soviet spooks packing. In the twelve months that followed, 135 Soviets were exposed as agents and forced to abandon their overseas assignments – more than ever before in the history of East-West relations.

On 22 April 1983, the US Department of State summoned Soviet Ambassador Anatoly Dobrynin and announced that three high-level diplomats were being declared *persona non grata*. Oleg Konstantinov, Third Secretary at the United Nations in New York, had been apprehended by FBI agents in Manhasset trying to obtain secret aerospace documents; his 44-year-old UN colleague, KGB spy Aleksandr Mikheyev, was snared seeking confidential foreign policy papers from a Congressional aide in Washington, while Yevgeny Barmyantsev, identified as a lieutenant-colonel of the GRU, was caught trying to gain access to military laser secrets in Maryland. That same day, on the other side of the world, authorities in Canberra, Australia, exposed and extradited Embassy secretary Valeri Ivanov for KGB activities.

On 26 April Stockholm recalled its Ambassador from Moscow. The Swedes, still fuming about a recent espionage affair at the Göteborg shipyards, were provoked by persistent Soviet submarine violations of their territorial waters.

Vice-Consul Vladislav Istomin was caught red-handed with NATO know-how in Geneva, while Danish counter-intelligence exposed Yevgeniy Motorov, chief of Line 'X' in Copenhagen. In Belgium, authorities halted the efforts of Soviet 'businessman' Yevgeniy Mikhailov to harvest computer hardware; in Tokyo an associate named Arkhadii Vinogradov was caught stalking Hitachi software.

Within six months London had expelled eight Soviets, Madrid four, Bonn three, Berne two, and Ottawa one. Even the Iranian regime of Ayatollah Khomeini took action, throwing out eighteen Soviet diplomats on 4 May. They were seen off at Tehran airport by

## Soviets exposed as spies in the West between December 1982 and February 1984

| | | | |
|---|---|---|---|
| 17/12/82 | Rome | 1 | Lt.-Col. Ivan Heliog (Military Attaché/GRU) |
| 23/12/82 | Stockholm | 2 | Yuri Averine (Soviet Consul-General/GRU); Lt-Gol. Piotr Skiroky (Military Attaché/GRU) |
| 12/2/83 | Rome | - | Viktor Pronin (Aeroflot/KGB): arrested |
| 17/2/83 | Cologne | 1 | Guennadi Batachev (STM/KGB): convicted, later extradited |
| 24/2/83 | Rome | - | Konayev (Soviet chemical company): arrested |
| 3/83 | London | 8 | Names not known |
| 3/83 | Toronto | 1 | Names not known |
| 3/83 | Madrid | 4 | Soviet shipping agency employees, names not known |
| 3/83 | Bonn | 2 | Names not known |
| 3/83 | Amsterdam | 1 | Names not known |
| 5/4/83 | Paris | 47 | Nikolai Chetverikov (First Secretary/KGB Station Chief); Vasily Golitsyn (Naval Attaché /GRU Chief); Oleg Shirokov (Tass); Vladimir Kulikovskykh (Tass)) Edward Sokolov (Marseille Consulate); J. Krivtzov (UNESCO); J. Matveyev (UNESCO); S. Yakubenko (UNESCO); and others |
| 22/4/83 | Washington | 3 | Yevgeny Barmyantsev (Military Attaché/ GRU); Oleg Konstantinov (UN/KGB); Aleksandr Mikheyev (UN/KGB) |
| 22/4/83 | Canberra | 1 | Valeri Ivanov (First Secretary/KGB) |
| 26/4/83 | Stockholm | - | Ambassador ordered home by Moscow |
| 29/4/83 | Berne | 1 | Aleksei Dumov (Novosti/KGB) |
| 30/4/83 | Copenhagen | 1 | Yevgeny Motorov (First Secretary/KGB) |
| 4/5/83 | Tehran | 18 | Diplomats, names not known |
| 13/5/83 | Brussels | 1 | Yevgeny Mikhailov(Elorg Manager/KGB) |
| 19/5/83 | Bangkok | 1 | Viktor Barychev (Soviet Trade Mission/GRU) |
| 17/6/83 | Tokyo | 1 | Arkhadii Vinogradov (First Secretary/KGB) |
| 7/83 | Geneva | 1 | Vladislav Istomin (Vice-Consul) |
| 8/83 | Washington | 2 | Anatoly Skripko (First Secretary); Yuri Leonov (First Secretary) |
| 8/83 | Brussels | 1 | Gruchine (First Secretary/KGB) |
| 21/12/83 | Bangladesh | 9 | Names not known |
| 28/12/83 | Bangladesh | 6 | Names not known |
| 2/2/84 | Oslo | 5 | Leonid A. Makarov (First Secretary/KGB); Stanislav Tsyibotok (First Secretary/KGB); Yuri A. Anisimov (First Secretary/GRU), Mikhail Ovtkin (Soviet Trade Mission/KGB); Anatoli A. Artamonov (Soviet Trade Mission/GRU) |

## 186 Consequences

an angry mob of Islamic revolutionaries hooting: 'Death to the traitors.'

A NATO report of the time estimated that 70 per cent of the staff at Eastern diplomatic missions were employed by intelligence services. The mass expulsions of 1983 were the first serious countermeasures taken by the West. It was not surprising that the initiative originated in Paris. Curious was the fact that it came from the Mitterrand Government. There was another organization based in the French capital which bore prime responsibility for such matters. It should have been in the forefront. But it wasn't.

The organization is called CoCom – the Co-ordinating Committee for East-West Trade. Housed in an inconspicuous backdoor building (Annexe D) of the United States Embassy, its international staff has the job of identifying militarily applicable technology, drawing up embargo lists and setting down policy guidelines for strategic trade. Its charter provides that member states develop and co-ordinate enforcement strategies. In short, the Co-ordinating Committee is designed to define, expose and combat technical and scientific espionage; it is the watchdog of the West.

CoCom was created on 22 November 1949, just a few months after the Atlantic Alliance was founded. The seven signatory nations were France, Great Britain, Italy, Belgium, Luxembourg, Holland and the United States. The Federal Republic of Germany, Denmark, Norway, Canada and Portugal soon joined, followed later by Japan, Greece and Turkey. Today, all NATO states, with the exception of Iceland and Spain, are members.

CoCom was conceived as a child of the Cold War. Shortly before its inception, the Soviets had imposed a Stalinist government on Czechoslovakia and blocked access routes to West Berlin. Continuing East-West tension promoted CoCom's early development. China invaded Korea and Moscow continued its brutal Sovietization of Eastern Europe, quashing popular uprisings in East Germany (1953) and Hungary (1956) with Red Army tanks. With the detonation of the first Soviet atomic bomb, nuclear energy became the first dual-purpose technology to be embargoed by the West.

Today, the internal workings of CoCom are carefully screened from public view. There is no sign on the entrance to Annexe D to indicate its existence, no entry in the local telephone directory. Information about its meetings is limited, if available at all. Camera crews or photographers trying to sneak an exposure of the grey façade in the Rue de La Boetie are rudely shoved aside by guards. Reporters' questions are traditionally greeted with a terse 'No comment'. On occasion, participants at CoCom meetings have been known to deny that a meeting had taken place at all.

The secrecy shrouding the Paris panel, however, is not so much motivated by fear of Eastern espionage. Rather, it is the squabbles among friends that are CoCom's best-kept secret. Ever since the onset of the 1980s, a heated controversy has been raging among member states about the wisdom of NATO embargo measures.

The conflict began with the invasion of Afghanistan. In response to it, President Carter imposed a grain embargo on the USSR, also announcing a sweeping review of all American East-West trade policies. Over 400 US export licences were cancelled overnight, among them those for computer components for the Kama River project. But when the American President turned to his allies hoping to find support for his sanctions, he discovered that CoCom had deteriorated from a hard-hitting regulatory arm of NATO to an indifferent provincial outpost dozing in a dull bureaucratic sleep. CoCom, watchdog of the West, had long since lost its bite.

Staff members – all fourteen of them – were ill-equipped for the job at hand. They could judge neither the significance of the new equipment they were supposed to be protecting nor the global scope of scientific smuggling they were supposed to be preventing. Most of them were trade diplomats. Technical and military expertise, absolutely essential to serious regulatory supervision, was available only during rare high-level meetings. There were no simultaneous translators, no stenographers, not even a functioning Mimeograph machine. The total annual budget was a laughable $500,000.

The embargo lists of CoCom were thus hopelessly obsolete long before they went into effect. Innovations in Silicon Valley and elsewhere moved far faster than the bureaucratic machinery designed to protect them. 'If CoCom had existed in the year 3000 BC,' lamented one member, 'the wheel would still be in the export lists.'

Efforts on the enforcement front were hardly more spectacular. The Paris organization had no computers to record and correlate data on the scores of phoney companies, front organizations and middlemen being used; no data-processing equipment to trace the tangled shipping routes and unravel the twisted corporate networks. What little intelligence CoCom did have had simply been jotted onto hand-written file-cards. Employees spent their working days transcribing, translating and distributing minutes of the countless, but generally unproductive, CoCom meetings.

All this, of course, was fine with most Europeans. Interest in

## 188 Consequences

trade restrictions had waned in the fat years of détente between 1972 and 1980. The number of strategic products on the embargo list declined from 300 in 1951 to less than half that number by 1980. Remaining restrictions were shot full of holes by constant exemptions. In 1979 alone, member states submitted over 1,500 applications for exemptions to CoCom rules. Many countries on the Continent, in particular the Federal Republic of Germany, had become embarrassingly dependent on Eastern trade. All were earning billions of dollars in their dealings with Warsaw Pact countries. A weak embargo policy, they thought, was in their very best interest.

In the wake of the Afghanistan shock, President Carter began planning new priorities for CoCom. He wanted its facilities modernized, its personnel bolstered, its general mandate expanded. He suggested CoCom move from the Embassy annexe building to its own headquarters. He wanted a secretary-general and a staff of full-time military advisers installed. He now regarded Soviet theft of military know-how as a matter of grave concern and he appealed to Western nations to support his initiative.

German Chancellor Helmut Schmidt, whose low opinion of Jimmy Carter is now a matter of public record, regarded the US embargo initiatives as ill-conceived, untimely, even stupid. His criticism was shared in London and Paris. European governments were familiar with the appeals of the Carter Administration. And they distrusted them. All too often they had grudgingly consented to co-operate with an unpopular Washington policy – only to see the Americans themselves abandon it without notice.

Carter was in a weak bargaining position. His own Administration had contributed substantially to the relaxation of strategic trade policies. American companies had joined the Eastern business boom with verve, quickly overtaking the Europeans in both the quantity and the quality of their high-tech sales. They had led the way in computers, peddling more electronic wares to Moscow than all other CoCom countries combined. Over half of the 1,500 applications for exemptions in 1979 had been submitted by Washington. US companies had supplied over $200 million worth of top technology to the USSR. And now, suddenly, Washington was arguing that such transactions posed a fundamental threat to Western security.

European leaders wouldn't support Carter. If Washington wanted to restrict its own exports, they argued, let it. But they were quite unwilling to follow the lead taken by a wavering White House. Detached, disinterested, on occasion even bemused, they sat back to watch the first insecure steps of Jimmy Carter on the slippery stage of embargo politics.

**US Computer Exports to the USSR (in computer units)**

| 1972 | USA | 20 |
|---|---|---|
| | other Cocom nations | 18 |
| | total | 38 |
| 1973 | USA | 41 |
| | other Cocom nations | 36 |
| | total | 77 |
| 1974 | USA | 57 |
| | Other Cocom nations | 37 |
| | total | 94 |
| 1975 | USA | 162 |
| | other Cocom nations | 97 |
| | total | 259 |
| 1976 | USA | 122 |
| | other Cocom nations | 36 |
| | total | 158 |
| 1977 | USA | 67 |
| | other Cocom nations | 73 |
| | total | 140 |
| 1972-7 | USA | 469 |
| | other Cocom nations | 297 |
| | total | 766 |

*(Source:* Kenneth Tasky, 'Soviet Technology Gap', *Soviet Economy in a Time of Change*, US Joint Committee of Congress, Washington DC, 10 October 1979.)

The American President announced his decision in a televised speech on 4 January 1980. He wanted to hit the Soviets where it would hurt most: in the bread-basket. The main thrust of his action was an embargo on the sale of American grain, on which the USSR depended heavily. At the time, its own agricultural industry was in serious trouble. Although 1978 had produced a record grain crop (235 million tons), 1979 was a typical Soviet disaster (only 179 million tons).

Traditionally, the USSR writes off one quarter of its harvest to transportation losses, rodents, mildew and thievery. One million tons were already allocated to Vietnam and North Korea, another million to Afghanistan. The poor 1979 harvest meant that the Soviet Union would need to import 34 million tons. The lion's share – 25 million tons – was on order from the United States.

Bread is a mainstay of the Soviet diet. It is one of the few commodities in the USSR which is seldom scarce. Despite the squeeze,

## 190 Consequences

officials felt assured they could continue to meet demand for it. 'Rations for supplying the Soviet population with bread and grain products,' Leonid Brezhnev announced on 12 January 1980, 'will not be reduced by a single kilogram.' The problem, however, was not so much bread for the people as feed for livestock. The embargo would be felt first and foremost by the Kolkhozniks on Soviet agricultural collectives. Their herds were fattening on grain from the American Midwest. If the impending deficit forced a wholesale slaughter of cattle and swine, the situation could become explosive. Meat shortages in Eastern-bloc countries have been known to cause social upheavals and rioting.

Moscow apparently got wind of Jimmy Carter's intentions. The day before the presidential announcement, the Soviets placed a last-minute order for an additional 3 million tons of American grain. Soon Soviet purchasers swarmed out to scour world markets for alternative sources. Their biggest finds were made in Latin America. They closed long-term contracts with fifteen countries, importing coffee and sugar from Colombia, fruit conserves and spices from Mexico, rice from Costa Rica and bananas from Ecuador. In Argentina they purchased 83 per cent of the country's total grain exports. Canadian wheat sales rocketed. Europeans were soon buying huge quantities of American soya beans for reprocessing and resale to the Soviets. In the north German port of Hamburg, hordes of freighters brimming with Eastbound grain were sighted steaming out of the harbour. By October, *Pravda* was boasting that Jimmy Carter's grain embargo was a flop.

Many European heads nodded knowingly.

Circumvention of the American ban did, however, extract a high price from Kremlin treasuries. The Soviet Union spent a total of over 2 billion dollars on contracts with its new agricultural partners. Attempts to substitute barter arrangements for cash payments failed. The wares the Soviets had promised in exchange for grain – primarily farm machines – couldn't be delivered. They were badly needed at the agricultural collectives at home. Another problem was cash. Most of the grain-sellers demanded hard currency, and brokers at international commodity markets noticed a sudden surge of activity, as Moscow traders put millions of dollars' worth of oil, diamonds and precious metals up for sale. A year later, when commodity prices plummeted, they were forced to finance their continuing grain imports with short-term loans – at outrageous interest rates of up to 17 per cent.

Still, the Soviets seemed able to cope with the American embargo with relative ease – a fact that was interpreted in many circles as evidence of the futility of trade sanctions in general. One reason for the ineffectiveness of the embargo was revealed by Fred Asselin, a US Senate staff investigator at the Subcommittee on Investigations, when he testified at technology transfer hearings on 5 May 1982:

Mr Asselin: A former Compliance Division investigator told the subcommittee minority staff that he was given the assignment of investigating embargo violations. He said no other agents in the Compliance Division were assigned to assist him. . .

Senator Nunn: You mean to testify that based on your investigation, the grain embargo that resulted after the Soviets invaded Afghanistan had one person in the Commerce Department checking on violations?

Mr Asselin: Yes, sir. . .

Senator Nunn: So you have a major policy decision by the President of the United States on a major foreign-policy issue involving the United States and the Soviet Union, involving all our allies, involving the credibility of our foreign policy and economic policy around the world, and you have the Commerce Department assigning one individual for enforcement purposes?

Mr Asselin: That's right.

Most American farmers co-operated voluntarily with the regulations of their government. The problem was not compliance. The problem was that the grain embargo was ill-conceived.

It was designed to punish a nation for its foreign-policy behaviour. 'The Soviet Union,' Jimmy Carter said, 'must pay a concrete price for its aggression.' But no trade embargo will work unless it has widespread support from many countries, a factor which Jimmy Carter simply ignored. His action also failed to capture the public imagination. It was hard for people to understand what guns for the battlefields of Afghanistan had to do with bread for the tables of Soviet citizens. The brunt of the embargo was felt, as Soviet dissident Roy Medvedev pointed out, by 'the small man who had nothing to do with the invasion and little influence on those who did'.

Most seriously, the grain embargo failed to provide an adequate response to Moscow's mounting military aggression. Whatever the President's intention – whether to teach the Soviets a moral lesson or to whip them into submission – it was a delusion. Carter lacked a cohesive political strategy. Worse, his action helped unify the states

## 192  Consequences

of the Warsaw Pact while it brought divisiveness to the Atlantic Alliance.

From the Kremlin's perspective, the cost of the grain embargo had been an acceptable one. The problem could be reduced to a matter of mere money. High-tech trade restrictions, however, were more serious. The equipment needed for Warsaw Pact armaments programmes could not be purchased in Costa Rica, Argentina or even Canada. Carter's cancellation of computer exports thus caused grave concern in Moscow. While Soviet leaders hurriedly developed contingency plans, they continued their worried watch on the Western Alliance.

By mid-1980, with US elections approaching, Carter began to feel the hot breath of his Republican presidential rival. Ronald Reagan was a hardliner on Soviet affairs and political pressures on the White House started mounting. Carter pursued his CoCom initiatives with new vigour. He pressurized the member nations to abandon their policy of promoting Eastern trade with government subsidies. He presented a 'Military Critical Technologies List' of know-how that American defence officials considered vital to Western security. The Pentagon denial list, about the size of a large telephone directory, included all the bread-and-butter exports of Western Europe's burgeoning Eastern trade: advanced machine tools, modern steel plants, sophisticated oil-drilling and petro-chemical equipment, and technology for automobile production plants.

In Europe, resentment was growing. Leaders in London, Rome, Paris and Bonn sought to limit the damage done to East-West relations by Afghanistan. They wanted to salvage détente and they were enraged by the new American proposal, which went to the core of their vital economic interests. West Germany, for example, was selling 64 per cent of its total export of automated lathes to the USSR. France and Britain, too, were far more dependent on Eastern trade than the United States. They expressed willingness to accept modest changes in CoCom procedures. An across-the-board freeze on Eastern exports, however, was out of the question. The Europeans announced publicly that they would fight such a move tooth and nail.

The full extent of Alliance irritation surfaced on one occasion during the Washington visit of Count Otto Lambsdorff, West Germany's trade minister. His hosts were complaining about subsidies for Eastern trade in general and German government credits called Hermes Guarantees in particular, when Lambsdorff suddenly exploded. As one eye-witness reported, he screamed at startled

CoCom – The Toothless Watchdog    193

Commerce Department officials, ranting that Washington was try-ing to ruin the German economy, slamming his walking-cane on the conference table for emphasis. Upon his return home, the minister carried his criticism to the media – a reliable sign that allied relations have reached a new low – claiming the American initiatives were aimed at bringing German-Soviet trade to a standstill.

The US had its own reasons for anger. Several American com-panies had been forced to abandon lucrative contracts with the East, only to see European competitors snap them up – often with the explicit approval of their governments. The Armco Corpora-tion, for example, had been negotiating over the construction of an electric steel plant in Novolipetsk for four years. Most of the con-tracts had already been signed when President Carter suddenly can-celled the export licences. Helplessly, the American company was forced to look on while the state-owned heavy-equipment manufac-turer Creusot-Loire of France made a grab for the million-dollar deal. Armco's losses were $6.5 million on negotiating costs, $16.7 million on profits and an expected $20 to $30 million on service and spare parts orders. It was not an isolated incident.

Despite its influence as a superpower, the United States was un-able to coerce its partners into co-operation. CoCom is not a treaty organization. Its charter is not anchored in international law. All decisions are based on an informal 'gentlemen's agreement'.

And CoCom's members were not behaving like gentlemen.

Ronald Reagan fought and won his election campaign on the technology issue, and the team he took with him to Washington was well suited to the issue: Caspar Weinberger, Fred Iklé and Richard Perle, a trio of hardliners whose first stop was the Defense Depart-ment's technology office. They found it in a state of total chaos. Staffing was inadequate; files were either non-existent or unusable. The defence budget was boosted by $25 million, personnel increased by twenty-five, and a modern data-processing system cal-led Fordtis installed. Secretary Weinberger optimistically ear-marked $2 million for CoCom modernization and generously trans-ferred an additional $30 million from his budget to the Treasury Department to bolster enforcement efforts.

William von Raab, the new Chief of Customs, invested the new money in a special task force. At the time, only four agents were fer-reting out high-tech leaks at the nation's hundreds of airports and harbours. He trained over 400 new recruits and stationed them quickly at US embassies and consulates around the world. 'Opera-tion Exodus' was officially established in October 1981. Within the

## 194 Consequences

first twelve months of its existence, 2,330 illegal shipments were seized, valued at a total of $148.8 million.

Before Reagan's arrival, the US Department of Commerce bore primary responsibility for strategic trade violations. The new Administration transferred major authority to Customs and the Department of Defense. At Commerce, Reagan reinstated Lawrence Brady, who had departed during his dispute with Carter over the Kama River project, promoting him from a deputy department head to assistant secretary. As chief of his enforcement branch Brady chose the young California District Attorney Dr Theodore Wu, who had won national recognition for his handling of the Spawr and Bruchhausen cases.

One of Reagan's first presidential actions was to direct a Central Intelligence Agency investigation into the consequences of the technology haemorrhage. As a result of the devastating CIA report, discussed earlier in this book, a specialized team of intelligence experts was formed at the Langely headquarters under the name 'Technology Transfer Assessment Center'. Its members soon joined Customs and Commerce officials in an inter-agency task force, travelling to the capitals of the Western world to co-ordinate US enforcement activities with those of other governments.

Ronald Reagan hardly intended to limit his technology initiatives to a reshuffling of the Washington bureaucracy. He believed that Carter's problems with the Europeans resulted from a lack of resolve. After a mere six months in office, the new President broached the delicate subject of embargoes at his first summit meeting in Ottawa. He addressed a personal appeal to the leaders of the ten nations gathered there and one tangible result was the high-level meeting of CoCom which followed on 19 January 1982.

For the first time in over twenty-five years, cabinet-level representatives of the CoCom member states assembled in the Rue de La Boetie to discuss new strategic trade initiatives. They were greeted with a barrage of facts and figures from US security experts. With slides and charts, diagrams and satellite photographs Pentagon specialists briefed the assembled ministers on the awesome advances of the Warsaw Pact: SS-18 atomic missiles built with ball-bearing grinders from Vermont; SS-20 launch vehicles designed with IBM computers; nuclear warheads constructed with Western CAD/CAM systems. The subjects ranged from 'A' for aeronautics to 'Z' for ZIL trucks, and included over 160 Eastern weapons systems built with Western technology.

Customs officials elaborated on the complex corporate constructions and covert smuggling routes used by embargo-runners – nearly all passing through Europe – and named names of notorious high-tech pirates who, although exposed and indicted in the United States, were continuing their devious dealings from European soil. Intelligence experts detailed the inner workings of the KGB's Directorate 'T' – with over 20,000 agents rustling electronics manufacturing equipment at outposts all over the world.

'It was,' as one European participant concluded, 'an extremely impressive presentation. We were all simply floored.' The CoCom meeting could have ended as a resounding success. But things didn't turn out that way.

The technology issue, demonstrated so vividly in Paris, was blurred by other American initiatives which coincided with the CoCom meeting. At the same time, the Reagan Administration was pushing vigorously for a reduction in lending to the Soviet bloc which, in the Americans' view, was paying the way for KGB crooks to steal NATO know-how. US financial institutions, like the New York Chase Manhattan Bank, had long since reduced their Eastern holdings. But European banks continued to extend loans to the East. Poland's debts alone had risen from $1.2 billion in 1971 to the dizzying height of $26.5 billion by 1981. Chances were, they would never be able to pay even the interest.

White House efforts on the financial front focused on a mammoth natural-gas project then under negotiation between Moscow and several European countries. The West was to provide capital and technology to build a pipeline connecting the natural gas fields at Urengoy in Siberia with Western Europe. The Soviets were dependent on foreign know-how to tap their vast deep-earth Arctic reserves. But they were not the ones footing the bill. Contracts called for the Soviets to receive hard-currency earnings of between $35 billion and $70 billion. In return, Moscow promised to deliver cut-rate gas to the West Europeans – when construction of the pipeline was completed.

Not only were the profits way out of proportion; the transaction gave Moscow long-term leverage over the West. Once the pipeline was completed, there was no guarantee the Soviets would deliver their part of the bargain. The Europeans, Washington believed, were setting themselves up for blackmail. If the worst came to the worst, Moscow could threaten to cut energy supplies and the political pressure to consent to its demands would be enormous: from housekeepers who feared for their home heating, from bankers who feared for their loans and from industrialists who feared for their

## 196   Consequences

profits. The Soviets had been known to violate contracts on other occasions for reasons of political expediency. There were plenty of precedents: Yugoslavia in 1948, Israel in 1956, Finland in 1958, China in 1964 and Poland in 1981.

Export licences for the US companies participating in the pipeline deal, Caterpillar Tractor and General Electric, were cancelled by the Department of Commerce in December 1981. It would be advisable for the Europeans to follow suit with their companies, Washington warned. Otherwise, the US government might resort to steps of its own to punish Continental companies. It was an ominous economic threat. But President Reagan was firmly convinced that he would be spared the fate of his predecessor. Alliance politics, he believed, was merely a question of resolve.

The Europeans viewed the American announcement as an intolerable infringement of their national sovereignty. Once again, Washington was attacking a trade practice that went to the core of European economic interests. Besides, under European law valid private contracts could not be cancelled at will by the government merely because foreign policy had changed. European leaders would not contemplate violating agreements already signed and sealed with the Soviets. So they balked. Paris: 'In France, the French rule alone.' London: 'I do not believe it was right to do this.' Bonn: 'We will abide by all existing contracts.' Criticism of Washington came from the highest levels of government and was announced publicly.

For Ronald Reagan, things were getting slippery on the diplomatic stage. Six months had passed since the high-level CoCom meeting and no progress was being reported from Paris in high-tech negotiations. Delivery deadlines for the pipeline deal were approaching fast. Right to the finish, the White House hoped the Europeans would give in. To add weight to US demands, a special task force was formed under the leadership of Secretary of State George Shultz. It worked late into the night developing strategies to deal with companies which refused to co-operate with the American initiative. Options were discussed with President Reagan at his West Coast ranch, and approved by him.

On the morning of 27 August 1982, the French freighter *Borodine* heaved anchor and steamed out of the port of Le Havre. On board the ship, bound for Riga in the USSR, were three compressors for the Siberian pipeline in Urengoy. The *Borodine* sailed with the explicit approval of the Paris Government. Within thirty minutes of its departure, US sanctions took effect. Secretary of Commerce

Malcolm Baldrige immediately cancelled the American trading privileges with America enjoyed by two participating companies: Creusot-Loire, the state-owned French equipment manufacturer, and Dresser France, a Paris-based subsidiary of Dresser Industries of Dallas. The US company was ordered to end all technical communications with its French office. In Pittsburg, technicians flipped a switch at Dresser's computer base, cutting data lines between corporate headquarters in Texas and Paris. Terminals at Dresser France suddenly went blank.

For Commerce Secretary Baldrige the step was a 'moderate action'; for Dresser it meant near disaster. The computer connections to Texas main frames were a corporate lifeline. Electronic brains at the company's data base transmitted a torrent of up-to-the-minute information on latest technological innovations, international financial developments, designs and drilling results. Offices and construction sites in over 100 countries of the world were hooked into the Dresser network. Cut off from the rapid flow of vital data, the Paris subsidiary was virtually paralysed. Within days, an Australian account worth $3 million was cancelled.

Bonn was hoping its good relations with Washington would spare it from a similar fate. When West German Foreign Minister Hans-Dietrich Genscher breakfasted with George Shultz in New York on 4 October his optimism was growing. The German freighter *Horst Bischoff* had unloaded pipeline parts from AEG-Kanis in the Soviet port of Klaipeda the day before. The Secretary of State made no mention of it. But after the meeting, when Genscher looked through the *New York Times*, his spirits collapsed. Four German companies were on the US black list. Genscher's response: the Federal Republic of Germany would continue to abide by its contractual obligations.

Italian Prime Minister Spadolini suffered a similar fate when he travelled to Washington for talks two weeks later. Shortly before his arrival, US customs had seized thirty crates containing rotary blades in New York harbour. The shipment, addressed to the Italian company Nuovo Pignone, was also destined for the Siberian pipeline in Urengoy. But Spadolini, too, remained steadfast.

Ultimately, Ronald Reagan was forced to see the futility of American strong-arm tactics. His own efforts to enforce a pipeline embargo had proved just as ineffective as Jimmy Carter's sorry attempt at a grain ban. On 11 November 1982, a year after he launched his anti-pipeline programme, the President surrendered. When West Germany's newly elected (and staunchly pro-Ameri-

## 198   Consequences

can) Chancellor Helmut Kohl arrived in Washington a few days later, the agony was over. The two Western leaders could address other issues in a more relaxed climate. There were pressing political reasons for Reagan's retreat. The stationing of nuclear missiles was about to begin in Europe and anti-Americanism on the Continent was running high. The months ahead would be turbulent enough, and Reagan figured this was no time to be straining relations with loyal allies.

If West Germany's Christian Democratic Chancellor was treated to carrots in Washington, his Social Democratic counterpart in Austria got the stick.

Bruno Kreisky's visit to Washington was preceded by a barrage of interviews in Austrian newspapers. In the Vienna *Presse*, US Assistant of Defense Fred Iklé denounced the complacency of the neutral Alpine republic on the issue of illicit technology transfer. For years, Austria had served as a staging area and sanctuary for Moscow's most notorious high-tech traffickers. Scores of local companies, including a number of renowned corporations, were cooperating in the shady strategic scam. In fact, Iklé declared, over 100 of them had landed on American export denial lists.

While government circles in Vienna were pondering Iklé's statements, wondering which companies he meant, Richard Perle from the Pentagon appeared in the *Presse* with the answer: one of the companies involved was the Gesellschaft für Fertigungstechnik und Maschinenbau (GFM) in Styre. Rotary forges sold to the Soviets by the state-owned Austrian metals-manufacturer were being used to construct armour-piercing smooth-bore cannons for the Red Army. Perle: 'The cannon for the T-72 tank is produced on the Austrian equipment, and the same goes for a number of cannons of the Soviet navy.'

The press campaign came as an embarrassment, to say the very least. Kreisky was planning a good-will tour to the American capital. On his agenda were licences for a joint venture between the Vöst-Alpine steelworks and American Microsystems Inc. The crisis-ridden Austrian corporation planned a $46-million chip-making factory which, Kreisky hoped, could become the cornerstone of a European Silicon Valley in Austria. Export licensing applications, however, had got lost in the Washington bureaucracy. At least, that's what Kreisky thought. He wanted to expedite them.

In Washington, the Austrian leader learned differently. At a tête-à-tête in the White House, Kreisky was told no licences for the Vöst venture would be issued unless Vienna agreed to co-operate with

Western enforcement agencies. A detailed contract proposal had been drawn up. And Kreisky signed. 'We didn't get everything we wanted,' an American official later commented, 'but we got a lot.'

The tactics, described by *Business Week* as 'a little economic blackmail', proved effective. In October 1984 Austria announced it would tighten its export regulations and introduce stiff penalties for violators. They dubbed the law 'an autonomous Austrian solution' to the problem. *Izvestia* accused the US of gross interference in Austrian affairs.

The same squeeze was soon applied to the government of Sweden. Stockholm was reminded that crucial components for Swedish Air Force fighter-interceptors were made in the USA. If deliveries were to continue, co-operation was expected. In April 1984, after years of futile negotiations on the Datasaab case (described in Chapter Ten) the Swedish Foreign Office finally declared it was prepared to pay a civil penalty of $1 million. Datasaab was entered in the export denial lists.

Gentle persuasion of a similar nature was also used in Switzerland, Finland, India, even Hong Kong. In bilateral talks with the United States, a number of CoCom member states also demonstrated good will. European Customs authorities adopted the Pentagon's 'mushroom book' – a layman's guide to the complicated high-tech wares sought by KGB smugglers. Belgium tightened its export regulations, Japan introduced new travel restrictions for Soviet diplomats, and the British special task force 'Project Arrow' was scoring major successes. But the progress achieved in bilateral negotiations paled in comparison to the meagre results coming out of CoCom. New bite for the old dog was still nowhere in sight.

Neighbours near the Rue de La Boetie in Paris had noticed the lights at the CoCom headquarters burning until the small hours. National delegations had been meeting there regularly since October 1982. Washington presented proposals to update 160 items in ageing CoCom embargo lists, adding new technology and deleting those no longer regarded as strategically significant. The Europeans, however, consented only to the deletions. The introduction of new embargo restrictions in the fields of microelectronics and metallurgy, glass fibres and gas turbines continued to meet stiff resistance. An export ban on lasers, which obviously possessed great military potential and had already bolstered Warsaw Pact arsenals, was rejected. The American request for a technical advisory staff for CoCom was denied, as was the request for a new head-

## 200 Consequences

quarters and for modern computer equipment. In fact, even the request for US dollars, earmarked for the Paris organization by Defense Secretary Weinberger in 1981, was turned down. 'Many CoCom members,' a Commerce Department spokesman remarked in disgust, 'are hesitant partners, to put it mildly.'

While European negotiators stymied progress in Paris, European companies were making a killing on Eastern markets, snatching up the lucrative accounts their American competitors had been forced to abandon.

Caterpillar Tractor of Pretoria was forced to drop an $85 million contract for 200 pipe-layers. The profits were mopped up by the Japanese company Komatsu, which took an order for 1,500 similar machines.

At the 1983 spring trade fair in Leipzig, US exhibitors watched enviously from their nearly empty stands as the West Europeans closed contracts with eager Eastern buyers: Siemens of Germany sold microprocessing equipment; CIT-Alcatel of France sold digital lines; the Italians were peddling a new needle-matrix printer; the British offered a new word-processor.

In April 1983, Count Otto Lambsdorff was proclaiming that the Bonn government saw no need for any changes at CoCom. Six months later, the German minister travelled to Moscow with the German-Soviet trade commission, where Leonid Konstandov promised German contracts for large-scale Soviet industrial projects. Both announced their intention to expand bilateral trade. The USSR, Konstandov emphasized, was especially keen on the new technological innovations.

On 3 November 1983 Luxembourg representatives of fifteen Western banks, spearheaded by the Deutsche Bank of Germany, approved new multimillion-dollar loans for the destitute Nadlovy state bank of Poland.

Thus, it was the intelligence service which achieved the first major breakthrough. With the aid of a 36-year-old KGB defector named Vladimir Kuzishkin (who walked into the British Embassy in Tehran in June 1982), deciphering experts managed to crack several KGB secret codes. The incident provided Western eavesdroppers with their first reliable intelligence on the internal workings of Line 'X'. The British were extremely alarmed by what they heard, as were officials in other Western capitals who were told of the results. The French DST report and the murder of Bernard Nut did the rest.

By early 1983, things were moving. In addition to the Soviet spies

– now leaving Western capitals in droves – the discreet travels of US inter-agency teams, combined with mounting publicity on the technology issue, were prompting European enforcement officials to take action. The odyssey of the *Elgarin*, and the high-level attention given to the case in Washington, accelerated efforts. A number of important arrests followed. Werner Bruchhausen, who had been living free in and working with impunity from his German base of operations for years, was finally arrested and charged in Düsseldorf. Volker Nast and three other cohorts of the Richard Müller organization were imprisoned in the Baltic port of Lübeck on 13 December 1983. Megabuck himself, now on the run, found there were several countries where it was better not to show his face. In Britain, the new high-tech task force apprehended Bryan Williamson, who was to face trial for the first time.

But just as counter-intelligence and enforcement agencies began to make genuine progress, security experts were alarmed by another event that was soon causing deep concern and a high-level reassessment of the strategic significance of high-tech thievery. It happened on 3 June 1983 in the Memorial Sloan Kettering Hospital in New York City.

Systems manager Chen Chui spotted the problem immediately. Upon arrival at the hospital's data centre that Friday morning, his machines had informed him of a breakdown the night before. Apparently someone had been tampering with the computers. Five new files had been opened in the electronic brain, permitting access to its secret program. Portions of the memory had been erased. Mr Chui informed Sloan Kettering management. There was reason for alarm. The VAX-11/780 main-frame computer ran five X-ray machines at the New York cancer clinic, as well as answering inquiries from eight other hospitals across the country. An error in the computer program could lead to lethal overdoses of radiation in patient therapy.

After a crisis meeting, Mr Chui returned to the computer room, erased the new files and reprogrammed all access codes. With that, he thought, the problem was solved. But the following morning he discovered to his horror that the uninvited visitors had returned, this time planting a 'spy program' in the computer memory. It requested all users to reveal their secret passwords – a trick, Chen knew, that would enable the intruders to return to the computer's data base at will. They were dangerous. Kettering officials called the police. And the FBI.

By mid-August, the trail had been traced to West Allis, Wiscon-

## 202   Consequences

sin. From the attic of his home, 21-year-old Gerald R. Wondra had tapped into the Manhattan main frame and easily sliced through its layered electronic protection systems. His tool was a simple Apple-II home computer. Wondra belonged to a local group of young electronics enthusiasts who called themselves 'The 414s'. Cracking computer codes was their passion.

Investigators soon discovered that the youngsters had violated computers not only at the New York cancer clinic, but also at a consulting firm in Dallas, a bank in California and a cement company in Canada. Allegedly, they had even succeeded in penetrating non-confidential portions of a program at the Los Alamos National Laboratory – where atomic bombs were designed. They called their hobby 'hacking'. The phenomenon portrayed in the Hollywood thriller *War Games* (in which a teenage youngster cracks a Pentagon code and nearly provokes a Third World War) had been demonstrated in reality, at least on a modest scale.

Hacking immediately captured the public imagination. Dramatic cases of computer crime at financial institutions were already raising considerable doubts about the reliability of expensive data-protection systems. Now neighbourhood hot-shots were penetrating the fringes of national security computers. 'If this is what kids can do on a lark,' Adam Osborne of the Osborne Computer Corporation was soon asking, 'can you imagine what people are doing who are serious about this?'

Dr Willis H. Ware, data security expert at the Rand Corporation, welcomed the media commotion: 'The incident has been very useful in forcing the attention of the Federal Government to the issue.' For years Dr Ware and others had warned of the dangers of computer penetration. They were less worried, however, by the mildly malicious mischief of teenage tinkerers. Their concern was the KGB.

Early in the 1970s, Willis headed a government committee charged with investigating the vulnerability of high-security data systems. Its conclusion: advanced equipment was making computer-tapping easier by the day. Bugs could be attached to terminals or data lines; microwave transmissions intercepted in the air. Magnetic disks and tapes were easy to conceal and easy to steal. Protection programs, it was feared, could malfunction during a power failure, allowing open access to anyone. Computer switches and terminals – even electric typewriters – emitted weak radio-frequency radiation (known in the intelligence world as 'Tempest'), which could be intercepted and deciphered by finely tuned wireless receivers hidden in a nearby building, or parked in a nearby van.

Today, modern main-frame computers have become absolutely essential, monitoring nearly everything in the military that moves – from the supersonic speeds of bullets and ballistic missiles to the tedious tempo of slow ships and spare parts. Computers serve as guardians of the secret reconnaissance of Western spy-satellites, the unseen positions of NATO nuclear submarines or the identities of Western spies who have penetrated deep into Eastern establishments. They contain emergency plans for a nuclear alert, connect strategic bombers with one another and evaluate top-of-the-line weapons systems long before they are ever deployed. In short: computers know all the secrets that a superpower possesses.

The electronic share of total military intelligence is exploding. The Pentagon alone operates over 8,000 main-frame computers. Countless others are plugged into its international communications networks. Increasingly, top-secret information must be rapidly relayed from listening posts, reconnaissance satellites or military installations to evaluation centres at home, thus multiplying the opportunities for clandestine tapping. Sensitive magnetic tapes and disks, formerly locked away at night in high-security vaults, are now in constant use, never leaving their computers. There, of course, they are more vulnerable than ever. A computer is easier to crack than a safe.

In order to pinpoint potential soft spots in this network, 'Tiger Teams' were formed at the Pentagon and elsewhere – élite groups of specialists assigned to crack their country's computer codes. To the horror of their masters, they succeeded every time. Again and again, computers were reprogrammed or even completely redesigned, only to be penetrated once more. The Tigers tunnelled with relative impunity through layered software safeguards into the secret heart of strategic computer programs.

The tricks they played were simply fantastic. If an initial entry proved difficult, a 'trap door' was planted in the program which would flop open for the thief when he approached a second time. Tiger Teams revealed the darker side of the electronic revolution. 'They kept finding holes in what was already a Swiss cheese,' one official concluded. Particular concern was raised by their demonstrated ability to burgle a computer without leaving any telltale electronic traces. The National Security Agency had spent over $100 million to safeguard the top-of-the-line computer fleet at its cellar installation in Fort Meade, Maryland. Now, no one could guarantee that it had not already been compromised.

Computers, so vital to the defence of the West, had become one

## 204  Consequences

of its biggest security problems. Yet the most chilling conclusion was still to come.

In the world of electronic espionage and counter-espionage the Tiger Teams not only demonstrated that a trained expert was capable of cracking the most sophisticated of protection schemes – and living a superspy's dream inside the central nervous system of the NATO defence network; they were soon also exploring the menacing potential of software sabotage. Special programs, inserted clandestinely in the innermost data base of a strategic computer, could drive the entire system haywire.

A software saboteur could plant a program that would not be visible to even the most schooled eye. It could be designed to go unnoticed in routine situations. Triggered by a pre-set radio command or geographic parameter, the 'logic bomb' would be activated only in a real war, garbling communications, fogging radar or confusing navigational systems. Remote controls operated via satellite or internal timers accurate to a microsecond could also be used to switch them on. It was a military man's nightmare. Pentagon experts envisioned early-warning radar that refused to warn; helicopters and jet bombers that fell from the skies without apparent rhyme or reason; or ship-to-ship missiles that returned to sink the ship that had launched them.

Unlike hobby-hacking, serious military sabotage is no business for loners. A perpetrator would require a comprehensive knowledge of the system and its peripherals not available to single individuals, even those with high-security clearance. Equipped with the prepared program of an enemy intelligence service, however, a single agent – perhaps a trusted employee with an ID card dangling from his neck and a friendly wave for the security guard – could wreak absolute havoc.

The ugliest scenario is painted by Colonel Roger Shell of the National Security Agency: the re-targeting of nuclear missiles. A skilfully designed sabotage program, carefully concealed in the tangled bits and bytes of sophisticated software, could be programmed to ignore routine testing or electronic war-gaming, and thus go undetected for years in its quiet little corner of a missile guidance system. Not until the radio signals of a real alert reached it would the logic bomb go into action, commanding an atomic warhead to detonate in its home-town silo or reverse a trajectory, guiding a renegade rocket unerringly into Western targets in Washington, Paris, Hamburg or London. It is the ultimate threat of Moscow's computer spies: a lethal atomic boomerang soaring back to annihilate Western cities.

The Soviet military has neither the hardware nor the software capability to produce such systems. But there can be little doubt they are doing their very best to steal them.

## Source Notes

Bernard Nut: 'Real Goal', *US News and World Report*, 18/4/1983; *Chicago Tribune*, 10/4/1983; *New York Times*, 6/4/1983; 'Diplomatic Expulsions', *Guardian*, 17/4/1983.

Mass expulsions: *US News and World Report*, 9/5/1983; *Long Island Newsday*, 20/4/1983; *New York Times*, 7/4/1983, 24/4/1983 and 24/7/1983; *Christian Science Monitor*, 25/4/1983; UPI, 8/4/1983 and 24/4/1983; AP, 7/1/1983, 18/5/1983 and 23/12/1982; Reuters, 5/4/1983; *Frankfurter Allgemeine Zeitung*, 23/4/1983; *Financial Times*, 31/8/1983; *ARD Tagesthemen*, German television, 2/2/1984; Erwin Brunner, 'Dossier', *Die Zeit*, 3/2/1984.

CoCom (background): Rolf Hasse, *Theorie und Politik des Embargos*, Institut für Wirtschaftspolitik, University of Cologne, 1973, pp. 195-202; Ellen Frost and Angela Stent, 'NATO's Troubles with East-West Trade', *International Security*, summer 1983; *Frankfurter Allgemeine Zeitung*, 28/5/1983; *Frankfurter Allgemeine Zeitung*, 20/11/1982; *The Times*, 14/12/1983.

Grain embargo: *Der Spiegel*, 18/2/1980; *Baltimore Sun*, 6/3/1983; *Frankfurter Allgemeine Zeitung*, 9/2/1983; *Neue Züricher Zeitung*, 5/3/1982.

Quote from Assilin: US Senate Subcommittee on Investigations, Washington DC 1982, p. 92.

Quote from Jimmy Carter: *New York Times*, 5/1/1980.

Quote from Medvedev: 'Ein Frost so Kühl', *Der Spiegel*, March 1980.

Andropov's *Spiegel* Interview: *Der Spiegel*, 25/4/1983, p. 130.

Lambsdorff's explosion: Eyewitness interview with the author.

Armco: *Baltimore Sun*, 5/3/1982.

Pipeline embargo: *Frankfurter Allgemeine Zeitung*, 12/1/1982, 16/7/1982, 25/8/1982, 15/11/1982, 13/3/1983, 28/5/1983, and 28/8/1983; *New York Times*, 13/3/1983; *Der Spiegel*, 11/10/1982.

Dresser Industries: *Baltimore Sun*, 6/3/1983; *New York Times*, 13/3/1983.

Kreisky and the cannons: *Der Spiegel*, 3/1/1983; *Business Week*, 4/4/1983.

Sweden: *Die Welt*, 10/4/1984.

Operation Exodus: *Newsday*, 21/2/1983.

London: *Guardian*, 9/9/1983.

Hacker, Tiger Teams and software sabotage: Background to this section came from NATO high officials and other sources who offered information only on the condition they remain anonymous. The problem, however, has been discussed in public. See: *Newsweek*, 29/8/1983; *New York Times*, 25/9/1983; J. Goldstein, 'Technology Transfer from the Defence Perspective', *Signal*, August 1983; Tad Szulc, *Current News*, 5/11/1982; Tad Szulc, 'To Steal Our Secrets', *Parade*, 7/11/1982; Lawrence Meyer, 'Hey Ivan, say "Cheese"', *Washington Post Magazine*, 4/12/1983.

## 206   Consequences

Colonel Shell quote: 'Pentagon Computers: How vulnerable?' *US News and Report*, 31/10/1983; *USA Today*, 25/8/1983.

Excerpt from Tamir Agmon and Mary Ann Von Glinow (eds), *Technology Transfer in International Business*, 223–39.

# 12

# Pros and Cons of International Technology Transfer: A Developing Country's View

LINSU KIM

Recently, international technology transfer has been a subject of debate in both developed countries (DCs) and less developed countries (LDCs). Many argue that international technology transfer from DCs to LDCs should be promoted because it is mutually beneficial. They claim that international technology transfer to LDCs allows enterprises in DCs (1) to prolong the life cycle of products that are becoming obsolete in the home market, (2) to find new, growing markets, and (3) to ensure its own survival by relocating production segments to LDCs where labor costs are lower. Likewise, international technology transfer from DCs benefits enterprises in LDCs by enabling them to generate and improve products and processes and to gain export markets.

On the other hand, others argue that international technology transfer should be restricted to protect national interests; that its long-term effect is rather negative. Opponents in DCs argue that technology transfer to LDCs has "boomerang" effects, ultimately damaging their own industries and employment. Those opponents in LDCs argue that technology transfer results in an economic and technological dependence of LDCs on DCs.

This chapter assesses the pros and cons associated with international technology transfer from an LDC perspective. It presents a conceptual framework that may be useful to identify and assess different forms and channels of international technology transfer. With Korea as the case in point, it examines how an LDC acquires foreign technology from DCs and discusses implications for both DCs and LDCs.

## A CONCEPTUAL FRAMEWORK TO ASSESS TECHNOLOGY TRANSFER

Foreign technology transfer may be mediated by the market; the supplier and the buyer may negotiate payment for technology transfer, either em-

224     The Practice of International Technology Transfer in the Pacific Rim

**Table 12.1**   The Mode of Foreign Technology Transfer

| | | |
|---|---|---|
| Market mediated | Direct foreign investment, foreign licensing, turn-key plant, technical consultancy, made-to-order machinery<br>(Cell 1) | Standard (serial) machinery purchase[a]<br>(Cell 2) |
| Nonmarket mediated | Technical assistance by foreign buyers, technical assistance by foreign vendors[b]<br>(Cell 4)<br>Active | Imitation (reverse engineering) observation, trade journals, technical information service<br>(Cell 3)<br>Passive |

The Role of Foreign Suppliers

[a]Except for small, standard machinery, foreign suppliers send their engineers to assemble and test-run machinery sold. Often, they teach local personnel how to operate it and provide after-sale services. In this sense, the supplier's role is not passive, but compared with those mechanisms in cell 1, this mechanism can still be classified in cell 2.

[b]The vendor's service mentioned here refers to technical assistance not directly related to the operation of machinery sold; rather, the suppliers provide technical information and consultancy on operations not related to the machinery sold in exchange to a long-term purchase agreement.

bodied in or disembodied from the physical equipment. Foreign technology may also be transferred to local users without the mediation of the market. In this case the technology transfer usually takes place without formal agreements and payments. The foreign supplier can take either an active role, exercising significant control over the way in which technology is transferred to and used by the local recipient, or a passive role, having almost nothing to do with the way the user takes advantage of available technical know-how either embodied in or disembodied from the physical item. The recipient may be a producer, attempting to generate new products for import-substitution, or a user, deploying foreign technologies to improve productivity and product quality. These three variables—the mediation of the market, the role of foreign suppliers, and the role of recipients—offer a useful eight-cell matrix (Table 12.1) to identify and evaluate different mechanisms of international technology transfer and to assess its pros and cons.

Most studies of technology transfer have devoted their attention to the examination of those modes in cell 1 (market mediated with active suppliers). These include technology transfer accompanied by direct foreign investment (DFI), foreign licensing (FL), technical consultancy (TC), and made-to-order machinery. Recipients of foreign technology in this cell are mostly producers, with only a few exceptions.

Technology transfer in cell 2 is market mediated, but foreign suppliers play a relatively passive role, not exercising much control over the way in which technical know-how embodied in physical items is transferred to and used by the buyer. The purchase of standard serial machinery that embodies new technologies serves as the major means of transfer in cell 2. Recipients of foreign technology transfer in this cell are mostly users.

Technology transfers that take place in cell 3 are nonmarket mediated, in which the foreign suppliers play a relatively passive role. Reverse engineering done by local firms for their product and process development, reverse engineering by public research and development (R&D) centers for localization of technology beyond the capacity of local firms, and technical information services provided by various public agencies are good examples. Both producers and users could be in this cell.

In cell 4, technology transfer per se is not negotiated and priced in the market, but the foreign partners involved play a relatively active role in transferring technological know-how. There are at least two distinct cases. First, foreign buyers of locally produced goods under original equipment maker (OEM) arrangements deliberately provide technical know-how to ensure that local firms manufacture products according to the end buyer's stringent specifications. Second, foreign vendors of components and equipment provide local buyers with important technical services related to a system, for which they sell their products.

In conclusion, besides such formal mechanisms as direct foreign investment, foreign licensing, and technical consultancy, which have been studied at great length in the literature, technology may also be transferred across national boundaries through the various informal mechanisms outlined above. A discussion of how an LDC uses these mechanisms follows using Korea as the case in point.

## TECHNOLOGY TRANSFER IN PRACTICE: THE CASE OF KOREA

Korea is one of many LDCs that has developed comprehensive policies and measures to promote, enhance, and manage technology flows from DCs (Baranson and Roark, 1985). First, this section presents a brief discussion of Korea's public policies related to foreign technology transfer. Then Korea's experience in technology transfer under this public-policy environment is discussed. The Korea case illuminates further discussion of the pros and cons of international technology transfer.

### Technology Transfer in Cell 1

*Direct Foreign Investment*
The Korean government took initial steps to organize national policies for technology transfer in the 1960s. Korea's direct foreign investment (DFI) policy was quite free at that time, allowing any form of bona fide foreign capital, including fully owned subsidiaries, with extensive incentives. However, not much foreign investment came in during the 1960s, primarily due to questions about Korea's political stability and the uncertain economic outlook.

The government reversed its DFI policy in the 1970s, tightening controls, fearing that unlimited flow of foreign investment might adversely affect the domestic economy. Joint ventures received higher priority than wholly owned subsidiaries. A general guideline was adopted setting three criteria: eligibility, foreign ownership, and investment scale. Therefore, competition with domestic firms was seldom allowed in both domestic and international markets. Export requirements were forced on DFIs. Foreign participation ratios were basically limited to 50 percent, except for high technology and entirely export-oriented cases, rendering Korea one of a few countries with very restrictive foreign investment regulations.

In September 1980, the Korean government reversed its position, substantially liberalizing foreign investment guidelines. Another important reform was made effective from July 1984. The major change was a switch from the system of a "positive list" (listing which industries are open to DFI) to that of a "negative list" (listing those that are prohibited). The government's open policy for DFI is aimed at inducing the transfer of sophisticated new technologies and at promoting market competition for domestic firms, even within the domestic market, to intensify their innovation activities.

Under such a public policy environment, Korea has drawn an increasing amount of DFI through 1986 (Table 12.2), excluding the 1977–1981 period, which reflects an economic recession in Korea stemming partly from domestic political and social instability and partly from the global economic recession. Over 48 percent of the total DFI took place in the last period (1982–1986). Japan accounts for over 68 percent in the number of cases and 55 percent of the value, followed by the U.S. (20 and 26 percent, respectively). However, the average size of the American DFI is more than twice that of the Japanese. Chemicals, machinery, and electrical/electronics account for over 72 percent of the total DFI in the manufacturing sector and over 46 percent of the total DFI in Korea. Its proportion has steadily increased over time, as Korea has undergone a structural change from labor-intensive industries to relatively more knowledge-intensive industries.

The size of DFI and its proportion to total external borrowing are significantly lower in Korea compared with other newly industrializing countries (NICs) such as Taiwan, Singapore, and Brazil. One report (KEB, 1987) shows that Korea's stock of DFI in 1983 was only 7 percent of that in Brazil, 23 percent of that in Singapore, and less than half the size of that in Taiwan and Hong Kong. The proportion of DFI in Korea to the country's total external borrowing is only 6.1 percent, far below the levels of Singapore (91.9 percent), Taiwan (45 percent), and Brazil (21.8 percent). This figure appears to be the outcome of the nation's DFI strategy, under which Korea favored loans as a source of foreign capital and relied on other mechanisms to gain foreign technology.

Unlike other NICs, the contribution of DFI to economic development has not been significant in Korea. A recent study (Cha, 1983) shows that DFI's contribution to the growth of the gross national product (GNP) in Korea in the period from 1972 to 1980 amounted to only 1.3 percent, while its contribution to total and manufacturing value added was only

Pros and Cons of International Technology Transfer                    227

**Table 12.2**   Foreign Technology Transfer to Korea

| Source | 1962–1966 | 1967–1971 | 1972–1976 | 1977–1981 | (unit: U.S.$ million)<br>1982–1986 | Total |
|---|---|---|---|---|---|---|
| Japan | 8.3 | 89.7 | 627.1 | 300.9 | 875.2 | 1,901.2 |
| U.S. | 25.0 | 95.3 | 135.0 | 235.7 | 581.6 | 1,072.6 |
| Others | 12.1 | 33.6 | 117.3 | 184.0 | 309.7 | 658.7 |
| Total | 47.4 | 218.6 | 879.4 | 720.6 | 1,766.5 | 3,632.5 |
| **Foreign licensing** | | | | | | |
| Japan | — | 5.0 | 58.7 | 139.8 | 323.7 | 527.2 |
| U.S. | 0.6 | 7.8 | 21.3 | 159.2 | 602.7 | 791.6 |
| Others | 0.2 | 3.5 | 16.6 | 152.4 | 258.5 | 431.1 |
| Total | 0.8 | 16.3 | 96.5 | 451.4 | 1,184.9 | 1,749.9 |
| **Technical Consultancy** | | | | | | |
| Japan | — | 12.1 | 7.7 | 20.8 | 89.2 | 129.8 |
| U.S. | — | 3.1 | 6.0 | 16.7 | 159.1 | 184.9 |
| Others | — | 1.6 | 4.8 | 17.2 | 84.0 | 107.6 |
| Total | — | 16.8 | 18.5 | 54.7 | 332.3 | 422.3 |
| **Capital Goods Imports** | | | | | | |
| Japan | 148 | 1,292 | 4,423 | 14,269 | 20,986 | 41,118 |
| U.S. | 75 | 472 | 1,973 | 6,219 | 12,394 | 21,133 |
| Others | 93 | 777 | 2,445 | 7,490 | 53,338 | 64,143 |
| Total | 316 | 2,541 | 8,841 | 27,978 | 86,718 | 126,394 |

Sources: Ministry of Finance for data on direct foreign investment and foreign licensing, Ministry of Science and Technology for technical consultancy data, and Korea Society for Advancement of Machinery Industry for capital goods data.

1.1 percent and 4.8 percent, respectively, in 1971, and 4.5 percent and 14.2 percent, respectively, in 1980. Its contribution to employment was only 0.2 percent and 1.5 percent, respectively, for 1971 and 1980.

*Foreign Licensing*
Korea's policy on foreign licensing (FL) was quite restrictive in the 1960s. General guidelines stipulated that royalties should be set within 3 percent of sales and contract duration within 3 years for the manufacturing sector, and that no export restrictions were allowed. In the early years, such a restrictive policy on foreign licensings led some local licensees to bargain in purchasing generally known mature technologies. The restrictive guidelines were significantly relaxed in the 1970s to allow the transfer of more sophisticated technologies. The policy was further relaxed in 1984, opening for all industries and for all terms and conditions. The government also used public R&D centers to assist the private sector in identifying particular foreign technologies and their suppliers and in strengthening bargaining power in transfer negotiations.

Table 12.2 also presents basic statistics of FL to Korea. It shows that, as with DFI, over 67 percent of FL was signed in the last period (1982–1986) and that the United States leads Japan in the receipt of royalty pay-

228    The Practice of International Technology Transfer in the Pacific Rim

ments from the late 1970s, despite half the number of cases. These numbers indicate that the average royalty per case is far greater for the United States than for Japan, reflecting a significantly higher degree of sophistication in technology licensed from the United States. Kim's (1984) microeconomic study of 70 Korea firms confirms that American technologies are more capital-intensive, sophisticated, and complex than Japanese technologies.

*Technical Consultancy*
Table 12.2 shows that compared with other modes of technology transfer, technical consultancy is not significant in terms of value. The last period (1982–1986) accounts for over 78 percent of the total, and the United States leads Japan in supplying engineering consultancy services to Korea. The U. S. lead is attributed to the sharp increase of its share in the last period.

*Turnkey Plants and Machinery*
While Korea was restrictive on DFI and FL through the 1970s, it relied heavily on turnkey plants and made-to-order machinery for technology transfer. Firms in the process-based industries established in the 1960s and early 1970s, such as chemicals, cement, steel, and paper, resorted to the turnkey mode. In these industries, investment size was relatively large and proprietary engineering know-how was critical, although equipment was generally known. For these reasons, the best alternative for local firms without engineering capabilities was to rely completely on experienced foreign firms, to minimize the risks involved in a large investment and the time required to achieve normal operations. However, no quantifiable data is available to assess the quantity of technology transfer associated with turnkey plants' transfers.

**Technology Transfer in Cell 2**

The Korean government played a crucial role in transferring foreign technology embodied in standard serial machinery. The rapid growth of the Korean economy has called for paramount investment in production facilities. For example, the index of gross domestic fixed capital formation has risen rapidly from 50.7 in 1953 to 662.7 in 1970 and to 3,546.2 in 1984, with 1960 set equal to 100. Capital formation as a percentage of gross national product has risen from 5.5 percent to 25.3 percent to 33.4 percent during the same period. However, government policy has been biased in favor of foreign capital goods as a way to strengthen the international competitiveness of capital goods user industries. Such a policy led to massive imports of foreign capital goods, transferring new foreign technologies embodied therein, at the cost of retarding the development of the local capital goods industry (Kim, 1987a). The import liberalization rate in the machinery industry was relatively high until the first half of 1971, giving capital-goods users almost free access to foreign capital goods. Moreover, the slight overvaluation of the local currency, the tariff exemptions on imported capital goods, and the financing of purchases by supplier's credits, which carried low interest rates relative to those on the domestic market, all worked to increase the attractiveness of capital goods imports (Rhee and Westphal, 1977).

Table 12.2 also present data on capital goods imports. Japan leads the United States and the other countries in supplying capital goods to Korea. The U.S. proportion has steadily increased over time, but its share remained about half of the Japanese share in the whole period. Furthermore, of the four categories of technology transfer listed in Table 12.2, capital goods imports far surpass other means of technology transfers in terms of value. Capital goods imports were worth 34 times the value of DFI, 72 times the value of FL, and almost 300 times the value of technical consultancies. In sum, the total value of capital goods imports is 21 times that of all other categories combined. The trend is similar for individual source countries. For example, the total value of capital goods imports from the United States is 10.3 times that of all other categories combined from the United States. Thus, Korea may have acquired more technology from DCs through the importation of capital goods than through any other means. Compared to NICs, such as Argentina, Brazil, India, and Mexico, Korea has relied more on the import of capital goods for foreign technology transfer than on DFI and FL (Westphal et al., 1985).

## Technology Transfer in Cell 3

Technology transfers that take place in cell 3 are nonmarket mediated, in which foreign suppliers play a relatively passive role. Reverse engineering done by local firms for their product and process development is the most important mode in this cell; both producers and users take advantage of this mode. Reverse engineering by public R&D centers for localization of technology beyond the capacity of local firms and technical information services provided by various public agencies may also be good examples. Since technology transfers in this cell cannot be quantified, one has to be satisfied with fragmented pieces of evidence from empirical studies and other anecdotes.

### Reverse Engineering by Private Firms

The Korean government also has measures to facilitate the transfer of foreign technologies through the mode in cell 3. For example, in 1979 the Korean government introduced a scheme to designate locally developed innovative products and offered preferential treatment, tax incentives, and 2-year protection from entry by foreign technology or local imitation as a way to foster local development efforts. In the case of the machinery industry, preferential financing is offered even to the purchaser of such products.

There appear to be four types of firms that undertake reverse engineering. The first type comprises producers. Kim's (1980) study of consumer electronics firms reveals that of 15 black/white TV assemblers in 1975, 11 entered the industry by reverse engineering done by experienced engineers poached from existing firms. A study of 42 innovative capital goods producers shows that these firms primarily deployed the forms and channels to technology transfer in cell 3 (see Kim and Kim, 1985). Bae and Lee's (1986) study of small and medium-sized firms in Korea also shows that reverse engineering was the major means of product development for these firms. A recent study on the transfer of factory automation (F) to Korea shows that the majority of 28 firms doing business in CAD/CAM, 11 in automatic warehousing, and 22 in robotics, used reverse engineering as a way to get FA technology (Kim, 1988a).

The second type of firm is users. As in von Hippel's (1976) concept of user-paradigm, FA users first developed equipment related to FA in Korea. As mentioned previously, they imported the first batch of FA equipment to improve their productivity, but they soon developed FA equipment by assimilating imported technology on the basis of their experience in using foreign equipment and their own R&D efforts to meet subsequent needs and needs within their *chaebol* (Korean version of Japanese zaibatsu) group. An electronics firm has reportedly developed 2,500 units of 45 different pieces of FA equipment ranging from part-inserting robotics and automatic printed circuit board inspectors to automatic wrapping and unmanned vehicles. Many chemical, cement, paper, and steel makers studied by Kim (1987b) largely resorted to turnkey plants for the initial setup, but used reverse engineering for subsequent expansions. All of these firms have progressively sequenced from the importation of foreign equipment and systems to the imitative production of their own models for import-substitution. These firms later became equipment producers to capitalize on their capabilities, as the local market slowly took shape.

The third type includes firms that entered their manufacturing business first as sales and service agents for foreign exporters. Such experience provided local agents with opportunities to assimilate foreign technologies. For example, most of the 28 firms in the CAD/CAM business started as local sales/after-service agents for foreign suppliers. Many of them are now investing for their own R&D for assimilation and import-substitution of foreign systems. A similar pattern is also observed in Korea's entry into the computer industry (Kim et al., 1987).

The fourth type of firm includes new technology-based small producers that are started by technical entrepreneurs who have recently spun-off from local universities, R&D centers, or existing firms. Given the initial capability in innovation, they began developing innovative equipment and systems through reverse engineering of sophisticated foreign products (see Kim, 1988b).

## Reverse Engineering by Public R&D Centers

Public R&D centers, 16 in number in 1981 and then reduced to 8 by mergers, have played a major role in developing new technologies through reverse engineering. Public R&D centers undertake joint research with the private sector to help local firms acquire new technology without collaborating with foreign sources or to help them gain better bargaining power in technology transfer negotiations. An example of the former would be the Korean reaction to the Japanese refusal to license technology for polyester film production to Korea, in fear of losing its market in Korea; a public R&D center in collaboration with a local chemical firm successfully undertook a reverse engineering task to develop the technology. Korea now is the world's major supplier for audio and video cassette tapes.

Public R&D centers also play an important role in developing new technologies that have large economic externalities. For example, the Korean government invested about $300 million in developing design and production capabilities for semiconductors and computers. The government established a public R&D institute several years before the private sector's entry into computers and semiconductors in an attempt to gain first-stage experience in R&D in the new technologies and to generate experienced researchers and production workers. This R&D center has been the spearhead and backbone of R&D in computer and semiconductor design.

In short, the major role of public R&D centers in a developing country, as discussed by Utterback (1975), is to facilitate and lubricate foreign technology transfer by assisting in the private sector's acquisition of foreign technology, formally or informally; to solve immediate or short-term problems; and to undertake their own R&D activities to pave the road for the private sector to have an advantageous position at a later point in time in acquiring newer technologies.

## Technical Information Service

The government also exerts efforts to facilitate the international transfer of scientific and technical (S&T) knowledge through its technical information services provided by various public agencies. Through several monthly periodicals and on-line computer systems, these agencies collect, process, and disseminate S&T information on a nonprofit basis.

## Technology Transfer in Cell 4

Technology transfer takes place in cell 4 where technology transfer per se is not negotiated and priced in the market, but foreign partners play a relatively active role in transferring technological know-how. Like those in cell 3, qualitative assessments have to be made, because no quantifiable information is available for the modes in cell 4.

There are at least two distinct cases. First, foreign buyers of locally produced goods under OEM arrangements deliberately provide technical knowhow to ensure that locally manufactured products meet buyers' stringent

232        The Practice of International Technology Transfer in the Pacific Rim

specifications. Second, foreign vendors of components and equipment provide local buyers with important technical services indirectly related to a system, for which they sell their products. The first mode of technology transfer, which was mentioned repeatedly by local firms in field interviews (e.g., Kim, 1987b; Westphal et al., 1981; Kim, 1980) is a by-product of the export-oriented industrialization strategy in Korea. Therefore, public policy for export promotion has had an important impact on foreign technology transfer.

In conclusion, the government has played an important role in managing and facilitating the transfer of foreign technology by regulating collaborative agreements concerning direct foreign investment and foreign licensing and by providing incentives and preferential financing to those who acquire foreign technologies through means other than collaborative agreements. The government has also contributed significantly to technology transfer through the development of an R&D infrastructure such as public R&D institutes and technical information centers, which have played an important role in helping local firms acquire better bargaining power in technology transfer negotiations and also in the reverse engineering of foreign technologies.

## IMPLICATIONS FOR DEVELOPED COUNTRIES

The conceptual framework and empirical evidence in Korea lead to several questions that DCs need to address to gain a better understanding of the pros and cons of technology transfer to LDCs. Should DCs try to stop transferring technology to LDCs in fear of boomerang effects? If yes, then can DCs stop transferring technology to LDCs? The evidence presented above suggests that the answer to both questions is "no."

Any attempt to restrict technology transfer to LDCs through the modes in cell 1 will jeopardize economic growth and yet cannot prevent LDCs from obtaining technology. The restriction of DFI to LDCs will jeopardize the global strategy of multinational firms in DCs, while that of FL will shorten the economic life cycle of their technologies and products. If one DC refuses to transfer a technology, a sophisticated technology buyer in the LDCs can turn to an alternative source. Examples abound. When the Japanese refused to license peroxide technology to Korea in fear of losing its market in the country (Korea imported 100 percent of its peroxide from Japan), Koreans turned to America, who was willing to license the technology to enter the new market. When large semiconductor firms in Japan and the United States were reluctant to transfer semiconductor technology to Korea, Korean firms were able to negotiate a dozen licensing agreements with small semiconductor firms in the United States between 1983 and 1985. These small firms were willing to sell the technology for a quick infusion of cash during the business slump in semiconductors (Kim, 1987c).

Alternatively, a sophisticated technology buyer can turn to another mode of technology transfer. Examples are again plentiful. When the Japanese refused to license video cassette recorder (VCR) technology to Korea, Korean firms employed reverse engineering (the mode in cell 3) of Japanese products (the mode in cell 2), augmented by technical consultancy (a mode in cell 1) provided by foreign engineers recruited on a short-term basis by the Korean firms. Now Korea challenges Japan in the world market.

DCs cannot stop transferring technology to LDCs through the mode in cell 2, because the transfer of standardized machinery is merely a pure form of commodity trade. Any restriction on trade will hamper economic growth of the individual firms and, in turn, of the entire nation. The empirical data (the value of Korea's imports of capital goods was 21 times those of DFI, FL, and TC combined) and impressions gained in field interviews with Korean firms suggest that Koreans appear to have learned more from imported capital goods than from other modes of technology transfer. Kim and Kim's (1985) study offers useful insights as to how Koreans innovate by taking advantage of locally available foreign products.

DCs can do little to stop technology transfer through the modes in cell 3. Rather, it is the outcome of the absorptive capacity and entrepreneurship of firms and R&D centers in LDCs that take advantage of these modes. Foreign technologies transferred through other means serve as important inputs for reverse engineering. The most innovative and profitable papermaker in Korea, for example, never entered collaborative agreements with foreign partners in the form of either DFI or FL but progressively reverse-engineered foreign models in developing the most sophisticated, computerized paper-making line (Kim, 1987b).

Technology transfer modes in cell 4 are also important means of economic activity for multinationals from DCs. OEM manufacturing in LDCs is a strategy of international sourcing of components and end-products by firms in DCs to sustain price competitiveness in both domestic and international markets. Informal technical assistance provided by foreign component suppliers is simply an extra service needed to conclude sales and to enable successful implementation of their components by the buyer.

None of the technology transfer modes in the four cells can be restricted by DCs to reduce the boomerang effects from LDCs. All modes, except for those in cell 3, about which DCs can do little, are, in fact, vital economic means for the growth of firms in DCs.

A second question arises: If technology transfer to LDCs cannot and should not be restricted, then does technology transfer from DCs to LDCs directly result in boomerang effects in DCs? Data show otherwise. Figure 12.1 presents a market-mediated total technology transfer to Korea in dollar value from Japan and the United States, together with the two sets of trade data between Korea and these two countries. The figure also shows an inverse relation between technology transfer to and exports from Korea. The total value of technology transfer from Japan is almost double that from the United States. Korea's exports to the United States, however, are almost double those to Japan. Furthermore, in 1986, for example, the

234        The Practice of International Technology Transfer in the Pacific Rim

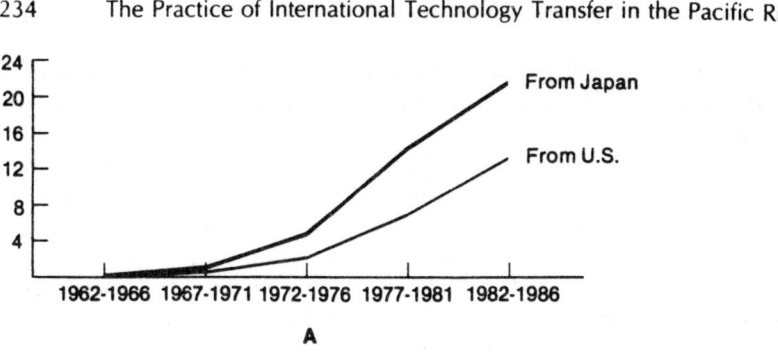

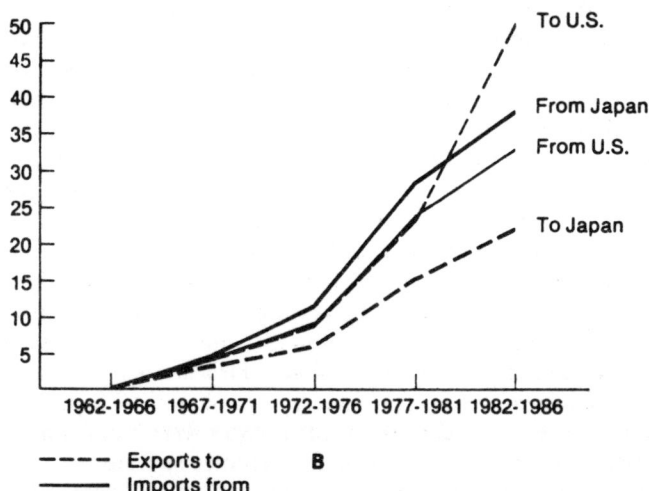

**Figure 12.1**  Technology and commodity trade between Korea and its two leading partners (1962–1986).

trade imbalance with the United States was almost $7 billion in favor of Korea, whereas that with Japan is $5.5 billion in favor of Japan. That is, Korea obtained foreign technology primarily from Japan but exported its products mainly to the United States. Therefore, Korea's export performance to the United States is not the direct outcome of the U. S. technology transfer to Korea. An Organization for Economic Cooperation and Development (OECD) study (1981) also concluded that trade competition came from countries that were not the biggest customers for technology and capital goods. Then, what is the source of the trade competitiveness of the fast growing LDCs? The international competitiveness of LDCs vis-à-vis a particular DC stemmed not so much from that DC's technology transfer to the country as from its absorptive capacity to assimilate, adapt, and improve foreign technologies. The internal weaknesses of DCs in industrial structure and innovation also account for the rapid export expansion of LDCs. Japan's postwar industrial history and the recent growth of Asian NICs (e.g., Taiwan, Korea, Hong Kong, and Singapore) demonstrate this point.

The third question emerges: What then is the net effect of the technology transfer of DCs to LDCs? Technology transfer definitely benefits recipient countries because it can be a vital source of technological change, leading to productivity improvement or to new products, processes and industries (see Kim, 1980). Technology transfer also benefits supplier countries. The OECD study (1981) concluded that the net return from technology transfers by DCs has been positive overall in terms of trade, employment, and consumer benefits in DCs. However, gross effects have fallen unequally on different industries. Engineering and machinery industries have benefited from increasing demand in LDCs during the period of low demand in other DCs. Some light manufacturing industries, however, had their problems compounded by the increased competition from LDCs. Kim's (1985) study of the United States–Korea link in technology transfer also concluded that although technology transfers from the United States to Korea have been accompanied by an increase in the volume of light commodity exports from Korea, its capital goods imports from the United States have increased dramatically to $6.2 billion in the 1977–81 period and doubled again to $12.4 billion in the 1982–1986 period. One might argue that the United States also faces increasing capital goods imports from Korea. American multinational corporations, however, comprise a significant part of the machinery industry in Korea. Helleiner (1979) shows that 64.2 percent of Korean general machinery exports and 67.3 percent of its electrical machinery exports to the United States were from American multinational corporations with production facilities in Korea.

Finally, what should DCs do to maintain technological leadership? No nation can maintain technological leadership through stringent controls of technology outflows: Witness the United Kingdom in the nineteenth century. Transfers are likely to continue at a significant rate, because on the supply side alternative sources of technology are increasing and the firms in possession of the technology may have to transfer it, as discussed previously, to expand sales and to extend the economic life of their technologies and products, maximizing the return from the technologies they possess. On the demand side, industries in some LDCs have developed increasing capabilities to master received technologies, to exert strong bargaining power in acquiring foreign technologies, and to undertake R&D to create their own innovations. Only through continuous innovation can DCs maintain their current position in technology leadership.

## IMPLICATIONS FOR LESS DEVELOPED COUNTRIES

The conceptual framework and Korea's experience as presented above also offer useful implications for other LDCs that attempt to maximize benefits from technologies available in DCs. Three issues may be addressed in discussing the implications. The first issue is related to market-mediated technology transfers in cells 1 and 2. That is, LDCs should not view technology transfer from DCs to be a source of their foreign dependency and for this reason restrict it. The experiences of LDCs, like India, that adopted restrictive policies on foreign technology transfer and that stuck to policies of self-reliance, like the People's Republic of China in the 1960s and 1970s, have demonstrated that the preclusion of foreign technology can retard long-term economic growth. Rather, they should view technology transfer as a catalytic source of technological change, leading to the international competitiveness and economic growth of their countries. The experiences of other countries, like the Asian NICs, demonstrate that it is not technology transfer that leads to technological dependency on multinationals. Rather, it is the lack of local absorptive capacity to assimilate, adapt, and improve imported technologies that leads to dependency on foreign suppliers.

Nevertheless, heavy reliance on DFI as a means of technology transfer may, to some extent, lead to foreign dependency. DFI definitely transfers production capability (capability to operate and maintain a production system) but hardly investment capability (capability to set up or expand new production systems) or innovation capability (capability to innovate new products and processes), particularly when the parent company undertakes DFI to exploit the local market in LDCs. Kim's (1987c) study of microelectronics in Korea concluded that semiconductor manufacturing by multinational firms in Korea before 1984 only transferred simple packaging technology but fostered neither the skills, knowledge, and learning capabilities of the production workers nor the engineering capabilities of the domestic economy. Ernst (1985) reached a similar conclusion in his study of automation and the Third World. Amsden and Kim's (1985) comparative study of Hyundai and Daewoo (a joint venture with GM) in the automobile industry in Korea shows that the local firm outperformed the multinational subsidiary in product development and market performance, as the parent company constrained the latter's investment in R&D and GM was inactive in transferring technology transfer to its subsidiary.

The second issue is related to nonmarket-mediated technology transfer modes in cells 3 and 4. That is, LDCs can benefit greatly from informal technology transfers. It cannot be said in precise quantitative terms how important informal technology transfers have been in Korea, but earlier discussions, together with other studies (e.g., Kim, 1982; Kim and Kim, 1985) clearly indicate that they have been very important in Korea's acquisition of technological capabilities. This mode of technology transfer has clearly prevailed in innovative small firms and for a long time has been significant in broadening all exporters' capabilities (Westphal et al. 1985). In short, Korea's experience indicates that the majority of important or crucial information needed to solve technical problems can be obtained, free of charge, through nonmarket-mediated mechanisms, if LDCs have the local capability to undertake reverse engineering tasks.

The final issue concerns the views on foreign technology transfer and indigenous efforts in LDCs. Foreign technology transfer should not be viewed as a substitute for indigenous technological efforts or vice versa. Rather, they should be viewed as complementary. By providing new dimensions in technology, foreign technology transfer enables LDCs to make a quantum jump in indigenous technological efforts. Similarly, indigenous technological efforts lead to local ability to identify appropriate technology and to bargain better in technology transfer negotiations. Once imported, capabilities accumulated through indigenous efforts enable local firms to quickly assimilate, adapt, and improve transferred foreign technology.

## SUMMARY AND CONCLUSIONS

The process of technology transfer is so diverse and complex that it defies a simple analysis. But the conceptal framework presented here enables one to identify various mechanisms of international technology transfer and to examine Korea's public policies promoting the inflow of foreign technologies and Korea's actual experience in the international transfer of foreign technology. The analysis led to the conclusion that informal mechanisms (i.e., those in cells 2, 3, and 4) are as important, if not more, as formal mechanisms (i.e., DFIs, FLs) in international technology transfer, particularly when the recipient country has absorptive capacity. This chapter then discussed the various implications of the conceptual framework and Korea's experience for both DCs and LDCs.

Table 12.3 schematically presents the discussions so far and their conclusions. In quadrant 2, where DCs transfer technologies but LDCs do not have absorptive capacity, DCs will enjoy their technological monopoly power and LDCs will suffer from the syndromes of technological dependency, a

**Table 12.3**    Assessment of Pros and Cons of International Technology Transfer

| | | The Existence of Absorptive Capacity in LDCs | |
|---|---|---|---|
| | | Yes | No |
| Willingness of DC to transfer technologies through formal mechanisms to LDCs | Yes | Transfer takes place. Both DCs and LDCs gain (1) | Transfer takes place. DCs gain but LDCs become dependent (2) |
| | No | Transfer takes place. DCs lose but LDCs gain (3) | Transfer does not take place. Both DCs and LDCs gain little (4) |

238     The Practice of International Technology Transfer in the Pacific Rim

typical outcome of a zero-sum game. Neither DCs nor LDCs gain in quadrant 3, where DCs are not willing to transfer their technology LDCs and LDCs do not have the capability to benefit from existing technology in DCs. It is in quadrants 1 and 4, where LDCs have absorptive capacity, that DCs worry about the boomerang effects of technology transfer. But regardless of whether or not a particular DC is willing to transfer technologies through formal mechanisms, LDCs in quadrants 1 and 4 would be able to acquire foreign technologies, either through alternative sources or through alternative (informal) mechanisms. Why, then, do DCs not take advantage of opportunities to expand their market through DFI or to extend the economic life of technologies through FL (quadrant 1)? It is also in this quadrant 1 that LDCs will not end up being technologically dependent on DCs. That is, quadrant 1 is the best alternative for both DCs and LDCs. Here international technology transfer benefits both the supplier and recipient countries.

## REFERENCES

Amsden, Alice H., and Kim, Linsu. (1985). "The role of transnational corporation in the Republic of Korea's production and exports of automobile," HBS Working Paper Cambridge, Mass.: Harvard Business School.

Bae, Zong-tae, and Lee, Jinjoo. (1986). "Technology development patterns of small and medium sized companies in the Korean machinery industry," *Technovation*, 4:279–296.

Baranson, Jack, and Roark, Robin. (1985). "Trends in north–south transfer of high technology," in Nathan Rosenberg and Claudio Frischtak (eds.), *International Technology Transfer: Concepts, Measures, and Comparisons*, New York: Praeger Press, pp. 24–42.

Cha, Dong-Se. (1983). *Weja Doipeo Hyogwa Boonsuk* (An analysis of the effects of direct foreign investment), Seoul: Korea Institute for Economics and Technology.

Ernst, Dieter. (1985). *Automation, Employment and the Third World - The Case of the Electronics Industry*, The Hague: IDPAD, Institute of Social Studies.

Helleiner, G. D. (1979). "Transnational corporations and trade structure," University of Toronto. Mimeo.

KEB (Korea Exchange Bank). (1987). "Direct foreign investment in Korea," *Monthly Review*, October, pp. 3–13.

Kim, Kee Young. (1984). "American technology and Korea's technological development" in Karl Moskowitz (ed.), *From Patron to Partner*, Lexington, Mass.: Lexington Book, D.C. Heath Co.

Kim, Linsu. (1980). "Stages of development of industrial technology in a LDC: A model," *Research Policy*, 9:254–277.

Kim, Linsu. (1982). "Technological innovation in the capital goods industry in Korea: A micro analysis," Working Paper, International Labor Office in Geneva.

Kim, Linsu. (1985). "Technology transfer and R&D in Korea: national policies and the U.S.–Korea Lin," *Korea's Economy*, 1(1): April.

Kim, Linsu. (1987a). *Technological Transformation in Korea: Progress Achieved and Problems Ahead*, a paper prepared for the World Institute for Development Economic Research, the United Nations University, Helsinki, Finland.

Kim, Linsu. (1987b). *Imitating and Apprentice: How Korea Acquired Technological Capability Fast*, College of Business Administration, Korea University.

Kim, Linsu. (1987c). *The Generation and Diffusion of Microelectronics: Local Capability, Employment Effects, and Public Policies in Korea*, a paper presented at a U.N. University New Technology Conference at the University of Limburg, Maastricht.

Kim, Linsu. (1988a). "The transfer of programmable automation to a rapidly developing country: An initial assessment," forthcoming in *International Economic Journal*.

Kim, Linsu. (1988b). "Entrepreneurship and innovation in a rapidly developing country," *Journal of Development Planning*, 18.

Kim, Linsu, and Kim, Youngbae. (1985). "Innovation in newly industrializing country: A multiple discriminant analysis," *Management Science*, 31(3):312–322.

Kim, Linsu, Lee, Jangwoo, and Lee, Jinjoo. (1987). "Korea's entry into the computer industry and its acquisition of technological capability," *Technovation*, 6:277–293.

OECD (Organization for Economic Cooperation and Development). (1981). *North/South Technology Transfer: The Adjustment Ahead*, Paris: OECD.

Rhee, Yung, W., and Westphal, Larry E. (September 1977). "A micro econometric investigation of choice of technology," *Journal of Development Economics*, 4:205–237.

Utterback, James M. (1975). "The role of applied research institutes in the transfer of technology in Latin America," *World Development*, 3(9):665–673.

Von Hippel, Eric. (1976). "The dominant role of users in the scientific instrument linovation process," *Research Policy*, 5 (3):212–239.

Westphal, Larry E., Kim, Linsu, and Dahlman, Carl J. (1985). "Reflections on the Republic of Korea's acquisition of technological capability" in Nathan Rosenberg and Claudio Frischtak (eds.) *International Technology Transfer: Concepts, Measures, and Comparisons*, New York: Praeger, pp. 167–221.

Westphal, Larry E., Rhee, Yung W., and Pursell, Garry (1981). "Korean industrial competence: Where it came from," World Bank Staff, Working Paper No. 469, Washington, D.C.: World Bank.

# C
# Private Business and Multinational Enterprise

# [21]

*Economic History Review*, XLV, 2(1992), pp. 285–307

# Strategies for innovation: the diffusion of new technology in nineteenth-century British industry[1]

## By CHRISTINE MACLEOD

In so far as it has been studied at all, the British mechanical engineering industry in the nineteenth century has generally been treated as just one more industry among many—if, by the end of that century, one of the most crucial ones in the story of 'decline and fall'.[2] Historians have largely taken for granted the industry's role in supplying the machinery and power sources needed by manufacturers of consumer goods, promoters of railways, shipowners, farmers, and others, in the same way that a ritualistic recital of a few major inventions has often been substituted for any serious consideration of their productive consequences.[3] With the exception of the machine-tools branch, whose role in promoting unprecedented accuracy and speed of production is recognized, the industry's structural function and the relationship between suppliers of capital goods and their customers have been largely ignored.[4] This is curious, since the mechanical engineering industry constituted the major part of a burgeoning new sector, specializing in the production of capital goods, with a vested interest in their adoption, their standardization and, arguably, in their rapid obsolescence.[5] The present article attempts to remedy this omission through an exploration of the role of capital-goods suppliers in the innovation and diffusion of technical change.

## I

Many users of capital equipment were able to construct it themselves. For those who could not, there had long been craftsmen—wheelwrights,

[1] The research for this paper was funded by an E.S.R.C. postdoctoral fellowship, 'Capital goods industries and technical innovation in Britain, 1780-1880' (1986-8). My thanks to Kristine Bruland, D. C. Coleman, Wendy Faulkner, and Mark Goldie, who have read through and improved earlier drafts of this article, and to participants in the Economic History Seminar, University of Oxford, the History and Philosophy of Science Seminar, University of Cambridge, and the Edinburgh Seminar, Dept of Economic and Social History, University of Edinburgh, for their helpful comments, criticisms, and suggestions.
[2] Burgess, thesis; Burstall, *Mechanical engineering*; Coe, *Engineering industry*; Jefferys, *Story of the engineers*; Musson and Robinson, 'Origins'; Musson, 'Engineering industry'; Rolt, *Victorian engineering*; Saul, 'Engineering industry'; *idem*, 'The market'.
[3] Mathias, 'The machine', pp. 11-4.
[4] Floud, *Machine tool industry*; Musson, 'Growth'; Rosenberg, *Perspectives*, pp. 9-31; Saul, 'Machine tool industry'; but see Tann, *Factory*; *idem*, 'Textile millwright'; and, for the later nineteenth century, Kirk, thesis.
[5] 'Capital goods' is used here to denote machinery and equipment used largely by manufacturing industries. This is to restrict it to a part of a much broader sector of the economy that also includes, for example, the shipbuilding and construction industries.

blacksmiths, carpenters, for example—making and repairing the carts, ploughs, spinning wheels, presses, and other tools, bespoken by their customers, usually to traditional, local patterns; such items represented, however, but a part of their more generalized output. A few trades specialized in the production of industrial equipment at an early date: the framesmiths of Nottingham (who numbered 14 by 1739);[6] the wool-card makers and loom builders of the textile regions;[7] the millwrights, who constructed the wind, water, and horse mills, and later, atmospheric engines. What marks out the second half of the eighteenth century is the unprecedented expansion and specialization of the production of industrial and agricultural equipment— itself slight in comparison with the rate of growth of the mechanical engineering industry during the next century.[8]

At the same time, producers of capital goods also began to feature prominently in the patent records. Rare before 1750, by the 1790s they were obtaining approximately one-eighth of all English patents; their number had risen steeply both relatively and absolutely over that half century. They gave their occupations as 'millwrights', and the more novel trades of 'engine makers' and 'machine makers'. Not only did they ascribe to themselves new specialisms in trade, but they were also specialists in their patenting: nine out of ten of their patents covered capital goods, whether prime movers (such as steam engines), industrial production machinery (for example, spinning mules, or carding engines), agricultural implements, or mining equipment. Their number could be increased by a further 25 per cent by including men from the older, more general 'engineering' trades mentioned above (such as carpenters, or clockmakers), who were patenting similar items.[9] This prompts one to ask whether the capital-goods makers' share of patents continued to rise exponentially after 1800. Did machine makers assume the burden of invention, lifting it from the shoulders of machine users? If so, what difference did it make? Would machine makers manage their intellectual property differently, and how would this affect the diffusion of innovation?[10]

Research over the last two decades has provided us with a picture of diffusion that emphasizes the interpersonal demonstration of technical know-how among users of capital goods. The skilled personnel who constructed and operated equipment played a crucial role as the principal agents of technology transfer. Influential in the development of this model has been Harris's work on the problems of transferring coal and iron technologies from England to France in the eighteenth century, and Jeremy's on the transatlantic exchange of textile technology, largely prior to 1830. Jeremy has explicitly minimized the role of machine makers in the export of British

---

[6] Deering, *Nottingham*, pp. 94-5.
[7] Kerridge, *Textile manufactures*, pp. 174-5.
[8] Jefferys, *Story of the engineers*, ch. 1; Tann, *Factory*, ch. 6; Gauldie, *Scottish country miller*, ch. 8; Hartridge, thesis, pp. 4-20.
[9] MacLeod, *Inventing*, pp. 136-9.
[10] I owe the concept of management of intellectual property to Carolyn C. Cooper. For her pioneering study in this field, see Cooper, thesis.

technology.[11] Two other studies of international technology transfer in the second half of the nineteenth century, however, while not denying the skills embodied in human agents, emphasize the part played by machine-making firms in the export of British technology.[12] This difference in findings has a firm empirical basis: prior to 1843 it was illegal for British machine makers to export many types of machinery, and most of them found sufficient business in the domestic market not to run the risk of smuggling.[13] The export market grew rapidly after 1843 and flourished whenever machine makers and engineers were faced with domestic recession in their respective user industries.[14] Concentration on the international transfer of technology implies perhaps that its domestic diffusion was non-problematic, even routine. The present study suggests, however, that to the extent that this was so, it was largely owing to the unsung role played by manufacturers of capital goods.

If historians have tended to overlook the structural function of the mechanical engineering industry and minimized its role in innovation, their omission pales against that of the theorists of technological change. They, while assuming the capital-goods sector to be the fount of invention, have largely ignored its role in diffusion, erecting grand explanatory models on the basis of users alone. This doubtless arises from the domination of such model building by demand factors; a common explanandum is the lag in diffusion within a single culture—why users adopt new technologies at differential rates. Epidemic theories of diffusion, for example, assume that users 'catch' a new technology from one another.[15] Yet, in nineteenth-century Britain, as we shall see, many users anxiously imposed 'quarantine' measures on themselves to prevent others 'catching' their invention; it was often only through the medium of their capital-goods suppliers that information about a new technology was passed back and forth among users. In drawing attention to the influence of supply side factors on the pace of diffusion, however, Rosenberg has also emphasized the importance of the capital-goods industries: 'Creating a capital goods industry is, in effect, a major way of *institutionalizing* internal pressures for the adoption of new technology.'[16] Taking his cue from Marx's observation that the fundamental technical characteristic of modern industry was 'to construct machines by machines', Rosenberg explores the implications of innovation within the capital-goods sector for capital saving, mass production, standardization, and technology transfer.[17]

The distinction between users and makers merits attention because there is reason to think that the strategies of the two groups for managing their

---

[11] Harris, 'Skills'; *idem*, 'Attempts'; Jeremy, *Transatlantic industrial revolution*, esp. pp. 54-7. See also Mathias, 'Skills', and, for a relatively rare study of the means of transfer within a single culture at this period, Smith and Hall, 'Simeon North'.

[12] Bruland, *British technology*; Saxonhouse, 'Japanese technological diffusion'.

[13] See, however, Chaloner, 'Richard Roberts'.

[14] Bruland, *British technology*, pp. 148-52.

[15] For valuable critiques of diffusion models, see Stoneman, *Economic analysis*, pp. 71-2, and *passim*; Metcalfe, 'Impulse', pp. 350, 355; Gold, 'Adoption'.

[16] Rosenberg, *Perspectives*, p. 164.

[17] Ibid., pp. 151-72; *idem*, *Inside the black box*, pp. 43-51.

intellectual property should be distinctive and should affect the pace at which their respective inventions were diffused. To the user or consumer-goods manufacturer any new machine that he invented was a *process* invention. He could keep the invention for his own use, perhaps working it secretly, and could expect to profit by producing goods that were cheaper or of better quality than those of his competitors. To the capital-goods manufacturer, however, a new machine was a *product* invention, and he could only expect to capture the profits of his invention by selling it to his customers, the machine users. Secrecy was out of the question; his profits depended on his product's diffusion.[18]

My basic hypothesis, therefore, is that user-inventors would tend to restrict diffusion and maker-inventors would tend to promote it. It is similar to the model recently employed by von Hippel, and found by him to be applicable to the innovation strategies operative in late twentieth-century North American industry. Von Hippel recognizes that the two weak points in his model's structure are patent licensing and fluidity between the two sectors, but discounts their empirical importance in his study.[19] It is not possible to ignore them so easily here: the opportunities created by the patent system either for monopoly or for licensing extended the range of strategies available, particularly to the user-inventor. This will require the hypothesis to be modified. I shall examine, first, the sources of innovation—which sector was the more productive of them—and then, in the light of this, consider the strategies actually employed, and their implications for both individual firms and industries.

As a preliminary step, one needs to recognize the costly and uncertain nature of patent protection in this period, since this was relevant in determining the management of intellectual property.[20] Until 1852 patents for invention were very expensive—approximately £100 for England and Wales, £350 to include Scotland and Ireland. They were also insecure: they were based on the minimum of statute law, and case law precedents, becoming more common after 1750, were frequently contradictory. Litigation was costly, slow, and hazardous, so that many patentees preferred to tolerate infringements rather than risk going to court—where until the 1830s judges and juries appear to have shared a prejudice that patentees were parasitic monopolists, to be unsuited at every opportunity.[21] Furthermore, patents were not as open to inspection as their name implied. The English system was one of registration, not of examination. Patentees were obliged to file a specification describing their invention in sufficient detail for 'one skilled in the art' to imitate it. However, there was no quality control at the point of enrolment, so that the only time the sufficiency of a specification was tested was when and if there were court proceedings. Neither were the specifications normally published; several journals, such as the *Repertory of Arts and*

[18] Freeman, *Economics*, pp. 150, 162.

[19] Von Hippel, *Sources*, pp. 44-51. He finds that much contemporary inventive activity is conducted in the user sector; see also Freeman, *Economics*, p. 41.

[20] This question is examined at greater length in MacLeod, 'Paradoxes'.

[21] Dutton, *Patent system*, pp. 70-81; review of C. Babbage, *Reflexions on the decline of science in England*, in *Qu. Rev.*, XLIII (1830), pp. 333-41.

## STRATEGIES FOR INNOVATION                    289

*Manufactures*, began to print a selection of them in the 1790s, but otherwise a specification had to be tracked down to the one of three possible offices where it had been registered, and then to be copied for a fee. It was quite feasible in these circumstances for a patentee to attempt to work his patent in secrecy. Many inventors probably never went to the expense (and possible exposure) of a patent, but for those who feared their invention might be stolen by an unscrupulous employee or rival, who could turn it against them, the establishment of priority, if nothing else, was important. Much patenting was defensive.[22]

The Patent Amendment Act of 1852 altered the situation markedly. Patents were cheapened: £25 secured provisional protection for the whole of the British Isles, extendible to 3, 7, and then 14 years by further payments.[23] The system remained one of registration, but specifications were all published in full and distributed by the newly established Patent Office to the major public libraries, along with a series of indexes, to assist inventors in discovering whether their idea really was new. Litigation remained expensive and slow but at least there was now a substantial body of case law on which to draw and judges were more consistent and likely to share the revised view of inventors as national benefactors. Furthermore, since the early nineteenth century a professional body of patent agents had emerged, on whose experience and information an inventor could draw.[24] In summary, it became harder to operate a patent in secret, but, since it was also easier to defend it at law and therefore to enforce it, the need for secrecy diminished. The reduction of costs, combined with the greater amount of information available, also made it more feasible for a capitalist to buy up all the patents in his sphere—a major element in the new strategic game.[25] Criticism of the patent system was by no means silenced by these reforms, and Britain was not exempt from the Europe-wide agitation over the next 20 years for the total abolition of patents, but the system worked much more smoothly than it had done, which was particularly to the advantage of less wealthy inventors. The law was altered again in 1883, to introduce a degree of examination and to reduce the cost (to £4 for the first four years' protection). The first effect, as it had been in 1852, was dramatically to increase the number of applications.[26]

## II

Did machine makers strengthen their hold on the patent system after 1800? The simple answer is that they did, but their efforts were not evenly spread. Analysis of the patent records shows machine makers dominant in heavy engineering, taking out the majority of patents for steam engineering

---

[22] MacLeod, *Inventing*, pp. 40-96.

[23] The total cost was £170. Approximately 10 per cent of sealed patents were kept in force for the full 14 years, by payment of the final instalment of £100: *Commissioners of patents for invention for the year 1880* (P.P. 1881, XXXVII), p. 824.

[24] Dutton, *Patent system*, p. 86 ff.

[25] See below, pp. 294-5, 297-8.

[26] Boehm and Silberston, *British patent system*, pp. 29-37; Coulter, thesis.

and machine tools (though the latter were scarce before 1840). They also took a large and constant share (never less than 40 per cent) of patents for agricultural implements and machinery from the 1790s. However, in the field of production machinery—for example, textile, papermaking, and printing machinery—they continued to be outnumbered (in some industries, overwhelmed) by machine users until at least the 1830s. From that time, patents were more evenly distributed between the two sectors in most industries, but it was rare for machine makers to take the lead before the final quarter of the nineteenth century. (It is hard to be precise because of the increasing numbers of patentees who gave no indication of their occupation.) We may take, however, the example of the paper industry, which occupied an intermediate position in the timing of these developments— in line with printing and cotton spinning, later than flax spinning, but earlier than weaving, cloth finishing, lace and hosiery, or wool and worsted spinning. Among (native) British patentees of papermaking machinery, there were approximately six times as many users as makers in the period 1800-19, and nearly three times as many in 1820-49; the position was reversed around 1860, there being twice as many makers as users obtaining patents by 1870.

But concentration on the number of patents obscures the picture. For one thing, patents are more indicative of invention than of commercial innovation; for another, patents cover both major inventions and minor improvements indiscriminately. Closer scrutiny reveals that machine makers were rarely responsible for radical inventions in production machinery; these were the province of users and of independent or 'professional' inventors otherwise unconnected with the industry.[27] The contribution of machine makers was made overwhelmingly through incremental inventions—though not, of course, to the exclusion of users or independent inventors—and this helps to explain the chronology just noted. Mechanization of most textile processes, first achieved between 1780 and 1830 when the machine-making industry was in its infancy, was accomplished by users (and outsiders). They were still responsible for major breakthroughs in the 1840s and 1850s, when there were many more machine-making firms in business, but their patents were gradually outnumbered by the incremental inventions of the machine-makers, as mechanisms were refined and improved. Moreover, before 1830, artisans who made radical innovations tended to use, rather than make and sell, them. The continuing poverty of John Leavers of Nottingham, for instance, was ascribed by contemporaries to his departure from this practice, in producing his bobbin-net frame for sale; it certainly makes a sharp contrast with the fortune of his contemporary, John Heathcoat, who employed his frames to manufacture lace.[28]

Typically, an independent inventor or user made the initial breakthrough, and then a series of users and their employees produced major modifications

---

[27] It was commonly said, in the nineteenth century, that major inventions were all the work of 'outsiders'. Recent research by Dutton, illuminating the role of 'the quasi-professional' inventor, appears to confirm this. However, Dutton included machine makers in his figures for non-users (i.e. outsiders), and I would query whether they should be regarded as independent in this way: Dutton, *Patent system*, pp. 122-4, and *passim*; MacLeod, *Inventing*, pp. 139-43.

[28] Felkin, 'Hosiery and lace', p. 75.

STRATEGIES FOR INNOVATION                              291

and complementary inventions that rendered the new machine viable in the workplace. For example, the Rev. Edmund Cartwright is credited with inventing both the power loom and the woolcombing machine between 1785 and 1792; both proved disappointing in initial use. The first commercially viable power looms were made by William Horrocks of Stockport, a cotton manufacturer, and Robert Miller, a Dumbarton calico printer, in the first decade of the next century; the power loom was of little use until warp dressing was also mechanized—a feat achieved by William Radcliffe, another Stockport cotton manufacturer.[29] Automatic features, which allowed looms to be run much faster, were added from the 1840s, most of them again by users, led by the Blackburn firm of Hornby & Kenworthy. Woolcombing took much longer to mechanize satisfactorily. Cartwright's machine and its derivatives were used on coarse wool, but it was half a century after Cartwright's initial breakthrough that Samuel Cunliffe Lister and his associates (Isaac Holden and G. E. Donisthorpe), all of them worsted spinners keen to release the woolcombing bottleneck, invented machines that were universally applicable. Flax-spinning presents a similar picture. Mechanized initially in the 1780s by John Kendrew, an optical-glass grinder, and Thomas Porthouse, a clockmaker, it was improved considerably by one of their manufacturing licensees, John Marshall of Leeds, with his mechanic Matthew Murray. Wet spinning was introduced by James Kay in 1825, and important preparatory machines a few years later by W. K. Westley and Peter Carmichael, all of them spinners. Analogous examples could be given from a host of other technologies, including cloth finishing, calico printing, papermaking and printing machinery, glassmaking equipment, preparatory machinery for textile-spinning, and hosiery and carpet looms.

The one major exception to this pattern—of radical innovations produced by users or independent inventors—is instructive. The spinning mule, invented by Samuel Crompton, a weaver, and made partially self-acting by William Kelly, the manager of New Lanark spinning mill, *c.* 1790, follows thus far the usual pattern. Forty years later, however, two fully self-acting mules were patented, one by James Smith of Deanston, a cotton spinner, the other (ultimately the more successful) by Richard Roberts, engineer and partner in Sharp Roberts, the Manchester machine makers. The account of his invention that Roberts gave in 1851 stressed his great reluctance to undertake the task. He had been pressured by a group of Lancashire spinners, eager to undermine the mule operators, whose skills were their strongest wage-bargaining counter. He had refused the employers' commission twice before accepting the challenge and felt his caution justified when, once the self-acting mule was invented, the spinners had lost interest in it, their employees having been subdued by the slump in trade. It was ten years before Roberts began to make a profit from it.[30]

Roberts's attitude provides the clue to a major reason for this pattern—a differential in the risk factor. The risks incurred by a user were considerably less than those taken by a maker. Both faced the hazard of technical failure,

[29] Ure, *Cotton*, II, pp. 287-9; Fairbairn, 'Rise', p. ccxi; Hills, *Power*, pp. 220, 224.
[30] *S. C. on Letters Patent* (P.P. 1851, XVIII), pp. 426, 429-32.

292                              CHRISTINE MACLEOD

but the maker incurred the additional, commercial uncertainty involved in selling his invention. On the one hand, if the maker misjudged his market, the trade might deem the new machinery inappropriate and ignore it; on the other, if he put on sale a radical invention that threatened its extant capital stock with obsolescence, the trade might try to resist it and thereby at least delay commercial success.[31] Market uncertainty suggested that machine makers would be wiser to devote their inventive efforts to innovation in their own processes, not their customers'.[32]

A second, facilitating, reason for the pattern was that the necessary mechanical ability existed in both sectors. Mechanics were retained by users to repair and service their machinery, and in some cases also to build it. A large cotton or flax spinning mill in the mid nineteenth century would employ approximately 20 men and boys just to maintain the machinery.[33] Managers with technical skills were preferred over those without them: Peter Carmichael of Dundee recalled that since he intended to be a flax-spinner, 'it was therefore decided that I should begin by serving my apprenticeship as a mechanic', a decision that took him to the workshops of both William and Peter Fairbairn—the prelude to an innovative career with Baxters, the flax spinners.[34] Building one's own machinery offered the opportunity for experimental activity and, if an invention were made, for its development and perhaps deployment in secret.[35] There might be a further benefit. Any new machine invented and constructed by the user firm would presumably receive more expert maintenance, repair, and adjustment than one that depended on the suppliers' after-sales service or the user firm's mechanics' less intimate knowledge of its design. It might therefore enjoy a higher level of performance, as well as a more sympathetic trial, encouraging its adoption and further investment in its development.

Radical invention in industrial production machinery appears then to have been in the hands of users, who, as my working hypothesis suggested, might opt for secrecy and monopoly over dissemination. I shall first explore the implications of this, and then suggest some modifications.

The hypothesis may be illustrated by the contrasting management of the two papermaking machines patented in England in the first decade of the nineteenth century, one of which was treated as a product invention, the other purely as a process invention. The former was the Fourdrinier machine, invented in France and produced in London by the engineer Bryan Donkin for sale to paper makers. By 1851 Donkin's firm had made 191 machines, 83 of them for British paper mills; once the patent had expired, in 1821, several other engineering firms began to produce them.[36] Secret working was, of course, not an option for Donkin, as a machine maker, and he placed no restrictions on anyone viewing his products. (His machine tools

[31] For an example of trade resistance delaying innovation, see *S. C. on Fourdrinier's Patent* (P.P. 1837, xx), p. 54. See also Freeman, *Economics*, pp. 150-63 and *passim*; von Hippel, *Sources*, pp. 72-3.
[32] See below, pp. 300-1.
[33] *S. C. on Exportation of Machinery* (P.P. 1841, VII), pp. 217, 320.
[34] Gauldie, ed., *Dundee*, p. 32.
[35] The pros and cons of building one's own machinery were considered by William Brown, the Dundee flax spinner, in the early 1820s; Dundee University Library (hereafter D.U.L.), MS 15/26, fos. 208-15.
[36] Clapperton, *Paper-making machine*, pp. 171, 226.

were, however, kept secret, neither patented nor for sale.) It was in his interest that the invention should be diffused as widely as possible, under his aegis, which could have involved licensing other machine makers, had the initial demand been sufficiently strong—which it was not. The other machine was that of John Dickinson, invented for his own use and patented in 1809. Dickinson had no intention of selling the machines but only of using them in his own paper mills. His methods contrast with Donkin's: he was notoriously secretive, banning strangers from his Hertfordshire mills and making the machinery on site.[37] Dickinson's machinery was the equal of Donkin's, in both technical and economic terms. This is demonstrated not only by its widespread adoption in the US, but also by Dickinson's reputation for high-quality products.[38] In what other ways, defensive or offensive, Dickinson might have managed his patent if the Fourdrinier machine had not been available to rival paper makers is a matter for speculation.

Strategies similar to Dickinson's are not hard to find, though not all were equally successful. In-house machine building and innovation were common practices in the Nottingham lace trade in the first half of the nineteenth century.[39] Lace manufacturers vied to employ the most ingenious mechanics, whose function was to devise adjustments to the bobbin-net frames that would produce new patterns of figured lace. One of the more successful manufacturers was James Fisher, who secured the services of a series of inventive mechanics, not least William Crofts, who obtained and assigned to Fisher 18 patents for such adjustments between 1831 and 1842, including those that allowed Fisher to monopolize the production of fashionably spotted lace for many years.[40] This practice was undermined by Deverill's invention in 1841 of a rotary leavers frame whose versatility in producing lace patterns made the mechanic's role in design obsolete. The fancy-lace frame could now be made to a standard design—and, incidentally, be steam-powered. The 1850s saw the increasing separation of the machine-building from the lace-making industry in Nottingham, with the growth of independent machine-making firms, beholden to none.[41]

One industry where a specialized machine-making sector never emerged was (thrown) silk. Several producers of spinning and weaving machinery for the cotton, linen, and woollen industries also built silk machinery to order, but none of them specialized and none of them invented for the trade. A partial explanation lies in the scattered location of this relatively small manufacture, but we should look also to the tradition of in-house machine making and innovation, practised by the leading firms. Silk crape-making in particular was shrouded in mystery and virtually absent from the patent

[37] Ferguson, ed., *Engineering reminiscences*, pp. 118-30.

[38] Ironically, it became the standard type used in the United States thanks to a classic piece of transatlantic industrial espionage, when his foreman was lured to Philadelphia and divulged the design. By 1882 61 per cent of all US papermaking machines were of the cylinder-mould type: McGaw, *Most wonderful machine*, pp. 101-2. For Dickinson's paper, see *S. C. on Fourdrinier's Patent* (P.P. 1837, XX), p. 70; Evans, *Endless web*, pp. 11, 21.

[39] *S. C. on Exportation of Machinery* (P.P. 1841, VII), pp. 187, 189.

[40] Church, *Economic and social change*, p. 71.

[41] Varley, *History*, pp. 22-9, 48-51.

294                              CHRISTINE MACLEOD

records. George Courtauld II commented in 1848 that 'our machinery being peculiar to ourselves . . . we should feel no doubt willing to give a good deal to keep this machinery from the possession of other parties';[42] it was all designed and built in the firm's workshops at Bocking in Essex.

User-inventors were not always so machiavellian: an invention marooned behind factory walls probably represented a wasted opportunity more often than an attempted monopoly. A user may have invented something that answered his own needs but have been unwilling to undertake the expense and effort of diffusion, or to risk the charge of infringing someone else's patent with consequent prosecution or royalty payments. The insecurity of patents under the unreformed system offered considerable incentives to secret working. Alternatively, the new machine may have worked less than perfectly, requiring frequent 'tinkering' to get anything at all out of it; a technical dead-end that might have been reopened had an outsider been consulted. Andrew Ure was able to point to four other cotton spinners who had achieved some measure of success in making their mules self-acting and setting them to work in their own factories, by the time that Roberts and Smith respectively were beginning to offer self-actors for sale.[43]

All the examples I have given so far represent what might be termed a weak form of the model of users' behaviour—process inventions, patented or unpatented, being worked in secret. However, in a stronger version of it, secret working might be supplemented by the purchase of rival patents and a regular resort to litigation, in order to exclude most or all competitors from the advantages offered by a new technology.[44] Such practices with a view to establish a monopoly or oligopoly appear to have been relatively rare in Britain. So far, I have identified only four instances.[45] The first case is that of James Fisher, who supplemented his own collection of patents for lace machinery with the purchase of others' until the trade united against his prosecution of infringements in 1847.[46] Secondly, in 1851 five of the largest Lancashire cotton spinners combined to purchase Heilmann's combing patent for cotton, and commissioned John Hetherington & Sons of Manchester to produce it for them. This oligopoly was short-lived: once the five had 'obtained a command of the market', Hetherington was licensed to build the machines for others, the syndicate collecting three-fifths of the £500

---

[42] Coleman, *Courtaulds*, I, p. 116. And see the revealing later incident, in Essex County Record Office (hereafter E.R.O.), D/F3/2/37, p. 19.

[43] Ure, *Cotton*, II, pp. 193-6.

[44] Two recent pieces of research have shown how influential this could be: Honeyman and Goodman, *Technology*, p. 91, and *passim*; Noble, *America*, pp. 85-109.

[45] It should have been possible to chart these strategies very accurately for at least part of the period, since between 1852 and 1883 all transactions involving patents were registered at the Patent Office. However, this entire class of records was destroyed by the Public Record Office in the 1960s, apart from two tantalizing samples and the indexes (now class BT900). By scanning these indexes it is possible to draw up a rough sketch of who was dealing extensively in patents, but rarely possible to reconstruct what type of transaction was involved—licensing, assignment, etc. Among users, the names that stand out are Crossleys of Halifax and S. C. Lister; among makers, Platt Brothers, and Curtis, Parr & Madeley, both textile-machine makers, and John Fowler, the steam-plough pioneer. These results are not surprising and, in the case of Platts and Fowler, the original documents remain among their archives, supplying a much fuller picture of their activities.

[46] Chapman, ed., *Felkin's history*, pp. 328-9.

purchase price as royalties.[47] Around the same time, Chances of Birmingham and Pilkingtons of St Helens persuaded James Hartley to license them under his lucrative patent of 1847 for rolled plate glass. Hartley explained to Pilkingtons:

> I have no desire to license the patent to anyone, it is no advantage to me to receive £500 a year [royalty] . . . , I am out of pocket by what you and Messrs Chance pay me, as compared to what I should be if you did not make rolled glass, but it is not good policy for a patentee to grasp too much.

However, if they wished him to grant no more licences, they should agree to share the burden of litigation costs—which they did.[48] Two syndicates of six and four manufacturers respectively, headed by the firm of Tomkinson & Adam, similarly monopolized the production of Axminster carpets through patents for power looms in the 1880s and 1890s.[49] One might add Isaac Holden to this list, since he kept control of his woolcombing machine in England much as he had in France, through in-house development and close patent management. But by the time he was operating in England, alternative machines were well established and his influence was consequently limited— not least by his ex-partner, S. C. Lister. In this he was more akin, say, to John Dickinson, the paper maker, operating in splendid isolation.

### III

And so, having failed to strengthen the hypothesis, I shall now relax it. For, as it stands, it leaves no space for a more liberal strategy by machine users, who might opt instead to license competitors and collect royalty payments from them. This strategy also might involve extensive purchasing of patents and their ruthless enforcement. Paradoxically, the most effective dissemination of radical innovations was, I shall argue, achieved by user-inventors employing this more liberal strategy. User-inventors, therefore, were able to offer both the best and the worst diffusion rates. Typically, the liberal strategy was feasible where the patentee was confident of a large, assured market for fairly standard goods, such as bulk textiles, paper, print, or grain, which he could never hope to monopolize. It was also likely where there was a machine-making sector, on the one hand able to offer technical help and marketing services to inventive users, and, on the other, ready to undermine secrecy and provide (help to invent, if necessary) a feasible alternative to a valuable monopoly. The most extreme case is to be found in agriculture: no matter how far any one inventive farmer was able to reduce his own costs through innovation, he could never expect to control the market, and was limited by the extent of his own land; it was rational for him to maximize his return on the invention by extracting a premium on its dissemination, most conveniently by licensing one or more implement makers to produce it.

User-inventors who planned to collect patent royalties had to decide how

[47] Leigh, *Science*, II, p. 162.
[48] Chance, *History*, pp. 78-9.
[49] Bartlett, 'Michael Tomkinson', p. 537.

296                          CHRISTINE MACLEOD

close an interest they wished to maintain in their invention. At one end of the spectrum, they could diversify into machine making, seeking their return in sales of machinery (as products) as well as in royalties and their own use of the process invention. This was the route followed to great effect by John Crossley and Sons, the carpet makers of Halifax. It was also the course taken by James Smith of Deanston, a partner in the great Scottish cotton-spinning firm of James Finlay and Sons.[50]

At the other end of the spectrum, they might simply license other users to employ their invention, supplying the blueprints and leaving them to make their own arrangements with a machine maker. This appears to have been the means whereby Arkwright reluctantly disseminated the water-frame, and James Kay the wet-spinning of flax.[51] Again, they might opt out completely and sell the patent rights to a machine maker outright.

Between these two extremes lay the option of some sort of arrangement with one or more machine makers, who would be authorized to produce the invention, remitting a premium and, perhaps, a share of the profits on sales. Thus the Fourdriniers engaged Bryan Donkin, and Donkin in turn incorporated the improvements proposed to him by various customers.[52] And thus S. C. Lister made his fortune from woolcombing machinery, receiving £1,000 for every machine made and sold for him by Taylor, Wordsworth & Co. of Leeds. It provided an opportunity, in particular, for the employees of user firms, lacking capital of their own, to retain a stake in the sales of their inventions.[53]

Far from user-inventors always hindering the dissemination of radically novel techniques, they were often indispensable. First, they were able to demonstrate by their own practice the feasibility of a new machine, in a way that was not normally available to a machine maker. The relative risks have already been discussed. Cost benefits were rarely clear cut. Nobody wished to play the guinea pig, to discover on behalf of the trade the operating difficulties inherent in some machine maker's brainchild.[54] The editor of *The Textile Manufacturer* encapsulated this distrust when he wrote in 1878:

> As is well known, the improvements made in our machine shops, and which have their origin there, are frequently conjectural, and hence often valueless. But those which come from our spinning and weaving mills, where the difficulties are being constantly experienced, will be found to be of a very different nature.

From the managers and operatives of the latter one could 'expect the most useful inventions'.[55] Doubtless, here was more than a hint of flattering his readers' prejudices (at the expense of his advertisers' pride), and his statement had less empirical support than when Roberts was inventing his self-actor

[50] See below, pp. 297-9.
[51] Fitton, *Arkwrights*, p. 75; Dutton, *Patent system*, pp. 189-92.
[52] See, for example, Clapperton, *Paper-making machine*, pp. 114, 147-8, 226.
[53] See, for example, *S. C. on Letters Patent* (P.P. 1851, XVIII), pp. 250, 253, 324, 409, 422, 448, 524, 574.
[54] See, for example, Leigh, *Science*, I, p. 116; Montgomery, *Manual*, p. 81; Manchester Central Library (hereafter M.C.L.), John Mason & Co. MSS, Box 2, fo. 116; University of Glasgow Archives (hereafter U.G.A.), UGD 91/80, fos. 19-20.
[55] *Textile Manufacturer*, 3 (15 Jan. 1878), p. 1.

50 years earlier, but such prejudices had their own part to play in shaping decisions.[56] Some makers resorted to a degree of forward linkage in order to demonstrate their inventions to wary customers.[57] The Bradford machine makers, Joseph Jefferson & Brothers, advertised in 1888 that, since they had entered a partnership with a firm of commission woolcombers, 'they have found out a great many improvements in machinery . . . [they] try all their experiments and make them perfect before offering them to the public, and their friends can see the machines at work before purchasing'.[58] Manufacturers of agricultural machinery, faced with a traditionally more conservative market, paid much more attention to publicity from an early date, even institutionalizing the promotion of innovation in ploughing matches and other regular trials. Garretts of Leiston owned a farm, which they 'maintained for demonstration purposes'.[59]

Secondly, and more crucially, user-inventors could exert the pressure of competition.[60] The most striking examples postdate the reform of the patent system in 1852. Yorkshire woolcombers queued up to pay Lister £1,000 royalty above the £200 price of each machine; they sought cheaper alternatives, but as Lister closed these off by purchase and litigation, they swallowed hard and paid the price demanded to avoid being squeezed out of the newly mechanized trade and the profits it promised. In his memoirs, Lister claimed that 'for years I sold a large number', and boasted 'it was the largest patent right that was ever paid for a machine'. This was, to say the least, fortunate, since he had bought Heilmann's patent for £33,000, Noble's for £20,000, and Donisthorpe's half share in their joint patent for the 'nip' machine for £27,000, and there were other purchases besides.[61] His success was quickly achieved. Already in 1857 it was stated that 'by far the greater proportion of all wool, of whatever kind, now combed, is combed by Lister's machine'.[62] By 1867 there were 1,038 combing machines at work in the United Kingdom, a large proportion of them constructed under Lister's patents; by 1859 he was also said to have exported more than 1,000.[63]

The firm of John Crossley & Sons was similarly responsible for the rapid mechanization of the carpet-weaving industry in the 1850s. Carpet manufacturers had little choice but to buy the patented looms and pay for licences. Crossleys supplied the looms at what they claimed to be the cost price of £100 each (reduced to £60 in 1854) and charged an annual licence fee. Royalty receipts ranged from £24,000 to £33,000 per annum between 1857 and 1863, outstripping, in the second half of that period, the profits they derived from carpet manufacturing itself. Yet in 1863 Crossleys themselves produced approximately one-third of all Brussels and tapestry

[56] See below, pp. 302-3.
[57] Ward, thesis, pp. 422-3; Hodgson, *Keighley*, p. 256.
[58] *Industries of Yorkshire*, I (1888), p. 246.
[59] Whitehead, *Garretts*, pp. 32-3.
[60] Gold, 'Adoption', pp. 112-3.
[61] [Lister], *Lord Masham's inventions*, pp. 42-7.
[62] James, *History*, p. 577; see also Burnley, *History*, p. 403.
[63] Jenkins and Ponting, *Wool textile industry*, p. 108; *Schlumberger v. Lister*, in Hayward, ed., *Hayward's patent cases*, VII, p. 990.

carpets made in Britain. Kidderminster (Worcs.) was the major centre of carpet manufacture, and the 1851 census had recorded 26 handloom firms, of which only 2 remained ten years later. Both subsequently mechanized and by 1866 there were 15 firms in the area employing Brussels power looms, the smallest with nine looms, the largest with 85. The other hand-loom manufacturers had retired or become bankrupt. The survivors had invested in new factory buildings and engines, as well as looms, incurring an outlay of approximately £11,000 for a 40-loom establishment. Furthermore, they were forced repeatedly to cut their prices by Crossleys, who were anxious that demand should expand to absorb the massively increased supply.[64]

The mechanization of both woolcombing and carpet weaving was a brutally swift process, in which the chief sufferers were the many skilled operators of hand combs and looms. Dragooned by ambitious manufacturers who bought up patents and were prepared to spend large sums on both purchase and litigation to secure their monopolies, neither industry evinced the conservatism and hesitancy to innovate that is often imputed to British industry in the second half of the nineteenth century. Three experienced patent lawyers, in their testimony to the 1871 select committee, commented favourably on this phenomenon. William Grove, for instance, argued that

> Some of those great manufacturers . . . who buy up a large number of inventions, frequently do good, because they stimulate other manufacturers. People would, perhaps, go on in the dull, easy routine of making thousands a year, and would not adopt new patents unless some patentee or monopoliser of patents stirred them up.

Thomas Webster even described Crossleys as 'very benificent despots'.[65] It should be added that many witnesses disputed the virtue of patents at this same period, but not on these grounds. Both their supporters and detractors denied that consolidation of patents was a problem, and compulsory licensing was generally thought unnecessary, its introduction in 1883 being a response to fears concerning the intentions of the growing band of foreign patentees.[66]

## IV

To turn now to the other side of the equation. Is this to conclude that the machine-making industry was of no consequence in innovation and its diffusion? On the contrary, it played an important role, but rarely a pioneering one. To draw an oversharp distinction, machine makers specialized in diffusion and left users to assume the risks of invention and commercialization. Only a small minority of inventor-users, like Crossleys, were able and willing to extend into machine building. Most had to rely on a machine maker to implement their ideas, and to distribute them. Although he supplied some local spinners himself, James Smith of Deanston, for example, licensed J. C. Dyer of Manchester to make his patent mules and carding engines for

[64] Bartlett, 'Mechanisation', pp. 49-62; *idem, Carpeting*, pp. 22-3; P.R.O., C31/1556/2727.
[65] *S. C. on Letters Patent* (P.P. 1871, x), pp. 626, 661-2, 730.
[66] Boehm and Silberston, *British patent system*, p. 30.

STRATEGIES FOR INNOVATION 299

France, and a number of other Lancashire and Glasgow makers to help him supply the domestic market.[67] Machine makers thus provided the major channels through which innovation, whatever its source, could flow.

Their disseminating role did not stop there. Bruland has shown that British machine makers supplied not only machinery, but also a 'package' of consultancy services to Norwegian manufacturers.[68] These functions were not restricted, however, to the export market: machine-making firms were extending to their foreign customers a type of service that they had already developed, if less comprehensively, at home. It was normal to send skilled mechanics out to erect machinery on the customer's premises, and for them to stay for days or weeks to instruct the operatives and the mill mechanics— sometimes they were poached, to the natural annoyance of the machinist.[69] Machine makers were asked to look out for suitable managers, foremen, and mechanics, and for advice on new technical developments, the selection of compatible machinery, or the most efficient running speeds.[70] The existence of machine makers ready both to construct and to instruct must have facilitated entry to an industry, particularly as machinery became more complex.

Where a specialized machine-making industry existed, it was harder for secrecy to prevail. What is striking to the eyes of one used to nineteenth-century British industrialists' secretive practices is the readiness of Norwegian and Japanese manufacturers to share technical experience with their local competitors.[71] In contrast, many British manufacturers not only kept their rivals at arm's length, but also enjoyed a love-hate relationship with their machine makers, springing from the latters' different attitude to secrecy. John Marshall II, the Leeds flax-spinner, summed it up when he advised his partners, in 1825:

> Looking forward to the many improvements which we expect to make in the preparing processes, it does not seem right to have these improvements spread as soon as possible through the medium of the machine shops in which such machinery might be made. We should therefore establish one of our own as quickly as we can.[72]

In the event, Marshalls continued to buy in their machinery. They helped finance a newcomer to Leeds, Peter Fairbairn, to establish a machinist's business, and with him enjoyed a long period of cooperation and exchange

---

[67] G.U.A., UGD91/80, fos. 1, 34-5, 43, 60-1.
[68] Bruland, *British technology*, pp. 75-88, 108-36; Saxonhouse, 'Japanese technological diffusion', pp. 149-65.
[69] Fox, *Quaker homespun*, pp. 54-5; Riden, *Butterley Company*, p. 49; Bradford District Archives, George Hattersley & Sons Ltd MS 32D83/32/1 (letter to A. Tatton, 8/11/1853); Brotherton Library, Leeds (hereafter B.L.L.), Marshall MSS 15/14, 16/29; D.U.L., Baxter MS 11/5/14, fo. 38.
[70] See, for example, Fox, *Quaker homespun*, p. 53; B.L.L., Holden MSS 6/4, 7/1; John Rylands Library, Manchester, McConnell & Kennedy MSS, Correspondence 1795, Gill & Toldery to McConnell & Kennedy, 18/3/1796 [sic]; Letter Book (out) 1796-1805, McConnell & Kennedy to George Hannay, 18/4/1800; E.R.O., D/F 3/2/78, fo. 94; D/F 3/2/80, fos. 21-4; D/F 3/2/81, fo. 52; D.U.L., Douglas Fraser & Sons Ltd MS, 42/4/1, fo. 138.
[71] Bruland, *British technology*, pp. 137-46; Saxonhouse, 'Japanese technological diffusion', pp. 160-2.
[72] Quoted in Rimmer, *Marshalls*, p. 153. See also *S. C. on Letters Patent* (P.P. 1871, x), p. 752.

of ideas, learning to tolerate the 'leakage' of improvements as the price of receiving many others.[73]

By arranging for potential purchasers to view their products at work in a customer's factory or by impressing them with details of what they had supplied to whom, machine makers helped to open up some channels of information among users who had often previously had little idea of their competitors' activities.[74] This intrusiveness may have encouraged user-inventors to patent and license, rather than attempt the more risky strategy of secret working (with or without a patent). Probably more influential, however, was the growing security offered by patents from the 1830s, as the bias of judges and juries seems to have swung round to favour patentees at the expense of infringers, and further after 1852, when the Patent Amendment Act simplified and cheapened the system.[75] Inventors were thereby empowered to manage their intellectual property through more public transactions, with greater confidence in the availability of some legal sanction on enforcement.

Machine makers themselves became increasingly patent conscious. Their purpose was normally to protect improvements rather than radical inventions.[76] This points to another role played by machine makers, the importance of which should not be underrated—their interest in the refinement of techniques. If a user made the initial technical breakthrough, there was often a machine maker who either gave him some assistance at the time or made subsequent improvements to rub smooth the rougher, technical edges. For example, the woolcombing machine invented by Noble and Donisthorpe was developed under licence from Lister (the assignee) by the firm of Taylor Wordsworth, which held 16 current patents for improvements in 1888, on top of the many that had already expired.[77] Hetheringtons similarly developed the Heilmann cotton comber, and Donkin, the Fourdrinier machine. This was a relatively low-cost and low-risk strategy for a machine maker to adopt. For, once a radical innovation had been made and users had demonstrated their willingness to invest in the new machinery, they could be expected to favour its most advanced form. A string of patents for incremental inventions helped both to direct their choice and generally to advertise the makers as technically advanced machinists.

Machine makers made another vital contribution by inventing machine tools that permitted the more accurate measurements and precise finishes on which increasingly sophisticated production machinery and steam engines depended. Invented either by specialist machine-tool manufacturers or by machine makers—for their own use or to meet the requirements of their customers—machine tools also facilitated the switch from wooden to cast-iron machinery, and later to steel. Improvements in machine tools that cheapened the production of machinery and enhanced its quality were passed

[73] *S. C. on Exportation of Machinery* (P.P. 1841, VII), pp. 194, 200-1.
[74] B.L.L., Marshall MS, 37, fos. 49, 51, 62; D.U.L., Baxter MS 11/5/14, fo. 78; E.R.O., D/F 3/2/57, fo. 3; D/F 3/2/80, fos. 25-6, 35; Aberdeen University Library, Alexander Pirie & Sons MS, 2911/6/1.
[75] See above, pp. 288-9.
[76] See above, pp. 290-2.
[77] Burnley, *History*, p. 392.

on to the user industry as capital savings. Similarly, improvements made to steam engines and to gearing and shafting enabled textile manufacturers to run their machinery economically at higher speeds.[78]

It was unusual for a machine maker to attempt to control a patent as tightly as might a user, like Holden or Lister. The distinction between a radical and an incremental invention was doubtless crucial here: a monopoly over the latter offered little strategic advantage.[79] The records of Platt Brothers indicate a routine and uneventful market in invention among machinists: a willingness to sell, buy, and pool licences to use each other's patents, returning royalties on moderate terms.[80] This indicates that we should not judge capital-goods manufacturers as a whole by the example of James Watt, who issued no licences to other steam-engine builders and probably held back development. Watt's engine was a radical invention, and his position was unusual in a number of ways. Insecurity, engendered both by fear of putting his patent specification to the legal test and by the decentralized nature of steam-engine construction before the Soho foundry was opened in 1796, made him particularly defensive—and, once the specification, against all the odds, had finally been upheld, particularly aggressive in seeking out those he considered to be 'pirates'.[81] Nonetheless, because it represented a product invention, Watt's steam engine was diffused during the term of his extended patent—if not as widely nor in such an advanced form as it might have been.

## V

What follows from this pattern of innovation and diffusion? First, for the individual user-firm, exclusive reliance on the inventions of its own machine shop was a high-risk strategy. On the positive side, it offered the opportunity for a technological lead which, if managed astutely, might open the route to market domination, as it did for Crossleys and the glassmaking triopoly of Pilkingtons, Chances, and Hartleys. On the negative side, especially if pursued dogmatically, it could lead to expensive mistakes, technological dead-ends, inefficient and even obsolete machinery. This appears to have been a fate that threatened Samuel Courtauld & Co. in the early 1870s, averted, in part, by the renewal of contacts with machine makers in the north and, through them, by access to the modern looms developed for the younger and healthier spun-silk industry.[82] The railway companies pushed the policy of in-house building further than did most industries in the second half of the nineteenth century. On the one hand, this permitted, for example,

---

[78] Von Tunzelmann, *Steam power*, pp. 182-225, 240-51; Lyons, 'Powerloom profitability', suggests this was less important for power looms than for spinning machinery. For gearing and shafting, see Pole, ed., *Sir William Fairbairn*, pp. 114, 257-60.

[79] The dangers of radical inventions in the hands of ruthless machine makers were evinced later in the nineteenth century by the restrictive practices of the patentees of American boot-making machinery: Ravenshear, *Influence*, pp. 88-91.

[80] Lancashire County Record Office (hereafter L.R.O.), DDPSL 1/108/1-2.

[81] Tann, 'Mr Hornblower', pp. 95-105. This may be contrasted with the strategy of Robert Stephenson: Warren, *Century*, pp. 79-81, 95.

[82] Coleman, *Courtaulds*, 1, pp. 171-4, 181-91.

the Great Western to enjoy exclusively the technically sophisticated, and barely surpassed, locomotive designs of George Churchward; on the other, it condemned the London and North Western to the expense incurred by F. W. Webb's persistent refusals to accept technological defeat.[83] In effect, few firms were totally self-sufficient: most bought in some machinery, as well as components.

Buying in machinery was a more cautious strategy that suited most. It enabled a conscientious manufacturer to make an informed choice among techniques, and to patronize a range of machinists, each perhaps a specialist: the costs and risks of innovation were borne by others. In comparison the royalties he had to pay to use patented innovations were usually slight— even if they did elicit much grumbling. He could draw on the knowledge and advice of a number of machine makers, who might also keep him informed of his competitors' activities, at home and abroad, as well as assisting in any technical difficulties that arose. If he kept up to date in this way, he was unlikely either to be becalmed in a technical backwater or to be ruined by experimental schemes. This strategy did not, of course, preclude the possibility of dominating one's industry, as may be attested, for example, by Marshalls' pre-eminence in flax-spinning before 1870, but that dominance would not be the result of technological leadership.

As far as an industry as a whole was concerned, technological self-reliance combined with secrecy could be an unspoken suicide pact, especially when exposed to foreign competition. The sorry state of the British (thrown) silk industry by 1900 may be ascribed largely to changing fashions and the abolition of import tariffs in 1860, but the widespread practice of in-house machine building had discouraged new entrants and led to the technical isolation and obsolescence of the few remaining older firms. Parliament was told in 1912 that, between 1883 and 1902, 25 firms in the (thrown) silk industry had disappeared, chiefly through failure: 'The cause of failure was in every case clearly understood, for, in addition to mismanagement and unsound trading, a view of their machinery, at the subsequent auction, caused one to wonder how they had existed so long.'[84] Six years later, it was still reported that 'much of the plant is old-fashioned', and this was ascribed in part to the fact that 'the industry does not offer a sufficient market to machinists to induce them to incur large expenditure in experimenting in models specially adapted to the requirements of a very limited trade'.[85] In glassmaking also, where there were no specialist capital-goods manufacturers, free trade and Belgian competition saw the technically advanced few thrive at the expense of the many.[86]

There is, however, no simple dichotomy between the long-term prospects of industries that enjoyed the services of machine makers and those that did

---

[83] Westwood, *Locomotive design*, pp. 96-105, 130-1; Reed, *Crewe*, pp. 234-7. For a more constructive perspective on the lack of standardization in British steam locomotive design than is usual, see Kirby, 'Product proliferation'.

[84] *Report . . . industrial exhibitions* (P.P. 1912-3, XXII), p. 304.

[85] *Report . . . textile trades after the war* (P.P. 1918, XIII), p. 722.

[86] Barker, *Glassmakers*, pp. 99, 157-61.

not. It was hard to flourish without a capital-goods sector, but the existence of one did not guarantee rapid technical advance. What it would provide was at least steady progress within a technological paradigm. The Lancashire cotton industry (to take a well-worked example) may have rejected ring spinning, but it was equipped with mules that, in the 1890s, outperformed their predecessors of 60 years before by an estimated 60 per cent in output per spindle per hour, and with a considerable fall in capital costs. Much of this improvement could be credited to the machine makers.[87] Whether the majority of users were culpably conservative, as many contemporaries believed, remains a point for debate. Arguably, industrial machine makers did not put sufficient effort into persuasion and shared their customers' caution, but the difficulties of persuading their customers to be among the first to try innovatory machines must have played a part in discouraging machine makers from pursuing radical inventions. Yet, if we accept Sandberg's exculpation of cotton spinners from the charge of economic irrationality, we must also conclude that contemporary machine makers were wise to respect their customers' rejection of ring spinning and to pursue instead the perfection of mules, to the benefit of both sectors.[88]

There was clearly sufficient work normally available to machine makers in the equipment of an expanding market to make the pressure of obsolescence an unnecessary sales strategy, particularly once machinery exports were completely legalized in 1843. The ring spindles that most Lancashire spinners rejected were sold in great numbers, in Japan, India, and elsewhere, by British machinists under licence from their American inventors, and textile engineering, particularly, was becoming increasingly export-oriented.[89] With profitable outlets abroad, the machine makers had less incentive to bully manufacturers at home to update their techniques, let alone to invest in radically different machines whose economic advantages were not proven. And, if there were neither user-inventors forcing the pace nor direct competition from abroad, the whole user industry could lapse into a quiet middle age of perhaps steady improvement but no radical change. The major textile industries, apart from carpets and woolcombing, exhibit this pattern in the second half of the nineteenth century. Farnie argues that 'The [engineering] industry should not be blamed for any "failure" by Lancashire to adopt the ring frame or the automatic loom since it could supply its clients only with what they wanted', and shifts any criticism to the users for not generating 'a local demand . . . for a higher technology than the one in use'.[90] Yet any 'failure' lies as much in the supply of innovation as in the demand for it. The principal domestic source of radical innovations had been the user sector itself; if that source was drying up (and patent figures

---

[87] Nasmith, *Machinery*, pp. 5-7; von Tunzelmann, *Steam power*, pp. 224-5; and above, p. 300. As to whether productivity continued to grow between 1890 and 1913, see Mass and Lazonick, 'British cotton industry', pp. 28-9.

[88] Sandberg, *Lancashire*, pp. 44-8, and *passim*; Lazonick, 'Factor costs'.

[89] Saxonhouse, 'Japanese technological diffusion', *passim*; Farnie, 'Textile machine-making industry', pp. 152-5, 168; Kirk, thesis, pp. 196-7.

[90] Farnie, 'Textile machine-making industry', p. 161.

suggest it may have been), then the reasons must initially be sought there.[91] However, their avoidance of the risks of innovation leaves the textile engineers, like their clients, open to the charge of entrepreneurial failure, despite their short-term rationality. British machinists were accustomed to buying in radical innovations, supplementing those of domestic inventors with a growing number of foreign ones, particularly from the US. In this historical context it is unsurprising that 'all the major innovations made in the cotton industry in the twentieth century were made outside Britain', and that Textile Machinery Makers Ltd, the giant combine made up of the major Lancashire textile engineering firms, established a research section only in 1946.[92]

In nineteenth-century Britain, the most effective prescription for an innovative industry was an aggressively innovative market leader, willing to license competitors at reasonable rates, combined with a machine-making industry that would support it with a supply of equipment and expertise. This depended on the existence of a patent system that provided innovators with the security of being able to capture the profits of innovation. Occasionally, such security was used to benefit the individual user firm at the expense of the industry as a whole. But far more damaging had been the uncertainty of the unreformed patent system, particularly before 1830, that encouraged secret working and discouraged licensing. It is hard to disentangle the coincident effects, on the one hand, of patent reform and, on the other, of the growth of a capital-goods sector in breaking this down. Neither was alone sufficient, but both were necessary conditions for a more rapid and widespread diffusion of innovation. Unfortunately, by the time both were in place, the habits of secrecy had become so ingrained in many industries that there was little hope for cooperative advance among user firms, and machine makers had been deterred from the pursuit of radical innovation. However, this should not lead us to overlook a major determinant of the pace of diffusion arising on the supply side, that is the strategy adopted by the innovator for capturing the profits of his innovation: secret working or restrictive licensing by a user would limit and retard the supply; liberal licensing or sales of machinery by either a user or a maker would hasten its diffusion.

*University of Bristol*

[91] A plausible explanation of this might be the declining number of mechanically trained owners and managers, personally interested in the techniques used on the shop floor and able to appreciate suggestions for innovation arising there.
[92] Farnie, 'Textile machine-making industry', p. 163.

**Footnote references**

*Official publications*
Select Committee on Fourdrinier's Patent (P.P. 1837, XX).
Select Committee on Exportation of Machinery (P.P. 1841, VII).
Select Committee on Letters Patent for Invention (P.P. 1851, XVIII).
Select Committee on Letters Patent (P.P. 1871, X).
Report of the Commissioners of Patents for Invention for the year 1880 (P.P. 1881, XXXVII).

STRATEGIES FOR INNOVATION                   305

*Report of His Majesty's Commissioners for the industrial exhibitions at Brussels, Rome, Turin, 1910 and 1911* (P.P. 1912-3, XXII).
*Report . . . textile trades after the war* (P.P. 1918, XIII).

### Secondary sources

Barker, T. C., *The glassmakers. Pilkington: the rise of an international company, 1826-1976* (1977).

Bartlett, J. N., 'The mechanisation of the Kidderminster carpet industry', *Bus. Hist.*, IX (1967), pp. 49-62.

Bartlett, J. N., *Carpeting the millions: the growth of Britain's carpeting industry* (Edinburgh, 1978).

Bartlett, J. N., 'Michael Tomkinson (1841-1921)', in D. J. Jeremy, ed., *Dictionary of Business Biography*, vol. 5 (1986).

Boehm, K. and Silberston, A., *The British patent system: 1, administration* (Cambridge, 1967).

Bruland, K., *British technology and European industrialization: the Norwegian textile industry in the mid nineteenth century* (Cambridge, 1989).

Burgess, K.R., 'The influence of technological change on the social attitudes and trade union policies of workers in the British engineering industry' (unpub. Ph.D. thesis, Univ. of Leeds, 1970).

Burnley, J., *The history of wool and woolcombing* (1889).

Burstall, A. F., *A history of mechanical engineering* (1963).

Chaloner, W. H., 'New light on Richard Roberts, textile engineer (1789-1864)', *Trans. Newcomen Soc.*, 41 (1968-9), pp. 27-43.

Chance, J. F., *A history of the firm of Chance Brothers & Co.* (1919).

Chapman, S. D., ed., *Felkin's history of the machine-wrought hosiery and lace manufactures* (Newton Abbot, 1969).

Church, R. A., *Economic and social change in a midland town: Victorian Nottingham, 1815-1900* (1966).

Clapperton, R. H., *The paper-making machine: its invention, evolution and development* (Oxford, 1967).

Coe, W. E., *The engineering industry in the north of Ireland* (Newton Abbot, 1969).

Coleman, D. C., *Courtaulds: an economic and social history, 1: the nineteenth century, silk and crape* (Oxford, 1969).

Cooper, C. C., 'The roles of Thomas Blanchard's woodworking inventions in nineteenth-century American manufacturing technology' (unpub. Ph.D. thesis, Yale Univ., 1985).

Coulter, M., 'Property in ideas: the patent question in mid-Victorian Britain' (unpub. Ph.D. thesis, Indiana Univ., 1986).

Deering, C., *Nottingham vetus et nova* (Nottingham, 1761).

Dutton, H. I., *The patent system during the industrial revolution, 1750-1852* (Manchester, 1984).

Evans, J., *The endless webb: John Dickinson & Co. Ltd, 1804-1954* (1955).

Fairbairn, W., 'The rise and progress of manufacture and commerce', in T. Baines, *Lancashire and Cheshire past and present*, 2 (1871), pp. i-cclx.

Farnie, D. A., 'The textile machine-making industry and the world market, 1870-1960', *Bus. Hist.*, XXXII (1990), pp. 150-67.

Felkin, W., 'Hosiery and lace', in G. Phillips Brown, ed., *British manufacturing industries* (2nd edn. 1877), pp. 1-88.

Ferguson, E. S., ed., *Early engineering reminiscences (1815-40) of George Escol Sellers* (Washington, D.C., 1965).

Fitton, R. S., *The Arkwrights, spinners of fortune* (Manchester, 1989).

Floud, R., *The British machine tool industry, 1850-1914* (Cambridge, 1976).

Fox, H., *Quaker homespun: the life of Thomas Fox of Wellington, serge maker and banker, 1747-1821* (1958).

Freeman, C., *The economics of industrial innovation* (2nd edn. 1982).

Gauldie, E., ed., *The Dundee textile industry* (Edinburgh, 1969).

Gauldie, E., *The Scottish country miller* (Edinburgh, 1981).

Gold, B., 'On the adoption of technological innovations in industry: superficial models and complex decision processes', in S. MacDonald, D. McL. Lamberton, and T. Mandeville, eds., *The trouble with technology: explorations in the process of technological change* (1983), pp. 104-21.

Harris, J. R., 'Skills, coal and British industry in the eighteenth century', *Hist.*, 61 (1976), pp. 167-82.

Harris, J. R., 'Attempts to transfer English steel techniques to France in the eighteenth century', in S. Marriner, ed., *Business and businessmen: studies in business, economic, and accounting history* (Liverpool, 1978), pp. 199-233.

Hartridge, R. J., 'The development of industries in London south of the Thames, 1750 to 1850' (unpub. M.Sc. thesis, Univ. of London, 1955).

Hayward, P. A., ed., *Hayward's patent cases, 1600-1883*, 11 vols. (Abingdon, 1988).

Hills, R. L., *Power in the industrial revolution* (Manchester, 1970).

Hodgson, J., *Textile manufacture and other industries in Keighley* (Keighley, 1879).

Honeyman, K. and Goodman, J., *Technology and enterprise: Isaac Holden and the mechanisation of woolcombing in France, 1848-1914* (Aldershot, 1986).

306                         CHRISTINE MACLEOD

James, J., *History of the worsted manufacture* (1857).

Jefferys, J. B., *The story of the engineers* (1945).

Jenkins, D. T. and Ponting, K. G., *The British wool textile industry* (1982).

Jeremy, D. J., *Transatlantic industrial revolution: the diffusion of textile technology between Britain and America, 1790-1830* (Cambridge, Mass., 1981).

Kerridge, E., *Textile manufactures in early modern England* (Manchester, 1985).

Kirby, M. W., 'Product proliferation in the British locomotive building industry, 1850-1914: an engineer's paradise?', *Bus. Hist.*, XXX (1988), pp. 287-305.

Kirk, R. M., 'The economic development of the British textile machinery industry, c. 1850-1939' (unpub. Ph.D. thesis, Univ. of Salford, 1983).

Lazonick, W., 'Factor costs and the diffusion of ring spinning prior to World War I', *Qu. J. Econ.*, XCVI (1981), pp. 89-109.

Leigh, E., *The science of modern cotton spinning*, 2 vols. (Manchester, 2nd edn. 1873).

[Lister, S. C.], *Lord Masham's inventions, written by himself* (1905).

Lyons, J. S., 'Powerloom profitability and steam power costs: Britain in the 1830s', *Exp. Econ. Hist.*, 24 (1987), pp. 392-408.

McGaw, J. A., *Most wonderful machine: mechanization and social change in Berkshire paper making, 1801-1885* (Princeton, 1987).

MacLeod, C., *Inventing the industrial revolution: the English patent system, 1660-1800* (Cambridge, 1988).

MacLeod, C., 'The paradoxes of patenting: innovation and diffusion in nineteenth-century Britain, France, and North America', *Tech. & Cult.*, 32 (1991), pp. 885-910.

Mass, W. and Lazonick, W., 'The British cotton industry and international competitive advantage: the state of the debates', *Bus. Hist.*, XXXII (1990), pp. 885-910.

Mathias, P., 'Skills and the diffusion of innovation from Britain in the eighteenth century', *Trans. Roy. Hist. Soc.*, 5th ser., XXV (1977), pp. 90-113.

Mathias, P., 'The machine: icon of economic growth', in S. MacDonald, D. McL. Lamberton, and T. Mandeville, eds., *The trouble with technology: explorations in the process of technological change* (1983), pp. 11-22.

Metcalfe, J. S., 'Impulse and diffusion in the study of technical change', *Futures*, 13 (1981), pp. 347-59.

Montgomery, J., *The cotton spinner's manual* (Glasgow, 1850).

Musson, A. E., 'The growth of mass production engineering', in A. E. Musson and E. Robinson, eds., *Science and technology in the industrial revolution* (Manchester, 1969), pp. 473-509.

Musson, A. E., 'The engineering industry', in R. A. Church, ed., *The dynamics of Victorian business* (1980), pp. 87-106.

Musson, A. E. and Robinson, E., 'The origins of engineering in Lancashire', in A. E. Musson and E. Robinson, eds., *Science and technology in the industrial revolution* (Manchester, 1969), pp. 427-58.

Nasmith, J., *Modern cotton spinning machinery, its principles and construction* (1890).

Noble, D. F., *America by design: science, technology, and the rise of corporate capitalism* (New York, 1979).

Pole, W., ed., *The life of Sir William Fairbairn* (1876).

Ravenshear, A. F., *The industrial and commercial influence of the English patent system* (1908).

Reed, B., *Crewe locomotive works and its men* (Newton Abbot, 1982).

Riden, P. J., *The Butterley Company, 1790-1830* (Chesterfield, 1973).

Rimmer, W. G., *Marshalls of Leeds, flax-spinners, 1788-1886* (Cambridge, 1960).

Rolt, L. T. C., *Victorian engineering* (Harmondsworth, 1970).

Rosenberg, N., *Perspectives on technology* (Cambridge, 1976).

Rosenberg, N., *Inside the black box: technology and economics* (Cambridge, 1982).

Sandberg, L., *Lancashire in decline* (Columbus, Ohio, 1974).

Saul, S. B., 'The machine tool industry in Britain to 1914', *Bus. Hist.*, X (1968), pp. 22-43.

Saul, S. B., 'The engineering industry', in D. Aldcroft, ed., *The development of British industry and foreign competition, 1875-1914* (1968), pp. 186-237.

Saul, S. B., 'The market and development of the mechanical engineering industries in Britain, 1860-1914', in B. E. Supple, ed., *Essays in British business history* (Oxford, 1977), pp. 31-48.

Saxonhouse, G., 'A tale of Japanese technological diffusion in the Meiji period', *J. Econ. Hist.*, XXXIV (1974), pp. 149-65.

Smith, M.R. and Hall, J. H., 'Simeon North, and the nature of technological innovation among the antebellum arms makers', *Tech. & Cult.*, 14 (1973), pp. 573-91.

Stoneman, P., *The economic analysis of technological change* (Oxford, 1983).

Tann, J., *The development of the factory* (1970).

Tann, J., 'The textile millwright in the early industrial revolution', *Text. Hist.*, 5 (1974), pp. 80-9.

Tann, J., 'Mr Hornblower and his crew: Watt engine pirates at the end of the eighteenth century', *Trans. Newcomen Soc.*, 51 (1979-80), pp. 95-105.

Ure, A., *The cotton manufacture of Great Britain*, 2 vols. (London and Manchester, 1836).

Varley, D. E., *A history of the Midland Counties Lace Manufacturers' Association, 1915-1958* (Long Eaton, 1959).

Von Hippel, E., *The sources of innovation* (New York and Oxford, 1988).

Von Tunzelmann, G. N., *Steam power and British industrialization to 1860* (Oxford, 1978).

Ward, M. F., 'Industrial development and location in Leeds north of the River Aire, 1775 to 1914' (unpub. Ph.D. thesis, Univ. of Leeds, 1972).

Warren, J. G. H., *A century of locomotive building by Robert Stephenson and Co., 1823-1923* (Newton Abbot, repr. 1970).

Westwood, J. N., *Locomotive design in the age of steam* (1977).

Whitehead, R. A., *Garretts of Leiston* (1964).

[Yorkshire], *Industries of Yorkshire*, I (1888).

# [22]

## The Role of Private Business in the International Diffusion of Technology

CLEARLY, private business is but one agent for the diffusion of technology. Yet it is an important one. In the normal pursuit of business, technological knowledge and skills pass over political boundaries and private enterprise takes part in the international diffusion of technology. In this paper I want to try to delineate the means by which private companies have shared in the international diffusion of technology in the nineteenth and twentieth centuries. I will note briefly the "imitation lag" and then what I want to call the "absorption gap." From generalizations, I will turn to some explicit examples and analysis. Finally, in conclusion, I want to return to my concept of the absorption gap and the role of private enterprise in bridging that gap.[1]

### I

In theory at least there appear to be eight distinct ways by which a private company can act to transfer technology across political borders. The eight methods are broad and each contains sub-categories. First, a private concern can export for sale or for exhibition a new or improved product. If the exports are capital goods, they transfer technology directly when used in modernizing production processes. But any export, whether of a producer or a consumer good, may be imitated in foreign lands and by this means move technology from one country to another. Thus, the first manner by which a company transfers technology involves simply the *export of products*. Second, a private enterprise can take out patents in a foreign country, patents that may be worked in that nation. In reg-

This paper has been revised since its delivery on September 14, 1973, in light of comments made at the Economic History Association meeting by Professors Kozo Yamamura, Ralph Hidy, Stuart Bruchey, David Felix, and others.

1 Technology has been defined in narrow terms to comprise simply tools and machines. I prefer a more comprehensive definition, including in addition to tools and machines, product design, knowhow, and organizational ability, that is, concept along with technique. I am well aware that technological diffusion often takes place in "bits and pieces." Indeed, technology itself evolves in a complex fashion. While these points are not elaborated on herein, there is nothing in my paper to imply otherwise.

## Private Business 167

istering a patent, there is disclosure. There are opportunities for its sale or licensing, or for designing similar but not covered products. Thus, the second approach involves the *export of patents*. Third, a firm can make a range of different types of technical assistance agreements or provide technical aid to foreign companies or governments. Here there is the *export of technical knowledge and services*. And, fourth, a company can undertake direct foreign investments, that is, act as a multinational enterprise and transfer its technology abroad with its investment. This fourth approach involves an *export of, or rather an extension of, the firm itself abroad*. All these methods are those of enterprises that possess technology to transmit.[2] Note that a single company can participate in all four forms of transfer. The distinctions between these methods of transfer may be real or simply theoretical, depending on the particular circumstances.

The second four manners by which a company transfers technology are counterparts of the first group. These four are associated with the receipt of technology: One, a firm can *import* machines used in production processes new to its nation; an importer can sell or present any new product, which is then imitated within the recipient nation. Two, a company can commercialize the *patents* of a foreign enterprise in its domestic market. Three, it can make *technical assistance arrangements* from which it will benefit. Four, it can acquire technology from a *direct foreign investor*. All these last four approaches involve the utilization of foreign technology. Perhaps these four manners of receiving technology may be as much or more responsible for international technological diffusion as the four manners used by the holders of the technology.

Still, the second group of businessmen is often, although not always, reliant on the first group, since it is impossible to import if something is not exported; a patent must exist before any one can work or modify that particular patent; someone must have the technology before a technical assistance agreement can be made; for there to be technology derived from a direct foreign investor, there must be that investor. Note, however, that the holder of the technology—the exporter, the owner of the patent, the provider of tech-

[2] If the exporter of the product is independent of the producer of the product, it can be argued that the export firm does not have the technology of production; yet, it is still the holder of (owner of or agent for) the product that contains within it the technology.

## 168 *Wilkins*

nological assistance, or even the direct foreign investor—need not be a private company. In short, there need not be symmetry in the relationship between our two groups of private companies.[3] Note, too, the second group of businessmen may in certain instances be one and the same as the first group; exporter and importer may be part of one company; holder of the patent and exploiter of it abroad may be identical; and so forth.[4] Once again, the distinctions may be real or simply theoretical.

There are additional ways by which a technologically-advanced company participates in a passive manner in the transfer of technology.[5] The eight modes described above seem to be the *active* ways by which private companies take part in technological transfers.

## II

Before I elaborate on these eight modes of transfer, it is worth considering the difference between mere transfer and the absorption of technology within the host country. A company can export capital goods. In one country the machines installed might be allowed to break down and eventually fall into disrepair; in another country, the same machines might be used efficiently in modern industry, copied, adapted, and produced locally. A company can ex-

[3] The holder of the technology may be, for example, a national or international public agency, or an individual or individuals, a periodical, or a scholarly text. Technology obviously does not have to be received by private companies from other private companies, or alternatively, transmitted by private companies to other private companies.

[4] By identical I mean within the same corporate group—company and branch or foreign subsidiary of that company. I am in this case piercing the corporate arrangements to determine the actuality.

[5] An employee of the technologically-advanced firm may leave it and travel abroad, taking with him technological information or proficiency. Nathan Rosenberg, "Economic Development and the Transfer of Technology: Some Historical Perspectives," *Technology and Culture*, XI (July 1970), 553ff., points out that in the nineteenth century it was common for technological transfers to be made by the migration of trained personnel. See also William Woodruff, *The Impact of Western Man* (New York: St. Martin's Press, 1967), chap. 5. A technologically-advanced company may open its plants to visiting foreign technicians, who see processes they can imitate in their homelands. A firm may sell or exhibit its products domestically, which products may be seen by foreigners who may reproduce the innovations in their own countries. A man may conduct industrial espionage in a plant of a technologically-advanced company and then transmit the secrets across the border. In each of these cases, technological transfer occurs, but the technologically-advanced private firm is essentially passive. It does not send goods, patents, technology, capital, or men across borders; instead, the agency for the transfer is an individual, an ex-employee, or a visitor who may or may not be associated with a private business *abroad*.

## *Private Business* 169

port consumer goods to two countries. In one the product might continue as an import; while in the second, host-nation businesses might manufacture the product. A company may register patents in countries abroad. In one nation the patents may not be worked, or be worked by foreigners; in a second, the patents may be commercialized by nationals of that land. A firm may transfer its technology through a technical assistance arrangement and in one country no one may be able or willing to utilize the advanced methods, whereas by contrast the technical assistance in another country might effectively train nationals of the host country. Similarly, a direct foreign investment carries with it technology but the technologies transmitted may be confined to the foreign corporation, or alternatively, may be absorbed by enterprises within the host nation. In each of these paired cases there is a transfer of technology, but only in the second situation in the pair does absorption or true international diffusion occur.

These comments distinguishing transfer and absorption are obviously too black and white, since they do not take into account time lags.[6] The poles are lack of absorption and rapid absorption; between the poles, absorption may take years or even decades. There seems to be not only an "imitation lag," but an "absorption gap." The literature on technological transfers says a great deal about the international imitation lag, defined by others as the lapsed time between when a product is first produced in the innovator country and in each subsequent nation.[7] The imitation lag is relatively uncomplicated to determine, yet it seems to me inadequate, because it says nothing about absorption (or true international diffusion). It does not differentiate whether the product was produced by nationals of the "imitating" country on their own, or by such nationals with extensive foreign assistance or by subsidiaries of multinational enterprises.[8] It would seem that only when nationals on their own

[6] Everett Rogers, *Diffusion of Innovations* (New York: The Free Press, 1962), pp. 18-19, 79-120, and his second edition of the same book, *Communication of Innovations* (New York: The Free Press, 1971), pp. 128-132, are excellent on time lags between awareness of the innovation and adoption.

[7] To my knowledge, the term "international imitation lag" was first used by Michael Posner, "International Trade and Technical Change," *Oxford Economic Papers*, XIII (October 1961), 323-341. See also G. C. Hufbauer, *Synthetic Materials and the Theory of International Trade* (Cambridge, Mass.: Harvard University Press, 1966), chaps. 1 and 5, and Louis T. Wells, Jr., ed., *The Product Life Cycle and International Trade* (Boston, Mass.: Division of Research, Graduate School of Business Administration, Harvard University, 1972), pp. 23-25.

[8] John R. Tilton, *International Diffusion of Technology: The Case of Semi-Con-*

(or virtually on their own) are able to produce the product does true diffusion—in contrast with mere geographical transfer—of the technology occur. Tentatively, I will define the absorption gap as the lapsed time between the introduction of a new technology, process or product, into a nation and the point when that technology is used in processes of comparable or near comparable efficiency and the manufacture of products of comparable or near comparable quality under ownership and control (defined here as technological ability) of nationals of that country. "Near comparable" is probably a better formulation than "comparable," for with effective absorption there will be modification when appropriate and also improvement to fit national requirements. Note that I am referring here to the initial—original—absorption of the new technology within a recipient nation. In going beyond international diffusion and dealing with economic development, obviously one must consider two absorption gaps, one defined, as above, to indicate simple *international* diffusion of the new technology, and the second defined to indicate infusion (or successful *national* diffusion), that is when the new technology is not only adopted and adapted by nationals but also becomes the dominant technology of the host nation industry.[9] In

---

*ductors* (Washington, D.C.: Brookings Institution, 1971), p. 23, introduces four different "lags" (including the imitation lag) but none takes into account the question of national ownership and control, although Tilton's book has much of value to say on this matter.

[9] Thus, for example, absorption gap (1) would indicate the lapsed time between the first introduction of British-made cotton textiles into India and the efficient production of such machine-made cotton textiles in a plant owned by Indian capital, run by Indian management, and operated in the main by Indian technicians. Absorption gap (2) might indicate the lapsed time between the first introduction of British machine-made cotton textiles into India and the time when say 50 percent of the output of the Indian cotton textiles came from the modern Indian cotton textile industry. I find myself very much in agreement with Stuart Bruchey's statement that "it is not so much the first appearance of new techniques as their spread [within a nation] that matters in economic growth." *The Roots of American Economic Growth* (New York: Harper & Row, 1965), p. 139. Nonetheless, the entry of new techniques in a nation and their initial absorption are clearly a precondition for their spread. Professor Solomon Barkin of the Department of Economics at the University of Massachusetts, Amherst, has been helpful to me in stressing that in considering "absorption" of technological ability—as I have defined it—I should consider separately managerial and technical personnel. I find this idea both stimulating and troubling. For example, in the 1890's, The Royal Dutch Company in the Dutch East Indies used American drillers in its crude oil producing operations. Royal Dutch had complete ownership and management control. Was the foreign technology absorbed? It seems to me that it was under Dutch corporate control, and one can legitimately refer to true technological diffusion as having taken place. Perhaps the test should be: If the business would fail or be seriously disrupted were foreign technicians removed, the control of technology cannot be said to be in national

## Private Business                    171

this paper, I am considering only the first absorption gap involving diffusion over international boundaries. In short, the international imitation lag (as defined by others) covers the transfer of technology to a foreign nation and does not take into account the nationality of the producer in the host country. The absorption gap stresses the absorption of the new technology by nationals of the recipient country. As I have defined these terms, international transfers are a necessary but not a sufficient condition for absorption or true international diffusion.[10]

With the concept of the absorption gap we are brought squarely to the need to analyze the conditions under which the international technology is received—that is, the institutional structure prepared to accept the technology, and for purposes of this paper, specifically private companies within the host country that can digest the technology. Our second group of four modes by which technology is transferred deals with this matter. Only if the companies in the second group (the receivers of technology) are nationals of the host country, I suggest, does effective international diffusion take place.

Since I am arguing that transfer does not necessarily mean diffusion, this brings me to the point that there are barriers to effective diffusion of new technology that directly relate to the receivers of technology. These often co-existent and sometimes overlapping barriers include: (1) demand barriers (there may not be sufficient de-

---

hands; if, by contrast, the business would remain viable and can find substitutes for the foreign technicians then it may be that despite the presence of foreign technicians in the operations, the technology has been effectively assimilated.

[10] In the literature on international technological transfer and diffusion, definitions vary. Sometimes the ideas of transfer, diffusion, and adoption are used interchangeably. (See for instance Tilton, *International Diffusion*, pp. 2, 163.) Sometimes "transfer" refers to the crossing of borders and "diffusion" to the spread within the borders. See John Joseph Murray's article in Daniel L. Spencer and Alexander Woronick, eds., *The Transfer of Technology to Developing Countries* (New York: Praeger, 1967), p. 9. Rogers, *Diffusion of Innovation*, p. 76, differentiates diffusion from adoption in that diffusion for him involves the spread from source to user or adopter, while adoption is an "individual matter"—"the mental process through which an individual passes from first hearing about an innovation to final adoption." In *Communication of Innovations*, pp. 12, 26, 99ff., Rogers defines diffusion as a special type of communication—"the process by which innovations spread to the members of a social system" and changes his definition of adoption to a "decision to make full use of a new idea." For purposes of this paper, as my reader is now aware, I am distinguishing between mere international transfers and true international diffusion (that is absorption)—the first implying the physical, geographical transfer of an innovation (specifically new technology) over borders, and the second designating the spread of that new technology to nationals of the host country to the extent that they can and do utilize the new technology in production.

172 *Wilkins*

mand to warrant national production); (2) capital barriers (local producers may not have or be able to obtain the capital to utilize the technology)[11]; (3) natural resource barriers (a nation's commercially-developed natural resources may be inappropriate for the effective utilization of the technology)[12]; (4) labor-cost barriers (low labor costs relative to other costs may discourage the application of a particular technology); (5) technological barriers (local producers may not have the skills or education to absorb the incremental technological knowhow)[13]; (6) scale barriers (foreign producers may have economies of scale that cheapen costs vis-à-vis host nation producers; without government protection, national producers may have no possibility of meeting foreign competition); (7) infrastructure barriers (there may not be sufficient supporting services or complementary techniques to warrant diffusion); (8) cultural barriers (there must be values and norms of behavior conducive to the absorption of technology); and (9) most easily overcome, language barriers, which may slow absorption. There may also be "priority barriers" within a particular economy.[14] Herein, I do not intend to elaborate on these barriers, which are obviously of vast importance. The barrier, however, that directly concerns me is one that should be (but is often not) included on the above list—that of "business organization." There must be effective business organization (private or governmental) to absorb the technology.[15]

[11] As Joseph Bower points out in *Managing The Resource Allocation Process* (Boston: Division of Research, Graduate School of Business Administration, Harvard University, 1970), p. 39, "Studies of the research and development process indicate that expenditures rise exponentially as a product moves from the basic and applied research steps to development and production. It is factories, tools, and dies, trained labor, reoriented channels of distribution and promotion which are the truly expensive part of innovation."

[12] See Nathan Rosenberg, ed., *The Economics of Technological Change* (London: Penguin Books, 1971), pp. 210, 274-281.

[13] Professor Rosenberg has put it, in describing the United States in the nineteenth century, "it required considerable technical expertise to borrow and exploit a foreign industrial technology." *Technology and American Economic Growth* (New York: Harper & Row, 1972), p. 82.

[14] See note 54 below for an example of priority barriers. Priority barriers may be erected by governments as well as faced by private companies. Thus, the Soviet government may decide not to manufacture certain consumer goods, not because of demand, capital, natural resource, labor-cost, technological, scale, infrastructure, cultural, or language barriers, or even business organization barriers, but because of priority barriers.

[15] Effective business organization includes attitudes of management. G. F. Ray, "The Diffusion of New Technology—A Study of Ten Processes in Nine Industries," *National Institute Economic Review* (May 1969), 83, concludes that the attitude of management has the "greatest impact on the application of new techniques." I am

## III

With these general comments, I am now ready to examine the actual process of technological transfer by private companies. Regrettably, my examples are unsystematic. The difficulty lies in the shortness of the paper. The examples should, however, demonstrate forms of transfer and their relation to diffusion, as well as provoke thought about methodology in dealing with international technological diffusion by private business. Whether the classification scheme proves useful in studying the success of the particular type of international diffusion has to be tested in subsequent research.

All the eight modes of transfer that I have outlined in the early part of this paper have existed in the nineteenth and twentieth centuries. First, exports: The British, fearing the diffusion of technology in the early nineteenth century, barred the sale abroad of certain textile machinery.[16] Britishers bypassed the law and established manufacturing enterprises on the continent and *exported* from there, directly transferring and diffusing British manufacturing methods.[17] In the late 1820's and 1830's, English builders sold

---

far from alone in talking about the absorptive capabilities of recipient firms. See Jack Baranson, "Technology Transfer Through the International Firm," *American Economic Review*, LX (May 1970), 435-436. Baranson in this article is concerned with factors affecting transfer logistics of the international firm; he barely touches on the problems of transfer as distinct from diffusion but he does recognize the importance of absorptive capabilities. In my general analysis in this paper, I find myself influenced by the body of work that deals specifically with transfers without diffusion, for example, the seminal article by Hans Singer, "Distribution of Gains between Investing and Borrowing Countries" (1950), reprinted in Hans Singer, *International Development* (New York: McGraw-Hill, 1964), pp. 161-172; the concept of "a dual economy" that now appears in most textbooks on less developed countries; and statements such as the one that appeared in a 1956 National Planning Association Study. After noting that U.S. firms had been transferring technology to Latin America for years, this study concluded, "Unfortunately, however, only a low proportion of the many small firms which are still using primitive practices throughout Latin America have as yet been reached by the methods and techniques which are being introduced by U.S. firms and their affiliates." National Planning Association Special Committee on Technical Cooperation, *Technical Cooperation in Latin America—Recommendations for the Future* (Washington, D.C.: National Planning Association, 1956), p. 77. This will henceforth be cited as NPA Technical Cooperation Study. I have a number of reservations about the legitimacy of such views, but find it essential to take their premises into account in a consideration of the history of the international diffusion of technology. These views touch on the basic question of the abilities of recipients of technology to absorb the technology.

[16] Details on such British restraints appear in Great Britain, "First Report from Select Committee to Inquire into the Operation of the Existing Laws Affecting the Exportation of Machinery," *Parliamentary Papers*, Vol. 7 (1841).

[17] For example the Cockerill firm, using British methods and manufacturing in Belgium and Germany, sold its machines as far east as Poland; new textile enterprises developed, incorporating the new technology. David S. Landes, *The Unbound*

174                           *Wilkins*

their locomotives in the United States; these were copied and improved upon and "a locomotive-building industry sprang up in the United States almost at once."[18] In the 1840's, Stephen Moulton carried to (exported to) England samples of Charles Goodyear's vulcanized rubber, exhibiting the product to prospective manufacturers; these samples were seen by Englishman Thomas Hancock, who had worked on rubber manufacture for many years; not long after, Hancock took out his own patents that virtually duplicated Goodyear's process. He then proceeded to manufacture rubber goods.[19] The Singer records are full of data expressing concern about the imitation of Singer sewing machines in western Europe.[20] In the twentieth century, capital equipment exports were often a means of transferring technology abroad. Likewise, exported products of all sorts were imitated.

Second, registering of patents abroad served to transfer technology. In the nineteenth and twentieth centuries U.S. companies in Europe and European enterprises in the United States obtained patents. The patents were worked in the foreign country. Examples include manufacture of revolvers, aluminum, electrical equipment, and chemicals.[21]

---

*Prometheus* (Cambridge, Eng.: Cambridge University Press, 1969), pp. 150, 148. Bruchey notes that despite the ban on British machinery exports, a substantial number of British machines reached the United States to be copied and, more important, modified to meet U.S. requirements. Bruchey, *Roots of American Economic Growth*, p. 167.

[18] Eugene S. Ferguson, "The Steam Engine Before 1830," in *Technology in Western Civilization*, eds. Melvin Kranzberg and Carroll W. Pursell, Jr. (New York: Oxford University Press, 1967), I, p. 299.

[19] Moulton was British, emigrated to the United States, and established his business there. He took the rubber samples to England, hoping to sell "the inventor's secret." William Woodruff, "Origins of An Early English Rubber Manufactory," *Bulletin of the Business Historical Society*, XXV (March 1951), pp. 32-36. Hancock's lawyers denied that the latter had gained technological information directly from Goodyear's samples. Woodruff suggests, "Perhaps his [Hancock's] genius lay in appreciating what Goodyear had done. . . .There can be no doubt that Goodyear's discovery stimulated the English inventor to still further effort."

[20] Singer Manufacturing Co. records, State Historical Society of Wisconsin, Madison, Wisconsin.

[21] Thus Colt licensed companies on the European continent to make revolvers under Colt patents in the 1850's. See Mira Wilkins, *The Emergence of Multinational Enterprise: American Business Abroad from the Colonial Era to 1914* (Cambridge, Mass.: Harvard University Press, 1970), p. 30. The predecessor of Aluminum Company of America, The Pittsburgh Reduction Company, that acquired the Hall patents for making aluminum in 1888, granted in 1895 a license under these patents to a small French firm, rights that soon passed to d'Alais et Camargues, later Cie. Pechiney. See George W. Stocking and Myron W. Watkins, *Cartels in Action* (New York: Twentieth Century Fund, 1946), pp. 220, 227. Annual reports and company prospectuses in the Scudder Collection, Columbia University Library, reveal numerous licens-

## *Private Business* 175

Third, a range of technical assistance arrangements have been made by private firms to communicate technology. Often capital equipment exports were accompanied by a single mechanic or a group of technicians that installed the equipment and instructed the customer in its operation and maintenance.[22] When, for example, American elevators and electrical equipment were marketed abroad in the late nineteenth century, technicians frequently accompanied the export.[23] The German Von Kohorn Company sold machinery and technical aid for the establishment of the viscose rayon industry in Czechoslovakia (1919), Greece (1923), Turkey (1935), Rumania (1937) and then farther afield in Peru (1946) and Egypt (1948).[24] In more recent times, as well, this phenomenon of exporters sending technical knowhow with their exports has persisted.[25] Likewise, when patents were worked abroad, often the innovating firm would transfer technological information beyond what was in the patent registration. Thus, for instance, when in the 1850's, the Singer Company sold its French patent, it agreed to send to the purchaser an aide for his manufacturing department so that merchant could make "perfect machines."[26] Frequently, in the twentieth century, the licensing of patents and technical assistance accords went together.[27]

While associated with exports and patents, technical assistance arrangements may go far beyond the other two modes of transfer. There were patents included in the interchange of information between Standard Oil (N.J.) and I. G. Farben and between du Pont and the large European chemical companies before World War II, but the technological assistance transcended the mere licensing of

---

ing relationships. Sometimes patents were taken out abroad under the names of individuals on behalf of a company, sometimes by the company itself. U.S. Bureau of Census, *Historical Statistics of the United States* (Washington, D.C.: G.P.O., 1960), pp. 607-608, gives data on the number of patents issued in the United States to residents of foreign countries and foreign corporations. I know of no one who has attempted to use these data in considering problems of international technological transfer.

[22] See, for example, Landes, *Unbound Prometheus*, p. 150, on Cockerill exports.

[23] Data from company records of exporters of these products.

[24] Hufbauer, *Synthetic Materials*, p. 93.

[25] All through the twentieth century, American firms have sent technicians to install machinery in plants in Latin America and over time have trained local employees to operate and maintain the machinery. NPA Technical Cooperation Study, pp. 76-77.

[26] Wilkins, *Emergence of Multinational Enterprise*, p. 38.

[27] See Mira Wilkins, *The Maturing of Multinational Enterprise: American Business Abroad from 1914 to 1970* (Studies in Business History, Cambridge, Mass.: Harvard University Press, forthcoming).

## 176 *Wilkins*

patents.[28] In recent years, the many agreements between U.S. and Japanese enterprises for technological exchanges sometimes include patent exchanges yet they comprise far more than the licensing of such patents.[29] Some technical assistance accords may be entirely independent of patents. In 1908, Herbert Hoover organized an international mining consulting firm to sell U.S. technological services.[30] British Managing Agents in India transferred technological knowhow.[31] When management contracts are made between western companies and firms or governments in less developed countries there is a sale of organizational and technological skills.[32]

Private business enterprise has had experience with a particular technology that it has developed or used. It has trained individuals to work with the technology. It has knowledge of the problems and difficulties in commercializing the particular technology. It has organizational knowledge. It is in short in a unique position in the transfer of the specific technology. It seems clear that often the product, or the description in the patent, or mere drawings and instructions, are inadequate for transfers of technology; men are needed to carry, explain, and facilitate the introduction of the new processes or products. The private firm can provide the institutional framework whereby these men can transfer the technology.[33]

Four, technology also crosses boundaries through direct foreign investment. Closely associated with exporting, registering patents abroad, and technical assistance is direct foreign investment. Generally, the international business carries on all these functions and has done so since the nineteenth century, if not before.[34] As prac-

[28] *Ibid.*

[29] Data obtained in Japan from U.S. and Japanese companies and the Ministry of International Trade and Industry.

[30] Herbert Hoover, *Memoirs* (New York: Macmillan, 1952), I, pp. 28ff.

[31] P. S. Lokanathan, *Industrial Organization in India* (London: George Allen & Unwin, 1970), pp. 15-16.

[32] See Peter Gabriel, *The International Transfer of Corporate Skills* (Boston: Division of Research, Graduate School of Business Administration, Harvard University, 1967).

[33] These comments are especially true of the twentieth century as technology became more complex. But even in the nineteenth century, as others have noted, imitation of products and processes and development of patents often required foreign personnel familiar with the techniques. Such men in the nineteenth century were sometimes (and sometimes not) associated with private companies. In recent times, when a company has moved away from specific to "overall technology," it may fail completely. This was the case with Litton Industries' much discussed contract with Greece for the economic development of Crete and the Peloponnesus peninsula.

[34] There are numerous instances in the nineteenth century wherein private companies carried technology over borders through direct investments. For the activities

## Private Business 177

tically every writer on the subject has pointed out, direct investors communicate management, technology, and skills across national boundaries. Recent studies have indicated that the firm that invests abroad generally has an advantage, an advantage in technology, product design, marketing, or managerial expertise.[35] It has this advantage to communicate. Corporations that own foreign factories, mines, oil properties, and plantations transmit technology in various ways: (1) There is clearly a physical (geographical) transfer. Beyond that, the products made and the processes used abroad are there to be imitated within the host nation.[36] Also, there can be a shift of the technology of the direct investor to host country nationals should expropriation occur or should a national firm purchase the properties of the direct investor.[37] (2) Host nation workers and managers gain knowledge of products and processes. The training may be on the job, in the corporation's home operations, at local educational institutions supported by the company, or at foreign universities (subsidized by scholarships granted by the multinational business).[38] The training can range from that in simple skills to that in highly-sophisticated modern technology and business administration. (3) If the activity of the international en-

of American companies in this respect see Wilkins, *Emergence of Multinational Enterprise.* How many cases of technological transfer through direct investment one can find in the eighteenth century is unknown to the present author, but clearly European companies before 1800 through direct investment appear to have played a role in technological transfer. Thus, in 1770, a French company operated a coal mine at Hagenbach in Baden and appears to have transferred the more advanced French methods to Germany. See Rondo E. Cameron, *France and the Economic Development of Europe 1800-1914* (Princeton, N.J.: Princeton University Press, 1961), p. 372. Alexander Gerschenkron, *Economic Backwardness in Historical Perspective* (New York: Praeger, 1965), pp. 38-39, suggests such activities by the Fuggers in the fifteenth and sixteenth centuries.

[35] See for example Raymond Vernon, *Sovereignty at Bay* (New York: Basic Books, 1971); Wilkins, *Emergence of Multinational Enterprise;* and Wilkins, *Maturing of Multinational Enterprise.*

[36] On imitation abroad see Robert B. Stobaugh, "The Product Life Cycle, U.S. Exports, and International Investment," DBA dissertation, Graduate School of Business Administration, Harvard University, 1968, and Tilton, *International Diffusion,* p. 164.

[37] These two types of diffusion are seldom discussed, yet they are of some significance.

[38] Today multinational corporations boast of their contributions in this sphere. It wasn't always so: William Woodruff tells of how in the 1850's American investors in a rubber plant in Scotland imported skilled labor from the United States; part of the reason was the company's fear that skilled British rubber workers "might only stay long enough to make off with the firm's secrets." William Woodruff, "The American Origins of a Scottish Industry," *Scottish Journal of Political Economy,* II (February 1955), 28.

178                        *Wilkins*

terprise is a joint-venture, foreign technology is brought under partial host nation ownership.[39] (4) Suppliers of the direct foreign
investor frequently obtain significant technological assistance.[40] (5)
If technology is broadly defined to encompass marketing experience
(including the servicing of complex products), technology is often
transferred to dealers and distributors. (6) In addition, indirectly,
but of great significance, the multinational corporation acts to transfer technology by paying taxes and offering employment in the host
nation, which actions create capital resources and demands there.
The resources and demands often in turn become magnets that will
result in the emergence of agencies for the subsequent transfer of
technology not specifically required by the multinational corporation.[41] Examples of all these types of transfers of technology
abound.[42]

The fact of transfer by export, by patent, by technical assistance,
or by direct investment says nothing about the appropriateness of
the technology transferred for the host country. Some argue that
technology suitable in one country may be less suitable in a second
nation that has different relative costs of factors of production and
a different demand structure. Indeed, in the main (although far
from always) when a firm transfers technology, it does little to modify manufacturing methods; it transmits what it has developed at

[39] A number of governments in recent years have forced multinational enterprises
to have local partners in part in order to diffuse technology.

[40] There is marvelous, detailed material to illustrate this point that I have uncovered in the files of Ford-Werke, Cologne. When in the early 1930's, Ford began
to manufacture in Germany, it needed local suppliers. It made arrangements for
German suppliers to learn about U.S. technology so that its German subsidiary could
buy quality German-made parts. See Frederick C. Young to E. C. Heine, Dec. 19,
1934; Frederick C. Young, "Report on Cologne," Dec. 22, 1934; and T. F. Gehle to
A. M. Wibel, Jan. 31, 1935 and Mar. 14, 1935, Ford-Werke Archives, Cologne.
Moving to a totally different area, multinational corporations that are buyers of
rubber, cotton, and bananas have given technical assistance to small growers. See
NPA Technical Cooperation Study, p. 76.

[41] Thus, for example, the Kuwait Oil Company pays taxes to the Kuwaiti government. That government used part of its revenue to buy technology to build a water
desalination plant for Kuwait City. Using tax revenues, the host government becomes
the vehicle for technological transfers. Or, as a second example, employees of Kuwait
Oil Company have demands. Technologically-advanced goods are imported into
Kuwait to meet the new demands.

[42] The literature on the multinational corporation and its technological contributions is substantial. The National Planning Association published in the 1950's and
early 1960's a series of Case Studies on United States Business Performance Abroad,
many of which studies sought to reveal the technological contributions of multinational corporations. James Brian Quinn, "Technology Transfer by Multinational Companies," *Harvard Business Review*, XLVII (Nov.-Dec. 1969), 147-161, provides
numerous examples of transfers of technology.

## Private Business 179

home. In many industries, the high engineering costs of designing plants "strongly militate against redesigning [them] to employ more labour and less capital" or to take advantage in other ways of different resource availability in the recipient nation.[43] Brazilian economist Celso Furtado insists that the introduction of new technology in manufacturing in less developed countries by giant technologically-advanced international business creates "structural imbalances."[44] On the other hand, many feel that not only in industrial nations, where there is more comparability in factor costs, but also in less developed countries, the advanced technology is appropriate.[45] More research needs to be done and better tests of appropriateness developed. Yet I would suggest—as I have earlier—that it is more the receiver than the communicator of the technology that is responsible for diffusion and the most stringent tests lie in that arena.

This brings me to the second group of transmitters of technology. First, the importer: Within the host country, private companies (the importer or other firms) may undertake to manufacture an import locally. Such import substitution is far from automatic. There must be a demand for the new product or processes and also an in-

[43] Quotation is from Hufbauer, *Synthetic Materials*, p. 68, and applies to his work on the chemical industry; it is also applicable to other industries, although in some industries there is clear evidence that multinational corporations have adapted to foreign conditions—if not too frequently to differences of factor availability at least to diverse foreign demand. Recent research at the Harvard Business School by Professor Robert B. Stobaugh indicates that often adjustments of technology by multinational corporations occur in *material handling* and *packaging* rather than in the actual production activity.

[44] Celso Furtado, *Obstacles to Economic Development* (Garden City, N.Y.: Doubleday & Co., 1970). On the many advocates of some technological adjustment, see Louis T. Wells, Jr., "Economic Man and Engineering Man: Choice of Technology in a Low-Wage Country," *Public Policy*, XXI (Summer 1973), 319, n. 1. This article has fascinating data on the selection of technology in plants in contemporary Indonesia.

[45] Such is certainly the thrust of the N.P.A. studies, cited in note 42 above. It is the policy implicitly accepted by Brazilian *government* economists. Gerschenkron, *Economic Backwardness in Historical Perspective*, pp. 9, 26, argues that largely by the application of "modern and efficient techniques" can a backward country achieve success and that the advanced technology is the right one. From a different point of view, others agree that techniques to be appropriate should be modified and feel that some international firms are "more willing and able than others to adjust industrial transfers to the specialized needs of developing countries." Baranson, "Technology Transfer through the International Firm," p. 440. For an intelligent, although limited, consideration of the impact of the foreign investor's communication of technology on the Canadian economy, see Report of the Task Force on the Structure of Canadian Industry, *Foreign Ownership and the Structure of Canadian Industry* (Ottawa: Privy Council Office, 1968), pp. 56-60, 66-70.

## 180                         *Wilkins*

stitutional structure to undertake the import substitution. While a great deal has been written about import substitution by less developed countries in recent times, there appears to be a paucity of analysis on the pace and character of import substitution in the nineteenth and early twentieth centuries and the extent to which import substitution has involved "mere transfer" or full absorption. For example, there are figures available on the number of power looms in France and Germany in the 1860's and 1870's,[46] but not the number actually manufactured within those countries; we have figures on the capacity of steam engines worldwide in the late nineteenth century,[47] but not the breakdown on the steam engines from abroad and those produced within a host nation. We have inadequate data on the extent to which power looms or steam engines that were made within a particular country were manufactured by nationals of that country and the extent to which they were produced by foreign companies operating within that country.

From available information, it is clear that one aspect of technological diffusion in the nineteenth century lay in the significant import substitution in the United States and western Europe. There is substantial evidence that in certain products, American firms, for example, rapidly substituted locally-produced goods for imports. Likewise, in the late nineteenth century, when British shoe manufacturers met American competition, *they* replicated U.S. methods to meet competition.[48] On the other hand, comparable import substitution did not occur when U.S. shoe manufacturers, for instance, sold in Latin America—at least in the nineteenth century. Americans and Britishers had companies capable of—and determined to—copy and adapt the methods of foreigners. In the twentieth century numerous cases exist of imitation of imports, resulting in the diffusion of technology.[49] On the other hand, not every nation has

[46] Landes, *Unbound Prometheus*, p. 214.

[47] *Ibid.*, p. 221.

[48] Rosenberg, *Technology and American Economic Growth*, p. 45n. The British started to import American shoe-making machinery to obtain the appropriate technological processes. See John H. Dunning, *American Investment in British Manufacturing Industry* (London: George Allen & Unwin, 1958), pp. 31-32, and International Management Association, *Case Studies in Foreign Operations* (New York: International Management Association, 1957), pp. 77-78, on United Shoe Machinery's activities in Britain providing American machinery for the "modernized industry."

[49] Product cycle theory argues that products are exported, imitated abroad, and that this becomes a basis for direct investment in foreign countries by the exporter. See Raymond Vernon, "International Investment and International Trade in the Product Cycle," *Quarterly Journal of Economics*, LXXX (May 1966), 190-207.

## *Private Business*                                      181

companies able or ready to imitate or adapt the technology. It may be foreign and not national business that provides for the import substitution, closing the imitation lag but not the absorption gap.[50]

Two: Firms operating on the basis of foreign patents sometimes merely transfer as distinct from diffuse technology. This may be the case when the manufacturer abroad is part of a multinational enterprise. It was the case when the revolvers produced by Europeans under Colt license never reached the high standards of the American product.[51] By contrast, often, the exchange of patents between private firms has proved highly effective in the diffusion of technology. Data available to business historians reveal substantial evidence of domestic-incorporated companies that have obtained licenses to work foreign patents at home.[52] The German General Electric Company (Allgemeine Elektrizitäts Gesellschaft) started its business on the basis of American patents and a minimum of technical aid. The assimilation of technology was highly successful, and A.E.G. was soon innovating.[53] Diffusion was not always so rapid; German chemical patents were registered in the United States in the nineteenth and early twentieth centuries. Some of these patents were worked by German subsidiaries in the United States. There was a transfer of the technology to this side of the Atlantic. Then, with World War I, these patents were confiscated and made available to American companies; only at this time was there absorption of the technology. In 1917-1918 there existed in the United States companies capable of working these patents and a domestic demand for the output under them.[54]

[50] The automobile industry is a fine example of this proposition. The mass produced American automobile made with interchangeable parts was exported to Europe. William Richard Morris (later Lord Nuffield) set out to compete with the Model T, producing his first car, the Morris-Oxford in April 1913. Morris borrowed American technology. Mira Wilkins and Frank Ernest Hill, *American Business Abroad: Ford on Six Continents* (Detroit: Wayne State University Press, 1964), p. 51. On the other hand, when in the late 1950's and 1960's Brazil and Argentina determined to substitute domestic car and truck production for imports, it was in the main foreign rather than domestic capital that undertook to manufacture. *Ibid.*, pp. 416-119.

[51] Wilkins, *Emergence of Multinational Enterprise*, p. 30.

[52] Such data are highly miscellaneous, ranging from annual reports, company records, government hearings and reports, antitrust case materials, to business histories.

[53] Wilkins, *Emergence of Multinational Enterprise*, p. 54.

[54] Why were Americans so slow to imitate in this case? The reason seems to lie not in business organization, not in capital, natural resource, labor-cost, technological, scale, infrastructure, cultural, or language barriers, or even completely in demand factors but rather in priorities. The demand was small; the profit potentials did not seem great; and more important up until World War I the Germans adequately filled the existing demand. With the war, the demand structure changed and the

We turn next to item three in this group: the receivers of technical assistance. Technical assistance may be communicated to firms incapable of absorbing this aid. It may be communicated to foreign subsidiaries of the holder of the technology and thus kept within the holders' own family group—a geographical transfer. On the other hand, technical assistance from abroad may be *requested* by —sought out by—host nation companies, be effectively utilized, and serve as a highly viable means of both technological transfer and diffusion. Technical assistance obtained from abroad may be particular or general, informal or formal, short-term or long-term. A few examples will suffice. In the late 1860's or early 1870's, Henry Phipps, financial director of Union Iron Mills Company (one of the firms that would become part of Carnegie Steel Company) visited a mill in Germany and noticed that the piles made ready for the heating furnace, to be used for rolling "I" beams, contained more than double the amount of scrap iron rails employed in Pittsburgh. He sketched the pile and once home ordered a change in Union Mills' practice. We are told that "the cost of this trip to Europe was saved almost daily thereafter to his firm."[55] Here was a case of specific technological assistance, informally obtained, on the basis of one journey. Similarly, on a European trip in 1872, Andrew Carnegie studied Bessemer steel works, recognized the significance of the new technology, and on his return made plans based on what he saw in England.[56] Here, too, we have technical information, informally obtained, on a single trip, but in this case general technological know-how transmitted by the chief executive of the recipient firm. Earlier in American history, when in 1801-1802, Irénée du Pont planned a powder mill in the United States, he drew on French technical aid. He sought out and arranged that French government draftsmen would draw the plans for his company's machinery, that the machines would be constructed in France, and that if needed, the French would send technical aid. Soon, however, Irénée du Pont's powder plant had absorbed the French technology and was

---

former sources of supply were gone. Working chemical patents became of high priority. Diffusion occurred.

55 James Howard Bridge, *The Inside History of the Carnegie Steel Company* (New York: The Aldine Book Co., 1903), p. 35.

56 *Ibid.*, pp. 75, 86. It is important that Carnegie grasped the potentialities of the new technology. See Robert A. Solo, "Technology Transfer," in Robert A. Solo and Everett M. Rogers, eds., *Inducing Technological Change for Economic Growth and Development* (East Lansing: Michigan State University Press, 1972), p. 18.

## Private Business 183

on its own.[57] Here we have general technical assistance on a formal but short-term basis.

Sometimes technical assistance came from the men hired. Thus, British mining companies in South Africa sought out and employed American technicians.[58] When the Belgian firm SIDAC began producing cellophane in 1925, it did so with the aid of the chief engineer from La Cellophane (a French company that was the first producer of cellophane in 1917). The engineer brought to the Belgian company blueprints and complete data on the French firm's secret processes for cellophane manufacture.[59]

J. S. Fforde in his volume, *An International Trade in Managerial Skills*, tells of how Britishers "of the technical managerial type" went to the Indian sub-continent and Latin America for "career service" in one business and would be recruited by one enterprise after another (in jute manufacture, cotton spinning and weaving, paper manufacture, flour milling, and light engineering) as "a type of efficiency expert."[60] Unfortunately Fforde does not tell when this practice started; presumably it relates to the late nineteenth and twentieth centuries. He also implies but does not state that these men were hired by local capitalists as well as foreign enterprise. How effective this was in diffusion of technology still needs closer study.

On a more formal basis we have the technological assistance arrangements that existed between Standard Oil of New Jersey and I. G. Farben and between du Pont and the major European chemical companies. In the chemical industry in the first part of the twentieth century, European, especially German, technology greatly impressed Americans. "I was plunged into a world of research and development on a gigantic scale such as I had never seen," wrote Frank A. Howard of Standard Oil Development Company (a subsidiary of Jersey Standard) after a tour of the Badische Anilin und Soda research laboratories at Ludwigshaften in early 1926. What Howard saw was a pilot plant for the hydrogenation of oil. Badische was then being merged into the newly-formed I. G. Farben and Jersey Standard entered into arrangements with the giant Ger-

---

[57] William S. Dutton, *Du Pont* (New York: Scribners, 1942), p. 31 and *passim*.
[58] Interviews in South Africa.
[59] Hufbauer, *Synthetic Materials*, pp. 88-89, 131.
[60] J. S. Fforde, *An International Trade in Managerial Skills* (Oxford: Basil Blackwell, 1957).

184                    *Wilkins*

man firm to obtain technological knowledge. Jersey Standard's
historians have recorded that "with the help of I. G. Farben 'know-
how'" engineers of Jersey Standard's affiliates "mastered a new,
difficult, and promising process." The company's historians conclude
that from I. G. Farben, Jersey Standard gained "research concepts
and techniques" as well as "the stimulus that [in time] contributed
to the building up of a large research staff soundly trained in chemistry and chemical engineering."[61] In a similar vein, an internal
memo from du Pont's files, dated December 9, 1936, shows the impact of the international exchange of technical aid on that receiving
company. "It should be borne in mind that a number of the du
Pont Company's most important activities have originated from
technical information derived from European sources, examples being rayon, 'Cellophane,' ammonia, hydrogen peroxide, titanium di-
oxide, to mention only a few." The memo noted that as a result of
its technical agreements with European groups "the du Pont Company has been able to offer numerous products developed in Europe
in the American market."[62]

Perhaps the most impressive (and successful) technical assistance accords have been between Japanese and Western firms in
the post-World War II years. Here, too, in the main, Japanese companies seem to have taken the initiative in seeking out the technology. Over the years 1950 to 1970, the Japanese government
approved 8,324 contracts made by Japanese concerns involving the
purchase of technology from western enterprises.[63]

These are only a scattering of technical assistance arrangements
prompted by actions (desires) of *recipient* private enterprises.
Clearly, one needs more than assorted instances and a systematic
treatment by industry, as well as by country and region, of the effectiveness of the various types of technical assistance in technological diffusion. My point here is simply that often domestic—

[61] Henrietta M. Larson, Evelyn H. Knowlton, and Charles S. Popple, *New Horizons, 1927-1950* (New York: Harper & Row, 1971), pp. 153-159.
[62] J. K. Jenney, Foreign Relations Department to J. E. Crane, Dec. 9, 1936,
Eleutherian Mills Historical Library, Greenville, Wilmington, Dela., Accession 1231,
Box 2.
[63] See Terutomo Ozawa, "Should the United States Restrict the Technology
Trade," *MSU Business Topics*, XX (Autumn 1972), 35. An excellent piece on technology transfers to Japan is George Hall and Robert Johnson, "Transfer of U.S.
Aerospace Technology to Japan" in *Technology Factor in International Trade*, edited
by Raymond Vernon (Special Conference Series No. 22; New York: National Bureau
of Economic Research, 1970).

## *Private Business*                       185

receiving—firms took the initiative in obtaining technological assistance from abroad and that this type of initiative should be tested
as possibly one of the most effective forms of international technological diffusion. I might suggest that the reason for its effectiveness
was that when this occurred there was an existing private business
structure, an agency, that could absorb the technology. The defined
demand was determined by the recipient rather than by the donor.
The selection of the technology to be received was by the recipient,
who hoped to profit from its receipt.

The last of our four modes by which private companies obtain
technology involves the receipt by companies of technology from
direct foreign investors. As we have earlier noted, when a company
invests abroad, in a geographical sense it transfers technology; yet,
as we have also noted, it may not diffuse technology for the latter
may be contained within the corporation. Yet there do exist imitators (absorbers) of the processes and products introduced by multinational corporations. In developed countries, it is commonplace
that when a direct foreign investor undertakes operations, competitors using similar processes and making similar products emerge.
(Sometimes the direct investment is made because the imitation of
the export has taken place and the holder of the technology can
not maintain its market unless it manufactures nearby; often imitation seems likely and occurs *after* the direct investment has been
made.)[64] This type of absorption by private companies in the host
country is, however, more difficult in less developed countries,
where private companies have neither the organization nor the
capital to replicate the methods of the foreign investor.[65] Sometimes
technology is diffused when a direct investor sells out to a private
domestic firm. In England, for example, before World War I Westinghouse set up a foreign manufacturing subsidiary; this subsidiary
was sold to the British, Metropolitan-Vickers, which obtained Westinghouse's technology in the transaction.[66] U.S. direct investors
dominated the Cuban sugar industry in the 1920's, introducing new
technology; gradually, over time, a number of the properties were

[64] The formulation often made is that direct foreign investment takes place after
the market for exports is "threatened"—meaning by the existence of imitators or the
*potential* for competition.

[65] Here the government, which is beyond the subject limits of this paper, may
take over the technology.

[66] Wilkins, *Maturing of Multinational Enterprise.*

186 *Wilkins*

transferred to Cuban capital, and Cuban businessmen took the place of Americans.[67]

Employees and managers of foreign subsidiaries often have been hired by host nation enterprises and serve to transfer technology. In recent years, European companies have eagerly sought personnel who have worked for foreign subsidiaries of American firms.[68]

When the direct investor participates in a joint-venture with a host nation firm, that company obtains valuable technology. This has been the motive of a number of host-nation companies that have approached foreign firms, suggesting joint-venture relationships.[69]

Because a direct investor in a foreign country creates certain demands, there are linkage effects resulting from the direct investor's activities. Private companies in the host country often seek to fill the demands. Here the technological diffusion often will be *associative* rather than direct. For example, when a foreign company invests in an extractive industry in a less developed country, its employees probably need housing; local private companies often learned from the foreign company not the basic technology of the latter's industry but rather a new technology of home building.[70] So, too, when Sears, Roebuck opened a department store in Peru in 1955, it did not create other mass marketers in Peru. Rather, its technological diffusion was associative. Soon it was seeking local suppliers, and local suppliers started to seek out Sears. In 1959, the president of the Lima firm of Industrias Reunidas, S. A. asked Sears whether it would be interested in marketing a nationally-made refrigerator. Sears was interested; the Peruvian firm obtained with Sears' help a license from an American manufacturer. Two years

[67] *Ibid.* From 1934 to 1958 there had been a steady decline in U.S. influence in the Cuban sugar industry.

[68] J.-J. Servan-Schreiber, *The American Challenge* (New York: Atheneum, 1968), p. 4.

[69] For example, in 1904 Gordon M. McGregor, a Canadian wagon builder, visited Henry Ford and convinced him operations in Canada were desirable. Ford agreed that he and the American company would furnish the Canadian enterprise with patents, plans, drawings and specifications needed to build Ford cars, and Ford personally would give "such reasonable and sufficient oversight" as was required. In return for the technology, the stockholders in Ford Motor Company obtained a 51 percent interest in the Canadian firm and Ford was paid a fee for his services. (Wilkins and Hill, *American Business Abroad*, pp. 14-18). In the late 1920's and early 1930's, the Japanese wanted to build their own refining industry. One way was through a joint-venture with an American oil company. In this case the Japanese held control, while obtaining U.S. technology. Data from the Archives, Mitsubishi Oil Company, Tokyo, Japan.

[70] Based on my own visits to such enterprises in Latin America and the Middle East.

## Private Business                              187

later, using American technology, the first refrigerator was made in Peru. The demand created by the direct investor had been the stimulant for such production; Sears had been the catalyst for the transfer and diffusion of technology; the Peruvian firm, however, had initiated the suggestion.[71]

Sometimes the linkage effects are more general; thus, in Canada, employment offered by multinational corporations has contributed to a higher standard of living and the raising of the level of demand. Canadian firms have sought to attract new technology to meet the demands.[72]

These are but a few of the many instances wherein private companies in receiving countries have tried to obtain technology and have been successful in obtaining that technology from, through, or based on the presence of the direct foreign investor.

### IV

In conclusion, then, this paper has been a modest attempt to define various aspects of the role of private business as a vehicle for the diffusion of technology. It has been difficult to write because there are so many facets of this fascinating subject that seem to cry out for exploration. Because of space limitations, I have had to be highly selective. Among the numerous relevant topics not discussed or barely considered are: (1) the relations between the *motive* behind technological diffusion by private enterprise and the effectiveness of that diffusion; (2) the process whereby private enterprise changes its strategies through time and takes on over the years an altered role in the diffusion of technology; (3) the attributes of successful diffusers of technology (do such attributes exist in the abstract?); (4) the variation between and among industries and technologies in technological diffusion by private firms (are certain industries and technologies more amenable to technological transfers by private enterprise than others?); (5) the success of technological diffusion by private firms as compared with other agencies for diffusion (has this varied through time?); (6) a systematic look at differences in receptivity to international diffusion of technology

[71] William R. Fritsch, *Progress and Profits: The Sears, Roebuck Story in Peru* (Washington, D.C.: Action Committee for International Development, 1962), pp. 22, 50-51.

[72] Often in Canada, it was other multinational corporations that met the new demands. Yet, there were cases wherein Canadian private companies took on that role.

## 188                          *Wilkins*

by various nations; (7) a comparison of the demand structure and factor proportions within both the donor and recipient nations and the effects on international diffusion by private firms; (8), which is associated with point (7), the appropriateness of technology developed by private firms for international diffusion; (9) the distinctions between what is economically sound for the private enterprise and for the nations receiving the technology; and (10) an exploration of the types of measures that might be employed as indices of the effectiveness of private enterprise in technological diffusion (how much, for example, can productivity data be used as a measure of technological diffusion?). On each of these topics, and many others as well, there is a vast amount to be learned. Because of space constraints, I have, however, limited my content.

I have herein presented eight ways by which private enterprise in technologically advanced and in receiving countries acts to transfer technology. Clearly, private enterprise transfers technology in a variety of manners within these eight categories. I have tried to emphasize that for true international diffusion there must be more than simple geographical transfer of technology; there must be absorption of technology by national enterprises within the host country. While there are a number of factors affecting absorption by the host nation, ranking high among them is the existence or non-existence of agencies to receive the technology. One of the most significant of such agencies has, in the past, been private business. Thus, I have argued in this paper that one must not only study the holders of technology as vehicles of diffusion but also the receivers of technology. I have suggested that with the existence of international business, the concept of the "imitation lag" may cover more transfers and not diffusion per se. A more fruitful concept might be that of an "absorption gap," a notion that considers the time that true international diffusion takes, the time between the introduction of a new technology into a nation and that point when the innovation is utilized in processes of near comparable efficiency and in the production of products of near comparable quality under the ownership and control of nationals of that land. Using such a concept may offer a more meaningful guide to questions of international diffusion. I hope the distinctions made in this paper will provoke further research on the agencies for technological diffusion, particularly the role of private business.

MIRA WILKINS

# [23]

Excerpt from *World Investment Report 1992: Transnational Corporations as Engines of Growth*, Transnational Corporations and Management Division, Department of Economic and Social Development, 131–62.

# Chapter VI

## TRANSNATIONAL CORPORATIONS, TECHNOLOGY AND GROWTH

Technology plays an undisputed role in economic growth by increasing the productivity potential of all factors of production, both tangibles such as labour and capital and intangibles such as organization and quality control. As the economies of the world are becoming increasingly globalized, technology emerges as the most decisive factor in determining international competitiveness and hence growth prospects. The present chapter provides a brief overview of the relationship between technology and growth, evaluates the role of TNCs in that relationship and draws some policy implications relevant to strengthening the contribution of TNCs to growth through technology transfer.

## A. Technology as a determinant of growth

### 1. Linkage between technology and growth

The concepts of technology and technological change encompass many dimensions. Technological progress in some cases involves process innovation, implying that new ways are found to produce existing goods and services, often involving less use of resources. In others, it involves product innovation, implying the introduction of new products or the improvement of quality. Technology comprises more than machinery and other forms of hardware embodied in physical goods. It can be considered as "the stock of knowledge (technical or management)" [1] used in production and marketing. A part of that knowledge is embodied in machines, but much of it is also embodied in human skills, management methods, organizational structures and work routines. Technology, therefore, takes different forms: "hardware", such as machinery and equipment; "software", such as blue prints and process specifications; and the "services" of technicians and professionals for tasks such as quality improvements, management and marketing know-how and process and product design. The software and service components of technology are becoming increasingly important in the international economy, with the emergence of information technology as the central element in the production of many goods and services.

The pervasive nature of the concept of technology, as indicated above, raises several problems in relating technology to growth. There is no single measure of the level of technology and the rate of technological change; it is not easy to separate the independent contribution of technology from other factors of production, particularly capital and labour, in which technology often becomes embodied; [2] and the impact of technology on growth depends on complex interactions between technological change, the structure of incentives confronting enterprises to apply, adapt and innovate upon available technologies, as well as institutional arrangements regarding, among other things, the flow of information and the functioning of markets. [3]

Despite those problems, analysts generally agree on the importance of technology as a determinant of growth. At the conceptual level, technology is considered to promote growth in several ways. First, advancements in technology enable a country to obtain a greater output from any given combination of inputs, which means that the productivity of factors of production is enhanced by technology. Second, technology can promote and sustain growth through the production of new products (including qualitatively superior products), with higher value-added and greater income elasticity. Third (related to the above but deserving of special mention), technology can foster growth through improved export performance, which often requires a shift in the composition of exports from primary commodities to manufactures, and within manufactures to more technology-intensive products.

## 2. Some empirical evidence on technology and growth

A substantial body of empirical evidence drawn from developed countries provides empirical support for the conceptual links between technology and growth. Empirical evidence for the central importance of product innovations in long-term growth was provided by S. Kuznets as early as 1930. [4] Based on the premise that old consumer goods typically suffer from low long-term income and price elasticity, he argued that a cost-reducing impact of technological change in old goods would have a small aggregative impact on growth. The long-term growth impulse, therefore, came from new products. Similarly, J. Schumpeter emphasized the role of "creative destruction" of old products and their replacement by new ones in the dynamics of growth. [5] Many subsequent studies at both the aggregate and sectoral levels have provided empirical evidence for the beneficial impact of technology on growth through increased productivity of factors of production. [6]

**Transnational Corporations as Engines of Growth**

Recent empirical studies on developing countries also demonstrate a significant impact of technology on growth through higher factor productivity. According to one study on Latin America, for example, nearly 20 per cent of growth in output for that region for the period 1940-1970 was accounted for by growth in total factor productivity. [7] The findings of several studies on countries/territories in Asia are presented in table VI.1. Very recently, furthermore, one study based on a sample of 25 countries, comprising developed countries as well as six newly industrializing economies (Argentina, Brazil, Hong Kong, Mexico, Republic of Korea and Taiwan Province of China), has found that innovation and diffusion of technology exerted a significant impact on growth of GDP and productivity for the period 1960-1985. [8]

As was noted earlier, technology can promote growth through improved export performance, which often requires a change in the composition of exports in favour of manufactures. Available evidence shows that, within the manufactures group of exports, R&D intensive industries have been the most rapidly growing exporters. Thus, over the period 1980-1987, the rate of growth of imports into developed countries averaged 10 per cent for high R&D intensive industries, while that of low R&D intensive industries was only 5 per cent. [9] It follows that, in so far as exports exert an influence on growth, the technological content will determine the strength of (export-led) growth.

The studies cited above provide convincing empirical evidence of the significant contribution of technology to growth, which in recent decades appears to have assumed an even greater importance for growth. For example, from about the mid-1970s, per capita use of commodity materials (such as, energy, steel, copper, cement) declined or levelled off, while per capita world GDP continued its upward trend; the difference between the growth in the use of materials and the growth of GDP can be largely attributed to growth in the use of knowledge-intensive new technologies such as electronics, computers and new materials. [10] Consistent with that assertion, data show a generally declining trend in the intensity of raw materials per unit of GDP in developed countries (figures VI.1 and VI.2).

Table VI.1. **Selected developing economies in Asia: growth of output and contribution of total factor productivity**
(Percentage)

| Economy | Period | Growth rate in output | Contribution of total factor productivity |
|---|---|---|---|
| Hong Kong | 1955-70 | 9.3 | 46.5 |
| | 1970-80 | 9.6 | 21.3 |
| India | 1950-80 | 3.5 | 39.1 |
| | 1970-80 | 3.0 | 0.2 |
| Indonesia | 1970-80 | 7.7 | 31.5 |
| Korea, Republic of | 1955-70 | 8.8 | 56.4 |
| | 1970-80 | 8.5 | 41.2 |
| Malaysia | 1970-80 | 7.8 | 21.7 |
| Philippines | 1957-62 | 4.9 | 0.0 |
| | 1963-69 | 5.2 | 15.4 |
| | 1970-74 | 6.3 | 19.0 |
| | 1970-80 | 6.2 | 20.6 |
| Singapore | 1957-70 | 6.6 | 55.2 |
| | 1966-72 | 12.5 | 4.8 |
| | 1972-80 | 8.0 | -11.3 |
| | 1970-80 | 9.1 | 19.7 |
| Taiwan Province of China | 1955-77 | 8.0 | 53.6 |
| | 1970-80 | 8.5 | 50.0 |
| Thailand | 1970-80 | 6.9 | 19.7 |

*Source:* Yukio Ikemoto, "Technical progress and level of technology in 1970-1980: a translog index approach", *The Developing Economies*, vol. XXIV, No. 4 (December 1986), pp. 368-390.

**Figure VI.1. Intensity of use of selected agricultural raw materials in developed countries**

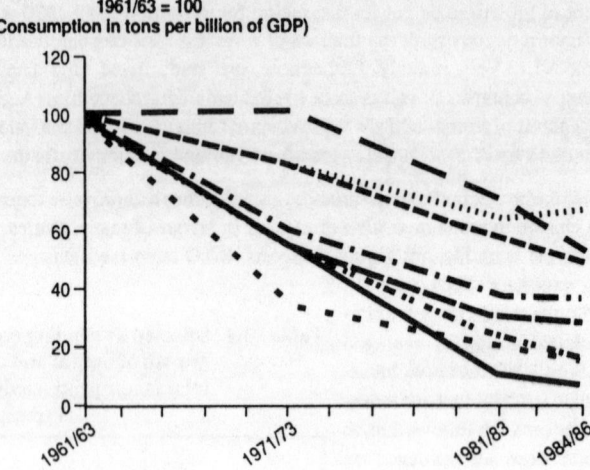

1961/63 = 100
(Consumption in tons per billion of GDP)

*Source:* UNCTAD, "Impact of technological change in patterns of international trade" TD/B(XXXV)/SC.I/CPR.2 (March 1989), p. 4.

*Key:*

| | | | | |
|---|---|---|---|---|
| Rubber | Cotton | Jute | Sisal | Abaca |
| Wool | Wood | Sugar | | |

## B. Transnational corporations and technology development

As mentioned earlier, the common denominator linking technology to growth is that it permits production of a greater amount of, or new output from, a given amount of resources. That production requires new technological development that normally involves R&D efforts. Results of R&D, in turn, are often reflected in patents. The present section, therefore, deals with the role of TNCs in technology development, as indicated by R&D and patents.

### 1. Overview

In recent years, a notable shift has occurred in the perception of the process of technology development. The conventional view was that scientific innovation and discoveries are largely made in academic institutions and research laboratories. Once these are in place, firms seek to commercialize them. This view of technology development is no longer appropriate. With the emergence of information based technologies, the increasing globalization of enterprises and growing competition in the interna-

**Transnational Corporations as Engines of Growth**

Figure VI.2. Intensity of use of selected minerals, ores and metals in developed countries

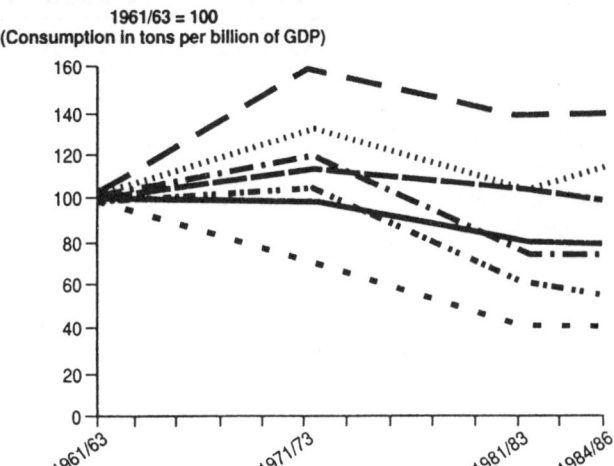

1961/63 = 100
(Consumption in tons per billion of GDP)

*Source:* UNCTAD, "Impact of technological change in patterns of international trade" TD/B(XXXV)/SC.I/CPR.2 (March 1989), p. 4.

*Key:*

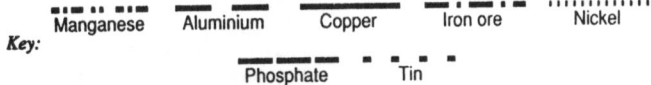

Manganese    Aluminium    Copper    Iron ore    Nickel

Phosphate    Tin

tional market place, science and technology have become more and more linked. It is being increasingly recognized that a large part of technological development occurs because of actions taken by enterprises. [11] Indeed, TNCs devote substantial resources to R&D, in addition to having a variety of institutional arrangements with universities, research institutions and other enterprises.

The role of private companies in R&D relative to overall national expenditure on R&D by major home countries of TNCs is illustrated in table VI.2. It is interesting to observe that the proportion of sales spent by these companies on R&D far exceeds the proportion of total national expenditure as a proportion of GDP. Furthermore, R&D expenditures of the limited number of companies for which data are presented in table VI.2 account for a significant share of total national expenditures. And most of these firms, in turn, are TNCs, which, recognizing the key importance of technologies, have undertaken extensive research programmes. Such TNCs as IBM, General Electric, Hitachi, General Motors and Siemens have allocated funds amounting to billions of dollars annually for R&D (table VI.3).

Research-and-development expenditures capture the resources devoted to technological development. They are, therefore, an *input* indicator. Results of R&D often find expression in patents, which can be viewed as an *output* indicator of technological development. Here, again, available data on patents registered in the United States clearly show the dominant role of corporations. Foreign-owned corporations together with those of United States origin account for over three-fourths of patents registered in the United States; the share of foreign-owned corporations has increased from the mid-1980s and is now larger than that of domestic firms (table VI.4). The top 50 TNCs accounted for more than one-fourth of all patents granted to corporations during the 1980s. Overall, the development of technologies appears to be increasingly undertaken by TNCs.

## 2. The internationalization of technological development and transnational corporations

Historically, TNCs have undertaken technological development mostly in their home countries. Foreign affiliates generally undertook modifications and adaptations to innovations, emanating mainly from the R&D establishments of their parent firms in home countries. That is still the predominant pattern. In recent times, however, there has been a marked growth in the internationalization of R&D. As TNCs become progressively more global and acquire a world orientation for their inputs, products and markets, a number of them are establishing

**Table VI.2. Research-and-development expenditure of selected countries and top companies [a]**

A. *National R&D expenditure, latest available years*

| Country | R&D expenditure (Million dollars) | Share of R&d expenditure in GDP | |
|---|---|---|---|
| | | Per cent | Year |
| Canada | 7 250.6 | 1.3 | 1989 |
| France | 22 241.0 | 2.3 | 1989 |
| Germany, Federal Republic of | 34 234.0 | 2.6 | 1989 |
| Italy | 14 189.6 | 1.3 | 1989 |
| Japan | 82 853.1 | 2.7 | 1989 |
| Netherlands | 4 792.3 | 2.1 | 1989 |
| Sweden | 5 459.5 | 2.9 | 1989 |
| Switzerland | 3 899.6 | 2.9 | 1986 |
| United Kingdom | 18 356.3 | 2.4 | 1988 |
| United States | 144 867.0 | 2.7 | 1989 |

B. *R&D expenditure of top companies*

| Country | R&D expenditure of top companies (Million dollars) | R&D expenditure as percentage of sales | Number of companies |
|---|---|---|---|
| Canada | 2 069 | 4.6 | 6 |
| France | 6 997 | 4.2 | 17 |
| Germany, Federal Republic of | 14 086 | 6.1 | 19 |
| Italy | 2 640 | 4.2 | 8 |
| Japan | 27 295 | 4.9 | 74 |
| Netherlands | 4 208 | 3.0 | 7 |
| Sweden | 3 454 | 6.5 | 10 |
| Switzerland | 4 426 | 5.9 | 10 |
| United Kingdom | 7 570 | 2.1 | 33 |
| United States | 37 569 | 4.7 | 28 |

*Sources:* Calculations of the Transnational Corporations and Management Division, based on OECD, *Basic Science and Technology Statistics* (Paris, 1991), table 3; United Nations, *National Accounts Statistics: Analysis of Main Aggregates*, various issues; and *Business Week;* Quality 1991, pp. 171-172 and 176-208.

[a] Companies with the highest absolute amount of R&D expenditures.

integrated R&D systems, with overseas laboratories playing a significant role. The increasing importance of economies of scope, shorter product cycles and rapid obsolescence, all of which require closer interaction with customers, have necessitated such an internationalization. In some cases, internationalization has been motivated by the desire to take advantage of scarce scientific and technical personnel, in which particular host countries possess a comparative advantage. At the same time, the process has been facilitated by the development of transnational computer-communication networks and on-line systems that permit the smooth flow of data and information among remote sites and, indeed, the on-line conduct of R&D. [12]

**Table VI.3. Research-and-development expenditure by top 20 transnational corporations, 1990**
(Millions of dollars)

| Ten non-United States TNCs | R&D expenditure | Ten United States TNCs | R&D expenditure |
|---|---|---|---|
| Siemens | 4 132 | General Motors | 5 342 |
| Hitachi | 3 011 | IBM | 4 914 |
| Matsushita | | Ford Motor | 3 558 |
|   Electrical Industrial | 2 423 | AT&T | 2 433 |
| Philips Electronics | 2 411 | Digital Equipment | 1 614 |
| Alcatel Alsthom | 2 237 | General Electric | 1 479 |
| Fujitsu | 2 097 | Du Pont | 1 428 |
| Toshiba | 1 864 | Hewlett-Packard | 1 367 |
| Nippon Telegraph | | Eastman Kodak | 1 329 |
|   & Telephone | 1 739 | Dow Chemical | 1 136 |
| NEC | 1 728 | | |
| Bayer | 1 699 | | |

*Source: Business Week,* Quality 1991, p. 176.

No comprehensive data exist on the geographical distribution of the R&D efforts of TNCs by country of origin. But the sketchy evidence available from limited survey data lends support to a growing internationalization. In the case of United States TNCs, the proportion of R&D expenditure accounted for by foreign affiliates increased to 10 per cent in 1989, from seven per cent in 1966. [13] The data seem to indicate that European TNCs have reached a considerably higher degree of internationalization of their R&D expenditures. Some 23 per cent of the R&D expenditures by 20 Swedish TNCs were undertaken abroad in 1987, compared with 21 per cent in 1980. [14] In the case of TNCs from the Federal Republic of Germany, it has been noted that the growth of R&D employment abroad has risen much faster than the growth of total employment abroad. A survey of 33 major firms showed that 18 per cent of their total R&D employees in 1989 were employed in affiliates abroad. [15] In contrast, a survey of 11 large TNCs from the Federal Republic of Germany at the end of the 1970s revealed that 15 per cent of their R&D personnel were employed abroad. [16] Some of the leading European TNCs, such as Ciba Geigy, Royal Dutch Shell, Bull, Philips, Olivetti, ABB and Norsk Hydro, each spend more than a third of their total R&D expenditure in foreign locations. [17]

Available data on patents also generally indicate a rising importance of R&D in foreign locations (table VI.5). Between the early 1970s and mid-1980s, the share of patents filed in the United States by TNCs that are credited to research undertaken outside the home country of the parent company has increased in 7 of the 11 countries included in the table.

Table VI.4. Number of United States patents, by type of grantee, [a] 1980-1991

(Thousands)

| | Year | | | | | | | | | | | |
|---|---|---|---|---|---|---|---|---|---|---|---|---|
| | *1980* | *1981* | *1982* | *1983* | *1984* | *1985* | *1986* | *1987* | *1988* | *1989* | *1990* | *1991* [b] |
| **United States-owned** | | | | | | | | | | | | |
| Individuals | 11.4 | 12.0 | 10.3 | 9.2 | 10.4 | 10.7 | 11.0 | 10.9 | 11.7 | 14.7 | 14.9 | 7.7 |
| Corporations [c] | 27.6 | 29.4 | 25.8 | 25.7 | 29.9 | 30.9 | 29.3 | 33.5 | 31.3 | 37.9 | 36.0 | 18.6 |
| Top 50 United States-based TNCs [d] | 9.2 | 10.1 | 8.7 | 8.8 | 10.1 | 10.3 | 9.6 | 10.8 | 9.5 | 11.2 | 10.5 | 9.2 |
| Government | 1.2 | 1.1 | 1.0 | 1.0 | 1.2 | 1.1 | 1.0 | 1.0 | 0.7 | 0.8 | 1.0 | 0.6 |
| **Foreign-owned** | | | | | | | | | | | | |
| Individuals | 4.1 | 4.2 | 3.4 | 3.3 | 3.7 | 4.0 | 4.3 | 4.9 | 4.8 | 5.4 | 5.3 | 2.7 |
| Corporations [e] | 19.2 | 21.3 | 19.6 | 19.9 | 24.0 | 26.8 | 27.6 | 33.5 | 32.1 | 38.7 | 37.3 | 19.1 |
| Top 50 non-United States-based TNCs [d] | 6.3 | 7.2 | 6.9 | 7.0 | 8.6 | 9.5 | 9.4 | 11.6 | 11.3 | 13.6 | 12.5 | 6.2 |
| Government | 0.3 | 0.3 | 0.4 | 0.3 | 0.4 | 0.5 | 0.5 | 0.6 | 0.5 | 0.4 | 0.4 | 0.8 |
| **All corporations c** | 46.8 | 50.7 | 45.4 | 45.6 | 53.9 | 57.7 | 56.9 | 67.0 | 63.4 | 76.7 | 73.3 | 37.7 |
| Top 50 TNCs [d] | 11.6 | 13.2 | 11.8 | 12.1 | 14.2 | 15.3 | 14.8 | 17.6 | 16.5 | 20.2 | 19.1 | 10.0 |
| All corporations as a percentage of total | 73.4 | 74.2 | 75.2 | 76.8 | 77.4 | 78.0 | 77.2 | 79.4 | 78.2 | 78.3 | 77.3 | 76.2 |
| Foreign corporations as a percentage of all corporations [e] | 41.0 | 42.0 | 43.2 | 43.6 | 44.5 | 46.4 | 48.5 | 50.0 | 50.6 | 50.5 | 50.9 | 50.7 |
| Top 50 TNCs [e] as a percentage of all corporations [e] | 24.7 | 26.0 | 26.0 | 26.5 | 26.3 | 26.5 | 26.0 | 26.3 | 26.0 | 26.3 | 26.0 | 26.5 |

*Sources:* United States Patent and Trademark Office, OEIPS/TAF Program within the Office of Information Systems, *All Technologies Report, January 1963-June 1991* (Washington, November 1991), *Design Patents Report, January 1977-June 1991* (Washington, November 1991) and OEIPS/TAF Program Database (Washington, June 1991).

a   Utility and design patents only.
b   Figures up to June 1991.
c   May also include non-corporate organizations.
d   Inventor patents only.
e   May exclude subsidiaries identified separately from parent organizations.

In sum, TNCs account for the bulk of R&D expenditures in their home countries, which, in turn, are the global leaders of technological development. A small but increasing share of those expenditures is being shifted to host countries, albeit primarily to developed ones.

Obviously, technology development is the pre-requisite for improved factor productivity and product innovations which, in turn, fuel growth. By virtue of their dominance in technological development, TNCs play a major role in the growth process, a role that is likely to become of greater importance in the future because of the increasing importance of technology as a determinant of growth.

## C. Transnational corporations and the transfer of technology to developing countries

The preceding section has demonstrated the importance of TNCs in the development of technology at the global level. The focus of analysis in the present section is on their role in technology transfer to developing countries. [18] The rationale is that, as demonstrated in the preceding section, technology development by TNCs mostly takes place in the home countries of those firms or in other developed host countries. [19] Therefore, access to technologies for developing countries is largely a matter of acquiring technologies from TNCs in developed countries. The impact of technology transfer from TNCs on the

Table VI.5.  **The share of United States patents of the largest firms world-wide attributable to research in foreign locations (outside the home country of the parent company), organized by the nationality of parent firms, 1969-1986**

(Percentage)

| Country | 1969-72 | 1973-77 | 1978-82 | 1983-86 |
|---|---|---|---|---|
| Belgium | 49.6 | 54.2 | 56.1 | 71.3 |
| Canada | 42.0 | 40.0 | 39.8 | 35.5 |
| France | 10.2 | 9.4 | 8.8 | 10.9 |
| Germany, Federal Republic of | 13.6 | 11.5 | 12.3 | 14.4 |
| Italy | 20.1 | 18.3 | 13.7 | 11.7 |
| Japan | 2.9 | | 1.9 | 1.3 |
| Netherlands | 63.9 | 68.8 | 64.1 | 70.0 |
| Sweden | 20.9 | 17.8 | 25.9 | 31.3 |
| Switzerland | 45.0 | 44.3 | 44.1 | 42.6 |
| United Kingdom | 43.3 | 40.5 | 38.7 | 45.0 |
| United States | 4.3 | 5.5 | 6.0 | 7.4 |

*Source:* John H. Dunning, "Multinational enterprises and the globalization of innovatory capacity", Rutgers University GSM working paper No. 91-03 (January 1991), table 7, p. 18.

growth of the host economy, however, depends on how the various modes of technology transfer interact with the local technological capabilities, incentive structures and institutional arrangements.

The principal sources of technology acquisition are scientific and technical publications (typically widely accessible at low costs); trade (through the import of machinery and equipment); FDI (through both wholly-owned foreign affiliates and joint ventures); and non-equity links with TNCs through mechanisms such as patents, licenses, technical assistance agreements and other contractual arrangements, as well as strategic alliances. Transnational corporations play a major role in all those modes of transferring technology, particularly so in the latter three.

## 1. Transnational corporations and the supply of capital goods

The import of capital goods is a prime determinant of the productive capacities of developing countries. As table VI.6 shows, developing countries in Africa recorded a significant decline in the absolute value of capital-goods imports during the decade of 1980s; those in Latin America and the Caribbean achieved only a marginal increase; and those in Asia and the Pacific raised their imports by nearly three-fourths. That is undoubtedly a significant explanatory variable in the differential growth performance of those groups of countries.

### Table VI.6.  Capital goods [a] imports by developing countries, 1980-1989

(Billions of dollars)

| Country group | 1980 | 1981 | 1982 | 1983 | 1984 | 1985 | 1986 | 1987 | 1988 | 1989 |
|---|---|---|---|---|---|---|---|---|---|---|
| All developing countries | 115.3 | 130.5 | 122.4 | 106.2 | 105.9 | 101.4 | 112.4 | 126.1 | 143.9 | 155.2 |
| Africa | 23.5 | 26.3 | 22.5 | 18.2 | 17.6 | 16.7 | 16.5 | 16.5 | 17.2 | 18.2 |
| Asia and the Pacific | 56.7 | 65.1 | 68.2 | 65.3 | 62.4 | 56.7 | 63.7 | 74.0 | 87.2 | 95.8 |
| Latin America and the Caribbean | 31.1 | 35.8 | 29.0 | 20.3 | 23.6 | 25.3 | 28.5 | 31.8 | 35.0 | 36.3 |

*Source:* UNCTAD Secretariat.

a Includes SITC Rev. 1, Section 7, machinery and transport equipment, except 7194 domestic appliances non-electrical; 7241 television receivers; 7242 radio receivers; 7250 domestic electrical equipment; 7321 passenger motor cars; 7326 chasis for passenger motor cars; 7329 motor cycles; and 7331 bicycles.

The key issue in the present context is the role of TNCs in capital-goods imports. The non-availability of data does not permit a disaggregation of imports of capital goods from TNCs as compared with imports from other entities. Indirect evidence suggests, however, a major role of TNCs in the supply of capital goods. For example, in 1989, at least 80 per cent of United States foreign trade was undertaken by those corporations, including parent companies in the United States, foreign affiliates of United States TNCs and United States affiliates of foreign TNCs. It is, therefore, not unrealistic to assume that the proportion of capital-goods imports of developing countries from the United States accounted for by TNCs is quite high. The importance of TNCs in the supply of capital goods is also underscored by the high proportion of intra-firm trade in some capital goods items. For example, exports of non-electrical machinery by United States TNCs are substantially intra-firm; in 1989, 60 per cent of such exports were represented by intra-firm transactions. [20] The importance of intra-affiliate transactions in total capital goods imports of developing countries, however, would be considerably less (see chapter VIII).

## 2. Technology transfer through foreign direct investment

A TNC normally undertakes FDI when it possesses certain technological or other economic advantages over its competitors, which it finds in its best interest to exploit internally from a foreign location. [21] Technology forms an important part of the competitive advantage of a TNC, and many firms choose to service their foreign markets through FDI, not only to exploit that advantage but also to retain company control over their technology. Transnational corporations generally transfer their most recent technology to their affiliates, while selling or licensing older technology to locally-owned firms and joint ventures. [22] Hence, FDI may be the only way for many developing countries to gain access to the latest technology and especially to certain key technologies.

Foreign direct investment can promote technological change in developing countries—and, as box VI.1 shows, in developed countries as well—in a number of ways. The direct impact may occur through its contribution to higher factor productivity, changes in product and export composition, R&D undertaken by foreign affiliates, the introduction of organizational innovation and improved management practices, and employment and training (the last of these aspects is being dealt with in chapter VII). Indirect impacts occur through collaboration with local R&D institutions, technology transfer to local downstream and upstream producers, the effects of the presence of foreign affiliates on competition and on the efficiency of local producers and the turn-over of trained personnel.

### (a)  Direct effects

#### (i)  Transnational corporations and factor productivity

An important contribution of technology to growth is through increased factor productivity. An evaluation of the contribution of TNCs to that process would require highly disaggregated data on the

---

### Box VI.1. Foreign direct investment in developed countries and technology transfer

A good part of the discussion of FDI and technology transfer in developed countries focuses on the possibility that FDI may, in fact, reduce the technological capacity of the host economy and, hence, impair its growth prospects. For example, it has been argued that Japanese TNCs in the United States cause a drain of United States technology to Japan. More specifically, it is feared that, when Japanese companies acquire United States firms, especially in high-technology industries, they do so in order to capture innovative products and their technologies for the parent firm in Japan. [1] It is argued that, in the longer run, foreign investors will shift the bulk of R&D activities from the United States to their home countries and denude the United States of its innovative capacity by making the results of R&D unavailable to firms in the United States.

It is very difficult to assess the longer run impact of foreign investors acquiring or displacing United States firms on the development of technological capacity in the United States. However, if the concern is that foreign investors will shift R&D activities from their United States affiliates back to headquarters, the data show that, in fact, foreign affiliates in all industries taken together in the United States perform twice as much R&D per worker than United States firms (table 1). In the manufacturing sector, however, the differential is less pronounced and the amount of company-funded R&D expenditure per worker for United States firms is marginally lower than for foreign affiliates. Those figures do not give any indication of the type or quality of R&D undertaken by the two categories of firms, but they do not support the view that foreign firms are transferring large amounts of R&D from their United States affiliates to headquarters. Similarly, a study of royalties and licence fees found that transfer of foreign technology into the United States by foreign affiliates was more than five times larger than technology transferred out by them. [2] In fact, measured by royalties and licence fees, the largest proportion of technology transfer from the United States was accounted for by United States parents of foreign affiliates.

Many foreign investors may locate their R&D activities in the United States in order to take advantage of the technology centres in that country. [3] A study of Japanese entries into 297 United States industries showed that Japanese FDI predominated in R&D intensive industries in respect of establishment of new plants, but there is no indication that Japanese acquisitions are more frequent in high-technology industries. [4] Many Japanese companies pursue a strategy of vertical integration for their overseas activities. Fujitsu, for example, has constructed a $100 million R&D, manufacturing and service facility in Texas for the development of fiber optic transmission systems for the United States market, jointly by United States and Japanese engineers. [5] Like-

**Table 1. Research and development by United States affiliates of foreign firms, 1988**

| | | United States firms | |
|---|---|---|---|
| | Foreign affiliates [a] | Total [b] | Company-funded |
| **All industries** | | | |
| R&D (millions of dollars) | 7 382 | 97 889 | 65 583 |
| Employment (thousands of workers) | 3 682 | 91 076 | ... |
| R&D per worker (thousands of dollars) | 2.00 | 1.07 | 0.72 |
| **Manufacturing** | | | |
| R&D (millions of dollars) | 6 402 | 89 776 | 60 223 |
| Employment (thousands of workers) | 1 762 | 19 341 | ... |
| R&D per worker (thousands of dollars) | 3.63 | 4.64 | 3.11 |

*Source:* Edward M. Graham and Paul R. Krugman, *Foreign Direct Investment in the United States* (Washington, D.C., Institute for International Economics, 1991), table 3.3, p. 73.

a Data are preliminary
b Includes federally funded as well as company-funded expenditure.

/.....

## Transnational Corporations as Engines of Growth

**(Box VI.1, continued)**

wise, in 1985, Honda established an R&D facility in Ohio, now employing 200 persons. Honda's goal is to employ 500 people in R&D centres in the United States by 1995. The strategy of the company is to develop and build cars in the areas in which they are sold. [6]

Foreign affiliates in the United States may also be helping to increase United States factor productivity by introducing new technology and management methods. This may particularly be the case in industries in which the United States is losing international competitiveness. In the automotive industry, the Japanese automakers in the United States are introducing new standards in the manufacturing and engineering of cars, which serve as models for United States automakers. For example, in 1990 the Ford Motor Company switched to the production of a new model without stopping the assembly line, by introducing reprogrammable machines that move on tracks. This is a practice Japanese automakers have developed, and used in the United States, to increase productivity. [7] In 1982, General Motors turned over a run down and inefficient auto plant in Fremont California to Toyota Motor Corp. as a part of a joint venture. By introducing new technology and a typical Toyota production system, with just-in-time delivery and a flexible assembly line, it only takes half of the previous work force to assemble the same number of cars. [8]

Japanese automakers in the United States also transfer technology indirectly, by providing technical assistance to United States suppliers. Since the surge of the yen made it more profitable for Japanese firms to source locally, they encourage United States car-part firms to adopt new methods to improve quality and lower production costs. [9]

In fact, Japanese FDI in the United States automotive industry has brought new investments and technology transfer to other declining industries in that country, like the steel and rubber industries. Faced with strong international competition and declining demand from the automotive industry, the United States steel industry verged on collapse at the beginning of the 1980s. However, many Japanese steel firms have invested heavily in the United States, building state-of-the-art plants for coating and preparing steel coils used by carmakers, and entering joint-ventures to modernize large integrated United States steel plants. [10] In this manner, the United States partners gain access to state-of-the-art Japanese technology, as well as the new market of Japanese automakers.

---

1 Marjorie Sun, "Investors' yen for U.S. technology", *Science*, vol. 246 (8 December 1988), pp. 1238-1241; Eduardo Lachicha, "Japanese firms the most active investors in U.S. high-tech concerns, study says", *The Wall Street Journal*, 14 May 1991; and Georgio Gilder, "American technology at fire-sale prices", *Forbes* (22 January 1990), pp. 60-64.

2 See Kan H. Young and Charles Steigerwald, "Is foreign investment in the U.S. transferring U.S. technology abroad?", *Business Economics*, vol. XXV, No. 4 (October 1990), pp. 28-30.

3 See Edward M. Graham and Paul R. Krugman, *Foreign Direct Investment in the United States* (Washington D.C., Institute for International Economics, 1991).

4 See Bruce Kogut and Sea Jin Chang, "Technological capabilities and Japanese foreign direct investment in the United States", *Review of Economics and Statistics*, vol. 73, No. 3 (August 1991), pp. 401-413.

5 In 1989, Fujitsu had seven R&D centres in the United States, mainly devoted to software and development of data storage equipment. See *Business International*, 23 October 1989.

6 See Martin Kenney and Richard Florida, "How Japanese industry is rebuilding the Rust Belt", *Technology Review*, vol. 94 (February/March 1991), pp. 24-33. Toyota also announced a major expansion of research facilities in the United States in 1991 (see *The Wall Street Journal*, 3 June 1991).

7 See *The New York Times*, 14 March 1990.

8 See *Business Week*, 14 July 1986.

9 See *The Wall Street Journal*, 12 April 1988.

10 See Kenney and Florida, op. cit.

use of factors of production and value-added, differentiated by ownership in different industries. Such data are not readily available.

There are case studies, however, that provide some evidence of the relative efficiency of the use of factors of production, as between foreign affiliates and domestic enterprises. For example, an analysis of 282 pairs of foreign and domestic firms of similar size drawn from 80 manufacturing industries in Brazil concluded that foreign firms have a significantly higher ratio of value-added to output than domestic ones. [23] A study on Thailand found that foreign firms had higher average productivity of both capital and labour in the manufacturing sector compared with domestic firms, and the difference was owing to the higher efficiency of foreign firms as measured by a technology co-efficient derived from production-function estimations. [24] Similarly, a study on the Republic of Korea observed that the marginal product of both capital and labour was higher in foreign firms compared with domestic firms, but the differential was much greater for capital than labour. [25]

All these studies, therefore, support the view that foreign firms can contribute to growth through the provision of technologies that make more efficient use of capital and labour.

### (ii) Transnational corporations and product composition

As noted earlier, the introduction of new products or qualitatively superior old products is one of the ways by which technology promotes growth. Transnational corporations can play a role in this process. One way of assessing the role is to examine the performance of TNCs in the production of relatively more research-intensive products (table VI.7). The table shows that, for United States

**Table VI.7. Shares of high and medium research-intensive industries [a] in total sales and manufacturing sales of foreign affiliates, 1982 and 1989**

(Percentage)

| Developing region | 1982 | | 1989 | |
|---|---|---|---|---|
| | Share in total sales | Share in manufacturing sales | Share in total sales | Share in manufacturing sales |
| **United States majority-owned affiliates** | | | | |
| Africa | 3.5 [b] | 59.2 [b] | 3.1 [b] | 23.0 [b] |
| Asia and the Pacific | 15.7 | .. | 50.7 [b] | .. |
| Latin America | 21.8 | 57.3 | 33.1 | 60.9 |
| **Japanese affiliates** | | | | |
| Africa [c] | 17.1 | 42.4 | 10.8 | 40.9 |
| Asia and the Pacific [d] | 29.0 | 74.5 | 25.9 | 79.8 |
| Latin America | 20.1 | 66.0 | 19.4 | 74.3 |

*Sources:* United States, Department of Commerce, *U.S. Direct Investment Abroad: 1982 Benchmark Survey*, (Washington, D.C., United States Government Printing Office, 1985), table III.D.3, and *1989 Benchmark Survey*, Preliminary Results (Washington, D.C., United States Government Printing Office, 1991), table 32; Japan, Ministry of International Trade and Industry, *The Fourth Basic Survey on Japanese Business Activities Abroad* (Tokyo, Okurasho Insatsu-Kyoku, 1991), p. 12, and *Survey on the Overseas Activities of Japanese Companies*, No. 12-13 (Tokyo, Toyo Hoki Shuppan, 1984), p. 43.

a High and medium research-intensive industries include chemicals, machinery (except electrical), electrical machinery and domestic equipment, and transportation equipment.
b Part of data are suppressed by the source to avoid disclosure.
c Includes South Africa.
d Includes Australia and New Zealand.

TNCs, the expansion of the share of sales of high and medium research-intensive industries primarily occurred in Asia and the Pacific. In that region, United States affiliates also had the largest increase of R&D expenditure as percentage of sales, as noted later. Latin America shows a much slower growth in sales of high and medium research-intensive industries and Africa shows a decline. For Japanese TNCs, the picture is similar as far as Africa is concerned. And, again, there has been an increase in the share of sales of high and medium research-intensive industries in manufacturing sales in Asia and the Pacific as well as Latin America, with a slightly more pronounced growth in the latter region in terms of the share of manufacturing sales.

The creation of production facilities by TNCs in high and medium research-intensive industries can imply technology transfer not merely through a changing product composition, but also through the training of host country personnel in new technical skills and the introduction of new management methods and new ways of organizing the production process. The impact of FDI on the transfer of skills from host-country personnel, however, does not depend only on the degree of complexity of the technology employed; it is also a function of the methods used for transferring skills, the quality of in-house training programmes, the promotion policy for nationals through exposure to progressively higher levels of responsibility and the provision of off-the-job training.

As a note of caution, it should also be mentioned that the data in table VI.7 do not provide information on the value-added activities of the foreign affiliates. It could be that some TNCs locate only relatively labour-intensive, low value-added operations of those research-intensive outputs in the host country, and that the high value-added operations are located in the home country.

### (iii) Transnational corporations and export composition

The technological content of exports can be an important determinant of growth performance. It is well known that R&D intensive exports generally have higher income elasticities; therefore, the growth of those exports is more sustainable over the long run. Besides, a rising share of such exports also carries the implication that the country concerned is in a position to take advantage of shifts in international demand (manifested in the growth of internationally competitive R&D intensive industries), rather than to rely exclusively on traditional exports based on natural-resource endowments or low labour costs. The role of TNCs in the export of R&D intensive products, therefore, deserves scrutiny.

The relevant data are presented in table VI.8. They show that, in the case of Japanese affiliates, the share of R&D intensive exports in total manufactured exports increased between 1982 and 1989 in Latin America and Asia, but declined in Africa, where an absolute decline of R&D intensive exports also occurred. In the case of United States affiliates, their share of R&D intensive exports increased somewhat in Latin America, declined slightly in Asia (though the share is still much higher than in Latin America) and remained very small in Africa. On the whole, affiliates have significantly increased R&D intensive exports.

Again, it is difficult to estimate the local value-added in the host country from export-oriented production. It should also be noted that the performance of TNCs in respect of R&D intensive exports is not necessarily better than that of local enterprises. In particular, local enterprises in certain Asian countries have clearly outperformed foreign affiliates. Total R&D intensive exports from Asia in 1989 were more than four times those recorded in 1982; [26] but the increase in R&D intensive exports by both United States and Japanese affiliates over the same period, though significant, did not reach a similar proportion. (See also the discussion on structural change of exports in chapter VIII.)

| *(iv) Research and development by affiliates* | Table VI.8. Manufactured exports and research-and-development intensive [a] exports of foreign affiliates, 1982 and 1989 |

The evidence that an overwhelming proportion of the *foreign* R&D of TNCs is located in developed countries does not necessarily imply that such R&D is insignificant from a host-country perspective. In countries such as India, the Republic of Korea and Singapore, the share of aggregate R&D expenditure attributable to foreign firms exceeded 15 per cent in the 1970s. [27] Moreover, some evidence indicates that foreign affiliates may now be devoting more of their resources than before to R&D. In the case of the majority-owned foreign affiliates of United States TNCs, there has been a noticeable increase in their R&D expenditures as a proportion of sales in a number of developing countries (table VI.9). But there are some noticeable regional differences. Research-and-development expenditure by United States affiliates as a percentage of sales increased four times between 1982 and 1989 in Asia and the Pacific, while it stagnated in Latin America and remained insignificant for the developing countries in Africa.

The location of R&D activities in developing countries can be explained by locational advantages and the corporate

(Millions of dollars)

| Developing region | United States majority- owned affiliates | | Japanese affiliates [b] | |
|---|---|---|---|---|
| | Manufactured exports | R&D intensive exports | Manufactured exports | R&D intensive exports |
| **Latin America** | | | | |
| 1982 | 4 692 | 2 908 | 971 | 84 |
| 1989 | 10 176 | 6 794 | 815 | 165 |
| Percentage increase | 117 | 134 | - 19 | 96 |
| **Asia** | | | | |
| 1982 | 5 954 c | 5 453 c | 5 950 | 3 027 |
| 1989 | 13 861 | 12 176 | 11 560 | 7 230 |
| Percentage increase | 133 | 123 | 94 | 139 |
| **Africa** | | | | |
| 1982 | 169 c | 3 c | 23 | 9 |
| 1989 | 566 | 9 c | 30 | 5 |
| Percentage increase | 235 | 200 | 30 | - 44 |

*Sources:* United States, Department of Commerce, *U.S. Direct Investment Abroad: 1982 Benchmark Survey,* op. cit., tables III.E.4. and III.E.5, and *1989 Benchmark Survey* tables 42 and 44; Japan, Ministry of International Trade and Industry, *Survey on the Overseas Activities of Japanese Companies,* No. 12-13, op. cit., pp. 90, 91 and 95 and No. 18-19, (Tokyo, Okurasho Insatsu-kyoku, March 1990), pp. 74-75, 78-79 and 82-83.

a Definition same as in table VI.7.

b The values may be substantially understated because of incomplete coverage of firms in the surveys.

c Part of the data is suppressed by the source to avoid disclosure.

strategies of TNCs. The decisions of corporations to locate R&D activities in certain host countries are very much dependent on factors, such as the availability of R&D facilities and of trained scientific and engineering personnel. Generally, countries with high expenditures for R&D are also the countries in which United States affiliates have a high proportion of R&D expenditure compared to sales.

Very little is known of the type of research undertaken by foreign affiliates. The R&D activities taking place within foreign affiliates are, most likely, typically confined to adapting the technology of the parent company to local conditions. In a sample of 218 Japanese TNCs, 57 per cent expressed the view that the main objective of their foreign R&D facilities was to develop products tailored to meet local demand. [28] The effect of TNCs on deeper indigenous research-and-innovation capabilities ("know-why") in developing countries is less evident. As TNCs can import all their "know-why" and need to perform only adaptive research in host countries, local firms may well conduct more research (as opposed to development) than do foreign affiliates.

**Table VI.9. Research and development expenditures of selected developing economies as a percentage of GNP and research and development expenditures for United States majority-owned affiliates as percentage of sales**

| Developing regions/economy | R&D expenditure of countries as percentage of GNP | | | | R&D expenditure of United States majority-owned affiliates as percentage of sales | |
|---|---|---|---|---|---|---|
| | Percentage | Year | Percentage | Year | 1982 | 1989 |
| **Latin America** | | | | | 0.2 | 0.2 |
| Argentina | 0.2 | 1982 | 0.5 | 1988 | 0.4 | 0.25 |
| Brazil | 0.7 | 1982 | 0.4 | 1985 | 0.4 | 0.3 |
| Mexico | .. | .. | 0.6 | 1984 | 0.3 | 0.2 |
| **Africa** | | | | | 0.01 | 0.02 |
| **Middle East** | | | | | 0.1 | 0.4 |
| **Asia and the Pacific** | | | | | 0.04 | 0.2 |
| Hong Kong | .. | .. | .. | .. | .. | 0.1 |
| India | 0.7 | 1982 | 0.9 | 1986 | 0.5 | 0.6 |
| Indonesia | 0.4 | 1983 | 0.2 | 1988 | 0.02 | 0.03 |
| Republic of Korea | 0.9 | 1982 | 1.9 | 1988 | .. | 0.3 |
| Malaysia | | | .. | .. | .. | 0.1 |
| Singapore | 0.3 | 1981 | 0.9 | 1987 | .. | 0.3 |
| Thailand | 0.3 | 1985 | 0.2 | 1987 | 0.03 | 0.02 |
| Taiwan Province of China | .. | .. | .. | .. | 0.3 | 0.4 |

*Source:* UNESCO, *Statistical Yearbook,* various issues; United States, Department of Commerce, *U.S. Direct Investment Abroad: 1982 Benchmark Survey,* op. cit., tables III.H.3 and III.E.1; and *1989 Benchmark Survey ,* op. cit., tables 40 and 76.

It may also be that a strong presence of TNCs can inhibit the development of an indigenous technological base beyond adaptive research. [29] When TNCs penetrate a host-country market, indigenous firms may be forced to cut back on research or to narrow their field of specialization, as they are confronted with declining market shares caused by competition with TNCs that possess much greater technological capacities. On the other hand, foreign competition could also induce domestic firms producing similar products to undertake R&D that otherwise would not have taken place, in order to improve their competitive advantage. In that case, FDI could advance local innovatory capacity in areas in which the host country and its firms are strongest and have a competitive market structure. In the case of greenfield investments, which do not compete with local industry, there is no displacement of local enterprises, and FDI will most likely lead to a net increase in the innovatory capacity of a host country, even through adaptive research. [30]

### (v)  Organizational innovation and management practices

Organizational innovation and improved managerial practices are being increasingly viewed as a major aspect of technological development for enhancing productivity and accelerating growth. The principal components of these aspects that have evolved over the last two decades or so can be summarized as follows: [31]

- The underlying philosophy of production has been altered: instead of producing to stock, goods are produced to order. That necessitates a demand-driven system capable of producing a variety of product types in much smaller volumes. Hence, lot sizes have been reduced dramatically.

- The efficient production of different products in small lot sizes requires minimizing downtime. That, in turn, requires quick line changeovers and tool setups. Machinery redesign becomes necessary but, more importantly, production-line workers must be trained to do changeovers rather than having them done by separate teams as in mass production.

- Production layouts need to be restructured, and changes made in the use and management of machines in order to create a smooth flow of smaller lot sizes.

- Inventories have to be reduced to a minimum "just-in-time" level rather than being stocked "just-in-case", so that the increased number of different product types can be accommodated without large carrying costs.

- Maintaining a smooth flow of production without inventories requires that components have zero defects or be of perfect quality, whether they come from suppliers or from in-house sources further back in the production line.

- Skill and craft demarcations among workers are eliminated and workers are trained to be multi-skilled; they are paid according to their skill level and the quality of their work.

The organizational changes involved extend throughout the firm: from design to marketing to production; from senior management to the shop floor; and from management's relations with its workforce to the firm's relations with its suppliers.

Transnational corporations from Japan, particularly those in the automobile industry, have been the pioneers of these developments. It was during the 1980s that these organizational techniques began to be introduced outside of Japan. In some cases this was a direct result of the operations of the Japanese affiliates themselves, especially in the electronics, automobiles, component and machine tool industries that had been established in North America and Europe. In other cases, non-Japanese suppliers of these Japanese foreign investors began to restructure to incorporate new patterns of organization in order to meet the requirements of their Japanese customers. A third source of innovation were the practices of those firms that had subsidiaries or joint ventures in Japan and which were learning through the operations of these subsidiaries - Bendix's production of auto components and Xerox's restructuring of the mid 1980s are cases in point. By the late 1990s, the central tenets of the new organizational paradigm had filtered through to the major non-Japanese TNCs and were being implemented at the plant level in various industrialized countries.

More recently, TNCs from Japan and elsewhere have started implementing organizational changes in developing countries. No systematic data are as yet available to document the extent of such technology transfer. However, available case studies show that some developing country firms have adopted these changes either as joint venture partners of TNCs or under licensing agreements (box VI.2); in other cases, similar changes have been introduced in TNC affiliates or subsidiaries in developing countries. Examples of the adoption of these technological changes can be found in such diverse countries as Brazil, the Dominican Republic, India, Mexico and Zimbabwe. [32]

## (b)  Indirect effects

Foreign direct investment can promote growth through several indirect mechanisms of technology transfer. For example, backward linkages to local firms, in the form of subcontracting the supply of parts, components and services, create additional demand for intermediate products. A supplier firm in a developing country that is in a subcontracting relationship with a foreign subsidiary can receive technical assistance to improve its product quality and production process or to undertake new product development. When upgrading the technological level of supplier industries, FDI often increases the local value-added and generates growth. The presence of foreign affiliates can increase competition and thereby force domestic enterprises to improve productive efficiency, which is growth-enhancing.

An earlier chapter has provided evidence that TNCs may be increasing their use of inputs from local sources. Local sourcing of inputs, particularly when done under subcontracting arrangements, is often associated with technological assistance to the local suppliers by TNCs. In a survey of the largest foreign affiliates operating in Mexico, for example, it was found that almost two thirds of them had local subcontracting relationships. Almost all of the foreign affiliates that subcontracted locally imparted some kind of training to their national subcontractors: 87 per cent provided training in quality control,

68 per cent gave technical assistance and 22 per cent offered financial assistance to their subcontractors. [33]

As to the spillover impact of TNCs on the technological capacity and productive efficiency of indigenous enterprises, several studies on developed countries provided mixed evidence. [34] The same is true of developing countries. A recent study on Mexico showed that the rate of productivity growth of local firms and their ability to reach the productivity standards of TNCs were positively related to the degree of foreign ownership of an industry. [35] That estimate was interpreted to imply that competition from foreign affiliates forced Mexican firms to increase productivity by investing in human capital and new technology. The study could not exclude, however, the possibility that the competitive pressure from foreign affiliates had simply forced out inefficient local firms, thus improving the average productivity performance of Mexican firms. In contrast, a study on Morocco did not provide any evidence that the presence of foreign firms resulted in increased productivity of domestically-owned firms. [36] Although

---

**Box VI. 2. Transfer of organizational technology: the case of Escorts Ltd. in India**

Escorts Ltd. is a large Indian firm which grew to prominence since the mid-1950s. It began by producing motorcycles, and diversified into tractors and automobile components. In 1985, following the general opening-up of the Indian automobile industry to TNCs, Escorts entered into a licensing agreement to manufacture Yamaha motorcycles in a new plant in Surajpur. This commenced production in 1986, manufacturing 100 cc motorcycles predominantly designed by Yamaha. Escorts' older Faridabad plant producing motorcycles of wholly Escorts design remained in operation. In early 1990, the two Escort plants accounted for 40 per cent of the Indian motorcycle market.

A key strategic decision was taken to build a new plant and to employ a young and skilled labour-force rather than to attempt a turnaround of the existing plant. The youth of the labour-force (average age of 25 years in 1990) was intended to facilitate training in radically new forms of work-organization; it was also designed to reduce pressure from workers and trades unions to "impose traditional workpractices".

Training has therefore been a priority for Escorts in its new plant. It began with senior managers, senior technical personnel and supervisors. Yamaha organized extensive training for these groups, including spells in Japan—from two weeks to six months, depending upon the tasks involved. Thereafter, training was extended to the direct work-force by teams of 10 Japanese and 10 Indian trainers. Workers received two weeks initial training before going on to the shop-floor. After approximately one month, they received training in new skills (off the shop-floor), with this cycle being repeated for a period of approximately six months, until workers were deemed to have reached a minimum acceptable standard. Thereafter, additional training was provided at regular intervals as the average skills of the labour-force were gradually increased, especially in the acquisition of multiple skills. This is reflected in the payments system, where basic wages are supplemented by increments for skill acquisition and are thus partly paid on the principle of what the workers can do, rather than what they actually do.

The plant is laid out on a cellular basis, with kanban carts moving work-in-progress between various stages of stamping, machining and assembly. Typically, each operator is responsible for a number of machines, unlike Faridabad where each machine tends to have a dedicated operator. Work-teams are responsible for each

/.....

---

## Transnational Corporations as Engines of Growth

foreign firms had higher levels of productivity, domestic firms showed faster productivity growth; but that could not be attributed to dynamic externalities from FDI.

In some cases, TNCs stimulate technology development by local R&D institutions. In India, for example, one TNC recently signed a letter of intent with a Government-funded telecommunications R&D facility—the Centre for the Development of Telematics—to use switches designed by the Centre in a new open-architecture cellular system. In addition, the TNC intends to sponsor research at nine leading engineering colleges. [37]

---

### (Box VI.2, continued)

cell, and workers within each team are generally cross-trained to perform all the tasks in the cell (as well, in many cases, as tasks in other cells). Just-in-time production on work-in-progress is carefully observed. The result, as can be seen from table 1, is that changeover time in the Surajpur press shop is eight to sixteen times quicker than at Faridabad. Batch-sizes were less than half, and inventory components and raw materials were generally six times greater than at the sister plant.

Table 1. Comparison between the Surajpur and Faridabad plants of Escorts Ltd., 1990

| Plant | Annual output (Number of motor-cycles) | Number of workers | Output per worker (Number) | Change-over time | Batch size | Inventory inputs |
|---|---|---|---|---|---|---|
| Surajpur | 77 500 | 625 | 124 | 30-60 minutes | 4 000 | 15-30 days |
| Faridabad | 96 000 | 4 000 | 24 | 8 hours | 8 000-10 000 | 3-6 months |

The Surajpur plant has thus experienced considerable progress and is considerably more efficient than its sister-plant at Faridabad. Labour productivity was almost five times higher, with 625 workers producing 77,500 motorcycles (124 motorcycles per worker), compared with the 4,000 workers manufacturing 96,000 units at Faridabad (24 motorcycles per worker). Most of this superiority in performance was due to organizational factors, although the product design by Yamaha also played a role. The decidedly superior performance of the Surajpur plant clearly illustrates how transfer of organizational technology by TNCs can bring about major improvements in productivity.

---

*Source:* Transnational Corporations and Management Division, *Transnational Corporations and the Transfer of New Management Practices to Developing Countries* (New York, United Nations, forthcoming).

### 3. Transnational corporations and technology transfer through non-equity forms

Apart from wholly- or majority-owned FDI (usually known as internalized forms of transfer), TNCs also transfer technologies through a variety of externalized (primarily non-equity) forms, which include minority joint ventures, licensing, management and marketing contracts and international subcontracting. (Box VI.3 discusses transfer of technology through joint ventures in developed countries.) But data on such technology transactions between developing countries and TNCs are sketchy and difficult to interpret. Few developed countries disaggregate their technology receipts by country, type of transaction (for example, licensing or management contracts) or relationship between the receiving and the paying enterprise (an affiliate or an unrelated enterprise). Similar comments also apply to developing country-data on technology payments. Hence, it is a formidable task to assess the role of TNCs in technology transfer through such forms, and it is harder still to assess their impact on growth.

An earlier study by United Nations Centre on Transnational Corporations concluded that, during the 1970s, and up until the mid-1980s, the incidence of externalized forms of technology transfer seems to have increased. [38] Data on United States TNCs suggest a weak corroboration of that trend (table VI.10). Attention should be drawn, however, to the fact that technology receipts from unaffiliated enterprises in developing countries account for a very small proportion of total receipts.

Data for host countries tell a similar story. In the Republic of Korea, for example, payments for foreign licensing and technology contracts increased their share in total technology-transfer transactions (defined as FDI inflows plus payments for foreign licensing and technical consultancy plus capital-goods imports) from 1.2 per cent in 1972-1976 to 1.6 per cent in 1982-1986. [39] In Thailand, technology payments (comprising payments for royalties, trademarks, technical and management fees) increased from around 0.1 per cent of GDP during the 1972-1976 period to 0.2 per cent during the 1984-1987 period. [40]

Several points need to be considered to put the pattern into perspective:

- The first concerns the potential of non-equity forms for the future. It is well known that many factors—such as the age and sophistication of a technology, industry characteristics, corporate strategies within particular industries as well as the level of host-country entrepreneurial, technological and human-resources development—affect the choice of particular TNCs regarding externalized technology transfer. That form of transfer, however, may be even less favoured in the future than it was in the past. For one, recent developments in information technologies tend to increase the internalization advantages of TNCs. Those developments facilitate and cheapen the cost of intra-firm communication, coordination and control. The high costs of development and rapid obsolescence are likely to reinforce efforts of TNCs to secure a quicker pay-back through internalization. Furthermore, the internalization of the R&D expenditure noted earlier and the trend towards strategic alliances among TNCs in respect of the development and transfer of technologies limit the plurality of sources in the technology market. The

**Transnational Corporations as Engines of Growth**

---

### Box VI.3. Joint ventures in developed countries and technology transfer

A study of Japanese FDI in the United States showed that joint ventures were the preferred entry mode for Japanese firms in industries with high United States R&D expenditure, while Japanese R&D expenditures had no significant influence on the mode of entry. [1] However, before any conclusions can be drawn on the role of Japanese-United States joint ventures in technology transfer from the United States, it is necessary to distinguish between different types of joint ventures. [2] One type is related to barriers in the form of Governments encouraging joint ventures in disadvantaged national industries to maximize the gains for domestic partners. A second type involves voluntary joint ventures between partners with mutually beneficial strengths. Again, voluntary joint ventures can occur between a strong firm in a national declining industry and a relatively weaker firm from a foreign firm with a strong competitive advantage or between equally strong partners with specialized advantages.

It can be assumed that technology transfer between the partners will be highest within voluntary joint ventures, and that a strong firm in a declining industry will receive relatively more technology than the partner with the competitive advantage. The well known joint ventures in the automotive industry, like Chrysler's joint venture with Mitsubishi, General Motors' investment in Isuzu and Ford's joint venture with Mazda, are examples of firms in a declining industry gaining access to new technology (among other things, the production of compact cars). The relatively smaller Japanese firms with the competitive advantage gain access to the United States market. The reverse example is Fujitsu's joint venture with Amdahl, a small and innovative computer firm. [3] Fujitsu gained United States technology while Amdahl received financial support.

Joint ventures of the type were the partners have mutual beneficial strengths are also numerous between United States and Japanese firms. One such joint venture is the Toshiba-Motorola venture to manufacture microprocessors (Motorola's competitive advantage) and large memory chips (Toshiba's competitive advantage). The joint venture with Toshiba has allowed Motorola to gain access to technology in order to compete in advanced dynamic random access memory (DRAM) chips. [4]

It is difficult to determine who has benefited the most from technology-related joint ventures between Japanese and United States firms. However, a study of the industry distribution of joint ventures between Japanese and United States firms in 1987 showed that industries in which Japan has a clear competitive advantage (measured by market share in the OECD countries) account for the largest share of joint venture assets. [5] This indicates that the existing joint ventures in manufacturing presumably provide more opportunities for the transfer of technology from Japan to the United States than from the United States to Japan.

---

1 Bruce Kogut and Sea Jin Chang, "Technological capabilities and Japanese foreign direct investment in the United States", *Review of Economics and Statistics*, vol. 73, No. 3 (August 1991), pp. 401-413.

2 See Dorothy B. Christelow, "U.S.-Japan joint ventures: who gains?", *Challenge*, vol. 32, No. 6 (November-December, 1989), pp. 29-38.

3. Ibid.

4 See *Asian Business*, January 1991.

5 See Christelow, op. cit.

deceleration in the growth of external resource inflows through official development assistance and private flows other than FDI would limit the ability of developing countries to acquire unpackaged technology. Finally, recent policy changes in developing countries in favour of FDI tend to reduce the cost of internalization. Those factors are likely to increase the importance of FDI as an instrument of technology transfer.

• The second observation relates to the interrelationship between FDI and externalized technology transfer. With the exceptions of India and the Republic of Korea, the bulk of technology receipts of United States TNCs from unaffiliated enterprises in developing economies originates precisely in economies such as Argentina, Brazil, Hong Kong, Indonesia, Malaysia, Mexico, Philippines, Singapore, Taiwan Province of China and Venezuela, which are also among the largest FDI host economies. Thus, familiarity with the enterprises of a country and their capabilities gained through FDI may be a precondition for (or at least a facilitator of) externalized forms of technology transfer. In the case of India and the Republic of Korea, a combination of restrictive policies towards FDI, the availability of a substantial pool of well-trained human resources and the large market size encouraged externalized forms of technology transfer despite comparatively less FDI.

**Table VI.10.  United States and the Federal Republic of Germany: technology receipts, 1986-1990**

(Millions of national currencies)

| Year | Receipts from all countries | Receipts from all unaffiliated enterprises | | Receipts from unaffiliated enterprises in developing countries | |
|---|---|---|---|---|---|
| | | Value | Percentage | Value | Percentage |
| **United States** [a] | | | | | |
| 1986 | 7 531 | 1 842 | 24.5 | 296 | 3.9 |
| 1987 | 9 419 | 2 171 | 23.1 | 364 | 3.9 |
| 1988 | 11 211 | 2 513 | 22.4 | 443 | 4.0 |
| 1989 | 12 404 | 2 814 | 22.7 | 563 | 4.5 |
| 1990 | 15 840 | 3 445 | 21.7 | 690 | 4.4 |
| **Germany, Federal Republic of** [b] | | | | | |
| 1986 | 1 690 | 134 | 7.9 | 10 | 0.6 |
| 1987 | 1 670 | 163 | 9.8 | 7 | 0.4 |
| 1988 | 1 892 | 129 | 6.8 | 9 | 0.5 |
| 1989 | 2 166 | 189 | 8.7 | 7 | 0.3 |
| 1990 | 2 360 | 157 | 6.7 | .. | .. |

*Sources:* United States, Department of Commerce, *Survey of Current Business*, vol. 71, No. 9 (September 1991), tables 4.1-4.5, pp. 74-78; and *Monthly Report of the Deutsche Bundesbank*, various issues.

a Includes royalties and licence fees.
b Includes receipts from patents, inventions and processes.

• Third, the growth impact of technologies transferred through externalized forms depends, as in the case of capital-goods imports, largely on the capacity of domestic entrepreneurs to make the right selection, use the acquired technologies effectively and adapt and innovate continuously. Relevant also to the growth-promoting impact of technologies acquired through externalized forms is the capacity to negotiate reasonable terms. The many imperfections in technology markets give rise to a considerable scope for bargaining. If technology purchasers are not equipped with adequate knowledge concerning the availability of alternative sources of the same or similar technologies and their costs, transactions are very likely to be settled in terms more favourable to the sellers. In the case of Thailand, for example, it has been observed that no correlation existed between licensing fees and the complexity of technologies; fees were paid for technologies no longer covered by patents; and technical fees paid for similar technologies in the pharmaceutical industry ranged from 0.4 per cent to 20 per cent of sales. Even though many of the buyers were aware that the terms were disadvantageous, they did not have adequate information on alternative sources. [41]

## 4. Strategic alliances and technology transfer

High risks and rising R&D costs (especially in the area of new technologies) and the rapid obsolescence of new products have forced many TNCs to form technology-related strategic alliances to share development costs, acquire new technologies and make better use of scarce qualified personnel. [42] The substantial number of strategic alliances in existence now is a relatively new phenomenon, and it is very difficult to obtain precise data on its frequency and purpose. There are indications, however, of an emerging trend towards a very high proportion of agreements involving the development of and access to technologies. [43] The alliances of IBM with several other corporations for the purpose of developing its personal computer are an example: the Lotus Corporation provided the application software, and Microsoft wrote the operating system, for a micro-processor that was produced by Intel. [44] IBM (traditionally reluctant to conclude alliances) has now created alliances with more than 40 partners around the world, pooling technology and customer bases in the telecommunications and related fields. As a response to competition from IBM, the Japanese computer firm Fujitsu formed alliances with Texas Instruments, Siemens and Hitachi. Such alliances are often undertaken for the joint development of new generations of products and to set industry standards. Table VI.11 illustrates the geographical and industry breakdown of technology alliances among TNCs from the Triad. Transnational corporations from the United States and Europe are clearly the most active participants in strategic alliances, most of which take place in information technologies.

Technological alliances can be viewed as a way of providing collective protection to technological advances among a few partners. The increasing incidence of such alliances combined with the current pace and cost of technological development makes it more difficult for developing countries to acquire technology through traditional non-equity arrangements. Many alliances also involve common actions for setting international standards that increase the barriers to entry (including, for new products from developing countries) in the international market. Some developing countries, particularly the newly industrializing ones, have the potential and capability, however, to become partners in technology alliances. In the information-technology industry, for example, Taiwan Province of China, has made extensive use of alliances with TNCs to acquire technological capabilities. A typical example of that use is in the area of computer software, where the Government has set up two software engineering firms in cooperation with IBM. [45] Taiwan Province of China, provides good quality engineers at a relatively low cost while IBM provides experience in software research and development. Similarly, the Sony Group is to transfer advanced technology to the electronics industry in Taiwan Province of China. Sony has announced that it has entered into alliances with 130 electronics companies from that country working with a "technology development centre" to create a production base for export to Japan and affiliated companies of Sony world-wide. [46] Similarly, several firms in the automobile industry in the Republic of Korea have entered into alliances with TNCs from the Triad. Examples are those of Hunday with Mitsubishi and Chrysler; Daewoo with General Motors, Suzuki and Isuzu; and Kia with Ford and Mazda. [47]

These examples, however, represent only a small number of alliances that include developing countries. Indeed, only 2 to 3 per cent of technology alliances in the 1980s were between companies from the Triad and firms from newly industrializing economies, and less than two per cent included firms from other developing countries. [48] For most developing countries, then, the acquisition of new technologies is likely to rely—at least for the present—on intra-firm transfers by TNCs, rather than on inter-firm alliances between independent firms.

**Table VI.11. International distribution of technology cooperation agreements in biotechnology, information technologies and new materials, cumulative 1989**

(Number and percentage)

| Area | Biotechnology | | Information technologies | | New materials | |
|---|---|---|---|---|---|---|
| | Number | Per cent | Number | Per cent | Number | Per cent |
| Japan | 58 | 5 | 95 | 4 | 88 | 13 |
| United States | 428 | 35 | 707 | 26 | 139 | 20 |
| United States-Japan | 155 | 13 | 406 | 15 | 94 | 14 |
| United States-Western Europe | 245 | 20 | 599 | 22 | 133 | 19 |
| Western Europe | 223 | 18 | 509 | 19 | 118 | 17 |
| Western Europe-Japan | 38 | 3 | 177 | 7 | 49 | 7 |
| Other | 66 | 5 | 225 | 8 | 67 | 10 |
| *Total* | 1 213 | 100 | 2 718 | 100 | 688 | 100 |

*Source:* John Hagedoorn and Luc Soete, "The internationalization of science and technology (policy): how do 'national' systems cope?" in H. Inose, M. Kawasaki and F. Kodama, eds., *Science and Technology Policy Research: What Should be Done? What Could be Done?* (Tokyo, Mita Press, 1991), pp. 201-216.

## D. Assessment

The present chapter has shown that technology is a key determinant of growth. It promotes growth by increasing factor productivity, enabling the introduction of new products with greater long-run income elasticities and bringing about shifts in the export composition in favour of research-intensive exports with higher growth potential. In recent decades, the importance of technology as a determinant of growth has been increasing.

Transnational corporations are responsible for the bulk of technological development. Therefore, in so far as growth is driven by technology, the growth and development of developing countries are closely linked to a variety of equity and non-equity links with TNCs that permit access to technologies.

One channel of access to technology is through the import of capital goods, in the supply of which TNCs play a dominant role. Because the choice of technologies acquired through that mechanism largely rests with domestic importers, the contribution of TNCs to growth through this mechanism is essentially indirect.

Foreign affiliates can promote technological change in developing countries—and thereby growth—through their own R&D. During the past decade, data for United States majority-owned affiliates in developing countries show that the share of R&D expenditures in sales, though small, has increased. The net impact on growth arising from increased R&D expenditure by foreign affiliates also depends on what effects such an increase has on the R&D capabilities of indigenous firms. In general, it appears that the effect is likely to be more beneficial where domestic firms are capable of undertaking R&D to meet the challenge of competition from foreign affiliates. Foreign affiliates generally appear to exhibit higher factor productivity, which contributes to growth. They also appear to have contributed to the growth of developing countries through increasing the share of R&D intensive products in their total sales and their manufactured exports, over the past decade. Domestic enterprises in Asia, however, outperformed TNCs in respect of R&D intensive manufactured exports.

Foreign affiliates also have contributed to the growth of developing countries indirectly by increasing their purchase of local inputs; but the level and nature of such purchases is conditioned by the level of the industrial development of the host country (see chapter V). In some cases, foreign affiliates have stimulated R&D by local institutions through collaborative arrangements. Sparse information on that aspect does not allow any conclusion of the overall effect of such a stimulation on growth.

In some cases, a growth stimulus has also been generated through significant technology transfer by TNCs via such non-equity channels as licensing and subcontracting. The countries that have benefited most from such transfers appear to be typically the largest host countries; but, sometimes, those forms involved unfavourable terms that imply an avoidable drain on domestic resources.

In recent times, there has been an upsurge of technological alliances among TNCs, particularly in respect of new technologies. The upsurge raises the concern that reduced competition in the international technology market and restrained access for developing countries could limit the contribution that the dissemination of new technologies can make to growth.

In sum, TNCs are making a worthwhile direct contribution to the growth and technological development of host developing countries through the R&D expenditure of affiliates, changes in their product and export composition and higher factor productivity. As to the indirect stimuli to growth through non-equity forms of technology transfer, integration with domestic economies, the stimulation of local R&D and technological alliances with developing country enterprises, the evidence is rather mixed. The beneficial impact of those mechanisms appears to be largely contingent on, among other things, the domestic capacity of the host country to generate and adapt acquired technologies; the competitive ability of domestic enterprises; and the availability of well-trained human resources. It can be concluded, therefore, that, as regards indirect stimuli to growth through a wider dissemination of technologies, TNCs can strengthen a national technological base where the above conditions already exist; but they are unlikely to create them.

# E. Some policy implications

Several policy implications emerge from the findings in the present chapter. For one, R&D by foreign affiliates in developing countries appears to be mainly located in those countries that already possess some domestic technological competence and a reasonable supply of trained scientific and technical personnel. Policy measures directed towards inducing TNCs to undertake greater R&D in host economies should, therefore, be conceived in the broader context of the indigenous technological development policy of a country, encompassing, among other things, the creation of an adequate human-resources base for technological activities.

As noted earlier, FDI has made a notable contribution to technology transfer and thereby growth through changes in the composition of products and exports in favour of greater technological intensity. Here, again, performance variations of TNCs appear to be related to indigenous technological capacities of host economies; hence, the conclusion noted above with regard to inducing TNCs to undertake greater R&D in host economies is applicable here as well. In addition, it should be pointed out that, while FDI may be a useful means of quickly benefiting from the results of new innovations abroad through the transfer of production, it does not necessarily imply a dissemination of technological knowledge to domestic producers. That raises the question of the choice of mode of transfer as between FDI and externalized forms. There are contrasting experiences in respect of that choice, even though national-growth performances have been quite comparable. Singapore can be easily cited as a case of high reliance on FDI, while the Republic of Korea represents a greater reliance on externalized forms. That was facilitated in the Republic of Korea both by the creation of an ample entrepreneurial skill base and by governmental assistance to local enterprises that provides information and support in bargaining. Any country seeking to pursue a similar strategy would be well advised to evaluate carefully the level of human resources development of the country, the entrepreneurial capacity of domestic producers and the ability of the Government to provide appropriate information and guidance to domestic firms. Besides, it should be pointed out that restrictive policies towards FDI may severely limit access to sophisticated technology, as amply demonstrated by the experience of India. Furthermore, even the Republic of Korea chose to liberalize its FDI policy progressively since the early 1980s, as more modern technologies were needed to sustain international competitiveness and, as noted in chapter V, FDI contributed almost one half of the new capital in technology-intensive industries such as electrical machinery and transportation equipment.

In the context of the varying performance of TNCs in respect of linkages with the domestic economy, the question of performance requirements assumes relevance. Quite obviously, for example, if the objective is to promote efficiency in the use of resources in order to promote faster growth, the imposition of a local content requirement, in the absence of an internationally competitive supplier industry, would be counterproductive, at least in the short run. There may be a case for a highly selective use of such requirements, however, in cases with a high probability that local producers would be able to achieve quickly international standards with assured demand for their products. [49]

Even where TNCs are willing to transfer technologies in externalized forms, the terms of transfer may leave something to be desired, as the experience of Thailand (cited earlier) demonstrates. Excessive payments imply a drain on domestic resources and thus may inhibit growth. Technology purchasers should, therefore, be provided with adequate information regarding available alternatives, to enable them to make informed choices. An arbitrary imposition of limits on royalty payments or licence fees is likely to limit access to desired technologies.

A correlation appears to exist between FDI and access to externalized forms of transfer in most cases. Hence, highly restrictive policies towards FDI may also limit the scope to acquire technology through other channels, unless the country concerned has a strong bargaining position because of its large market size or its capacity to develop technologies independent of an association with TNCs. The link between FDI and externalized forms also raises a formidable technology barrier for the vast majority of developing countries that attract little FDI because of their structural constraints, no matter how liberal their national policies towards such investment.

The present chapter has also demonstrated that there is a marked tendency among TNCs to hold new technologies closely among themselves through strategic alliances. There have been, at the same time, some instances of such alliances with developing-country enterprises. Local enterprises, therefore, deserve encouragement to enter into such arrangements with TNCs from advanced countries, wherever possible, in order to gain access to new technologies, or to be able to apply them more widely in the interest of sustaining competitiveness and growth.

Finally, it should be emphasized that the growth-promoting impact of technologies acquired through FDI as well as other forms of association with TNCs ultimately depends on the incentive structure faced by both foreign and domestic enterprises in acquiring, adapting, innovating upon and diffusing technologies. The incentive structure is conditioned by a host of public policies, concerning physical infrastructure, human resources development, R&D, technology and FDI, competition, international trade, factor pricing, venture capital, subsidies etc. The formulation of such a holistic approach is, no doubt, immensely complex; but without such an approach the contribution of TNCs to growth through technology transfer will fall short of its potential.

## Notes

[1]Francois Chesnais, "Science, technology and competitiveness", *STI Review*, No. 1 (Autumn 1986), pp. 85-129.

[2]OECD has published a major work dealing with these issues: *Technology and Productivity: The Challenge for Economic Policy* (Paris, OECD, 1991).

[3]Sanjaya Lall and George Kell "Industrial development in developing countries and the role of government intervention", *Banca Nazionale del Lavoro Quarterly Review*, No. 178 (September 1991), pp. 271-292.

[4]S. Kuznets, *Secular Movements in Production and Prices* (Boston, Houghton Mifflin, 1930).

[5]J. Schumpeter, *Capitalism, Socialism and Democracy* (New York, Harper and Row, 1942).

[6]See, for example, Robert M. Solow, "Technical change and the aggregate production function", *Review of Economics and Statistics*, vol. XXXIX, No. 3 (August 1957), pp. 312-320; Edward Denison, *The Sources of Economic Growth in the United States* (New York, Committee for Economic Development, 1962) and *Why Growth Rates Differ?* (Washington, D.C., The Brookings Institution, 1967); M. Abramowitz "Resource and output trends in the United States since 1870", *American Economic Review: Papers and Proceedings*, vol. XLVI, No. 2 (May, 1956), pp. 5-24; J. Kendrick, "Productivity trends: capital and labour", *Review of Economics and Statistics*, vol. XXXVIII, No. 3 (August 1956), pp. 248-257; Z. Griliches, "Hybrid corn: explanation in the economics of technological change, *Econometrica*, vol. 25, No. 4 (October 1957), pp. 501-522; W. Parker and J. Klein, "Productivity growth in grain production in the United States, 1840-60 and 1900-10", in Brady Dorothy, ed., *Output, Employment and Productivity in the United States after 1800* (New York, National Bureau of Economic Research, 1966).

[7]V. J. Elias, "Sources of economic growth in Latin American countries", *Review of Economics and Statistics*, vol. 60, No. 3 (August 1978), pp. 362-371.

[8]J. Fagerberg, "Innovation, catching up and growth", in OECD, *Technology and Productivity: The Challenge for Economic Policy* (Paris, OECD, 1991), pp. 37-43.

[9]UNCTAD, "Impact of technological change on patterns of international trade", TD/B(XXXV)/SC.1/CRP.2 (March 1989), table 3, pp. 8-9.

[10]William F. Miller, "Europe 1992: regionalism and globalism", *The International Executive*, vol. 33, No. 2 (September-October 1991), pp. 28-35.

[11]Paul M. Romer, "Endogenous technological change", *Journal of Political Economy*, vol. 98, No. 5 (1990) pp. S71-S102. For a state-of-the-art review of the role of TNCs in innovatory activity, and a collection of major writings in this respect, see J. Cantwell, ed., *Transnational Corporations and Innovatory Activities. United Nations Library on Transnational Corporations* (London, Routledge, forthcoming).

[12]Jeremy Howells, "The location and organization of research and development: new horizons", *Research Policy*, vol. 19, No. 2 (1990), pp. 133-146 and Karl P. Sauvant, *International Transactions in Services: The Policies of Transborder Data Flows* (Boulder, Colorado, Westview Press, 1986).

[13]John H. Dunning, "Multinational enterprises and the globalization of innovatory capacity", Rutgers University GSM working paper No. 91-03 (January 1991), mimeo., and United States Department of Commerce, *U.S. Direct Investment Abroad:*

## Transnational Corporations as Engines of Growth

*1989 Benchmark Survey Preliminary Results* (Washington, D.C., United States Government Printing Office, 1991), tables 76 and 91.

[14]Dunning, "Multinational enterprises", op. cit.

[15]Christoph Doerenbacher and Michael Wortmann, "The internationalization of corporate research and development", *Intereconomics*, vol. 26, No. 3 (May/June 1991), pp. 139-144.

[16]Quoted in Michael Wortmann, "Multinationals and the internationalization of R&D: new development in German companies", *Research Policy*, vol. 19, No. 2 (1990), pp. 175-183.

[17]John H. Dunning, *Multinational Enterprises and the Global Economy* (Reading, Massachusetts, Addison Wesley, forthcoming), chap. 11.

[18]For a recent state-of-the-art review of the subject and a collection of major writings in this respect, see E. Chen, ed., *Transnational Corporations and Technology Transfer. United Nations Library on Transnational Corporations* (London, Routledge, forthcoming).

[19]In this context, it should be noted that only 2 per cent of total R&D expenditures of 11 TNCs of the Federal Republic of Germany surveyed at the end of the 1970s were in developing countries (see Wortmann, op. cit.). In the case of United States TNCs, 5 per cent of foreign R&D expenditures were in developing countries in 1982, and that proportion declined to 4 per cent in 1989 (see United States, Department of Commerce, *U.S. Direct Investment Abroad: 1982 Benchmark Survey* (Washington D.C., United States Government Printing Office, 1985), tables III.H.3 and III.E.1, and *1989 Benchmark Survey*, op. cit., tables 40 and 76.

[20]United States, Department of Commerce, *U.S. Direct Investment Abroad: 1989 Benchmark Survey*, op. cit., table 85. For electrical machinery, 32 per cent of the exports were intra-firm. There are no separate data on the industrial distribution of exports by United States TNCs to developing countries. However, the intra-firm portion of exports to developing countries is generally lower than it is for developed countries.

[21]John H. Dunning, *Explaining International Production* (London, Unwin Hyman, 1988), p. 10.

[22]Magnus Blomström, "Host country benefits of foreign investment", Working Paper No. 3615 (Cambridge, National Bureau of Economic Research, 1991), mimeo.

[23]Larry N. Wilmore, "The comparative performance of foreign and domestic firms in Brazil", *World Development*, vol. 14, No. 4 (April 1986), pp. 489-502.

[24]Somsak Tambunlertchai and Eric D. Ramstetter, "Foreign firms in promoted industries and structural change in Thailand", in Eric D. Ramstetter, ed., *Direct Foreign Investment in Asia's Developing Economies and Structural Change in the Asia-Pacific Region* (Boulder, Colorado, Westview Press, 1991), pp. 65-102.

[25]Chung H. Lee and Eric D. Ramstetter, "Direct investment and structural change in Korean manufacturing", in ibid, pp. 105-141.

[26]Calculations of the Transnational Corporations and Management Division, based on United Nations Statistical Office, *Monthly Bulletin of Statistics*, various issues.

[27]John H. Dunning, "Multinational enterprises", op. cit., p. 16.

[28]Dunning, *Multinational Enterprises and the Global Economy*, op. cit., chap. 11.

[29]J. Cantwell, *Technological Innovation and Multinational Corporations* (London, Basil Blackwell, 1989), p. 178.

[30]John H. Dunning, "Multinational enterprises", op. cit.

[31]UNCTC, *New Approaches to Best-Practice Manufacturing: The Role of Transnational Corporations and Implications for Developing Countries* (United Nations publication, Sales No. E.90.II.A.13).

[32]Transnational Corporations and Management Division, *Transnational Corporations and the Transfer of New Management Practices to Developing Countries* (New York, United Nations, forthcoming).

[33]UNCTC, *Foreign Direct Investment and Industrial Restructuring in Mexico* (United Nations publication, Sales No. E.92.II.A.9).

[34]See, for example, Richard Caves, "Multinational firms, competition and productivity in host country markets", *Economica*, vol. 14, No. 162 (1974), pp. 176-183; S. Globerman, "Foreign direct investment and 'spill-over' efficiency benefits in Canadian manufacturing industries", *Canadian Journal of Economics*, vol. 12, No. 1 (1980), pp. 24-52.

[35]Magnus Blomström and Edward N. Wolf, "Multinational corporations and productivity convergence in Mexico" (1989), mimeo.

[36]Mona Haddad and Ann Harrison, "Are there dynamic externalities from foreign direct investment? Evidence from Morocco", in R. Newfarmer and C. Frischtak, eds., *Transnational Corporations, Market Structure and Industrial Performance. United Nations Library on Transnational Corporations* (London, Routledge, forthcoming).

[37]*Business Asia*, vol. XXIII No. 46 (18 November 1991), p. 14.

[38]UNCTC, *Transnational Corporations and Technology Transfer: Effects and Policy Issues* (United Nations publication, Sales No. E.87.11.A.4), p. 14.

[39]Calculated from table 2 in Lim Su Kim, "Technological transformation in Korea and its implications for other developing countries", *Development and South-South Cooperation*, vol. IV No. 7 (December 1988), pp. 19-29.

[40]The World Bank, Industry and Energy Department, *Technology Strategy and Policy for Industrial Competitiveness: A Case Study of Thailand* (Washington, D.C., The World Bank, April 1990), p. 16.

[41]Ibid.

[42]See Jonathan B. Tucker, "Partners and rivals: a model of international collaboration in advanced technology", *International Organization*, vol. 45, No. 1 (Winter 1991), pp. 83-120.

[43]See Lynn Krieger Mytelka, ed., *Strategic Partnerships: States Firms and International Competition* (London, Pinter Publishers, 1991), pp. 7-35.

[44]See Kenichi Ohmae, "The global logic of strategic alliances", *Harvard Business Review*, vol. 89, No. 2 (March-April 1989), pp. 143-154.

[45]See Gee San, "Technology, investment and trade under economic globalization: the case of Taiwan", in OECD, *Trade, Investment and Technology in the 1990s* (Paris, OECD, 1991), pp. 57-97.

[46]See *Financial Times*, 20 November 1991.

[47]*Asian Business* (January 1991), p. 26.

[48]The numbers are based on a total of 4,192 strategic technology alliances contained in the MERIT data bank, University of Limburg, Maastricht.

[49]For a discussion of issues related to performance requirements, see UNCTC and UNCTAD, *The Impact of Trade-related Investment Measures on Trade and Development* (United Nations publication, Sales No. E.91.II.A.19).

# [24]

# TECHNOLOGICAL CAPABILITIES AND JAPANESE FOREIGN DIRECT INVESTMENT IN THE UNITED STATES

Bruce Kogut and Sea Jin Chang*

*Abstract*—This article examines the effect of relative technological capabilities on Japanese direct investment into the United States by looking simultaneously at industry conditions in the two markets. A negative binomial regression model is specified to estimate the effects of R & D capability and industry structure on a count measure of Japanese entries across 297 industries. The results indicate that Japanese direct investment in the United States is drawn to industries intensive in R & D expenditures summed across both countries; voluntary restraints on Japanese exports encourage direct investment. When the entries are disaggregated by mode (e.g., new plant or acquisition), there is a significant indication that joint ventures are used for the sourcing and sharing of U.S. technological capabilities.

THERE is an on-going policy concern in the United States over the loss of the resident technological capabilities due to foreign direct investment. Popular attention has focused particularly on Japanese investments in high technology industries. Yet, little analysis has been given to whether these investments reflect the extension of Japanese firm advantages through foreign direct investment as opposed to the targeting of American technology.

These issues are also of interest to the broadening of theoretical and empirical studies on the relationship of R & D expenditures to foreign direct investment. Home technological capability has been found in numerous empirical studies to correlate significantly with outward exports and foreign direct investment.[1] The explanation for this finding has conventionally emphasized that foreign direct investment and exporting are jointly determined by a combination of factors reflecting

Received for publication December 27, 1989. Revision accepted for publication August 9, 1990.

* Wharton School of the University of Pennsylvania.

This paper is supported under a project grant from the German Marshall Fund; additional funding was given by AT & T under the auspices of the Reginald H. Jones Center. We acknowledge the research assistance of James Jurney and the comments of Erin Anderson, Paul Rosenbaum, Louis Wells, and the anonymous referees. The data for the study were collected by the International Trade Administration at the Department of Commerce under the guidance of the late Mr. Richard Apcar, to whom we would like to express our gratitude and appreciation for his dedicated public service.

[1] See the summaries by Deardorff (1984) for exports and by Caves (1982) for foreign direct investment.

dynamic comparative advantage and industry rivalry. R & D expenditures lead to the creation of valuable rent-yielding and intangible assets, such as the knowledge involving new product and process technologies. Home oligopolistic rivalry not only promotes the creation of these technologies, but leads to eventual investments in overseas markets.[2]

Frequently neglected in the empirical studies on the determinants of foreign direct investment are the conditions of the foreign market. This neglect is particularly important to identifying whether foreign direct investment is motivated by the home technological advantage or by the desire to source technology in the foreign market. Because of the high correlation of the industry distribution of R & D expenditures among countries, the positive relationship between home R & D and outward direct investment cannot itself distinguish between these two distinct motives.

For example, due to the paucity of non-American data on R & D expenditures, most empirical studies on foreign investment into the United States have relied on American industry data, assuming a high correlation of R & D expenditures among countries. But the positive relationship between U.S. R & D expenditures and foreign entry raises, in fact, a paradox that foreign investments into the United States favor industries intensive in domestic R & D expenditures.[3] Without observations on R & D intensity of the investing country, it has been impossible to sort out whether foreign direct investment in the United States is motivated by the exploitation of an initial technological advantage or by the seeking of new technologies resident in the United States.

[2] Caves (1971) and Magee (1976) note the relationship between foreign direct investment and oligopoly may be spurious, as both arise from the possession of entry deterring and intangible assets by a firm.
[3] See the reviews by Pugel (1981) and McClain (1983).

The importance of distinguishing between the relative technological capabilities of the source and recipient countries has recently been raised by the studies of Audretsch and Yamawaki (1988) and Yamawaki and Audretsch (1988) on the determinants of Japanese exports to the American market. In their work, Japanese exports have been shown to be sensitive to competitive conditions in the home and foreign market. Relative Japanese advantage in R & D is a significant determinant of exports to the American market.

The following article addresses the question of the effects of Japanese and U.S. technological capability on Japanese foreign direct investment into the United States. Two industry measures of technological capability are examined: R & D expenditures for each country and innovation frequency (though data for the latter exist only for the United States). These variables, along with other determinants of foreign direct investment, are then regressed against a count measure of the number of Japanese entries in 297 United States industries. Because the dependent variable is a count of entries, a negative binomial regression is specified, with heterogeneity being captured by an error term. The count measure allows for a lower level of aggregation (4-digit SIC) than previous studies and for the elimination of censoring bias if an industry has not evidenced an entry.

## I. Industry Influences on Foreign Entry

There is little theoretical controversy over the importance of the role of the competitive conditions of the foreign market in the theory of foreign direct investment. The incorporation of the effects of the competitive nature of the foreign market on inward direct investment poses, however, complications to the traditional explanation stressing intangible assets of the firm and oligopolistic rivalry of the home market. These complications are of two sorts. One is simply whether the foreign country is advantaged in location.[4] Lower factor costs of the foreign relative to the home country should increase the attractiveness of direct investment. There is some empirical evidence to support this commonsensical relationship, especially in the findings regarding the positive effect of tariffs on inward investment flows.[5]

Common sense notwithstanding, the high degree of flows of foreign direct investment among developed countries suggest that location advantages in factor costs are not pivotal. In fact, Swedenborg (1979) found that the country distribution of Swedish outward investment flows are correlated with high wages. One possible explanation for this anomaly derives from Caves' (1971) explanation that foreign direct investment stems from the public good character of the rent-yielding intangible asset and the necessity of local production to adapt products to the foreign market. Alternatively, Swedenborg's result could represent a confounding of horizontal and vertical investment. Vertical direct investment need not be restricted to raw material and low-cost labor sites, but also can occur for the sourcing of the technical capabilities (as embodied in skilled and professional workers) from foreign countries. Both Dunning (1988) and Cantwell (1989) hypothesize, with empirical support, that countries leading in technology will draw foreign direct investment.

The second complication posed by incorporating competitive conditions of the target country in the theory of foreign direct investment is that the foreign market may itself be oligopolistic. In the 1960s, when the overwhelming proportion of the world's foreign direct investment flows was American, the interaction between national oligopolies did not seem very consequential. Hence, the emphasis by Vernon (1966) solely on the influence of home rivalry in new product and process technologies on foreign direct investment was reasonable and supported by considerable empirical evidence.[6] Subsequent empirical work, especially of Flowers (1976) and Graham (1978), and the theoretical model of cross-hauling investments of Brander and Krugman (1983) indicate oligopolistic rivalry at the international, rather than national, level as influencing the foreign direct investment decision. This perspective coincides with the recent treatment of exports and international R & D rivalry, where export compe-

---

[4] The effect of location is best summarized in Dunning's "eclectic" approach. See Dunning (1981).

[5] See Horst (1972) and Caves (1982).
[6] See the volume of empirical studies edited by Wells (1972), as well as Knickerbocker (1973).

JAPANESE FOREIGN DIRECT INVESTMENT IN U.S.     403

TABLE 1.—JAPANESE ENTRY INTO 2-DIGIT SIC INDUSTRIES

| SIC Code | Industry Group | Entry Count | Value (million $) |
|---|---|---|---|
| 20 | Food and kindred products | 73 | 1298 |
| 21 | Tobacco products | 0 | 0 |
| 22 | Textile mill products | 13 | 130 |
| 23 | Apparel and related products | 7 | 40 |
| 24 | Lumber and related products | 10 | 0 |
| 25 | Furniture and fixture | 8 | 37 |
| 26 | Paper and allied products | 9 | 174 |
| 27 | Printing and publishing | 9 | 166 |
| 28 | Chemical and allied products | 100 | 2382 |
| 29 | Petroleum and coal products | 2 | −18 |
| 30 | Rubber and plastics, n.e.c. | 29 | 804 |
| 31 | Leather and leather products | 2 | 0 |
| 32 | Stone, clay, and glass products | 22 | 205 |
| 33 | Primary metal industries | 72 | 2195 |
| 34 | Fabricated metal products | 48 | 434 |
| 35 | Machinery, except electrical | 133 | 3210 |
| 36 | Electric and electronic equipment | 209 | 2334 |
| 37 | Transportation equipment | 114 | 3111 |
| 38 | Instruments and related products | 49 | 1389 |
| 39 | Miscellaneous manufacturing | 29 | 335 |
| | Total | 938 | 18,226 |

Note: The amount of Japanese investment in the United States is calculated for 1980–88. The entry count is for 1976–87.

tition (sometimes abetted by governments) is driven by investments in R & D.[7]

Though R & D influences jointly exports and outward direct investment, there is a subtle difference. By Vernon's product cycle argument, foreign direct investment should occur in industries that are maturing and, hence, declining in R & D expenditures. Similar implications are to be found in Dornbusch, Fischer, and Samuelson's (1977) Ricardian analysis of trade, where differences in relative productivity among two countries result in each country specializing in the industries where it maintains a relative technical advantage. Productivity gains localized to a few export industries force the marginal industries to withdraw from the market. Two implicit alternatives are for home firms to transfer production to the foreign market if they are at a home location disadvantage but otherwise maintain a competitive position in the foreign market, or to increase investments in R & D expenditures in order to renew the technological base of export or import-competing industries.

In summary, there exists three issues of theoretical and policy importance that are currently

[7] See Brander and Spencer (1983, 1985), as well as the empirical treatment by Scherer (1988).

unsettled: whether foreign direct investments into the United States respond to conditions of the home or foreign market structure; whether they are motivated by the exploitation of home technological advantage or by the sourcing of the U.S. locational advantages in technology; and whether exports is the preferred mode in those industries where home technological capabilities are relatively advantaged. The first two of these issues are examined below by looking jointly at the influence of Japanese and U.S. industry factors on Japanese direct investments in the United States; the last issue is only addressed by joint inference from our and other studies' results.

## II. Empirical Model

### Measurement of Foreign Entry

Sorting out explanations for these issues has always been difficult because foreign direct investment figures are prone to substantial measurement errors. (R & D measurement has also been a stumbling block, as discussed below.) Balance of payment records neglect the investment portion derived from local debt or the capitalized contribution of intangibles, such as technology or goodwill. Foreign sales in a country have been a preferred measurement, but are also vulnerable to error by neglecting local content in the value-added. Like R & D expenditures, most available data on foreign direct investment or foreign sales tend to be aggregated to the 3 or 2 digit SIC levels.

We employ a novel measure in the analysis below, that is, a count of the number of entries in each 4-digit industry made by Japanese firms between 1976 and 1987. (See the appendix on data sources.) An entry count has the methodological advantage in providing a disaggregated measurement of foreign direct investment, thus allowing for a more refined determination of the influence of industry variables. Table 1 provides the industry breakdown of Japanese entries into the United States and the value of these investments as provided by balance of payments statistics. (We give only the entry count and value for manufacturing industries; minority investment positions are excluded.[8]) The correlation between these two series is 0.89.

*Variable Measurement*

A major problem in analyzing the relative influence of industry and country variations in R & D expenditures on foreign entry is the paucity of data. When available, the data are highly aggregated. In addition, problems of collinearity are unavoidable, because industry R & D expenditures are correlated across countries.

Audretsch and Yamawaki (1988) and Yamawaki and Audretsch (1988) address these problems by using multiple measures of technological intensity, in addition to looking at the raw differences in industry R & D expenditures between the two countries. We follow this approach by gathering information on R & D expenditures as a percentage of sales and (for the United States only) on innovation frequencies. This latter measure is a count of the number of innovations reported at the four-digit industry level.[9] These measures have been used successfully in previous applied work.[10] All three variables (*Japan R & D,*

U.S. R & D, and *Innovation Frequency*) are expected to be positively related to Japanese entry.

The drawback of the R & D expenditures is the high correlation between the U.S. and Japanese series. We addressed this problem by both summing and subtracting the two countries' expenditures.[11] The R & D sum provides a measure of the overall effect of R & D on entry; it captures the dimension of international rivalry in R & D, as well as its properties as an intangible and, thus, internationally transferable asset. The R & D difference serves to indicate whether entry is motivated by a home (i.e., Japanese) technological capability or by the desire to source U.S. technology. The two variables are weakly correlated with each other.[12]

The remaining variables are designed to capture other conditions of rivalry in Japan and the United States, as well as barriers to entry into the United States. Following Hymer (1976), Knickerbocker (1973), and others, we use the industry concentration ratio as an index of rivalry (*U.S. 8-Firm Concentration* and *Japan 8-Firm Concentration*).[13] Because intermediate levels of concentration have been found empirically to lead to oligopolistic instability, 8-firm concentration indices for Japan and the United States were used. (Regression results with 4-firm concentration rates are similar, as to be expected given the high correlation between the two indices.)

As foreign entry shares many of the characteristics of entry into new product markets, we can rely on the results of the recent literature on entry to identify additional variables that influence foreign investment. One variable characterizing the attractiveness of an industry for entry is shipment growth (*Shipment Growth*).[14] A growing industry permits entry without necessarily displacing encumbents. Industry advertising and marketing expenditures (*US Advertising*) have, however,

---

[8] Because a few industries (especially auto-related) experienced an inordinately large number of entries, the regressions reported below were reestimated without them; no change in the results were found.

[9] See Acs and Audretsch (1987) for a description.

[10] See Audretsch and Yamawaki (1988) and Yamawaki and Audretsch (1988). An alternative strategy of using company data is provided by Goto and Suzuki (1989), but this information is available only for a fraction of the cases.

[11] We thank Paul Rosenbaum for this suggestion.

[12] Since the U.S. and Japanese R & D series are not identical in their aggregation, we also built the same sum and difference measures using the National Science Foundation R & D data, which aggregated similarly to the Japanese series. None of the results reported below were changed.

[13] Similar findings for the degree of concentration and outward U.S. direct investment were found by Baldwin (1979) and Pugel (1978) and by Caves, Porter, and Spence (1980) for U.S. investment in Canada.

[14] See Gorecki (1975), Duetsch (1984), and Khemani and Shapiro (1986). Shapiro (1983) also confirms this finding for foreign entry into Canada.

JAPANESE FOREIGN DIRECT INVESTMENT IN U.S.                405

TABLE 2.—DATA DESCRIPTIONS AND PREDICTED SIGNS FOR INDEPENDENT VARIABLES

| Variable | Definition | Predicted Sign |
|---|---|---|
| Japan R & D | Japanese R & D expenditure/sales, (%) eleven years average for 1976–85 | + |
| U.S. R & D | U.S. R & D expenditure/sales (%) 1977 | + |
| R & D Sum | Japanese R & D expenditure + U.S. R & D expenditure | + |
| R & D Difference | Japanese R & D expenditure − U.S. R & D expenditure | − |
| Innovation Frequency | Number of new innovations, collected by Small Business Administration, 1982 | + |
| Japan R & D Growth | Average growth rate of Japanese R & D expenditure/sales, 1975–85 | − |
| Japan 8-Firm Concentration | 8-firm concentration ratio (%), 1982 | + |
| U.S. 8-Firm Concentration | 8-firm concentration ratio (%), 1982 | − |
| Shipment | Eleven years average value of industry shipment, 1975–85 | + |
| Shipment Growth | Average growth rate of industry shipment, 1975–85 | + |
| Import | Import/Shipment (%), eleven years average 1975–85 | + |
| U.S. Advertising | Advertising expense/sales (%), 1977 | − |
| Export Restriction | Dummy variable noting the existence of quota and voluntary restraints for Japanese export | + |

been usually found to deter entry.[15] While advertising levels have been found to be correlated with outward flows of foreign direct investment, they should deter inward entry. To these two variables, we add the degree of import penetration (*Import*); industries with high penetration of imports should also be easier to enter by local operations. To control for industry size effects, we include shipments volume (*Shipment*) as a right-hand variable. Finally, in line with the findings of Horst (1972) that import tariffs encourage inward investment, a dummy variable indicating quotas (*Export Restriction*) is constructed.

The above arguments and variable definitions are given in table 2. (Sources for the data are given in the appendix.) Descriptive statistics are given in table 3. The correlation matrix of the variables (not reported here) indicates a substantial degree of collinearity exists for the Japanese and U.S. R & D expenditure variables. A low correlation between innovation frequency and Japanese R & D expenditures suggests one avenue to compare the relative effects of Japanese and U.S. technological investments without de-

[15] An exception is Highfield and Smiley (1987) who found no relationship between advertising and entry rates, whereas Kessides (1986) finds an overall positive impact of advertising on entry.

grading the standard errors. Some caution in interpretation is required, however, for, as discussed below, innovation frequency and R & D expenditures are imperfect substitutes for each other.

*Sample Size*

The Department of Commerce listing of foreign entries indicates 938 Japanese entries occurred in 218 industries in the period of 1976 through 1987; the remaining 245 industries, as identified by the U.S. Standard Industrial Classification, did not experience a Japanese entry during the period. Entries only into wholesale and retail distribution are not classified in the manufacturing industries and are not included in the analysis; industries experiencing no entries are included. Unfortunately, matching Japanese industry concentration data to the U.S. industries left 122 U.S. industries without a match; an additional 34 industries were lost due to missing values of other variables. We eliminated these industries in the runs. (Including these industries and eliminating the Japanese concentration variable does not change the results reported below.) This elimination left 825 entries in 297 industries, of which 165 industries experienced entries and 132 industries experienced none. These 297 in-

TABLE 3.—DESCRIPTIVE STATISTICS AND CORRELATION MATRIX

| Variable | Mean | Standard Deviation | Lowest | Highest |
|---|---|---|---|---|
| (1) *Entry Count* | 2.78 | 6.74 | 0.00 | 72.00 |
| (2) *Japan R & D* | 1.72 | 1.14 | 0.20 | 5.70 |
| (3) *U.S. R & D* | 1.35 | 1.46 | 0.00 | 10.20 |
| (4) *R & D Sum* | 3.07 | 2.29 | 0.40 | 15.90 |
| (5) *R & D Difference* | 0.36 | 1.25 | −6.76 | 3.98 |
| (6) *Innovation Frequency* | 11.44 | 56.78 | 0.00 | 860.00 |
| (7) *Japan R & D Growth* | 0.05 | 0.02 | −0.01 | 0.12 |
| (8) *Japan 8-Firm Concentration* | 76.04 | 19.26 | 17.50 | 100.00 |
| (9) *U.S. 8-Firm Concentration* | 54.27 | 22.70 | 5.00 | 100.00 |
| (10) *Import* | 0.12 | 0.27 | 0.00 | 2.98 |
| (11) *Shipment* | 5,503.90 | 14,689.00 | 65.34 | 209,924.27 |
| (12) *Shipment Growth* | 0.02 | 0.05 | −0.15 | 0.41 |
| (13) *U.S. Advertising* | 1.62 | 2.39 | 0.00 | 20.20 |
| (14) *Export Restriction* | 0.03 | 0.17 | 0.00 | 1.00 |

Note: $N = 297$.

dustries accounted for 88% of all manufacturing entries.

*Model Specification*

The use of a count of foreign entries presents econometric and measurement issues similar in the study of patents. We follow the approach of Hausman, Hall, and Griliches (1984) on patent counts by specifying a Poisson regression to model the probability that the number of entries will occur $n$ times (with $n = 0, 1, 2, \ldots$) as follows:

$$\text{Prob}(Y = yj) = e^{-\lambda_j}\lambda_j^{Y_j}/Y_j! \qquad (1)$$

with $Y_j$ being the count of relationships for the entries of the $j^{th}$ industry. To incorporate exogenous variables, lambda can be made a function of the covariates:

$$\lambda_j = \exp(\Sigma B_i X_{ij}), \qquad (2)$$

where $B$'s are the coefficients, $X$'s are the covariates (with $X1$ set to one), $i$ indicates the $i^{th}$ variable, and $j$ is the $j^{th}$ industry. The exponential function ensures non-negativity.

The Poisson distribution contains the strong assumption that the mean and variance are equal to lambda. Based on an OLS regression, an estimate of the Breusch-Pagan statistic for heteroscedasticity was found to be highly significant. Since some of the (heteroscedastic) variation can be attributed to the large differences in shipment

volume across industries, we calculated the Goldfeld-Quandt statistic, which indicated significant differences in variances due to industry size.

To eliminate some of this variation, the initial Poisson regression (using the same variables given in column 3 of table 4) was weighted by shipment volume (*Shipment*). As a heuristic diagnostic (described in Cameron and Trivedi (1986)), the squared residual (i.e. $(y_j - \lambda_j)^2$) was calculated from this initial regression and then regressed on the predicted count $(\lambda_j)$. The coefficient of 0.02 is significant at less than 0.001, indicating substantial overdispersion.

To address the persisting problem of heteroscedasticity, we incorporate explicitly the disturbance of industry heterogeneity by specifying a compound distribution through an addition of an error term. Equation (2) now becomes

$$\lambda_j = \exp(\Sigma B_i X_{ij})\exp(u_j); \qquad (3)$$

$\lambda_j$ is no longer determined but is itself a random variable. As $u_j$ is unobserved, it is integrated out of the expression by specifying a gamma distribution for the error term, whereupon the now compound Poisson reduces to the negative binomial model (Johnson and Kotz, 1970). Only the scale of the distribution is permitted to vary as a function of the covariates, with the variance of $Yj$ parameterized to equal $(1 + \alpha)E(Yj)$, that is, a

## JAPANESE FOREIGN DIRECT INVESTMENT IN U.S.

TABLE 4.—NEGATIVE BINOMIAL REGRESSION ESTIMATES ON ENTRY COUNTS

| Variable Name | Equations | | | | |
| --- | --- | --- | --- | --- | --- |
| | (1) | (2) | (3) | (4) | (5) |
| *Intercept* | 0.36 | 1.22 | 0.57 | 0.58 | 0.58 |
| | (0.90) | (3.12)[a] | (1.41) | (1.44) | (1.43) |
| *Japan R & D* | 0.47 | | 0.26 | | |
| | (6.04)[a] | | (2.67)[a] | | |
| *U.S. R & D* | | 0.29 | 0.20 | | |
| | | (6.39)[a] | (3.69)[a] | | |
| *R & D Sum* | | | | 0.23 | 0.22 |
| | | | | (5.78)[a] | (5.35)[a] |
| *R & D Difference* | | | | 0.02 | 0.04 |
| | | | | (0.36) | (0.60) |
| *Innovation Frequency* | | | | | $0.48 \times 10^{-3}$ |
| | | | | | (0.17) |
| *Japan R & D Growth* | −14.27 | −23.06 | −16.03 | −16.17 | −16.21 |
| | (−2.30)[b] | (−4.06)[a] | (−2.70)[a] | (−2.72)[a] | (−2.74)[a] |
| *Japan 8-Firm* | 0.02 | 0.02 | 0.02 | 0.02 | 0.02 |
| *Concentration* | (3.34)[a] | (4.39)[a] | (4.55)[a] | (4.56)[a] | (4.50)[a] |
| *U.S. 8-Firm* | −0.01 | −0.02 | −0.02 | −0.02 | −0.02 |
| *Concentration* | (−3.30)[a] | (−4.37)[a] | (−4.37)[a] | (−4.39)[a] | (−4.31)[a] |
| *Import* | 0.16 | 0.41 | 0.25 | 0.25 | 0.25 |
| | (0.25) | (0.62) | (0.40) | (0.40) | (0.40) |
| *Shipment* | $-0.54 \times 10^{-5}$ | $-0.47 \times 10^{-5}$ | $-0.49 \times 10^{-5}$ | $-0.49 \times 10^{-5}$ | $-0.49 \times 10^{-5}$ |
| | (−1.04) | (−0.89) | (−1.01) | (−1.01) | (−0.99) |
| *Shipment Growth* | 4.56 | 1.18 | 1.90 | 1.86 | 1.24 |
| | (3.11)[a] | (0.39) | (1.18) | (1.15) | (0.66) |
| *U.S. Advertising* | −0.03 | −0.04 | −0.04 | −0.04 | −0.03 |
| | (−0.72) | (−1.17) | (−0.39) | (−0.86) | (−0.82) |
| *Export Restriction* | 2.58 | 3.02 | 2.84 | 2.85 | 2.93 |
| | (6.99)[a] | (8.99)[a] | (8.43)[a] | (8.46)[a] | (8.26)[a] |
| *α* | 0.72 | 0.70 | 0.65 | 0.65 | 0.65 |
| | (7.73)[a] | (7.58)[a] | (7.47)[a] | (7.48)[a] | (7.11)[a] |
| *Log-Likelihood* | −714.84 | −712.10 | −707.08 | −707.12 | −707.01 |

Note: *t*-statistics are in parentheses.
[a] $p < 0.01$.
[b] $p < 0.05$.

constant variance-mean ratio.[16] Given this specification, it becomes straightforward to account for overdispersion.[17] The parameter estimate of $\alpha$ provides a measure of contribution of intrinsic randomness or (as is more probable) heterogeneity caused by omitted variable bias. One likely source of heterogeneity is the omission of the effects of time. Given the availability of a single panel for most of the variables, the data were pooled over time and analyzed by cross-section. Using 1981 and 1985 as breakpoints (which closely correspond to obvious disturbances caused by

changes in the value of the dollar/yen exchange rate), the data were partitioned into three samples. No important differences were found in the estimates of the coefficients.

### III. Statistical Results

In the results described below, we report only the negative binomial regression estimates. The Poisson estimations tend to give substantially higher *t*-statistics; there are few cases where a variable that is significant under the negative binomial specification is not also significant under the Poisson. All tests are two-tailed. We discuss first the R & D results across all the regressions and then turn to the other estimates.

In table 4, the first two columns report the results when R & D expenditures for Japanese and the United States are separately entered. The findings indicate strongly that the number of

---

[16] We verified the specification against an alternative by running the diagnostic recommended by Cameron and Trivedi (1986) to regress the squared residual (divided by the predicted count) on a constant and the predicted count. The fit is positive and only significant for the intercept, supporting the constant variance-mean parameterization.
[17] Greene's statistical package (LIMDEP) provides this test as a standard feature.

entries is related to Japanese and U.S. R & D expenditures, at least when they are separately entered. Interestingly, growth in Japanese R & D expenditures depresses the tendency to enter the United States. As explained further below, this result implies that in industries where Japanese corporations are increasing their technological capabilities, foreign direct investment in the United States is relatively less prevalent. The next regression, reported in column 3, enters U.S. and Japanese R & D expenditures simultaneously. Despite their collinearity, both variables are estimated to be positively and significantly related to entry.

Because of possible contamination of the standard errors due to collinearity, other specifications were explored. As explained earlier, we summed the Japanese and U.S. R & D expenditures (*R & D Sum*), as well as subtracted the U.S. expenditure from the Japanese (*R & D Difference*). The results, given in column 4, indicate that entry is more likely in high R & D expenditure industries; the relative technological advantage of either country does not itself appear to influence the likelihood. The finding points to the importance of the intangibility and transferability of R & D assets, as well as to international rivalry in high technology industries.

In column 5, *Innovation Frequency* is added to the regression. Though positive, its coefficient is not significant. This result is somewhat surprising given a relatively large bivariate correlation (0.4) between *Innovation Frequency* and entry. (Inspection of the correlation table suggested that this correlation is potentially spurious due to the multicollinearity of *Innovation Frequency*, *Shipment Growth*, and *U.S. R & D*.)

The coefficient signs and significance levels of the other variables are fairly robust over the estimated regressions. The coefficients to the 8-firm concentration measures indicate that market structure in both countries exerts a very significant effect on entry. High home concentration clearly encourages direct investment in the United States; high U.S. concentration deters Japanese entry. These results reinforce the traditional findings of the importance of oligopolistic conditions in the home market, but they also show that direct investment is sensitive to the market structure of the target country.

The only other variable that is consistently significant is *Export Restriction*. Because the auto-related industries were outliers in entries and also were characterized by export restrictions, we reestimated the regression in their absence. The results remained largely the same. The imposition of voluntary export restraints promotes Japanese entry into the United States through foreign direct investment. (A coefficient of 2.8 implies that entry should increase by 16, a significant amount!)

The estimates to the remaining variables are not impressive. High important penetration (*Import*) and domestic growth in shipments (*Shipment Growth*) are very weakly related to foreign entry. The coefficient to U.S. advertising and marketing expenditures is negative, indicating that entry is deterred, but the result is not significant. The absolute volume of shipments is also not significantly related to entry.

In all of the regressions, the estimates for unexplained heterogeneity ($\alpha$) are highly significant. Clearly, there is a significant proportion of inter-industry variation in entry that is not being explained by the included variables. One possibility is that investment motivations differ by the mode of entries (i.e., acquisition, new plant, or joint venture), which vary in frequency across industries. We address this issue below.

### Discussion

The regression analysis strongly confirms the traditional findings that foreign direct investment is related to the intensity of industry R & D expenditures. In fact, the strong correlation to both U.S. and Japanese R & D expenditures indicate substantial competition in technological intensive industries. These results are consistent with the recent findings of Scherer (1988), Cantwell (1989), and Yamawaki and Audretsch (1988).

The results regarding the negative effects of the growth of Japanese expenditures suggest that the relationship is more complicated than is frequently depicted. Part of the explanation may lie in a specification bias due to the omission of the export alternative. As Horst (1972) pointed out, exports and foreign direct investment are jointly determined and are likely to be influenced by many of the same factors. It appears that exports remain favored in those industries where

Japanese firms are proportionately expanding most rapidly their technological capabilities.

When analyzed in conjunction with Yamawaki and Audretsch's (1988) positive finding between exports and Japanese expenditures relative to the United States, the results on R & D point to the speculation that growth in Japanese R & D capabilities discourages foreign direct investment because exports persist as the optimal mode. Both modes require a domestic technological capability to be viable. But the switch point from exports to foreign direct investments appears to be driven by dynamic changes in comparative advantage in maintaining and developing the technological capability of export industries.

An examination of the data provides further insight into the analysis. Among the fast growing R & D industries are sectors which are known to be facing considerable import competition in Japan and which are difficult to transplant, e.g., metals (zinc, aluminum, iron and steel) and petroleum refining. The slowest R & D growth industries include television and radio, transportation (exclusive of aircraft), and chemicals. This pattern suggests the immobility of production in import-competing industries stimulates R & D expenditures; the causal path between the feasibility and desireability to shift production offshore and R & D expenditures is likely to flow in both directions. The data are consistent with the broad arguments of Vernon (1966) and Dornbusch, Fischer, and Samuelson (1977), but they also give the important reminder of the endogeneity of R & D and productivity expenditures.

Though not significant, *U.S. Advertising* is negatively signed, consistent with the literature on entry.[18] As the usual finding is that domestic advertising is associated with outward foreign direct investment flows, this negative relationship between advertising and inward entries suggests a potential qualification to past empirical work. In particular, one can speculate that advertising advantages are harder to transport across borders because it is not product embodied and is often tailored to national cultures (Caves, 1982). In addition, entry is deterred by the presence of encumbent brands.

It is of some considerable policy interest that export restrictions promote foreign entry, with a significance of less than 0.001. In light of Yamawaki's (1986) finding that quotas decrease Japanese exports, the effect of export restrictions is to discourage exports and increase foreign entry into the United States. This result is not surprising, but whether it is the intention of policymakers is dubious.

In summary, the statistical analysis generally confirmed the expected relationships between the independent variables and the count frequency of Japanese investment entry into U.S. manufacturing industries. The surprising findings that entries are less likely from Japanese industries showing growing technological capabilities can be reconciled with recent theories of export and production hysteresis, where production from marginally-viable industries are shifted to foreign sites (Kogut and Kulatilaka (1990)). Yet, the alternative explanation that Japanese investments are promoted to source the technological advantages resident in the United States is equally consistent with the results.

### IV. Entry by New Plants, Acquisition, or Joint Venture

Some insight into the relative merits of these arguments can be gleaned by partitioning the data by mode of entry.[19] In table 5, the regression results are provided for each entry mode. Columns 1 to 3 are the negative binomial regression estimates for entry with both Japan and U.S. R & D expenditures included; columns 4 to 6 give the estimates with the R & D expenditures summed and subtracted.

The results given in columns 1 to 3 indicate important differences from the results in table 4 and substantial variation across mode of entry. Of most importance, Japanese R & D expenditures have no significant influence on entry into the United States by acquisition or joint venture. For new plants, the effect of Japanese R & D investment is to increase the likelihood of entry. U.S. R & D expenditures are associated with higher entry rates by all three modes, but especially by

---

[18] See Shapiro and Khemani (1987).

[19] Because of its small sample size, entry by plant expansion is not examined.

Table 5.—Negative Binomial Regression Estimates by Type of Entry

| | Acquisition | Joint Venture | New Plant | Acquisition | Joint Venture | New Plant |
|---|---|---|---|---|---|---|
| *Intercept* | -0.14 | -0.94 | -0.50 | -0.10 | -0.91 | -0.50 |
| | (-0.20) | (-1.50) | (-0.88) | (-0.15) | (-1.44) | (-0.88) |
| *Japan R & D* | 0.05 | -0.11 | 0.41 | | | |
| | (0.44) | (-0.82) | (3.00)[a] | | | |
| *U.S. R & D* | 0.18 | 0.34 | 0.20 | | | |
| | (2.56)[b] | (3.92)[a] | (2.87)[a] | | | |
| *R & D Sum* | | | | 0.10 | 0.11 | 0.31 |
| | | | | (1.95)[c] | (2.12)[b] | (5.87)[a] |
| *R & D Difference* | | | | 0.03 | -0.25 | 0.08 |
| | | | | (0.35) | (-2.34)[b] | (0.81) |
| *Innovation Frequency* | | | | 0.003 | -0.001 | -0.001 |
| | | | | (1.04) | (-0.24) | (-0.26) |
| *Japan R & D Growth* | -14.80 | -2.08 | -21.99 | -13.50 | -2.26 | -22.25 |
| | (-1.89)[c] | (-0.27) | (-2.43)[b] | (-1.91)[c] | (-0.29) | (-2.48)[b] |
| *Japan 8-Firm Concentration* | 0.02 | 0.02 | 0.02 | 0.02 | 0.02 | 0.02 |
| | (3.13)[a] | (2.69)[a] | (3.20)[a] | (2.94)[a] | (2.70)[a] | (3.28)[a] |
| *U.S. 8-Firm Concentration* | -0.03 | -0.01 | -0.02 | -0.03 | -0.01 | -0.02 |
| | (-5.18)[a] | (-2.54)[b] | (-2.77)[a] | (-4.85)[a] | (-2.58)[a] | (-2.90)[a] |
| *Import* | 0.57 | -0.88 | -0.19 | 0.71 | -0.87 | -0.20 |
| | (0.61) | (-0.96) | (-0.24) | (0.85) | (-0.95) | (-0.25) |
| *Shipment* | $-0.28 \times 10^{-5}$ | $-0.94 \times 10^{-5}$ | $0.12 \times 10^{-5}$ | $-0.30 \times 10^{-5}$ | $-0.95 \times 10^{-5}$ | $0.12 \times 10^{-5}$ |
| | (-0.46) | (-1.33) | (0.20) | (-0.47) | (-1.25) | (0.20) |
| *Shipment Growth* | 2.75 | -3.64 | 1.52 | -0.69 | -3.19 | 2.51 |
| | (1.54) | (-1.75)[c] | (0.80) | (-0.28) | (-1.35) | (1.06) |
| *U.S. Advertising* | 0.02 | -0.21 | -0.05 | 0.03 | -0.22 | -0.05 |
| | (0.47) | (-2.75)[a] | (-0.77) | (0.74) | (-2.78)[a] | (-0.82) |
| *Export Restriction* | 1.67 | 3.33 | 2.69 | 1.53 | 3.37 | 2.71 |
| | (4.74)[a] | (8.40)[a] | (7.56)[a] | (4.09)[a] | (7.79)[a] | (7.53)[a] |
| *α* | 0.62 | 0.32 | 0.83 | 0.56 | 0.33 | 0.82 |
| | (3.25)[a] | (2.98)[a] | (5.54)[a] | (3.22)[a] | (2.73)[a] | (5.35)[a] |
| Log-Likelihood | -394.46 | -330.94 | -518.71 | -391.36 | -330.60 | -518.61 |

Note: *t*-statistics are in parentheses.
[a] $p < 0.01$.
[b] $p < 0.05$.
[c] $p < 0.10$.

joint ventures. These results suggest the use of cooperative ventures to source, or benefit from, U.S. technological capabilities in industries where the United States, but not Japan, are intensive in R & D expenditures. But it should be noted that the effect of the growth in Japanese R & D expenditures is to dissuade entry, no matter by what mode.

The remaining variables present some interesting contrasts. Acquisitions are strongly favored when the investment is from a concentrated Japanese industry but into a low concentrated U.S. industry. This relationship suggests that home rivalry might spill over into a rush to invest by acquisition. Sparsity of American targets appears to discourage acquisitions, presumably because the likelihood of a takeover is captured in the valuation of share prices.

Joint ventures favor industries where advertising expenditures are low. This finding reinforces

the explanation that joint ventures are oriented towards technological sourcing. All three modes are stimulated by the existence of export restraints.

Columns 4 to 6 report the estimates by replacing each country's R & D expenditure by the variables *R & D Sum*, *R & D Difference*, and *Innovation Frequency*. The results point to a striking difference. Unlike the other entry modes, joint ventures clearly favor industries where the United States has greater R & D expenditures. The results for entry by new plants resemble the findings for the overall sample, as given in table 4. Acquisitions are insensitive to differences in R & D expenditures and are only weakly related to the total expenditures in both countries.

The difference between new plants and the other two modes is echoed in the other results. Joint ventures are shown again to be less likely in high advertising expenditure industries. A posi-

## JAPANESE FOREIGN DIRECT INVESTMENT IN U.S. 411

tive but modest effect is evident for acquisitions; the effect is insignificant but it raises the speculation that acquisitions are also used for brand label and distribution channel access. The importance of home (i.e., Japanese) market concentration is confirmed for all three entry modes. The highly significant and negative coefficient to the measure of U.S. concentration indicates, however, that foreign market structure (and, possibly, the lack of acquisition targets) is an important deterrent to entry.

In summary, the results on entry mode choice provide an interesting blend of support for a Ricardian investment model of the product cycle and technological sourcing explanations.[20] Entries by new plants occur in industries where the total R & D expenditures are high and where Japanese R & D expenditures are not growing quickly. As predicted by Vernon's product cycle explanation or by recent arguments of direct investment under shifts in relative technical efficiency, industries with declining relative shares of R & D are eventually forced to withdraw from exports or to shift overseas. However, the industry distribution of entries by joint ventures points to the motivation to source U.S. technological capabilities.

### V. Conclusions

The above analysis of Japanese entries into 297 U.S. industries throws light on a number of areas of theoretical and policy importance. In large part, the findings are consistent with recent work on entry into U.S. industries, no matter if the entrant is U.S. or foreign. Industry growth and R & D expenditures attract entry; high levels of seller concentration deter entry.[21]

But unlike previous studies on domestic entry or on foreign direct investment, our findings look not only at the conditions of the target industry, but also permit an analysis of the effects of the industry structure of entering firms on the probability of entry. The implications of the findings are straightforward and appealing. Japanese di-

rect investments in new plants predominate in high R & D expenditure industries, but are not sensitive to a relative technological advantage in either market.

These issues are important to the on-going debate on the benefits of foreign direct investment in the United States. It has long been a contention in the popular press that foreign investments drain American technology. With the recent strengthening of the oversight provisions of the interagency Committee on Foreign Investment in the United States, the executive branch has the authority to block foreign acquisitions in violation of national security interests. Many of these interests have been defined as the safeguarding of U.S. technological strengths.

The results of this paper do not indicate that Japanese acquisitions are more frequent in high-technology industries. In fact, the entry count of acquisitions is related only with weak significance to any of the R & D measures. The results indicate, however, that Japanese–U.S. joint ventures appear to be motivated by the sourcing of U.S. technology. As recent work has shown a strong tendency for manufacturing joint ventures to terminate by acquisition (Kogut, 1991), attention should be directed—to the extent that policy makers should be concerned at all over direct investment—in areas of joint U.S. and Japanese cooperation.

There is an important and unobserved institutional explanation for the relationship of R & D to Japanese foreign entry, for it has been the policy of MITI to encourage the investment of both physical and technological capital in targeted export and import-competing industries. The strong effect of the variable *Japan R & D Growth* may be capturing this institutional effect. In part, this interpretation raises questions of the robustness of the findings to other countries. It also raises, more intriguingly, important issues of the relationship of government policy to the composition of trade and investment and the long-term evolution of a country's comparative advantage.

---

[20] The significance levels to the estimates of $\alpha$ are also lower compared to those in table 4, suggesting that some of the earlier heterogeneity stemmed from the pooling of counts across entry modes.

[21] See the review of Acs and Audretsch (1989). They qualify the R & D results by noting that high R & D expenditure industries deter small firm entry.

### REFERENCES

Acs, Zoltan, and David Audretsch. "Innovation, Market Structure, and Firm Size," this REVIEW 69 (Nov. 1987), 567–574.

———, "Small-firm Entry in US Manufacturing," *Economica* 56 (May 1989), 255–265.

Audretsch, David, and Hideki Yamawaki, "R & D Rivalry, Industrial Policy and U.S. Japanese Trade," this REVIEW 70 (Aug. 1988), 438–447.

Baldwin, Robert E., "Determinants of Trade and Foreign Investment: Further Evidence," this REVIEW 61 (Feb. 1979), 40–48.

Brander, James A., and Paul R. Krugman, "A 'Reciprocal Dumping' Model of International Trade," *Journal of International Economics* 15 (1983), 313–321.

Brander, James A., and Barbara Spencer, "International R & D Rivalry and Industrial Strategy," *Review of Economic Studies* 50 (Oct. 1983), 707–722.

———, "Export Subsidies and International Market Share Rivalry," *Journal of International Economics* 18 (Feb. 1985), 83–100.

Cameron, A. Colin, and Pravin Trivedi, "Econometric Models Based on Count Data: Comparisons and Applications of Some Estimators," *Journal of Applied Econometrics* 1 (Jan. 1986), 29–53.

Cantwell, John, *Technical Innovations in Multinational Corporations* (London: Basil Blackwell Inc., 1989).

Caves, Richard, "International Corporations: The Industrial Economics of Foreign Investment," *Economica* (Feb. 1971).

———, *Multinational Enterprise and Economic Analysis* (New York: Cambridge University Press, 1982).

Caves, Richard, Michael E. Porter, and A. Michael Spence, with John T. Scott, *Competition in the Open Economy: A Model Applied to Canada* (Cambridge: Harvard University Press, 1980).

Deardorff, A. A., "Testing Trade Theories and Predicting Trade Flows," in R. W. Jones and P. B. Kenen (eds.), *Handbook of International Economics* (Amsterdam: Elsevier Science Publishers, B.V., 1984).

Dornbusch, Rudiger, Stanley Fischer, and Paul Samuelson. "Comparative Advantage, Trade, and Payments in a Ricardian Model," *American Economic Review* 67 (1977), 823–839.

Deutsch, Larry, "Entry and the Extent of Multiplant Operations," *Journal of Industrial Economics* 32 (1984) 477–487.

Dunning, John H., "Explaining the International Direct Investment Position of Countries," *Weltwirtschaftliches Archiv* (1981), 30–64.

———, *Multinationals, Technology, and Competitiveness* (Boston: Unwin Hyman, 1988).

Flowers, Edward, "Oligopolistic Reactions in European and Canadian Direct Investment in the United States," *Journal of International Business Studies* 7 (1976), 43–55.

Gorecki, Paul, "The Determinants of Entry by New and Diversifying Enterprises in the UK Manufacturing Sector, 1958–1963: Some Tentative Results," *Applied Economics* 7 (1975), 139–147.

———, "The Determinants of Entry by Domestic and Foreign Enterprises in Canadian Manufacturing Industries," this REVIEW 58 (Nov. 1976), 485–488.

Goto, Akira, and Kazuyuki Suzuki, "R & D Capital, Rate of Return on R & D Investment and Spillover of R & D in Japanese Manufacturing Industries," this REVIEW 71 (Nov. 1989), 555–564.

Graham, Edward, "Transatlantic Investment by Multinational Firms: A Rivalristic Phenomenon?" *Journal of Post-Keynesian Economics* 1 (1978), 82–99.

Greene, William H., *LIMDEP* (Philadelphia: Social Science Data Center).

Hausman, Jerry, Bronwyn Hall, and Zvi Griliches. "Econometric Models for Count Data with an Application to

the Patents–R & D Relationship," *Econometrica* 52 (July 1984), 909–938.

Highfield, Robert, and Robert Smiley, "New Business Starts and Economic Activity: An Empirical Investigation," *International Journal of Industrial Organizations* 5 (1987), 51–66.

Horst, Thomas, "The Industrial Composition of U.S. Exports and Subsidiary Sales to the Canadian Market," *American Economic Review* (Mar. 1972), 37–45.

Hymer, Stephen, *International Operations of National Firms: A Study of Direct Foreign Investment* (Cambridge: MIT Press, 1976).

Johnson, Norman, and Samuel Kotz, *Discrete Distributions* (Boston: Houghton Mifflin, 1970).

Kessides, Ioannis, "Advertising, Sunk Costs, and Barriers to Entry," this REVIEW 68 (Feb. 1986), 84–95.

Khemani, R. S., and D. Shapiro, "The Determinants of New Plant Entry in Canada," *Applied Economics* 18 (1986), 1243–1257.

Kindleberger, Charles P., *American Business Abroad: Six Lectures on Direct Investment* (New Haven: Yale University Press, 1969).

Kogut, Bruce, "Joint Ventures and the Option to Expand and Acquire," *Management Science* 37 (1991), 19–33.

Kogut, Bruce, and Nalin Kulatilaka, "Direct Investment, Hysteresis, and Real Exchange Volatility," Working Paper, Reginald H. Jones Center for Corporate Strategy, The Wharton School (1990).

Knickerbocker, Federick, *Oligopolistic Reaction and Multinational Enterprise* (Boston: Division of Research, Graduate School of Business Administration, Harvard University, 1973).

Magee, Stephen, "Information and the Multinational Corporation: An Appropriability Theory of Direct Foreign Investment," in Jagdish Bhagwati (ed.), *The New International Economic Order* (Cambridge: MIT Press, 1976), 317–340.

McClain, David, "Foreign Direct Investment in the United States: Old Currents, 'New Waves,' and the Theory of Direct Investment," in C. Kindleberger and D. Audretsch (eds.), *The Multinational Corporation of the 1980's* (Cambridge, MA: MIT Press. 1983).

Pugel, Thomas, *The Industry Determinants of Foreign Direct Investment into the United States* (New York: Division of Research. Graduate School of Business Administration, New York University. July 1986).

———, "The Determinants of Foreign Direct Investment: An Analysis of U.S. Manufacturing Industries." *Managerial and Decision Economics* 2 (1981), 220–228.

———, *International Market Linkages and U.S. Manufacturing: Prices, Profits and Patterns* (Cambridge, MA: Ballinger, 1978).

Scherer, Frederick, "International R & D Races: Theory and Evidence." The Prince Bertil Symposium on "Corporate and Industry Strategies for Europe," Stockholm School of Economics, Nov. 9–11, 1988.

Shapiro, Daniel, "Entry, Exit and the Theory of the Multinational Corporation," in C. Kindleberger and D. Audretch (eds.), *The Multinational Corporation in the 1980's* (Cambridge, MA: MIT Press, 1983).

Shapiro, Daniel, and R. S. Khemani, "The Determinants of Entry and Exit Reconsidered," *International Journal of Industrial Organization* 2 (1987), 15–26.

Swedenborg, Birgitta, *Multinational Operations by Swedish Corporations* (Stockholm: Almqvist and Wiksell, 1979).

Vernon, Raymond, "International Investment and International Trade in the Product Life Cycle," *Quarterly Journal of Economics* 80 (May 1966), 190–207.

Wells, Louis (ed.), *The Product Life Cycle and International Trade* (Graduate School of Business, Harvard University, 1972).

Yamawaki, Hideki, "Exports, Foreign Market Structure and Profitability in Japanese and U.S. Manufacturing," this REVIEW 68 (Nov. 1986), 618–627.

Yamawaki, Hideki, and David B. Audretsch, "Import Share Under International Oligopoly with Differentiated Products: Japanese Imports in U.S. Manufacturing," this REVIEW 70 (Nov. 1988), 569–579.

## DATA APPENDIX

Since the entries are counted at the 4-digit SIC level, most of the independent variables were collected also at the 4-digit level, except for Japanese R & D expenditures and its growth rate. *ENTRY COUNT* was constructed by calculating total numbers of Japanese firms' entries into the U.S. manufacturing industries during the period 1976–87. The raw data on the Japanese entries into the United States were collected by the International Trade Administration at the Department of Commerce, as published annually in *Foreign Direct Investment in the United States*, Department of Commerce. ITA. *JAPAN R & D* was constructed from the *Kagaku Gizyuitsu Kenkyou Chousha Houkou* (Research Report on Science and Technology), based on the survey by Statistics Bureau, Office of Prime Minister, Japan. *JAPAN R & D* is for 1975–85 and is SIC aggregated to the equivalent 2 or 3 digit level. *JAPAN R & D GROWTH* is computed as the average of annual growth rates

of *JAPAN R & D*. *JAPAN CONCENTRATION* are 8-firm concentration ratios, constructed from *Syuyou Sangyoni Okeru Seisan Syouchyoudo to Herfindahl Index no Syui* (Trends in Production-based Concentration Ratios and Herfindahl Indices for Major Industries), Report #12-87-001, 262-00-A, Fair Trade Commission, Japan. 1987. The matching of Japanese industry concentration data to the U.S. industries left 122 out of 454 4-digit SIC manufacturing industries without a match.

*U.S. R & D* and *U.S. ADVERTISING* are constructed from the Federal Trade Commission Line of Business Report for 1977. *SHIPMENT*, eleven years' average value of industry shipment, 1975–85, was obtained from unpublished reports from the Department of Commerce. *SHIPMENT GROWTH* is the average of annual growth rates of *SHIPMENT*. *IMPORT* is import value divided by industry shipment, eleven year average 1975–85, from the same source as *SHIPMENT*. *US CONCENTRATION* are 8-firm concentration ratios for 1982, as listed in the *1982 Census of Manufacturing*, Department of Commerce. *INNOVATION FREQUENCY* measures the numbers of new innovations for each industry, which are collected by Small Business Administration for 1982. *EXPORT RESTRICTION* is a dummy variable noting the existence of U.S. quota or voluntary export restraints by Japanese companies. Those industries include steel industries (3312, 3313, 3315, 3316, and 3317), television and radios (3651), semiconductors (3674), automobiles and parts (3711 and 3714). Textile industries were not included in the sample because of the missing data in other variables.

# [25]

Excerpt from David J. Jeremy (ed.), *The Transfer of International Technology: Europe, Japan and the USA in the Twentieth Century*, 167–90.

## 8. Japanese Motor Vehicle Technologies Abroad in the 1980s

### Tetsuo Abo

---

It was during the 1980s that Japanese motor-car manufacturers began in earnest to transplant their production technology abroad, particularly in developed countries, having exported their motor vehicle products on a massive scale since the late 1960s. From the early 1960s, they operated small local plants or had technological assistance contracts with local governments or private firms in developing countries. However the scale of local production and the level of technology brought in were limited to the narrow range that is possible under an import substitution policy in a closed economy. This reluctance to embark on foreign direct investment by the Japanese motor industry, in sharp contrast to its very strong inclination to export, is related to the general nature of Japanese-style production technology itself, which is in turn influenced to a great extent by the sociocultural background in Japan.

In the following study I will first analyse the features of Japanese motor vehicle technologies and outline their international transfers. Then the focus will shift to the activities of Japanese motor firms in the USA, using the materials and data collected and assessed by the joint research team, Japanese Multinational Enterprise Study Group (JMNESG).[1]

## FEATURES OF JAPANESE MOTOR VEHICLE TECHNOLOGIES

As Hymer's thesis on multinational enterprise shows,[2] one essential condition for a technology to be transferred abroad is that an industry or a firm have some technological advantage. What kind of advantages are there when Japanese motor-car companies set up their local productions in foreign countries or have licensing contracts with foreign companies? Following Abernathy,[3] it is helpful to divide the question into two aspects – product and process technologies – since a sharp contrast in the different levels of advantage in Japanese industries between these two can be considered as one of the most distinguishing characteristics of Japanese-type technology or know-how. I will take, to begin with, the common features of Japanese industries in terms of product and process technologies and then those of the motor industry.

## Advantages of Japanese-type Production Technologies

Generally speaking, Japanese industries have had stronger competitive advantages in process technologies than in product ones. This situation is fundamentally due to the fact that Japan is a latecomer as an industrialized country. Japanese industries and firms have therefore traditionally concentrated their research and development (R&D) efforts on the introduction of theories, ideas and designs related to product innovations invented in more developed, Western countries and on the manufacturing of cost-efficient products by taking advantage of lower wages and a very diligent working people. Since the mid-1960s, especially after the oil crisis of 1973–4, Japanese industries have been greatly concerned with quality of products as well, and with reducing costs of materials and parts by implementing quality circle (QC) activities and just-in-time (JIT) methods (described later), which primarily apply to practices or engineering technologies in manufacturing processes.[4]

What is most significant in the product technologies of Japanese industry is the objective of modifying or improving Western products to suit Japanese market conditions, which are deeply influenced by the natural, economic and sociocultural environment. One typical example of such a product technology is *Kei-Haku-Tan-Sho* (KHTS, light–thin–short–small). The finished goods – from transistor radios, pocket-sized calculators or portable tape recorders, to various kinds of smaller-sized durable goods, such as refrigerators, TVs, vacuum cleaners or motor vehicles – as well as semi-manufactured products, not only parts and components but also materials (such as, typically, steel) have been made lighter, thinner, smaller and so on. These characteristics are profoundly influenced by geographical, social and economic conditions in Japan: small islands, small houses, narrow roads, a traditionally lower level of income, relatively plentiful and high-quality labour, scarce natural resources and so forth. By adeptly responding to such circumstances, Japanese firms have created a modified and even improved product technology as compared with the Western, especially American, models. In this respect it is useful to point out that such a Japanese-type product technology can be an 'interceptor'[5] and transferred to a considerable extent to developing regions or economies, such as many of the Asian countries where the environmental conditions mentioned above are similar to those of Japan. The common feature in this type of product technology or know-how is the steady incremental modifications and improvements of established products to fit changes in market conditions. This produces large product variations and, as also pointed out by Abernathy,[6] is usually seen as a stage of diffusion for a radical product innovation. The main factor in innovation is thus process technology.

In the motor industry, this type of product technology has produced small, sub-compact and even 550 cc mini-cars (called light cars in Japan) as well as many sizes of motor-cycle. In this respect the situation is similar to that of other small countries in Europe. Compared to Japan, Britain has much more road space. In part this is owing to the higher ratio of flat to hilly land, despite the smaller total land area; also the population has been smaller, and the income level of the people has been higher. The Japanese motor industry has therefore had to tailor its product to the needs of each group of customers. It has done so by making the best use of labour-intensive practices, including the subcontract system, and by developing more specific types of motor vehicle, not only smaller in size and lower in price, but also with a larger variety in size, model and cost, on the basis of smaller volume production, than in Britain. It will not be difficult, then, to understand that this traditional background is one of the most important origins in terms of product technology of the earlier and easier development of the recent 'larger variety in smaller volume' production system in Japan.[7]

However it was not until the bombshell of the first oil crisis that these product technologies in the Japanese motor industry acquired a powerful international comparative advantage. Sharp increases in petrol prices and the general atmosphere of resource shortage have made KHTS-type, efficient fuel usage and resource-saving technologies the most favoured internationally, even in US markets. As the air pollution problem is in the same environmental context as the resource shortage one, engine design technology for anti-exhaust emission devices in Japan, such as Honda's CVCC engine and Toyo Kogyo's (Mazda's) rotary engine, have also enjoyed notable superiority over other foreign motor manufacturers. This became apparent after the Clean Air Act (known as the Muskie Act) in the USA (1970) and the standards controlling emissions in Japan (1973) were established. Other Japanese motor manufacturers, Toyota, for example, had to develop comparable anti-pollution engine technology under political pressure from the National Diet and the Environment Agency (created in 1971) against the background of rising public concern about environmental issues.[8] The Japanese situation should be contrasted to that in the USA, where the enforcement of the Muskie Act was postponed and in European countries, where no legislative regulation has been established.

## Advantages of Japanese-type Process Technologies

It has been very significant in process technology that, since the 1970s, the Japanese-type production system achieved its comparative advantage over those of other industrialized countries. The level of product quality and efficiency in production at Japanese automotive firms has been far ahead of foreign rivals working to produce a greater variety of cars in smaller volumes with less defects and more resource- and cost-saving methods. Here flexibility emerges as the essential element, but, to put it more strictly, a compatible combination of flexibility and cost-minimizing practices in process technology has become the most decisive and necessary condition for car makers to survive.

A flexible manufacture (or production) system (FMS) has been essential for any manufacturers wishing to adapt to changing market needs in this period. For car makers it is particularly necessary to be able to respond to new tastes for diversification in car models and options in a mature market with limited growth. Two types of FMS are seen among world motor vehicle manufacturers: the Western and the Japanese. Generally speaking, most European and American firms have tried to adapt to this situation primarily by introducing machine- and computer-oriented methods, typically represented by CAD (computer-aided designing), CAM (computer-aided manufacturing), CIM (computer-integrated manufacturing) and so forth.[9] Such methods, mainly supported by numerically controlled (NC) machine tools and robots based on microelectronics (ME) technologies, are of course effective in manufacturing a large variety of models, types and options in smaller or even variable volumes on the same line. This system, however, has soon come up against serious problems in its operation. First of all, it has encountered the problem of 'qualification'; that is, inadaptability of skilled workers. Especially in European countries there are long historical traditions of the apprenticeship which trains skilled workers and gives them qualifications according to their jobs. The system is so firmly established that it is not easy to adapt to the changes of skills reflecting the new FMS. In the USA there has also been a rigid job classification system based on American-style mass-production techniques (Fordism or Taylorism). It is interesting to note that both of these job classification systems have been supported and strengthened by trade unions. The systems therefore cannot adjust to the new process technologies. Often they become one of the major obstacles to introduction of FMS owing to union resistance. A second problem for Western-type FMS is the enormous cost of capital investment. Methods principally relying on capital equipment in order to create flexibility in production processes need huge amounts of money if they are to operate successfully, causing a heavy burden of depreciation costs long after settlement terms.[10]

The Japanese-type FMS, by contrast, essentially relies on the human element. It goes without saying that Japanese car makers have been much ahead of American and European makers in introducing robots and microelectronic-controlled machine tools, though original R&D and the pioneering introduction of robots, CAD and CAM systems and so on were led by American companies.[11] However in Japanese car plants people are still playing key roles at many levels. The plant-level people in the automated FMS are expected, first of all, to be 'multifunctional' or 'intellectual'[12]; that is, having the ability to understand more of the linked processes before and after their own jobs and to be involved in attending the machines as well as the products handled by them. The critical role of shop-floor managers and workers here is to check and respond to usual and unusual changes in the operation of numerically controlled equipment and robots as well as to spot product defects. In the processes of LVSV (large variety in smaller volume) production, various levels of 'usual' changes, such as of models, types and options according to market conditions, would induce more frequent unusual changes, such as missed supply of parts and components, problems or missed adjustment of robots and other ME-controlled machines. In order to cope with these situations, it would be much more costly to bring in more sophisticated machines and more machine coordinators, maintenance specialists and 'checkmen' (quality inspectors) on the shop floor. In Japanese motor plants ordinary workers, besides doing their own regular work, also monitor the proper functioning of the machines they attend and respond to defects in products. As they are more or less 'multiskilled' many of them can be transferred to other work groups or shops as 'temporary helpers' for unusual changes and even to a different plant when their current plant is shut down. As the result of such combinations of machines and shop-floor workers, Japanese plants are being operated very flexibly and incur much lower costs, both for capital equipment and employees, compared to Western car companies.

This raises the question, how can Japanese motor firms train such unique multifunctional workers and managers? An answer is connected to the 'universality–particularity' issue of Japanese-type skill formation. To begin with, this type of skill should be distinguished from a traditional one – the ability to do a specific job, such as shaping, drilling, welding and painting, with the skill of a craftsman–artist. These new types of multiple or versatile skills need not be as intensive as traditional skills because many complex tasks can now be performed by NC machine tools or robots. A less intensive but broader range of skills is required here.

What should be emphasized, in my opinion, is that the Japanese-style workplace-oriented team concept is the basic background for skill formation. It is true, indeed, as K. Koike stresses,[13] that skill formation is a matter of techniques for training workers, but, at the same time, the extent to which the technique is effective largely depends on the various conditions surrounding plants. According to Koike, 'intellectual skills' require shop-floor workers to handle flexibly various kinds of unusual situations during the production process; these skills result from both horizontal and vertical 'career' extension, primarily through 'on-the-job training' (OJT) at the workplace. He also emphasizes that, though developed in Japan, such methods can be generalized as objective and universally applicable production management techniques. It is doubtless useful to see these techniques as the core of Japanese-style skill formation, but I also believe that we should take into consideration the limitations and costs of the international application of such methods. This is closely related to the problem of team concept.

It is my principal understanding that the multifunctional sense and behaviour of employees in Japanese companies cannot be sufficiently acquired through technical training alone because they are deep-rooted in the 'my company' consciousness of almost all Japanese employees. Such behaviour is thought to be, to a considerable extent, the function of employees' loyalty and expectations of the improving performance of their company or shop. Such a strong sense of participation in their company leads employees to pay closer attention to their workplace. And, as is often pointed out, since the 'my company' consciousness in Japan is greatly influenced by the 'high context' of society's group orientation, with people good at communication informally as well as through the more formal structures of companies, we have to take into account here the sociocultural factors.[14] If we neglect these dimensions, we would be unable to explain (a) why Japan alone has diffused such management practices, without being definitely aware of their real meaning,[15] into almost all large firms; and (b) why such a strong spirit of involvement can be easily created at almost every level of staff in a company, as well as in a broad range of subcontractors.

It is well known that the team-oriented centripetal force in Japanese firms is also critical in achieving the incremental accumulation of cost-saving measures, such as bit-by-bit avoidance of waste and defects, or improvements (*Kaizen*) in production know-how.[16] Needless to say, these measures are supplemented by 'voluntary' small group activities with 100 per cent participation, such as QCC (quality control circles) or suggestion systems. The JIT (just-in-time) system,[17] for instance, is one of the typical information-sharing measures, both inside a plant and between an assembler and materials or components vendors, made possible by group orientation. This system was developed mainly for assemblers to minimize inventory stock at the cost of parts suppliers, but now, in the era of LVSV, it has been playing a more important role, not to mention its flexible adjustment of inventory stock, in order to control all the production processes from the vendors to the final assembly lines.

# INTERNATIONAL TRANSFER OF JAPANESE MOTOR VEHICLE TECHNOLOGIES

Considering these salient features of Japanese-type culture-embodied production technologies, it is easy to understand why Japanese firms, when applying their technological advantages to foreign countries, would have to face problems in adapting to local environments. Japanese motor technology is heavily culture-specific. We in the JMNESG have therefore created a working model, called the 'application-adaptation dilemma model', which indicates the difficult or even trade-off relationship between both aspects, and have also developed a hybrid ratio (HR, to be mentioned later) to test the correlation between the degrees of application of Japanese management techniques at the plants abroad and of adaptation to the local climates. Hereafter, in accordance with the above point of view, I will draw first an overview on the overseas production activities including licensing agreements of Japanese motor firms and their impact on the technological transfer to local motor industry. Then I will focus on the findings and analysis regarding the degree and situation of application–adaptation of Japanese production and management methods in the USA.

## An Overview on the Overseas Production Activities of Japanese Motor Firms

One of the fundamental historical features in the overseas business of Japanese firms is, as mentioned earlier, that leading large manufacturing firms traditionally tried to expand abroad mainly through the export of their products. Further, they were rather reluctant to opt for offshore production through foreign direct investment, opposing both base and host positions. By sharp contrast, Western, and especially American, enterprises were very active in transplanting their production and management methods almost immediately after establishing their US businesses.[18] Recently Japanese firms have begun to change this pattern under the threat of trade friction and the drastic appreciation of the yen. This is also the case with the Japanese motor industry. It is clear that the uniqueness of Japanese-style production technology described above is the principal reason why Japanese motor manufacturers were so reluctant to manufacture abroad. And this situation is closely connected to the manner, degree and regional distribution of technological transfer by Japanese car makers.

Historically, although the start-up of Honda's passenger car production in 1982 in Marysville, Ohio, was so epoch-making, 'grass-roots' small-scale local manufacturing by other Japanese makers had begun in the mid-1960s. Figure 8.1 shows the distribution of the manufacturing and KD (knock down) plants of the Japanese motor vehicle industry, by company and by country, as of the end of March 1989. Many of those plants which were set up before the 1970s are very small-scale assembly facilities in developing countries, mainly in Asia, Oceania and Africa. As KD export-related plants, which range from 'semi-knock down' (bolt and nut completing) to 'complete knock down' (the value of export is less than 60 per cent of a complete car), are major forms, these are classified as export rather than local manufacturing. Indeed, in the regions of Asia, Oceania and Africa, about one-third of total assembly facilities are among 'manufacturing', but most of them are operating under a far lower level of normal economies of scale, which have been imposed by the import substitution policies of the nationalistic local governments. Therefore we cannot simply see them from the usual economical or efficiency point of view; still they should be considered as one limited route of technological transfer from the developed countries to the latecomers.

In these developing regions, the second or third ranking group motor companies are distinguished in local production. Mitsubishi Motor Co. is the most outstanding, in that its number of local manufacturing plants (eight), as well as its total (11), is remarkably large. The company, ranked third to fifth in domestic production, has apparently sought comparative advantage over big competitors in those countries where a combination of minute techniques and know-how in manufacturing, marketing and even political negotiation is required in the tiny markets protected by local regulations policies.

Mitsubishi has taken advantage of its special management resources as a division of the leading corporate group (pre-war *zaibatsu*) by making the best use of its related *sogo shosha* (general trading company), such as Mitsubishi Shoji and Nisho Iwai, and the political influence of Mitsubishi Heavy Industry. The national car project in Malaysia is representative of Mitsubishi Motor's activities. The Proton Corporation, a joint venture of the Malaysia Heavy Industry Development Corporation (70 per cent), Mitsubishi Motor (15 per cent) and Mitsubishi Shoji (15 per cent), started in 1985 to produce two models of passenger cars (1.3–1.5 litre) and jumped up from number four to number one in 1986, producing around 66 000 units in 1989, followed by Nissan (KD plant, 100 per cent local ownership), Toyota (KD plants, 15 per cent) and so on. Mitsubishi Motor also has two joint venture (JV) plants in Thailand which are for KD and engine assembly. And it is interesting to note that these two or three Mitsubishi plants in the South East Asian region have been trying between them to implement an international division of labour in their press parts, engine and so on, though it has been very limited. Judging from the fact that in these markets an 80 to 90 per cent share is occupied by Japanese cars and trucks, we can safely say that the Japanese-type KHTS product technology and human-oriented flexible management and production methods, with Mitsubishi as a typical example, are more applicable there than the Western type. The extent to which the production technology has really been transferred to the local industry and people, however, is not easy to report definitely because there are few and limited field researches on this,[19] and there is another problem – how to evaluate international competitiveness in this kind of closed economy.

In this respect, however, Mitsubishi Motor has also provided good evidence in Korea. Although the Korean motor market has been closed to foreign competitors the sharp increase in production in the Korean motor industry since around 1983 has been supported primarily by export to developed markets, particularly to Canada and the USA. Mitsubishi Motor had an agreement with the Hyundai Motor Co. regarding technology licensing (since 1974) and Joint Venture (JV) (since 1982) in order to produce the Pony. It is clear that these agreements and the related technological cooperation between the two companies must have played a critical role in production areas. Hyundai Motor started to manufacture cars in 1967, as a very late entrant in the motor manufacturing industry, with the assistance of Ford. The technological tie with Ford was abandoned in 1985, when Mitsubishi Motor and Mitsubishi Shoji raised their ownership interests in Hyundai Motor from 10 per cent to 15 per cent (7.5 per cent each). Hyundai has since become an exporter of small cars to the US market. This indicates that Japanese product and process technologies have been substituted for American ones.

In terms of process technology, Mitsubishi must have helped Hyundai learn the fundamental engineering methods and know-how for small-car production which Japanese companies had accumulated. For that purpose Mitsubishi has retained several dozen expatriates from its Japanese operations at the Hyundai plant. This is important, especially for Japanese-style manufacturing practices, since the training system is principally OJT, based on workplace-oriented human relations. It is also interesting to learn that a retired high-level Mitsubishi engineer has been working as a production advisor at the Hyundai plant in Korea. Needless to say, Hyundai, on the other hand, has made the best use of the strong support of the nationalistic government, its organizing power as a newly developed *zaibatsu*, as well as the cheaper labour costs, longer working hours and the undervalued won. Judging from the short plant tour observations made by myself in the spring of 1989,[20] however, Hyundai is now at a turning-point. It will have to improve its techniques at the plant level: for example by mastering the ability to manufacture a larger variety of cars with a lower defect ratio; a smaller volume of inventory stock; a lesser number of workers; and so forth, in order to meet the standards of quality and efficiency in the US market and also in the domestic market where the rapid progress of Kia Motors Corporation is remarkable.

Kia, ranked third, just below Daewoo Motor Co. in 1988 production, was allowed by the Korean government to reopen the production of passenger

*Figure 8.1   Distribution of overseas production plants of Japanese motor firms, by country and company, 1989*

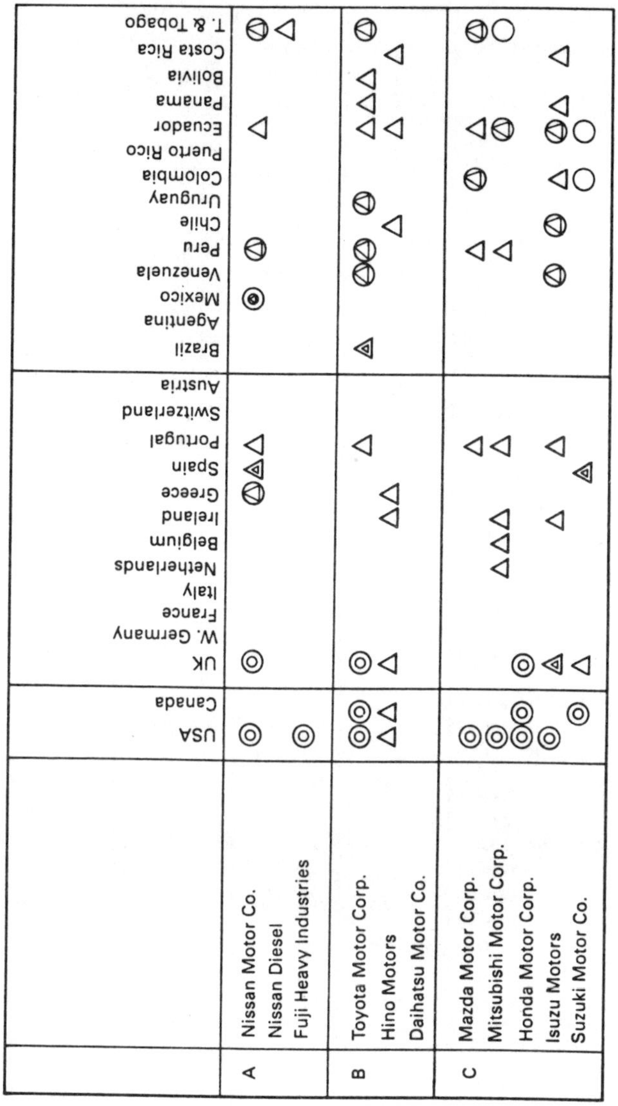

| | USA | Canada | UK | W. Germany | France | Italy | Netherlands | Belgium | Ireland | Greece | Spain | Portugal | Switzerland | Austria | Brazil | Argentina | Mexico | Venezuela | Peru | Chile | Uruguay | Colombia | Puerto Rico | Ecuador | Panama | Bolivia | Costa Rica | T. & Tobago |
|---|---|---|---|---|---|---|---|---|---|---|---|---|---|---|---|---|---|---|---|---|---|---|---|---|---|---|---|---|
| **A** Nissan Motor Co. | ◎ | | ◎ | | | | | | | ◎ | ◎ | △ | | | | | ◎ | | ⊘ | | | | | △ | | | | ⊘ △ |
| Nissan Diesel | ◎ | | | | | | | | | | | △ | | | | | | | | | | | | | | | | |
| Fuji Heavy Industries | | | | | | | | | | | | | | | | | | | | | | | | | | | | |
| **B** Toyota Motor Corp. | ◎ △ | ◎ △ | ◎ | | | | | | | | | △ | | | ◁ | | | ⊘ | ⊘ | △ | ⊘ | | | △ | △ | △ | △ | ⊘ |
| Hino Motors | | | △ | | | | | | | △ | | | | | | | | ⊘ | ⊘ | | | | | △ | △ | △ | | |
| Daihatsu Motor Co. | | | | | | | | | | △ | | | | | | | | | | | | | | | △ | | △ | |
| **C** Mazda Motor Corp. | ◎ | ◎ | ◎ | | | | | | | | | △ | | | | | | | △ | | | ⊘ | | ⊘ | | | | ⊘ |
| Mitsubishi Motor Corp. | ◎ | ◎ | △ | | | | △ | | △ | | | △ | | | | | | | △ | | | | | ⊘ | | | | ○ |
| Honda Motor Corp. | ◎ | ◎ | △ | | | | △ | | △ | | | △ | | | | | | ⊘ | | ⊘ | | | | ○ | | | | |
| Isuzu Motors | ◎ | ◎ | | | | | △ | | | | | ◁ | | | | | | | | △ | | △ | | ○ | △ | | △ | ○ |
| Suzuki Motor Co. | | ◎ | | | | | | | | | | | | | | | | | | | | | | | | | | |

| | | Zaire | Morocco | S. Africa | Algeria | Liberia | Nigeria | Ivory Coast | Congo | Tunisia | Togo | Cameroon | Kenya | Sierra Leone | Egypt | Malawi | Zimbabwe | Turkey | Iran | Iraq | U.A. Emerates | Saudi Arabia | Taiwan | Korea | China | Hong Kong | India | Malaysia | Philippines | Indonesia | Thailand | Singapore | Burma | Bangladesh | Pakistan | Australasia | New Zealand | Micronesia |
|---|---|---|---|---|---|---|---|---|---|---|---|---|---|---|---|---|---|---|---|---|---|---|---|---|---|---|---|---|---|---|---|---|---|---|---|---|---|---|
| **A** | Nissan Motor Co. | | | | | | | | | | | | | | | | | | | | | | | | | | | | | | | | | | | | | |
| | Nissan Diesel | | | | | | | | | | | | | | | | | | | | | | | | | | | | | | | | | | | | | |
| | Fuji Heavy Industries | | | | | | | | | | | | | | | | | | | | | | | | | | | | | | | | | | | | | |
| **B** | Toyota Motor Corp. | | | | | | | | | | | | | | | | | | | | | | | | | | | | | | | | | | | | | |
| | Hino Motors | | | | | | | | | | | | | | | | | | | | | | | | | | | | | | | | | | | | | |
| | Daihatsu Motor Co. | | | | | | | | | | | | | | | | | | | | | | | | | | | | | | | | | | | | | |
| **C** | Mazda Motor Corp. | | | | | | | | | | | | | | | | | | | | | | | | | | | | | | | | | | | | | |
| | Mitsubishi Motor Corp. | | | | | | | | | | | | | | | | | | | | | | | | | | | | | | | | | | | | | |
| | Honda Motor Corp. | | | | | | | | | | | | | | | | | | | | | | | | | | | | | | | | | | | | | |
| | Isuzu Motors | | | | | | | | | | | | | | | | | | | | | | | | | | | | | | | | | | | | | |
| | Suzuki Motor Co. | | | | | | | | | | | | | | | | | | | | | | | | | | | | | | | | | | | | | |

*Notes*
1. Planned plants are partially included.
2. A: Nissan group B: Toyota group C: Others
3. ⊚ Passenger car manufacturing; ○ Passenger car KD; ◮ Truck manufacturing; △ Truck KD

*Sources:*   Nissan Motor Co. (ed.), *Jidosha Sangyo Handobukko (Automobile Industry Handbook); Kinokuniya shoten (1988) (in Japanese) and others.*

177

cars in 1987, with the technological assistance of Mazda Motor Corporation, which in 1983 participated in the capital ownership (8 per cent) of Kia with Itoh Chu (2 per cent), a Japanese general trading company. Kia has been quick to introduce Japanese-style flexible management and production methods, from a sort of LVSV and JIT practices at the plant to organizing a cooperative association of affiliated suppliers. There is only space here to mention the well-known problems which the Korean motor industry has recently faced: from bottlenecks due to insufficient development of the technology, to a lack of key components in sufficient quantities, such as engines and transmission, and of parts suppliers, and the shortage of workers (especially skilled workers), to the drastically appreciated won. It would be tempting to examine the extent to which Korean firms or society are similar to the Japanese in terms of multifunctional flexibility (or individualistic demarcation), team or group orientation and so on, compared with Western countries. However owing to the limitation of data and space, I can only suggest here that there seem to be important similarities between Japanese and Korean societies. At the same time we should not overlook certain differences, a delicate divergence of implications arising from family orientation in Korea and group orientation in Japan, for example.

In Asia, and to a considerable degree in other developing areas as well, as is seen in Figure 8.1, Isuzu Motors Ltd and Hino Motors Ltd are distinguished in their overseas production activities, though both are mainly concentrated in trucks and buses. Isuzu, the world's leading small–medium-sized truck maker, is especially noticeable in that the company has taken advantage of its special connection with GM (General Motors), which has a 40.2 per cent ownership interest in Isuzu. Consequently it has exploited the huge international network of GM management assets. The Isuzu–GM JV truck and commercial car plants in Egypt, the Philippines and the UK are among former GM facilities in which the actual responsibilities of production have been taken over by Isuzu. IBC Vehicles Ltd (former Bedford Motors facility, the UK commercial car maker) is a similar case in a developed country. Isuzu is also an OEM (Original Equipment Manufacturer) supplier for the US market for small-sized passenger cars as well as trucks and transaxles to GM.

Suzuki, the leading mini-car maker in the Japanese domestic market, has a very similar relationship to Isuzu in terms of a business alliance with GM, which owns a 4.95 per cent equity share of Suzuki. At the Toronto plant in Canada, a 50–50 JV with GM, Suzuki started in the spring of 1989 to produce small cars both for GM (OEM) and itself for Canadian and US markets. On the other hand, Suzuki is also characterized as one of the 'grass-root'-type multinational producers in many developing countries. Its activities as a leading car maker in India are of especial interest. Toyota, the overwhelming market leader in Japan, has had relatively wide and deep commitments, particularly in Asia and Oceania, although it has been generally slow in direct foreign investment. In Thailand the company has, since 1962, been eagerly stamping major parts, such as body presses, assembling engines and transmissions and producing finished cars under the closed economy.

In comparison with the above Japanese motor companies, Nissan has been the most internationalized all-round player in terms of the regional distribution of offshore production.[21] It has started up several important foreign manufacturing plants since the mid-1960s, both in developing and developed countries. In developing regions Nissan's salient feature has been the setting up of production facilities, not only in Asian and Oceanian countries, such as Taiwan, Thailand and Australia, but also in Mexico and Spain, where no other Japanese car maker (apart from Suzuki's jeep plant in Spain) operates. Most Japanese makers have been particularly reluctant to manufacture motor-cars in these developing Latin countries, since sociocultural and economic environments there are extremely different from those of Japan. Nissan has somehow managed to develop its manufacturing plants against extremely different cultural backgrounds in these Latin countries for a long period without making substantial profits. In several recent years, however, both these Latin subsidiaries have been making money and are looking forward to better performance.

Nissan Mexico, in particular, is outstanding in that, since it began to assemble motor-cars in 1966, it has been competing in the closed domestic market against the subsidiary plants of major world car makers such as Chrysler, Ford, GM and Volkswagen (VW), all of which were set up before or just after the First World War. Nissan at last took the place of number one producer (market share of 29 per cent, 1988) from VW in 1987. Nissan Mexico has two contrasting plants as well as a small engine casting plant. One is the oldest in Cuauhnabuac, and has been assembling small cars and trucks including some for export (20 per cent of product units) by actively adapting to the various aspects of local environment. It is procuring locally a large variety of parts and components, employing relatively larger numbers of employees (4 860 in 1989) in order to respond to much less flexibility of job rotation (partly because of the policy of traditional unions), a lower level of product quality and so on. It has developed a special engine which is suited to low atmospheric pressure and steep gradients on roads at high altitudes, and it imports a small number of key parts (a company's imports cannot exceed its exports in a given year). Another new plant in Aguascalientes (1 865 employees in 1989) was built in 1983 to manufacture the functional components (engine, transaxle and so on) both for Cuauhnabuac plant and the Smyrna plant in the USA (mentioned later). At this plant, Nissan Mexico took advantage of its location where the influence of unions is weaker. Here Nissan has tried to introduce more aggressively Japanese-style management practices such as a smaller number of job classifications and an individual merit system based on more innovative hardware production technology. Also it has decided to build, in the near future, a new and very advanced assembly facility for small-car manufacture for export to the US market. The combination of these two types of Nissan Mexico plants is one of the most suggestive models of a Japanese motor manufacturer that applies its managerial advantages to developing countries, simultaneously making considerable efforts to adapt to local conditions.

The main overseas production activities of Nissan, however, have been pursued in developed countries. Nissan Motor Manufacturing (UK) Ltd (NMUK) is one example, along with Nissan US, which will be described later. The NMUK plant at Sunderland started to produce cars in mid-1986, far earlier than other Japanese motor manufacturers in the UK (Isuzu at West Bromwich, West Midlands in 1987, mentioned above; the Honda engine plant at Swindon in 1989; the Toyota plant at Derby will be opened in 1992). The NMUK plant seemed, judging from my own observation at the end of 1986, to be eagerly pursuing the transplantation of Japan-style practices at the shop-floor level, as compared with other Nissan plants abroad. The training of multifunctional workers must be among its top priorities because I saw every production worker or 'technician' (with no job classification and covering a very wide range of job elements on a station of the assembly line – a range comparable to that found at its parent plant in Japan) working at an extremely slow speed, at least during the initial phase. Most of the workers trained as technicians would be candidates for team leader and maintenance people in the next phase of plant expansion. The 'single union agreement' is of special importance for these practices in the UK where traditionally a company has to have many contracts with many unions at its plants, resulting in solid walls of demarcation between jobs. NMUK has been able to introduce a sort of individual merit system which evaluates the performance of each worker: consequently the amount of wages paid can be different, from worker to worker, within a small margin, even in the same job classification. This individual evaluation is taken into account in promotion, which obviously affects the work attitude of workers.[22] Notably there is a relatively larger number and a higher level of Japanese expatriates in the UK: as of mid-1989, the ratio of Japanese to total employees (2.6 per cent) is higher than in the US plant (0.6 per cent) and those in Mexico (Aguascalientes, 2.5 per cent and Cuauhnabuac plant, 1.2 per cent respectively). The NMUK has thus learned from experiences derived from Nissan's many plants abroad.

It is interesting that such group-oriented types of production methods have been accepted by NMUK's blue-collar workers (though not necessarily by white-collar ones) in Britain, which is regarded as the mother country of individualism. At the same time, however, we have to take into account the

*Technology Transfer and Business Enterprise*

performance of time-consuming and costly practices, productivity and profitability as well as quality, which largely depend on the scale of production. With fewer than 30 000 units produced in 1986 (77 000 in 1989) it is not easy to assess such a level of performance.

### Application and Adaptation of Japanese Motor Vehicle Technologies in the USA

Here I will introduce the outline of the interim report of the joint field research project of JMNESG (mentioned at the beginning of the chapter) carried out in 1986 on Japanese motor and electronics plants in the USA, and add to it some new information which was derived from a second plant tour by the same project team during the summer and autumn of 1989.[23] In Table 8.1, we see profiles of major Japanese motor plants in the USA and Canada, as of autumn 1989. Our project team visited the plants in 1986 (plants A–E) and 1989 (plants A–K). It is our policy not to publish the names of the companies which own them. Eight assembly plants of eight makers (three plants for one company and one JV plant for two Japanese companies) and two plants of parts suppliers are included.

The plants A–E in the table are the earliest of these plants. We tried to apply a working model called the 'application and adaptation model' to their local production activities in order to illuminate the level and features of transfer of Japanese-style production systems in the motor industry to the USA. 'Application' is the aspect of an MNE's (multinational enterprise) activities which retain the comparative advantages by introducing its methods to local production facilities. In MNE theory it is referred to as a 'firm-specific factor'. 'Adaptation' is the other aspect of the MNE's activities which involves modifying or adjusting parent practices to the various kinds of local environment. In MNE theory it is described as the 'location-specific factor'.[24] It will be clear, judging from the characteristic features of Japanese-style management described in the first section of the chapter, that it is not easy for Japanese car makers to balance these two types of activity. We thus regard these two aspects as presenting a trade-off relationship or dilemma. What interests us most is, therefore, the degree of the mixture or 'hybrid ratio', in other words the extent to which the system of the Japanese motor industry is either applied or adapted in the USA. In assessing this hybrid ratio we identified over 20 characteristics of Japanese management practices and grouped them into six categories, as shown in Table 8.2. Data were gathered from plant observations and interviews at the above Japanese motor plants in Japan (ten) and the USA (five), and at one other US plant. In order to quantify and illustrate major findings from this study, we then created a five-point grading technique for each category. Within this ranking

*Table 8.1   Profile of Japanese motor plants in the USA, autumn 1989*

| Company | Opening year | Main products | Production capacity (000 units) | Number of employees | Location | Mode of entry |
|---|---|---|---|---|---|---|
| A | 1982[1] | passenger cars, motor-cycles | 510(360)[2] | 6 000(3 900)[2] | Mid-west | wholly owned |
| B | 1983 | passenger cars, pickup trucks | 265(240)[2] | 3 300(3 200)[2] | Mid-south | wholly owned |
| C | 1984 | passenger cars | 250(250)[2] | 2 500(2 500)[2] | West coast | JV(US maker) |
| D | 1984 | air conditioners, radiators, etc. | | 670(420)[2] | Mid-south | wholly owned |
| E | 1985 | plastic parts, instruments | | 410(130)[2] | Mid-south | wholly owned |
| F | 1987 | passenger cars | 240 | 3 500 | Mid-west | wholly owned |
| G | 1988 | passenger cars | 200 | 2 950(3 500)[3] | Mid-south | wholly owned |
| H | 1988 | passenger cars | 240(100)[4] | 2 800(2 900)[3] | Mid-west | JV(US maker) |
| I | 1988 | passenger cars | 50(25)[4] | 710 | Canada | wholly owned |
| J | 1989 | passenger cars, pickup trucks | 60 | 550(1 700)[3] | Mid-west | JV(JPN makers) |
| K | 1989 | passenger cars | 200(40)[4] | 1 000 | Canada | JV(US maker) |

*Notes*
[1] Production of motor-cycles started in this plant in 1979.
[2] Numbers in parentheses are those as of autumn 1986.
[3] Numbers in parentheses are those planned for employees in full production.
[4] Numbers in parentheses are the actual ones produced in 1989.

182

a five-point score indicates the highest possible degree of 'application' (and, consequently, the lowest level of 'adaptation' – such a score would be given to a Japanese plant operating in Japan). A one-point ranking indicates the highest possible degree of 'adaptation' (and would be given to an American plant operating in the USA). Although these rankings are by no means exact,[25] with these we can easily estimate 'application–adaptation' levels. In Table 8.2, the main findings are as follows.

1.  Regarding the meanings of the interrelationships between the six groups, first of all, I (job organization and operation) and II (production control) are the core parts of the Japanese production system at the shop-floor level. III (sense of togetherness) and IV (employment situation) are human-related factors supporting the core parts. V (procurement) involves material factors which have a great influence on efficiency and quality in the operation of the core parts. VI (parent–subsidiary relations) is the upper framework of administrative organization which decides and controls the methods and directions of the core parts.

2.  Group I (GI), the organizational framework of the flexible production system, is assessed, on average, as the next highest (3.7) to GIV (4.1). Japanese car makers show a strong inclination to transfer the integral aspects of Japanese production techniques in GI to their plants in the USA in terms of 'method', compared with GII, which reveals, on the other hand, a clear inclination to bring in the Japanese techniques in terms of 'outcome' or 'result'. 'Method' means ideas or know-how about practices regarding process technology which can be transferred only when such practices are actually perceived and realized by local employees. Job classification (JC) will necessarily provide the basis for the implementation of other practices, especially within those in GI. JC is assessed very highly (4.0) because all the Japanese car plants in the USA (and also the Nissan plant in the UK) have adopted an extremely simplified JC system, with essentially only one classification, as contrasted with more than 200 JCs at a normal US motor plant. Even 40 JCs at the US plant which we visited at the same time is exceptional by American standards (this plant was shut down in the summer of 1988).[26] It is interesting to note that such a simplification is not to be seen even in Japan! This is a modification, indeed, but in which direction? I will come back to this point later. As was explained in the second section of this chapter, with such a simplified JC system an equally high score of job rotation (JR, 4.0), training (4.0) and role of supervisor (3.8) were made possible. On-the-job training (OJT) through JR was being tried in order to widen the employee's workplace horizons and his or her multifunctional skills. However the range of JR was primarily limited to a team and wage system (3.0), so was rigidly connected to a simple JC system. A Japanese-style promotion system based on seniority and individual merit evaluation

*Table 8.2   Hybrid ratio of Japanese motor plants in the USA, 1986*

| | | | |
|---|---|---|---|
| I | | Work organ. & its operation | 3.7 |
| | 1 | Job classification | 4.0 |
| | 2 | Wage system | 3.0 |
| | 3 | Job rotation | 4.0 |
| | 4 | Training | 4.0 |
| | 5 | Promotion | 3.5 |
| | 6 | Role of supervisor | 3.8 |
| II | | Production control | 3.5 |
| | 7 | Process technology | 4.6 |
| | 8 | Quality control | 3.2 |
| | 9 | Maintenance | 2.6 |
| III | | Sense of togetherness | 3.6 |
| | 10 | Employment security | 4.0 |
| | 11 | Small-group activity | 2.8 |
| | 12 | Open-style office & dining hall | 3.4 |
| | 13 | Uniform | 4.0 |
| | 14 | Socializing | 4.0 |
| | 15 | Company meeting | 3.5 |
| IV | | Employment situation | 4.1 |
| | 16 | Homogeneity of employees | 3.4 |
| | 17 | Turnover rate | 4.6 |
| | 18 | Union & labour relations | 4.4 |
| V | | Procurement | 3.3 |
| | 19 | Local contents | 3.2 |
| | 20 | Suppliers | 3.4 |
| VI | | Parent–subsidiary relations | 3.5 |
| | 21 | Ratio of JPN expatriates | 2.8 |
| | 22 | Decision making | 3.8 |
| | 23 | Status of American managers | 3.8 |
| | | Average | 3.6 |

*Note*:   5=the highest degree of application (the lowest degree of adaptation); 1=the lowest degree of application (the highest adaptation).

*Source*:   Japanese Multinational Enterprise Study Group.

was working only to a certain extent (so 'promotion' is assessed as 3.5). On the other hand, the wage system among production workers who belong to a JC is similar to the American one (so the score is 3.0). This restricts the scope for working a Japanese seniority-related system linked to a correspondingly increasing wage level, which helps to accumulate broader skills based on long-term experience. (In Japan it is an institutional condition that supervisors and technicians, including even maintenance people, should be trained inside the plants by production workers.) Also this wage system makes it difficult to introduce individual merit awards, which motivate workers to increase their income and, even more importantly, acts as a system to evaluate workers for promotion. In this sense, the scores for promotion (3.5) and supervisors (3.8) are not very high at these plants in the USA. In short, Japanese car plants in the USA were especially eager to transfer the 'method' of human-related production techniques, but with some limitations.

3.  The extremely high score of process technology (4.6) in GII meant, by contrast with the above, that the visible 'outcome' of Japanese techniques such as machine, equipment and layout of manufacturing lines, was directly transferred into the Japanese plants in the USA, along with imported parts and components, ensuring the principal competitive advantage of these plants against US makers. On the other hand, quality control (QC, 3.2) and maintenance (2.6) in GII were assessed as among the lowest because the 'method' of QC and maintenance used there was rather similar to the American one: specialist engineers or technicians in the QC and maintenance sections, trained in colleges or vocational schools, were primarily responsible for this work. Although Japanese car plants in the USA emphasize high quality of product and well-maintained automation processes, the ways of implementing the same goal as in Japan have to be much more 'adaptation'-oriented, and therefore are more costly and less flexible.

4.  Among various kinds of human-related elements in GIII and IV, which support GI and II, higher scores of employment security (4.0), turn-over rate (4.6) and union and labour relations (4.4) and the low score of small-group activity (2.8) are particularly noticeable. All Japanese car plants in the USA emphasized a policy (written or unwritten) of avoiding lay-offs for as long as possible. Turnover rate was assessed as very high in terms of 'hybrid ratio' since the actual turnover rates at these plants (around 4 per cent per year) were much lower than those (10 per cent) at the plants of US car makers, though higher than at the Japanese parent plants. These kinds of long-term employment practices, needless to say, are the crucial conditions for the Japanese-style training–promotion system. Four of the five Japanese plants are non-union and the remaining one has accepted UAW (Union of Automobile Workers) under an exceptionally modified agreement in which the drastic reduction of the number of JCs and flexible operation of JR and so on, mentioned above, are possible. With a distinctive difference between the functions of the UAW (United Auto Workers) and the Japanese Federation of Auto Workers in terms of flexibility concerning GI, II and III, it is no wonder that all the Japanese

motor companies prefer non-union plant in the USA. QC (quality circle) activity, one of the typical small-group activities, is among the most difficult of practices for Japanese firms to introduce into the USA. Of course most Japanese plants there, except for one parts maker, were trying to implement some sort of QC but differences from Japanese-style QC were quite clear in many respects: (a) the main goal in the Japanese plants in the USA was just to enhance team and quality consciousness as far as possible. By contrast, in Japan, QC is a team activity for all the production members, intended to improve quality and efficiency (*Kaizen*) in production processes; (b) 'voluntary' participation rates in the USA were 20–30 per cent, compared with the 100 per cent (!) usual in Japan; (c) as a rule, QC activities in the USA were held during company time and paid, while in Japan there is no payment and they are pursued during the workers' own time.

5.   The local contents of three Japanese assembly plants were around 50 per cent and those of two parts plants were 60–70 per cent (score 3.2). In other words, almost half of the parts and components used in major plants, which consisted of key components such as engines, transmissions and axles, were imported mainly from Japan. These were the critical material condition, the 'outcome' of Japanese techniques, which compensated for the insufficient transplant of the core elements. The important difference between this and the above process technology (introduction of machine and equipment produced in Japan) is that imported parts by themselves do not effect technology transfer, whilst the machine and equipment can be a part of it if combined with 'method'.

6.   The ratio of Japanese expatriates in GVI is the key human element which expresses the most significant characteristic of Japanese manufacturing plants in foreign countries. Expatriates play the decisive role in the overall performance of the plants. This is not a problem of system or 'method' but, here also, one of 'outcome' in the form of human beings – although Japanese expatriates can either train or substitute for local employees. The very low score of this ratio (2.8) is a little misleading, because two extreme groups of plants were mixed here. One is a group where the ratios were higher than 5 per cent (score of 4.0) and the ratios in another group were around 1 per cent (1.0–2.0). Either way these ratios must be much higher than in the case of Western MNEs. My special emphasis is that it would be difficult for local Japanese plants to survive (as of autumn 1989) without a considerable number of lower- to middle-level Japanese managers (section managers or managers), called 'coordinators' or 'advisors', who, like the shadows of American or Japanese general managers, take charge of almost all practices, particularly at the shop-floor level. This means that the role of these 'coordinator'-type Japanese expatriates may be even more crucial than that of top managers.

## CONCLUSION

Historically, Japanese motor companies have been very reluctant to transplant their production technologies to foreign countries because of the culture-embodied nature of their process technologies which are their main
comparative advantage compared with their product technologies. Japanese-
type product technologies such as KHTS, however, found their advantage
and began to be transferred, first to developing countries and, after the oil
shocks, even to developed countries. In Asian countries, Japanese motor
vehicle production technologies began to be transferred as an import-substitution measure in closed economies. Later they were transferred to support
the export-oriented developments of local motor industries. Here the sociocultural similarity seems to be one of the important factors for successful
transfer. It is also interesting to note that Japanese motor companies in the
second and third ranking group have been more aggressive in local production (direct foreign investment) in developing regions and have had various
kinds of tie-up relations with local governments or private firms, including
US car makers.

In developed countries, according to our joint researches, we can offer
the following conclusions regarding the present stage in the transfer of
Japanese motor vehicle technologies, particularly to the USA. We have
found a strong orientation towards the application of Japanese management
practices. In order to apply the human aspect of Japanese 'methods' at the
shop-floor level of plants in the USA, Japanese companies, to a considerable
extent, have been introducing the core part of practices such as simplified
JC, OJT, and JR and the supporting subsystem or conditions such as social-
izing activities, longer-term employment and cooperative labour relations.
At the same time there is particular emphasis on the introduction of hardware or 'outcome' of technologies, such as process technology and key
parts, from Japan, which play critical roles in achieving high product quality
and efficiency in the production processes. On the other hand, however,
Japanese plants in the USA have had to adapt to local conditions and to
modify to a greater or lesser extent the original methods, such as, in particular, QC, maintenance and promotion, which should be closely connected
with the above core part and subsystem, and are integral aspects of the
Japanese production system. As a whole, it should be expressly noted that
all the above transplanting processes of Japanese methods are actually enabled
to materialize by a considerable number of Japanese expatriates, particularly
those called 'coordinators'. In relation to this, one of the most impressive
findings during our second-round 1989 plant tour in the USA was that many
Japanese coordinators were actively trying to develop the ideas of modifications that cannot yet be seen, either in plants in Japan or in the US plants of

American companies. A good example, as well as the simplified JC, is the 'Skilled Trade Training Program', which has just been tried at plant G. In this programme to select and train maintenance technicians, qualified employees who have passed a certain screening test take a series of training courses that would enable them to learn the essence of multifunctional maintenance skills. At the Japanese parent plants, instead, OJT methods play the primary role in this process, as described earlier. Japanese 'coordinators' in the USA are also energetically contributing to analyse the parent system and to translate it into a 'third form' adaptable to American conditions. Of course, so far, the result is not certain.

To sum up, the overall level of technology transfer of Japanese motor vehicle firms to the USA, maintaining the same level of product quality as in Japan and reducing the productivity gap (productivity is usually said to be 10–20 per cent less than in Japan) entails a sizeable cost both in terms of manpower expense and hardware technologies. The transfer of 'outcomes' of technologies is by no means an actual transplanting of technologies, although Japanese 'coordinators' could train local people to perform Japanese 'methods' so as to reduce costs. At any rate, as the cost performance of the Japanese car plants in the USA has been somewhat better than that of the plants of US motor manufacturers, almost all of those Japanese plants will still be able to survive, so long as 'market discrimination', such as voluntary restraint agreements, tariffs and appreciation of the yen, remain fairly effective in preventing import competition from Japan.

## NOTES

1.  Japanese Multinational Enterprise Study Group, directed by Professor Tetsuo Abo, University of Tokyo, has been organized since 1983 by Japanese and American researchers and awarded grants from the Toyota Foundation in 1985 and 1987–88 for the researches under the title of 'Japan–US Joint Research on Problems with Local Production by Japanese Manufacturing Firms in the United States: An Assessment of Japanese-Style Management Transferability in the Automobile, Consumer Electronics and Semiconductor Industries'. We undertook field studies in the summer/autumn of 1986 and 1989 in the USA and have published in 1990 the English version of the interim paper for the preliminary research submitted to the Toyota Foundation and published by the Japanese in 1988: The Institute of Social Science, University of Tokyo, *Local Production of Japanese Automobile and Electronics Firms in the United States: The 'Application' and 'Adaptation' of Japanese Style Management* (Research Report No. 23, 1990).
2.  S.H. Hymer, 'The International Operations of National Firms: A Study of Direct Foreign Investment', MIT, PhD thesis, 1960.
3.  W.J. Abernathy, *The Productivity Dilemma: Roadblock to Innovation in the Automobile Industry* (Baltimore, 1978), chs 1–4.
4.  The emphasis of Japanese manufacturing firms on process or engineering technologies can be shown, for example, in the international comparison of the number of graduates of BEng and BSc in the following table:

| | Japan | | USA | | UK | | W. Germany | |
|---|---|---|---|---|---|---|---|---|
| | 1975 | 1987 | 1975 | 1983 | 1974 | 1985 | 1975 | 1985 |
| B Eng | 65 422 | 75 843 | 53 520 | 128 195 | 10 374 | 16 600 | 4 344 | 7 869 |
| BSc | 9 504 | 13 389 | 88 990 | 75 522 | 15 479 | 25 100 | 4 656 | 8 184 |

*Source*:   Japan Ministry of Education, *International Comparison of Indicators on Education* (Tokyo, 1988) (in Japanese).

5.  T. Ozawa, *Multinationalism, Japanese Style* (Princeton, 1979), pp. 206–11.
6.  Abernathy, *Productivity Dilemma*, especially ch. 2.
7.  T. Ohno, *Toyota Seisan Hoshiki (Toyota Production System)* (Daiyamondo Sha, 1978), ch. 1, *passim*; M. Cusumano, *The Japanese Automobile Industry* (Cambridge, Mass., 1986), ch. 5.
8.  Toyota Motor Corporation, *Toyota: A History of the First 50 Years* (Toyota City, 1988), pp. 203–9.
9.  US Dept. of Commerce, *The US Automobile Industry, 1984* (Washington, DC, 1985); *Ward's Automotive Yearbook* (Detroit, various issues, especially 1984–7).
10. 'Make-or-Break Time', *Fortune*, 15 February 1988; T. Abo, 'The Capital Expenditures of US Manufacturing Industries in the 1980s', *Sekai Keizai Hyoron* (*World Economic Survey*) (November 1988).
11. See T. Abo, 'New Technology and Manpower Utilization in Japanese Automobile Firms in Japan and Their Plants in the United States', in Sung-Jo Park (ed.), *Technology and Labor in the Automotive Industry* (Frankfurt/New York, 1991). Papers of the Seoul International Symposium sponsored by the Korean Automobile Manufacturers Association.
12. K. Koike, *Understanding Industrial Relations in Modern Japan* (London, 1988), pp. xiii–xv and 266 ff.
13. Ibid.
14. E.T. Hall, *Beyond Culture* (New York, 1976), p. 39 and chs 6–8; also see T. Abo, 'The Emergence of Japanese Multinational Enterprise and the Theory of Foreign Direct Investment', in T. Shibagaki, M. Trevor and T. Abo (eds), *Japanese and European Management: Their International Adaptability* (Tokyo, 1989), pp. 11–12.
15. It is interesting to learn the following stories regarding this. One of the members of the board of directors in the Japanese headquarters of plant B, who had been an advisor to the first American president at that plant, told us that Japanese people could not answer immediately after they were asked by the American president staying at a Japanese plant how Japanese practices were created, because they had not necessarily developed all such practices on purpose. The active Japanese coordinators at plant G told us that, when they tried to develop modified training methods for the American plant of the Japanese parent plant, described on page 87–8, they had to reinterpret for themselves the real meaning or 'logic' of what they are practising at their Japanese plants, which is now a world-famous production system.
16. *Kaizen* essentially means, said a Japanese vice-president at plant H, 'that people do more than what is decided to do'.
17. Regarding JIT, see Y. Monden, *Toyota Production System* (Atlanta, 1983), chs 2–3.
18. Regarding the reluctant nature of Japanese multinationals, see M. Trevor, *Japan's Reluctant Multinationals* (London, 1983); M. Yoshino, *Japan's Multinational Enterprises* (Cambridge, Mass., 1976). As for some examples of the active American multinationals, see T. Abo, 'American Automobile Enterprises Abroad during the Interwar Period Case Studies on Ford and General Motors with Emphasis on the Process of Their Multinational Adaptation to Local Climates', *Annals of the Institute of Social Science* (University of Tokyo) no. 22, (1981); also Mira Wilkins, *The Maturing of Multinational Enterprise: American Business Abroad from 1914 to 1970* (Cambridge, Mass., Harvard University Press, 1974).
19. International joint research directed by Professor Shoichi Yamashita, Hiroshima Uni-

versity, (University of Tokyo Press, 1990), is among the most useful regarding this, but not specifically for the motor industry. Professor Hiroshi Itagaki, Saitama University and a member of our JMNESG, made several on-site researches on Japanese car plants in ASEAN countries in the autumn of 1989 and kindly provided me with some recent information about those plants. Also Nissan Motor Corporation, *The Handbook of Motor Vehicle Industry* (1988) and *Overseas Productions of Nissan Motor* various issues (both in Japanese), are very useful sources and have data for other motor companies besides Nissan.

20. The major Korean motor manufacturers invited foreign researchers including myself to their plants when they sponsored an international workshop in the spring of 1989 (see note 11).

21. People at Nissan Motor have been very cooperative in arranging my plant tours and interviews in many countries, such as the USA (1986, 1989), the UK (1990), Spain (1987), Mexico (1989) and Japan, which has been a great help for the present chapter.

22. P. Wickens, *The Road to Nissan* (London, 1987), ch. 8 is also very useful. (The author accepted me as one of the interviewees at the UK plant in 1986.)

23. Regarding the joint field researches of JMNESG, besides the forthcoming interim report and my papers (see notes 11, 26), see the several outcomes in English by its members: D. Kujawa and M. Yoshida, 'Cross-Cultural Transfers of Management Practices: Japanese Manufacturing Plants in the United States', paper presented at the 1987 Annual Meeting of the Academy of International Business in Chicago; H. Itagaki, 'Application–Adaptation Problems in Japanese Automobile and Electronics Plants in the USA', in Shibagaki *et al.*, *Japanese and European Management*; R. Grosse and D. Kujawa, *International Business* (Irwin, 1988), pp. 322–7.

24. J.H. Dunning, *International Production and Multinational Enterprise* (George Allen & Unwin, 1981) Part One. N. Hood and S. Young, *The Economics of Multinational Enterprise* (London, 1979), ch.2.

25. We decided an evaluation criterion for each item in Table 8.2 according to Japanese model (5) and American model (1). But for the items for which we could not find any quantitative index we had to judge the scores from qualitative information through very intensive discussion.

26. For this American plant, also see T. Abo, 'The Application of Japanese-Style Management Concepts in Japanese Automobile Plants in the United States', in B. Dankbaar, U. Jurgens and T. Malsch (eds), *Die Zukunft der Arbeit in der Automobilindustrie* (Berlin, 1988), pp. 333–5.

# [26]

Excerpt from Sanjaya Lall, *Multinationals, Technology and Exports: Selected Papers*, 114–30.

# 7 Multinationals and Technology Development in Host Countries[1]

## I INTRODUCTION

The promotion of domestic technological capabilities is becoming a growing concern for the more industrialised of the developing countries (LDCs). Among the many issues that have arisen in this area, some of the most complex and controversial have concerned the proper balance and the nature of the interaction between foreign and indigenous technologies.[2] In an activity like technological development, where dynamic learning effects and externalities are rife, many analysts have argued that there is a strong need for government intervention: untrammeled market forces by themselves would not lead to the optimal level of technological development in LDCs, and some deliberate exclusion of foreign technologies may be necessary to protect the learning process there.[3]

Multinational corporations (MNCs) from the developed world appear as an important topic in this debate. In many developing countries, MNCs are major producers of advanced industrial products and among the most important conduits for the transfer of modern technologies from abroad. They are often in the forefront of export activity and, in larger countries with relatively developed industrial structures, of local technological activity. It is the very fact of their technological strength, based upon 'frontier' innovation in the industrialised countries, and their competitive prowess in host LDCs which raises fears that they may stifle indigenous capabilities. In the language of the more polemical critics of MNCs, they act as 'agents of technological dependence' of host LDCs.

The case against MNCs in this guise is, however, neither coherent

*Multinationals and Technology Development in Host Countries* 115

nor well articulated. There are different versions of the criticisms, some of which contradict the others. This paper is an attempt to separate out some of the strands of the arguments advanced and to evaluate them against some (regrettably scanty) evidence. It will focus on the experience of India, where I have been conducting some research on the process of technological development. It may be said, with some justice, that India is an inappropriate example. It has severely restricted the inflow of foreign direct investment in the last two decades and has sought, behind formidable protective barriers and with widespread inefficiencies, to promote not just self-reliance in production but also self-reliance in technology.[4] Existing foreign investments in the country have been forced to dilute the extent of foreign ownership, and their growth and diversification in response to their competitive strengths in open markets have been tightly controlled. Nevertheless, most of the world's leading MNCs still operate in India, and are particularly active in high-technology industries.[5] India's attempt to develop its technological capabilities itself makes the role of MNCs an interesting one. Thus, the use of Indian examples is still of relevance to this discussion.

A primary source of confusion in this area is the lack of a clear definition of what 'technology development' means. Part II of this chapter discusses this problem. Another source of confusion is that the impact of MNCs on different levels of technology development can be assessed in several different contexts: technology development *within* the LDC affiliate of the MNC; technology development of the affiliate *relative to* similar local firms in the host country; and the *direct or indirect effects* of the activities of MNC affiliates upon technology development in related (vertically linked or competitive) firms in the host country. Parts III–IV of the chapter deal with these questions in turn. The main conclusions are drawn in part VI.

Two other introductory points: first, this chapter does not deal with such difficult issues as the 'appropriateness' of products or processes transferred to host countries by MNCs, and, second, it does not define MNC presence *per se* as a bad thing (because it 'denationalises' industry) or as a symptom of indigenous technological 'dependence'. On the first point, it is assumed that these are broader questions of distribution and factor-pricing policy which have to be taken as given, and that foreign and local firms in similar activities do not differ much in their behaviour once the larger parameters are set. On the second, it is assumed that the same technological activity undertaken in a host country yields the same benefit whether it is by a foreign affiliate or a local firm: benefits differ only if behaviour patterns are different.

116    *Multinationals, Technology and Exports*

## II   WHAT IS 'TECHNOLOGY DEVELOPMENT'?

This chapter is concerned only with manufacturing industry. In industry, technology development can be defined to encompass an enormous variety of activities.[6] We can simplify these into four categories.

In the sense that technology is simply the knowledge of how to carry out manufacturing activity, even the setting-up of a new assembly activity in a developing country can be said to transfer some new knowledge and so contribute to the technology of the host country. From this elementary stage, technology development can progress to the knowledge of operating imported technologies in increasingly sophisticated industries. The gaining of such operating knowledge – for which we use the convenient label of *'know-how'* – will include not only the assimilation of imported techniques (which can itself be a lengthy and active learning process) but also quality control (which also involves active technical effort), improved plant layout and production practices, slight modifications to equipment and tooling, trouble-shooting, the use of different raw materials, and so on. It is well accepted now that the acquisition of know-how, even in the context of imported technologies, is a real and significant source of technological progress in LDCs.[7]

The next stage of technology development involves the understanding of the nature of the underlying process and product technologies, and leads to their substantial adaptation, improvement and even replacement by new processes or products. We may call this the development of *'know-why'* capabilities. Such technology development arises partly as a natural extension and deepening of know-how capabilities, and partly as a result of conscious efforts to develop design, testing, pilot-plant and similar activities (the 'development' part of research and development, R & D, in industrial enterprises).

Know-why development may be followed by *applied research,* the application of given scientific knowledge to the process of commercial innovation. Again, the line between the acquisition of know-why and its extension into genuine innovation is very unclear, but we mention it because of the possibility (real in many LDCs) that the understanding of why technologies work the way they do may not lead on to risky and costly efforts to undertake innovation. The final stage of technology development within industry comprises the ability to undertake *basic scientific research,* pushing back the frontiers of knowledge without regard to specific commercial application.

The contribution of MNCs to technology development in host LDCs can take place at any or all of these levels. A positive effect at one level may co-exist with a negative effect at another. Moreover, at any given level, a positive contribution to technology development within the MNC affiliate may be associated with a negative effect on such development in other enterprises in the host country. The range of potential permutations is enormous, and we clearly cannot analyse them in any detail here. Certainly the lack of sufficient empirical evidence limits the scope for meaningful discussion. For this reason, we will simplify further and refer in the following analysis *only to know-how and know-why development* (the latter comprising all the advanced stages of technological activity from design and development onwards).

A very important point to bear in mind about the different levels of technology development in LDCs concerns its *net social value.* All technology development entails a cost, and each successive stage probably entails higher costs. The development of indigenous technological capability is not an end in itself: technology is simply one input into the process of industrial production, and the efficiency of the production process which results from the application of technology is the 'bottom line' which the country is concerned with. True, technology development has various cumulative dynamic and external effects, but the social cost entailed must always be compared with the cost of importing it from abroad.

These points are relevant here because India seems to have pushed technological self-reliance almost as hard as inward-looking industrialisation. Of all the industrially advanced LDCs, it is the one with the least relative reliance on foreign technology in any form (MNCs, capital goods imports, licensing, foreign engineering) in the past two decades. As I have argued elsewhere (my paper in Fransman and King), this policy has been pushed to counter-productive extremes, reinforcing the inefficiencies engendered by industrial policy. Not only is the extent of local technology generation by MNC affiliates and local firms highly conditioned by this framework, the broader economic implications of the efforts also have to be borne in mind.

Let us turn now to the three forms in which the effects of MNCs on technology development are evaluated.


## III  TECHNOLOGY DEVELOPMENT WITHIN MNC AFFILIATES

There is a widespread, but mistaken, impression that technologies developed by MNCs in their home countries are transferred costlessly and without adaptation to affiliates in LDCs. Teece (1976) shows in great detail that the transfer of any technology to a new location – developed or less-developed – entails substantial transfer and assimilation costs. These costs arise, of course, whether the recipient is an affiliate or an independent firm, and range from pre-transfer engineering, equipment specification, adaptive R & D (most of these are undertaken in the technology exporting country) to detailed engineering, local design adaptation, start-up, trouble-shooting, and training (which are mostly undertaken in the recipient). After a plant is running, subsequent technology creation in the form of know-how (production engineering and minor adaptation) occurs over time, reaching a peak soon after technology introduction and tailing off gradually until a new technology is transferred or new problems encountered (Davies, 1979).

In the know-how phase, there seems no *a priori* reason to differentiate between the performance of foreign and local firms. As noted earlier, know-how accumulates as a result of problem-solving, and foreign and local firms encounter identical problems in given locations. Katz's (1978) studies show that foreign affiliates undertake minor innovation-type engineering just as successfully as local firms, while in terms of adaptation to local factor prices there is no sustained difference between the two groups (Lall, 1978). It is sometimes suggested that MNCs may perform better because of their higher-quality technical manpower, or because of their ready access to the accumulated know-how of the parent company. In India, there is little evidence to support or refute this at the level of production engineering.

It is when we come to the know-why and more advanced research phases that the literature leads us to expect MNCs to be relatively backward in local technology development. The economics of R & D location (see Lall, 1979, and Hirschey and Caves, 1981) dictate that the bulk of basic design and development work by MNCs be highly centralised; the need for high-level scientific manpower of several kinds and a highly developed technological infrastructure dictates that this centralisation occur in the developed world; and the need for close interaction, especially in engineering industries, between innovation, management and production entails that R & D be placed near the head office or the largest markets.

This does not mean that MNC affiliates do not perform basic design or development work in host LDCs. There has been a growing tendency among MNCs (if US data can be extrapolated to other home countries) to place R & D facilities in *some* LDCs. These are the larger,

## Multinationals and Technology Development in Host Countries 119

industrially more advanced and technically better endowed host countries. There are several sound economic reasons why R & D facilities are set up there: first, the transfer of increasingly more complex know-how is itself facilitated by the existence of local know-why capabilities (Teece, 1976); second, in certain products the needs of the local market differ from those of developed countries (e.g. automobiles, food products, toiletries) and large local markets justify the investment in R & D facilities; third, the use of some LDC facilities for export to the rest of the Third World leads both to long production runs and the need for product adaptation, strengthening the previous tendency (e.g. VW in Brazil); fourth, the use of local raw materials or the need to test certain products locally induces local investigative R & D activity (pharmaceuticals and food products); and, finally, in exceptional cases LDC R & D facilities can be utilised for the MNCs' global needs of innovation (but this generally occurs *after* local R & D is well established).

Thus, the facts that a relatively small proportion of MNCs' total R & D spending is allocated to the Third World and that most LDCs host no R & D by MNCs are quite compatible with the fact that large countries like Brazil or India today support very active R & D by MNCs. In India there are some 500 officially recognised R & D establishments in private manufacturing industry. Firms with foreign equity (including those which have diluted the foreign share to 40 per cent because of government policy) account for over half these firms; in fact, nearly all foreign affiliates of any size have set up recognised R & D facilities (the next section will present an explicit comparison of R & D by local and foreign firms). Some of these, like the one set up by the Unilever affiliate, have gone in for basic scientific research, contributed major product innovations and led to the utilisation and export of formerly unused local materials.[8] A number of affiliates earn royalties and technical fees from abroad by the export of their know-how. It is often suggested that their access to the parent company enables them to utilise 'frontier' technology to enhance the value of their Indian R & D.

All this suggests that, *up to a certain level*, MNCs do not lag in innovative activity in large LDCs with a pool of skilled manpower. The question then arises: is this level the socially optimal one? Clearly, local R & D in LDCs is much smaller, more application-oriented and less 'deep' than R & D in the home countries of MNCs.[9] To a large extent this is inevitable, given the relative sizes of operations, the level of technological sophistication and the competitive needs of the two locations. However, many host countries (including relatively advanced ones like Canada) feel that MNC R & D is 'truncated' to an undesirable extent, with the truly innovative functions kept at home even when they could be efficiently located abroad.[10] The allegation is difficult to prove or disprove empirically: the social and private perceptions of the net long-term benefits of establishing R & D in a newly industrialising country can be widely different, depending upon the relative assessments of research capabilities, minimum scale requirements of research, dynamic learning effects and externalities. In their survey of the literature, Kamien and Schwartz (1982) demonstrate clearly how fragile is the empirical basis for generalising about the determinants of innovation even in the highly industrialised countries. In LDCs, where much less is known on the economics of industrial progress, the area of ignorance is far larger. Non-quantifiable and dynamic factors further compound the analytical difficulties.

Since firm conclusions are not feasible, let me round off this section with my own impressions on technology development within MNCs. First, to the extent that MNCs are the most efficient international transmitters of industrial know-how, their presence in LDCs promotes technology development *at the production level* in those sectors where the speed of technical change and the complexity of technology makes other methods of transfer relatively inefficient or costly.[11] By implication, their direct participation may be less necessary in activities with 'lower' or more stable technologies. It may be argued that newly industrialising countries have their comparative advantage in acquiring know-how, rather than spreading their limited technological resources thinly over 'deeper' technical learning. Certainly the most successful NICs, the Gang of Four in East Asia, have benefited enormously from this strategy of efficiently utilising imported technologies without forcing the pace of know-why accumulation. Judged by this criterion, the extent of technology development by MNCs is probably optimal.

Second, to the extent that countries wish to go beyond the know-how stage into *know-why and basic research,* and have the wherewithal to do so efficiently, MNCs may go to the desired extent only in a few instances. Some government intervention may then be necessary to bring private costs and benefits of local R & D into line with social ones. Two outcomes are possible: the MNC is induced to launch local R & D to the extent desired (and is able to do it very efficiently because of the backing of the parent company), or it refuses to undertake 'deeper' R & D in the given host country (or conducts it inefficiently), and the latter's objectives are only achieved by setting up independent R & D facilities. I suspect the actual outcome lies somewhere between these extremes. A case may then be made that in some cases MNCs do

*Multinationals and Technology Development in Host Countries* 121

retard 'deep' technology learning. But it is a qualified case, and not as obvious as may appear at first sight.

## IV TECHNOLOGY DEVELOPMENT IN MNCs RELATIVE TO LOCAL FIRMS

An earlier survey of the literature (Lall, 1978) had suggested that there was no marked or consistent difference between MNCs and local firms as far as the adaptation of imported technologies was concerned. Given the specific needs of the host country, its technical capabilities and the inherent adaptability of the technologies in question, both sets of firms seemed to respond similarly to the economic environment. The available evidence on the relative efficiency of their responses is very mixed and of uneven quality, so we can deduce little from it.

What, then, of their relative propensities to undertake 'deeper' technological effort? Since no detailed empirical studies exist on the issue of successful innovation by foreign and local firms in LDCs, we may resort to two sets of rather crude indicators: exports of technology by foreign and local firms, and formal R & D spending by them.

Exports of technology have been used as a rough indicator of the level of technological competence reached by a developing country (Lall, 1982). The available evidence suggests that the bulk of technology export activity is undertaken by local firms in LDCs[12] If this is an indicator of 'deeper' technological learning, we may infer that MNCs lag in the accumulation of know-why in these countries.

We must, however, note two important qualifications. First, many important forms of innovation do not show up in the foreign activities of industrial enterprises, and, if they do, they tend to affect the export of products rather than the export of technologies. Second, MNC affiliates may transfer their technology abroad *via* the parent company network (for royalties or in exchange for other technologies[13]) rather than in contractual forms which are caught by the data. In the Indian case, we have already noted that MNC affiliates are actively selling technologies in return for royalties. The Unilever affiliate exchanges technical information freely with its related firms overseas, and also exports its expertise by posting Indian engineers in other LDCs. Many of the innovations made in India are utilised in affiliates in S.E. Asia and Brazil. Certainly its impressive export effort (Hindustan Lever is the largest single private sector exporter in India) is based partly on the exploitation of its local research effort. Moreover, given the very

122          *Multinationals, Technology and Exports*

restricted scope permitted to MNCs in Indian industry, it would be premature to infer from technology export data that foreign affiliates are less capable of know-why accumulation than local firms.

The use of R & D data may appear to be more reliable in this context. There are, however, problems here also. A number of other factors which influence firm-level R & D performance (see Kamien and Schwartz, 1982) must be controlled for before the influence of foreign ownership is assessed; R & D expenditures are an indicator of research input, not its output (innovation or learning); different firms may define it differently, and the composition of R & D between basic research, applied research and imitation may differ;[14] and a comparison of firms in different industries may be misleading if the inherent research-intensity of the industry is not taken into account. Some data I have analysed on the R & D performance in India, which deal with some of these problems, may be instructive.[15]

These data pertain to 145 firms in only two sectors, engineering and chemicals, minimising the differences in 'technological opportunity' which crucially affect R & D performance. The 100 engineering firms are also assigned a dummy variable distinguishing those engaged in complex (machine building) activities from those in simpler metal shaping activities. The two sectors chosen account for 80 per cent of India's foreign collaborations, so are a fair representation of MNC activity there. A number of control variables are included besides the degree of foreign ownership. The shortage of space precludes a detailed analysis of their rationale and expected signs. Some of the results for each of the sectors – we did not aggregate them for this exercise because of the differences in their technological characteristics – are shown in the Appendix Table.

As far as the impact of foreign ownership (variable $x_1$) is concerned, the two sectors show different tendencies. Taking a large number of regressions (not shown in the table) into account, engineering has a positive and consistently significant sign, while chemicals has a negative and only occasionally significant sign.[16] The two sectors also show interesting differences in the impact on R & D of firm size, age, export performance, and the proportion of highly paid employees: clearly, generalisations across different sectors cannot be drawn. Foreign ownership has a positive relationship with R & D in one major sector. *There is no support for the case that MNCs in general are less R & D intensive than local firms.*

In sum, these two indicators do not, in the final analysis, enable us to conclude that there are significant differences between the relative technological performance of foreign and local firms in India. Given the stringent controls on MNCs in India and the consequent low appropriability of the returns from their research, it is perhaps surprising that the evidence does not show a more negative result for foreign affiliates. A truer test would ideally cover a more open economy with a more competitive industrial structure: but we have to make do with the little information that we do possess.

## V EFFECTS OF MNCs ON RELATED FIRMS' TECHNOLOGY DEVELOPMENT

There are two sets of firms 'related' to MNCs: those that supply inputs to MNC affiliates and those that compete with them in their product markets.

As far as vertically linked firms are concerned, the limited Indian evidence suggests that MNCs are as active in transferring skills, know-how and product design to suppliers as local firms.[17] The exact pattern of inter-firm linkages depends upon the technical characteristics of the products and the relative sizes of the firms concerned. The highly inward-looking Indian regime has induced all firms, regardless of ownership, to set up extensive local linkages: many of these linkages may, however, be socially inefficient and the resulting production facilities internationally uncompetitive. However, it is likely that the basic propensities to create linkages remains the same in more dynamic and efficient economies.

As far as competing firms are concerned, the debate is still conducted at a very general level. It is argued, on the one hand, that competition from technologically efficient MNCs induces local firms to improve their own technology,[18] and that the presence of MNCs leads to much more rapid diffusion of technology by imitation and 'contagion'.[19] On the other, it is argued that a strong foreign presence inhibits local firms from undertaking risky and costly research activity.[20] The two apparently contradictory views can be reconciled, of course: local firms may well be induced to upgrade their production technology by a greater reliance on licensing when faced with competition from MNCs operating with 'frontier' technologies, and, in the process, their independent know-why development may suffer. A strong MNC presence may, in other words, be associated both with better local know-how and 'shallower' local know-why (needless to say, this is on the assumption that the indigenous industrial sector survives MNC competition on the

124       *Multinationals, Technology and Exports*

basis of licensed technology – the phenomenon of 'denationalisation' is assumed to be contained within socially acceptable limits).

We should, however, note some important qualifications to the argument that MNCs inhibit know-why development in competing firms. First, it assumes a specific relationship between licensing and in-house research, treating them as substitutes at 'deeper' levels of R & D; this may not be the case as firms grow more sophisticated and larger. Thus, licensing may become a *necessary input* into in-house technological development;[21] this is certainly the case with Japan, which for decades was the world's largest importer of licensed technology. If local enterprises in the NICs have reached this stage (and I believe many have), increased MNC competition will have net technological benefits. Second, we must be careful not to generalise to all industries: a number of basic sectors (textiles, steel, cement, machine tools) are not important stamping-ground for MNCs, and in these the inhibiting effects do not really arise. Third, the Indian case illustrates the dangers of closing the economy too much to foreign competition. Not only has industrial efficiency suffered and damaging technological lags developed (it is becoming increasingly apparent that independent R & D efforts in LDCs simply cannot keep up with world frontiers in many industries, even if they are mere imitators), but a great deal of technological effort has gone into socially unproductive uses like finding high-cost local substitutes for imported materials and equipment. Finally, some advanced technologies may be available *only* via MNC entry, so the question of inhibition is redundant.

These qualifications still leave a grey area where genuine concern can be voiced about the effects of MNCs on firms in lesser industrialised countries. Moreover, to the extent that know-why development does require large, sheltered markets in the initial, high-risk phases, a powerful foreign presence may increase the risk to unacceptable levels. The Japanese experience lies behind many arguments that local technology development can proceed very successfully without direct MNC participation. Unfortunately, it is not clear how far Japan can be emulated by the LDCs of today; and, given Japan's human resources, its actual experience does not prove that it may not have done *even better* with a more liberal policy on MNC entry.

## VI  CONCLUSIONS

The analysis of the nature and determinants of technology develop-

*Multinationals and Technology Development in Host Countries* 125

ment in LDCs is in its infancy. We know little about its different phases and about the true comparative advantage of LDCs in the scale of technology 'deepening'. Not surprisingly, therefore, the relationship between MNCs and local technology development is fraught with confusion and needless polemic.

This chapter has sought to clarify some of the elementary concepts and separate out the different strands of the debate. It has distinguished between various sequential stages of technology development, concentrating on the basic categories of 'know-how' and 'know-why' acquisition. For each of these, it has discussed the impact of MNCs on technology development within the firm, in relation to local firms and in linkage or competition with local firms.

It would appear that the effect of MNCs on know-how development is positive in all its manifestations. Given their undoubted technological superiority in most areas of modern industry, and given also the inherent need to adapt technology to each particular host environment, MNCs contribute to know-how acquisition within their affiliates and, by linkages and competition, to related local firms. As far as know-why development is concerned, the picture is more mixed. MNCs have good economic reasons to keep basic innovative effort centralised in the developed world (though exceptions exist), and generally transfer less to LDCs than many host countries would like. Their presence may also inhibit local firms investing in their own know-why development.

These conclusions need two major qualifications. First, the comparative advantage of LDCs may in fact lie in know-how development, and the correct strategy for them may be to efficiently exploit imported know-why. If know-why develops naturally over time with increased sophistication of production, a reliance on MNCs (in sectors where they are predominant) may well be the best long-term policy. Second, even the development of know-why capabilities in competing local firms may be stimulated by MNC competition once these firms have a certain size and competence. Thus, MNC entry would hasten the technological maturity of 'infant' industries which may otherwise simply grow into slothful, uncompetitive adolescents.

## NOTES

1. This paper was presented at the Seventh World Congress of the International Economics Association held at Madrid in September 1983.
2. For concise statements see Katz (1982) and Blumenthal (1979). Also see

126 *Multinationals, Technology and Exports*

the various papers in Fransman and King (1984), and in particular the introductory chapter by Fransman.

3. See ibid, Dahlman and Westphal (1982), and Lall (1982).
4. See my paper in Fransman and King (1984).
5. For a statistical analysis of the industrial characteristics of MNC investment in India, see Lall and Mohammad (1983).
6. The typology set out below draws upon, but is not identical to, that of Dahlman and Westphal (1982). Also see the classification in my 1982 book, pp. 66–8.
7. For a survey of the results of pioneering research in Latin America into plant-level development of know-how capabilities, see Katz (1978).
8. I am grateful to the executives of Hindustan Lever for providing me with detailed information on their R & D in India. This was used for a case study prepared for the ILO on technological change and employment generation by MNCs in developing countries (Lall, 1983a).
9. For a review of the literature see Caves (1982).
10. On Canada, conflicting views on 'truncation' are presented by McFetridge (1977) and Globerman (1976), the former suggesting that truncation presents real problems for Canada, the latter that it is the small scale of protected Canadian operations which restricts the generation of technology and also its efficient application (by licensing). Globerman concludes that economic nationalism has led to a misallocation of scientific resources in Canada.
11. On this point see Casson (1979).
12. This is the finding of an inter-country study of technology exports by the World Bank and the Inter-American Development Bank in which I am participating.
13. In Australia, for instance, Parry and Watson (1979) show that MNC affiliates often have technology swap arrangements with the parent companies.
14. For an examination of the impact of firm size and market structure on the composition of R & D expenditures, see Mansfield (1981).
15. I am very grateful to Mohammad Saqib for his help in collecting and processing this data. Some of the findings, pertaining to the engineering industry, are discussed in Lall (1983b).
16. In a similar exercise for Australia, Parry and Watson (1979) found a negative impact for the extent of foreign ownership on R & D within a sample of firms which all had some foreign shareholding.
17. See Lall (1980) for an empirical study of the truck manufacturing industry in India.
18. See Caves (1982), chapter 4.
19. The 'contagion' idea is developed by Findlay (1978), who argues, on this assumption, that LDCs' technical change is a direct function of the extent of foreign presence in the local economy.
20. See Lall (1982).
21. See Link (1983).

Multinationals and Technology Development in Host Countries 127

# REFERENCES

Blumenthal, T. (1979) 'A Note on the Relationship between Domestic Research and Development and Imports of Technology', *Economic Development and Cultural Change,* vol. 27, no. 2.

Casson, M. (1979) *Alternatives to the Multinational Enterprise* (London: Macmillan).

Caves, R. E. (1982) *Multinational Enterprise and Economic Analysis* (Cambridge: Cambridge University Press).

Dahlman, C. and Westphal, L. E. (1982) 'Technological Effort in Industrial Development – an Interpretative Survey of Recent Research', in F. Stewart and J. James (eds) *The Economics of New Technology in Developing Countries* (London: Frances Pinter).

Davies, S. (1979) *The Diffusion of Process Innovations* (Cambridge: Cambridge University Press).

Findlay, R. (1978) 'Relative Backwardness, Direct Foreign Investment, and the Transfer of Technology: A Simple Dynamic Model', *Quarterly Journal of Economics,* vol. XCII, no. 1.

Fransman, M. and King, K. (eds) (1984) *Technological Capability in the Third World* (London: Macmillan).

Globerman, S. (1976) 'Canadian Science Policy and Economic Nationalism', *Minerva,* vol. 14, no. 2.

Hirschey, R. C. and Caves, R. E. (1981) 'Internationalization of Research and Transfer of Technology by Multinational Enterprises', *Oxford Bulletin of Economics and Statistics,* vol. 43, no. 2.

Kamien, M. I. and Schwartz, N. L. (1982) *Market Structure and Innovation* (Cambridge: Cambridge University Press).

Katz, J. (1978) 'Technological Change, Economic Development and Intra and Extra Regional Relations in Latin America', Buenos Aires: IDB/ECLA, Working Paper no. 30.

Katz, J. (1982) 'Technological Change and Development in Latin America', in R. Ffrench-Davis and E. Tironi (eds) *Latin America and the New International Economic Order* (London: Macmillan).

Lall, S. (1978) 'Transnationals, Domestic Enterprises and Industrial Structure in Host LDCs: A Survey', *Oxford Economic Papers,* vol. 30, no. 2.

Lall, S. (1979) 'The International Allocation of Research Activity by U.S. Multinationals', *Oxford Bulletin of Economics and Statistics,* vol. 41, no. 4, and chapter 3 of this volume.

Lall, S. (1980) 'Vertical Inter-Firm Linkages in LDCs: An Empirical Study', *Oxford Bulletin of Economics and Statistics,* vol. 42, no. 3, and chapter 12 of this volume.

Lall, S. (1982) *Developing Countries as Exporters of Technology* (London: Macmillan).

Lall, S. (1983a) 'Technological Change, Employment Generation and Multinationals: A Case Study of a Foreign Firm and a Local Multinational in India', Geneva: International Labour Office, Multinational Enterprises Programme, Working Paper no. 27.

Lall, S. (1983b) 'Determinants of R & D in an LDC: The Indian Engineering Industry', *Economics Letters,* vol. 13, no. 4.

128            *Multinationals, Technology and Exports*

Lall, S. and Streeten, P. P. (1977) *Foreign Investment, Transnationals and Developing Countries* (London: Macmillan).

Lall, S. and Mohammad, S. (1983) 'Multinationals in Indian Big Business: Industrial Characteristics of Foreign Investment in a Heavily Regulated Economy', *Journal of Development Economics*, vol. 13, no. 1, and chapter 5 of this volume.

Link, A. N. (1983) 'Inter-Firm Technology Flows and Productivity Growth', *Economics Letters*, vol. 11, nos. 1–2.

McFetridge, D. G. (1977) *Government Support for Scientific Research and Development: An Economic Analysis* (Toronto: University of Toronto Press).

Mansfield, E. (1981) 'Composition of R & D Expenditures: Relationship to Size of Firm, Concentration and Innovative Output', *Review of Economics and Statistics*, vol. 63, no. 4.

Parry, T. G. and Watson, J. F. (1979) 'Technology Flows and Foreign Investment in the Australian Manufacturing Sector', *Australian Economic Papers*, vol. 18, no. 32.

Teece, D. J. (1976) *The Multinational Corporation and the Resource Cost of Technology Transfer* (Cambridge, Mass.: Ballinger).

APPENDIX TABLE  Regression analysis (log linear) of determinants of R & D performance in India (1978–9)

| Independent variables | Industry and Equation (Numbers) | | | | |
| | Engineering (n = 100) | | Chemicals (n = 45) | | |
| | 1 | 2 | 3 | 4 | 5 |
|---|---|---|---|---|---|
| $x_1$ | | 0.095[b] (1.711) | −0.120[c] (1.329) | −0.070 (0.701) | 0.287 (0.761) |
| $x_2$ | 0.903[b] (2.661) | | −0.017 (0.047) | | |
| $x_3$ | 0.574 (1.164) | 0.694[c] (1.359) | | | −1.810[b] (2.456) |
| $x_4$ | 0.900 (0.815) | 0.176[c] (1.531) | | | 0.207 (0.752) |
| $x_5$ | 0.369[a] (2.811) | | 0.116 (0.616) | | |
| $x_6$ | −0.238[b] (2.298) | −0.172[c] (1.549) | 0.605[a] (3.879) | 0.480[a] (3.250) | |
| $x_7$ | 1.322[c] (1.628) | 1.596[b] (1.853) | | −0.086 (0.424) | |
| $x_8$ | | | | 0.222[c] (1.650) | 0.300[b] (2.231) |

APPENDIX TABLE    *Regression analysis (log linear) of determinants of R & D performance in India (1978–9)*

|          |                    |                    |          |          |          |
| -------- | ------------------ | ------------------ | -------- | -------- | -------- |
| $x_9$    | 1.069[a] (4.487)   | 1.061[a] (4.133)   |          |          |          |
| $x_{10}$ |                    | −0.718 (1.202)     |          |          |          |
| $R^2$    | 0.414              | 0.356              | 0.268    | 0.279    | 0.208    |
| $F$      | 8.79[a]            | 5.95[a]            | 3.168[b] | 3.407[b] | 2.821[b] |

Notes:    $t$ statistics in parentheses

Significance levels: a – 99%; b – 95%; c – 90%

Sources:   Company balance sheets, Department of Science and Technology, Directorate General of Technical Development.

Symbols and Definitions:

Dependent variable: R & D expenditures in 1978 as a percentage of sales in that year.

Independent variables:

$x_1$  —   share of foreign equity in total equity

$x_2$  —   size of firm as measured by sales

$x_3$  —   age of firm in India

$x_4$  —   number of foreign licensing agreements

$x_5$  —   royalties paid as a percentage of sales

$x_6$  —   exports as a percentage of sales, a measure of competitiveness

$x_7$  —   percentage of total wages accounted for by a highly paid technical and managerial employees, a measure of top-level skills

$x_8$  —   selling expenses as percentage of sales

$x_9$  —   dummy variable for engineering firms, with 1 = firms engaged in machinery manufacture and other sophisticated activies and 0 = others

$x_{10}$  —   average wage level, an index of gneral production skills.

# Name Index